2023 International Conference on Microelectronics (ICM 2023)

Abu Dhabi, United Arab Emirates
17-20 December 2023

IEEE Catalog Number: CFP23473-POD
ISBN: 979-8-3503-8083-5

**Copyright © 2023 by the Institute of Electrical and Electronics Engineers, Inc.
All Rights Reserved**

Copyright and Reprint Permissions: Abstracting is permitted with credit to the source. Libraries are permitted to photocopy beyond the limit of U.S. copyright law for private use of patrons those articles in this volume that carry a code at the bottom of the first page, provided the per-copy fee indicated in the code is paid through Copyright Clearance Center, 222 Rosewood Drive, Danvers, MA 01923.

For other copying, reprint or republication permission, write to IEEE Copyrights Manager, IEEE Service Center, 445 Hoes Lane, Piscataway, NJ 08854. All rights reserved.

*** *This is a print representation of what appears in the IEEE Digital Library. Some format issues inherent in the e-media version may also appear in this print version.*

IEEE Catalog Number: CFP23473-POD
ISBN (Print-On-Demand):   979-8-3503-8083-5
ISBN (Online):   979-8-3503-8082-8
ISSN: 2332-7014

Additional Copies of This Publication Are Available From:

Curran Associates, Inc
57 Morehouse Lane
Red Hook, NY 12571 USA
Phone: (845) 758-0400
Fax: (845) 758-2633
E-mail: curran@proceedings.com
Web: www.proceedings.com

2023 International Conference on Microelectronics (ICM 2023)

Abu Dhabi, United Arab Emirates
17-20 December 2023

IEEE Catalog Number: CFP23473-POD
ISBN: 979-8-3503-8083-5

TABLE OF CONTENTS

Handwritten Signature Recognition Using Deep Learning 1
Basmala Mustafa, Radwa Taha, Omar M. Fahmy, Shereen Moataz Afifi

Efficient CNN Hardware Architecture Based on Linear Approximation and Computation Reuse
Technique 7
Mohammed F. Tolba, Hani Saleh, Mahmoud Al-Qutayri, Ayman Hroub, Thanos Stouraitis

A CMOS Analog Neuron Circuit with a Multi-Level Memory 11
Melvin D. Edwards, Nabil J. Sarhan, Mohammad Alhawari

Deep Image Deraining 16
Laila Hegazy, Shereen Afifi, Omar Fahmy

Acoustic Device for Detecting Red Palm Weevil Using Deep Learning and IoT 21
Maryam M. Atia, Amr I. Gouda, Mahmoud G. Ismail, Mohammed A.-M Salem, Mohamed A.
Abd Elghany

Enhancing Traffic Management with Embedded Machine Learning for Vehicle Detection 26
Mohamad Sofian Abu Talip, Mohd Zulhakimi Ab Razak, Mahazani Mohamad, Anis Salwa
Mohd Khairuddin, Tengku Faiz Tengku Mohmed Noor Izam, Azizul Azizan

Modified Arnold Transform and DNA Manipulation for Chaos-Based RGB Image Encryption 32
Marwan A. Fetteha, Wafaa S. Sayed, Lobna A. Said, Ahmed H. Madian

Aloe Vera Tissue Modeling and Parameter Identification Using Meta-Heuristic Optimization
Algorithm 36
Mohamed S. Ghoneim, Dalia A. Fathi, Lobna A. Said, Ahmed H. Madian, Magdy A. Bayoumi

Convolutional Autoencoder for Real-Time PPG Based Blood Pressure Monitoring Using TinyML 41
Noor Faris Ali, Mousa Hussein, Falah Awwad, Mohamed Atef

Multi-Class Classification of Melanoma on an Edge Device 46
Aser Ashraf Ali, Radwa Essam Taha, Ranpreet Kaur, Shereen Moataz Afifi

Deep/Federated Learning Algorithms for Ultrasound Breast Cancer Image Enhancement 52
Sarah M. Waly, Radwa Taha, Mohamed A. Abd Elghany, Mohammed A.-M Salem

AS-LR Emergency Detection Scheme for Biomedical Applications 58
Nadine Boudargham, Abdallah Kassem, Mustapha Hamad

Computational-Based Advanced Encryption Standard (AES) Accelerator 64
Enas Abulibdeh, Hani Saleh, Baker Mohammad, Mahmoud Alqutayri

A SDR Transmitter Baseband-To-IF IC with Digital Out-phasing Exponential Modulation in a
65nm CMOS LPE 70
Mihai Sanduleanu, Ahmed Mahdy

A Review on Hyperdimensional Computing 74
Maram Abdulrahman, Sandy Wasif, Miran Wael, Eman Azab, Maggie Mashaly, Mohamed A.
Abd El Ghany

Deep Neural Network Inference Processor 80
Ali Hegazy, Khaled Salah

A State-Of-the-Art Design: Applying Forward Kinematics to Improve Patient Positioning in Radiosurgery 86
Alaa Saadah, Donald Medlin, Jad Saud, Xiao Ran Zheng, Géza Husi

A Self-Aware Power Management Model for Epileptic Seizure Systems Based on Patient-Specific Daily Seizure Pattern 91
Shiva Maleki Varnosfaderani, Rihat Rahman, Nabil J. Sarhan, Mohammad Alhawari

Parkinson's Disease Detection Using Voice Features and Machine Learning Algorithms 96
Rumana Islam, Esam Abdel-Raheem, Mohammed Tarique

EEG Experiment Flow Control Using FPGA as an Alternative to Commercial Devices 101
Cristian Y. Olivares, Norelli Schettini

An ECG-Based Blood Pressure Estimation Using U-Net Auto-encoder and Random Forest Regressor 107
Elham Alaa Aldein, Mohamed Abdleraheem, Usama Sayed Mohamed, Mohamed Atef

Examining the Performance of Melanoma Classification Using Superpixel Segmentation: A Comparative Analysis 113
Faezeh Mohammadi Aydoghmishi, Sudipta Modak, Esam Abdel-Raheem, Luis Rueda

Spatial Multiplexing MIMO for Remote Areas Employing MMSE Parallel Interference Cancellation for Non-Orthogonal GFDM 119
Danilo Gaspar, Vanessa Mendes Rennó, Luciano Leonel Mendes, Tales Pimenta, Shahab Ehsanfar

Comparison of ANFIS and ANN for Small-Signal Modelling of GaN HEMT Up to 40 GHz 125
Bagylan Kadirbay, Saddam Husain, Anwar Jarndal, Mohammad Hashmi

Efficient Implementation of a 4x4 Enhanced Pipeline Multiplier Using Electric EDA Tool 131
Khader Mohammad, Nirmeen Al-Sheikh

Design and Simulation of Dual-Metal-Gate Tunnel Field Effect Transistor with Biomolecule Sensing Applications. 137
Mohammad Salim Wani, Hend I. Alkhammash, M. Shiblee, Sajad A. Loan

Design and Fabrication of Nanofibrous Membrane and Microelectrodes for Highly Robust Biocompatible Humidity Sensing 141
Afaque Manzoor Soomro, Faheem Ahmed, Muhammad Waqas

InGaAs Self-Switching Diode with Suppressed Harmonics for High Frequency Applications 145
B. Sharma, S. Garg, P. Singh, Supriya Garg, G. Das, D. K. Sharma, N. Gupta, A. K. Singh, S. Kumar, S. R. Kasjoo

Secured and Optimized Hardware Accelerators Using Key-Controlled Encoded Hash Slices and Firefly Algorithm Based Exploration 149
Anirban Sengupta, Aditya Anshul, Chirag Kothari, Sumer Thakur

On Structure Design Optimization of GaN Based Semiconductor Device for Reduced Trapping 153
Arivazhagan L, Anwar Jarndal

Fault Simulation Framework Using PyUVM 158
Mina Hanna Fayez, Mohamed Ahmed Eladawy, Micheal Safwat Sahyon, Islam Osama Ahmed, Omar Hossam El-Din, Mohamed Ahmed Elshafie, Mohamed Ayman Taha, Mohamed Gamal Talaat

AUTG: An Automatic UVM-Based TestBench Generator for VLSI Chip Design Verification 162
Mohammad Ismael, Ayman Hroub, Abdellatif Abu-Issa

A Sub-1dB Noise Figure Ku Band GaN Low Noise Amplifier for Space Applications 168
Husna Hamza, Anwar Jarndal

A Low-Power Analog Integrated Gaussian-based Neural Network Classifier with Application to
Hepatitis Disease Recognition 172
Vassilis Alimisis, Nikolaos P. Eleftheriou, Paul P. Sotiriadis

An Adaptive Analytical FPGA Placement Flow Based on Reinforcement Learning 178
C. Barn, S. Vermeulen, S. Areibi, G. Grewal

Supporting Dynamic Control-Flow Execution for Runtime Reconfigurable Processors 184
Hassan Nassar, Rafik Youssef, Lars Bauer, Jörg Henkel

Fast Parallel Multiple Access Distributed Arithmetic (FPMA-DA) Reconfigurable FIR Filter 190
Ahmed H. A. Bayoumi, Sameh A. Ibrahim, Hossam A. H. Fahmy

Hardware Acceleration for Deep Learning Model 196
Shereen Afifi, Abdelrahman Amgad Abdallah, Radwa Taha

Efficient Mux-Based Multiplier for MAC Unit 202
Huruy Tekle Tesfai, Hani Saleh, Mahmoud Meribout, Mahmoud Al-Qutayri, Thanos Stouraitis

Development and Optimization of a Planar Wideband Ultrathin Absorber Based on Equivalent
Circuit Model Analysis 206
Yasmine Abdalla Zaghloul, Hany Hammad

Augmented Reality as an Educational Enrichment Tool: Integrating the Virtual with the Real 210
*Eder Costa Maciel, Miller Henrique Lúcio Fernandes, Tales Cleber Pimenta, Jaqueline
Corrêa Silva De Carvalho, Marcos Alberto De Carvalho*

The Influence of Artificial Intelligence in Society 214
Milene Santos Moreira, Douglas De Tarso Da Silva, Tales C Pimenta

Fusing IP Vendor Palmprint Biometric with Encoded Hash for Hardware IP Core Protection of
Image Processing Filters 218
Anirban Sengupta, Aditya Anshul, Sumer Thakur, Chirag Kothari

Emotion Recognition Based on Electroencephalogram (EEG) Signals 222
Khader Mohammad, Saleem Hamo, Mohammad Abbas, Maen Mohammad

Hyperdimensional Computing Versus Convolutional Neural Network: Architecture, Performance
Analysis, and Hardware Complexity 228
Eman Hassan, Meriem Bettayeb, Baker Mohammad, Yahya Zweiri, Hani Saleh

PSO-GA-based Federated Learning for Predicting Energy Consumption in Smart Buildings 234
Nader Bakir, Ahmad Samrouth, Khouloud Samrouth

Optimizing Charging Schedules for WRSNs: A Multi-Criteria Decision-Making Approach with
Multiple Charger Vehicles 239
*Samah Abdel Aziz, Ammar Hawbani, Xingfu Wang, A. S. Ismail, Nasir Saeed, Saeed H.
Alsamhi, Liang Zhao, Ahmed Al-Dubai*

Exploring H_2S Gas Sensing with Graphene Nanoribbon Field Effect Transistors: A Semi-Empirical Simulation Approach 243
Asma Wasfi, Mohamed Atef, Falah Awwad

DRAM Bitline as a Delay Path for Potential PUF 248
Enas Abulibdeh, Leen Younes, Baker Mohammad, Hani Saleh, Mahmoud Alqutayri, Khaled Humood

Real-Time Switched Capacitor Based Power Side-Channel Attack Detection 253
Leen Younes, Baker Mohammad, Mahmoud Al-Qutayri, Hani Saleh, Dima Kilani

An Embedded Real-Time Driver Monitoring System Based on AI and Computer Vision 258
Leila Sharara, Alexandros Politis, Lubna Alazzawi, Mohammed Ismail

Development, Optimization, and Application of ML Based Modeling of Printed VO_2 RF Switch 264
Ahmad Khusro, Mohammad Hashmi, Muhammed Akmal Chaudhary

2.45GHz Low-Power Diode Bridge Rectifier Design 268
Koubar Gabriel, Haddad Fayrouz, Bouchra Nessakh, Rahajandraibe Wenceslas, Sadek Sawsan

A 0.3-V 10-nW CMOS OTA with Feedforward Body-Driven Structure 272
Hirokazu Yoshizawa, Naoki Inoue

A High PSRR CMOS Voltage and Current Reference in One Circuit Without Amplifier for Low Power Applications 276
Ashutosh Pathy, Andleeb Zahra, Amir Ahmad, Zia Abbas

A Low-Power Analog Integrated Deep Spatio-Temporal Inference Network with Application to Digit Classification 280
Vassilis Alimisis, Nikolaos P. Eleftheriou, Paul P. Sotiriadis

Regional CubeSat Communication and Constellation Design Evaluation 284
Khaled Mohammed, Hamzeh Abu Qamar, Ruhul Amin Khalil, Nasir Saeed

Innovative Hardware Architecture for Zero Emission Sea Drones 288
Ilya Kavalchuk, Saad Zafar, Shahid Islam

Layout-Based Reliability Analysis of openMSP430 Register File Under External Radiations 294
Vivek Bansal, Otmane Ait Mohamed, Fakhreddine Ghaffari

A Novel Architecture of CXL Protocol Data Link Layer for Low Latency Memory Access 298
Basma H. Mohamed, Esmail Hany, Mahmoud El-Tahawy, Mohamed Abdel-Salam, M. Watheq El-Kharashi, Mona Safar

The Electric Crisis of Brazil in 2021 Origins and a Solution Proposal 304
Milene Santos Moreira, Tales C Pimenta

A Fully-Differential Low-Noise Instrumentation Amplifier for Electrical Impedance Tomography 308
Ibrahim Alkhalifa, Yaqub Mahnashi

Radiation-Hardened Stabilized Power Supply Unit Based on Bipolar Transistors 313
Takato Tanizawa, Minoru Watanabe

An Analog Integrated, Low-Power, Area-Efficient, Gilbert, Modulo-based Classifier with Application to Lung-Cancer Classification 317
Vassilis Alimisis, Nikolaos P. Eleftheriou, Savvas Leventikidis, Paul P. Sotiriadis

Ultra-Low Power Self-polarized Dynamic Threshold Telescopic OTAs Circuits for Biomedical Applications.. 321
Dalila Laouej, Houda Daoud, Mourad Loulou

Energy-Efficient Computation-In-Memory Architecture Using Emerging Technologies 325
*Rajendra Bishnoi, Sumit Diware, Anteneh Gebregiorgis, Simon Thomann, Sara Mannaa,
Bastien Deveautour, Cédric Marchand, Alberto Bosio, Damien Deleruyelle, Ian O'Connor,
Hussam Amrouch, Said Hamdioui*

Author Index

2023 International Conference on Microelectronics (ICM)

2023 International Conference on Microelectronics (ICM) took place December 17-20, 2023 in Abu Dhabi, United Arab Emirates.

Committees

General co-Chairs

Mohamad Sawan (Westlake University, China)

General Local co-Chairs

Mahmoud Al-Qutayri (Khalifa University, United Arab Emirates)

Ghada Hussain Alsuhli (Khalifa University, United Arab Emirates)

Hani Saleh (Khalifa University, United Arab Emirates)

TPC co-Chairs

Abdallah Kassem (Notre Dame University, Lebanon)

Ahmed Madian (Nile University, Egypt)

Baker Mohammad (Khalifa University, United Arab Emirates)

Falah Awwad (UAE University, United Arab Emirates)

TPC Members

Najah A. Abu Ali (UAEU, United Arab Emirates)

Mahmoud Al Ahmad (United Arab Emirates University, United Arab Emirates)

Yousef Al Hammadi (UAE University, United Arab Emirates)

Amine El Moutaouakil (United Arab Emirates University, United Arab Emirates)

Mohammad Hayajneh (UAEU, United Arab Emirates)

Mahmoud Meribout (Khalifa University, United Arab Emirates)

Adel Najar (UAEU, United Arab Emirates)

Vinod Pangracious (American University in Dubai, United Arab Emirates)

Publications Committee Co-Chairs

Mohamed Tabaa (EMSI Casablanca, Morocco)

Financial Co-Chairs

Nawaf Almoosa (Khalifa University, United Arab Emirates)

Abdulhadi Shoufan (Khalifa University, United Arab Emirates)

Handwritten Signature Recognition using Deep Learning

Basmala Mustafa
Media Engineering and Technology
German University in Cairo
Cairo, Egypt
basmala.mahmoud@student.guc.edu.eg

Radwa Taha
Media Engineering and Technology
German University in Cairo
Cairo, Egypt
radwataha1999@gmail.com

Omar M. Fahmy
Electrical Engineering Dept.
Badr University in Cairo
Cairo, Egypt
omar.fahmy@buc.edu.eg

Shereen Moataz Afifi
Media Engineering and Technology
German University in Cairo
Cairo, Egypt
shereen.moataz@guc.edu.eg

Abstract—Handwritten signature recognition plays a crucial role in verifying document authenticity and preventing fraudulent activities. That's why this paper focuses on the development of a deep learning-based system for recognizing handwritten signatures. The main objectives include creating a diverse dataset of signatures, implementing a deep learning architecture for accurate signature recognition, and evaluating the system's performance using various metrics.The VGG16 architecture was chosen due to its effectiveness compared to other methods, and it served as the framework for further enhancements. The results demonstrate the model's outstanding accuracy. Specifically, the proposed model, trained on the merged dataset, achieved remarkable performance with a training accuracy of 99.78% and a validation accuracy of 99.75%. During testing, the model exhibited an impressive accuracy of 98.96%, confirming its effectiveness in identifying genuine signatures. Furthermore, the model trained on the collected dataset had shown an accuracy of 98.9% which ensures an efficient handwritten signature recognition.

Index Terms—Deep Learning, Handwritten Signature, Signature Recognition, Preprocessing, CNN, VGG16

I. INTRODUCTION

The handwritten signature is a crucial biometric trait for verifying identities in legal, financial, and administrative contexts [1], [2]. Manual authentication can be time-consuming and error-prone. Recent advances in deep learning and computer vision have paved the way for more accurate and efficient automated signature recognition systems, with potential applications in banking, law enforcement, and governmental organizations.

Despite their promise, these systems face challenges in accuracy and reliability, especially in detecting forged signatures due to variability in styles, pen pressure, and angles. To address this, recent research has turned to Convolutional Neural Networks (CNNs), achieving up to 98.8% accuracy in signature recognition and 89% in forgery detection [3], [4]. Architectures like GoogLeNet's Inception-v1 and Inception-v3, employing CNN models, have also shown promise, with validations of 83% and 75%, respectively [5]. This paper

leverages CNNs to enhance the accuracy and reliability of the proposed signature recognition system.

The manuscript aims to develop a deep learning-based system that not only identifies genuine signatures but also detects forgeries, reducing the need for manual intervention. This approach saves time and costs associated with traditional methods.

To achieve these goals, the manuscript focuses on collecting and preprocessing a comprehensive dataset of signatures [1]. Preprocessing includes noise removal to aid in implementing a deep learning architecture for signature recognition. Evaluation metrics like accuracy, precision, recall, and F1 score are used, and the system's performance is compared with other state-of-the-art methods.

In summary, this paper offers valuable insights into model performance, dataset requirements, and potential areas for improvement in the field of signature recognition. The findings emphasize the importance of diverse dataset training and highlight the capabilities of the VGG16 architecture in achieving high accuracy for handwritten signature recognition tasks.

II. LITERATURE REVIEW

Since the early 1990s, researchers have been investigating offline Handwritten Signature Recognition due to its importance as a behavioral biometric used for identification and authentication purposes. However, forged signatures can lose their key features, making signature verification a critical security aspect. Over the years, several solutions have been developed to address this challenge as shown in Fig. 1.

One such approach by Tarek Atia [6] involves classifying binary images using Convolutional Neural Networks (CNNs), specifically utilizing models like VGG-16, ResNet-50, Inception-v3, Xception, and a Custom CNN Model. The proposed system preprocesses signatures and uses binary image classification models to authenticate scanned signature

979-8-3503-8083-5/23 $31.00 © 2023 IEEE

images. The results showed promising accuracy, with the ResNet-50 model achieving 82.3% accuracy.

Other studies have explored different methods and architectures for signature recognition. Kancharla et al. [1] employed Convolution Neural Networks (CNNs) for offline signature recognition, using ADAM and RMSprop as adaptive learning rate methods. While ADAM required fewer epochs to achieve optimal performance, its accuracy varied with different datasets. Noor et al. [7] used preprocessed signature images to train CNNs and achieved high accuracy when the number of individuals was low and the number of signature samples was high.

Miaba et al. [8] explored online signature verification using hybrid transform features gathered from dynamic signature signals. They used different transforms like DFT, DCT, and DWT, achieving the best accuracy by combining the most effective signals from each transform.

Sam et al. [5] compared Inception-v1 and Inception-v3 deep convolutional neural network architectures for GPDS Synthetic Signature Database classification, finding Inception-v1 more suitable for low-resolution inputs. Yapici et al. [9] and Bonde et al. [10] experimented with CNN models and fine-tuning VGG16, achieving improved results.

Gupta Y et al. [11] conducted a study using various pretrained models and optimizers, and the approach utilizing VGG16 with the Adam optimizer for feature extraction showed the highest accuracy.

Considering the extensive research conducted on signature forgery detection and the promising results reported in the literature, the approach utilizing VGG16 with the Adam optimizer appears to be a compelling choice for further investigation in this paper. This approach combines a well-established model architecture and an effective optimizer, potentially enhancing the accuracy and reliability of signature forgery detection systems.

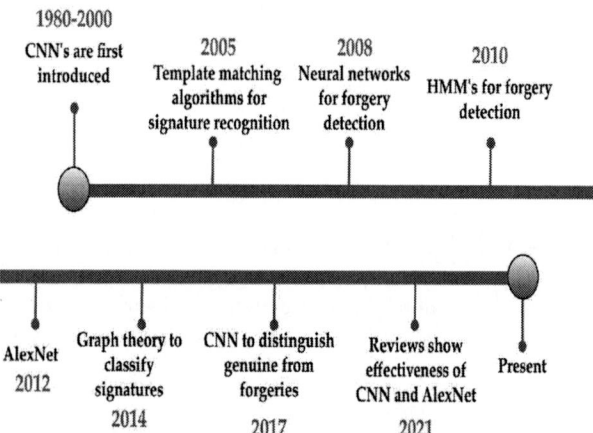

Fig. 1. Historical timeline of research on the problem of signature recognition [12].

III. METHODOLOGY

A. Methodology Overview

The methodology employed in this study adopts a supervised machine learning approach for signature verification.

The dataset used in this study consists of two main types: first dataset is a customized dataset created by merging two existing datasets (ICDAR 2011 [13] and CEDAR [14]) and the second dataset used is a collected dataset containing signatures from 17 individuals with diverse writing styles and formats, providing a broader representation of real-world signatures. The diversity within our dataset allows for a comprehensive exploration of signature verification across a wide spectrum of writing styles and individual characteristics. This inclusion in dataset composition ensures that our proposed methodology is robust and adaptable to various real-world scenarios.

The preprocessing steps involve resizing the images to a fixed size of 224x224 pixels, converting them to RGB format, and normalizing the pixel values. The chosen model architecture is a convolutional neural network (CNN), specifically the VGG16 model with transfer learning based on findings from the literature review. Several deep learning architectures for signature recognition were compared, including VGG16, Inception v-3, ResNet 50, and Xception, with each architecture tested using four optimizers: Adam, RMSprop, SGD, and Adagrad. VGG16 with the Adam optimizer demonstrated superior training and validation accuracy compared to the other architectures.

VGG16 is a widely recognized CNN architecture with 16 layers, including 13 convolutional layers and 3 fully connected layers. It excels at extracting hierarchical features from input images through convolutional and max pooling layers. The fully connected layers with ReLU activation help map learned features to output classes, and the softmax activation layer generates a probability distribution across the classes. Based on comparative accuracy analysis, VGG16 was selected as the primary architecture for the signature recognition system, benefiting from its pre-trained model on a large-scale image dataset.

The model is trained on a subset of the dataset using the Adam optimizer and categorical cross-entropy loss function. Evaluation is performed on a separate test set, and various performance metrics, such as accuracy, precision, and F1 score, are calculated. Overall, the methodology aims to deliver a robust and accurate approach to signature verification, as depicted in the block diagram shown in Fig. 2.

B. Data Preprocessing

1) Collected Dataset Preprocessing: The image preprocessing for signature verification starts with organizing the forged and genuine signatures data into training and testing sets. The data is split randomly, with 80% of training set and the remaining 20% to the testing set. The training set images undergo several preprocessing steps, including converting them to grayscale, enhancing brightness, and normalizing pixel values to improve contrast and consistency.

After preprocessing, the images are converted back to RGB format and resized to 224x224 pixels. The resulting images are added to the training data, along with labels indicating that they are forged signatures (label 1). The same preprocessing

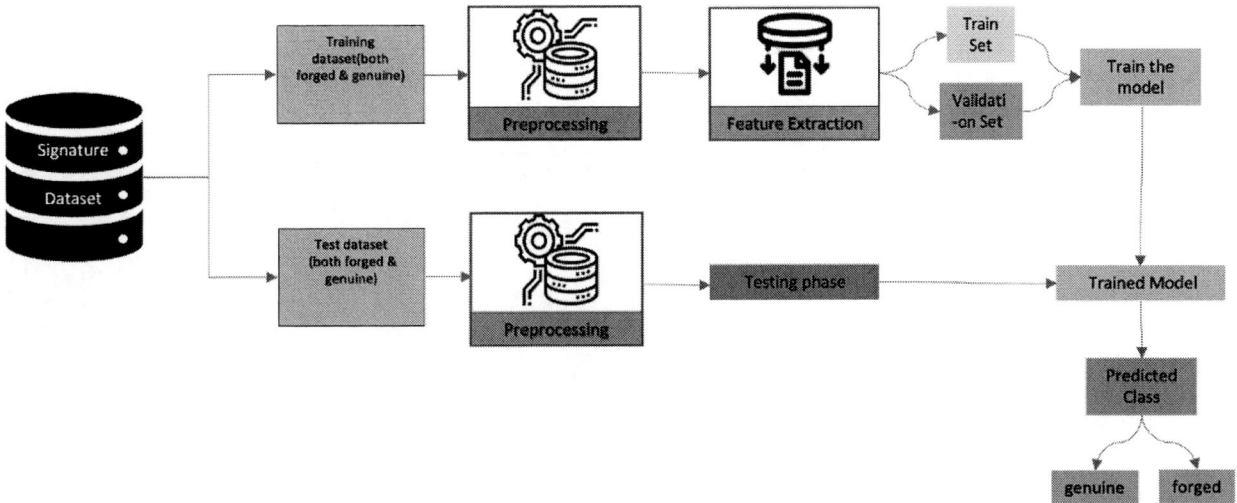

Fig. 2. Block Diagram of the Proposed Models.

steps are applied to the test data, and the preprocessed images are added to the test data with corresponding labels.

To facilitate further processing and training, the images and labels are converted into NumPy arrays, and the pixel values are normalized to the range of [0, 1]. This normalization ensures that the input data is within a suitable range for the machine learning algorithm.

In summary, the image preprocessing involves loading, splitting, and enhancing the signatures data for training and testing. The preprocessed images, along with their labels, are then converted into NumPy arrays and normalized to facilitate efficient processing and training of the signature verification model.Fig. 3 illustrates the signature before and after preprocessing.

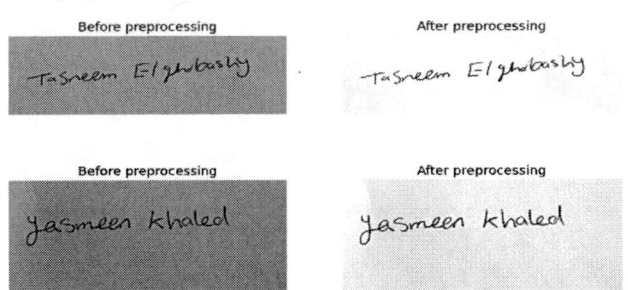

Fig. 3. Preprocessing Of Signatures From Collected Dataset.

2) Customized Dataset Preprocessing: In the preprocessing steps for signature recognition, the dataset undergoes various transformations to prepare it for training and testing. The first step involves resizing all images to a standardized size of 224 pixels, ensuring uniform dimensions for the deep learning model and reducing computational complexity.

The dataset is divided into two classes: "Forged Signatures" and "Genuine Signatures," with images for each class organized into separate directories. Random splitting is applied to create training and testing sets, with 80% of forged and genuine signatures allocated for training, and the remaining 20% used for testing.

After loading the images using the OpenCV library, they undergo preprocessing steps. Each image is converted to the RGB color space for consistency, followed by resizing to a uniform size suitable for the deep learning model.

The test data is also preprocessed in the same manner as the training data, and the preprocessed images and corresponding labels are added to the test data.

The images are stored in separate lists based on their respective classes, and the image data and labels are converted into NumPy arrays as previously described.

Overall, these preprocessing steps ensure the dataset is properly prepared for training and evaluation, with resizing, splitting, and normalization playing crucial roles in achieving optimal model performance. Fig. 4 illustrates the signature before and after preprocessing.

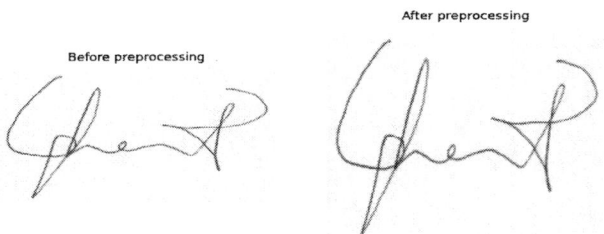

Fig. 4. Preprocessing Of Signatures From Merged Dataset.

C. Feature Extraction

The feature extraction stage in the handwritten signature verification system is crucial for capturing discriminative information from the input images. For this purpose, the widely recognized CNN architecture VGG16 is employed. VGG16 consists of multiple convolutional layers that systematically extract hierarchical features from the signatures. These convolutional layers use filters to highlight specific visual patterns, gradually building a comprehensive representation of the signature. Max-pooling layers down-sample the feature maps, emphasizing prominent features and enhancing the model's robustness to variations in position and scale. By removing the fully connected layers, the focus remains solely on extracting discriminative features from the signatures. Utilizing the VGG16 architecture enables the model to capture relevant and distinctive characteristics, facilitating accurate classification of genuine and forged signatures. Despite its high parameter count, VGG16 is widely used and well-suited for feature extraction in various applications.

D. Intial Implemented Models

In this study, three datasets were used for training and evaluating various models in the handwritten signature verification system. The initial experiments without randomization resulted in low test accuracy and precision. Several trials were conducted, each employing different architectures and hyperparameters based on the VGG16 model. Data augmentation techniques were used in most trials to improve model generalization.

To enhance the models, the ICDAR 2011 and CEDAR datasets were merged, resulting in a larger dataset of 4290 samples. Three models were then evaluated on this dataset, utilizing the VGG16 architecture with different regularization techniques and activation functions.

Two successful models were identified, demonstrating the effectiveness of the VGG16 architecture for signature verification. Building on these successes, a new model is proposed to further enhance the system's performance. The proposed model aims to achieve higher accuracy and lower error rates through innovative architectural designs and state-of-the-art techniques.

The subsequent sections of the paper will delve into the details of the proposed model, highlighting its novel features and discussing its potential for surpassing previous results. Through rigorous evaluation and comparison, the study aims to demonstrate the superiority of the proposed model in achieving enhanced performance in handwritten signature verification.

E. The Proposed Models

1) The First Proposed Model: The first proposed model is designed to enhance the performance of the handwritten signature verification system. It is trained on the customized dataset of 4290 signatures mentioned in section III-B2 and based on the VGG16 architecture. Data augmentation techniques, such as rotation, shifting, shearing, and zooming, are employed to diversify the training dataset and improve generalization.

The model consists of additional layers following the base model, including a flatten layer, two dense layers with ReLU activation, and dropout layers to prevent overfitting. The final dense layer uses the softmax activation function for classification probabilities.

During training, a categorical cross-entropy loss function and the Adam optimizer with a learning rate of 1e-5 are used. Early stopping with a patience of 5 helps prevent overfitting. The model is trained on augmented training data with a batch size of 32 for 30 epochs. The dataset is split into 80% for training and 20% for testing, with further splitting of training data into 70% for training and 30% for validation.

Fig. 5 shows snippet of the code implementing the model.

```
# Build the model
model = Sequential()
model.add(base_model)
model.add(Flatten(input_shape=base_model.output_shape[1:]))
model.add(Dense(256, activation='relu', kernel_regularizer=regularizers.l2(0.001)))
model.add(layers.Dropout(0.5))
model.add(Dense(128, activation='relu', kernel_regularizer=regularizers.l2(0.001)))
model.add(layers.Dropout(0.5))
model.add(Dense(output_, activation='softmax'))

model = Model(inputs=model.input, outputs=model.output)
model.compile(loss='categorical_crossentropy', optimizer=optimizers.Adam(learning_rate=1e-5), metrics=['accuracy'])

model.summary()

earlyStopping = EarlyStopping(monitor='val_loss', mode='max', verbose=1, patience=5)
callbacks = [earlyStopping]

# Train the model with data augmentation
history = model.fit(train_data_gen,batch_size=32, epochs=30, callbacks=callbacks, validation_data=val_data_gen)
```

Fig. 5. Snippet From The Code Of The First Proposed Model.

2) The Second Proposed Model: The second proposed model utilizes the collected dataset mentioned in section III-B1, consisting of signatures from 17 individuals with diverse signature styles and formats. The dataset aims to encompass a wider range of signature variations compared to the original dataset. The model uses transfer learning with the pre-trained VGG16 architecture, loading pre-trained weights from ImageNet and making the last few layers trainable. Data augmentation techniques, such as rotation, shifting, shearing, and zooming, are applied during training to enhance generalization.

The collected dataset is split into training and testing sets (80% and 20%, respectively), with the training set further divided into training and validation sets (80% for training, 20% for validation). The model includes a flatten layer and three dense layers with ReLU activation. L2 regularization is employed to prevent overfitting. The model is compiled with categorical cross-entropy loss and optimized using the Adam optimizer with a learning rate of 1e-4.

To prevent overfitting during training, early stopping is implemented with a patience of 10. The model is trained for 40 epochs with a batch size of 32. Training and validation accuracy, as well as training and validation loss, are monitored throughout the process.

Fig. 6 showcases snippet from the code of the model trained on the collected dataset.

```
# Split data into training and validation sets
split_idx = int(len(train_data) * 0.8)  # 80% for training, 20% for validation
train_data_gen = data_augmentation.flow(train_data[:split_idx], train_labels[:split_idx], batch_size=BS)
val_data_gen = data_augmentation.flow(train_data[split_idx:], train_labels[split_idx:], batch_size=BS)

# Build the model
model = Sequential()
model.add(base_model)
model.add(Flatten(input_shape=base_model.output_shape[1:]))
#model.add(Dense(128, activation='relu'))
#model.add(layers.Dropout(0.5))
model.add(Dense(256, activation='relu', kernel_regularizer=regularizers.l2(0.1)))
model.add(Dense(128, activation='relu', kernel_regularizer=regularizers.l2(0.1)))
model.add(Dense(output_, activation='softmax'))

model = Model(inputs=model.input, outputs=model.output)
model.compile(loss='categorical_crossentropy', optimizer=optimizers.Adam(learning_rate=1e-4), metrics=['accuracy'])
```

Fig. 6. Snippet From The Code Of The Second Proposed Model.

IV. RESULTS

A. The First Proposed Model Results

The first proposed model significantly enhances the hand-written signature verification system's performance. It is based on the VGG16 architecture and utilizes a pre-trained base model with trainable last few layers. Data augmentation techniques, such as rotation, shifting, shearing, and zooming, are incorporated to diversify the training dataset and improve generalization capabilities. The training and validation accuracy reach impressive values of 99.78% and 99.75%, respectively, while the test accuracy is 98.96%. The model achieves a false acceptance rate (FAR) of 0.0, indicating accurate identification of genuine signatures, and a false rejection rate (FRR) of 2.77%, indicating effective detection of impostor signatures. The precision of the model is 98.9%, and the F1 score is 0.986, reflecting its overall performance.

These results demonstrate the model's effectiveness in accurately classifying and verifying handwritten signatures. Its low false acceptance rate and reasonable false rejection rate contribute to its reliability and trustworthiness. The model shows great promise for real-world applications where precise signature verification is crucial. It represents a significant advancement and holds potential to enhance the reliability and robustness of signature verification systems.

Fig. 7 illustrates the training and validation accuracy, while Fig 8 showcases the training and validation loss, providing insights into the model's learning progress.

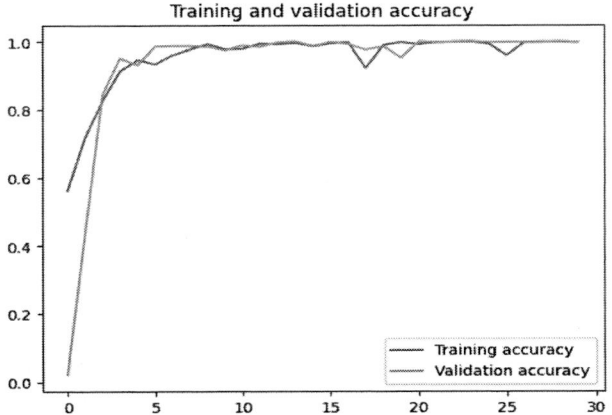

Fig. 7. Training and Validation Accuracy of The First Proposed Model.

B. The Second Proposed Model Results

The model is the model used for the collected dataset is a modified VGG16 architecture.

The training process yields promising results, with 98.9% training accuracy and 96.36% validation accuracy, indicating good generalization. The loss curves show effective learning during training.

The model's performance is evaluated on a separate test dataset, achieving 98.9% test accuracy, 0% False Acceptance Rate (FAR), and 0% False Rejection Rate (FRR), demonstrating accurate classification of genuine and forged signatures.

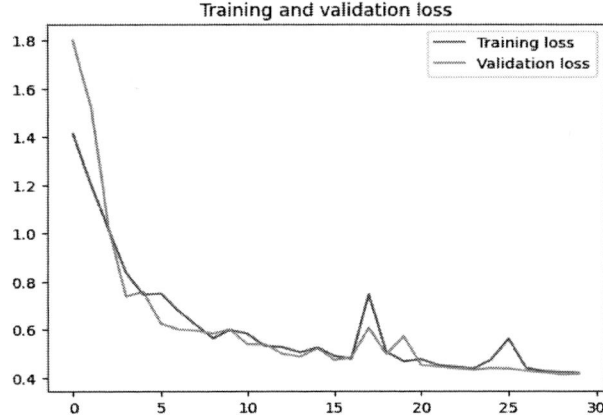

Fig. 8. Training and Validation Loss of The First Proposed Model.

The precision of 98.9% further confirms the model's reliability in distinguishing between genuine and forged signatures.

These results underscore the model's effectiveness in recognizing and verifying handwritten signatures, making it suitable for practical applications in authentication and security systems. Fig. 9 illustrates the training and validation accuracy, while Fig 10 showcases the training and validation loss, providing insights into the model's learning progress.

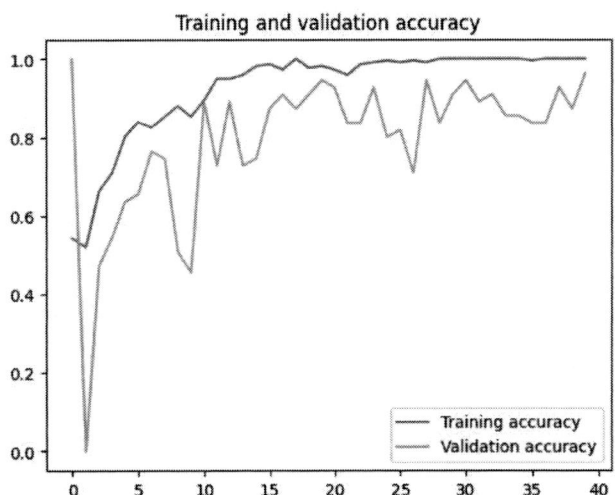

Fig. 9. Training and Validation Accuracy of The Second Proposed Model.

C. Discussion

The developed models in this study, based on the proposed VGG16 architecture and incorporating data augmentation, outperformed existing approaches for handwritten signature recognition in terms of accuracy. The model trained on the customized merged dataset as discussed in section III-E1 achieved a test accuracy of 98.96% with negligible false acceptance rate (FAR) and false rejection rate (FRR). The model trained on the collected dataset which discussed in section III-E2 achieved 98.9% training accuracy, 96.36% validation accuracy, and 98.9% test accuracy, with a FAR and FRR of 0, and precision of 98.9%.

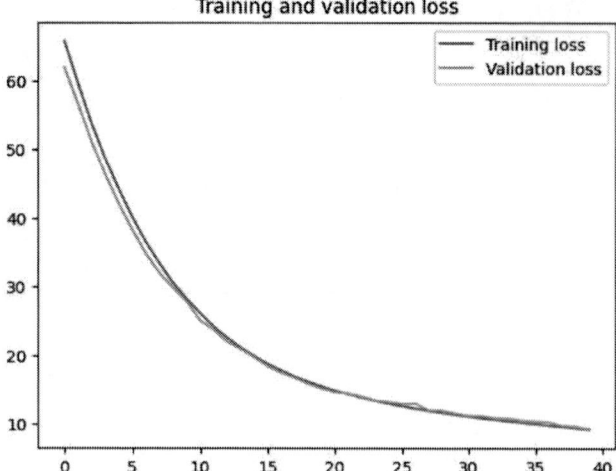

Fig. 10. Training and Validation Loss of The Second Proposed Model.

Comparing these results to related works, the proposed models demonstrated superior performance. For example, Kancharla et al. [1] achieved 98.9% accuracy with one dataset but lower accuracy with another dataset. Atia et al. [6] reported a maximum accuracy of 82.33% with the ResNet-50 model. Bonde et al. [10] achieved an accuracy of 92.03% with an FAR of 12.60% and an FRR of 3.35%. The results highlight the superiority of the proposed models in accurately recognizing handwritten signatures. Table I shows the comparison of accuracy between proposed models in this study with Gupta Y et al. [11] in the context of VGG16 Adam model.

TABLE I
COMPARISON OF PROPOSED MODELS WITH VGG16 ADAM MODEL
PRESENTED IN [11]

Model	Training Accuracy	Validation Accuracy	Test Accuracy
Gupta Y et al. Model [11]	95.84%	95.56%	-
The First Proposed Model	99.78%	99.75%	98.96%
The Second Proposed Model	98.9%	96.36%	98.9%

V. CONCLUSION

The conclusion of this paper highlights the progress in handwritten signature recognition and the goal of creating a deep learning-based system for distinguishing genuine from forged signatures. The project collected and preprocessed a dataset, implemented the VGG16 model with transfer learning for signature recognition, and evaluated the system's performance using various metrics.

The main findings emphasize the importance of training on larger and diverse datasets for greater robustness and generalization. Two proposed models were trained, one on merged data and the other on the collected dataset, achieving high training and validation accuracy. These models outperformed Gupta Y et al.'s VGG16 model in terms of accuracy with an accuracy of 98.96% on the customized dataset and and accuracy of 98.9% on the collected dataset.

However, the project faced limitations due to GPU availability, resulting in longer training times and limited diversity in the collected dataset. Future work should consider local machine setups, gather a more diverse dataset with various signature styles, and expand the dataset for better model robustness. Additionally, creating a user-friendly interface for the system would enhance its accessibility and usability.

Overall, addressing these areas of improvement would optimize the system's performance, flexibility, and user experience, advancing automated signature recognition technology in different applications.

REFERENCES

[1] K. Kancharla, V. Kamble, and M. Kapoor, "Handwritten signature recognition: a convolutional neural network approach," in 2018 International Conference on Ad- vanced Computation and Telecommunication (ICACAT), pp. 1–5, IEEE, 2018.

[2] F. Noor, A. E. Mohamed, F. A. Ahmed, and S. K. Taha, "Offline handwritten signa- ture recognition using convolutional neural network approach," in 2020 International Conference on Computing, Networking, Telecommunications Engineering Sciences Applications (CoNTESA), pp. 51–57, IEEE, 2020.

[3] J. Poddar, V. Parikh, and S. K. Bharti, "Offline signature recognition and forgery detection using deep learning," Procedia Computer Science, vol. 170, pp. 610–617, 2020.

[4] V. L. Souza, A. L. Oliveira, and R. Sabourin, "A writer-independent approach for offline signature verification using deep convolutional neural networks features," in 2018 7th Brazilian Conference on Intelligent Systems (BRACIS), pp. 212–217, IEEE, 2018.

[5] S. M. Sam, K. Kamardin, N. N. A. Sjarif, N. Mohamed, et al., "Offline signa- ture verification using deep learning convolutional neural network (cnn) architec- tures googlenet inception-v1 and inception-v3," Procedia Computer Science, vol. 161, pp. 475–483, 2019.

[6] O. Tarek and A. Atia, "Forensic handwritten signature identification using deep learning," in 2022 IEEE 9th International Conference on Sciences of Electronics, Technologies of Information and Telecommunications (SETIT), pp. 185–190, IEEE, 2022.

[7] F. Noor, A. E. Mohamed, F. A. Ahmed, and S. K. Taha, "Offline handwritten signa- ture recognition using convolutional neural network approach," in 2020 International Conference on Computing, Networking, Telecommunications & Engineering Sciences Applications (CoNTESA), pp. 51–57, IEEE, 2020.

[8] A. Miaba, M. Gwetu, and S. Viriri, "Online signature verification using hybrid trans- form features," in 2018 Conference on Information Communications Technology and Society (ICTAS), pp. 1–5, IEEE, 2018.

[9] M. M. Yapici, A. Tekerek, and N. Topaloglu, "Convolutional neural network based offline signature verification application," in 2018 International Congress on Big Data, Deep Learning and Fighting Cyber Terrorism (IBIGDELFT), pp. 30–34, IEEE, 2018.

[10] "S. Bonde, P. Narwade, and R. Sawant, "Offline signature verification using convo- lutional neural network," in 2020 6th International Conference on Signal Processing and Communication (ICSC), pp. 119–127, IEEE, 2020.

[11] Y. Gupta, S. Kulkarni, and P. Jain, "Handwritten signature verification using trans- fer learning and data augmentation," in Proceedings of International Conference on Intelligent Cyber-Physical Systems: ICPS 2021, pp. 233–245, Springer, 2022.

[12] J. A. Lopes, B. Baptista, N. Lavado, and M. Mendes, "Offline handwritten signature verification using deep neural networks," Energies, vol. 15, no. 20, p. 7611, 2022.

[13] ICDAR-2011-Signature-Verification-Competition "http://www.iaprtc11.org/mediawiki/index.php/ICDAR_2011_Signature _Verification_Competition_(SigComp2011)."

[14] "CEDAR-Dataset — kaggle.com." https://www.kaggle.com/datasets/ shreelakshmigp/cedardataset?rvi=1."

[15] Understanding VGG16: Concepts, Architecture, and Performance — data- gen.tech." https://datagen.tech/guides/computer-vision/vgg16/."

979-8-3503-8083-5/23 $31.00 © 2023 IEEE

Efficient CNN Hardware Architecture Based on Linear Approximation and Computation Reuse Technique

Mohammed F. Tolba*, Hani Saleh*, Mahmoud Al-Qutayri*, Ayman Hroub [†], and Thanos Stouraitis *

* SoC Center, Khalifa University, Abu Dhabi, UAE, P.O. Box 127788, Abu Dhabi, UAE
[†]Department of Electrical and Computer Engineering, Birzeit University, Birzeit, Ramallah, Palestine

Abstract—Large deep neural network (DNN) models pose significant computational and memory challenges, particularly when deploying them on edge devices. To address this, techniques such as pruning, quantization, data sparsity, and data reuse have been applied to DNNs, mitigating memory and computational complexity at the cost of some accuracy loss. This paper introduces an efficient hardware accelerator tailored for Convolutional Neural Networks (CNNs). The proposed architecture is the result of a co-optimized approach encompassing both algorithms and hardware. It leverages linear approximation of pre-trained network weights with minimal accuracy loss. A novel computational reuse method is presented to curtail the number of multiplication and addition operations and memory accesses, seamlessly integrated into the dedicated elements within the CNN design. To validate the effectiveness of this architecture, we conducted experiments on a gem5-based RISCV simulator, employing the VGG16 model for the CIFAR 100 dataset and the AlexNet model for the TinyImageNet dataset. The results showcased an impressive speedup of approximately $2\times$ on AlexNet compared to the reference model. Additionally, our proposed CNN design was successfully implemented on the Xilinx Kintex 7 Field Programmable Gate Array (FPGA), achieving a notable reduction in hardware resource utilization compared to prior research efforts. This work serves as a versatile framework for evaluating diverse trade-offs involving accuracy, latency, power consumption, and cost across different CNN architectures.

Index Terms—Deep neural network, Hardware acceleration, computational reuse, approximate computing, AI accelerator

I. INTRODUCTION

Convolutional Neural Networks (CNNs) play a pivotal role in various domains, including image recognition [1], speech recognition [2], and gaming [3]. However, their power comes at a cost, demanding billions of operations for high accuracy. This poses challenges for computing systems, particularly in edge computing, where speed and energy efficiency matter most [4]. For instance, VGGNet [5] uses about 46 billion multiply-accumulate (MAC) operations, typically run on high-capacity GPUs. These computational demands strain hardware resources, making it tough to maintain accuracy and real-time performance, notably on edge devices [6]–[8]. CNNs have transformed various fields, but their resource-heavy nature requires innovative solutions to balance accuracy with edge computing constraints.

Numerous processors have been specifically designed to accelerate CNN processing. Many of these processors focus on architectural innovations aimed at reducing data movement [9], [10], while others employ low-power design techniques to minimize power consumption [11]. There are also processors that leverage weight sparsity to optimize the computation of nonzero weights and activations, thereby speeding up CNN operations [12]. However, a significant opportunity for optimization lies in exploiting the large number of repetitive multiplications inherent in approximate computing techniques, which has not been fully tapped into by these processors. Moreover, it's worth noting that most processors dedicated to CNNs adhere to the layer-by-layer approach, which necessitates off-chip memory access. For larger CNN models, this reliance on off-chip memory access results in substantially higher energy consumption compared to arithmetic or SRAM operations, often exceeding a factor of 1000 [4]. Consequently, the compression of CNN models has emerged as a valuable strategy to reduce their size. This compression can be achieved through various techniques, including weight quantization [13], weight pruning [4], sparsity regularization [12], weight clustering [14], low-rank approximation [15], and weights approximation [16], [17], among others.

The linear approximation method stands out as a valuable approach for reducing CNN weights while minimally impacting accuracy. This method is applied during backpropagation for each batch, helping to recover any accuracy loss incurred. In the linear approximation method, all weights are approximated in the form of a first-degree polynomial:

$$f(x) = c_0 + c_1 x \qquad (1)$$

Here, c_0 and c_1 represent the polynomial coefficients. It's worth noting that increasing the degree of the polynomial can decrease the error between the actual and approximated values but comes at the cost of heightened hardware complexity. In the work presented in [16], the authors introduced a method that linearly approximate each filter row and replacing it with linear weights (LW). This approach offers the advantage of reusing sub-computations within the same filter and input feature map (ifmap), albeit without hardware implementation. Therefore, this work proposes a design and a hardware architecture that supports the linear approximation which requires less hardware utilization. In particular, the architecture can efficiently eliminate the redundant operations caused by the

979-8-3503-8083-5/23 $31.00 © 2023 IEEE

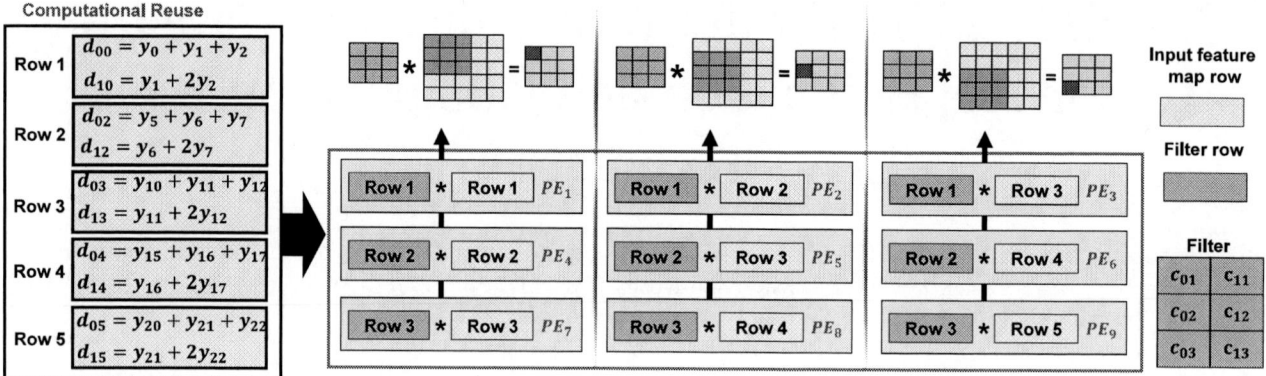

Fig. 1: An example for the proposed CNNs architecture, where filter reuse, input feature map reuse and computational reuse are achieved. $(y_0, y_1, ...)$ represent the input feature map and $d_{00}, d_{01}, ...$ depict the subcomputation that reused across the Processing Element Array (PEA). Each PE requires two adders and two multipliers.

repeated linear coefficients and exploit the computation reuse technique.

The design of efficient hardware for CNNs has always been a big challenge for resource-constrained devices such as Internet-of-Things (IoT) [8], [18], [19]. This is due to the cost of the high computational complexity of CNNs. Accordingly, different methods are used to improve energy efficiency and throughput with minimum impact on the accuracy or increasing hardware cost. This work proposes an Efficient CNN Accelerator designed to support the linear approximation technique, which requires less hardware utilization. In particular, the architecture can efficiently eliminate the redundancy operations caused by repeated linear coefficients and use the computation reuse technique.

The remainder of this paper unfolds as follows: In Section II, we delve into the intricacies of the proposed hardware architecture. Section III offers a comprehensive demonstration of the effectiveness of our architecture, supported by various experiments. Finally, Section IV concludes the paper.

II. PROPOSED HARDWARE ARCHITECTURE

Approximating the CNN weight lineally inside the filter causes a repetition of the multiplication for each multiply and accumulates in the same convolution operation. The algorithm was first introduced in [16] without a hardware implementation. Figure 1 illustrates an example of generating output partial sums for a CNN architecture featuring a 5×5 input feature map, a 3×3 filter, and PEAs (Processing Element Arrays) of size 3×3 based on [16]. The left-hand side of Figure 1 depicts the computational reuse block, which outlines the operations required for each row of the input feature map. These operations are diagonally reused across the PEA. Similar to Eyeriss [9], all weights are reused across PEs horizontally, and the partial sums are accumulated vertically. The number of PE columns and rows equal the number of output feature map rows and the number of filter rows, respectively.

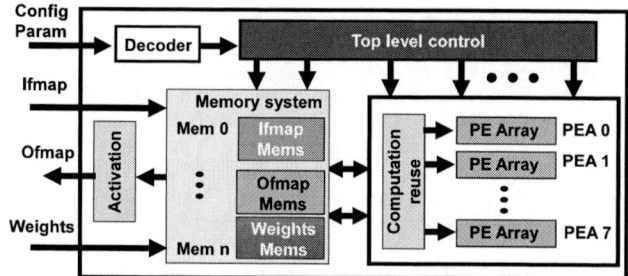

Fig. 2: Proposed CNN design top-level architecture, which consists of 8 PEAs, each 3×6 and each PE has two multipliers and two adders. In addition to 12 memory banks for the weights (128 KB), 32 memory banks (192 KB) are used for ifmaps and ofmaps.

The top-level architecture of the proposed CNN design is shown in Fig. 2, which is built for approximating the CNN linearly within the filter. The architecture mainly consists of the memory system, control block, activation, and PEAs, as shown in Figure 1. Configuration parameters are first loaded into the decoder during initialization to define the access pattern in the memory system and computation in the PEAs. The bit width of ifmaps and weights is chosen as 8 bits to maintain similar accuracy to the original one. The FSM (finite state machine) control block generates the PEAs' control signals and addresses needed for ifmaps, ofmaps, and weights memories. The designed PEAs support CNN models with 3×3 kernel sizes and relu activation. The PEAs comprise 144 PEs, divided into 8 PE array groups (PEA 0-7), each containing 3×6 PEs (18 PEs). To reduce memory access, row stationery (RS) dataflow is used [9]. The weights are reused over PEs horizontally, and input feature map values are reused diagonally. To reduce the data movement of Psums, all Psums are accumulated inside each PE and then accumulated vertically to compute the final Psums.

979-8-3503-8083-5/23 $31.00 © 2023 IEEE

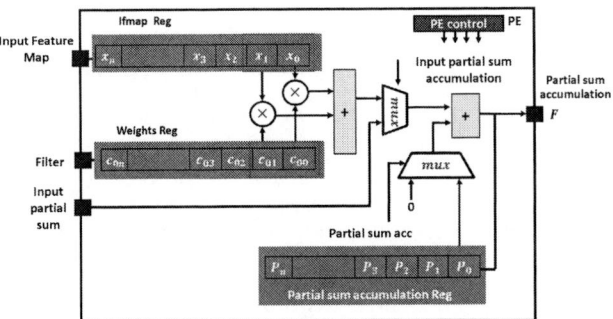

Fig. 3: Processing Element (PE) architecture, which consists of three registers, three multipliers, three adders, and two multiplexers.

A computation reuse block on the left-hand side of the PEA contains the operations needed for each input feature map row. This operation is a common factor among all the PEs that share the same ifmap row, and these operations are reused diagonally for multiple PEs. Each PE requires two multipliers and two adders as shown in Fig. 3. Moreover, 12 memory banks for the weights (128 KB) and 32 memory banks (192 KB) of ofmap and ifmap are used. The weights memory is reduced, as the number of weights in this design equals $2/3$ of the baseline weights.

Initially, all input feature maps (ifmaps) and weights are loaded and stored within dedicated ifmap and weight memories. Once this loading phase is completed, the necessary ifmaps and weights are fetched from the primary memory and subsequently stored within the registers of Processing Element Arrays (PEAs), as depicted in Figure 3. Each Processing Element (PE) comprises three registers: ifmap, Psum, and weight registers, accompanied by two multipliers and two adders. With all ifmaps and weights now residing in these registers, all PEs are activated to perform the element-wise multiplication of ifmaps with weights, with the results being stored in the Psum register. Upon completion of the computation for the first ifmap, the subsequent ifmap is loaded, and its convolutional output is accumulated with the previous Psum, subsequently stored in the Psum register. This sequential process is applied iteratively for all ifmaps, resulting in the accumulation of Psum values within each PE.

III. RESULTS AND DISCUSSION

To validate the efficacy of the proposed architecture, we selected two prominent CNN models, namely AlexNet and VGG16, as our benchmark standards. These models were trained and quantized in TensorFlow using diverse datasets, including TinyImagNet and CIFAR 100. VGG16 is celebrated for its architectural simplicity and consistency, relying on 3×3 convolutional filters across the network. In contrast, AlexNet comprises eight layers, featuring five convolutional layers and three fully connected layers, resulting in a substantial number of parameters.

TABLE I: Gem5 RISCV comparison for VGG16 on CIFAR 100, where G = Giga = 1 billion=10^9 and J is Joule.

	Execution time (s)	Energy (J)	CPU cycles (G)	Number of bytes read from Memory (G)	accuracy loss (%)
Ref	10.12	3.9	20.3	89.9	0
Proposed work	5.93	2.3	11.9	58	0.9

To assess the efficiency of our proposed architecture, we undertook C-based implementations for both VGG16 and AlexNet. Additionally, we employed fixed-point quantization to further curtail memory storage requirements. In this work, we adopted an 8-bit fixed-point representation for both the proposed architecture and the reference models. It is important to note that all benchmark models were consistently implemented using the same C-based architecture.

Given the increasing popularity of the open-source RISC-V instruction set architecture, particularly in deep learning applications due to its vector extensions, we cross-compiled the C code for our proposed algorithm into the RISC-V architecture using the GCC cross-compiler. Subsequently, we leveraged the Gem5 simulator for code execution [20]. In our specific setup, we configured Gem5 for the RISC-V architecture, establishing a straightforward single-core machine configuration comprising a 64 cache line size, AtomicSimpleCPU, 64 KB L1 cache, 1 MB L2 cache, 1 GB off-chip memory, and a CPU clock frequency of 1 GHz.

The proposed architecture was subjected to a comprehensive evaluation using standard CNN benchmark models, considering key performance metrics such as execution time, speedup, energy consumption, and CPU cycles. In Table I, the performance of the VGG16 model on CIFAR 100 is compared, revealing a significant reduction in the number of CPU cycles from 20.3 billion to 11.9 billion compared to the reference model. This improvement results in a noteworthy $1.7\times$ speedup, albeit with a minor accuracy loss of 0.9%. Furthermore, the proposed approach substantially reduces the volume of bytes read from memory, decreasing it from 98.9 billion to 58 billion, while also lowering the overall energy consumption from 3.9 Joules to 2.3 Joules.

In a similar vein, the comparison of the AlexNet model on TinyImagNet is presented in Table II. Notably, the proposed architecture achieves a substantial reduction in total energy consumption, decreasing it from 16.5 Joules to 8.1 Joules, alongside a notable reduction in CPU cycles with around $2\times$ speed-up, from 85.8 billion to 42.1 billion. These improvements are achieved with a modest 3% accuracy loss.

To further demonstrate the performance and efficacy of our proposed CNN designs, we took the initiative to implement them on Xilinx FPGA device. The implementation of this CNN design was carried out utilizing the Verilog Hardware Description Language (HDL) and subjected to simulation using Xilinx Vivado. Initially, we obtained results through MATLAB and Python software, and subsequently, we compared

979-8-3503-8083-5/23 $31.00 © 2023 IEEE

TABLE II: Gem5 RISCV comparison for AlexNet on TinyImagNet, where G = Giga = 1 billion=10^9 and J is Joule.

	Execution time (s)	Energy (J)	CPU cycles (G)	Number of bytes read from Memory (G)	accuracy loss (%)
Ref	42.9	16.5	85.8	424.6	0
Proposed work	21.1	8.1	42.1	204.9	3

TABLE III: FPGA hardware resources comparison between proposed CNN design and previous works.

	FPGA Platform	Frequency (MHz)	LUTs	BRAM	Power (W)
Proposed design	Xilinx Kintex 7	171	144.8 k	512	2.6
[21]	VX980T	150	NA	1492	12.39

these outcomes with those acquired from the FPGA implementation. In Table III, we present a comparative analysis between our proposed CNN design and previous implementations on FPGA. It's worth acknowledging that variations may exist in the FPGA platform and operating frequencies for each accelerator. However, the table unmistakably showcases that our CNN design stands out with the lowest power consumption and the least demand for FPGA resources.

IV. CONCLUSIONS

In summary, this paper presents a compelling solution in the form of an efficient hardware accelerator tailored specifically for Convolutional Neural Networks (CNNs). Our approach stands out as a testament to the synergy achieved through co-optimizing both algorithms and hardware components. To empirically validate the prowess of our architecture, a series of experiments were meticulously conducted on a gem5-based RISCV simulator. These experiments entailed the deployment of the VGG16 model with the CIFAR 100 dataset and the AlexNet model with the TinyImageNet dataset. The results are nothing short of impressive, with AlexNet exhibiting a remarkable speedup of approximately $2\times$ when compared to the reference model. In addition to simulation-based validation, we translated our proposed CNN design into practical implementation on the Xilinx Kintex 7 Field Programmable Gate Array (FPGA). This endeavor yielded tangible results, as we achieved a reduction in hardware resource utilization compared to previous research efforts.

V. ACKNOWLEDGMENT

This publication is based upon work supported by the [CIRA-2020-053] fund and the System-On-Chip center at Khalifa University of Science, Technology under Award No. [RC2-2018-020] and SRC project AIHW-2983.

REFERENCES

[1] K. He, X. Zhang, S. Ren, and J. Sun, "Deep residual learning for image recognition," in *Proceedings of the IEEE conference on computer vision and pattern recognition*, 2016, pp. 770–778.

[2] G. E. Dahl, D. Yu, L. Deng, and A. Acero, "Context-dependent pretrained deep neural networks for large-vocabulary speech recognition," *IEEE Transactions on audio, speech, and language processing*, vol. 20, no. 1, pp. 30–42, 2011.

[3] V. Mnih, K. Kavukcuoglu, D. Silver, A. Graves, I. Antonoglou, D. Wierstra, and M. Riedmiller, "Playing atari with deep reinforcement learning," *arXiv preprint arXiv:1312.5602*, 2013.

[4] X. Ma, S. Lin, S. Ye, Z. He, L. Zhang, G. Yuan, S. H. Tan, Z. Li, D. Fan, X. Qian *et al.*, "Non-structured dnn weight pruning–is it beneficial in any platform?" *IEEE Transactions on Neural Networks and Learning Systems*, 2021.

[5] K. Simonyan and A. Zisserman, "Very deep convolutional networks for large-scale image recognition," *arXiv preprint arXiv:1409.1556*, 2014.

[6] H. Mo, W. Zhu, W. Hu, G. Wang, Q. Li, A. Li, S. Yin, S. Wei, and L. Liu, "9.2 a 28nm 12.1 tops/w dual-mode cnn processor using effective-weight-based convolution and error-compensation-based prediction," in *2021 IEEE International Solid-State Circuits Conference (ISSCC)*, vol. 64. IEEE, 2021, pp. 146–148.

[7] V. Sakellariou, V. Paliouras, I. Kouretas, H. Saleh, and T. Stouraitis, "A multiplier-free rns-based cnn accelerator exploiting bit-level sparsity," *IEEE Transactions on Emerging Topics in Computing*, 2023.

[8] H. Mo, W. Zhu, W. Hu, Q. Li, A. Li, S. Yin, S. Wei, and L. Liu, "A 12.1 tops/w quantized network acceleration processor with effective-weight-based convolution and error-compensation-based prediction," *IEEE Journal of Solid-State Circuits*, vol. 57, no. 5, pp. 1542–1557, 2021.

[9] Y.-H. Chen, T. Krishna, J. S. Emer, and V. Sze, "Eyeriss: An energy-efficient reconfigurable accelerator for deep convolutional neural networks," *IEEE journal of solid-state circuits*, vol. 52, no. 1, pp. 127–138, 2016.

[10] W. Lu, G. Yan, J. Li, S. Gong, Y. Han, and X. Li, "Flexflow: A flexible dataflow accelerator architecture for convolutional neural networks," in *2017 IEEE International Symposium on High Performance Computer Architecture (HPCA)*. IEEE, 2017, pp. 553–564.

[11] S. Choi, J. Lee, K. Lee, and H.-J. Yoo, "A 9.02 mw cnn-stereo-based real-time 3d hand-gesture recognition processor for smart mobile devices," in *2018 IEEE International Solid-State Circuits Conference-(ISSCC)*. IEEE, 2018, pp. 220–222.

[12] W. Wen, C. Wu, Y. Wang, Y. Chen, and H. Li, "Learning structured sparsity in deep neural networks," *Advances in neural information processing systems*, vol. 29, pp. 2074–2082, 2016.

[13] M. F. Tolba, H. Saleh, M. Al-Qutayri, and B. Mohammad, "Reduce computing complexity of deep neural networks through weight scaling," in *2022 IEEE International Symposium on Circuits and Systems (ISCAS)*. IEEE, 2022, pp. 1249–1253.

[14] E. Park, J. Ahn, and S. Yoo, "Weighted-entropy-based quantization for deep neural networks," in *Proceedings of the IEEE Conference on Computer Vision and Pattern Recognition*, 2017, pp. 5456–5464.

[15] E. L. Denton, W. Zaremba, J. Bruna, Y. LeCun, and R. Fergus, "Exploiting linear structure within convolutional networks for efficient evaluation," in *Advances in neural information processing systems*, 2014, pp. 1269–1277.

[16] M. F. Tolba, H. T. Tesfai, H. Saleh, B. Mohammad, and M. Al-Qutayri, "Deep neural networks-based weight approximation and computation reuse for 2-d image classification," *IEEE Access*, vol. 10, pp. 41 551–41 563, 2022.

[17] M. F. Tolba, H. Saleh, B. Mohammad, M. Al-Qutayri, and T. Stouraitis, "Eacnn: Efficient cnn accelerator utilizing linear approximation and computation reuse," in *2023 IEEE International Symposium on Circuits and Systems (ISCAS)*. IEEE, 2023, pp. 1–5.

[18] E. Hassan, Y. Halawani, B. Mohammad, and H. Saleh, "Hyperdimensional computing challenges and opportunities for ai applications," *IEEE Access*, vol. 10, pp. 97 651–97 664, 2021.

[19] S. Yin, P. Ouyang, J. Yang, T. Lu, X. Li, L. Liu, and S. Wei, "An energy-efficient reconfigurable processor for binary-and ternary-weight neural networks with flexible data bit width," *IEEE Journal of Solid-State Circuits*, vol. 54, no. 4, pp. 1120–1136, 2018.

[20] N. Binkert, B. Beckmann, G. Black, S. Reinhardt, A. Saidi, A. Basu, J. Hestness, D. Hower, T. Krishna, S. Sardashti *et al.*, "The gem5 simulator. acm sigarch comput. archit. news 39 (2), 1 (2011)," 2020.

[21] W. Huang, H. Wu, Q. Chen, C. Luo, S. Zeng, T. Li, and Y. Huang, "Fpga-based high-throughput cnn hardware accelerator with high computing resource utilization ratio," *IEEE Transactions on Neural Networks and Learning Systems*, vol. 33, no. 8, pp. 4069–4083, 2021.

A CMOS Analog Neuron Circuit with A Multi-Level Memory

Melvin D. Edwards II, *Student Member, IEEE*, Nabil J. Sarhan, *Member, IEEE*, Mohammad Alhawari, *Member, IEEE*

Abstract—This paper presents a CMOS-based analog neuron circuit that utilizes a multi-level analog memory that is useful for mixed signal neural networks. The implementation of neural networks in the analog/mixed signal domain is crucial for edge applications of Artificial Intelligence (AI). The proposed circuit is able to generate both positive and negative output Multiply and Accumulate (MAC) currents by utilizing a bipolar supply. The multi-level analog memory is able to generate eight distinct analog voltages to drive the analog neuron circuit. Simulation results show that the analog neuron circuit is able to perform linear addition over the entire MAC current range. The MAC current per input-weight pair is between $\pm 25\mu A$ to $\pm 56\mu A$. The analog neuron circuit is able to generate 0A for zero-weight or zero-input. Each input-weight pairs consumes $170\mu W$. The circuit is designed and simulated in the 65 nm technology node.

Index Terms—Analog Neuron, Two-Quadrant Analog Neuron, Positive and Negative MAC (Multiply and Accumulate) Currents, Bipolar Supply, Analog Memory Circuit, 3-Bit Memory, Neuron Circuit, Edge Computing, 65 nm technology

I. INTRODUCTION

The integration of Artificial Intelligence (AI) into various devices and systems, particularly through the use of Deep Neural Networks (DNNs), has led to significant advancements in computer vision [1]–[3] and natural language processing [4]–[6]. However, as the Internet of Things (IoT) continues to advance, the need for devices that integrate AI models while optimizing energy usage is growing. Consequently, alternative computing approaches such as analog computation are being investigated to address this demand [7], [8].

Analog computation is particularly well-suited for low-energy and low-latency applications [9], [10], as it allows for energy-efficient Vector Matrix Multiplication (VMM) through Kirchoff's Current Law (KCL) and Ohm's Law. Additionally, In-Memory Computing (IMC) systems utilizing non-conventional memory technologies, such as analog memory or RRAM, can potentially store entire models on-chip [11], [12], eliminating the need for off-chip memory accesses. However, the use of these non-conventional memory technologies introduce new issues that must be taken into account such as CMOS compatibility, durability, sneak path [13], and variation [14] for successful application to the field of DNNs.

The authors are with the Electrical and Computer Engineering Department, Wayne State University, Detroit, USA (email: ev7854@wayne.edu, nabil.sarhan@wayne.edu, alhawari@wayne.edu).

Corresponding author is Melvin D. Edwards II.

This work was supported by the National Science Foundation (NSF) under grant number 2221753.

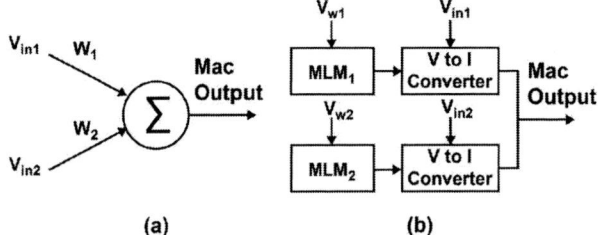

Fig. 1. Perceptron architecture. (a) High-level architecture. (b) Architecture of the proposed perceptron.

The signed MAC operation in the crossbar architecture is generally accomplished with one of two methods. The first method involves performing the signed operation digitally and assigning a negative sign in the digital domain to certain columns. This approach is referred to as the dual array approach to signed MAC operations. The second method, known as the dual row approach, which is implemented in this work, relies on a bipolar supply to perform signed operations in the analog domain. Although the dual array approach can reduce the latency of performing more operations in the analog domain, it is more sensitive to variation in the memory technology.

This paper presents a new design for an analog neuron circuit that utilizes a two-quadrant operation. The proposed circuit uses a recently introduced analog memory technology to store weights [15]–[17]. This memory has been shown to have low-power consumption and robustness to process variations (PVT).

The contribution of this paper is as follows:

- A low-power, analog perceptron with an all-CMOS implementation.
- Dual row implementation of the signed MAC operation.
- Usage of the same analog memory for both the positive and negative weights.
- Simulation results showcasing the operation of the analog neuron.

The remainder of this paper is structured as follows. Section II discusses the perceptron circuit design and the I/O signals from the perceptron. Section III presents the analytic results of the proposed neuron. Section IV discusses the simulation results of the proposed neuron. Section V concludes the paper.

979-8-3503-8083-5/23 $31.00 © 2023 IEEE

II. PROPOSED ANALOG NEURON CIRCUIT DESIGN

The focus of this work is on inference operations. The overall perceptron architecture is shown in Figure 1. This work utilizes a $\pm 1.2V$ supply. The architecture of this circuit is modular and extra input-weight pairs can be added as needed for any particular layer in the overall DNN by adding another Multi-Level Memory (MLM) and V-to-I converter as shown in Figure 1(b). Each input-weight pair consists of an Analog Multi-Level Memory (MLM) and a current summing architecture (V to I Converter). The current summer is connected to the shared MAC node which could be fed into a trans-impedance amplifier in system-level implementations of this neuron.

A. Analog Multi-Level Memory

Figure 2 shows the analog memory in this work, Figure 2(a), takes an analog input and quantizes that input to one of eight distinct memory levels which produces a stable voltage that drives the current summer. The analog MLM has two input signals and one output signal as shown in Figure 2(b). This memory is described in detail in [15]–[17].

B. Neuron Architecture

As shown in Figure 3, the neuron uses a bipolar supply $+V_{in}$ and $-V_{in}$ to generate the signed MAC current. The MAC node is grounded. This circuit has two pairs of transmission gates: TG1/TG2 which control zero versus non-zero weights and TG3/TG4 which control the sign of the MAC operation. There is a read transistor, M_{read}, which controls the V-to-I operation. There is a current mirroring structure that allows the current to be steered into the MAC node or out of the MAC node depending on the state of the V_{sel} pin.

The circuit in Figure 3 has two inputs $\pm V_{in}$ and $V_{w,q}$. $\pm V_{in}$ is the input activations in 1-bit bipolar binary form and $V_{w,q}$ is the weight which is coming from the Analog Multi-Level Memory. TG1 and TG2 control a zero weight versus non-zero weight. If V_{zs} is HIGH, TG1 is on and TG2 is off, voltage $V_{w,q}$ is allowed to through to the gate of M_{read} for a non-zero weight. If V_{zs} is LOW, TG1 is off, TG2 is on, the gate of M_{read} is connected to V_{in}, effectively making the gate-source voltage of M_{read} zero resulting in a zero weight.

The transistor M_{read} acts as a current source for the current steering architecture. M_{mirror} acts as a mirror for both M_1 and M_2. M_1 is the current source for the negative MAC current path while M_2 is the current source for the positive MAC current path. The positive MAC current path includes M_2, TG4, M_3, and M_4. The negative MAC current path includes M_1 and TG3. The positive MAC current path is enabled when TG3 is on and TG4 is off, this occurs when V_{sel} is HIGH. The negative MAC current path is enabled when TG3 is off and TG4 is on, this occurs when V_{sel} is LOW. M_3 and M_4 are added to allow for current to be sourced by M_4 for the positive MAC while M_1 acts as a current sink for the negative MAC.

Matching between positive and negative currents is a key design parameter. The unsuitability of NMOS and PMOS

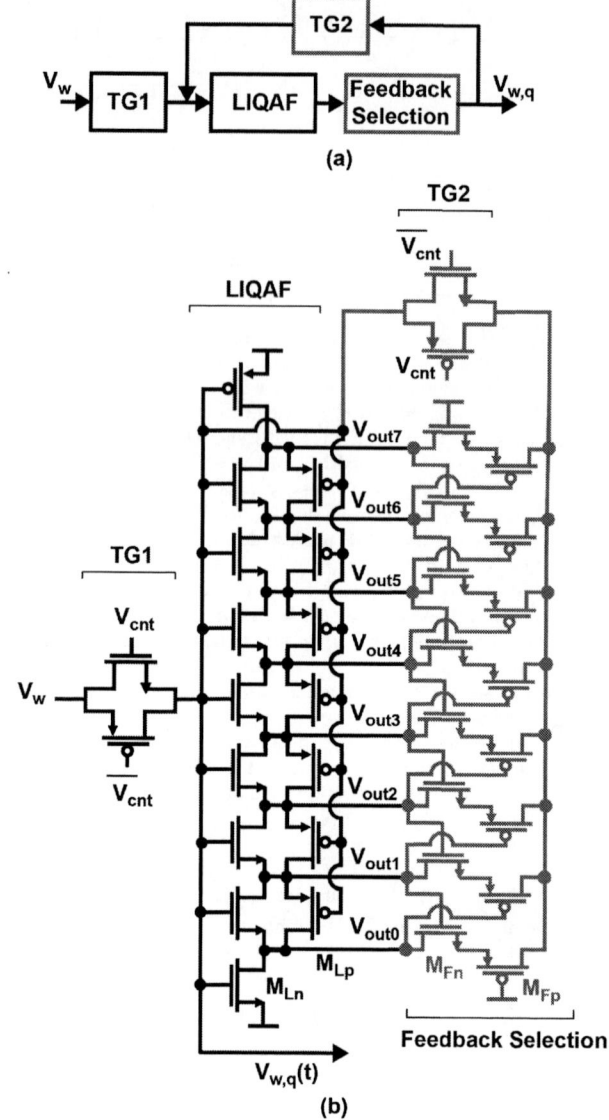

Fig. 2. (a). High-level block diagram of the analog memory. (b). transistor-level schematic of the analog memory.

transistors for the roles of current sinking and sourcing, respectively, arises from their opposite response to gate voltage modulation. Specifically, the overdrive of a PMOS transistor exhibits a decreasing trend as the gate voltage is increased. In contrast, an NMOS transistor demonstrates an increasing trend as the gate voltage is increased. Ideally, the current will be equal in magnitude and have opposite signs for all voltages of $V_{w,q}$. There is some mismatch in the currents from the positive and negative MAC operation. This non-ideality comes from the difference in current gain from the positive and negative current sources in the architecture. More details will be provided in the next section.

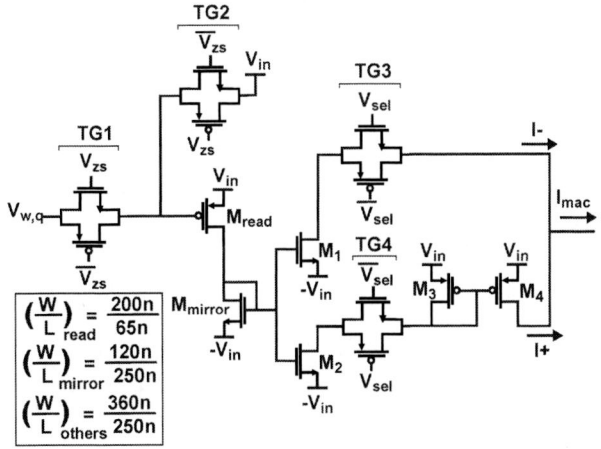

Fig. 3. Current Summing Architecture.

III. ANALYTIC RESULTS

The read transistor, M_{read}, converts the weight voltage into a current. The read transistor is biased to operate in the saturation region over the range of voltages that the analog multi-level memory outputs. The transconductance of the read transistor roughly follows that of the following equation.

$$g_m = \mu_n C_{ox} \frac{W}{L}(V_{SG} - |V_{th}|) \qquad (1)$$

The read transistor has a different gain for different weight values due to the over-drive voltage being modulated by the output of the analog memory's output. This is not an issue for two reasons. First, the architecture does not change weight values in run-time, weights are configured at the beginning of operation and remain until new weights are loaded into the inference machine. Second, this architecture is designed for 1-bit input activation which results in a discrete output of eight different levels or an output of zero in the case of zero input or zero weight.

The read transistor forms a current source for the central biasing branch of the current steering architecture shown in Figure 3. The read transistor expresses differences in weight voltages as a change in the over-drive voltage of the read transistor. Due to velocity saturation, there is a linear relationship between gate-source voltage and drain current in the saturation region. This results in a variable current source that is dependent on the weight voltage while the input voltage is binary. The current in the read transistor roughly follows the following equation.

$$I_D = \frac{1}{2}\mu_n C_{ox} \frac{W}{L}(V_{SG} - |V_{th}|)^2 \qquad (2)$$

The read transistor, M_{read}, is a current source to the current mirroring structure. The current mirroring structure takes this current and replicates it in the two signed branches connected to the drains of M_1 and M_2. Equation 2 explains the downward trend in the MAC current as the weight voltage is increased. As the weight voltage decreases, the voltage V_{SG} increases since

the source is fixed at V_{DD}. As the gate voltage decreases the over-drive increases which results in a larger current at the central biasing branch.

The current at the signed branches are roughly the following equations for transistors M_1 and M_2.

$$I_- = \frac{W_1}{W_{mirror}} I_{read} \qquad (3)$$

$$I_{M2} = \frac{W_2}{W_{mirror}} I_{read} \qquad (4)$$

In equation 4, I_{read} is being generated from M_{read}. I_- is being fed directly to the MAC node since it is sinking current from the node. The positive MAC node must undergo another conversion to convert from sinking current to sourcing current into the node. This equation is defined by the equation shown below.

$$I_+ = \frac{W_2 W_4}{W_{mirror} W_3} I_{read} \qquad (5)$$

Based on equations 3 and 5 sizing of transistors W_1 and W_2 must be equal and transistors W_3 and W_4 must be equal to ensure matching between the negative and positive MAC operation. Another consideration is channel length modulation which needs to be taken into account in design with cascoding techniques or increasing the transistor channel length. The length for this design was chosen to be 250 nm to minimize the effect of channel modulation on the mirroring operations.

IV. SIMULATION RESULTS

This section goes over the simulation results of the dual row MAC operation. Figure 4 shows the MAC current when $V_{in,1}$ is positive and $V_{in,2}$ is zero. The maximum current is 56 μA and the minimum current is 25 μA. The stairwell characteristic comes from the analog memory which is driving the current steering architecture. The analog memory outputs a voltage in the 0 to V_{DD} range with the analog outputs centered on $0.5V_{DD}$.

Figure 5 shows the output characteristic stairwell when both $V_{in,1}$ and $V_{in,2}$ are positive. The current summing action of the MAC architecture can be seen in this figure, where the maximum MAC current is 112 μA and the minimum current is 50 μA. This is due to both the inputs being positive and having the same weight applied to them. In actual operation, the two inputs do not need to have the same weight as each input/weight pair has a different analog memory to drive the current steering architecture.

Figure 6 depicts the MAC current when $V_{in,1}$ is negative and $V_{in,2}$ is zero. The maximum current is -25 μA and the minimum current is -58 μA. It should be noted that Figures 4 and 6 should have equal magnitude and opposite sign if perfect matching between the positive and negative path is accomplished based on section III. The difference between Figures 4 and 6 leads to Figure 7. Figure 7 quantifies the mismatch between the positive and negative paths. The maximum mismatch is at the two endpoint weight values which corresponds to 1.5 μA in each case. The mismatch of 1.5 μA is very small compared to the minimum non-zero

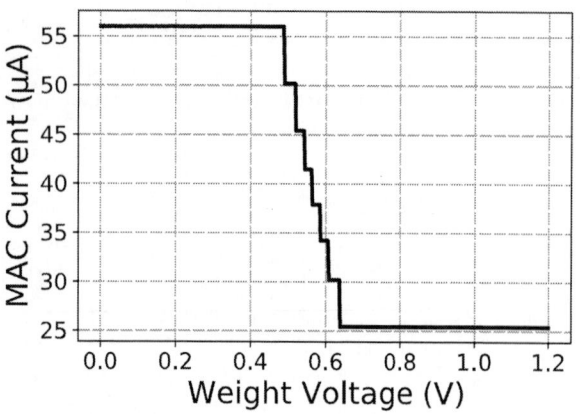

Fig. 4. Output MAC Currents when $V_{in,1}$ is positive and $V_{in,2}$ is zero.

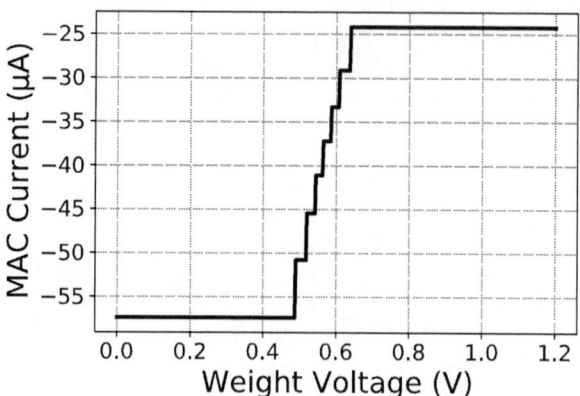

Fig. 6. Output MAC Currents when $V_{in,1}$ is negative and $V_{in,2}$ is zero.

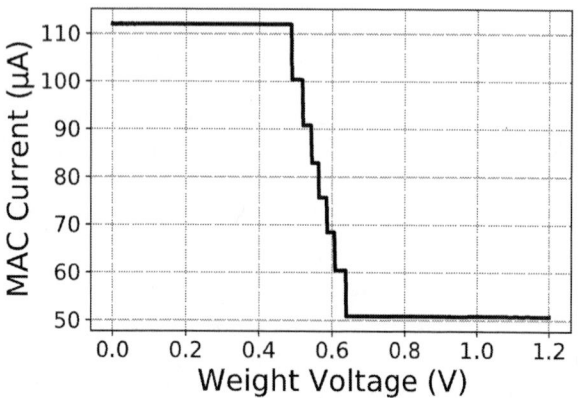

Fig. 5. Output MAC Currents when $V_{in,1}$ is positive and $V_{in,2}$ is positive.

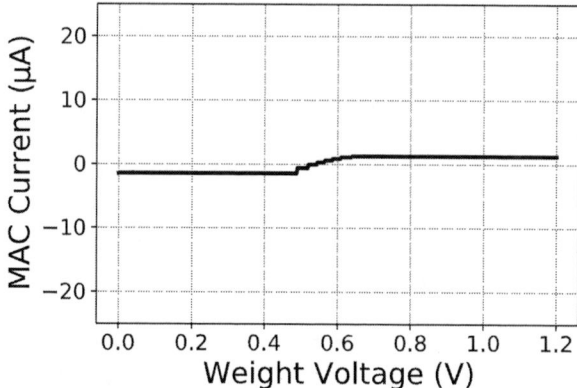

Fig. 7. Output MAC Currents when $V_{in,1}$ is positive and $V_{in,2}$ is negative with equal magnitude weights on both inputs.

current of 25 μA. This can effectively be considered zero in downstream operations where a trans-impedance amplifier (TIA) would amplify the current signal generated by this stage.

Figure 8 represents the actual MAC current in the time domain as a weight is programmed into the memory and held. In Figure 8 the programming of the memory happens from 0 s to 50 μs, this is outside the scope of this paper so the programming stage will not be discussed as the focus is on the inference stage of the MAC operation. Once the weight is chosen and the input voltage is input to the current summing circuit, the current is held indefinitely. The input to this architecture is one-bit, so it is either zero or a non-zero value determined by the analog memory.

Each input-weight pair consumes 170 μW. The circuit shown in this work consumes a maximum of 340 μW when both input-weight pairs are active.

V. CONCLUSION

The modular nature of the proposed analog neuron circuit enables the creation of neural networks with layers of varying

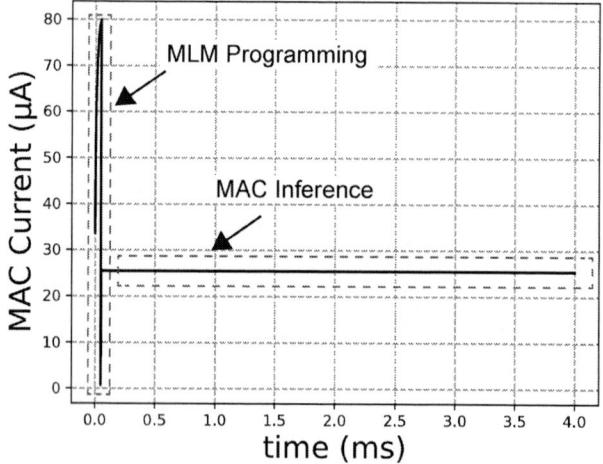

Fig. 8. Output MAC Current when $V_{in,1}$ is positive and $V_{in,2}$ is zero. A 1.2V weight is applied.

sizes. If a larger or smaller neural network is needed, input-weight pairs can be added or removed from the MAC node in the same manner as the first two input-weight pairs shown in this work. This level of flexibility enables this architecture to be used in neural networks that require varying numbers of connections for neurons, depending on the specific layer of the network. The modular and flexible nature of this block is a key feature that makes it an extremely versatile and valuable tool for building neural networks of all types and sizes.

REFERENCES

[1] Y. Liu, J. He, J. Gu, X. Kong, Y. Qiao, and C. Dong, "Degae: A new pre-training paradigm for low-level vision," in *2023 IEEE/CVF Conference on Computer Vision and Pattern Recognition (CVPR)*, pp. 23292–23303, 2023.

[2] F. Sammani, T. Mukherjee, and N. Deligiannis, "Nlx-gpt: A model for natural language explanations in vision and vision-language tasks," in *2022 IEEE/CVF Conference on Computer Vision and Pattern Recognition (CVPR)*, pp. 8312–8322, 2022.

[3] Y. Wang, X. Chen, L. Cao, W. Huang, F. Sun, and Y. Wang, "Multimodal token fusion for vision transformers," in *2022 IEEE/CVF Conference on Computer Vision and Pattern Recognition (CVPR)*, pp. 12176–12185, 2022.

[4] A. Soni, B. Amrhein, M. Baucum, E. J. Paek, and A. Khojandi, "Using verb fluency, natural language processing, and machine learning to detect alzheimer's disease," in *2021 43rd Annual International Conference of the IEEE Engineering in Medicine Biology Society (EMBC)*, pp. 2282–2285, 2021.

[5] M. Guo, Y. Chen, J. Xu, and Y. Zhang, "Dynamic knowledge integration for natural language inference," in *2022 4th International Conference on Natural Language Processing (ICNLP)*, pp. 360–364, 2022.

[6] S. Das, M. Ashrafuzzaman, F. T. Sheldon, and S. Shiva, "Network intrusion detection using natural language processing and ensemble machine learning," in *2020 IEEE Symposium Series on Computational Intelligence (SSCI)*, pp. 829–835, 2020.

[7] H. Tsai, P. Narayanan, S. Jain, S. Ambrogio, K. Hosokawa, M. Ishii, C. Mackin, C.-T. Chen, A. Okazaki, A. Nomura, I. Boybat, R. Muralidhar, M. M. Frank, T. Yasuda, A. Friz, Y. Kohda, A. Chen, A. Fasoli, M. J. Rasch, S. Woźniak, J. Luquin, V. Narayanan, and G. W. Burr, "Architectures and circuits for analog-memory-based hardware accelerators for deep neural networks (invited)," in *2023 IEEE International Symposium on Circuits and Systems (ISCAS)*, pp. 1–5, 2023.

[8] N. Udayanga, S. I. Hariharan, S. Mandal, L. Belostotski, L. T. Bruton, and A. Madanayake, "Continuous-time algorithms for solving maxwell's equations using analog circuits," in *2020 IEEE International Symposium on Circuits and Systems (ISCAS)*, pp. 1–1, 2020.

[9] M. de Prado, M. Rusci, R. Donze, A. Capotondi, S. Monnerat, L. Benini, and N. Pazos, "Robustifying the deployment of tinyml models for autonomous mini-vehicles," in *2021 IEEE International Symposium on Circuits and Systems (ISCAS)*, pp. 1–5, 2021.

[10] K. Matsubara, L. Hanno, M. Kimura, A. Nakamura, M. Koike, K. Terashima, S. Morikawa, Y. Hotta, T. Irita, S. Mochizuki, H. Hamasaki, and T. Kamei, "4.2 a 12nm autonomous-driving processor with 60.4tops, 13.8tops/w cnn executed by task-separated asil d control," in *2021 IEEE International Solid- State Circuits Conference (ISSCC)*, vol. 64, pp. 56–58, 2021.

[11] Y. He, Y. Huang, J. Yue, W. Sun, L. Zhang, and Y. Liu, "C-rram: A fully input parallel charge-domain rram-based computing-in-memory design with high tolerance for rram variations," in *2022 IEEE International Symposium on Circuits and Systems (ISCAS)*, pp. 3279–3283, 2022.

[12] J. Read, W. Li, and S. Yu, "Enabling long-term robustness in rram-based compute-in-memory edge devices," in *2023 IEEE International Symposium on Circuits and Systems (ISCAS)*, pp. 1–5, 2023.

[13] S. K. Kingra, V. Parmar, S. Negi, S. Khan, B. Hudec, T.-H. Hou, and M. Suri, "Methodology for realizing vmm with binary rram arrays: Experimental demonstration of binarized-adaline using oxram crossbar," in *2020 IEEE International Symposium on Circuits and Systems (ISCAS)*, pp. 1–5, 2020.

[14] J. Yang, X. Xue, X. Xu, Q. Wang, H. Jiang, J. Yu, D. Dong, F. Zhang, H. Lv, and M. Liu, "24.2 a 14nm-finfet 1mb embedded 1t1r rram with a 0.022μm2 cell size using self-adaptive delayed termination and multi-cell reference," in *2021 IEEE International Solid- State Circuits Conference (ISSCC)*, vol. 64, pp. 336–338, 2021.

[15] M. Alhawari, N. Albelooshi, and M. H. Perrott, "A 0.5v ¡4μw cmos photoplethysmographic heart-rate sensor ic based on a non-uniform quantizer," in *2013 IEEE International Solid-State Circuits Conference Digest of Technical Papers*, pp. 384–385, 2013.

[16] M. Alhawari, N. A. Albelooshi, and M. H. Perrott, "A 0.5 v < 4 μw cmos light-to-digital converter based on a nonuniform quantizer for a photoplethysmographic heart-rate sensor," *IEEE Journal of Solid-State Circuits*, vol. 49, no. 1, pp. 271–288, 2014.

[17] M. D. Edwards, H. Al Maharmeh, N. J. Sarhan, M. Ismail, and M. Alhawari, "A low-power, digitally-controlled, multi-stable, cmos analog memory circuit," in *2020 IEEE 63rd International Midwest Symposium on Circuits and Systems (MWSCAS)*, pp. 872–875, 2020.

Deep Image Deraining

Laila Hegazy
Media Engineering and Technology
German University in Cairo
Cairo, Egypt
Laylahegazyy@gmail.com

Shereen Afifi
Media Engineering and Technology
German University in Cairo
Cairo, Egypt
Shereen.moataz@guc.edu.eg

Omar Fahmy
Electrical Engineering Department
Badr University in Cairo
Cairo, Egypt
Omar.fahmy@buc.edu.eg

Abstract—**Image restoration is a crucial area within computer vision, with the goal of recovering high-quality images from degraded ones. To achieve the best outcomes, a delicate balance between intricate spatial details and comprehensive contextual information is essential. This research focuses on image deraining, which is a critical domain in computer vision for recovering high-quality images from degraded observations. The study explores various deraining methods through a comprehensive literature review, highlighting their strengths and weaknesses. The centerpiece is the MPRNet architecture, an innovative approach for image deraining, which is thoroughly studied and examined. The research also proposes a methodology to optimize the MPRNet model's performance and improve the quality of restored images resulting in better results both qualitatively and quantitatively. Overall, the research contributes to the field by introducing a novel approach and optimizing the MPRNet model's performance, paving the way for future advancements in image restoration.**

I. INTRODUCTION

The quality of digital images can be severely impacted by various environmental factors, with rain being one of the most troublesome elements. Rain streaks in images can significantly reduce visibility, making it challenging for both humans and computer vision systems to interpret the scene accurately. This degradation becomes particularly problematic in critical applications such as video surveillance, autonomous driving, and outdoor visual recognition systems, where clear and precise image information is essential for decision-making processes. As a result, there is an urgent need to develop effective image deraining techniques that can address these issues and ensure the reliability and accuracy of visual data.

Despite efforts in the field of image deraining, existing methods [1]–[4] still face limitations in their ability to remove rain streaks while preserving the crucial underlying image details. Some techniques might inadvertently introduce unwanted artifacts or fail to entirely eliminate rain streaks, leading to subpar results. Furthermore, a significant challenge lies in handling false positives during rain streak detection, as many current methods do not adequately address this issue. Consequently, unnecessary alterations in image regions can occur, which further complicates the deraining process and hampers the overall performance of these techniques.

To overcome these challenges and enhance the deraining process, this research turns its attention to the promising MPRNet (Multi-Stage Progressive Image Restoration network) [5].The MPRNet has shown great potential in various image restoration tasks and is now being explored for its effectiveness in image deraining. Leveraging its multi-stage progressive approach,the MPRNet offers a robust framework that aims to tackle rain streak removal while preserving the essential image characteristics. By employing advanced deep learning techniques,the MPRNet can potentially achieve more accurate and reliable results compared to traditional deraining methods. The hope is that by using MPRNet, we can pave the way for improved image deraining techniques that will significantly enhance the performance of computer vision systems in adverse weather conditions and real-world applications.

II. LITERATURE REVIEW

For several years, the field of computer vision and image processing has extensively studied image de-raining techniques. Rain streaks in images can significantly lower image quality and obscure crucial visual details, posing challenges for automated object detection and recognition systems. Consequently, a multitude of de-raining methods have emerged, spanning from traditional signal processing approaches to deep learning-based solutions. In this literature review, we will examine the diverse de-raining techniques proposed in previous research, assess their strengths and limitations, and offer potential insights into future advancements in this area.

CNN-based methods outperform other approaches in image restoration mainly due to their well-designed models. Various network modules and functional units have been developed for this purpose, such as recursive residual learning [6], dilated convolutions [6], attention mechanisms [8]–[10], dense connections [11]–[13], encoder-decoders [14]–[17], and generative models [18]–[23]. However, most of these models for low-level vision problems are designed as single-stage architectures. On the other hand, multi-stage networks [5] have shown better effectiveness in high-level vision tasks.

Recently, there have been attempts to apply the multi-stage design to image deblurring [24]–[26] and image deraining [27], [28]. However, these approaches face certain architectural limitations that hinder their performance. Firstly, existing multi-stage techniques often use the encoder-decoder architecture [25], [26], which effectively encodes broad contextual information but may fail to preserve spatial image details. Alternatively, some models use a single-scale pipeline [28], providing spatially accurate outputs but with less semantic

979-8-3503-8083-5/23 $31.00 © 2023 IEEE

reliability. An effective image restoration requires a combination of both design choices in a multi-stage architecture. Secondly, merely passing the output of one stage to the next stage without careful consideration yields suboptimal results. Thirdly, unlike some approaches [26], it is crucial to provide ground-truth supervision at each stage for progressive restoration. Finally, during multi-stage processing, a mechanism to propagate intermediate features from earlier to later stages is necessary to preserve contextualized features from the encoder-decoder branches.

A. Research Gaps

Deraining images still face several notable research gaps that require attention. Firstly, existing deraining methods often struggle to completely remove all rain streaks from images, leading to residual artifacts that can affect the overall visual quality and usability of the processed images. Secondly, during the rain removal process, there is a risk of losing essential background information, which is particularly critical for outdoor scenes or complex environments. Preserving such details while effectively removing rain remains a challenging task. Thirdly, the availability and quality of deraining datasets pose significant limitations. Insufficient diversity and volume of training data can hinder the generalization and robustness of deraining models to real-world scenarios. Lastly, another critical issue is the generation of blurred output images after rain removal. This occurs due to over-smoothing or the loss of high-frequency information during the deraining process, impacting the sharpness and clarity of the final results. Addressing these research gaps is crucial for advancing deraining techniques and achieving more accurate and visually pleasing results in various applications and environments.

B. Existing MPRNet Model

The MPRNet model was chosen to be enhanced [5] for image restoration due to its impressive performance and potential to address the limitations found in existing multi-stage techniques. The MPRNet has already demonstrated remarkable results in image restoration tasks, leveraging its CNN-based design and incorporating modules like recursive residual learning [6], dilated convolutions [6], attention mechanisms [8]–[10], dense connections [11]–[13], encoder-decoders [14]–[17], and generative models [18]–[22]. By building upon the MPRNet's foundation, I aim to introduce enhancements that overcome the architectural bottlenecks observed in current multi-stage approaches. With its capacity to combine both broad contextual information and preserve spatial image details, the MPRNet provides an excellent starting point to further optimize the restoration process. By implementing ground-truth supervision at each stage and ensuring proper feature propagation between stages, the enhanced MPRNet model will be better equipped to tackle complex image deraining challenges while achieving superior results in comparison to other methods such as DerainNet [3], SEMI [4], DIDMDN [7], UMRL [29], RESCAN [30], PreNet [31] and MSPFN [32].

III. METHODOLOGY

The Multi-Stage Progressive Image Restoration Network (MPRNet) utilized for image deraining used in this research is a comprehensive framework employing a combination of three stages for progressively restoring images, especially focusing on image deraining tasks. This chapter elaborates on the methodology involved in implementing the MPRNet for image deraining, which incorporates Encoder-Decoder Subnetworks, Original Resolution Subnetworks, Cross-Stage Feature Fusion (CSFF), and the Supervised Attention Module (SAM). Fig.1 shows the overall architecture of the MPRNet framework.

A. Proposed Methodology

The MPRNet framework shown in Fig.1 has shown impressive performance in image deraining tasks. However, there is always room for improvement and potential enhancements could further augment its performance.

1) Preprocessing: Preprocessing is a vital step in machine learning pipelines, including for the MPRNet deraining code. To prepare the raw data for input into the model, several preprocessing steps can be applied. First, images should be rescaled to a standard size of 512x512 to ensure consistent input dimensions, but caution should be taken to avoid information loss when downsizing high-resolution images. Normalization is crucial to standardize pixel intensity values, and for MPRNet, the mean and standard deviation values of [0.485, 0.456, 0.406] and [0.229, 0.224, 0.225], respectively, are commonly used. Channel splitting is recommended for color images to process individual color channels separately, potentially aiding the model's learning. Additionally, patch extraction can be used during training to increase the amount of training data and enhance the model's robustness to features' location in the input. These preprocessing techniques contribute to better model performance and stability in the deraining process.

2) Data Augmentation: Data augmentation is a useful technique for increasing the diversity of training data available for model development without the need for additional data collection. Various methods are employed, including random cropping to enhance robustness, flipping and rotation for better generalization, and brightness/contrast adjustment to recognize features under different conditions. Although increasing contrast improved rain detection, increasing saturation and hue did not impact model performance. Adding noise to the images did not show an improvement in performance, indicating the need for careful selection of augmentation methods based on specific task requirements.

3) Loss Functions: We use the Charbonnier loss and Edge loss. While these loss functions have proven effective, integrating other forms of loss could potentially optimize the model's learning process. The addition of a Laplacian gradient loss could be beneficial. This loss function penalizes the deviation of the gradient of the predicted and the target images, thereby preserving the edges and textures in the derained images. Such a loss could complement the existing ones and further improve the quality of the restored images. In addition, the total variation loss could be added as a regularization term to encourage spatial smoothness in the generated derained images

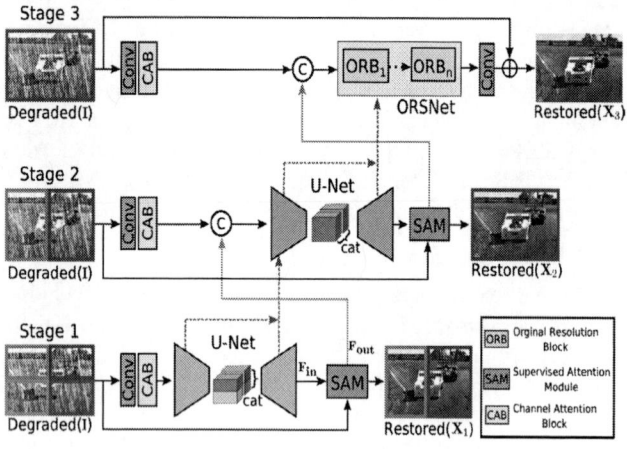

Fig. 1. Overall Framework of MPRNet [5]

IV. RESULTS

The MPRNet model is trained end-to-end, meaning all stages are simultaneously trained without the need for pretraining individual stages. This approach optimizes all stages together, resulting in improved overall performance. The training dataset contains 13,712 clean-rain image pairs collected from diverse datasets, enhancing the model's ability to generalize to unseen data. The Adam optimizer, a popular deep learning optimization algorithm, is used with an initial learning rate of 2×10^{-4}. The learning rate gradually decreases to 1×10^{-6} using cosine annealing, a strategy that smoothly reduces the learning rate across epochs for better model convergence. The Gradient Laplacian Loss was introduced during the model's training to preserve high-frequency details in restored images. This loss function is based on the Laplacian operator, which captures edges and textures in images, crucial for maintaining important high-frequency details.

To implement the Gradient Laplacian Loss, a function was defined within the train.py file. A Laplacian filter (3x3 matrix) was used on the model's output using PyTorch's functional API (F.conv2d function). The absolute difference between the filtered output and the original output was then computed and averaged to obtain the loss value.

In the first training trial, the Gradient Laplacian Loss was directly added to other components of the loss function without any weighting. However, this approach was ineffective as the lack of a balancing factor diminished the impact of the Gradient Laplacian Loss in reducing overall loss. In the second training trial, weights were assigned to the gradient loss, yielding better results than the first trial but still fell short compared to the loss without the gradient loss. Regarding qualitative results, the MPRNet demonstrated remarkable proficiency in handling different forms of degradation. It efficiently eliminated rain streaks with varying

angles and intensities, producing visually appealing images that closely matched the ground truth. This marked a clear distinction from other methods that compromised structural content, introduced artifacts, or struggled to fully remove rain streaks. Our enhanced Multi-stage Progressive Image Restoration Network (MPRNet) underwent rigorous evaluation, with a comprehensive analysis conducted using the PSNR (Peak Signal-to-Noise Ratio) and SSIM (Structural Similarity Index) metrics. These metrics served as crucial benchmarks, allowing us to quantitatively measure the performance of our enhanced MPRNet against existing state-of-the-art methods. The comparison was conducted across various datasets, and the results were meticulously documented in tables, as seen in Table I and Table II which show a comparison of PSNR and SSIM metrics results on both Rain100L and Test1200 datasets respectively. This extensive evaluation not only showcased the effectiveness of our enhanced MPRNet but also positioned it within the context of the latest advancements in the field, providing valuable insights into its superior performance and its potential applications in real-world scenarios. Fig.2. and Fig.3. show the qualitative results of the enhanced MPRNet model on both Rain100L and Test1200 datasets respectively.

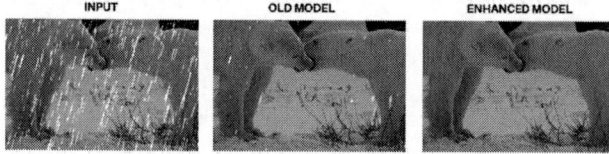

Fig. 2. Derained Image Example on Rain100L Dataset

TABLE I
COMPARISON OF PSNR AND SSIM RESULTS ON RAIN100L DATASET

	Metrics	
	PSNR	SSIM
DerainNet [3]	27.03	0.884
SEMI [4]	25.03	0.842
DIDMDN [7]	25.23	0.741
UMRL [29]	29.18	0.923
MPRNet [5]	36.40	0.965
Enhanced MPRNet	38.83	0.975

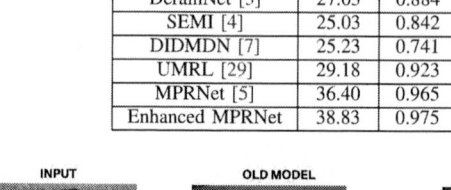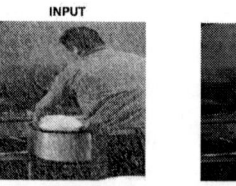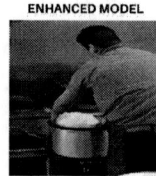

Fig. 3. Derained Image Example on Test1200 Dataset

TABLE II
COMPARISON OF PSNR AND SSIM RESULTS ON TEST1200 DATASET

	Metrics	
	PSNR	SSIM
DerainNet [3]	23.38	0.835
SEMI [4]	26.05	0.822
DIDMDN [7]	29.65	0.901
UMRL [29]	30.55	0.910
MPRNet [5]	32.91	0.916
Enhanced MPRNet	33.58	0.924

V. DISCUSSION

The MPRNet (Multi-stage Progressive Image Restoration Network) [5] stands out among various image restoration models due to its unique multi-stage progressive approach, offering versatile and refined results for multiple restoration tasks. In comparison to DerainNet [3], which focuses on single-image rain streak removal, MPRNet's adaptability and robustness make it a superior choice. Unlike SEMI [4], MPRNet employs a multi-stage progressive method, leading to gradual image refinement and improved outcomes. While DIDMDN [7] is effective for rain removal, MPRNet's iterative restoration approach sets it apart. In contrast to UMRL's [29] fixed structure, MPRNet's adaptability allows it to address diverse restoration challenges effectively. Compared to RESCAN's [30] spatial attentiveness, MPRNet balances both local and global features for comprehensive restoration. Finally, MPRNet's streamlined design eliminates the need for a separate pre-processing step, making it efficient for real-time applications, in contrast to PreNet. Furthermore, MPRNet's data augmentation during training contributes to improved generalization and reduced overfitting, enhancing its performance compared to MSPFN [32].

VI. CONCLUSION

In conclusion, the enhanced MPRNet's multi-stage progressive image restoration technique has proven to be a highly effective solution for image deraining tasks. Its performance, as gauged by both quantitative measures (PSNR and SSIM) and qualitative results, outperformed state-of-the-art methods. Its versatility in handling various types of degradation and compatibility with multiple tasks (deraining, deblurring, and denoising) mark it as a promising method for future research and development in the domain of image restoration.

The enhanced MPRNet model for deraining has shown promising results, but like all models, there is always room for improvement and exploration. Here are some potential directions for future work:

1) Incorporating Temporal Information for Video Deraining: Extending the model to handle video deraining could be an interesting direction. This would involve incorporating temporal information from multiple frames to improve the deraining performance.

2) Real-world Testing: The model could be tested on more diverse and challenging real- world datasets. This could provide a better understanding of the model's strengths and weaknesses and guide further improvements.

REFERENCES

[1] H. Zhang and V. M. Patel, "Density-aware single image de-raining using a multi- stream dense network," in Proceedings of the IEEE conference on computer vision and pattern recognition, pp. 695–704, 2018.

[2] X. Fu, J. Huang, X. Ding, Y. Liao, and J. Paisley, "Clearing the skies: A deep network architecture for single-image rain removal," IEEE Transactions on Image Processing, vol. 26, no. 6, pp. 2944–2956, 2017.

[3] W. Wei, D. Meng, Q. Zhao, Z. Xu, and Y. Wu, "Semi-supervised transfer learning for image rain removal," in Proceedings of the IEEE/CVF conference on computer vision and pattern recognition, pp. 3877–3886, 2019.

[4] R. Yasarla and V. M. Patel, "Uncertainty guided multi-scale residual learning-using a cycle spinning cnn for single image de-raining," in Proceedings of the IEEE/CVF conference on computer vision and pattern recognition, pp. 8405–8414, 2019.

[5] Zamir, Syed Waqas, et al. "Multi-stage progressive image restoration." Proceedings of the IEEE/CVF conference on computer vision and pattern recognition. 2021.

[6] Saeed Anwar and Nick Barnes. Real image denoising with feature attention. ICCV, 2019.

[7] Li, Yawei, et al. "Video deraining and desnowing using temporal correlation and low-rank matrix completion." Proceedings of the IEEE Conference on Computer Vision and Pattern Recognition (CVPR). 2017.

[8] Tao Dai, Jianrui Cai, Yongbing Zhang, Shu-Tao Xia, and Lei Zhang. Second-order attention network for single image super-resolution. In CVPR, 2019.

[9] Syed Waqas Zamir, Aditya Arora, Salman Khan, Munawar Hayat, Fahad Shahbaz Khan, Ming-Hsuan Yang, and Ling Shao. CycleISP: Real image restoration via improved data synthesis. In CVPR, 2020.

[10] Yulun Zhang, Kunpeng Li, Kai Li, Bineng Zhong, and Yun Fu. Residual non-local attention networks for image restoration. In ICLR, 2019.

[11] Tong Tong, Gen Li, Xiejie Liu, and Qinquan Gao. Image super-resolution using dense skip connections. In ICCV, 2017.

[12] Xintao Wang, Ke Yu, Shixiang Wu, Jinjin Gu, Yihao Liu, Chao Dong, Yu Qiao, and Chen Change Loy. ESRGAN: 14830 enhanced super-resolution generative adversarial networks. In ECCVW, 2018.

[13] Yulun Zhang, Yapeng Tian, Yu Kong, Bineng Zhong, and Yun Fu. Residual dense network for image restoration. TPAMI, 2020.

[14] Tim Brooks, Ben Mildenhall, Tianfan Xue, Jiawen Chen, Dillon Sharlet, and Jonathan T Barron. Unprocessing images for learned raw denoising. In CVPR, 2019.

[15] Chen Chen, Qifeng Chen, Jia Xu, and Vladlen Koltun. Learning to see in the dark. In CVPR, 2018.

[16] Orest Kupyn, Tetiana Martyniuk, Junru Wu, and Zhangyang Wang. DeblurGAN-v2: Deblurring (orders-ofmagnitude) faster and better. In ICCV, 2019.

[17] Olaf Ronneberger, Philipp Fischer, and Thomas Brox. UNet: convolutional networks for biomedical image segmentation. In MICCAI, 2015.

[18] P. K. Sharma, P. Jain, and A. Sur, "Dual-domain single image de-raining using conditional generative adversarial network," in 2019 IEEE International Conference on Image Processing (ICIP), pp. 2796–2800, 2019.

[19] Z. Guo, M. Hou, M. Sima, and Z. Feng, "Derainattentiongan: Unsupervised single-image deraining using attention-guided generative adversarial networks," Signal, Im- age and Video Processing, vol. 16, no. 1, pp. 185–192, 2022.

[20] Y. Guo, Z. Ma, Z. Song, R. Tang, and L. Liu, "Cycle-derain: Enhanced cyclegan for single image deraining," in Big Data and Security: Second International Conference, ICBDS 2020, Singapore, Singapore, December 20–22, 2020, Revised Selected Papers 2, pp. 497–509, Springer, 2021.

[21] P. Xiang, L. Wang, F. Wu, J. Cheng, and M. Zhou, "Single-image deraining with feature-supervised generative adversarial network," IEEE Signal Processing Letters, vol. 26, no. 5, pp. 650–654, 2019.

[22] Y. Ren, M. Nie, S. Li, and C. Li, "Single image de-raining via improved generative adversarial nets," Sensors, vol. 20, no. 6, p. 1591, 2020.

[23] N. K. ElFaramawy, O. M. Fahmy and S. M. Afifi, "Image De-Raining Enhancement Tool," 2023 International Mobile, Intelligent, and Ubiquitous Computing Conference (MIUCC), Cairo, Egypt, 2023, pp. 3-8, doi: 10.1109/MIUCC58832.2023.10278381.

[24] Maitreya Suin, Kuldeep Purohit, and A. N. Rajagopalan. Spatially-attentive patch-hierarchical network for adaptive motion deblurring. In CVPR, 2020.

[25] Xin Tao, Hongyun Gao, Xiaoyong Shen, Jue Wang, and Jiaya Jia. Scale-recurrent network for deep image deblurring. In CVPR, 2018.

[26] Hongguang Zhang, Yuchao Dai, Hongdong Li, and Piotr Koniusz. Deep stacked hierarchical multi-patch network for image deblurring. In CVPR, 2019.

[27] Xia Li, Jianlong Wu, Zhouchen Lin, Hong Liu, and Hongbin Zha. Recurrent squeeze-and-excitation context aggregation net for single image deraining. In ECCV, 2018.

[28] Dongwei Ren, Wangmeng Zuo, Qinghua Hu, Pengfei Zhu, and Deyu Meng. Progressive image deraining networks: A better and simpler baseline. In CVPR, 2019.

[29] Wang, Tianyu, et al. "Spatial attentive single-image deraining with a high quality real rain dataset." Proceedings of the IEEE Conference on Computer Vision and Pattern Recognition (CVPR). 2019.

[30] Luo, Weixin, et al. "Removing rain from single images via a deep detail network." Proceedings of the IEEE Conference on Computer Vision and Pattern Recognition (CVPR). 2015.

[31] Kim, Jiwoon, et al. "PReNet: Proposal rain removal network." Proceedings of the European Conference on Computer Vision (ECCV). 2018.

[32] Fu, Xueyang, et al. "Removing rain from a single image via discriminative sparse coding." Proceedings of the IEEE Conference on Computer Vision and Pattern Recognition (CVPR). 2017.

2023 International Conference on Microelectronics (ICM)

Acoustic Device for Detecting Red Palm Weevil Using Deep Learning and IoT

Maryam M. Atia[1], Amr I. Gouda[1], Mahmoud G. Ismail[2], Mohammed A.-M Salem[2], Mohamed A. Abd ElGhany[1],[3]

Electronics Engineering Dept., German University in Cairo, Egypt[1].

Computer Engineering Dept., German University in Cairo, Egypt[2].

Integrated Electronic Systems Lab, TU Darmstadt, Germany[3].

E-MAILS: maryam.atia@student.guc.edu.eg, amr.ibrahimgouda@student.guc.edu.eg, mahmoud.gamal@alumni2020.guc.edu.eg, mohammed.salem@guc.edu.eg, mohamed.abdel-ghany@guc.edu.eg

Abstract— **The Red Palm Weevil (RPW) is a major threat to the date palm industry, requiring early detection strategies. This paper presents a comprehensive exploration of two bioacoustic sensor prototypes tailored for the early detection of RPW. The first prototype employs a Raspberry Pi 4 coupled with a Convolutional Neural Network (CNN), achieving a detection accuracy of 99.02%. The second device, built with an ESP32-WROVER module, integrates a Simple Neural Network (SNN) for audio classification and attains an accuracy of 98.79%. This design's affordability and low power consumption make it particularly suited for large-scale deployment in agricultural settings. This paper delves deeper into the bioacoustic sensor technology used, the dataset chosen for CNN and SNN models training, and provides a comparison between the two device designs in terms of cost and efficiency. Our project's commitment to iterative improvement aims to provide an effective solution to counteract RPW's destructive effects on date palm cultivation.**

Keywords— Red Palm Weevil, Bioacoustic Sensor, Raspberry Pi4, ESP32-WROVER Module, Neural Network, Audio Classification, Cost-effective, Machine Learning.

I. INTRODUCTION

Since the 1980s, the Red Palm Weevil (RPW) has become a significant global threat to the date palm industry, rapidly expanding westwards from Saudi Arabia and the UAE through the Middle East and into parts of Europe [1]. The RPW larvae cause significant damage to the palm trees, often undetected until it's too late. Current traditional detection methods, like pheromone traps [2], are proficient but fail to detect the pests at an early stage, often leading to the transportation of infested material to new locations [3].

Market-available technologies like Saudi Arabia's Palmear and Israel's Agrint have been developed for early RPW detection. Palmear employs acoustic AI technologies, though its reliance on Bluetooth and human intervention limits scalability [4]. Agrint's IoT system also detects larval activity acoustically but is hindered by its need for continuous internet access, posing a challenge in remote areas [5].

Addressing the shortcomings of existing technologies, our project introduces two innovative bioacoustic sensor designs for detecting RPW activity using the sound of feeding larvae inside the palm trunk [6,7]. The general design, as illustrated in Fig. 1, provides a block diagram illustrating the overall functionality of both sensor devices.

The first approach harnesses the computational power of the Raspberry Pi 4, detailed in Section II, which is capable of handling complex data analysis essential for accurate RPW detection. Our second approach, described in Section III, explores the viability of the ESP32-WROVER module. This cost-effective, low-power module is embedded with machine learning algorithms, offering an economically viable solution for large-scale application in date palm cultivation.

Our systems are designed to overcome the limitations of Palmear and Agrint by providing a versatile, efficient, and scalable solution for RPW management. With this dual-approach system, we aim to deliver a proactive tool for early detection, thereby mitigating the RPW threat and protecting the date palm industry.

Figure 1. Block diagram of our proposed system

After the introduction, Section II details the design and technical intricacies of the Raspberry Pi-based solution. Section III elaborates on the ESP32-WROVER module-based design and its features. Section IV delves into the integration of machine/deep learning techniques. Section V discloses the results and performance metrics of both designs, emphasizing detection accuracy and deployment efficiency. The final section, Section VI, concludes the paper, emphasizing the implications of this research for the global date palm industry and outlining potential avenues for future work.

II. DESIGN OF THE RASPBERRY PI-BASED DEVICE

Figure 2 provides an overview of the system's components and their interconnections, depicting the overall process from sound capture by the Sensing Node to data processing and analysis by the AI Node, facilitated by a Raspberry Pi 4, to determine the presence of RPW infestations, incorporating a Sensing Node to capture the unique sounds of the RPW and an AI Node to process this data. Aimed at being both portable and cost-effective, the Sensing Node utilizes an audio probe—a 10cm long, hollow

979-8-3503-8083-5/23 $31.00 © 2023 IEEE

21

cylinder with a 10mm diameter—fitted with a MAX9418 microphone on its side to amplify the sounds using the probe's resonant properties. This setup is intended to be mounted on the palm tree at a height between 1 and 1.5 meters. It also includes a MAX9814 microphone amplifier, selected for its low noise and automatic gain control, to ensure superior sound quality. The AI Node, on the other hand, is the brains of the operation, comprising a Raspberry Pi 4 that handles sophisticated machine learning algorithms to analyze the audio data. A Bluetooth module facilitates the transfer of audio files in .wav format from the Sensing Node, while an Arduino Nano controls the recording process. Additionally, a Micro SD Module serves as the initial storage point for the files before processing. After analysis, the data is uploaded to the cloud, contributing to ongoing improvements in the device's detection capabilities and providing valuable insights into pest infestation patterns.

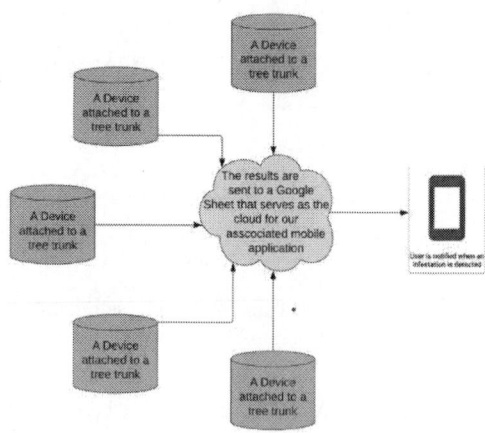

Figure 3. Block diagram of the ESP32-WROVER-based device

This design promises better efficiency in operation without hiking up the costs. It offers a pragmatic and easy-to-use solution for early RPW detection, aiming to be a top choice in the market. Figure 4 shows you what this fully assembled device looks like, ready to be put to the test in the field.

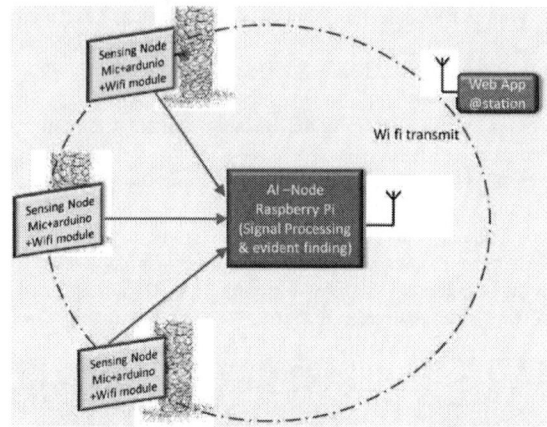

Figure 2. Block diagram of the Raspberry Pi 4-based device

III. DESIGN OF THE ESP23-WROVER-BASED DEVICE

Figure 3 highlights a refined device blueprint that marries advanced bioacoustic sensing with cost-effective computing power to detect the Red Palm Weevil (RPW). At the core of this setup is the Sensing Node, which houses an audio probe fitted with the INMP441 sensor—a budget-friendly microphone that accurately picks up the unique sounds made by RPWs inside palm trees. This sensor ensures that the sounds of the weevils are captured in high quality for processing.

The AI Node, which is essentially the brain of the operation, leverages the ESP32-WROVER microcontroller. This component is not only cost-effective but also powerful enough to run the necessary detection algorithms. It comes with a dual-core processor and Wi-Fi capabilities, enabling it to process the audio data efficiently and send alerts to a centralized system when it confirms an RPW presence, ensuring quick action can be taken.

The entire system is powered by a power bank, which is a crucial choice in the design as it needs to sustain the energy demands of both the Sensing and AI Nodes reliably. This power bank is selected based on its output, longevity, and ability to withstand various environmental factors, ensuring that the nodes are operating at their best.

Figure 4. Final Proposed device

IV. MACHINE/DEEP LEARNING MODEL

Machine learning has been highly effective in environment evaluation, notably through image analysis. Nevertheless, the field of audio classification, with its distinct potential, has not been explored to its full extent [8]. The persistent sound characteristics, not reliant on external factors such as light conditions, offer distinct advantages over visual data. Our approach encompasses two neural network models: A Simple Neural Network (SNN) for the resource-constrained ESP32 microcontroller, offering on-site RPW detection with minimal power, and a Convolutional Neural Network (CNN) for the Raspberry Pi 4, which affords complex audio pattern analysis thanks to its superior computational ability. These models reflect our commitment to adaptable and efficient RPW monitoring across varying palm tree farms.

A. Dataset and Data Collection

In our study, the dataset—featuring the distinctive sounds of the Red Palm Weevil (RPW) and a wide array of environmental noises—was comprehensively compiled from the Agricultural Research Center at Cairo University Road in Giza. We undertook an extensive field recording initiative, capturing high-quality and diverse audio samples at different times and locations within the farm.

For the Raspberry Pi4-based device, geared towards a Convolutional Neural Network (CNN) model [9] running on a Raspberry Pi, we secured 47 original recordings. These encompassed RPW sounds and an array of negative samples, including birdsong, human conversations, traffic noise, and miscellaneous ambient sounds with spectrogram features similar to those of RPWs. Each recording was segmented into one-second intervals, yielding 814 RPW samples and 905 samples of negative sounds.

To accommodate the Simple Neural Network (SNN) model for the ESP32—a simpler architecture requiring a larger dataset for effective learning—we further expanded our data collection. We integrated more recordings, culminating in a more extensive set of 3561 RPW files and 3390 negative samples. This larger dataset aimed to provide the SNN with a more comprehensive range of examples to enhance the learning process. To bolster the dataset and counter any class imbalance, we augmented these samples by repeating each RPW segment 33 times and each negative segment 35 times.

Both the CNN and SNN models utilized the same data division approach, with an 80% allocation for training and equal splits of 10% for both validation and testing. This consistent division ensured that each model was evaluated under uniform conditions, allowing for an equitable comparison of their respective performances.

B. CNN Model for Raspberry Pi4-based Device

For audio classification on the Raspberry Pi 4, we leveraged the full capabilities of the Librosa library for advanced feature extraction. Librosa enabled us to extract Mel-Frequency Cepstral Coefficients (MFCC) from audio signals [10]. This technique encapsulates the spectral properties fundamental for distinguishing unique sounds such as those made by the Red Palm Weevil (RPW).

Taking advantage of the Raspberry Pi 4's more robust computational resources, we constructed a Convolutional Neural Network (CNN) with a sequential four-layer architecture utilizing the TensorFlow and Keras libraries. Our CNN commenced with an input layer designed to handle the two-dimensional structure of MFCCs, followed by convolutional and pooling layers to detect and emphasize salient features in the frequency domain. We also incorporated dropout layers to reduce overfitting and dense layers that culminated in a softmax activation function in the output layer to classify between RPW and other environmental sounds.

C. SNN Model for ESP32-WROVER-based Device

In audio classification, we address the challenges posed by the limited resources of the ESP32-WROVER microcontroller. Initially, we attempted to implement the Mel-Frequency Cepstral Coefficients (MFCC) feature extraction part of the Librosa library in C/C++. However, the ESP32's memory limitations precluded the use of such a tool. Consequently, we developed a lightweight library tailored to the ESP32, which could perform Fast Fourier Transform (FFT) [11] to extract frequency domain features directly. The FFT approach proved to be a memory-efficient alternative, adeptly converting audio frames from the time domain to the frequency domain, thus capturing the essential frequency and time characteristics of the sounds required for our classification task.

For the ESP32, we needed a simple but effective neural network. We built a Simple Neural Network (SNN) with four layers using the tools Keras and TensorFlow. The network starts with an input layer to handle the FFT data. Then it has three hidden layers with 128, 64, and 32 neurons. Each hidden layer uses a ReLU activation, which helps the network learn complex patterns from the sound data. The last layer uses softmax activation to tell apart RPW from other sounds.

For training our network, we selected the Adam optimizer and categorical cross-entropy as our loss function, with accuracy as our chief performance metric. This led to an SNN with 555,426 trainable parameters, adept at detecting nuanced acoustic signatures within our dataset.

After training, we changed the model into a TensorFlow Lite [12] format and then into a hex array with the .xxd command in Linux. This was important so we could use the SNN on the ESP32.

V. Results and Comparisons

The implementation of both the Raspberry Pi 4-based and ESP32-WROVER-based devices showed promising results in the detection of Red Palm Weevil (RPW) infestations. The following sections present the results of both devices and a comparison of their performances, shedding light on their potential benefits over existing market solutions.

A. Raspberry Pi4-based Device Results

The Raspberry Pi 4-based device delivered exceptional performance, achieving an accuracy of 99.02% after 50 epochs, precision of 99%, and recall of 98%. This high accuracy indicates the device's success in analyzing the audio signals from the palm trees, identifying the patterns of RPW infestations, and predicting the presence of these pests with remarkable reliability. Fig. 5 shows the accuracy obtained vs the number of epochs for the Raspberry Pi 4-based device.

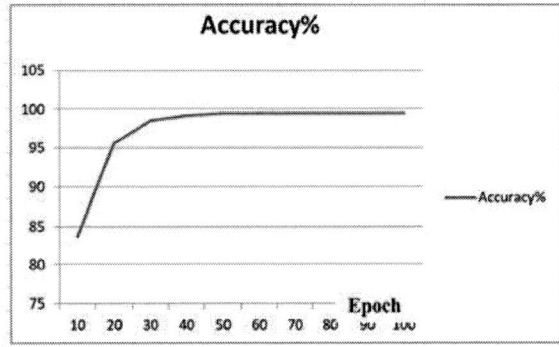

Figure 5. Accuracy vs. Epoch (Raspberry Pi 4-based device)

B. ESP32-WROVER-based Device Results

The ESP32-WROVER-based device, despite the introduction of FFT for feature extraction, displayed commendable performance. The model trained using FFT-derived features predicted RPW infestations with an accuracy of 98.79%, precision of 97.23%, and recall of 97.1%. While this accuracy is slightly lower than that of the Raspberry Pi 4-based device, it is important to note that both devices produced identical results in practical field testing, indicating the ESP32-WROVER-based device's reliable RPW detection based on audio signals. Fig. 6 shows the confusion matrix for the SNN model which we can obtain the accuracy from by a simple equation.

$$((TP+TN)/(TP+FP+TN+FN)) \qquad (1)$$

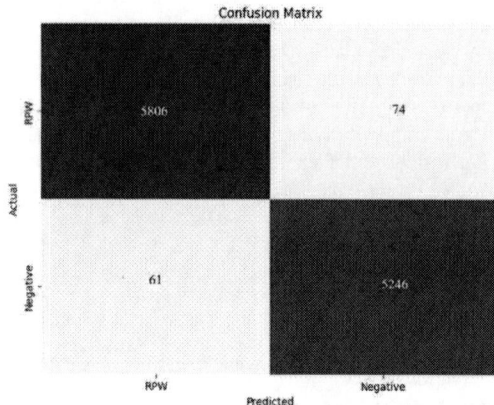

Figure 6. Confusion matrix (ESP32-WROVER-based device)

C. Results Comparison

Comparing the two devices' results, there is a clear trade-off between accuracy and cost. While the Raspberry Pi 4-based device achieved slightly higher accuracy, the ESP32-WROVER-based device offered a more cost-effective solution. Both devices demonstrated consistent performance in practical field tests, reinforcing their reliability in real-world scenarios.

The difference in the accuracy of the machine learning models – 99.02% for the Raspberry Pi 4-based device and 98.79% for the ESP32-WROVER-based device - did not affect the outcomes of practical applications. This observation highlights the practicality of both designs in real-world conditions.

D. Cost and Efficiency Comparison

In terms of cost-efficiency, the ESP32-WROVER-based device was a clear winner. A detailed cost breakdown is provided in Table I, showing the financial considerations for both the ESP32-WROVER and Raspberry Pi 4-based devices.

TABLE I. FINANCIAL MATRIX

Component	ESP32 Device	Raspberry Pi Device
AI-Node (per unit)	$4-6	$45-55
Audio Probe (per unit)	$1.2-2	$4-6
Power Supply (per unit)	$15-20	$20-25
Casing Production Cost (per unit)	$2-3	$4-6
Overall Cost (per unit)	$22.2-31	$73-92

The overall implementation cost of the ESP32-WROVER-based device is significantly lower than that of the Raspberry Pi 4-based device, making it a more affordable solution without compromising effectiveness. The ESP32-WROVER's design not only achieved similar performance to the Raspberry Pi 4-based device but did so while reducing costs, power consumption, and device size, as evidenced by the financial matrix.

Additionally, the ESP32-WROVER-based device is complemented by an associated mobile application, presented in Figure 7, which offers users real-time feedback, quick setup, and the convenience of monitoring the Red Palm Weevil infestations on the go.

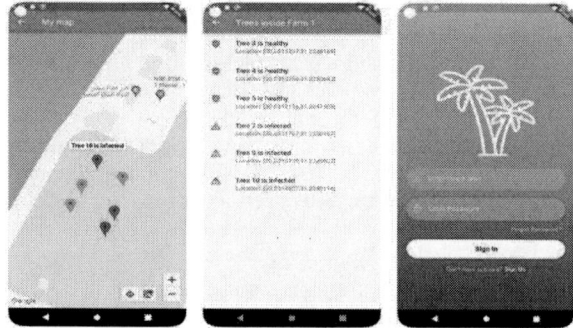

Figure 7. Our associated mobile application

Conversely, the Raspberry Pi 4-based device boasts superior accuracy, distinguishing it from the ESP32-WROVER device. This heightened accuracy is a great advantage, especially for this application. Furthermore, the Raspberry Pi 4-based device has the added capability of transmitting audio recordings to the cloud. This not only aids in data storage and accessibility but also provides an avenue for continuous model improvement by using this cloud-stored data for further training, ensuring that the machine learning model evolves with the new data over time.

E. Market Solutions Comparison

To provide a clearer perspective, Table II contrasts the two devices with existing market solutions, underlining the competitive edge of our devices in terms of both accuracy and cost.

979-8-3503-8083-5/23 $31.00 © 2023 IEEE

TABLE II. COMPARISON TABLE

Component	ESP32 Device (Egypt)	Raspberry Pi Device (Egypt)	Palmear (KSA)	Agrint (Israel)
Cost/Unit	-	-	$73-92	$22.2-33
Data Transfer	Bluetooth	Cloud (Wi-Fi)	Cloud (Wi-Fi)	Cloud (Wi-Fi)
Scalability	Human-dependent	High	High	High
Deployment	Human-operated per tree	Automated per tree	Automated per tree	Automated per tree
Interface	Manual input required	Mobile app, IoT system	Indicating LEDs, standalone or IoT system	Indicating LEDs, standalone or IoT system, mobile app

The comparison illustrates that both our devices introduce a level of autonomy that the Palmear system lacks. The Palmear device necessitates manual intervention, where a person must physically interact with each tree, inserting their phone into the device to get readings. This process is labor-intensive and may lead to variability in results due to human error.

In contrast, our Raspberry Pi 4-based and ESP32-WROVER-based devices are designed for autonomous operation, equipped with indicating LEDs that provide immediate visual status indication. They can function as standalone units or be integrated into an IoT system, allowing for scalability and ease of deployment. Each tree is individually monitored, and data is automatically transferred to the cloud, which is especially beneficial in extensive palm groves where manual checking would be impractical.

While Agrint's device shares cloud connectivity and IoT integration, it lacks the indicating LEDs found in our designs. These LEDs are crucial for immediate, on-site troubleshooting and status checks without the need to consult the mobile app. Moreover, our ESP32-WROVER-based device further enhances user convenience with its associated mobile application, offering a direct and user-friendly interface for real-time feedback and monitoring, a feature that aligns with Agrint's user interaction but at a significantly lower cost.

In conclusion, the Raspberry Pi 4-based and ESP32-WROVER-based devices present a cost-effective, accurate, and scalable alternative to current market offerings. With their advanced features and user-friendly design, they are poised to transform the efficiency and effectiveness of RPW detection and management.

VI. CONCLUSION

The bioacoustic devices developed in this study represent significant advances in combating the RPW threat. The results obtained from both the Raspberry Pi 4-based and ESP32-WROVER-based devices, as well as their comparative analysis, underscore the potential of machine learning and audio classification in pest control. While the ESP32-WROVER-based device is more cost-efficient and offers an integrated mobile experience, the Raspberry Pi 4-based device's superior accuracy and cloud integration make it an equally compelling choice, depending on the specific needs of the application. The success of these devices hints at the potential for further improvements and wider applications of this technology in the pest control industry and beyond.

Future work should explore improvements in detection accuracy and energy efficiency, possibly exploring renewable energy sources. Extensive field trials in varied regions will refine the device's performance, potentially leading to more comprehensive and adaptable pest detection solutions.

REFERENCES

[1] A. Elsabea, J. R. Faleiro, and M. Abo-El-Saad, "The threat of red palm weevil Rhynchophorus ferrugineus to date plantations of the Gulf region in the Middle-East: An economic perspective," Outlooks on Pest Management, vol. 20, no. 3, pp. 131–134, June 2009.

[2] Grand View Research, "Palm Oil Market Size, Share & Trends Analysis Report By Product (Crude Palm Oil, Palm Kernel Oil), By Application (Edible Oil, Cosmetics, Bio-diesel, Lubricants, Surfactants), By Region, And Segment Forecasts, 2021-2028," 2021.

[3] S. T. Murphy and B. R. Briscoe, "The red palm weevil as an alien invasive: biology and the prospects for biological control as a component of IPM," Biocontrol, vol. 20, pp. 45–46, 1999.

[4] Palmear, (n.d.), [Online]. Available: https://www.palmear.ai/.

[5] AgrInt, (n.d.), [Online]. Available: https://www.agrint.net/.

[6] W. Hussien, M. Hussien, and T. Becker, "Detection of the Red Palm Weevil Rhynchophorus ferrugineus Using its Bioacoustics Features," University of Munich, Freising, Germany, 2010.

[7] A. Hetzroni, V. Soroker, and Y. Cohen, "Toward practical acoustic red palm weevil detection," Computers and Electronics in Agriculture, vol. 124, pp. 100–106, 2016.

[8] Imran, M. Safwat; Rahman, A. Fahmida; Tanvir, Sifat; Kadir, Hamim Hassan; Iqbal, Junaid; Mostakim, Moin. "An Analysis of Audio Classification Techniques using Deep Learning Architectures." IEEE Conference Publication | IEEE Xplore, Jan. 20, 2021.

[9] K. Palanisamy, D. Singhania, and A. Yao, "Rethinking CNN Models for Audio Classification," 2020

[10] Kamarulafizam, I. & Salleh, Shussain & Jamaludin, Mohd Najeb & Ariff, A. & Chowdhury, Abdur. (2007). Heart Sound Analysis Using MFCC and Time Frequency Distribution. 10.1007/978-3-540-68017-8_102.

[11] Oberst, Ulrich. (2007). The Fast Fourier Transform. SIAM J. Control and Optimization. 46. 496-540. 10.1137/060658242.

[12] R. David, J. Duke, A. Jain, V. J. Reddi, N. Jeffries, J. Li, N. Kreeger, I. Nappier, M. Natraj, S. Regev, R. Rhodes, T. Wang, and P. Warden, "TensorFlow Lite Micro: Embedded Machine Learning on TinyML Systems," 2020

[13] Verdouw, Cor & Wolfert, Sjaak & Tekinerdogan, Bedir. (2016). Internet of Things in agriculture. CAB Reviews. 11. 1-12. 10.1079/PAVSNNR201611035.

Enhancing Traffic Management with Embedded Machine Learning for Vehicle Detection

Mohamad Sofian Abu Talip
Department of Electrical Engineering
Faculty of Engineering
University of Malaya
50603 Kuala Lumpur, Malaysia
sofian_abutalip@um.edu.my

Mohd Zulhakimi Ab Razak
Institute of Microengineering and
Nanoelectronics
Universiti Kebangsaan Malaysia
43600 Bangi, Selangor, Malaysia
zul.hakimi@ukm.edu.my

Mahazani Mohamad
Department of Electrical Engineering
Faculty of Engineering
University of Malaya
50603 Kuala Lumpur, Malaysia
mahazani@um.edu.my

Anis Salwa Mohd Khairuddin
Department of Electrical Engineering
Faculty of Engineering
University of Malaya
50603 Kuala Lumpur, Malaysia
anissalwa@um.edu.my

Tengku Faiz Tengku Mohmed Noor
Izam
Department of Electrical Engineering
Faculty of Engineering
University of Malaya
50603 Kuala Lumpur, Malaysia
tengkufaiz@um.edu.my

Azizul Azizan
Advanced Informatics Department
Razak Faculty of Technology and
Informatics
Universiti Teknologi Malaysia
54100 Kuala Lumpur, Malaysia
azizulazizan@utm.my

Abstract—In recent years, vehicle detection has become vital for applications ranging from autonomous driving to traffic control, surveillance, and monitoring. The demand for efficient real-time detection systems has surged, prompting the integration of machine learning algorithms into embedded platforms as a promising approach. This paper focuses on developing a robust and efficient system deployable on the NVIDIA Jetson Nano 2GB Developer Kit. The system harnesses machine learning algorithms tailored for resource-constrained embedded systems to achieve high detection accuracy in real-time. The process encompasses data preparation, preprocessing, feature extraction, and classification. Deep learning models used are You Only Look Once (YOLO) algorithm, YOLOv5n, and YOLOv7-tiny, trained on labeled datasets to classify regions of interest based on unique vehicle attributes. For inference on the Jetson Nano, both are chosen for their real-time capabilities and high object detection accuracy, are employed. To leverage the Jetson Nano's GPU power, NVIDIA's Compute Unified Device Architecture (CUDA) toolkit is installed, enabling parallel computing and deep learning model optimization. Results indicate that YOLOv7-tiny achieves the better precision-confidence at 92.7% and a recall-confidence of 94% compared to YOLOv5n. This study also uses various other evaluation metrics such as accuracy, precision, recall, confusion matrix, and F1 score to measure system performance and examine computational efficiency, to help in the selection of appropriate models for embedded systems.

Keywords—traffic management, vehicle detection, embedded systems, machine learning, deep learning

I. INTRODUCTION

Despite the recent virus outbreak, technological advancements, particularly in Artificial Intelligence (AI), continue to progress unabated. The current state of AI is widely recognized as a potent practical tool capable of simplifying a multitude of tasks across various professions worldwide. It is increasingly apparent that AI's rapid technological evolution will inevitably lead to its surpassing human intelligence in executing complex tasks with exceptional efficiency. These characterizations underscore the significant role that AI plays in contemporary computing technology [1].

Furthermore, in recent years, there has been a proliferation of applications encompassing various subfields of AI, including Machine Learning (ML), Computer Vision, Natural Language Processing (NLP), Deep Learning (DL), Neural Networks, and Cognitive Computing [2]. These applications have not only had a profound impact on human endeavours but have also contributed to the advancement of AI itself.

Traffic management using AI is a transformative approach to addressing the growing challenges of urban congestion and transportation efficiency. AI technologies, including machine learning and computer vision, are being harnessed to revolutionize how traffic is monitored, analysed, and controlled in cities worldwide [3]. By deploying AI-powered traffic management systems, cities can achieve several key objectives.

Moreover, vehicle detection is a critical component of modern transportation and surveillance systems. It involves the identification and tracking of vehicles within a given area, typically using various sensor technologies and computer vision techniques. The primary objectives of vehicle detection include traffic management, security monitoring, and data collection for smart city applications. This paper proposed vehicle detection systems using embedded machine learning to help optimize traffic flow by adjusting traffic signals in real-time, reducing congestion, and improving transportation efficiency.

The rest of this paper is organized as follows. Section II discussed the related work of this project. Section III explained the methodology development and implementation. Section IV presented the results and discussion followed by Section V as a conclusion to conclude the work and some suggestions for future works.

II. RELATED WORK

A. Overview of Machine Learning via Embedded System

Machine Learning is the field of study that focuses on designing and developing algorithms and statistical models that enable computer systems to execute tasks without the need for explicit instructions. Some of the platforms used for embedded ML are Jetson Nano and Raspberry Pi. ML is one of the branches for AI that utilizes data as well as statistical techniques that will allow machines to learn and improve their performance gradually. Machine Learning techniques can be divided into four types; supervised, unsupervised, semi-supervised and reinforcement [4].

979-8-3503-8083-5/23 $31.00 © 2023 IEEE

As embedded systems technology is developing rapidly due to recent advancements in computer architecture and ML. This resulted in embedded machine learning (EML) becoming more important in a variety of applications such as robust speech recognition, innovative healthcare, robotics, and advanced computer vision schemes [5]. Despite the potential advantages, effectively implementing machine learning algorithms in embedded systems is a significant challenge. These algorithms frequently have high computational and memory requirements, which makes it challenging to use them on devices with limited resources like embedded systems and mobile phones [6]. At both the algorithmic and hardware levels, new optimization techniques are required to get around this restriction.

Real-time intelligent solutions that are based on deep learning need to have certain key features such as being energy-efficient, cost-effective, and compact. They also need to strike a balance between accuracy and power consumption [7]. Typically, deep learning architectures are deployed in centralized cloud computing environments. However, there are challenges such as high network latency, energy consumption, and financial costs that can negatively impact system performance [8]. To address these limitations, a new approach called "edge AI" or edge computing has emerged, which involves performing computations locally using data from various devices or sensors [9]. The major challenge in implementing edge AI is to achieve high-accuracy results from the algorithms while keeping power consumption low. However, recent advancements in hardware options such as central processing units (CPUs), graphics processing units (GPUs), application-specific integrated circuits (ASICs), and system-on-a-chip (SoC) accelerators have made it possible to achieve these goals and thus making edge AI a possibility [10].

B. Embedded Machine Learning Platforms

Currently there are already a variety of platforms that contribute to the development of EML which are led by brands such as NVIDIA, Intel, and Qualcomm. Among all these brands, Intel's Movidius Neural Computing Stick (NCS) is the cheapest option for running complex algorithms with multiple layers of Convolutional Neural Networks (CNN). Another popular embedded hardware for machine learning is NVIDIA's Jetson, which is widely used as an accelerator for various algorithms. One of the key features of the Jetson is its small size, lightweight, and low power consumption. However, to fully utilize the capabilities of Jetson and achieve real-time performance, it is necessary to optimize both the Jetson hardware and the neural network algorithms. Variants of the Jetson such as TK1, TX1, TX2, Xavier NX and AGX Xavier have been widely used in recent years [10]. In this project we chose NVIDIA Jetson Nano 2GB as our embedded platform.

C. Machine Learning Models for Vehicle Detection

The reason why vehicle classification system and image classification system needed to be chosen wisely is because different methods proposed for vehicle classification and each one of them results in varying accuracy rates. The accuracy of the systems depends on the method used [11]. For example, some methods only classify vehicles based on their front and rear views, which results in some vehicles not being detected and classified. Therefore, it is necessary to use techniques that detect the vehicle from multiple viewpoints. In this work, images of vehicles captured from multiple viewpoints are used

as input for training. Different data annotation and data augmentation techniques are applied to produce a more comprehensive database from the acquired images.

III. METHODOLOGY AND IMPLEMENTATION

In this section, we outline the systematic approach employed to integrate embedded machine learning techniques into our traffic management system. Leveraging advanced algorithms, we detail the process of vehicle detection and the key steps taken to enhance the overall efficiency of traffic control. Subsequently, we provide a comprehensive overview of the practical deployment of these methodologies.

A. Data Acquisition

Before proceeding with the development and implementation of the system, the data acquisition process needed to be done first to initiate the ML training. For the proposed system for vehicle detection and classification, the training, testing and evaluation will be based on internet dataset and self-built dataset. The dataset that will be used consists of multiple types of vehicle class, hence, to approach this issue, multiclass classification may be considered. To have a reliable dataset, the images that are collected must be on a bright day and the images are not blurry and have high quality condition. But for some ML techniques, the images used may need to be changed in their dimensions, which significantly reduced the quality of the datasets, and to solve this, a huge number of datasets needed to be considered. Images of poor quality may affect the ML training and may directly reduce the efficiency and performance of the proposed vehicle detection and classification system.

For internet dataset, the images obtained from Kaggle.com, and it contains a huge number of images of vehicles and non-vehicles that can be used for ML training. For self-built dataset, the images obtained are captured at around Universiti Malaya, Selangor and Kuala Lumpur. The dataset is manually acquired by using Samsung Galaxy S22 camera and built in 50MP main camera to ensure the images captured are crystal clear. The data taken both as images and videos, to be extracted and used in ML training. Currently, the total images collected is approximately 100 images caught manually via smartphone, and for online datasets, almost 30,000 images are used. For both internet and self-built dataset, the images collected consists of various types of vehicles, car, bus, lorry and motorcycle.

To be able to achieve satisfactory results, machine learning techniques rely heavily on the dataset that is used to perform the learning process. Therefore, the selection of datasets that are used in this work is closely observed and tracked to avoid any 'bad data', which surely can affect the results of the research.

B. You Only Look Once (YOLO)

For this project, YOLO will be used for vehicle detection and classification algorithms. YOLO is made to locate objects quickly and precisely within an image and identify them. To accomplish this, a single neural network is used to analyse the entire image after using bounding boxes to separate objects. Instead of analysing the image in sections, this method has the advantage of analysing the entire image at once. It also improves object detection by using the predicted areas [7].

The first single-stage object detection algorithm Redmon J. suggested is the YOLO algorithm. The bounding box and

979-8-3503-8083-5/23 $31.00 © 2023 IEEE

classification problems are combined into a regression problem, which eliminates the candidate box extraction step from the two-stage algorithm. The process of YOLO algorithm is as follows:

1. Image is divided into S ×S meshes, where each grid is responsible for predicting where the target's actual box will fall in the centre of the grid (Jain & Nandy, 2019).

2. S ×S ×B of bounding boxes generated.

3. Each bounding box has five parameters: targets' width and height dimensions (x, y, w, h), target centre point coordinates, and confidence of the target contained.

4. The category predicted of the target in the grid is predicted by the S ×S grids.

5. The category score for each prediction box is then calculated by multiplying the category probability by the prediction bounding box confidence.

6. The final prediction results are obtained by filtering these prediction boxes using non-maximum suppression (NMS).

The YOLO model architecture includes several convolutional layers followed by fully connected layers to predict bounding boxes and class probabilities. It takes an input image and produces a set of bounding box predictions along with corresponding class probabilities for each predicted object. A deep convolutional neural network (CNN) that completes end-to-end object detection makes up the YOLO model architecture.

C. YOLOv5

For YOLOv5, there are multiple object detection models that can be utilized. The model can be divided into five different network structures depending on their sizes YOLOv5n, YOLOv5s, YOLOv5m, YOLOv5l and YOLOv5x, sorted ascendingly based on their models' sizes, from smallest to largest.

In this project, since the platform where this Machine Learning model will be deployed which is the NVIDIA Jetson Nano 2GB Developer Kit, YOLOv5n will be used as it is the smallest in the family and meant for the edge, IoT devices, and with OpenCV DNN support as well. It is less than 2.5 MB in INT8 format and around 4 MB in FP32 format. It is ideal for mobile and embedded solutions.

D. YOLOv7

Like YOLOv5, YOLOv7 also consists of multiple models depending on their sizes. In this work, YOLOv7-tiny will be used among all YOLOv7 models. The reason is because it is a basic model tailor-made for edge GPU. Computer vision models with the suffix "tiny" in YOLOv7-tiny, means that they are lighter to run ML on mobile computing devices or distributed edge servers and devices. They are also optimized for Edge AI and deep learning workloads. For distributed real-world computer vision applications, this model is crucial. The edge-optimized YOLOv7-tiny differs from the other versions in that it uses leaky Rectified Linear Activation function (ReLU) as the activation function as opposed to Sigmoid Linear Unit (SiLU) in the other models. Theoretically, YOLOv7-tiny will perform better than YOLOv5n and will be decided based on the result. Compared to YOLOv5n, YOLOv7-tiny is 127 FPS faster and 10.7% more accurate on average precision [1].

IV. RESULTS AND DISCUSSIONS

A. Dataset for YOLOv5 and YOLOv7

In this research, deep learning methods, specifically YOLOv5 and YOLOv7, have been employed. These models utilize Convolutional Neural Networks (CNNs) as their foundational machine learning technique, requiring a distinct approach to image representation within the dataset. Deep learning models typically demand substantial amounts of data due to their intricate architecture. Given the heightened complexity of deep learning techniques, increased data volume can significantly enhance their performance, as these models excel at extracting intricate patterns and representations from the data.

i. Data Preparation

As deep learning techniques do not rely on handcrafted features extraction algorithms like traditional machine learning, the preparation of dataset for both YOLOv5 and YOLOv7 is much simpler but instead it takes much more disk space due to the raw data in nature. In this research, the website Roboflow.com has been used as a platform to store all the datasets that have been used to simplify the process as it comes with multiple functionalities that can be utilized. The dataset is uploaded and then split up into 80% training set, 10% validation set and 10% testing set.

ii. Dataset Preprocessing

Similar to conventional machine learning, data preprocessing plays a crucial role in optimizing the computational efficiency and enhancing the performance of deep learning approaches. This involves applying image transformations to all dataset images. Since the dataset comprises images with varying dimensions, it's imperative to ensure uniformity in input image size during both training and inference. However, in deep learning, the adjustments made to dimensions are not as extensive as in traditional techniques, as it's essential to preserve the original aspect ratio of the images. As previously mentioned, due to the heightened complexity of deep learning model architectures, it becomes critical to maintain the visual integrity of objects within the images while simultaneously increasing the precision of the object detection model. In this project, the images in the dataset were resized to a consistent dimension of 416×416 pixels to facilitate these objectives.

iii. Data Augmentation

As the traditional machine learning dataset used consists of 21,000 images while the deep learning dataset is around 9,000 images, to balance it, data augmentation is done by using the functionality in Roboflow.com. In order to add new variations and more images to the dataset, data augmentation performs transformations on the current images. This will result in the deployed models to be more accurate across a broader range of use cases. For vehicle detection and classification, one of the most important aspects to improve the accuracy of the deep learning algorithms is to have a huge number of images in a dataset. The reason is the features for each classes; car, bus, lorry and motorcycle, it can be seen that there are apparent differences in the features. Therefore, there are many data augmentation processes that can be deployed to increase the number of variations and the amounts of images used in the dataset. Since vehicle detection and classification is insensitive in the way images are represented, for example whether the images are inverted or not, the colors of the images used whether it is grayscaled or have different amounts of exposures, or even the way the image in the dataset is rotated to a specific degree of rotations, it would not significantly affect the accuracy of the deployed models.

Hence, for the data augmentation of the images in the dataset, the augmentations that have been done are the flipping of the images horizontally only to avoid weirdly flipped images from being used. With these augmentations, the dataset can now be generated and can be used for the training of YOLOv5 and YOLOv7. The new current amounts of images after augmentation are at 22,971 images which is approximately the same as the value of images in the dataset for traditional machine learning. Hence, the data augmentation helped simplify the task of having more data without directly increasing the current numbers of images in the dataset which simplified the tedious task of finding more images and then annotated it.

B. Implementation Using YOLOv5n

In this project, for the training of the YOLOv5n, the training has been done with 20 epochs, and to complete the training, it took around 48 hours approximately. Based on the results of the training, the information obtained is shown in Table I.

YOLOv5n have been trained, based on the normalized results, for bus class the correctly predicted is about 65% which is the highest correct class prediction, cars class with 45% which is the lowest class prediction, lorry with 63% and motorcycle 50%. For all the classes except the cars class, they have higher correct predictions than the background false negative and false positive, but the results are not significant due to the model type used which sacrifice the overall performance to have much more faster train time, faster speed with much higher efficiency in computational cost and is much more feasible to be deployed on is Jetson Nano 2GB Developer Kit.

Based on the precision-confidence curve as shown in Fig. 1, for bus class, the precision is at 1.00 at approximately 0.7 confidence, for cars class, the precision is at 1.00 at approximately 0.85 confidence, lorry class with precision at 1.00 at approximately 0.9 confidence and motorcycle class at precision 1.00 at approximately 0.64 confidence. For all classes, the precision is at 1.00 at approximately 0.902 confidence. Based on the recall-confidence curve as shown in Fig. 2, all classes are at 0.00 recall at 0.88 confidence.

Based on Fig. 3, for precision-recall curve, for all classes, given that the IoU threshold is at 0.5, it indicates that the model can accurately detect objects with a precision of 0.666 while still achieving a reasonable recall rate. Hence, with both precision-confidence, recall-confidence and precision-recall as reference, the f1-score-confidence curve is as shown in Fig.

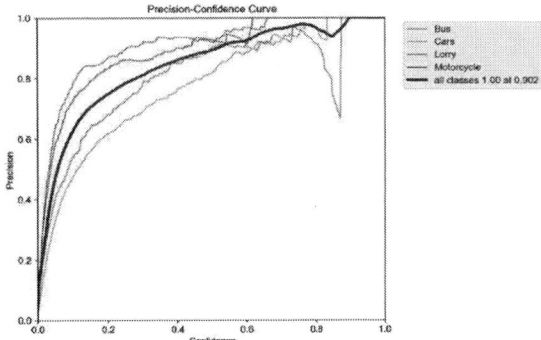

Fig. 1. Graph of precision-confidence curve for YOLOv5n

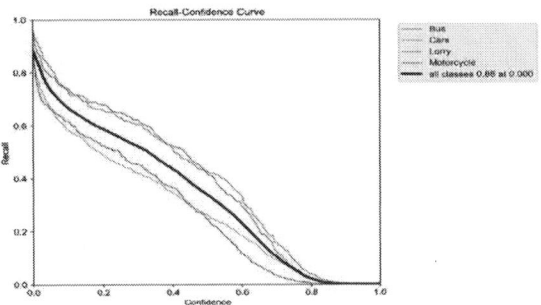

Fig. 2. Graph of recall-confidence curve for YOLOv5n

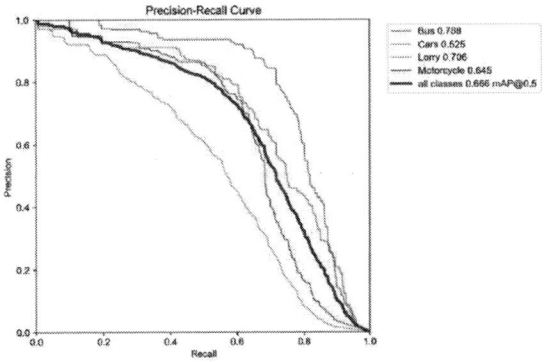

Fig. 3. Graph of precision-recall curve for YOLOv5n

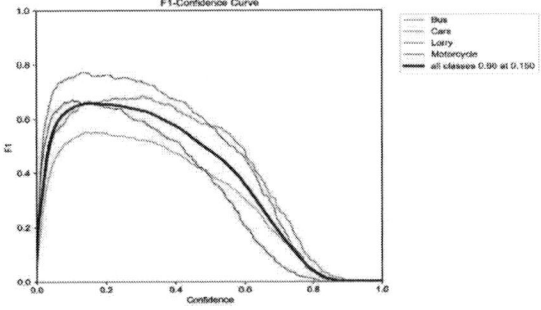

Fig. 4. Graph of F1-score-confidence curve for YOLOv5n

TABLE I. TRUE POSITIVE (TP), FALSE POSITIVE (FP) AND FALSE NEGATIVE (FN) FOR EACH CLASS, YOLOV5N

Class\Metrics	True Positive (TP)	False Positive (FP)	False Negative (FN)
Bus	0.65	0.06	0.34
Cars	0.45	0.77	0.53
Lorry	0.63	0.11	0.28
Motorcycle	0.50	0.07	0.49

C. Implementation Using YOLOv7-tiny

Again, for the training of the YOLOv7-tiny, the training has been done with 20 epochs, and to complete the training, it took around 24 hours approximately, which is significantly faster than the YOLOv5n model, around 100% faster for the same number of epochs. Based on the results of the training, the information obtained is shown in Table II.

YOLOv7-tiny that have been trained, based on the normalized results, for bus the correctly predicted is about 73% which is the highest correct class prediction, cars with 52% which is the lowest class prediction, lorry with 70% and motorcycle 59%. Overall, the accuracy of YOLOv7-tiny model is much higher than YOLOv5n, which approved the theory where YOLOv7 models perform slightly better than the YOLOv5 models. For all the classes except the cars class, they have higher correct predictions than the background false negative and false positive, but the results are not significant due to the model type used which sacrifice the overall performance to have much more faster train time, faster speed with much higher efficiency in computational cost and is much more feasible to be deployed on is Jetson Nano 2GB Developer Kit.

Based on the precision-confidence curve as shown in Fig. 5, for bus class, the precision is at 1.00 at approximately 0.85 confidence, for cars class, the precision is at 1.00 at approximately 0.78 confidence, lorry class with precision at 1.00 at approximately 0.92 confidence and motorcycle class at precision 1.00 at approximately 0.75 confidence. For all classes, the precision is at 1.00 at approximately 0.927 confidence. Next performance metric is the recall-confidence curve as shown in Fig. 6. Based on the recall-confidence curve, all classes are at 0.00 recall at 0.94 confidence. Next performance metric is the precision-recall curve as shown in Fig. 7. Based on results for all classes, given that the IoU threshold is at 0.5, it indicates that the model can accurately detect objects with a precision of 0.653 while still achieving a reasonable recall rate. Hence, with both precision-confidence, recall-confidence and precision-recall as reference, the f1-score-confidence curve as shown in Fig. 8.

D. Inference for YOLOv5n and YOLOv7-tiny (Image Input)

For these inferences, the intersection over union (IoU) threshold is at 0.45 and for the confidence, both models used 0.3 confidence. We used same inference image for both model as shown in Fig. 9 and Fg. 10. The results are shown in Table III. Based on the results, YOLOv7-tiny outperforms YOLOv5n in terms of inference speed, frames per second, Giga Floating-Point Operations Per Seconds (GFLOPs), Non-Max Suppression (NMS) time, and layer complexity, which is most of the performance evaluation metrics.

TABLE II. TRUE POSITIVE (TP), FALSE POSITIVE (FP) AND FALSE NEGATIVE (FN) FOR EACH CLASS, YOLOV7-TINY

Class\Metrics	True Positive (TP)	False Positive (FP)	False Negative (FN)
Bus	0.73	0.09	0.23
Cars	0.52	0.62	0.45
Lorry	0.70	0.13	0.21
Motorcycle	0.59	0.15	0.39

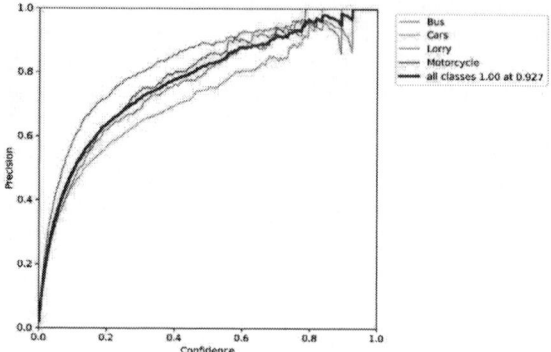

Fig. 5. Graph of precision-confidence curve for YOLOv7-tiny

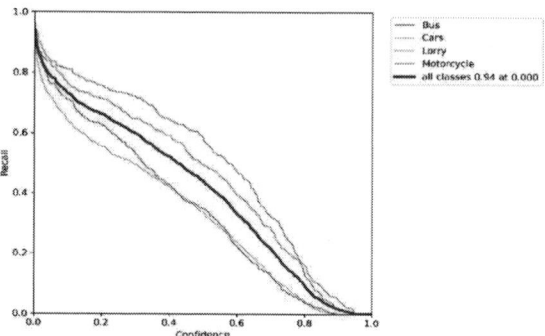

Fig. 6. Graph of recall-confidence curve for YOLOv7-tiny

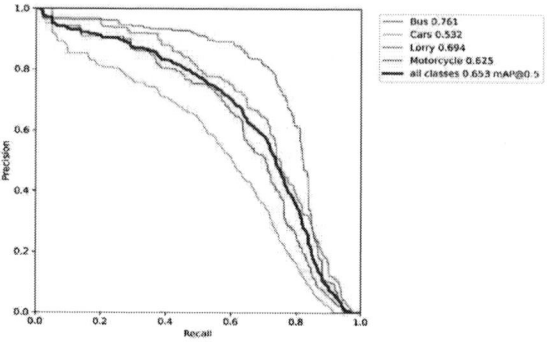

Fig. 7. Graph of precision-recall curve for YOLOv7-tiny

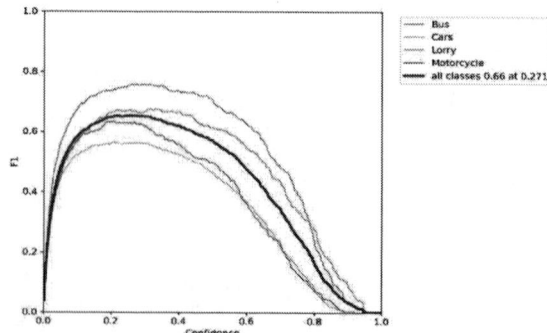

Fig. 8. Graph of F1-score-confidence curve for YOLOv7-tiny

Fig. 9. Result for inference (YOLOv5n)

Fig. 10. Result for inference (YOLOv7-tiny)

TABLE III. PERFORMANCE EVALUATION BASED ON INFERENCE FOR YOLOV5N AND YOLOV7-TINY

	YOLOv5n	YOLOv7-tiny
Inference computational time (ms)	89.9	44.9
Frames per seconds (fps)	11.1	22.3
Giga Floating-Point Operations Per Seconds (GFLOPs)	4.1	13.0
Parameters	1,764,577	6,015,714
Non-Max Suppression (NMS) per image (ms)	862.6	550.2
Layers	157	208

V. CONCLUSION

In conclusion, this work focuses on the development of vehicle detection and classification using embedded machine learning, which the edge device used for the deployment of the machine learning models is NVIDIA Jetson Nano 2GB Developer Kit. We have successfully designed efficient and accurate vehicle detection system for embedded systems via machine learning techniques, this has been achieved by deploying multiple machine learning algorithms to be used in Jetson Nano, currently, only deep learning algorithms are the best in order to obtain robust models, with the ability to do real-time detection.

Next, we implement a vehicle detection system tailored for embedded platforms. Deep learning remains the most efficient method for achieving strong overall performance metrics. Given the Jetson Nano's capacity to run deep learning algorithms, it's the optimal choice for achieving high accuracy while maintaining real-time performance. We also assess the performance of vehicle detection and classification in embedded platforms in two stages. First, we select machine learning algorithms based on performance metrics like accuracy, precision, recall, F1-score, and confusion matrix. Then, we evaluate the computational performance of the chosen models during inference using the Jetson Nano.

For future works, it will be the expansion of the dataset and more diverse samples to further enhance the system's performance as deep learning requires a huge amount of dataset to be able to detect with high levels of accuracy. Finally, integration with real world industry, as in Malaysia, there are still less usage of real-time vehicle detection and classification. It is worth noting that this project can be deployed as in the future, more and more cutting-edge devices will be introduced which allow us to have much more robust detection systems that can be integrated with real life situations.

REFERENCES

[1] Oluwaseyi, O., Irhebhude, M., & Evwiekpaefe, A. (2023). A Comparative Study of YOLOv5 and YOLOv7 Object Detection Algorithms. Journal of Computing and Social Informatics, 2, 1-12. https://doi.org/10.33736/jcsi.5070.2023

[2] Ajani, T. S., Imoize, A. L., & Atayero, A. A. (2021). An Overview of Machine Learning within Embedded and Mobile Devices–Optimizations and Applications. Sensors, 21(13).

[3] Joshi, P., Hasanuzzaman, M., Thapa, C., Afli, H., & Scully, T. (2023). Enabling All In-Edge Deep Learning: A Literature Review. IEEE Access, 11, 3431-3460. https://doi.org/10.1109/ACCESS.2023.3234761

[4] Mazzia, V., Khaliq, A., Salvetti, F., & Chiaberge, M. (2020). Real-Time Apple Detection System Using Embedded Systems With Hardware Accelerators: An Edge AI Application. IEEE Access, 8, 9102-9114. https://doi.org/10.1109/ACCESS.2020.2964608

[5] Neupane, B., Horanont, T., & Aryal, J. (2022). Real-Time Vehicle Classification and Tracking Using a Transfer Learning-Improved Deep Learning Network. Sensors, 22(10).

[6] Ong, K. W., & Loh, S. L. (2022). Vehicle Classification Using Neural Networks and Image Processing. International Journal of Electrical Engineering and Applied Sciences (IJEEAS), 5(2). https://ijeeas.utem.edu.my/ijeeas/article/view/6144

[7] Süzen, A. A., Duman, B., & Şen, B. (2020, 26-28 June 2020). Benchmark Analysis of Jetson TX2, Jetson Nano and Raspberry PI using Deep-CNN. 2020 International Congress on Human-Computer Interaction, Optimization and Robotic Applications (HORA),

[8] Alippi, C., Disabato, S., & Roveri, M. (2018, 11-13 April 2018). Moving Convolutional Neural Networks to Embedded Systems: The AlexNet and VGG-16 Case. 2018 17th ACM/IEEE International Conference on Information Processing in Sensor Networks (IPSN),

[9] Balid, W., Tafish, H., & Refai, H. (2017). Intelligent Vehicle Counting and Classification Sensor for Real-Time Traffic Surveillance. IEEE Transactions on Intelligent Transportation Systems, PP, 1-11. https://doi.org/10.1109/TITS.2017.2741507

[10] Blanco-Filgueira, B., García-Lesta, D., Fernández-Sanjurjo, M., Brea, V. M., & López, P. (2019). Deep Learning-Based Multiple Object Visual Tracking on Embedded System for IoT and Mobile Edge Computing Applications. IEEE Internet of Things Journal, 6(3), 5423-5431. https://doi.org/10.1109/JIOT.2019.2902141

[11] Carrasco, D. P., Rashwan, H. A., García, M. Á., & Puig, D. (2023). T-YOLO: Tiny Vehicle Detection Based on YOLO and Multi-Scale Convolutional Neural Networks. IEEE Access, 11, 22430-22440. https://doi.org/10.1109/ACCESS.2021.3137638

Modified Arnold Transform and DNA Manipulation for Chaos-Based RGB Image Encryption

Marwan A. Fetteha*, Wafaa S. Sayed†§, Lobna A. Said*, Ahmed H. Madian*‡

*Nanoelectronics Integrated Systems Center (NISC), Nile University, Giza 12588, Egypt. †Engineering
Mathematics Department, Faculty of Engineering, Cairo University, Giza 12613, Egypt.
‡Radiation Engineering Department, NCRRT, Egyptian Atomic Energy Authority, 29 Nasr City, Cairo, Egypt.
§Corresponding author e-mail:wafaa.s.sayed@eng.cu.edu.eg

Abstract—**Multimedia applications use image encryption algorithms extensively to safeguard and authenticate digital images. This paper presents an RGB image encryption method, which uses Chaos, DNA, pixel sum, and modified Arnold transform. The suggested algorithm is validated to be robust and resistant to visual, statistical, differential, and brute-force attacks. Additionally, the resulting encrypted images pass all tests of the NIST SP 800-22 test suite.**

Index Terms—**Arnold transform, Chaos, DNA, Image encryption.**

I. INTRODUCTION

The proliferation of digital communication networks and big data applications has made data protection an urgent concern. Data encryption is the primary method used to safeguard sensitive data. Therefore, cryptography is the most efficient and essential technology for information security [1].

A chaos system is a good tool for producing large amounts of pseudo-random sequences and building nonlinear encryption mechanisms. Chaotic systems can produce a vast number of keys rapidly and efficiently [2].

Many researches used chaotic systems as pseudo-randomness sources for their encryption algorithms. Double chaotic systems were employed in [3], where the Baker chaotic map determines the logistic chaotic map's state variables and parameters. A memristive chaotic system with transcendental nonlinearities was used in [4] to encrypt RGB images. In [3], the grayscale images were encrypted using a true random number generator and a chaotic system, which results in the growth of the system's complexity. Based on two chaotic systems and DNA manipulation, a plaintext-related encryption technique was described in [5].

This work proposes a novel technique to encrypt RGB images using the Lorenz hyperchaotic system, modified Arnold transform, and DNA manipulation. The methodologies used are briefly described in Section II. The presented encryption and decryption techniques are explained in Section III. Their performance is evaluated in Section IV. The paper is then wrapped up in Section V.

II. PRELIMINARIES

Randomness sources that can be replicated in the decryption process are needed for encryption systems. The primary sources of randomization used in the suggested system are described in this section.

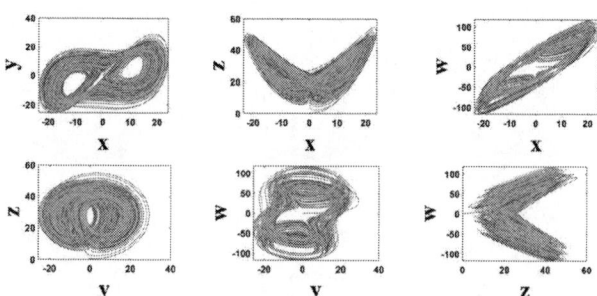

Fig. 1. The results of Lorenz hyperchaotic system.

A. Hyperchaotic Lorenz

The required randomness for encryption is provided by hyperchaotic Lorenz system [6], which is solved using Euler technique as follows:

$$x_{i+1} = h(\alpha(y_i - x_i) + w_i) + x_i, \tag{1a}$$
$$y_{i+1} = h(\theta x_i - y_i - x_i z_i) + y_i, \tag{1b}$$
$$z_{i+1} = h(x_i y_i - \beta z_i) + z_i, \tag{1c}$$
$$w_{i+1} = h(y_i z_i + \rho w_i) + w_i, \tag{1d}$$

where $\alpha = 10$, $\theta = 28$, $\rho = -1$, $\beta = 8/3$, and $h = 0.01$.

Considering the initial conditions x_0, y_0, z_0 and w_0 = 0.23, 0, 0.7, and 0.11 respectively. The outputs are shown in Fig. 1.

B. DNA Coding

The bit values are changed using DNA coding [7]. This is done to increase the algorithm's security. The four nucleotides that make up DNA are Cytosine (C), Thymine (T), Adenine (A), and Guanine (G). They are related in such a way that "A" complements "T" and "G" complements "C". Each DNA base corresponds to a binary code as shown in Table I.

If the relationship between these bases does not change, we can apply rules to edit the data based on these connections. Table I shows all potential rules that may be used for the encryption technique using a random integer in the rule choice. The associated DNA base is then substituted for the two input bits. For example, if the input is 'A' and the rule of choice is 4, the output will be 'T' equivalent to '10'.

Table II represents the DNA subtraction and addition results. These results could be performed by straightforward operations on DNA bases if Table I is used as a representation.

979-8-3503-8083-5/23 $31.00 © 2023 IEEE

TABLE I
DNA BINARY CODES, AND THERE ENCODING AND DECODING RULES

DNA base	Binary code	Rules							
			8	7	6	5	4	3	2	1
C	11	C	A	G	T	G	C	T	A	C
T	10	T	C	T	C	A	A	G	G	T
A	01	A	G	A	G	T	T	C	C	A
G	00	G	T	C	A	C	G	A	T	G

TABLE II
THE SUBTRACTION AND ADDITION RULES OF DNA

+	G	A	T	C	−	G	A	T	C
C	C	G	A	T	C	C	T	A	G
T	T	C	G	A	T	T	A	G	C
A	A	T	C	G	A	A	G	C	T
G	G	A	T	C	G	G	C	T	A

Every base in the DNA sequence is repeated four times, resulting in the pattern (G, A, T, C, G ...). So, it is possible to execute "DNA cycling" by using $mod4$ on the number of cycles.

C. Modified Arnold Transform

In [8], the Arnold transform is a periodic transformation, and after a certain number of iterations or cycles, the transformed or permuted picture becomes identical to the original image [8].

Suppose the number of Arnold transform cycles is determined at random. If it is either P or 0, the image will not be permuted, considering P is the period of the transformation for an image with $M \times M$ dimensions.

A modified Arnold transform was suggested in [9], which gets around this periodicity and permute the picture across any number of selected cycles (Cyc) by applying $G = 1 + mod(Cyc, P-2)$. This will prevent these two scenarios 0 and P and completely rule out the possibility of periodicity by lowering the adequate number of cycles G to a range between 1 and $(P-1)$.

III. PROPOSED ALGORITHM

1) In the hyperchaotic Lorenz system (1), each variable, x_0, y_0, z_0, and w_0, has its starting state defined by the hexadecimal representation of the four input sub keys (K_1, K_2, K_3, and K_4). The following are the calculated initial conditions that will keep this chaotic system inside its attraction basin:

$$x_0 = \left(\frac{K_1}{A/40}\right) - 20, \quad y_0 = \left(\frac{K_2}{A/40}\right) - 20, \quad (2a)$$

$$z_0 = \left(\frac{K_3}{A/50}\right), \quad w_0 = \left(\frac{K_4}{A/200}\right) - 100, \quad (2b)$$

where $A = 2^{52}$. So, the four chaotic sequences w, y, z and x generated with length $= M^2 + 1000$.

2) To produce X_h, Y_h, Z_h, and W_h, for each of the four chaotic sequences, the first 1000 iterations are

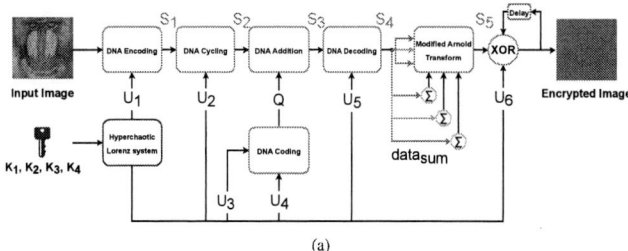

(a)

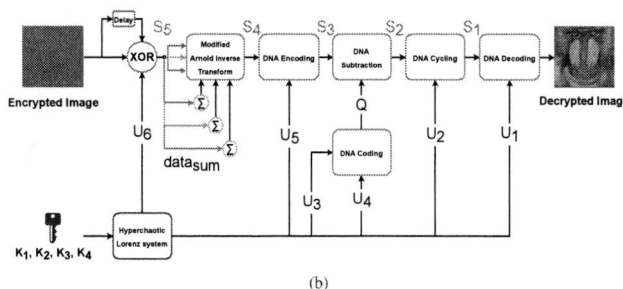

(b)

Fig. 2. Block diagram of (a) Encryption and (b) decryption processes.

eliminated. Next, the vectors U_1, U_2, U_3, U_4, U_5, and U_6 are created using:

$$U_1 = 1 + mod(\lceil X_h \times 10^{13}\rceil, 8), \quad (3a)$$

$$U_2 = mod(\lceil (U - \lceil U\rceil) \times 10^{13}\rceil, M^2), \quad (3b)$$

$$U_3 = 1 + mod(\lceil W_h \times 10^{13}\rceil, 8), \quad (3c)$$

$$U_4 = 1 + mod(\lceil (X_h + Y_h) \times 10^{13}\rceil, 256), \quad (3d)$$

$$U_5 = 1 + mod(\lceil Y_h \times 10^{13}\rceil, 8), \quad (3e)$$

$$U_6 = mod(\lceil (W_h + Z_h) \times 10^{13}\rceil, 256), \quad (3f)$$

where $\lceil\ \rceil$ is the ceiling operator, and $U = [X_h, Y_h, Z_h, W_h]$.

3) The rule of DNA that encodes the input image in accordance with Table I is chosen Using U_1.

4) DNA cycling is done on S_1 using $mod(U_2, 4)$.

5) DNA encodes U_4 using U_3 to produce Q. Following that, S_2 is subjected to the following equations in accordance with Table II:

$$q = Q(1) - Q(M^2), \quad (4a)$$

$$S_3(1) = Q(1) + S_2(1) + q, \quad (4b)$$

$$S_3(i) = S_2(i) + S_2(i-1) + Q(i). \quad (4c)$$

6) U_5 is used for the rule selection of DNA decoding of S_3 according to Table I.
The DNA decoding rule for S_3 is chosen using U_5 according to Table I.

7) The color channels (G, B and R) of S_4 are separated. For each channel, the "$data_{sum}$" is computed independently. Each channel is subjected to the suggested modified Arnold transform individually to produce S_5, where $Cyc = data_{sum}$.

8) To produce the encrypted image, S_5 is XORed with U_6.

A. Decryption Process

1) The input encrypted image is XORed with U_6.

TABLE III
STATISTICAL ANALYSIS RESULTS FOR THE PROPOSED SYSTEM.

Image	Channel	Encryption Quality Metrics			Correlation($\times 10^2$)			Entropy	Robustness Against Differintial Attacks	
		MSE	PSNR	SSIM	H	V	D		NPCR	UACI
House (256×256)	R	6837	9.7820	0.0111	-0.38	0.30	0.05	7.9969	99.5487	33.3784
	G	8661	8.7550	0.0105	0.42	-0.40	0.53	7.9969	99.6165	33.4696
	B	9665	8.2784	0.0101	0.18	-0.06	-0.78	7.9971	99.6035	33.4700
Baboon (512×512)	R	8650	8.7606	0.0111	0.02	-0.22	0.19	7.9993	99.6117	33.4618
	G	7743	9.2413	0.0098	0.02	-0.17	-0.14	7.9993	99.6045	33.5048
	B	9490	8.3581	0.0084	-0.10	0.22	-0.16	7.9993	99.6072	33.4568
Pepper (512×512)	R	8000	9.0993	0.0110	0.21	0.00	0.15	7.9992	99.6070	33.4700
	G	11259	7.6155	0.0082	0.21	0.08	-0.16	7.9993	99.6071	33.4313
	B	11123	7.6681	0.0072	-0.05	-0.04	0.13	7.9993	99.6134	33.4680
Lake(512×512)	R	7301	9.4969	0.0101	-0.11	-0.08	-0.12	7.9994	99.6121	33.4617
	G	11448	7.5434	0.0086	0.17	0.23	0.02	7.9992	99.6121	33.4799
	B	11537	7.5096	0.0070	-0.21	-0.12	-0.13	7.9993	99.6116	33.4397

2) The decryption process in this step is the same as step 7 in the encryption process. The only difference is the use of the inverse transform of Arnold. We take the "$data_{sum}$" before the inverse transform, which is not symmetric with the encryption process. The transformation of Arnold does not change the pixel's values; it changes their positions only. This property makes the decryption possible.

3) U_5 is used to select the DNA coding rule for S_4.

4) U_3 is used to DNA encode U_4 to generate Q. The following equations are applied according to Table II

$$q = Q(1) - Q(M^2), \tag{5a}$$
$$S_2(1) = S_3(1) - Q(1) - q, \tag{5b}$$
$$S_2(i) = S_3(i) - Q(i) - S_3(i-1). \tag{5c}$$

5) U_2 is used for cyclic shift S_2. It is done by checking the $mod(U_2, 4)$ result to choose the number of times the data is shifted.

6) U_1 is used for the selection of DNA decoding rule for S_1 to restore original image.

IV. EVALUATION OF PERFORMANCE

A. Key Space

A key length of 128 bits or more is recommended for security against brute-force attacks [10]. The suggested approach employs four 52-bit sub-keys. This accumulates to a total key space of 2^{208}.

B. Statistical Analysis

1) Histogram, MSE and PSNR: Figure 3 demonstrates the RGB histograms for the original and encrypted images. This figure shows a flat and uniform distribution of the results. The metrics used to evaluate the effectiveness of encryption are the Peak Signal-to-Noise Ratio (PSNR) [11] and the Mean Square Error (MSE) [12]. The values of High MSE and low PSNR indicate a significant difference between both the original and encrypted images. Table III gives the MSE and PSNR results.

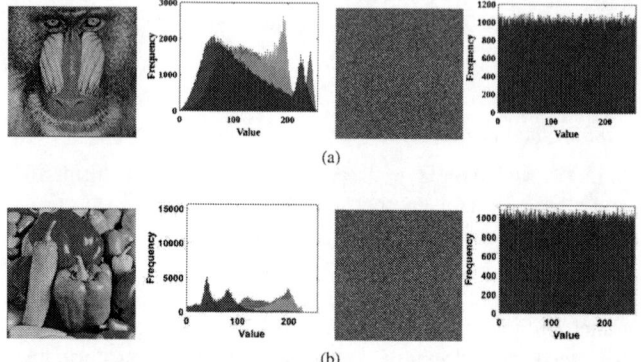

Fig. 3. RGB histogram for the original and encrypted (a) Baboon and (b) Pepper images.

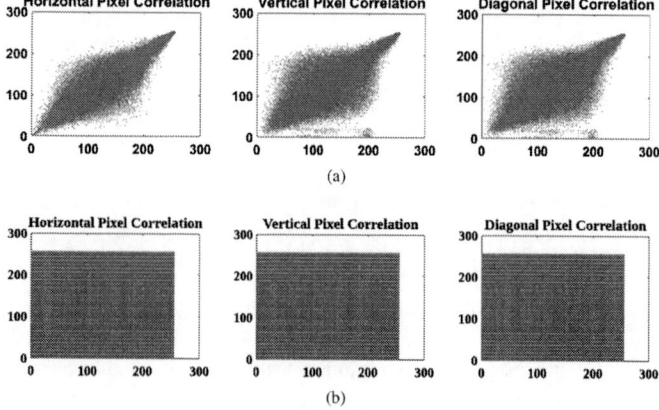

Fig. 4. Baboon's red channel's image correlation.

2) The Correlation Analysis: For encrypted images, the correlation coefficient values [11] must be near 0, which indicates that even adjacent pixels lack correlation. From Table III findings, the correlation coefficients are almost zero. Additionally, Figure 4 illustrates that the pixel values of the original image are clustered in an area, demonstrating their correlation. The pixel values of the encrypted images, however,

TABLE IV
COMPARISON OF THE STATISTICAL RESULTS WITH OTHER WORKS

	Proposed	[16]	[17]	[18]
MSE	8387.66	8395.53	8427.04	8351.64
PSNR	8.93846	8.89032	8.87405	8.91309
Entropy	7.99696	7.99729	7.9577	7.999
NPCR	99.5895	99.5972	99.6063	-
UACI	33.4393	29.6072	29.3973	-

TABLE V
BABOON (1024×1024) NIST RESULTS

Test name	P-VALUE	Result	PROPORTION	Result
BlockFrequency	0.025193	✓	1.000	✓
Runs	0.162606	✓	1.000	✓
OverlappingTemplate	0.534146	✓	1.000	✓
LongestRun	0.534146	✓	1.000	✓
Rank	0.534146	✓	1.000	✓
CumulativeSums	0.674344	✓	1.000	✓
Frequency	0.048716	✓	1.000	✓
FFT	0.162606	✓	0.958	✓
NonOverlappingTemplate	0.319325	✓	0.990	✓
Universal	0.534146	✓	1.000	✓
RandomExcursions	0.118159	✓	0.991	✓
Serial	0.187958	✓	1.000	✓
RandomExcursionsVariant	0.178160	✓	0.996	✓
ApproximateEntropy	0.637119	✓	1.000	✓
LinearComplexity	0.350485	✓	1.000	✓

are dispersed throughout.

3) The Information Entropy:: The average quantity of information that the pixels convey is known as information entropy [11]. The optimal value for an 8-bit value per channel is 8 to ensure the information is dispersed evenly across pixel values. In Table III, the findings show that the entropy successfully approaches 8 for each channel of the encrypted images.

4) Differential Attack Robustness: The least significant bit of a random pixel in the original image is changed for this test. The newly encrypted image is compared with the original image encrypted by using the Unified Average Changing Intensity (UACI) and the Pixel Number Change Rate (NPCR) [13]. The average values of 20 iterations were used to generate the NPCR and UACI values given in Table III. They are near the optimal levels, 99.61 and 33.46 percent, respectively [14], [15]. The statistical results are compared to other works in Table IV. The results show that the system is secure.

C. NIST SP 800–22 Test

The NIST SP 800–22 tests [19] are used to evaluate the sensitivity and unpredictability of the random sequence. The suite of tests evaluates the minimal requirements for a random-number stream via fifteen tests. The system successfully passes all tests using the Baboon image with a resolution of 1024×1024. Test results are shown in Table V.

V. CONCLUSION

This paper introduced an RGB image encryption system that uses hyperchaotic Lorenz system, DNA manipulation, and modified Arnold transform-based. The cases when pixel permutation cancels the modified Arnold transform are eliminated. The performance assessment of the proposed system demonstrates its capability of encrypting RGB images and suitability for security applications.

ACKNOWLEDGMENT

This paper is based upon work supported by Science, Technology, and Innovation Funding Authority (STIFA) under grant number (#38161).

REFERENCES

[1] Q. Lu, C. Zhu, and X. Deng, "An efficient image encryption scheme based on the lss chaotic map and single s-box," *IEEE Access*, vol. 8, pp. 25 664–25 678, 2020.

[2] S. M. Mohamed, W. S. Sayed, L. A. Said, and A. G. Radwan, "Reconfigurable fpga realization of fractional-order chaotic systems," *IEEE Access*, vol. 9, pp. 89 376–89 389, 2021.

[3] S. Zhou, X. Wang, Y. Zhang, B. Ge, M. Wang, and S. Gao, "A novel image encryption cryptosystem based on true random numbers and chaotic systems," *Multimedia Systems*, vol. 28, no. 1, pp. 95–112, 2022.

[4] S. M. Mohamed, W. S. Sayed, A. H. Madian, A. G. Radwan, and L. A. Said, "An encryption application and fpga realization of a fractional memristive chaotic system," *Electronics*, vol. 12, no. 5, p. 1219, 2023.

[5] M. Li, M. Wang, H. Fan, K. An, and G. Liu, "A novel plaintext-related chaotic image encryption scheme with no additional plaintext information," *Chaos, Solitons & Fractals*, vol. 158, p. 111989, 2022.

[6] X. Wang and M. Wang, "A hyperchaos generated from lorenz system," *Physica A: Statistical Mechanics and its Applications*, vol. 387, no. 14, pp. 3751–3758, 2008.

[7] J. Wu, X. Liao, and B. Yang, "Image encryption using 2d Hénon-sine map and DNA approach," *Signal processing*, vol. 153, pp. 11–23, 2018.

[8] L. Wu, J. Zhang, W. Deng, and D. He, "Arnold transformation algorithm and anti-Arnold transformation algorithm," in *2009 first international conference on information science and engineering*. IEEE, 2009, pp. 1164–1167.

[9] M. A. Fetteha, W. S. Sayed, L. A. Said, and A. G. Radwan, "Chaos-based image encryption using dna manipulation and a modified arnold transform," in *Model and Data Engineering: 11th International Conference, MEDI 2022, Cairo, Egypt, November 21–24, 2022, Proceedings*. Springer, 2022, pp. 3–15.

[10] S. Lian, *Multimedia content encryption: techniques and applications*. Auerbach Publications, 2008.

[11] M. Kaur and V. Kumar, "A comprehensive review on image encryption techniques," *Archives of Computational Methods in Engineering*, vol. 27, no. 1, pp. 15–43, 2020.

[12] I. Mehra and N. K. Nishchal, "Optical asymmetric image encryption using gyrator wavelet transform," *Optics Communications*, vol. 354, pp. 344–352, 2015.

[13] Y. Pourasad, R. Ranjbarzadeh, and A. Mardani, "A new algorithm for digital image encryption based on chaos theory," *Entropy*, vol. 23, no. 3, p. 341, 2021.

[14] A. Alghafis, N. Munir, M. Khan, and I. Hussain, "An encryption scheme based on discrete quantum map and continuous chaotic system," *International Journal of theoretical physics*, vol. 59, no. 4, pp. 1227–1240, 2020.

[15] F. Yu, X. Kong, A. A. M. Mokbel, W. Yao, and S. Cai, "Complex dynamics, hardware implementation and image encryption application of multiscroll memeristive hopfield neural network with a novel local active memeristor," *IEEE Transactions on Circuits and Systems II: Express Briefs*, vol. 70, no. 1, pp. 326–330, 2022.

[16] W. Alexan, N. Alexan, and M. Gabr, "Multiple-layer image encryption utilizing fractional-order chen hyperchaotic map and cryptographically secure prngs," *Fractal and Fractional*, vol. 7, no. 4, p. 287, 2023.

[17] M. Gabr, H. Younis, M. Ibrahim, S. Alajmy, I. Khalid, E. Azab, R. Elias, and W. Alexan, "Application of dna coding, the lorenz differential equations and a variation of the logistic map in a multi-stage cryptosystem," *Symmetry*, vol. 14, no. 12, p. 2559, 2022.

[18] W. Alexan, M. ElBeltagy, and A. Aboshousha, "Image encryption through lucas sequence, s-box and chaos theory," in *2021 8th NAFOSTED Conference on Information and Computer Science (NICS)*. IEEE, 2021, pp. 77–83.

[19] A. Rukhin, J. Soto, J. Nechvatal, M. Smid, and E. Barker, "A statistical test suite for random and pseudorandom number generators for cryptographic applications," Booz-allen and hamilton inc mclean va, Tech. Rep., 2001.

979-8-3503-8083-5/23 $31.00 © 2023 IEEE

Aloe Vera Tissue Modeling and Parameter Identification Using Meta-heuristic Optimization Algorithm

Mohamed S. Ghoneim[a], Dalia A. Fathi[b], Lobna A. Said[b], Ahmed H. Madian[b,c], and Magdy A. Bayoumi[a,d]

[a]Department of Electrical and Computer Engineering, University of Louisiana at Lafayette, LA, USA.
[b]Nanoelectronics Integrated Systems Center (NISC), Nile University, Cairo, Egypt.
[c]Radiation Engineering Dept., NCRRT, Egyptian Atomic Energy, Cairo, Egypt.
[d]Center for Advanced Computer Studies, University of Louisiana at Lafayette, LA, USA.

Abstract—The agricultural industry's use of non-invasive bio-impedance monitoring methods is expanding quickly. These measured impedance fluctuations reflect imperceptible biophysical and biochemical changes in living and non-living tissues. Bio-impedance circuit modeling is a valuable method for fitting the measured impedance in biology and medicine. A study on two samples of Aloe Vera leaves is conducted to identify the best model representing Aloe Vera leaves, and two different inter-electrode spacing distances are used to measure each sample. An electrochemical station (SP150) is used to detect bio-impedance at frequencies between 80 Hz and 200 kHz. The effectiveness of the employed models is evaluated by fitting them to the measured data. Among all models used, the medical plant stem and simplified plant stem models show the best results for all inter-electrode spacing distances.

Index Terms—Bio-impedance, Fractional-order circuits, Aloe Vera, Metaheuristic optimization.

I. INTRODUCTION

Aloe Vera is a cactus-like plant that can be found in dry hot areas around the world and lack of water places. It is a stemless or very short-stemmed plant that belongs to the Lily family [1]. Decades ago, Aloe Vera has been widely used in medicine as it has many antibacterial and antioxidant properties. It consists of 75 different potentially active components like vitamins, sugars, enzymes, lignin, minerals, and amino acids [2]. Aloe Vera unlike any other plant is used externally in treating burns, radiation effects on skin, incisions, and inflammation. In other words, Aloe Vera is essential to get Clammy Skin and speed up wound healing [3]. It was also used to prevent cancer and diabetic activities by extracting and ingesting its juice [4]. Aloe Vera can be used as a mouthwash to treat a sore mouth produced by radiotherapy treatment for cancer. Several medical studies have proven that Aloe Vera is very useful for patients with ulcerative colitis. In addition, Aloe Vera is a very effective treatment for those suffering from stomach colitis [5], and decreases the gastric mucosal lesions caused by 0.6 M HCl acid [6]. Moreover, Aloe Vera is effective for organ transplant patients as it can help in severe oral mucositis treatment [7]. Aloe Vera tissue has closed bio-impedance electrical circuits in its leaves [8]. Several species of aloe

Vera were studied in [9] to figure out some of the different biological properties. Additionally, it can be used to study electrochemical circuits and the responses of the biological closed circuits in different environments [10], [11]. In [10], the electrical signals transmissions over a long distance which were relevant to the biological closed circuits were studied. By using DC charged capacitor method (CSM), they found that Aloe Vera has a strong electrical anisotropy in its leaves.

Bio-impedance is defined as the reaction of biological cells to stimuli by an AC signal (voltage or current) [12], [13]. It is a safe technique used to figure out tissue characterization and disease diagnosis, and it changes according to size, health status, shape, type, and chemical composition [13], [14]. Although bio-impedance is applied in several human health and medical applications, it can be used to monitor plants in order to minimize damage and reduce the spread of plant diseases under several environmental conditions. In addition, Bio-impedance is used to study the effects of environmental changes on plants. On the other hand, Bio-impedance portrays the plants' behavior under special conditions such as plant maturity, heating and freezing. Besides, Bio-impedance is used to determine root growth water content [14], [15].

Bio-impedance circuit models are considered crucial aspects in biomedicine, biology, and plant physiology. Such models have been investigated for many purposes like characterizing the tissues of different fruits and vegetables to detect their physiological changes [13], [16], and to study the effects of drying and freezing-thawing treatments on eggplant [14], [17]. Several bio-impedance models have been proposed throughout the previous decades in order to accurately figure out the frequency-dependent impedance spectrum [16]. Bio-impedance circuit modeling is a significant tool to fit the measured impedance with biological data [15], [18].

Several bio-impedance models have been proposed throughout the previous decades in order to accurately figure out the frequency-dependent impedance spectrum and to reflect the electrical properties of plant biological cells [16]. The single-Cole dispersion model was the first model introduced to all bio-impedance circuit models. It was introduced in 1940

979-8-3503-8083-5/23 $31.00 © 2023 IEEE

by Kenneth and became the simplest commonly used model with good accuracy [19]. Later in 1990, a new model called double shell was presented which also gave good accuracy results at high frequencies [12]. Afterward, the Double-Cole circuit model was introduced to be an updated version of the single-Cole circuit model but with better accuracy in a wide frequency range [20]. Later in 2022, the medical plant stem model was introduced in [13]. This model was inspired by the stem of the plant and represents the stem tissue in its circuit model. After that, this model was simplified (simplified plant stem model) to reduce the vascular bundle elements and group them in one fractional-order capacitor and resistor.

Regardless of the model utilized, metaheuristic optimization approaches are typically applied to identify the model parameters [16]. Metaheuristic optimization is inspired by biological activity from nature. Water cycle algorithm (WCA) which is well known as the hydrological cycle is a powerful technique that can mimic the observation of the river flow and streams toward the sea [21]. Recently, numerous WCA applications have been introduced to several optimization fields [13], [22]. WCA affords high-accuracy solutions in solving optimization problems compared to other optimization techniques [13], [15], [23]. In other words, WCA is utilized in several areas to solve large-scale optimization problems with high performance. Motivated by this, we use the WCA optimization technique in this work to obtain high-accuracy results.

In this paper, Five fractional-order electrical impedance models are tested to represent Aloe Vera tissue. Using the SP150, the impedance of the two Aloe Vera samples is measured. The five fractional-order bio-impedance models are fitted with the measured impedance data, and the parameters of the models are then obtained using the metaheuristic optimization algorithm (WCA). Nyquist plot shows the fitted data compared to the measured data for all models. To determine the optimal model, the difference between the fitted data and measured data is calculated.

The structure of this paper is as follows: The analysis and representation of the impedance model circuit are briefly discussed in Section II. The formulation of the problem is shown in Section III. The experimental findings and discussion are presented in Section IV. Section V provides the paper's conclusion.

II. ALOE VERA MODELING

When used on various tissues, the Cole model, shown in Fig. 1(a), produced a satisfactory fit to experimental impedance data. It was first introduced as an all-encompassing model. Later, it was used more precisely to characterize the state of plants. It was presented as follows:

$$Z(s) = R_\infty + \frac{R_o - R_\infty}{1 + S^\alpha C_\alpha (R_o - R_\infty)}, \quad (1)$$

where R_o represents the resistance at low frequency, R_∞ represents the high-frequency resistance and α represents the Constant phase element order [24].

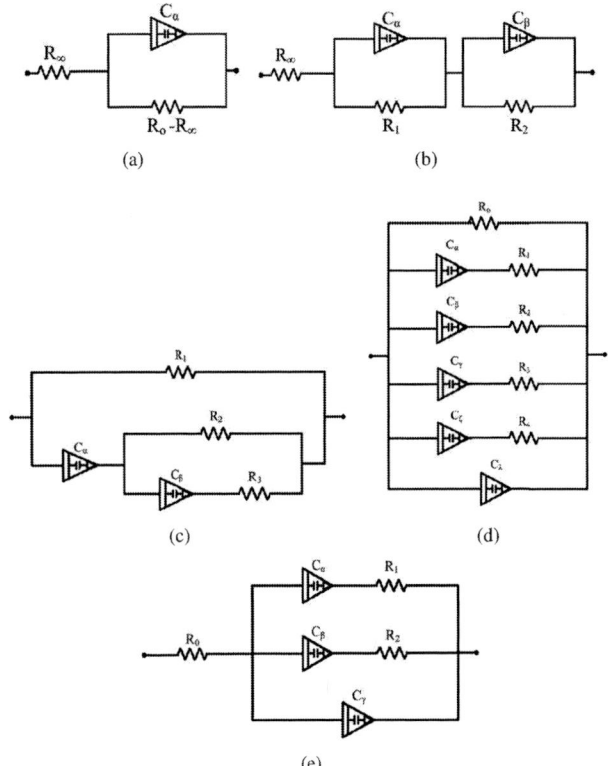

Fig. 1. Bio-impedance electrical circuit models (a) Single-Cole model (b) Double-Cole model (c) FO Double shell model (d) medical plant stem model and (e) simplified plant stem model.

Double-Cole is the expanded version of the single-Cole dispersion model. It was introduced to give complex materials better results and representation. The double-Cole is presented as follows:

$$Z(s) = R_\infty + \frac{R_1}{1 + S^\alpha R_1 C_\alpha} + \frac{R_2}{1 + S^\beta R_2 C_\beta}. \quad (2)$$

The double-Cole model is used to represent different plant and fruit tissues. The double dispersion Cole model has shown better results in [20] compared to the single-Cole dispersion model when both are applied on apples.

The fractional order double shell model was also introduced to represent plant cells in a better way and to give higher accuracy by adding the representation of the Vacuole inside the cell in 1990 [12]. The model is described as follows:

$$Z(s) = \frac{R_1 \left(S^{\alpha+\beta} C_\beta C_\alpha R_2 R_3 + S^\alpha C_\alpha R_2 + S^\beta C_\beta (R_2 + R_3) + 1 \right)}{S^{\alpha+\beta} C_\beta C_\alpha (R_2 R_3 + (R_2 + R_3) R_1)} \\ + S^\beta C_\beta (R_2 + R_3) + S^\alpha C_\alpha (R_1 + R_2) + 1}, \quad (3)$$

where C_β represents the plasma membrane capacitance, C_α and R_3 represent the Vacuole resistance and capacitance respectively, R_1 represents the extracellular resistance, and R_2 represents the intracellular resistance. It was proved that the fractional-order double shell model gives better results in plants' ripening period [16].

979-8-3503-8083-5/23 $31.00 © 2023 IEEE

The plant stem is a vital part of the plant. It is responsible for the plant's photosynthesis process, helps in water transportation, and protects the plant. Layers of the Plant stem can be divided into ground, vascular, and Epidermis systems mainly [13]. So, a plant stem model was introduced to present plant stem as follows:

$$Z(S) = \frac{R_o(1 + S^\alpha R_1 C_\alpha)(1 + S^\beta R_2 d_\beta)}{Z_b + Z_c + Z_r + Z_n}, \quad (4a)$$

$$Z_b(S) = \frac{(S^\alpha R_1 C_\alpha + S^\alpha R_o C_\alpha + 1)(S^\beta R_2 C_\beta + 1)}{(S^\gamma R_3 C_\gamma + 1)(S^\zeta R_4 C_\zeta + 1),} \quad (4b)$$

$$Z_d(S) = \frac{R_o(1 + S^\alpha R_1 C_\alpha)(S^{\gamma+\beta} R_3 C_\gamma C_\beta + S^{\gamma+\beta} R_2}{C_\beta C_\gamma + S^\beta C_\beta + S^\gamma C_\gamma)(S^\zeta R_4 C_\zeta + 1),} \quad (4c)$$

$$Z_r(S) = \frac{(1 + S^\beta R_2 C_\beta)(1 + S^\gamma R_3 C_\gamma)(1 + S^\alpha R_1 C_\alpha)}{S^\zeta C_\zeta R_o,} \quad (4d)$$

$$Z_n(S) = \frac{S^\lambda C_\lambda R_o(1 + S^\alpha R_1 C_\alpha)(1 + S^\beta R_2 C_\beta)}{(1 + S^\gamma R_3 C_\gamma)(1 + S^\zeta R_4 C_\zeta).} \quad (4e)$$

The stem circuit model was introduced to represent plant stems in 2022. This model was used to match the measurement data of Salvia officinalis L., Origanum majorana, and Lavandula species from the Lamiaceae family [13]. The plant stem model produced high-accuracy results compared to the Cole, double-Cole, and double-shell models. It was proposed mainly for medical plant stem [13].

The simplified stem model was introduced right after the plant stem model in [13]. It simplifies the plant stem model by grouping the Xylem and Phloem tissues as a vascular bundle in one branch. This model is represented as follows:

$$Z(s) = R_o + \frac{(S^\beta C_\beta R_2 + 1)(S^\alpha C_\alpha R_1 + 1)}{Z_k + Z_m}, \quad (5a)$$

$$Z_k(S) = S^{\alpha+\beta} C_\alpha C_\beta (R_1 + R_2) + S^\beta C_\beta + S^\alpha C_\alpha, \quad (5b)$$

$$Z_m(S) = \frac{S^\gamma C_\gamma (S^{\alpha+\beta} C_\alpha C_\beta R_1 R_2 + S^\beta C_\beta R_2}{+ S^\alpha C_\alpha R_1 + 1).} \quad (5c)$$

III. Problem Definition

Due to the notable medical uses and benefits of Aloe Vera and its availability, two samples are chosen and bought from

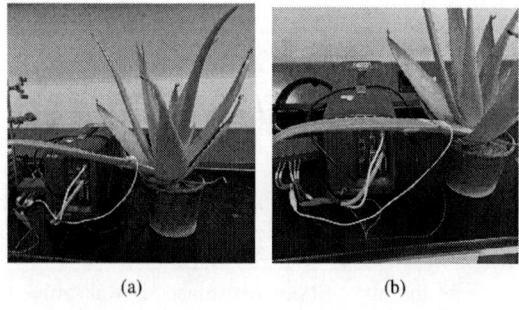

(a)   (b)

Fig. 2. Experimental samples setup for the Aloe Vera. Two electrodes are placed on the employed sample's leaves a) cross and b) at a distance of 5 cm.

the market. The experiment makes use of the two Aloe Vera plant samples. The impedance of the plants is measured, using an electrochemical workstation (SP150) that is frequently used in impedance analysis, at room temperature of 25°C for a frequency range from 80 Hz to 200 kHz. As demonstrated in Fig. 2 electrodes are inserted non-invasively across the plant leaves and 5 cm apart from one another for each sample. To prevent any changes in the impedance, the two trials are conducted back-to-back on both plants. There is no DC offset and the applied sinusoidal voltage excitation is $V_{rms} = 20mV$. 80 points per decade is the number used to measure the points. The log is then imported into MATLAB for post-processing to take place.

The measured impedance values are post-processed using WCA, a metaheuristic optimization algorithm, to determine the models' parameters that were employed. According to the literature, the implemented algorithm was selected since it produced results that were satisfactory. To achieve the optimum bio-impedance models' parameters, it is crucial to properly identify the variables impacting the optimization algorithm's outcome. Among these factors are the number of search agents, the objective function, the upper and lower bounds, runs and iterations, and the vector of optimized variables.

1) The sum of the absolute errors between the estimated impedance from the model and the measured impedance of the sample for each frequency point is the objective function ($f(x)$) described in Eqn. (6).

$$f(x) = min \sum_{i=1}^{n} \left| \frac{Zi_{model} - Zi_{measured}}{Zi_{measured}} \right|, \quad (6)$$

where impedance parameters (x) of each model depend on the problem size, $Z_{measured}$ is the measured response of the sample, while Z_{model} is the impedance equation of the models. n is the total number of the measured points.

2) The number of runs is 50 independent runs, 60 search agents used in the optimization, and through 1500 iterations for all the tested samples.

3) In a zone defined between a lower (LB) and an upper (UB) boundary that is defined independently for each model, the search agents look for the optimal answer, are shown in Table I.

4) a) the impedance parameters for Cole-impedance model are $[\alpha, R_\infty, R_o, C]$.

   b) for Double Cole-impedance model are $[\alpha, \beta, R_\infty, R_1, R_2, C_\alpha, C_\beta]$.

   c) for FO Double shell model are $[\alpha, \beta, R_1, R_2, R_3, C_\alpha, C_\beta]$

   d) The medical plant stem model parameters are $[\alpha, \beta, \gamma, \zeta, \lambda, R_o, R_1, R_2, R_3, R_4, C_\alpha, C_\beta, C_\gamma, C_\zeta, C_\lambda]$.

   e) The impedance parameters for the simplified plant stem model are $[\alpha, \beta, \gamma, R_o, R_1, R_2, C_\alpha, C_\beta, C_\gamma]$.

IV. Results and Discussion

The measured data across and at a distance = 5 cm of the studied samples are fitted on five different models (Cole,

TABLE I
THE LOWER BOUNDARY (LB) AND THE UPPER BOUNDARY (UB) FOR EACH MODEL.

Models / Parameters	Cole		Double Cole		Double shell		Medical Plant Stem model		Simplified Plant Stem model	
	LB	UB	LB	UB	LB	UB	LB	UB	LB	UB
R_∞	0	100KΩ	0	5KΩ	-	-	-	-	-	-
R_o	100Ω	1MΩ	-	-	-	-	0	1GΩ	0	100MΩ
R_1	-	-	0	1MΩ	1kΩ	1MΩ	0	1GΩ	0	100MΩ
R_2	-	-	0	10MΩ	1kΩ	1MΩ	0	1GΩ	0	100MΩ
R_3	-	-	-	-	0	10KΩ	0	1GΩ	-	-
R_4	-	-	-	-	-	-	0	1GΩ	-	-
C_α	10pF	10μF	10pF	9μF	0	3μF	0	100μF	0	10μF
C_β	-	-	10pF	9μF	0	3μF	0	100μF	0	10μF
C_γ	-	-	-	-	-	-	0	100μF	0	10μF
C_ζ	-	-	-	-	-	-	0	100μF	-	-
C_λ	-	-	-	-	-	-	0	100μF	-	-
α	0	1	0	1	0	1	0	1	0	1
β	-	-	0	1	0	1	0	1	0	1
γ	-	-	-	-	-	-	0	1	0	1
ζ	-	-	-	-	-	-	0	1	-	-
λ	-	-	-	-	-	-	0	1	-	-

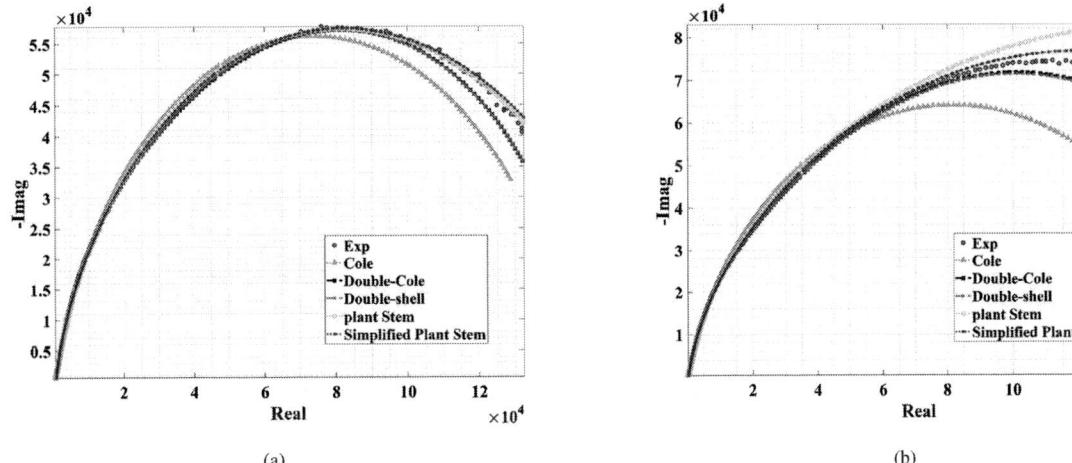

(a)   (b)

Fig. 3. Nyquist plots for all used models at (a) 5cm distance, and (b) across the sample.

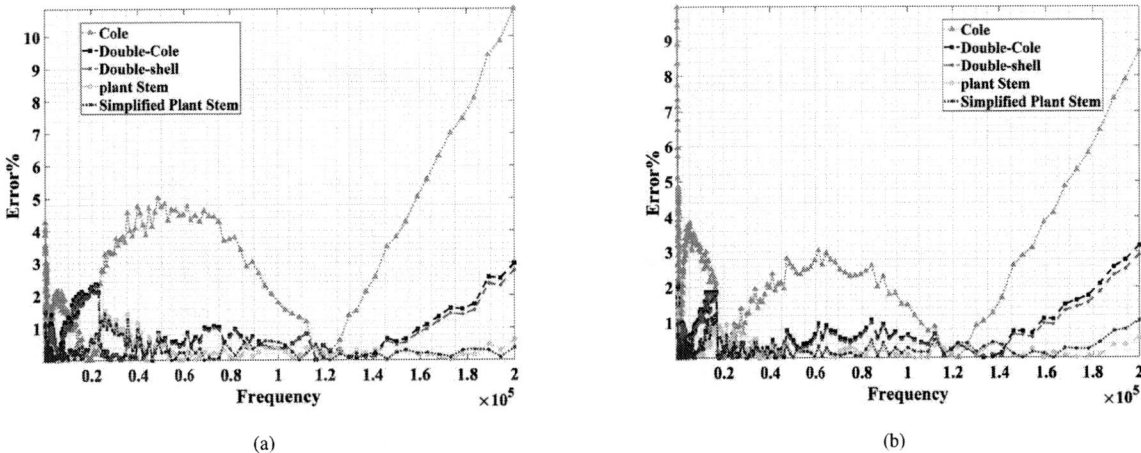

(a)   (b)

Fig. 4. Error plots for all used models at (a) 5cm distance, and (b) across the sample.

Double-Cole, Fractional-Order Double shell, medical plant stem, and simplified plant stem models) using WCA optimization technique to extract the models' parameters. As mentioned in the literature, WCA showed better performance compared with other optimization techniques [13], [15].

For the sample across the Aloe Vera leaves, the experimental data and the five applied models' extracted parameters are plotted on the Nyquist plot as seen in Fig.3. The maximum error calculated for the Cole model is around 10%, for the Double-Cole and Double shell are less than 3%, and for the medical plant stem model and the simplified plant stem model are around 1% as shown in Fig. 4(a).

For the sample at a distance of 5 cm, the experimental data and the five applied models' extracted parameters are plotted on the Nyquist plot as seen in Fig. 3. The maximum error calculated for the Cole model is around 9%, for the Double-Cole and Double shell are around 3%, and for the medical plant stem model and the simplified plant stem model are less than 1% as shown in Fig. 4(b).

The final outcome is that the simplified plant stem model showed a remarkable performance over the other employed models, followed by the medical plant stem model in representing Aloe Vera leaf tissue.

V. CONCLUSION

Aloe Vera is a very precious plant rich in its components, and it endures difficult conditions of scorching weather and a lack of water. Five bio-impedance models (Cole, Double-Cole, FO double shell, medical plant stem, and simplified plant stem) were compared to represent Aloe Vera leaf tissue. Nyquist plots showed the difference between the mentioned models at a 5 cm distance and across the sample. The simplified and the medical plant stem models gave 1% error, which is better than other employed models. It proved that the medical and the simplified plant stem models are the most suitable for representing Aloe Vera tissue.

REFERENCES

[1] D. Rakel and V. Minichiello, *Integrative Medicine, E-Book.* Elsevier health sciences, 2022.

[2] A. Surjushe, R. Vasani, and D. Saple, "Aloe vera: a short review," *Indian journal of dermatology*, vol. 53, no. 4, p. 163, 2008.

[3] D. Hekmatpou, F. Mehrabi, K. Rahzani, and A. Aminiyan, "The effect of aloe vera clinical trials on prevention and healing of skin wound: A systematic review," *Iranian journal of medical sciences*, vol. 44, no. 1, p. 1, 2019.

[4] T. Reynolds and A. Dweck, "Aloe vera leaf gel: a review update," *Journal of ethnopharmacology*, vol. 68, no. 1-3, pp. 3–37, 1999.

[5] K. Eamlamnam, S. Patumraj, N. Visedopas, and D. Thong-Ngam, "Effects of aloe vera and sucralfate on gastric microcirculatory changes, cytokine levels and gastric ulcer healing in rats," *World Journal of Gastroenterology: WJG*, vol. 12, no. 13, p. 2034, 2006.

[6] S. Yusuf, A. Agunu, and M. Diana, "The effect of aloe vera a. berger (liliaceae) on gastric acid secretion and acute gastric mucosal injury in rats," *Journal of ethnopharmacology*, vol. 93, no. 1, pp. 33–37, 2004.

[7] M. Guberti, S. Botti, C. Caffarri, S. Cavuto, L. Savoldi, A. Fusco, F. Merli, M. Piredda, and M. G. De Marinis, "Efficacy and safety of a colostrum-and aloe vera-based oral care protocol to prevent and treat severe oral mucositis in patients undergoing hematopoietic stem cell transplantation: a single-arm phase ii study," *Annals of Hematology*, vol. 101, no. 10, pp. 2325–2336, 2022.

[8] M. A. Mousa, M. Soliman, M. A. Saleh, and A. G. Radwan, "Tactile sensing biohybrid soft e-skin based on bioimpedance using aloe vera pulp tissues," *Scientific Reports*, vol. 11, no. 1, p. 3054, 2021.

[9] M. H. Radha and N. P. Laxmipriya, "Evaluation of biological properties and clinical effectiveness of aloe vera: A systematic review," *Journal of traditional and complementary medicine*, vol. 5, no. 1, pp. 21–26, 2015.

[10] A. G. Volkov, J. C. Foster, E. Jovanov, and V. S. Markin, "Anisotropy and nonlinear properties of electrochemical circuits in leaves of aloe vera l." *Bioelectrochemistry*, vol. 81, no. 1, pp. 4–9, 2011.

[11] K. Trebacz, H. Dziubinska, and E. Krol, "Electrical signals in long-distance communication in plants," *Communication in plants: Neuronal aspects of plant life*, pp. 277–290, 2006.

[12] M. Zhang, D. Stout, and J. Willison, "Electrical impedance analysis in plant tissues3." *Journal of experimental botany*, vol. 41, no. 3, pp. 371–380, 1990.

[13] M. S. Ghoneim, S. I. Gadallah, L. A. Said, A. M. Eltawil, A. G. Radwan, and A. H. Madian, "Plant stem tissue modeling and parameter identification using metaheuristic optimization algorithms," *Scientific Reports*, vol. 12, no. 1, p. 3992, 2022.

[14] B. M. Aboalnaga, L. A. Said, A. H. Madian, A. S. Elwakil, and A. G. Radwan, "Cole bio-impedance model variations in *daucus carota sativus* under heating and freezing conditions," *IEEE Access*, vol. 7, pp. 113 254–113 263, 2019.

[15] S. I. Gadallah, M. S. Ghoneim, A. S. Elwakil, L. A. Said, A. H. Madian, and A. G. Radwan, "Plant tissue modelling using power-law filters," *Sensors*, vol. 22, no. 15, p. 5659, 2022.

[16] M. Mohsen, L. A. Said, A. H. Madian, A. G. Radwan, and A. S. Elwakil, "Fractional-order bio-impedance modeling for interdisciplinary applications: A review," *IEEE Access*, vol. 9, pp. 33 158–33 168, 2021.

[17] M. Zhang and J. Willison, "Electrical impedance analysis in plant tissues: in vivo detection of freezing injury," *Canadian journal of botany*, vol. 70, no. 11, pp. 2254–2258, 1992.

[18] M. S. Ghoneim, A. A. Mohammaden, M. Mohsen, L. A. Said, and A. G. Radwan, "A modified differentiator circuit for extracting cole-impedance model parameters using meta-heuristic optimization algorithms," *Arabian Journal for Science and Engineering*, vol. 46, pp. 9945–9951, 2021.

[19] K. S. Cole, "Permeability and impermeability of cell membranes for ions," in *Cold Spring Harbor symposia on quantitative biology*, vol. 8. Cold Spring Harbor Laboratory Press, 1940, pp. 110–122.

[20] T. J. Freeborn, B. Maundy, and A. S. Elwakil, "Extracting the parameters of the double-dispersion cole bioimpedance model from magnitude response measurements," *Medical & biological engineering & computing*, vol. 52, pp. 749–758, 2014.

[21] A. Sadollah, H. Eskandar, H. M. Lee, J. H. Kim *et al.*, "Water cycle algorithm: a detailed standard code," *SoftwareX*, vol. 5, pp. 37–43, 2016.

[22] A. A. Heidari, R. Ali Abbaspour, and A. Rezaee Jordehi, "An efficient chaotic water cycle algorithm for optimization tasks," *Neural Computing and Applications*, vol. 28, pp. 57–85, 2017.

[23] A. JABBAR and S. ZAINUDIN, "Water cycle algorithm for attribute reduction problems in rough set theory." *Journal of Theoretical & Applied Information Technology*, vol. 61, no. 1, 2014.

[24] S. Mancuso, "Seasonal dynamics of electrical impedance parameters in shoots and leaves related to rooting ability of olive (olea europea) cuttings," *Tree physiology*, vol. 19, no. 2, pp. 95–101, 1999.

Convolutional Autoencoder for Real-Time PPG Based Blood Pressure Monitoring Using TinyML

Noor Faris Ali, Mousa Hussein, Falah Awwad, Mohamed Atef*

Electrical and Communication Engineering Department, College of Engineering, United Arab Emirates University, 15551
Al Ain, Abu Dhabi, United Arab Emirates, e-mail: moh_atef@uaeu.ac.ae

Abstract—In this paper, we propose an efficient and robust convolutional autoencoder (CAE) model for continuous real-time blood pressure (BP) monitoring. The proposed model was implemented on a resource-constrained edge device. The model was built to capture the hidden patterns among successive segments and alleviate the effects of momentary glitches and outliers. The model was deployed and assessed on the Arduino Nano 33 BLE Sense in a real-time environment by means of Tiny Machine Learning (TinyML). Extensive results revealed that the proposed model improved BP prediction accuracy on both offline and real-time experiments. With 4 features, the model achieved a mean absolute error±standard deviation (MAE±SD) of 2.81±2.84 and 1.51±1.85 mmHg for systolic BP (SBP) and diastolic BP (DBP), respectively, on a dataset of 40 subjects. Whereas microcontroller unit (MCU) based real-time continuous predictions attained 2.25±2.82 for SBP and 5.01±2.10 mmHg for DBP, on 8 volunteers. Compared to the state-of-the-art models implemented on tiny devices, our model showed superior robustness and accuracy. Overall, the study offered some important insights into the significance of compact and impactful feature set and the effectiveness of the proposed model in a real-time setting.

Keywords— Blood Pressure, PPG, TinyML, MCU, Convolutional Autoencoder, Arduino Nano

I. INTRODUCTION

Hypertension is increasingly recognized as an alarmingly worldwide public health concern [1]. In the last decade, researchers have devoted tremendous efforts to developing cuffless blood pressure (BP) monitors using Photoplethysmography (PPG) signals [2].

There exists a considerable body of literature on different machine learning (ML) algorithms and architectures to provide PPG based BP monitoring systems. For example, a multi-stage Convolutional Neural Network- Long Short-Term Memory (CNN-LSTM) model by [3] achieved mean absolute error±standard deviation (MAE±SD) of 3.97±5.55 mmHg for systolic BP (SBP) and 2.10±2.84 mmHg for diastolic BP (DBP). Prior work [4] developed a bidirectional LSTM model using PPG and electrocardiography (ECG) signals and achieved an MAE±SD of 0.73±0.95 for SBP and 0.55±0.51 for DBP. The work presented in [5] involved real-time acquisition of dual-site PPG signals to train and test an artificial neural network (ANN) model. The model achieved 0.29±4.49 mmHg for SBP and 0.5±2.4 mmHg for DBP.

There is a small number of research implemented real-time ML models on tiny devices for BP prediction. A major challenge with this kind of application is the constraints imposed by tiny devices including memory and energy limitations [6]. ML algorithms must be optimized to be more hardware-friendly while maintaining minimal error. This could be possible by lowering the input dimensionality and reducing the number of operations and computations

required, without a massive scarification in accuracy. Hence, data flow and memory necessities will be reduced [7].

In the last few years, Tiny Machine Learning (TinyML) has gained huge traction and has been rapidly evolving since then [8]. This quantum leap makes it possible to turn AI-based healthcare advances into applicable, wearable, and mobile products by deploying them in compact devices.

In the last couple of years, few papers have emerged to test the feasibility of running ML models on edge devices for real-time implementation. For example, in study [9] ANN model was implemented on STM32L4 microcontroller unit (MCU) for the real-time prediction of SBP using PPG signals. When the deployed model was tested on 8 subjects, it achieved an MAE±SD of 3.85±4.29 mmHg for SBP. Another study [10] intended to acquire synchronized PPG and ECG to extract hemodynamic parameters for BP monitoring. In this study, NRF52832 MCU was used for real-time implementation. 22 subjects were included in this experiment. A multiple linear regression (MLR) model was developed, and the average error was 0.002±8.544 mmHg for SBP, 0.005±6.690 mmHg for DBP.

PPG-based BP monitoring system was implemented in [11]. Four features were extracted for efficiency purposes from PPG and ECG. The study tested multiple models on three different edge devices. Briefly, the best results obtained, on 10 subjects, were by decision tree (DT) implementation on Arduino. The MAE±SD obtained was 6.85±9.16 and 14.08±17.82 for DBP and SBP, respectively.

From the short review above, research pertaining to offline testing showed outstanding results using a wealth of models that showed robust performance. The research that implemented models on tiny devices exhibited mixed results and reduced accuracy due to memory constraints and design restrictions to reduce complexity.

This work seeks to illuminate this uncharted area and remedy the problems in the extant literature. Accordingly, we designed a high-performing, interpretable, and efficient convolutional autoencoder (CAE) model and embedded it in a low-resources MCU using TinyML. The proposed model outperformed existing BP predictive models implemented on tiny edge devices in real-time operation.

II. METHODOLOGY

The methodological approach taken in this study proceeds in two main stages. First, the development and evaluation of the proposed convolutional autoencoder model, which involves database selection and processing of the PPG data to ensure model's validity for hardware implementation. Second, model deployment in a tiny device, namely Arduino Nano 33 BLE Sense, employing TinyML approach. Subsequently, we assess the performance of the model

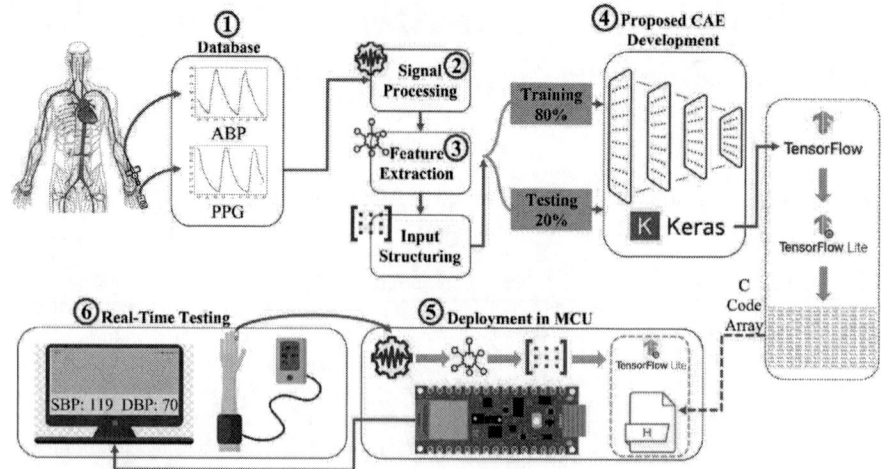

Fig. 1. The Overall Procedure of Model Development, TinyML Deployment, and Testing.

embedded in the hardware. In a real-time environment, features were extracted from a streamline of PPG signals acquired from new subjects (unseen data). Fig. 1 provides a summary of the steps taken to complete this work.

1. Dataset

The performance of the model was evaluated on a publicly available dataset, Physionet's MIMIC II [12]. Datasets of 40 individuals were selected considering three different BP categories: Normotension (40%), Hypertension (40%), and Hypotension (20%). Two-minute PPG and the corresponding arterial blood pressure (ABP) recordings were available for each subject. The SBP and DBP were extracted from the ABP signals.

2. PPG Signal Processing

PPG signals are prone to motion artifacts and other noise sources which interfere with the original signal bandwidth. Therefore, to mitigate motion artifacts, we harnessed a Butterworth 2nd order digital Low-Pass Filter (LPF) designed with a cut-off frequency of 6 Hz. Moreover, a digital High-Pass Filter (HPF) with a cut-off frequency of 0.5 Hz was employed for DC offset and baseline wander removal.

The amplitude of the PPG signal was normalized to a predefined range (from 0 to 1) to simplify the segmentation and features extraction procedures. The normalization was accomplished using the min-max formula expressed by (1).

$$Xn = \frac{X - \min(X)}{\max(X) - \min(X)} \tag{1}$$

Each PPG segment represents a cardiac cycle containing two valleys and one systolic peak. Peaks and valleys were detected using adaptive thresholding and comparison of neighboring samples techniques. After segmentation, the overall number of segments resulted was 7186 segments along with the corresponding SBP and DPB values extracted from the ABP signal.

3. Feature Extraction

From the empirical findings of our previous work [13], we highlighted some PPG features that have a significant influence on BP prediction accuracy and model interpretability. Accordingly, the following four features were selected to train and test our model: heart rate (HR), stroke volume (SV), peak width at half amplitude (PW), and diastolic time (peak-to-valley) (PVI). These four powerful features aid in developing an efficient, interpretable, and

effective model. The stroke volume quantity corresponds to the area under the curve (AUC) of one cardiac cycle or PPG segment.

The AUC was computed in valley-to-valley temporal window using Simpson integration rule (approximated by second order polynomial quadratic interpolant) using eq. (2). Where $\Delta t = (t_n - t_0)/n$, t_0= valleys[i] (beginning of a segment), t_n= valleys[i + 1] (end of a segment), and $t_0,...,t_n$ are the ends of the n subintervals.

$$SV = \int_{valleys[i]}^{valleys[i+1]} PPG(t)dt = \frac{\Delta t}{3}\left(PPG(t_0) + 4PPG(t_1) + 2PPG(t_2) + \cdots \; 2PPG(t_{n-2}) + 4PPG(t_{n-1}) + PPG(t_n)\right) \tag{2}$$

The rest of steps 4, 5, and 6 in Fig. 1: (4) development of proposed CAE model, (5) deployment in MCU, and (6) real-time testing are explained in next section III.

III. PROPOSED CONVOLUTIONAL AUTOENCODER DEVELOPMENT AND DEPLOYMENT

4. Proposed CAE Development

Despite the satisfactory performance of typical CNN, it is susceptible to noise and might be impractical due to the extraction of redundant uninterpretable features, resulting in a complex and inefficient model. To overcome this problem, we adopted the Convolutional Autoencoder (CAE) model because it is able to filter noise and compress information into a latent and refined representation while preserving the spatial coordinates processed by the CNN. The CAE topology guarantees a robust representation of features and thus stable predictions [14]. It is worth noting that convolutional autoencoder models are not as popular as typical CNNs in this research area, but they have been widely used in other fields [14-15].

The proposed model was deliberately designed to be efficient, interpretable, and powerful. Instead of conventionally feeding the raw PPG signals to the model, we structured the data as a matrix of successive segments and the corresponding interpretable features to make it suitable for CNN processing and to strike a trade-off between resources, accuracy, and interpretability of the model.

As depicted in Fig. 2, the model's first layer accepts 8*4 (segments*features) input matrix. The matrix enters two

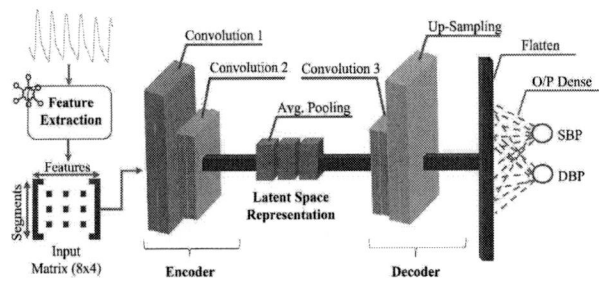

Fig. 2. Proposed Convolutional Autoencoder Topology.

Fig. 3. Deployment and Real-Time Implementation Procedure.

successive convolutional layers and gets convolved with the kernels to extract abstracted features maps at different levels. Average Pooling is applied to reduce the dimensionality (down sampling process) of produced features maps vectors Then, another convolutional layer followed by an up-sampling layer were set to increase the dimensionality of the features map, i.e., equalizing the spatial dimensions of the third convolutional layer to the first convolutional layer. The up-sampling layer carries the fine-grained details of feature maps generated by the last convolutional layer. Finally, a flatten layer followed by an output dense layer is assigned to perform the prediction of SBP and DBP.

In this scenario, average pooling layer serves as the bottleneck of the encoder. The coder or bottleneck acts as an intermediate medium holding a compressed and abstracted version of the data. Following, the decoder part starts with a convolutional layer that accepts the compacted high-level features representation and reconstructs the activation map.

Our goal was not to reconstruct the input, but to obtain detailed, high-level features and to capture the underlying complexities and hidden patterns among successive segments and alleviate the effects of outliers among the segments' features by smoothing and filtering through the model's bottleneck.

TABLE I shows the results obtained by the proposed CAE and a conventional CNN architecture with the same number of convolutional layers and parameters. All models were trained on the same datasets. The number of epochs was 200, learning rate was 0.007, and the batch size was 32 for all the models. It is explicitly observed that the proposed model delivered the best results; MAE±SD of 2.81±2.84 and 1.51±1.85 mmHg for SBP and DBP, respectively. This can be attributed to the CAE capabilities of mitigating noise and redundancy, learning richer representations, and capturing temporal relations among successive segments.

5. Model Deployment in MCU

To test our developed model for real-time inference, we selected the Arduino Nano 33 BLE sense MCU, with embedded artificial intelligence feature. It offers running edge computing applications using TinyML. It has a 1 MB CPU Flash Memory and 256 KB SRAM and

TABLE I. Typical CNN Vs Proposed CAE Model Results.

Model	Layers	MAE±SD (SBP) mmHg	MAE±SD (DBP) mmHg
CNN	3 Convolution, 1 Average Pooling	3.72±4.01	2.27±2.56
Proposed CAE	3 Convolution, 1 Average Pooling, 1 Up-sampling	2.81±2.84	1.51±1.85

runs operations with a clock speed of 64 MHZ [16]. As shown in Fig. 1, the proposed model was converted from Keras to TensorFlow Lite (TFL) format. Afterwards, in the Arduino sketch, we included all required libraries of TFL for MCU. We initialized TFL global variables and input (I/P) and output (O/P) tensors, pulled TFL model required operations, set memory buffer, allocated memory for the tensors, obtained TFL model representation in hexadecimal array (stored in the header file) and created interpreter to run the model.

As shown in Fig. 3, to obtain a BP prediction, a streamlined raw input from a PPG sensor was recorded. Then, raw signals were transformed into the same input shape that our model was trained on. All the processing steps accomplished offline were identically achieved by the MCU in real-time. Next, we passed this transformed data into our model input tensor and ran inference through invoking the deployed model. The output predictions of BP were elicited from the output tensor. Accounting for transient outliers and temporary glitches, we averaged BP predictions for both SBP and DBP, aiming for consistent behavior.

6. Real-Time Testing and Experimental Protocol

The proposed model was tested for its performance on new subjects in real-time. 8 subjects were recruited: 3 of them were clinically diagnosed with hypertension, while the other 5 were healthy. The following steps were accomplished to test the model on each subject, see Fig. 4:

1. Two Reference BP values measured using OMRON BP monitor were averaged to ensure reliable measurements.
2. The PPG sensor was attached to the subject's finger.
3. The TinyML-based program developed was executed on the Arduino IDE.
4. The features matrix (8*4) calculated in real time was passed to the input tensor.
5. Five continuous consecutive SBP and DBP predictions by the model were recorded and averaged.
6. The average predictions of the model were compared with the average reference BP to calculate the absolute error (AE).

IV. RESULTS AND DISCUSSIONS

1. Real-Time Testing Results

In this section, we present the experimental results. TABLE II demonstrates the averaged continuous real-time predictions of SBP and DBP obtained from the deployed convolutional autoencoder model in Arduino Nano 33 BLE Sense MCU and the measured reference values. Overall, the model achieved MAE±SD of 2.25±2.82 and 5.01±2.10 mmHg for SBP and DBP, respectively.

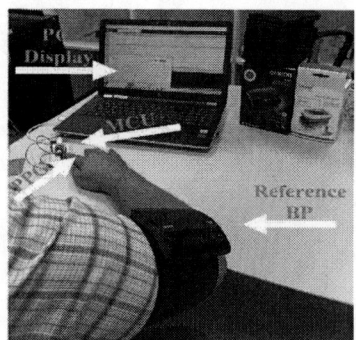

Fig. 4. Real-Time Testing for BP Prediction.

The empirical results confirm the validity and reliability of our proposed algorithm according to AAMI standards.

Another promising finding was the strong effect of the proposed convolutional autoencoder model in hardware implementation. In real-time evaluation, the model showed predominantly persistent behavior even when transitory outliers occur. This can be attributed to the robustness of the model and its filtering capabilities in real-time setting.

Regarding the limitations, it could be argued that the error in DBP predictions was higher compared to that produced on an offline basis (database). Surprisingly, the model performance on the database test set was 1.51±1.85, whereas, on the real-time setting, it obtained 5.01±2.10. We speculate that there is a possible confound in the database where the DBP values distribution was imbalanced. Accordingly, the accuracy of the DBP prediction could be enhanced by providing additional DBP data with a more symmetric distribution for training. In contrast, SBP predictions were mostly accurate. The results in real-time settings (2.25±2.82) are comparable to that in offline (2.81±2.84). Further investigations with larger data sizes are needed to validate the obtained results.

The proposed model exhibited outstanding performance with an efficient design and interpretable and compelling features set in both offline setting and hardware real-time testing. Together these results cast a new light on the combination of the proposed efficient model solidity and the compact features set sufficiency.

TABLE II. Real-Time Testing on MCU Results.

Subject	Averaged Reference SBP mmHg	Averaged Reference DBP mmHg	Averaged SBP Prediction mmHg	Averaged DBP Prediction mmHg
#1	116	71.5	116.959	65.424
AE			0.959	6.076
#2	125	75.5	125.618	71.658
AE			0.618	3.842
#3	163.5	75.5	161.904	71.306
AE			1.596	4.194
#4	121	79	122.046	72.57
AE			1.046	6.43
#5	122.5	74	122.022	75.13
AE			0.478	1.13
#6	167	81.5	158.632	73.57
AE			8.368	7.93
#7	112.5	75.5	112.16	71.26
AE			0.34	4.24
#8	151.5	76	156.12	82.242
AE			4.62	6.242
MAE±SD			2.25±2.82	5.01±2.10

2. Comparison with Other Models Implemented on MCU

Here we compare the results of the proposed method with those of traditional methods implemented in hardware [9-11] and other ones tested models on a PC [17-18]. As illustrated in TABLE III, compared to other models implemented on MCU using PPG with or without ECG signals, our model yielded the most accurate and stable predictions. In [10], both PPG and ECG signals were used to train an MLR model. The obtained average error is low, but high error variability is observed. In [11], PPG and ECG signal features were used, and a decision tree model was developed. The registered MAE±SD was extremely high. Both models implemented on edge devices in [10-11] did not meet AAMI standards. In study [9], PPG signals were used to train an ANN model and tested on STM32L4+ hardware for SBP prediction; the results were acceptable. In contrast, our model showed the lowest prediction error for both SBP and DBP, with only 4 features extracted from one signal (PPG).

Models introduced in [17-18] where tested on a PC. In [17], a convolutional autoencoder was built to receive raw PPG and ECG signals, the model resulted in 9.61±7.75 and 6.73±5.13 for SBP and DBP, respectively. On the other hand, the CNN-LSTM model implemented in [18] presented superior performance using raw PPG and ECG signals (2.57±2.39 and 3.44±2.63), nevertheless, our model achieved better DBP accuracy when tested on a PC (1.51±1.85) and comparable performance on SBP prediction (2.81±2.84). Overall, our method was the one that provided the most stable results in both real-time and offline settings.

TABLE III. Comparison with Other Models.

Study	Data Size	Signal	MCU	Model	MAE± SD (SBP) mmHg	MAE± SD (DBP) mmHg
[10]	22	PPG & ECG	NRF52832	MLR	0.002± 8.54	0.005± 6.690
[9]	8	PPG	STM32L4+	ANN	3.85± 4.29	N/A
[11]	10	PPG & ECG	Arduino Uno	DT	14.08± 17.8	6.85± 9.16
[17]	62	PPG & ECG	PC	CAE	9.61± 7.75	6.73± 5.13
[18]	30	PPG & ECG	PC	CNN-LSTM	2.57± 2.39	3.44± 2.63
This Work	40	PPG	PC	CAE	2.81± 2.84	1.51± 1.85
This Work	8	PPG	Arduino Nano 33 BLE Sense	CAE	2.25± 2.82	5.01± 2.10

CONCLUSION

This study provided the application of TinyML to implement a high-performance BP predictive model on a resource-limited edge device. It shows the compelling nature of the selected powerful features and the developed convolutional autoencoder's superior performance on both a PC and an edge device (Arduino Nano 33 BLE Sense). The model achieved 2.81±2.84 and 1.51±1.85 mmHg on a PC and 2.25±2.82 and 5.01±2.10 mmHg on the MCU for SBP and DBP, respectively. The results showed supportive evidence that our proposed CAE outperformed other models deployed on tiny devices. Further attempts could be quite beneficial to the literature to validate the conclusions drawn by this study.

References

[1] F. D. Fuchs and P. K. Whelton, "High Blood Pressure and Cardiovascular Disease," Hypertension, vol. 75, no. 2, pp. 285–292, Feb. 2020.

[2] C. El-Hajj and P. A. Kyriacou, "Cuffless blood pressure estimation from PPG signals and its derivatives using deep learning models," Biomedical Signal Processing and Control, vol. 70, p. 102984, Sep. 2021.

[3] J. Esmaelpoor, M. H. Moradi, and A. Kadkhodamohammadi, "A multistage deep neural network model for blood pressure estimation using photoplethysmogram signals," Computers in Biology and Medicine, vol. 120, p. 103719, May 2020.

[4] Y.-H. Li, L. N. Harfiya, K. Purwandari, and Y.-D. Lin, "Real-Time Cuffless Continuous Blood Pressure Estimation Using Deep Learning Model," Sensors, vol. 20, no. 19, p. 5606, Sep. 2020.

[5] A. Rababah, M. A. Khan, F. Awwad and M. Atef, "Dual-site Photoplethysmography Sensing for Noninvasive Continuous-time Blood Pressure Monitoring Using Artificial Neural Network," 2022 IEEE Biomedical Circuits and Systems Conference (BioCAS), Taipei, Taiwan, 2022, pp. 462-466

[6] A. Elsts and R. McConville, "Are Microcontrollers Ready for Deep Learning-Based Human Activity Recognition?," Electronics, vol. 10, no. 21, p. 2640, Oct. 2021.

[7] V. Sze, Y. -H. Chen, J. Emer, A. Suleiman and Z. Zhang, "Hardware for machine learning: Challenges and opportunities," 2017 IEEE Custom Integrated Circuits Conference (CICC), Austin, TX, USA, 2017, pp. 1-8.

[8] P. P. Ray, "A review on TinyML: State-of-the-art and prospects," Journal of King Saud University - Computer and Information Sciences, vol. 34, no. 4, pp. 1595–1623, Apr. 2022.

[9] B. C. Casadei, A. Gumiero, G. Tantillo, L. Della Torre, and G. Olmo, "Systolic Blood Pressure Estimation from PPG Signal Using ANN," Electronics, vol. 11, no. 18, p. 2909, Sep. 2022.

[10] Y. Zhang, C. Zhou, Z. Huang, and X. Ye, "Development of a Continuous Blood Pressure Monitoring System based on Pulse Transit Time and Hemodynamic Covariates," Proceedings of the 13th International Joint Conference on Biomedical Engineering Systems and Technologies, 2020.

[11] K. Ahmed and M. Hassan, "tinyCare: A tinyML-based Low-Cost Continuous Blood Pressure Estimation on the Extreme Edge," 2022 IEEE 10th International Conference on Healthcare Informatics (ICHI), Rochester, MN, USA, 2022, pp. 264-275.

[12] "PhysioNet," physionet.org. https://physionet.org/

[13] N. F. Ali and M. Atef, "LSTM Multi-Stage Transfer Learning for Blood Pressure Estimation Using Photoplethysmography," Electronics, vol. 11, no. 22, p. 3749, Nov. 2022.

[14] E. Pintelas, I. E. Livieris, and P. E. Pintelas, "A Convolutional Autoencoder Topology for Classification in High-Dimensional Noisy Image Datasets," Sensors, vol. 21, no. 22, p. 7731, Nov. 2021.

[15] K. Rai, F. Hojatpanah, F. Badrkhani Ajaei, and K. Grolinger, "Deep Learning for High-Impedance Fault Detection: Convolutional Autoencoders," Energies, vol. 14, no. 12, p. 3623, Jun. 2021.

[16] "Nano 33 BLE Sense | Arduino Documentation," docs.arduino.cc. https://docs.arduino.cc/hardware/nano-33-ble-sense.

[17] J. Zhang, D. Wu and Y. Li, "Cuff-less and Calibration-free Blood Pressure Estimation Using Convolutional Autoencoder with Unsupervised Feature Extraction," 2019 41st Annual International Conference of the IEEE Engineering in Medicine and Biology Society (EMBC), Berlin, Germany, 2019, pp. 3323-3326.

[18] P. Nandi and M. Rao, "A Novel CNN-LSTM Model Based Non-Invasive Cuff-Less Blood Pressure Estimation System," 2022 44th Annual International Conference of the IEEE Engineering in Medicine & Biology Society (EMBC), Glasgow, Scotland, United Kingdom, 2022, pp. 832-836.

Multi-class classification of melanoma on an edge device

Aser Ashraf Ali
Media Engineering and Technology
German University in Cairo
Cairo, Egypt
aser.ali@student.guc.edu.eg

Radwa Essam Taha
Media Engineering and Technology
German University in Cairo
Cairo, Egypt
radwataha1999@gmail.com

Ranpreet Kaur
Software Engineering and AI
Media Design School
Auckland, New Zealand
ranpreet.kaur@mediadesignschool.com

Shereen Moataz Afifi
Media Engineering and Technology
German University in Cairo
Cairo, Egypt
shereen.moataz@guc.edu.eg

Abstract—Melanoma is a dangerous skin cancer that requires early detection for successful treatment. This study presents an implantable diagnostic device that points to revolutionize early detection of melanoma. The device combines the most recent designs with re-configurable computing strategies, permitting for precise diagnosis and analysis of potential melanoma pictures. Furthermore, it utilizes advanced hardware and software components. Utilizing the VGG-16 profound learning model and augmented HAM10000 dataset, this study accomplishes a staggering accuracy rate. It is critical that a Raspberry Pi was utilized to test the usefulness of this model. Thorough tests and experiments have affirmed that this device can enormously improve the speed and accuracy of melanoma detection, potentially making a life-saving contribution within the field of early detection of melanoma. This study highlights the significance of utilizing state-of-the-art innovation in the battle against skin cancer and offers a promising arrangement for real-time detection of melanoma.

Index Terms—Melanoma, early detection, medical device, deep learning, skin lesion images, VGG-16,ResNet-50,multi-class.

I. Introduction

Skin cancer, including fatal melanoma, could be a worldwide public health issue with increasing rate. Early detection is critical to progress patient results and decrease mortality. Conventional discovery strategies based on visual inspection by dermatologists are subjective and time consuming. This study uses progressed technologies such as computer vision and deep learning [1] to develop a cost-effective and viable medical device for early detection of melanoma employing a Raspberry Pi-based platform. The device highlights fine-tuned VGG-16 CNN model trained on the HAM10000 dataset [2] to precisely classify skin lesions, including melanoma, nevus, and other conditions. The objective is to supply real-time predictions and empower fast and accessible melanoma screening in a variety of clinical settings, in this manner tending to a major healthcare challenge.

Later advances in computer-aided determination have showed the potential of profound learning algorithms, particularly convolutional neural networks (CNNs), in medical imaging applications. CNN exceeds expectations at image recognition assignments, making it reasonable for analyzing dermoscopy pictures of skin injuries [3]. This study builds on these accomplishments and leverages the control of huge datasets such as HAM10000 [2] to highlight the utilization of CNNs in melanoma detection. The proposed device not only represents a progressive application of this technology in medicine, but also will transform the landscape of melanoma diagnosis, eventually contributing to improved quiet results and decreased melanoma-related mortality.

This research is significant because it pioneers the combination of computer vision and deep learning, on a Raspberry Pi based platform to detect melanoma at a stage. The study demonstrates how a adjusted VGG 16 CNN model, trained using the HAM10000 dataset can accurately classify types of skin lesions. By utilizing the capabilities of networks in medical imaging specifically in dermoscopy analysis this research not only represents a noteworthy advancement, in healthcare but also has the potential to revolutionize melanoma diagnosis worldwide.

II. Related Work

Afifi et al. [4]–[12] proposed different designs for implementing a support vector machine (SVM) classifier implemented on a field programmable gate array (FPGA) targeting melanoma detection. The proposed SoC [4] was implemented on a Zynq-7000 FPGA, achieved classification accuracy from 97% to 98% for detecting melanoma and non melanoma images, while achieved low power consumption. The papers also proposed a dynamic hardware system for SVM classification of melanoma [5]. The dynamic hardware system achieved classification accuracy of 98.4%, which is higher than the accuracy of the static hardware system proposed in previous papers, and proposed a new medical device for early detection melanoma. This device has been implemented on a low-cost FPGA platform, making it suitable for mobile devices.

Agarwal et al. [13] proposed a convolutional neural network (CNN)-based approach for the classification of skin cancer

979-8-3503-8083-5/23 $31.00 © 2023 IEEE

images. They trained their CNN model on a dataset of 1,000 skin cancer images and achieved a classification accuracy of 98.2%. Their model was able to accurately classify benign and malignant skin lesions, demonstrating the potential of CNNs for skin cancer diagnosis.

Existing studies have demonstrated the effectiveness of deep learning models, such as ResNet-50, in melanoma detection, achieving high accuracies. Durães et al. [14] proposed a smart embedded system for skin cancer classification using a cascaded deep learning model. The model was trained on the HAM10000 dataset, which consisted of 10,000 dermoscopic images of melanoma and non-melanoma skin lesions. The system was implemented on a ZYNQ UltraScale+ MPSoC ZCU104 evaluation kit with a ZU7EV device. The system achieved a classification accuracy of 87%, with a frame rate of 13.5 FPS. The system was implemented on a low-cost FPGA platform, making it suitable for portable devices.

Jaisakthi et al. [15] proposed a deep learning model for skin cancer classification from dermoscopic images. The model was trained on the ISIC 2018 dataset, which consisted of 2,000 dermoscopic images of melanoma and non-melanoma skin lesions. The model was implemented on a NVIDIA GeForce GTX 1080 Ti GPU. The model achieved a classification accuracy of 95.2%. Bibi et al. [16] proposed a framework for skin lesion segmentation and classification using a combination of conventional and deep learning methods. It was evaluated on the ISIC 2017 dataset, which consisted of 2,000 dermoscopic images of melanoma and non-melanoma skin lesions, achieving a segmentation accuracy of 94.5% and a classification accuracy of 93.1%. It was implemented on a desktop computer with an NVIDIA GeForce GTX 1080 Ti GPU.

Sabouri et al. [17] proposed a cascade classifier for the diagnosis of melanoma in clinical images. The classifier was trained on a dataset of 1,000 melanoma and non-melanoma images, achieved a classification accuracy of 92.4%. The classifier was implemented on a personal computer with an Intel Core i7-4770 CPU.

However, challenges remain in terms of small datasets and manual feature extraction. Our research addresses a gap in the existing literature by focusing on the development of a resource-efficient and real-time melanoma detection device. Unlike previous studies, our contribution extends beyond algorithmic advancements, as we have implemented our melanoma diagnosis approach on hardware. This hardware implementation aims to facilitate real-time applications, providing a significant leap forward in the field. By bridging the gap between algorithmic research and practical hardware deployment, our work would add substantial value to skin melanoma detection in real-world scenarios.

III. METHODOLOGY

Melanoma is a dangerous skin cancer that has been on the rise in recent years. Early detection is critical for effective treatment. However, it can be troublesome for a dermatologist to distinguish a suspicious mole without performing a biopsy.

To address this issue, attempts are being made to create medical devices based on convolutional neural networks (CNNs) that might help in early detection of melanoma. The CNN model utilized is based on the VGG-16 architecture [18], trained on a dataset of around 32,000 skin lesion pictures, including 10,000 pictures from the HAM10000 dataset [2], and subjected to data augmentation techniques to progress class balance. Other CNN models such as a custom-built model, a Resnet-50 model, and VGG-16 modified model were tested too to compare performance. The VGG-16 model demonstrated the highest accuracy among the tested models, achieving a score of 76%. In comparison, the Resnet-50 model scored 70%, the modified VGG-16 model scored 61%, and the custom-built model scored 66%. This indicates that the VGG-16 model performed the best in terms of accuracy, outperforming the other models in the evaluation. This device is executed on a Raspberry Pi 3 Model B for real-time low power detection. This section covers data augmentation and preparation, the CNN model design, and the Raspberry Pi implementation.

A. Data Collection

The dataset used in this research is the HAM10000 dataset as shown in Fig. 1, a comprehensive collection of 10,000 dermoscopic pictures of skin lesions. The dataset includes seven particular classes, specifically Actinic keratoses and intraepithelial carcinoma (AKIEC), Basal cell carcinoma (BCC), Benign keratosis-like lesions (BKL), Dermatofibroma (DF), Melanocytic nevi (NV), Melanoma (MEL), and Vascular injuries (VASC). Each class captures particular characteristics and highlights relevant to melanoma detection. [2].

B. Pre-processing and Data Augmentation

Earlier to model training, the HAM10000 dataset underwent preprocessing and data augmentation to improve the model's execution and generalization capabilities. The pictures were resized to a standardized input size of 224 x 224 pixels, ensuring compatibility with the VGG-16 architecture. Different techniques have been utilized to expand the dataset, enormously increasing its quality and diversity. One of the main improvement strategies utilized was contrast alteration, which successfully changes the qualification between bright and dim areas in an image. This alteration expands the model's capacity to capture a more extensive range of highlights and varieties show in skin lesion pictures. Also, the dataset experienced a rotational change, presenting different rotation points such as 30°, 60°, 240°, and 270°. the dataset has been augmented by including images of skin lesions captured from different angles, and this diversity is illustrated in Fig. 2. The augmentation process yielded a considerable increase in the dataset size, resulting in a comprehensive collection of 32,000 pictures. This augmented dataset highlighted a wide cluster of melanoma and non-melanoma skin lesions, offering a more comprehensive and representative dataset for both training and evaluating the melanoma detection model.

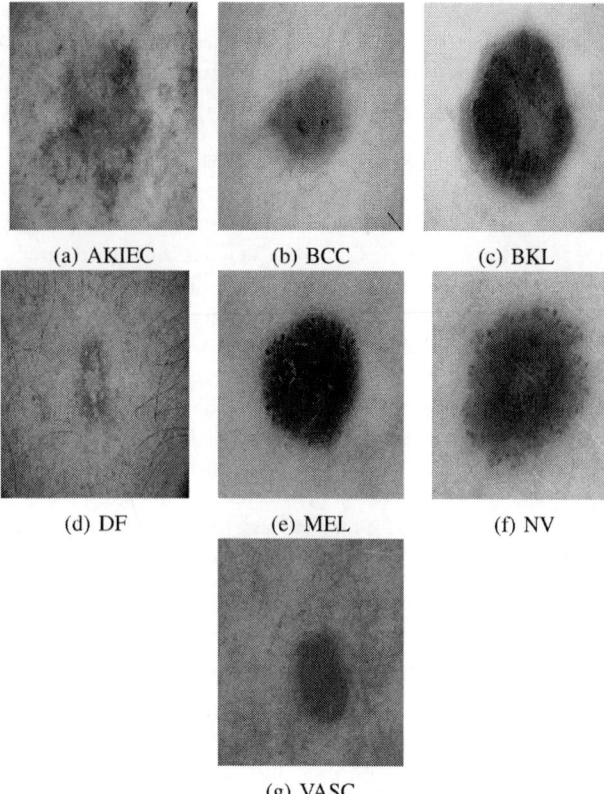

(a) AKIEC   (b) BCC   (c) BKL

(d) DF   (e) MEL   (f) NV

(g) VASC

Fig. 1: Different Types of Skin Lesions from the collected dataset.

Besides, the dataset enlargement techniques, especially contrast alterations and image rotation, altogether contributed to improving the model's overall performance. By growing the dataset and introducing varieties, the model acquired enhanced generalization capabilities. The contrast alterations improved the visibility of complicated points of interest inside the images, whereas the rotational changes presented varieties in orientation, improving the model's robustness to different perspectives. These results emphasize the considerable impact of data augmentation methods in supporting the dataset and improving the model's precision and reliability in classifying melanoma and other skin lesions, eventually leading to more exact melanoma detection results.

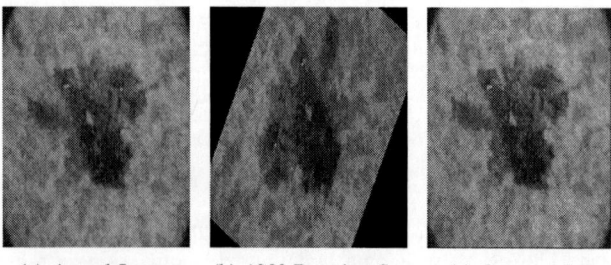

(a) Actual Image   (b) 120° Rotation Set   (c) Contrast Set

Fig. 2: Dataset symptoms after Applying Augmentation.

C. Model Development

The VGG-16 [18] (Visual Geometry Group 16-layer) architecture is a deep convolutional neural network (CNN) designed for image classification. It was introduced by the Visual Geometry Group at the University of Oxford and has become a widely used model in computer vision tasks. VGG-16 is characterized by its depth, consisting of 16 layers, and its simple and uniform architecture.

The architecture is composed of convolutional layers with small 3x3 filters, followed by max-pooling layers to reduce spatial dimensions. The use of small filters allows the network to capture intricate patterns in the data. The deeper layers of VGG-16 are designed to learn more complex and abstract features. The VGG-16 model [19] was chosen as the base architecture for this study due to its proven effectiveness in image classification tasks. This model consists of 16 convolutional layers followed by three fully connected layers and is equipped with a large number of trainable parameters, making it capable of obtaining complex features from dermoscopic images. The VGG-16 model was initialized with pre-trained weights on the ImageNet dataset and refined on the HAM10000 dataset using transformational learning techniques.

D. Tools used

During the implementation, various software tools were used. Google Colab was initially used for model training, then the Kaggle platform to leverage more powerful GPU resources. MATLAB and its related add-ons, including the Deep Network Toolbox and the Deep Network Quantizer, are used to support simulation and quantization tasks after training the model. Libraries like Keras, TensorFlow, Matplotlib, and NumPy in Python have been used for model development, data visualization, and numerical computation.

E. Training and Validation

The dataset was divided into training and validation sets. The model was trained on 29775 images belonging to 7 classes and validation was performed on 1985 images belonging to the same 7 classes. Its performance was continuously evaluated using the validation set to ensure optimal training.

F. Evaluation

The trained model was evaluated on a separate testing dataset, consisting of about 2000 unseen skin lesion images. Performance metrics such as accuracy and loss were calculated to assess the model's effectiveness. The methodology employed in this research ensured robust data collection, rigorous preprocessing, and the development of an accurate deep learning model for melanoma detection.

G. Dermoscopic Dataset

An additional dataset comprising 10,000 dermoscopic images was incorporated into this study [20]. This dataset was configured as a binary classification problem, categorizing images as either Benign or Malignant. Within this dataset, 7,500 images were allocated for training, while 1,500 images served as the validation set, and a further 1,000 images were reserved for testing. Although the deep learning model employed, ResNet50, achieved an impressive accuracy rate of 89%, it's essential to note that the primary focus of the study was training a model on a dataset encompassing seven distinct classes, making it distinct from the binary classification results.

H. Clinical Dataset

A clinical dataset of approximately 500 images was collected for potential integration into the medical device [21]. The VGG-16 model was trained for 50 epochs. The model's accuracy and loss values fluctuated, but it consistently achieved a validation accuracy of 75.9%. The results indicate that the VGG-16 model trained on the clinical images dataset had limited improvement in accuracy across the epochs. However, due to various considerations, the clinical dataset was not utilized in the training or evaluation of the deployed melanoma detection model. The decision to focus solely on the dermoscopic images from the HAM10000 dataset was made to maintain consistency and compatibility with the primary training dataset.

I. Raspberry Pi implementation

- Compact and Portable: The Raspberry Pi 3 Model B's small size allows for easy deployment in various settings, including clinics and remote areas. - Cost-Effective: The affordability of the Raspberry Pi makes the medical device accessible, particularly in resource-constrained environments. - Conversion to TensorFlow Saved-Model: The trained CNN model was converted to the TensorFlow Saved-Model format for seamless deployment on the Raspberry Pi. - Library Installation: Required libraries such as cv2, numpy, and tensorflow-runtime were installed on the Raspberry Pi. - System Updates: The Raspberry Pi system was updated to ensure compatibility and optimal performance. - Integration with LCD: The device was integrated with an LCD for a comprehensive implementation of early melanoma detection.

The use of Raspberry Pi 3 Model B in the implementation of the medical device offers portability, cost-effectiveness, and seamless integration with the trained CNN model. This enables widespread accessibility and practical deployment for early melanoma detection.

J. Proposed Device

The implementation of the device included a few steps to ensure real-time detection of melanoma using a Raspberry Pi Model 3 [22]. The Raspberry Pi is a user friendly computing platform. One of its advantages is that it's cost effective when compared to FPGAs which can come with higher costs and additional expenses, for development tools. While it may not be able to achieve the level of hardware optimization as FPGAs the Raspberry Pi strikes a balance between performance and simplicity making it an excellent choice for various projects, with different computational needs. As shown in Fig. 3, a camera was associated with the Raspberry Pi, permitting for real-time capture of pictures. The camera served as the input source for the detection process. Moreover, an LCD was associated with the Raspberry Pi, which would show the results of melanoma detection. During the device operation, the captured image was preprocessed to ensure compatibility with the VGG-16 model's input requirements. The resizing of the image to 224 x 224 pixels guaranteed a consistent input size for the model. This preprocessing step was crucial to enable seamless communication between the Raspberry Pi and the deployed CNN model, facilitating real-time melanoma detection.

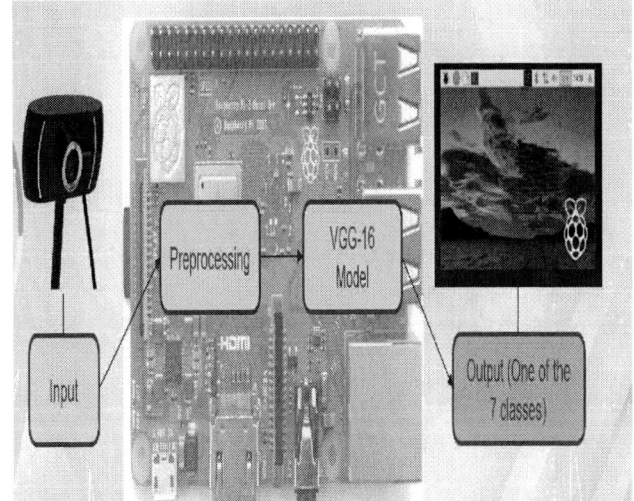

Fig. 3: Proposed System on Raspberry Pi for Melanoma Detection

IV. EXPERIMENTAL RESULTS

A. Performance Evaluation

The obtained results were analyzed to determine the model's performance in melanoma detection. The accuracy metric indicates the percentage of correctly classified samples, while the loss metric measures the model's learning progress during training. The model achieved an accuracy of 76% on the testing dataset. This indicates the percentage of correctly classified skin lesion images, reflecting the model's ability to differentiate between melanoma and other skin conditions. As shown in Fig. 4, the snippet showcases the device accurately predicting the presence of a melanoma lesion in the image.

Some other type of skin lesions were detected too such as nv as shown in Fig. 5

In certain cases, as shown in Fig. 6, it was observed that slight variations in camera positioning or changes in lighting conditions could lead to different predictions for the same captured image. This highlights the sensitivity of the device to external factors and the importance of consistent capture conditions for accurate predictions.

979-8-3503-8083-5/23 $31.00 © 2023 IEEE

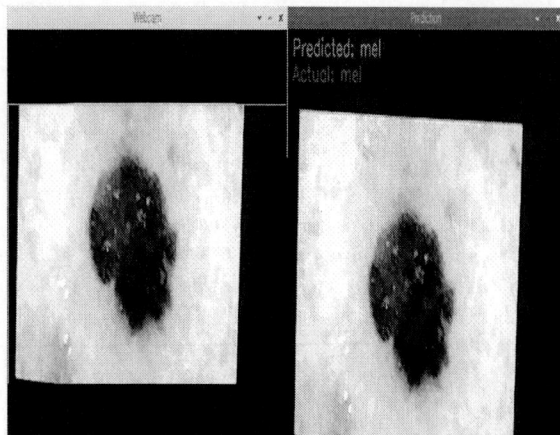

Fig. 4: Melanoma Prediction through Proposed Device

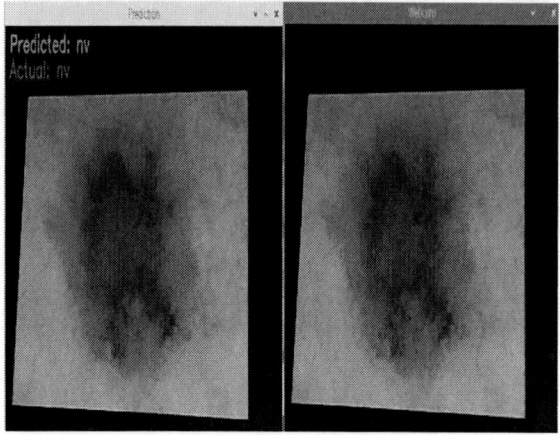

Fig. 5: Melanocytic Vs Nevi Prediction through Proposed Device

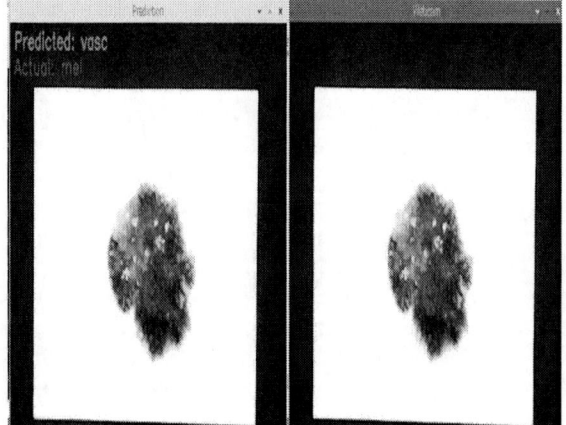

Fig. 6: Melanoma False case Prediction through Proposed Device

B. Device Efficiency

The efficiency of the proposed device was assessed in terms of frames per second (FPS) and execution time. On average, the device accomplished 12 FPS during real-time image processing, permitting for quick and consistent melanoma detection. The execution time for handling a single picture, including picture capture, preprocessing, highlight extraction, and prediction, averaged 8 seconds as shown in Fig. 7. This illustrates the device's ability to provide opportune results, encouraging fast diagnosis and decision-making.

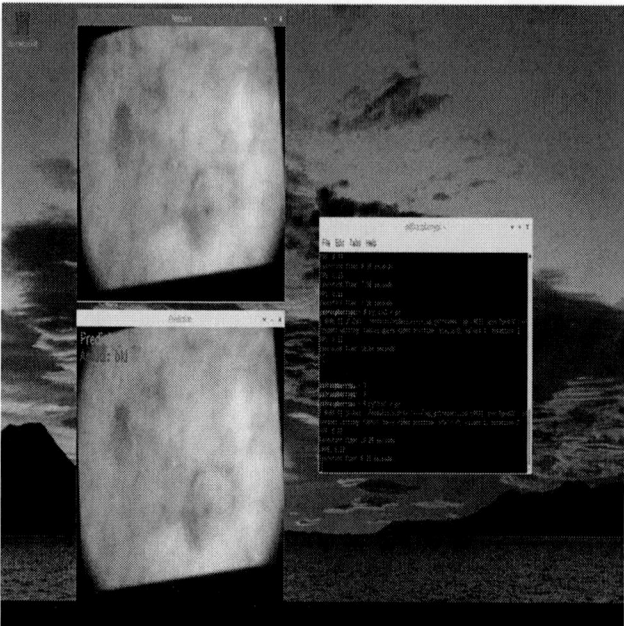

Fig. 7: Benign Keratosis Prediction through Proposed Device

Nevertheless, challenges continued, especially in managing with limited datasets. It is worth emphasizing that previous research basically centered around training models for binary classification scenarios, while our work addresses this gap by centering on the improvement of a resource-efficient and real-time melanoma detection device custom fitted for multiclass classification, enveloping all seven distinct classes. This contribution not only expands the scope of melanoma detection but also guarantees accessibility and reasonableness through its deployment on a low-cost device and targeting primary healthcare. Eventually, our research points to improve persistent results by facilitating early diagnosis over the different range of skin lesions.

In addition, it should be noted that our approach differs from some previous work, which used FPGA-based solutions [14]. On the other hand, we have deliberately chosen the Raspberry Pi platform to ensure cost-effectiveness. Despite these platform differences, our searches yielded significantly comparable results in terms of frames per second (fps). While they report an impressive 13.5fps, our implementation trails behind at a commendable 12fps.

V. CONCLUSION

This study illustrates the potential of deep learning procedures for melanoma detection, yielding promising accuracy levels. The proposed medical device offers noteworthy guarantee in advancing early melanoma diagnosis, and giving helpful and exact results. Its portability and reasonableness make it accessible in different healthcare settings, including under served-areas. Future bearings should investigate advanced designs, optimization strategies, and clinical information integration to upgrade framework precision and clinical utility.

The study effectively implemented a deep learning model prepared on the HAM10000 dataset for precise skin lesion picture classification. The developed device accomplished a satisfactory 76% accuracy on the testing dataset, appearing potential for melanoma detection. The contributions incorporate a novel approach to real-time melanoma detection and practical implementation on a Raspberry Pi.

Developing the Raspberry Pi-based medical device confronted limitations and challenges. The require for a capable GPU for training extended the training process and restricted exploration of different models and hyper-parameters. Equipment setup required cautious configuration to guarantee consistent communication, but variations in camera positioning and lighting conditions influenced prediction accuracy. The clinical dataset, collected for real-world scenarios, wasn't utilized due to compatibility and size restrictions. Over-fitting during training and FPGA deployment challenges obliged in general device performance.

In future work, a few avenues for improvement can be pursued. Firstly, improving GPU resources should be prioritized to speed up the training process and empower the investigation of advanced deep learning models, including transfer learning. Furthermore, exploring FPGA deployment and optimizing hardware is significant to achieve higher FPS rates and decrease deduction times, upgrading the device's reasonableness for clinical settings. Finally, integrating clinical information and embracing multi-modal analysis, incorporating patient-specific data and dermoscopic images, offers the potential to promote diagnostic capabilities, improving precision and reliability. Collaborations with healthcare specialists will give important experiences for refining device performance and fitting it to the particular needs of clinical practice.

REFERENCES

[1] Yann LeCun, Yoshua Bengio, and Geoffrey Hinton, "Deep learning," *Nature*, vol. 521, no. 7553, pp. 436-444, 2015. Publisher: Nature Publishing Group.

[2] Tschandl, P. (2018). The HAM10000 dataset, a large collection of multi-source dermatoscopic images of common pigmented skin lesions. Harvard Dataverse. DOI: 10.7910/DVN/DBW86T. URL: https://doi.org/10.7910/DVN/DBW86T

[3] Kai Arulkumaran, Marc Peter Deisenroth, Miles Brundage, and Anil Anthony Bharath. "A Brief Survey of Deep Learning." *IEEE Signal Processing Magazine*, vol. 34, no. 6, pp. 11-19, 2017.

[4] Shereen Afifi, Hamid GholamHosseini, and Roopak Sinha. "A system on chip for melanoma detection using FPGA-based SVM classifier." *Microprocessors and Microsystems*, vol. 65, pp. 57-68, 2019. ISSN: 0141-9331. DOI: https://doi.org/10.1016/j.micpro.2018.12.005.

[5] S. Afifi, H. GholamHosseini, and R. Sinha, "Dynamic hardware system for cascade SVM classification of melanoma," *Neural Computing and Applications*, vol. 32, pp. 1777-1788, 2020. DOI: https://doi.org/10.1007/s00521-018-3656-1.

[6] Shereen Afifi, Hamid Gholamhosseini, Roopak Sinha, and Maria Lindén, "A Novel Medical Device for Early Detection of Melanoma," *Studies in Health Technology and Informatics*, vol. 261, pp. 122-127, January 2019.

[7] Shereen Afifi, Hamid GholamHosseini, and Roopak Sinha, "FPGA Implementations of SVM Classifiers: A Review," *SN Computer Science*, vol. 1, Article number: 133, April 2020.

[8] Shereen Afifi, Roopak Sinha, and Hamid Gholamhosseini, "An Optimized Hardware System on Chip for a Support Vector Machine Classifier: a Case Study on Melanoma Detection," 2018. https://openrepository.aut.ac.nz/handle/10292/11913

[9] Shereen Afifi, Hamid Gholamhosseini, and Roopak Sinha, "FPGA Implementations of SVM Classifiers: A Review," *SN Computer Science*, vol. 1, Article number: 133, April 2020.

[10] Shereen Afifi, Hamid GholamHosseini, and Roopak Sinha, "SVM Classifier on Chip for Melanoma Detection," In 2017 39th Annual International Conference of the IEEE Engineering in Medicine and Biology Society (EMBC'17), Jeju Island, Korea, July 2017.

[11] Shereen Afifi, Hamid GholamHosseini, and Roopak Sinha, "A Low-Cost FPGA-based SVM Classifier for Melanoma Detection," In 2016 IEEE EMBS Conference on Biomedical Engineering and Sciences (IECBES), pp. 631-636, Malaysia, December 2016.

[12] Shereen Afifi, Hamid GholamHosseini, and Roopak Sinha, "Hardware Acceleration of SVM-based Classifier for Melanoma Images," In Image and Video Technology–PSIVT 2015 Workshops: RV 2015, GPID 2013, VG 2015, EO4AS 2015, MCBMIIA 2015, and VSWS 2015, Auckland, New Zealand, November 23-27, 2015. Revised Selected Papers, F. Huang and A. Sugimoto, Eds., ed Cham: Springer International Publishing, 2016, pp. 235-245.

[13] K. Agarwal and T. Singh, "Classification of skin cancer images using convolutional neural networks," *arXiv preprint arXiv:2202.00678*, 2022.

[14] Durães, P. F., Véstias, M. P. (2023). Smart Embedded System for Skin Cancer Classification. Future Internet, 15(2), 52. DOI: 10.3390/fi15020052

[15] Jaisakthi, S. M., Mirunalini, P., Chandrabose, Aravindan,Rajagopal, Appavu. (2023). Classification of skin cancer from dermoscopic images using deep neural network architectures. Multimedia Tools and Applications, 82, 15763-15778. DOI: 10.1007/s11042-022-13847-3

[16] Amina Bibi, Muhammad Khan, Muhammad Javed, Usman Tariq, Byeong-Gwon Kang, Yunyoung Nam, Reham Mostafa, and Rasha Sakr, "Skin Lesion Segmentation and Classification Using Conventional and Deep Learning Based Framework," *Computers, Materials and Continua*, vol. 71, February 2022. DOI: https://doi.org/10.32604/cmc.2022.018917.

[17] P. Sabouri, H. GholamHosseini, T. Larsson, and J. Collins, "A cascade classifier for diagnosis of melanoma in clinical images," , vol. 2014, pp. 6748-6751, 2014. ISSN: 2375-7477. DOI: https://doi.org/10.1109/embc.2014.6945177.

[18] Varshney, P. (2020). VGGNet-16 Architecture: A Complete Guide. Available online: https://www.kaggle.com/code/blurredmachine/vggnet-16-architecture-a-complete-guide/notebook

[19] S. Mascarenhas and M. Agarwal, "A comparison between VGG16, VGG19 and ResNet50 architecture frameworks for Image Classification," in *2021 International Conference on Disruptive Technologies for Multi-Disciplinary Research and Applications (CENTCON)*, pp. 96-99, 2021. DOI: https://doi.org/10.1109/CENTCON52345.2021.9687944.

[20] Muhammad Hasnain Javid, "Melanoma Skin Cancer Dataset of 10,000 Images," Available online at: https://www.kaggle.com/dsv/3376422. DOI: https://doi.org/10.34740/KAGGLE/DSV/3376422. Publisher: Kaggle, 2022.

[21] ISIC Archive, Available online at: https://www.isic-archive.com/. Accessed on May 31, 2023.

[22] Ghael, H. (2020, January). A Review Paper on Raspberry Pi and its Applications. DOI: 10.35629/5252-0212225227.

Deep/Federated Learning Algorithms for Ultrasound Breast Cancer Image Enhancement

Sarah M. Waly[1], Radwa Taha[1], Mohamed A. Abd ElGhany[2,3], and Mohammed A.-M Salem[1]

[1]Faculty of Media Engineering & Technology, German University in Cairo, Egypt
[2]Faculty of Information Engineering & Technology, German University in Cairo, Egypt
[3]Integrated Electronics Systems Lab, TU Darmstadt, Germany
[1]Emails: sarah.waly@student.guc.edu.eg, radwataha1999@gmail.com and mohammed.salem@guc.edu.eg
[2]Email: mohamed.abdel-ghany@guc.edu.eg

Abstract—Breast cancer stands as a leading cause of global cancer-related deaths among women, emphasizing the urgency for advancements in early detection and precise diagnosis to enhance patient outcomes. This study addresses the pressing need for improved breast cancer imaging through the integration of deep learning algorithms.

This paper specifically investigates the application of U-Net architecture and Federated Averaging (FedAvg) techniques for enhancing ultrasound breast cancer images thereby assisting radiologists in accurate cancer diagnosis. Leveraging two authentic datasets, we conduct experiments varying the number of epochs (50, 100, and 200) for FedAvg with U-Net. Our findings demonstrate promising accuracy values: 89.4716, 89.48, and 89.48, showcasing the effectiveness of our approach.

Additionally, we explore the impact of U-Net alone and in conjunction with FedAvg on the segmentation process of ultrasound breast cancer images. This dual-pronged investigation offers insights into the synergistic potential of these methodologies for image enhancement.

In conclusion, our study contributes to the ongoing efforts for improved breast cancer imaging. Through the fusion of U-Net and FedAvg, our approach demonstrates significant promise, providing a valuable avenue for advancing medical image analysis in the context of breast cancer diagnosis and treatment planning.

Index Terms—Federated Learning, Deep Learning, Ultrasound Breast cancer.

I. INTRODUCTION

Breast cancer, a critical global health issue, accounts for nearly 700,000 deaths annually. It develops from breast tissue [1], where unhealthy cells are generated rapidly instead of healthy cells. The lack of a permanent cure highlights the importance of early detection, primarily through mammography, which is adept at identifying deep tissue abnormalities and microcalcifications indicative of tumors. However, the effectiveness of mammography is lessened in cases of dense breast tissues, presenting a significant challenge in early cancer detection and accurate diagnosis. Thereby comes the role of ultrasound which is used in the cases of dense breast tissues to detect cancer.

This challenge underscores the need for advanced methods in breast cancer imaging, where Federated Learning [2] offers a promising solution. Traditional machine learning models, often centralized, raise concerns about data privacy, especially with sensitive health information. Federated Learning, as an innovative approach, enables a collaborative yet private enhancement of machine learning models. Processing data locally and using Federated Averaging (FedAvg), allows for the creation of a powerful, collective model without compromising individual data privacy. This methodology has potential applications beyond healthcare, including smartphone technology and autonomous vehicles.

The integration of the U-Net [3] architecture into Federated Learning frameworks significantly augments its potential in medical imaging applications, a deep learning model renowned for its effectiveness in semantic image segmentation. U-Net's unique structure—consisting of an encoder for high-resolution feature extraction, a bottleneck for crucial feature encapsulation, and a decoder for integrating these features—makes it particularly suitable for detailed analysis in medical imaging, such as breast cancer detection in ultrasound images. By integrating U-Net within a Federated Learning framework, we aim to address the limitations of traditional ultrasound, thereby improving early detection and diagnostic accuracy.

This paper is structured to provide a comprehensive overview of these interconnected themes: Section II discusses previous work in this area. Section III elaborates on our methodology, leveraging Federated Learning and U-Net for breast cancer imaging. Section IV presents the datasets used and the results obtained. Finally, Sections V and VI conclude the paper, discussing its broader implications and limitations.

II. LITERATURE REVIEW

This section delves deeper into the specifics of breast cancer imaging, the challenges faced in early detection, and the current state of technological advancements in this domain. We review the role of federated learning and deep learning in medical imaging, particularly focusing on their application in breast cancer detection. We also examine previous studies that have utilized these technologies, highlighting their strengths, limitations, and potential for future applications in medical imaging.

H. et al. [4] discussed the performance verification of a federated learning algorithm in a real-world healthcare environment. They used 5 deep learning algorithms that are

979-8-3503-8083-5/23 $31.00 © 2023 IEEE

trained with a federated learning system and compared against centralized techniques. It was indicated that federated learning has a similar performance to that of centralized deep learning. Holger et al. [5] used FedAvg to train the DenseNet-121 deep-learning model. The model is improved by an average of 6.3 percent using federated learning instead of training the models only on an institute's local data. Elshabrawy et al. [6] discussed the implementation of an Ensemble federated learning algorithm with a deep learning classification model to handle Non-IID when detecting COVID-19 and tested Non-IId skewed data on 5 FL algorithms. All FL models almost had the same results, except FedBN which gave the best result. Chai et al. [7] proposed FedAt: federated learning system with asynchronous tiers. Li et al. [8] proposed Federated Learning on non-IID Features via Local Batch Normalization (FedBN). It was indicated that it mitigates the issue of non-IID data across clients. It improves the convergence and the performance of the model. It outperforms the state-of-the-art methods of FedProx and FedAvg. It has achieved the highest accuracy and improvements. Mohamed et al. [9] discussed the implementation of a classification and detection system for breast cancer. They designed a deep learning-based system that combines a U-Net network and a two-class CNN-based deep learning model. They used (DMR-IR) database. It was indicated that the performance of the proposed model is superior to any other CNN model's performance on the same dataset except VGG16Net. Uysal et al. [10] aimed to make the process of computer-aided diagnosis more effective and easier and to classify the types of breast cancer with a deep learning model. They used a two-class classification with deep learning models. They used the BUSI dataset. Despite that ResNeXt50 gives better accuracy, ResNet50 has better stability and class-based results. Jabeen. et al. [11] proposed a system of 5 phases to classify breast cancer in ultrasound images using a deep learning model and fusion of the best-selected features. They used the BUSI dataset. The system has achieved 99.1 percent accuracy which is the best value obtained for accuracy with the best computational time that equals 13.599 (s). Balkenende et al. [12] provides an overview of deep learning research in breast cancer imaging, highlighting its applications in many fields. Studies show promising results with deep learning algorithms outperforming radiologists in some tasks. However, further large trials are needed, especially for ultrasound and magnetic resonance imaging, to determine the full potential of deep learning in breast cancer imaging. Debe et al. [13] explores deep learning techniques across various cancer types, with a focus on imaging for breast and cervical cancers, and MRIs for brain tumors. It examines methods like starting from scratch, transfer learning with layer freezing, and architecture optimization. Nazir et al. [14] reviews federated learning's role in medical imaging with deep neural networks (DNNs), highlighting its benefit in preserving data privacy. It describes a process involving local model training, gradient sharing, and global model refinement through algorithms like FedAvg.

III. METHODOLOGY

This research utilizes a two-part approach combining a custom U-Net model with Federated Averaging (FedAvg) for enhanced segmentation of breast cancer tumors in ultrasound images. First, we will start with training and testing the U-Net model directly on the datasets we have, then we will calculate the performance metrics and the accuracy and we will save the results. Second, when implementing the FL system, each client will have a U-Net model of its own to be trained and tested on its data. Then after training, each client would send its model updates to the global server to aggregate these updates and generate another global model that would be sent again to each client to repeat the process of FL until we have a converged global model. Then after the FL system finishes, we calculate the performance metrics and the accuracy of the resulting global model and compare it to the first U-Net model that was trained on the datasets directly.

We will be tackling first the deep learning part in Section 3.2 and then we will be tackling the Federated Learning part in Section 3.3.

A. The U-Net model

The custom U-Net model is implemented using PyTorch. It is a Supervised machine learning model [15], as it will be trained on ultrasound images and their masks, then it will be tested on a dataset with the same features as the one used for training.

The objective of using U-Net is due to its widespread use in medical imaging and its efficiency in computational resources. These qualities make U-Net an ideal candidate for processing medical images, where precision and resource management are paramount.

To streamline the training process within this framework, we consolidated the Baheya dataset by integrating both benign and malignant images, along with their respective masks. An identical procedure was applied to dataset B. This consolidation was key to constructing two distinct data loaders—one for each client—which simplifies the implementation and facilitates the subsequent steps of the machine learning workflow.

Using the previous datasets after handling them, we wanted to enhance model performance, particularly given the inherent challenges of medical image quality, so we applied several pre-processing and post-processing techniques:

1) Histogram Equalization: which is applied first to the dataset images and is used to boost contrast and enhance an image's overall appearance. It reorganizes an image's pixel intensity values to produce a more uniform histogram, which improves the contrast of the picture.

2) Sobel Edge Detection: which is applied after the histogram equalization to the dataset images and is used to detect the regions or boundaries of objects in an image, by computing the gradient magnitude of the pixel intensities, it can identify edges and highlight regions with sudden changes in intensity.

979-8-3503-8083-5/23 $31.00 © 2023 IEEE

3) *Connected Component Analysis*: which is applied to the predicted masks and is used to identify and label distinct connected regions or components within an image. It assigns a unique label or identifier to each connected region, enabling further analysis and manipulation of individual components. It was done on the predicted masks.

Figure 1 below illustrates the impact of these pre-processing steps on a sample from the Baheya dataset, showcasing the image quality improvement, where the image on the left is the image before and the image on the right is the image after applying the pre-processing.

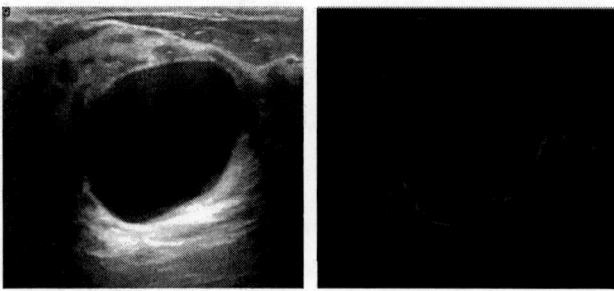

Fig. 1. A visualization of an image from the first dataset before and after any pre-processing.

Anticipating the risk of overfitting—a prevalent issue in machine learning when limited data are available—we strategically employ the Early Stopping technique during our training process. This not only prevents the model from learning noise but also optimizes the training duration.

To further ensure the model's generalizability and performance, we have meticulously selected a set of hyperparameters: Number of Channels: set to 1, indicating that the input images are grayscale. This is typical for medical images like X-rays or ultrasound images where color is not a factor. Number of Classes: set to 1, which suggests that the model performs binary segmentation (e.g., tumor vs. non-tumor). Number of Levels: set to 3, This likely refers to the depth of the U-Net architecture, meaning that there are three levels of downsampling in the encoder and upsampling in the decoder. Threshold: set to 0.4 based on empirical results to maximize the precision and recall of the segmentation. It is used to filter out weak predictions in the output segmentation map. It helps in differentiating between the foreground (e.g., tumor) and the background. A batch size of 16 balances the computational demand against the gradient descent's efficiency and strikes a balance between GPU resource utilization and gradient descent stochasticity, a learning rate of 0.001 encourages steady convergence and is selected for optimal performance after testing various values, and the image dimension of 256x256 provides a suitable trade-off between detail preservation and computational feasibility.

Delving into the network's architecture is essential for understanding its segmentation capabilities. As shown in figure 2, our network comprises 7 blocks and is illustrated below.

The encoder part consists of 3 blocks, each consisting of 3 convolutional layers with kernel size 3x3 for each followed by a ReLU activation function. Then, a max-pooling layer after each block to downsample the feature maps and capture more abstract features.

The decoder part, which comes directly after the encoder blocks and is connected to it, consists of 3 blocks, each consisting of 3 upsampling (convolutional transpose) layers with kernel size 2x2 for each to increase the spatial resolution of the feature maps followed by a convolutional layer with kernel size 3x3 followed by ReLU activation function. The decoder blocks also perform skip connections by concatenating the upsampled feature maps with the corresponding feature maps from the encoder.

The last layer, the Regression head, is a convolutional layer with kernel size 1x1 that maps the features from the decoder to the desired number of output classes to generate the final segmentation mask.

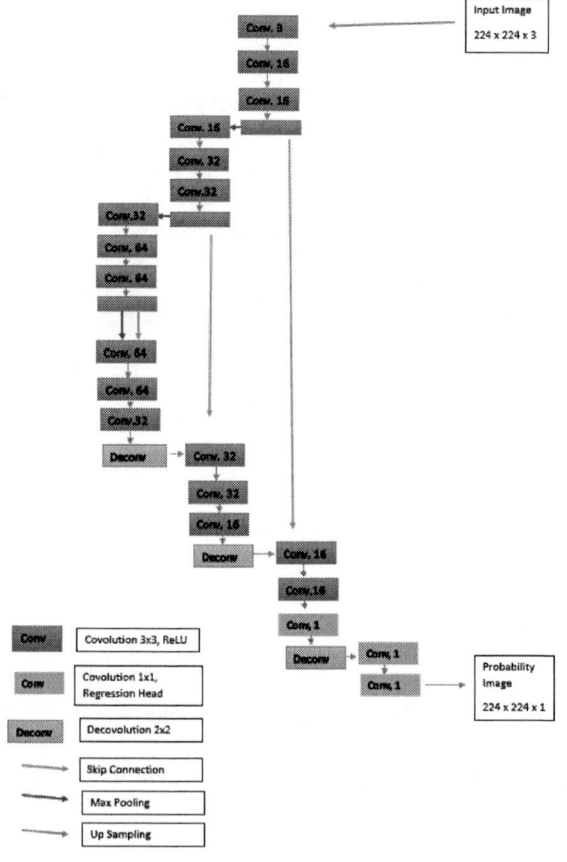

Fig. 2. The custom U-Net Architecture.

B. The Federated Learning part

In the Federated learning part, we used Federated Averaging (FedAvg) with the custom U-Net. Federated Averaging enables collaborative model training across numerous remote devices or clients while upholding data privacy. It is intended to deal with the problems brought on by decentralized data sources, such as mobile or edge devices, where direct sharing or centralization of data is not possible because of privacy issues or data ownership. The impetus for employing FedAvg stems from the need to optimize computational resources and to effectively utilize our limited dataset.

The key idea behind FedAvg is that each client trains the model using its local data, which helps capture unique patterns or characteristics specific to that client's data distribution. By averaging the models' parameters across multiple clients, the algorithm leverages the collective knowledge from diverse data sources without directly accessing or exposing the raw data. This approach ensures data privacy and security while still allowing the global model to benefit from the aggregated knowledge of all participants.

For the implementation of FedAvg, we utilized the Flower library [16], which provides a robust baseline for such distributed machine learning tasks. Consistency in data handling was maintained between the individual deep learning and federated learning components; both the Baheya dataset and dataset B were utilized as separate clients in the FL process, each contributing to the aggregated model learning.

IV. EXPERIMENT

A. DATASETS

- Dataset used: We used 2 datasets in this paper, the first one was the Baheya dataset [17] which was collected in 2018. It consists of 1578 images and their corresponding masks. It has 3 classes which are benign, malignant, and normal. The second is dataset B [18] which is a dataset from Spain. It has 2 classes which are benign and malignant. It consists of 163 images and their respective masks. Dataset B is too small to be used but these 2 datasets are the only ones publicly available for usage. We tried to contact many sources to provide us with ultrasound breast cancer datasets but our request was rejected due to privacy concerns. So used these 2 datasets to be able to work on the federated learning with 2 clients.

B. RESULTS

- The results of the U-Net model
We conducted two experimental trials with the U-Net model, utilizing different epochs to evaluate performance. The trials were designed to assess the impact of our preprocessing, post-processing, and early stopping techniques on the model's accuracy and loss metrics. The number of epochs are as follows:

The results are in tables I and II where Dataset 1 is the Baheya dataset, and Dataset 2 is the Dataset B. We will show the results before and after applying the preprocessing, post-processing, and early stopping techniques.

- The 1st trial: we had epochs= 100 and the result is as in figure 3 for the 1st dataset and in figure 4 for the 2nd dataset.

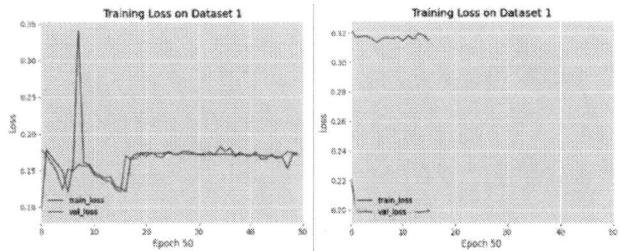

Fig. 3. The average loss of training the model in the first dataset for epoch=100 before and after applying the pre-processings and post-processings and the early stopping, where before is the image on the left side and after is the image on the right side.

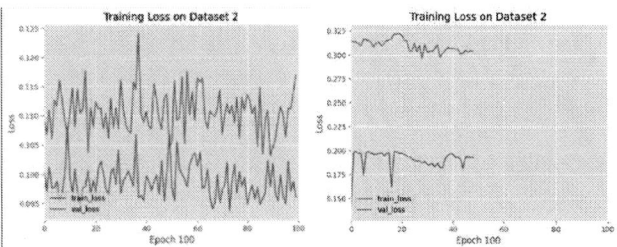

Fig. 4. The average loss of training the model in the second dataset for epoch=100 before and after applying the pre-processings and post-processings and the early stopping, where before is the image on the left side and after is the image on the right side.

- The 2nd trial: we had epochs= 200 and the result is as in figure 5 for the 1st dataset and in figure 6 for the 2nd dataset.

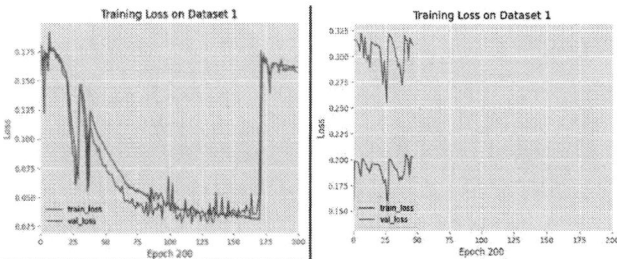

Fig. 5. The average loss of training the model in the first dataset for epoch=200 before and after applying the pre-processings and post-processings and the early stopping, where before is the image on the left side and after is the image on the right side.

From the previous results, we can see a big gap between the training loss and the validation loss in all figures 3, 4, 5 and 6 which means that the model suffers from overfitting which is due to insufficient data.

Furthermore, in the first trial with 100 epochs, our analysis revealed that the model's performance closely mirrored the ground truth for both the Baheya and Dataset B, as depicted in Figures 3 and 4, respectively. The results indicated that 100 epochs were optimal as extending the training to 200 epochs did not yield significant improvements and in some aspects, the performance was reduced as depicted in Figures 5 and 6.

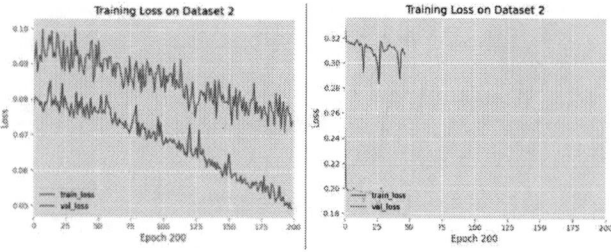

Fig. 6. The average loss of training the model in the second dataset for epoch=200 before and after applying the pre- and post-processings and the early stopping, where before is the image on the left side and after is the image on the right side.

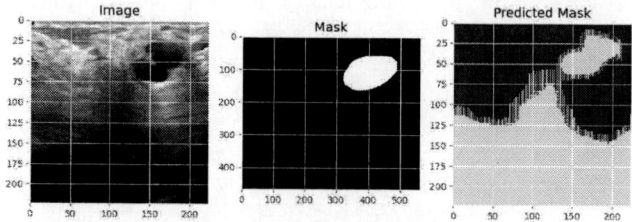

Fig. 7. The original image, the original ground truth, and the Predicted mask before applying the pre- and post-processing, respectively.

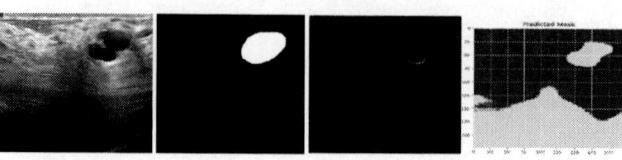

Fig. 8. The original image, The original Mask, The image after pre-processing, and The predicted mask after post-processing, respectively.

Our findings, as detailed in Table I, demonstrate promising accuracy and precision for both datasets before applying any pre-processings as shown in figure 7. Notably, the application of our pre-processing techniques led to varied outcomes as detailed in Table II. While Histogram Equalization and Connected Component Analysis improved the model's clarity as shown in figure 8, the inclusion of Sobel Edge Detection inadvertently omitted critical features, though reducing noise, resulting in the model's tendency to detect edges rather than the tumors themselves. This is evidenced by the increased noise in Figure ?? compared to the cleaner results in Figure ??. Nonetheless, the model's performance could potentially improve with additional epochs, allowing for better feature learning beyond edge detection.

TABLE I
SOME METRICS' RESULTS OF THE U-NET MODEL BEFORE APPLYING THE PRE- AND POST-PROCESSINGS AND THE EARLY STOPPING.

Trials	Dataset 1				Dataset 2			
	Accuracy	Loss	Recall	Precision	Accuracy	Loss	Recall	Precision
100 epochs	91.44	0.1755	0.0051	0.9497	96.75	0.1107	0.0	1.0
200 epochs	91.42	0.0809	0.2409	0.3488	96.76	0.0851	0.2137	0.2806

TABLE II
SOME METRICS' RESULTS OF THE U-NET MODEL AFTER APPLYING THE PRE- AND POST-PROCESSINGS AND THE EARLY STOPPING.

Trials	Dataset 1				Dataset 2			
	Accuracy	Loss	Recall	Precision	Accuracy	Loss	Recall	Precision
100 epochs	89.48	0.3055	0.1201	0.8722	89.4656	0.0634	0.2755	0.1755
200 epochs	89.48	0.3093	0.1492	0.8063	89.4832	0.3033	0.0440	0.7049

1) A sample of the results when applying the U-Net model before applying the pre- and post-processings and the early stopping in the following figure 7
2) A sample of the results when applying the U-Net model after applying the pre- and post-processings and the early stopping in the following figure 8

- **The results of the FedAvg**

  For the Federated Learning aspect, we utilized the FedAvg approach with two clients corresponding to our two datasets. The federated learning was executed over two rounds, each with 100 and 200 epoch trials. The FedAvg results, presented in Table III, indicated consistent accuracy across both trials. Noteworthy is the improvement in loss, precision, and recall, suggesting that FedAvg contributed to a more generalized model by effectively leveraging diverse data while maintaining data privacy.

TABLE III
SOME METRICS RESULTS OF THE FEDAVG APPROACH.

Trials	Accuracy	Loss	Precision	Recall
100 epoch	89.52	0.0542	0.5395	0.1060
200 epoch	89.42	0.0519	0.5787	0.1110

One of the most significant limitations when it comes to training deep learning models is the lack of computational resources. The training of such models requires GPUs so that the training can be completed efficiently and within a considerable amount of time. As for the datasets, due to privacy constraints, it is challenging to get access to large medical datasets as they require ethical permissions from the patients and the doctors as they are private data that belong to the patients.

C. Comaprison with previous work

In this paper, the FedAvg approach was applied to two datasets to enhance breast cancer segmentation in ultrasound images and underscore the potential of federated learning when integrated with U-Net. Our method achieved a consistent accuracy of approximately 89.5 percent across both 100 and 200 epochs trials, with noted improvements in loss, precision, and recall, indicating effective leveraging of diverse data while maintaining privacy. Contrastingly, Elshabrawy et al. [6] investigates the impact of non-IID data in healthcare, using four chest X-ray datasets to conduct federated learning experiments with five different algorithms: FedAvg, FedProx, FedNova, SCAFFOLD, and FedBN, simulating a multi-hospital scenario.

TABLE IV
A COMPARISON BETWEEN THE RESULTS OF INTEGRATING FEDERATED AVERAGING WITH U-NET ON BUS AND OTHER FEDERATED LEARNING ALGORITHMS ON CHEST X-RAY BY [6]

FL Algorithms	Datasets	Accuracy (%)
FedAvg with U-Net	BUS [17]	89.52
FedAvg [6]	Chest X-ray	79.58
FedProx [6]	Chest X-ray	76.92
FedNova [6]	Chest X-ray	5.57
SCAFFOLD [6]	Chest X-ray	79.18
FedBN [6]	Chest X-ray	84.4

A comparison between the results of integrating federated averaging with u-net and the other different algorithms by [6], where we can detect that integrating fedAvg with U-Net achieved the highest results as shown in the below table IV.

V. CONCLUSION

In conclusion, a comparative analysis of the U-Net model and the FedAvg approach highlights the advantages of federated learning in our context. FedAvg demonstrated an ability to enhance the U-Net's segmentation results significantly. It is also pertinent to mention that despite the limitations posed by the size and quality of the datasets and computational resources, the federated learning system, especially when integrated with the U-Net, showed the potential to improve medical image analysis while preserving data confidentiality.

Moreover, It is worth mentioning that the pre-processings, especially the sobel edge detection, were unsuitable to be used on medical images as they removed some important resilient features that the model should have learned to detect the tumor instead of only detecting the edges.

Looking forward, it is imperative to extend our research to larger and higher-quality datasets to fully capitalize on the benefits of federated learning. Exploring other FL algorithms and more advanced U-Net architectures could provide further enhancements. Additionally, a systematic approach to hyperparameter tuning is recommended to refine model performance.

REFERENCES

[1] M. Fouad, M. A. A. E. Ghany, and G. Schmitz, "A single-shot harmonic imaging approach utilizing deep learning for medical ultrasound," *IEEE Transactions on Ultrasonics, Ferroelectrics, and Frequency Control*, vol. 70, no. 3, pp. 237–252, 2023.

[2] T. Sun, D. Li, and B. Wang, "Decentralized federated averaging," *pubmed*, 2021.

[3] O. Ronneberger, P. Fischer, and T. Brox, "U-net: Convolutional networks for biomedical image segmentation," 2015.

[4] J. H. L. K. H. J. K. S. K. K. N. I. C. J. Y. H. L. M. M. H. M. A. L. K. K. S. K. H. Lee H, Chai Y, "Federated learning for thyroid ultrasound image analysis to protect personal information: Validation study in a real health care environment," pp. 1–9, 09 2021.

[5] H. R. Roth, K. Chang, P. Singh, N. Neumark, and W. Li, "Federated learning for breast density classification: A real-world implementation," *Springer International Publishing*, 2020.

[6] K. M. Elshabrawy, M. M. Alfares, and M. A.-M. Salem, "Ensemble federated learning for non-ii d covid-19 detection," pp. 057–063, 2022.

[7] Z. Chai, Y. Chen, A. Anwar, L. Zhao, Y. Cheng, and H. Rangwala, "Fedat: A high-performance and communication-efficient federated learning system with asynchronous tiers," *arXiv*, 2021.

[8] X. Li, M. Jiang, X. Zhang, M. Kamp, and Q. Dou, "Fed{bn}: Federated learning on non-{iid} features via local batch normalization," 2021.

[9] G. T. K. O. Mohamed EA, Rashed EA, "Deep learning model for fully automated breast cancer detection system from thermograms," pp. 1–20, 2022.

[10] M. Uysal, F.; Köse, "Classification of breast cancer ultrasound images with deep learning-based models," *Eng. Proc*, 08 2023.

[11] A. M. T. U. Z. Y. H. A. M. A. D. R. Jabeen K, Khan MA, "Breast cancer classification from ultrasound images using probability-based optimal deep learning feature fusion," *Sensors (Basel)*, 2022.

[12] L. Balkenende, J. Teuwen, and R. M. Mann, "Application of deep learning in breast cancer imaging," *Seminars in Nuclear Medicine*, vol. 52, no. 5, pp. 584–596, 2022. Breast Cancer.

[13] T. G. Debelee, S. R. Kebede, F. Schwenker, and Z. M. Shewarega, "Deep learning in selected cancers' image analysis—a survey," *Journal of Imaging*, vol. 6, no. 11, 2020.

[14] S. Nazir and M. Kaleem, "Federated learning for medical image analysis with deep neural networks," *Diagnostics*, vol. 13, no. 9, 2023.

[15] harshavardhan, "Fetal-ultrasound-image-segmentation-using-u-net," 2023.

[16] D. J. Beutel, T. Topal, A. Mathur, X. Qiu, J. Fernandez-Marques, Y. Gao, L. Sani, H. L. Kwing, T. Parcollet, P. P. d. Gusmão, and N. D. Lane, "Flower: A friendly federated learning research framework," *arXiv preprint arXiv:2007.14390*, 2020.

[17] W. Al-Dhabyani, M. Gomaa, H. Khaled, and A. Fahmy, "Dataset of breast ultrasound images," *pubmed*, vol. 28, p. 104863, 2020.

[18] M. R. Y. M. Z. R. Thomas C, Byra M, "Bus-set: A benchmark for quantitative evaluation of breast ultrasound segmentation networks with public datasets," 2023.

AS-LR Emergency Detection Scheme for Biomedical Applications

Nadine Boudargham
Faculty of Engineering
Notre Dame University
Deir El Kamar, Lebanon
nboudargham@ndu.edu.lb

Abdallah Kassem
Faculty of Engineering
Notre Dame University
Zouk Mosbeh, Lebanon
akassem@ndu.edu.lb

Mustapha Hamad
Faculty of Engineering
Notre Dame University
Zouk Mosbeh, Lebanon
mhamad@ndu.edu.lb

Abstract—The fast development of medical sensors allowed the emergence of a contemporary biomedical application called Body Sensor Network (BSN). BSN allows medical personnel to monitor the status of the person, through analyzing physiological parameters (Heart Rate (HR), Blood Pressure (BP), body temperature ($T°$), etc.) collected by miniature medical sensors placed on the human body. The main aim of BSN is to enhance people's life through detecting emergency cases as fast as possible and taking proper actions before it is too late. Since sensors in BSN have very limited energy that should be preserved, the challenge arises in designing schemes that ensure correct and fast emergency detection while maintaining low energy consumption of nodes. Machine Learning (ML) techniques are vastly adopted nowadays for anomaly detection due to the large amount of data collected by the sensors. Even though these techniques provide high percentage of accuracy in detecting emergencies, they operate over the whole dataset which quickly drains the energy of nodes due to the large amount of data that should be transmitted. Thus, in this article, we propose an algorithm that efficiently decreases the amount of transmitted data through dynamically adapting the sampling rate of every sensor based on the risk of emergency detected by the Linear Regression (LR) ML model. Adaptive sampling is used to reduce the energy consumption of nodes and to accelerate the detection of emergencies, while LR is used to identify true urgent cases. Several simulations were conducted to evaluate the performance of the proposed scheme. Results show that the suggested algorithm outperforms other ML based anomaly detection schemes presented in the literature and succeeds in addressing the BSN challenges.

Index Terms—Biomedical, BSN, anomaly detection, sampling algorithms, LR

I. INTRODUCTION

Body Sensor Network (BSN) is an emerging field of biomedical applications that is gaining a lot of research interest in the last few years. BSN consists of small wireless sensors that are placed on or implanted inside the human body. Their main role is to sense specific physiological parameters (such as Heart Rate (HR), body Temperature ($T°$), mean values of Blood Pressure (BPmean), respiration rate (RESP), etc.), and to transmit them to a coordinator node, who will process the data and send it to medical personnel for correct assessment of the current state of the person.

In BSN, adopting reliable anomaly detection schemes is important to identify emergency cases and take actions before it is too late. But at the same time, it is essential to preserve the limited energy of the sensors to prolong their lifetime since

it is hard to replace the batteries of the nodes, specially the implanted ones. This challenge can be addressed by adopting proper data reduction algorithms i.e. sampling algorithms, to reduce the amount of data transmitted by the sensors, since the majority of the energy consumption of nodes is caused by radio communication that depends on the number of bits that are sent within the network [1].

The great increase of the amount of data generated by sensors in BSN raised the need to automate data analysis using Machine Learning (ML) techniques [2], allowing BSN to better detect, monitor, and prevent possible fault assessment. Classification algorithms such as Decision Trees (DT) and Support Vector Machines (SVM), regression models such as Linear Regression (LR) and Random Forest (RF), and clustering techniques including K-means clustering and Artificial Neural Networks (ANN) [2] [3] are currently deployed for anomaly detection in BSN.

Even though many ML based anomaly detection schemes are found in the literature, they do not consider sampling or reducing the data before detecting anomalies. In most of the proposed schemes, anomaly detection is performed by the coordinator who has more energy capacity than the sensors; however, the amount of data transmitted by the sensors is very large which will quickly drain their energy from one side, and will delay the detection of emergency cases from the other side [4].

For this reason, we propose an efficient Adaptive Sampling approach based on Linear Regression (AS-LR) for emergency cases detection in BSN. The idea of the proposed scheme is to adapt the sampling rate of each sensor based on the difference between the actual and the predicted value computed by the LR prediction model. This combination of both data sampling and anomaly detection approaches guarantees fast detection of anomalies and ensures low energy consumption of nodes. Adaptive deterministic sampling is used for data reduction, while LR is used for correct detection of emergency cases. Deterministic sampling is chosen since it performs better than other sampling techniques in BSN [4], and LR is chosen since it is a technique of supervised ML widely used in BSN [5] [3] [6] and proven to be an efficient prediction model [7]. In our previous work [4], we applied the adaptive sampling approach to control charts. Since ML is more efficient when dealing with

979-8-3503-8083-5/23 $31.00 © 2023 IEEE

large amount of data, in this article, we test the performance of the adaptive sampling approach when applied on LR with respect to other ML schemes used for anomaly detection.

The rest of this paper is organized as follows. The related work is presented in Section II. The theoretical background is explained in Section III. The proposed scheme is elaborated in Section IV. The experiments and results are discussed in Section V, to conclude in Section VI.

II. RELATED WORK

Many anomaly detection schemes based on ML techniques were proposed for BSN in the literature. For instance, authors in [3] proposed using J48 decision tree along with LR model. J48 is used to classify normal and abnormal measurements, whereas LR model is used to identify true anomalies and sensor faults. In [5], the authors suggested using Support Vector Machine (SVM) with LR for anomaly detection in BSN, where measurements are classified based on SVM, and anomalies are identified through LR. The authors in [6] compared the performance of several ML algorithms to detect anomalies in BSNs. Three classification algorithms used to identify sensors generating abnormal values were compared, namely J48, k-Nearest Neighbors, and Random Forests, while two prediction models specifically LR and Additive Regression (AR) were used to locate anomalies. Results showed that using Random Forests with AR outperforms the others. In [8], researchers presented an anomaly detection technique through integrating the Artificial Neural Network (ANN) with LR. Also, in [9], the authors proposed adopting Gaussian regression along with majority voting for detecting anomalies.

Even though the proposed schemes in the literature aim to accurately identify emergency cases in BSN, to the best of our knowledge, none of these schemes propose sampling or reducing the amount of data prior to detecting anomalies, which exhausts the energy of the sensors and slows the detection operation.

From another side, few adaptive sampling approaches were suggested in the literature. These schemes propose adapting the sampling rate of sensors according to various criteria such as level of criticality of sensed values [10], or variance of vital signs [11], or based on the classification results of decision trees [12]. The main drawback of the proposed schemes is that they are implemented at the node level, which increases the energy consumption of these nodes due to the high computational requirements.

Therefore in this article, we propose a reliable and efficient adaptive sampling approach that adjusts the sampling rate based on LR prediction model outcome. The proposed scheme is implemented at the coordinator level to ensure maximum energy efficiency of nodes.

III. BACKGROUND

A. Linear Regression

LR presents a relation between an independent variable x_k (predictor) and a dependent variable y_k (response) [13]. The model has the following equation:

$$y_k = C_0 + C_1 x_k$$

C_0 and C_1 are computed in the training phase where C_0 is the intercept, i.e. the value of y when x is zero, and C_1 is the slope of the line, computed via the following formula:

$$C_1 = \frac{Cov(X,Y)}{VAR(X)} = \frac{\sum (x_k - X_{mean})(y_k - Y_{mean})}{\sum (x_k - X_{mean})^2}$$

In LR, the Root Mean Square (RMS) error can be used to detect anomalous values [14]. the RMS is the square root of the Mean Square Error (MSE), which is based on distance between the actual value y_i and the predicted value on the regression line $\hat{y}_i$ and is computed as follows:

$$MSE = \frac{\sum_{i=1}^{n}(y_i - \hat{y}_i)^2}{n}$$

Where n is the total number of samples. In general, 95% of the dataset values lie within two RMS, and 99.7% lie within three RMS error. For this reason, a sample is considered anomalous and an alarm is raised when the distance between the actual value and the predicted one, i.e. the computed error, is greater than three RMS which is the threshold limit of normal measurements [14].

B. Deterministic Sampling

Deterministic sampling selects the data values in a systematic way every predetermined interval. There is no randomness in the selection of samples. Deterministic sampling is faster and easier to execute than random sampling techniques such as Simple Random Sampling (SRS) [4]. It also provides better representation of the dataset, which is essential to detect anomalous data.

IV. AS-LR ALGORITHM FOR EMERGENCY CASES DETECTION

Our proposed AS-LR scheme is formed of three main phases:
1) LR modeling
2) Adaptive sampling based on LR model
3) Emergency cases detection

A. LR Modeling

The first phase consists in building the LR models. For every sensor attribute, the LR is trained using 1000 non-anomalous measurements to build accurate prediction models [15]. The obtained trained models will be used to estimate the value of the measurement y_k at every k^{th} instance, allowing the computation of the distance between the actual sensed value and the predicted one.

B. Adaptive Data Sampling Using LR

The main aim of the second phase is to adapt the sampling rate of sensors using LR. For this purpose, a jumping window is used. Over every window, and for each sensor, the coordinator computes the maximum distance between the samples' sensed and predicted value in the LR model, and selects the proper Sampling Rate (SR) to be used in the next window. SR is

979-8-3503-8083-5/23 $31.00 © 2023 IEEE

therefore adapted to the variation of data values. The sensors' role will be to select and transmit data based on the computed SR.

The rationale behind this idea is illustrated in Fig. 1. This figure represents an example of the LR model for a blood pressure subset. The actual sensed value, the prediction line and the threshold limits are plotted. Samples that lie between the threshold limits are considered normal measurements, and those beyond these limits are considered anomalous. As explained in Section III-A, the threshold limit line is three RMS away from the prediction line.

When the sensed values are close to the predicted ones, they will be far away from the threshold limits as shown in Fig. 1a. Therefore, the probability of emergency is low, and thus, it is not necessary to take all the samples in the next jumping window. A low SR of the sensor can therefore be used without compromising the outcome of the anomaly detection scheme.

However, when the difference between the sensed and predicted values increases as shown in Fig. 1b, the sensed values become closer to the threshold limits; thus more samples should be collected to better assess the situation, and therefore, the SR should be increased in the next window.

Further increase in the distance between the sensed and predicted values is shown in Fig. 1c. In this case, the coordinator might need to set the sampling rate to 100% in order to receive every data measurement in the next window as the risk of emergency is too high.

Once the difference between the sensed and predicted values decreases again below the threshold as shown in Fig. 1d, SR can be decreased, since the risk of emergency is reduced again.

The proposed algorithm operates as follows: For every sensor, the coordinator starts by setting SR to 1 in order to receive all the samples of the first window. It then applies LR on the window samples to find the maximum difference between the corresponding actual values and the predicted ones. The percentage of the maximum difference with respect to the threshold value is used to specify the sampling rate in the next window. In our algorithm, the maximum difference is considered in order to make sure that enough samples are collected in the next window for correct anomaly detection. For example, if the maximum difference is found to be one RMS, the coordinator will set the SR to 33.33% since the risk of emergency is very low, and the sensor will only collect 33.33% of the samples in the next window. The coordinator sends the evaluated SR to the sensor who will select the proper number of samples. The coordinator will then re-adapt the SR based on the new received data.

Algorithms 1 and 2 explain the proposed adaptive sampling approach.

C. Emergency Detection Approach

In the third phase, LR is used to identify true emergency cases from the selected samples. For each sensor, an anomaly is detected when the distance between the sensed and predicted value exceeds the threshold limit which is $3 \times RMS$. Since the coordinator adapts the SR of every sensor independently, the

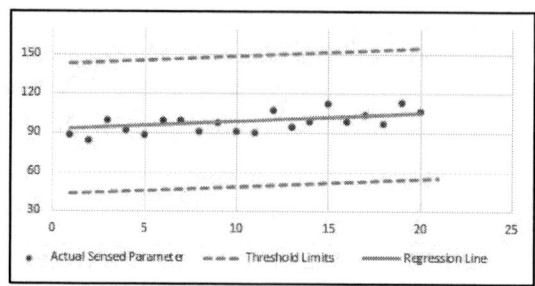

(a) Sensed Values Away from Threshold

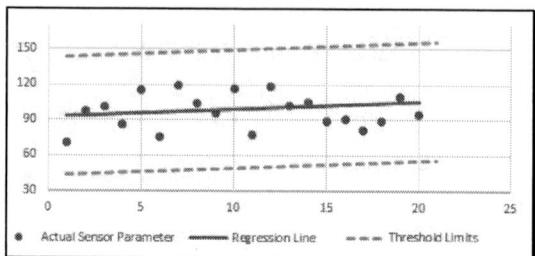

(b) Sensed Values Closer to Threshold

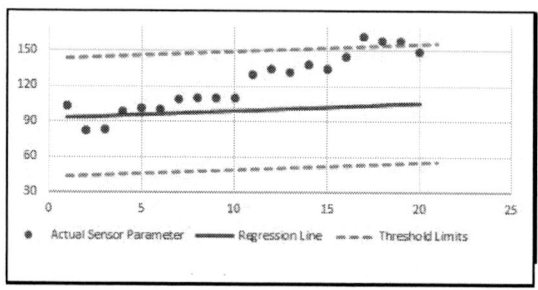

(c) Sensed Values Reaching Threshold

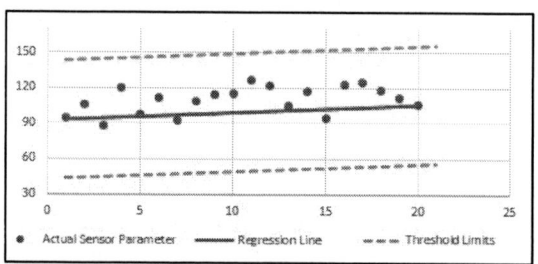

(d) Sensed Values Moving Away from Threshold

Fig. 1: Example of LR model

number of received samples can vary from one sensor to the other; therefore, in order to specify when the anomaly occurred in each sensor, the Start Time (ST) and End Time (ET) of the detected anomaly are recorded. ST of the anomaly is set to the beginning of the increase of the distance between the actual and predicted values, and ET of the anomaly is set after several decreases in the distance, which is equal to the number of successive decreases in the training window of each sensor [4].

Since the physiological parameters are highly correlated, deviations usually occur in at least two sensors; for example, the Heart Rate (HR) and the temperature sensor values increase simultaneously. Therefore to identify true emergency cases, a spatial correlation test is performed between the different sensors, where STs and ETs of the detected anomalies in different sensors are compared. If they overlap, the anomaly is considered as a true emergency and an alarm is raised. If not, then the anomaly is detected in only one sensor and is therefore classified as sensor fault.

The flow chart of the emergency detection approach is illustrated in Fig. 2

All the computations are performed by the coordinator and not the nodes to preserve their energy. The coordinator builds the LR model, computes the SR of each sensor, checks for anomalies, and raises alarms. The nodes' role is just to select the samples based on the received SR and send them to the coordinator.

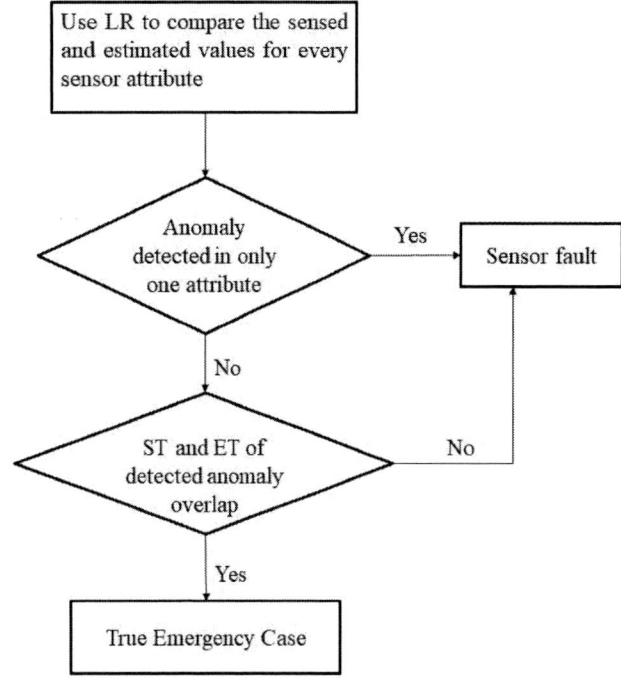

Fig. 2: Emergency Detection Algorithm using LR

ALGORITHM 1. AS-LR at the Coordinator Side

1: S_i: Sensor parameter
2: Th_i: Threshold value   ▷ Th_i= 3×RMS
3: SRw_0: Initial sampling rate of S_i
4: SRw_i: Computed sampling rate of S_i
5: n: Size of window
6: **for each** sensor S_i **do**
7:     $SRw_0 = 1$
8:     Send SRw_0 to S_i ▷ SR is 100% for the first window
9:     Receive samples from S_i
10:     **for** j=1:n **do**
11:         d_{imax}= $max[sensed(j) - predicted(j)]$;
12:     **end for**
13:     $SRw_i = min[1, d_{imax}/Th_i]$;   ▷ Compute the SR to be used by sensors in next window
14:     Send SRw_i to S_i
15: **end for**

ALGORITHM 2. AS-LR at the Sensor Side

1: SR_w: Sampling rate in window w
2: S_w: number of samples collected from the window
3: n: Size of window
4: **if** $window = 1$ **then**   ▷ for the first window
5:     $S_w = n$;   ▷ collect all samples of window
6:     **if** $window != 1$ **then**   ▷ for other windows
7:         Get SR_w from coordinator
8:         $S_w = SR_w \times n$;
9:     **end if**
10: **end if**
11: Select S_w samples from $window$
12: Send samples to coordinator

V. SIMULATION OF THE PROPOSED SCHEME

A. Simulation Parameters

To assess the performance of AS-LR, real medical data from Physionet database is used [16]. Four correlated attributes were considered : Heart Rate (HR), body temperature (T°), mean values of Blood Pressure (BPmean), and respiration rate (RESP). Each attribute contains 30000 measurement. Python programming language was used. Simulation parameters are presented in Table I.

Simulations are divided into two parts. In the first part, we compare the proposed AS-LR scheme to LR without sampling in order to evaluate the effect of using adaptive sampling on data prior to performing anomaly detection. In this part, the accuracy, energy consumption of nodes, and the execution time are compared for both AS-LR and LR without sampling. The accuracy rate demonstrates the ability of the algorithm to detect

TABLE I: Simulation Parameters

Parameter	Value
Number of Sensors	4
Number of Samples per Sensor	30000
Window Type	Jumping Window
Sample Size	2 Bytes
Channel Rate	250 Kbps
Decision Threshold for anomalies	$3 \times RMS$

979-8-3503-8083-5/23 $31.00 © 2023 IEEE

correct cases, whether the situation is a true emergency or a sensor fault:

$$Accuracy = \frac{TP + TN}{P + N}$$

Where TP and TN are the number of True Positives and True Negatives respectively; P is the total number of Positives and N is the total number of Negatives.

The nodes' total energy consumption is the energy consumed by sensors on sensing, processing, and transmitting the data to the coordinator; it shows the energy efficiency of various algorithms. Whereas the execution time reflects the coordinator's ability for fast detection of anomalies when different schemes are used; it encloses the sensors' data collection and transmission time, as well as the time needed to detect anomalies and raise alarms.

In the second part, AS-LR is compared to two anomaly detection schemes proposed in the literature, specifically J48 decision tree with LR [3], and SVM with LR [5]. These schemes are chosen since they use LR as a prediction model to detect anomalies, which is close to our proposed scheme. The three algorithms are compared with respect to accuracy rates, execution time, and sensors' energy consumption, as well as the precision, recall and specificity. The precision shows how strong the algorithms are in identifying true emergency cases from sensor faults. It is computed as follows:

$$Precision = \frac{TP}{TP + FP}$$

Where TP is the number of True Positives and FP is the number of False Positives.

Recall is the True Positive Rate (TPR). It demonstrates the ability of the schemes to detect the emergency cases. It is computed as:

$$Recall(TPR) = \frac{TP}{TP + FN}$$

Where TP is the number of True Positives and FN is the number of False Negatives.

Specificity shows the ability of different schemes to detect false alarms and is computed as follows:

$$Specificity = 1 - FPR = \frac{TN}{TN + FP}$$

Where FPR is the False Positive Rate, TN is the number of True Negatives and FP is the number of False Positives.

B. Results and Interpretation

1) Comparison Between AS-LR and LR Without Sampling: The accuracy, energy consumption of nodes, and execution time of both AS-LR and LR without sampling were simulated for different window sizes. Results are shown in Fig.3, Fig. 4 and Fig. 5 respectively.

Fig. 3 shows that AS-LR provides the same accuracy as LR without sampling (87.7%) for window sizes $w \leq 25$. The accuracy of AS-LR decreases with larger window sizes, since smaller windows permit more frequent adaptation of the SR

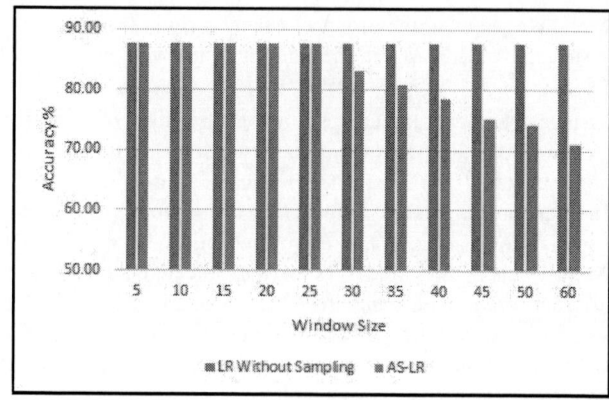

Fig. 3: Impact of the window size on the accuracy

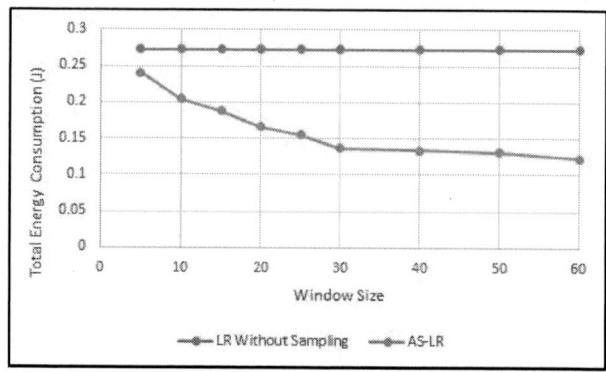

Fig. 4: Impact of the window size on the energy consumption

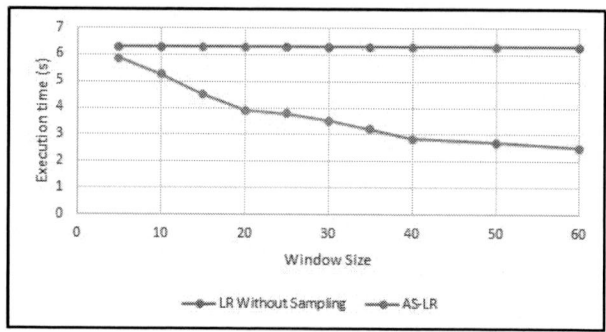

Fig. 5: Impact of the window size on the execution time

and thus better selection of samples in the next window. Fig. 4 shows that AS-LR significantly reduces the energy consumption of sensors; this is because AS-LR reduces the number of transmitted samples and the main reason for the sensors' energy consumption is data transmission which depends on the number of sent bits. As for the execution time, simulation results illustrated in Fig. 5 show that AS-LR decreases the emergency detection time compared to LR without sampling. The reason is that in AS-LR, less number of samples are collected by the sensors, leading to faster transmission of data from the nodes and faster computations by the sink.

TABLE II: Performance Comparison of Various Schemes

Algorithm	Precision (%)	Recall (%)	Specificity (%)	Accuracy (%)	Execution Time (s)	Energy Consumption (J)
AS-LR	87.1	86.18	85	87.7	3.78	0.155
SVM with LR	85.5	84.35	83.1	84.5	7.1	0.31
j48 with LR	84.2	81.5	82	83.42	6.22	0.28

The above simulation results prove that using AS-LR with window size $w \leq 25$ addresses BSN requirements better than LR without sampling, as it reduces the energy consumption of nodes and accelerates the detection of emergencies, while offering the same accuracy rate.

2) Comparison Between AS-LR and Anomaly Detection Schemes in the Literature: The simulation results of the comparison between AS-LR, J48 with LR, and SVM with LR with respect to the prediction, recall, specificity, accuracy, execution time and energy consumption of nodes are presented in Table II.

Results show that the proposed AS-LR outperforms the other two schemes in terms of precision, recall, specificity and accuracy. The reason is that the sensed data is classified using J48 and SVM prior to applying LR in the compared models, which increases the probability of erroneous detection due to misclassified instances. Whereas in the proposed scheme, LR is applied to the proper number of samples and provides a close look to the samples when there is a risk of emergency, which leads to better detection of anomalies. Also, results show that the proposed scheme provides a much better performance than the others in terms of time of execution and energy of nodes. In fact, AS-LR is 46% faster than SVM with LR and 39% faster than J48 with LR; also, the nodes in AS-LR consume 50% less energy than SVM with LR and 45% less energy than J48 with LR. The reason is that the reduced amount of the transmitted samples in AS-LR decreases the time to transmit data, accelerates the detection of emergencies, and preserves the energy of nodes. Whereas all the samples are transmitted in the other compared schemes, and classification and LR are applied on the whole dataset, which increases both the execution time and the nodes' energy consumption.

The obtained performance comparison results prove that AS-LR is an efficient scheme that is able to address various BSN challenges.

VI. CONCLUSION

In this article, an adaptive sampling approach is proposed based on LR model. The main aim of the proposed scheme is to accelerate the detection of true emergency cases and to decrease the energy consumption of nodes. Simulations showed that the proposed AS-LR scheme achieves the same accuracy as LR without sampling for window sizes less than 25, while reducing both the detection time of anomalies and the energy consumption of nodes. Simulations also showed that AS-LR outperforms other anomaly detection schemes in terms of precision, recall, specificity, and accuracy, while achieving a much lower execution time and energy consumption. This proves that the proposed scheme satisfies the requirements of BSN and succeeds in addressing its challenges. Future work includes proposing a LR re-modeling technique to enhance the performance of the prediction model, as well as evaluating the proposed adaptive sampling approach when applied to other ML techniques such as decision trees and artificial neural networks.

REFERENCES

[1] A. Kumar, M. Amarlingam, and P. Rajalakshmi. Random node sampling approach for energy efficient data gathering in wireless sensor networks. In *Proceedings of the IEEE Region 10 Symposium (TENSYMP)*, pages 1–5. IEEE, 2017.

[2] A. Albattah and M. A. Rassam. A correlation-based anomaly detection model for wireless body area networks using convolutional long short-term memory neural network. *Sensors*, 22(5):1951, 2022.

[3] O. Salem, Y. Liu, and A. Mehaoua. Anomaly detection in medical wireless sensor networks. *Journal of Computing Science and Engineering*, 7(4):272–284, 2013.

[4] N. Boudargham, R. El Sibai, J. Bou Abdo, J. Demerjian, C. Guyeux, and A. Makhoul. Toward fast and accurate emergency cases detection in bsns. *IET Wireless Sensor Systems*, 10(1):47–60, 2020.

[5] O. Salem, A. Guerassimov, A. Mehaoua, A. Marcus, and B. Furht. Anomaly detection in medical wireless sensor networks using svm and linear regression models. *International Journal of E-Health and Medical Communications (IJEHMC)*, 5(1):20–45, 2014.

[6] G. Pachauri and S. Sharma. Anomaly detection in medical wireless sensor networks using machine learning algorithms. *Procedia Computer Science*, 70:325–333, 2015.

[7] S. Habib. Machine learning to identify aberrant energy use to detect property failures, 2020.

[8] S. K. Nagdeo and J. Mahapatro. Wireless body area network sensor faults and anomalous data detection and classification using machine learning. In *2019 IEEE Bombay Section Signature Conference (IBSSC)*, pages 1–6. IEEE, 2019.

[9] M. U. H. Al Rasyid, I. U. Nadhori F. Setiawan, A. Sudarsonc, and N. Tamami. Anomalous data detection in wban measurements. In *2018 International Electronics Symposium on Knowledge Creation and Intelligent Computing (IES-KCIC)*, pages 303–309. IEEE, 2018.

[10] C. Habib, A. Makhoul, R. Darazi, and C. Salim. Self-adaptive data collection and fusion for health monitoring based on body sensor networks. *IEEE transactions on Industrial Informatics*, 12(6):2342–2352, 2016.

[11] M. Mohammad, I. Attarzadeh, and M. Hosseinzadeh. Adaptive sampling rate determination for energy efficiency of wireless body area networks. *Journal of Soft Computing and Information Technology (JSCIT)*, 8(1):1–13, 2019.

[12] A. Y. Benbasat and J. A. Paradiso. Groggy wakeup-automated generation of power-efficient detection hierarchies for wearable sensors. In *Proceedings of the 4th International Workshop on Wearable and Implantable Body Sensor Networks (BSN)*, pages 59–64. Springer, 2007.

[13] A. Abou Jaoude. The analysis of selected algorithms for the statistical paradigm. *The Republic of Moldova: Generis Publishing*, 2020.

[14] M. Rafferty, P. Brogan, J. Hastings, D. Laverty, X. A. Liu, and R. Khan. Local anomaly detection by application of regression analysis on pmu data. In *2018 IEEE Power & Energy Society General Meeting (PESGM)*, pages 1–5. IEEE, 2018.

[15] R. D. Riley et. al. Calculating the sample size required for developing a clinical prediction model. *Bmj*, 368, 2020.

[16] Physionet Database, URL: http://www.physionet.org/cgi-bin/atm/ATM., 2023.

Computational-Based Advanced Encryption Standard (AES) Accelerator

Enas Abulibdeh, Hani Saleh, Baker Mohammad, Mahmoud Alqutayri

System on Chip Lab (SoC), Department of Electrical and Computer Engineering

Khalifa University, Abu Dhabi, UAE

100059804,hani.saleh,baker.mohammad,mahmoud.alqutayri} @ku.ac.ae

Abstract—**The demands of high-level security and performance for resource-constrained SoC represent real challenges. Consequently, the dedicated accelerators are designed to deliver a high-quality function with minimal costs. This paper introduces a high-performance Advanced Encryption Standard (AES) accelerator that minimizes the area and power overhead. The suggested design replaces the LUT-based implementation of the substitution block (SBox) with a combinational circuit to break the fixed memory accesses and reduce power consumption. The proposed architecture reduces the hardware complexity by integrating the transformation and its inverse in one block and utilizes one key expansion block. The suggested accelerator outperforms the standard implementation of encryption by 25% and takes the benefits of the design aspects that are utilized in it.**

Index Terms—**AES, SoC, SBox, pipeline, hybrid block, ShiftRow, MixColumn, area, power, performance**

I. INTRODUCTION

[1] The Rapidly-growing of Internet-of-Things (IoT) and the ease of accessing their devices make data security essential and uncompromisable. Thus, cryptography algorithms are built and integrated on SoC. However, providing a high level of security for real-stream applications degrades the overall performance. The security and performance can be satisfied by sacrificing the area and energy. The limited space and battery life of SoC drop the traditional implementation of the related approaches and open the door to design efficient and dedicated accelerators.

AES accelerator is a cryptography accelerator that performs iterative symmetric key encryption and decryption according to Federal Information Processing Standard (FIPS) Publication 197. The accelerator implements three main processes: key expansion, plaintext encryption, and ciphertext decryption. The flow of these processes is shown in Fig. 1. The key expansion generates Nr round keys, which are used through encryption or decryption iterations. When the initial key is 128/192/256-bit, the algorithm applies 10/12/14 rounds either for encryption or decryption. The plaintext encryption includes four transformations: SubByte, ShiftRow, MixColumn (except the last iteration), and AddRoundKey. The ciphertext decryption applies the inverse of the four transformations of the decryption process in the opposite order. Given that all transformations are performed in the finite field GF (2^8).

The key expansion, encryption, and decryption use SBox, and its usually built using look-up table. Hence, 10 rounds of AES encryption involves 160 memory accesses for state

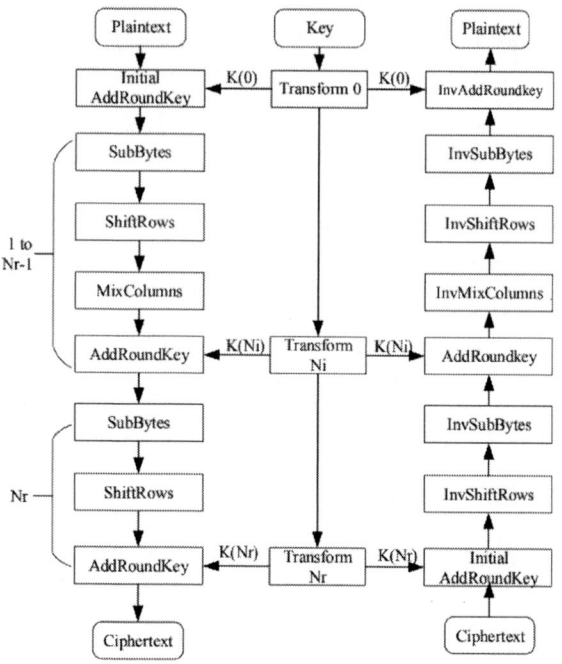

Fig. 1: The data flow and key generation inside AES accelerator [2]

substitution and 40 memory accesses to generate rounds' key, which in total 200 accesses [3]. On the other side, the same values can be generated using proper combinational circuit that built using a set of logical functions [4]. Analyzing the power consumed by each [3], the LUT-based implementation consumes 50 to 6K times of the combinational logic as shown in Fig. 2, which makes Sbox the most power-hunger unit in AES. Also, using one block for both key expansion and encryption/decryption, their application can't be parallel, which declines the performance. In addition to the SBox, key expansion, ShiftRow, MixColumn, and AddRoundKey are the encryption's transformations and their inverses are applied on the decryption. Since the implementation of each transformation and it's inverse are very similar, they can be integrated in one block to reduce the hardware complexity [2] [5]. Integrating the encryption and decryption impacts the overall performance [6] especially when the block is utilized

979-8-3503-8083-5/23 $31.00 © 2023 IEEE

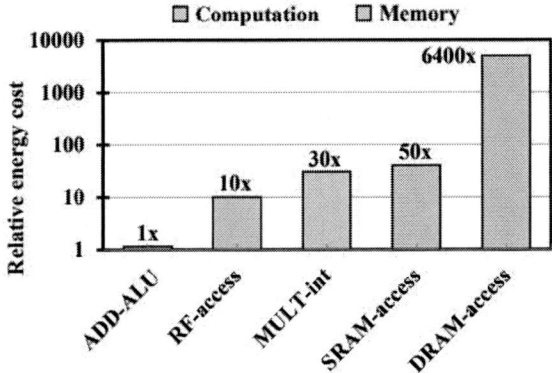

Fig. 2: Relative energy cost of combination's elements and memory accesses [3]

more than once during the execution such as SBox [2]. Also, on-fly operations demand a high generation rate of output. Various mechanisms are applied to maintain and enhance the execution time including breaking the critical path, pipelining the flow, and multi-core implementation. Additionally, an efficient combination of different approaches produces a satisfied design.

II. PROBLEM DESCRIPTION

Energy, area, and performance are three critical aspects of SoC design, and the dominant factor negatively affects the other two parameters. This work upgrades the performances of the AES accelerator to match the requirements of real-stream applications and minimizes the additional cost on both area and power aspects. The proposed architecture unrolls AES rounds into the hardware-based sequential rounds to reduce the number of execution cycles per operation [7] [8], supports two types of pipeline: out-round and in-round to overlap operations execution and increase throughput [9] [10], and breaks down the execution time of each round to enhance the frequency [10]. Additionally, the power overhead is minimized by building SBox using combinational logic. Also, the hardware complexity is reduced by using one block of KeyExpantion units along with all rounds and combining the implementation of each transformation and its inverse in one unit.

III. RELATED WORK

Duplicating hardware blocks, reducing critical paths, pipelining the data flow, and parallel execution paths are different mechanisms that are applied to enhance the design performance. To speed up the execution, [11] and [12] utilizes multiple SBoxs that enable parallel operating paths. 128-AES accelerator [13] achieves a high throughput of 30 to 70 Gbps by pipelining the internal and external data flow. The outer pipeline considers each AES round as a single stage, while the round's transformations are the stages of the inner pipeline.

This structure consumes 41 cycles to encrypt (or decrypt) a 128-bit block. P. Dong et al. [12] propose a high throughput AES core for real-time applications. The proposed architecture is 11 pipeline stages, where each stage presents one round. Also, a parallel path of the key expansion is proposed of 11 stages. Inside each round, 16 blocks of S-Box are utilized, ShiftRow is implicitly implemented through wiring, and microarchitectures are proposed for MixColumn and Key Expansion. The design is verified using Synopsys tools with COMS 45nm technology and achieved a throughput of 111 Gbps, and latency of 12.6 ns. However, the decryption process is not supported, and pipelining the round's transformations is discarded to reduce the latency. The AES core [12] is utilized in [14] to build a multi-core accelerator that achieves 853 Gbps at 667 MHz. The architecture uses ten cores with a key expansion unit shared among all cores. The suggested AES in [15] reduces both area and power consumption. They replace flip-flops with latches to reduce the hardware complexity and break the S-Box into two stages to shorten the critical path and decrease the cycle time by 28%. The design achieves 46.2 Mbps throughput at 0.47V. However, this throughput can't support the real-time application. Replacing LUT-SBox with the associated circuit has become common in recent research [2], [3], [5], [16], [17]. The SBox Acceleration Scheme [3] introduces a custom instruction in general purpose processor, and instead of retrieving the S-value from memory, the instruction calculates the substituting values using a combinational circuit. The proposed scheme takes up to 15 iterations for 8-bit entry and enables four simultaneous applications due to the pipelined architecture. The low-cost AES [2] implements a SubByte module without a look-up table to maximize the utility of the block and reduce the hardware complexity. Atomic-AES [6] provides dual functionality of encryption and decryption in 8-bit implementation. The design occupies 2227 GE and has an encryption/decryption latency of 246/326 cycles. The latency difference arises from reusing the forward transformations without optimizing them to support the forward and the inverse operations. AES crypto-engine [18] is an integrated design of AES encryptor and decryptor for hardware-critical applications. The architecture implements the function and its inverse in one clock for each, but the design is not pipelined. The low-cost AES [2] reuses the round unit for both encryption and decryption. The round unit includes four blocks on sequence, and each implements the transformation and its inverse, which is defined as a hybrid block. AddRoundKey is repeated at the beginning and the end of the round unit. The data path bypasses MixColumn at the final round, and the final output feeds AES's input for multi-round execution. The architecture consumes 92/110/128 cycles to perform AES-128/192/256 respectively. Nano-AES [5] enhances the area and the delay of the encryption process using an 8-bit datapath, and the MixColumn is reconfigured to 8-bit width using four internal registers. The Ultra-low power ultra-low energy AES [11] is a 32-bit architecture and delivers three levels of security: 128, 192, and 256-bit. The accelerator in [5] and [11] use Two register banks each with 16 x 8-bit

979-8-3503-8083-5/23 $31.00 © 2023 IEEE

registers. The State and Key registers store the round key and the intermediate results, and the implementation of ShiftRow is embedded in the State register. AES-128 consumes 527 cycles by Nano-AES [5] and 44 cycles by Ultra-low power AES [11]. The pipelined 128-AES encryption [19] pipelines the initial, nine intermediate, and final rounds to enhance the execution time, even though the design adds extra area overhead. SIMON Crypto-Core [20] saves both energy and area for low-level security SoC. The core uses a 64-bit key on 32/64 plaintext, and the architecture can be scaled up using 4 cores to support full security when needed.

IV. THE PROPOSED ARCHITECTURE

This work suggests a high-performance architecture of AES with minimal cost in both area and power. The proposed accelerator re-configures different implementations from literature, and the implementation to match design requirements. These aspects include:

1) Replacing LUT-based SubByte with the equivalent combinational circuit.
2) Aggregating the implementation of the transformation and its inverse in one unit.
3) Pipelining of AES's rounds and transformations.

Also, this work contributes to AES design by utilizing one block of the keyExpansion among all AES rounds, breaking Sbox access time into three stages, which forms with the MixColumn the four stages of the inner pipeline.

A. SubByte Implementation

A substitution byte is a set of permutations at the bit level and is applied on one byte to generate one byte. These permutations can be modeled mathematically using a set of equations that are applied in a finite field, where the output of any operation always fits in 8 bits. The look-up table is the classical implementation of this unit, where the 8-bit input represents the address of the table and the selected byte is the output. While the traditional implementation consumes power [21] [16] and area [2] [11], the associated combinational circuit [4] saves both. The logical implementation involves two main operations: multiplicative inverse and Affine transformation. The multiplicative inverse is the core of the block and is shown in Fig. 3. The design mainly uses XOR and offers minor usage of AND and NOT logic gates. δ and δ^{-1} are the isomorphic mapping to composite field $GF(2^8)$ and it's inverse respectively, x^2 is the squarer in $GF(2^8)$, $x\lambda$ and $x\phi$ are multiplication with constant in $GF(2^4)$ and $GF(2^2)$ respectively, and x^{-1} is the multiplicative inversion in $GF(2^4)$.

B. Hybrid Blocks

The hybrid block is the logic circuit that supports the implementation of the function and its inverse in one unit [6] [2]. Since the AddRoundKey operation is the same in both encryption and decryption, the combination of the function and its inverse is applicable for SubByte, ShiftRow, and MixColumn transformations. Affine transformation with

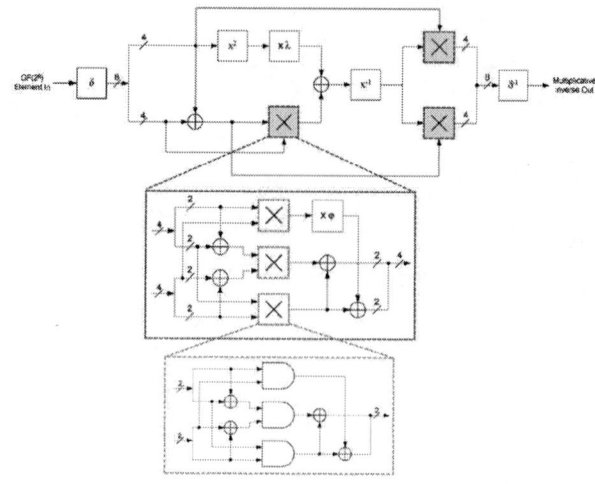

Fig. 3: The hardware implementation of the multiplicative inverse [4]

the multiplicative inverse delivers the full implementation of SubByte. When Affine transformation is applied after δ^{-1}, the circuit operates as the forward SBox. While applying Inverse Affine on the input of δ supports the reverse function. Fig. 4 shows the hybrid design of SubByte. The forward ShiftRow

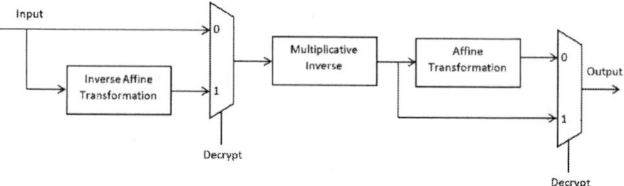

Fig. 4: The block diagram of hybrid SubByte

rotates the row's bytes to the left by the index of the row, while the inverse rotates the bytes right by the same amount for each row. Thus, the hybrid ShiftRow block attaches a multiplexer at the input of each byte, which selects between the circulated-left byte or the circulated-right byte. Since the 0th and 2nd rows are the same for both transformations, the Muxes for the 0th and 2nd rows are absent [6]. Because this block is embedded in the pipeline state register (as will be described later), the Muxs' output will be stored in the intermediate register. The MixColumn and its inverse can be combined with one circuit as shown in Fig. 5. a, b, c, and d are the bytes of the same column starting from the top (a) to down (d). xout and yout are 8-bit outputs of the encryption and decryption respectively. All column bytes are involved to generate the output byte as shown in the green box Fig. 5. Connecting four green blocks generates a 32-bit Hybrid MixColumn. Similarly, the red box can be duplicated to create a 128-bit Hybrid MixColumn unit. This work does not investigate the hybrid design of key expansion units. Therefore, the design has two

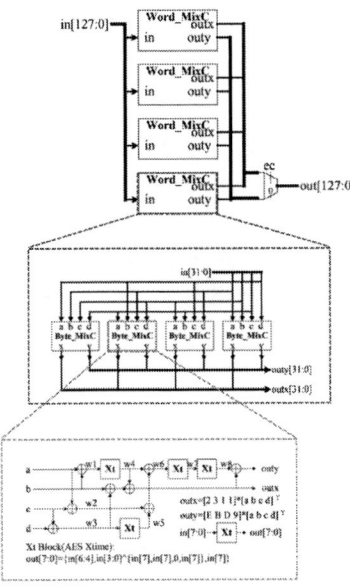

Fig. 5: 128-bit Hybrid MixColumn block [18]

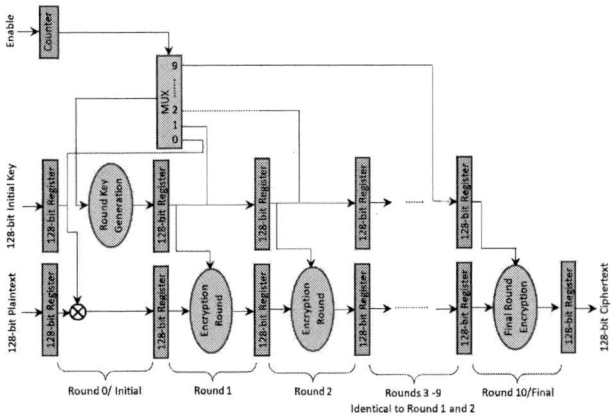

Fig. 6: The pipeline architecture of the proposed AES

dependent Sbox and MixColumn. The ShiftRow is embedded

key expansion units, and the input of AddRoundKey will be selected based on the type of operation at that stage.

C. The Efficient Pipeline

The pipeline is suggested to enhance AES's throughput. To enable the overlap execution, the architecture duplicates the round blocks into 11 blocks connected in sequence, and supports parallel independent paths of the data flow, inside each round. The top view of the proposed architecture is shown in Fig. 6. Two parallel paths flow concurrently to generate an intermediate state and sub-key for each round. The inputs are: 128-bit plaintext/ciphertext, 128- initial key, and the enable signal, and the output is 128-bit ciphertext/plaintext. The sub-key and the intermediate result of the round are latched through two 128-bit registers. Including the input and output registers, the top-level design utilizes 23 128-bit registers. Since the key (and sub-keys) will be the same for the whole message, the upper path only acts at the beginning of the process or when the key is updated externally. Therefore, the upper utilizes one KeyExpansion block, which generates the sub-keys for all rounds and latch them through the round's register. The counter output determines which input should be selected to KeyExpansion and enable the writing on the target register. The counter resets then counts when the enable signal asserts (a new key is counted). When the counter reaches 10, it stops. The lower path represents the encryption/decryption process. The initial round applies XOR between the initial key and input, the following nine rounds use the same logic, and the last round utilizes the same rounds logic except MixColumn is removed.

Each round encapsulates 16 blocks of SBoxs and 16 blocks of MixColumn as shown in Fig. 7. 128-bit of plaintext is divided into 16 concurrent paths, where each flows through in-

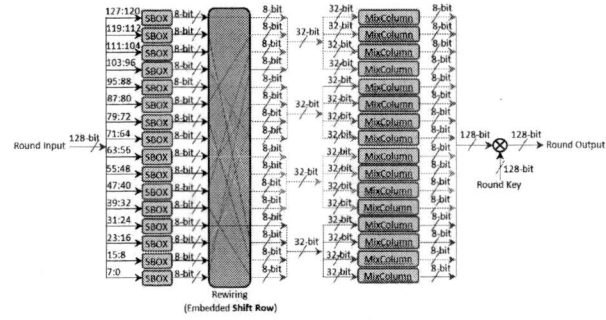

Fig. 7: The data flow and the architecture of the encryption round

by reordering the connection between SBox and MixColumn, and this order is visually described in Fig. 8. The colors represent the order before the shifting and the labels represent the order after the shifting. For instance, the first column before the ordering (represented in green) was in R00, R01, R02, and R03 due to the shifting on the second row by one byte, R01 previously becomes R05, and the two shifts on the third row put R02 previously in R10 and so on. The pipeline

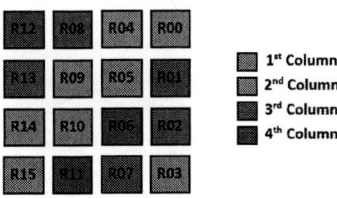

Fig. 8: The visual presentation of the embedded ShiftRow [11]

stages are the internal stages of the unrolled 11 rounds. Since the transformations' delay is intensively imbalanced, the stages of the round are selected to minimize the critical delay. Thus, combinational SBox is divided into three stages as shown in Fig. 9. The red lines represent the stage registers. The first stage includes inverse Affine transformation, isomorphic map, squaring, and multiplication in $GF(2^4)$. The second stage includes the multiplicative inverse in $GF(2^4)$. The third stage includes the multiplication in $GF(2^4)$, inverse isomorphic map and Affine transformation. The fourth stage is MixColumn with the AddRoundKey. The length (in bits) of the stage registers are 16x (16, 12, 8, and 8) respectively.

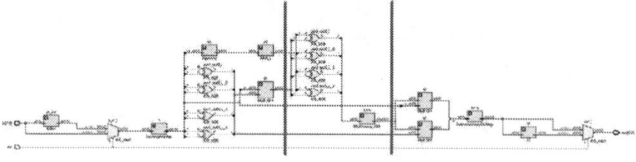

Fig. 9: The pipeline architecture of SBox

The initial round doesn't contain round block. However, the four stages are required for the KeyExpansion block, which block utilizes four parallel SBoxs as shown in Fig. 10.

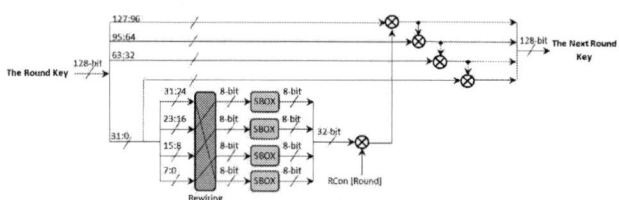

Fig. 10: Internal design of KeyExpansion

V. RESULTS

This work is implemented using the Vivado 19.2 tool and Synthesised on Artix-7. To analyze the design, it is compared with the unrolled standard AES implementation, which supports the encryption process only, and two variations of the proposed architecture. The first variation (I) supports a KeyExpansion unit for each round, which means 10 KeyExpantion units are utilized, and the round is designed without the internal four stages. The second variation (II) replaces the combinational SBoxs inside KeyExpansion with the traditional implementation of SBoxs.

To ensure the functionality of the design, a set of plaintext and ciphertext pairs are taken from the standard document and operate on the design through the simulation phase. Fig. 11 shows the initial flow of the signals when AES starts work. After the reset, the plaintext is latched by the data path every clock cycle, and the key is latched every four cycles. Since one key is utilized per message, key expansion blocks are reduced into one. As a result, all the key registers latch the new value (if available) every four cycles. Once the sub-keys are generated and latched by the designated register, the value will not change until a new key is inserted. The overhead of filling the key's pipeline will be the same as filling the data pipeline, therefore no extra overhead is added.

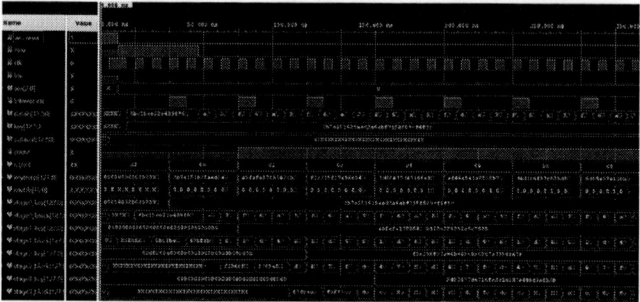

Fig. 11: The signals flow at the beginning of AES operation

After filling the pipeline, the ciphertext is generated one block per cycle (128-bit block) as shown in Fig. 12. The first output block is generated after 41 cycles of the first input, which are four stages per round, and one cycle to latch the first input. This work targets the throughput by reducing cycle

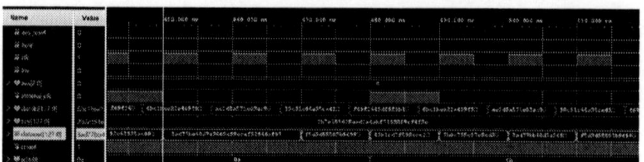

Fig. 12: The proposed AES outputs after first 41 cycles

time. Implementing SBoxs as a combinational logic increases the delay as shown in Table I. The standard AES implements 11 stages of the pipeline, where each stage is a single round and utilizes LUT-based SBox. While variation (I) is also an 11-stage pipeline, the round utilizes the combinational circuit. The cycle time for the second architecture is 50% greater, but it's breakable. Thus, the round is divided into four stages. However, the synthesis results of the proposed AES show that the cycle time remains the same (8 ns). The selection of the input to the KeyExpantion unit among all sub-key registers through the Mux and based on counter value are extra time overhead that is added to the first stage. The first stage of key expansion takes a minimum of 8 ns, which makes our enhancements pointless. Since the sub-keys are computed prior to the round by four stages, and the four stages are unnecessarily in the KeyExpansion block, LUT-based SBox is utilized in the KeyExpansion block, because its access time is less and not required to be pipelined. This modification in

TABLE I: The performance measurements of the fourth architectures

Architecture	Cycle Time (ns)	Throughput (Gbps)	Latency (cycles)
Standard AES	8	14.2	11
Proposed AES	8	14.2	41
Variation I	12	10.7	11
Variation II	6	21.3	41

the architecture is labeled as Variation II, which reduces the cycle time by 25% as shown in Table I. The latency pays the overhead of this enhancement. The plaintext is encrypted in 88 ns by standard AES, while it takes 246 ns in Variation II. The power and area are the compromised factors to enhance the performance as shown in Table II and Table III. Standard AES is minimal in both the area and power. The power consumed by Variation II is duplicated because of SBox utilization, while the hardware complexity of Variation II is reduced due to utilizing one KeyExpansion block.

TABLE II: The power consumption of the compared fourth architectures

Architecture	Total	Static	Dynamic
Standard AES	0.633	0.11	0.523
Proposed AES	0.917	0.11	0.807
Variation I	0.922	0.132	0.789
Variation II	1.45	0.111	1.34

TABLE III: The hardware utilization of the compared architectures

Architecture	LUTs	Registers	F7 Muxes	F8 Muxes	IO
Standard AES	9626	2560	3812	1473	385
Proposed AES	12580	8666	114	-	391
Variation I	12994	5900	118	-	391
Variation II	12377	8559	190	-	391

VI. CONCLUSION

The proposed high-performance AES for SoC applies a set of performance-based design aspects trying to increase the throughput of the accelerator to support real-time applications. The proposed architecture unrolls AES rounds into 11 sequential units and breaks SBox's access time into three stages to reduce the cycle time. To minimize the power and hardware overheads, encryption and decryption transformations integrate withing one block, only one KeyExpansion block is utilized and SBox is implemented using the equivalent logic.

REFERENCES

[1] M. D. P. E. Nitin Dahad. (2020) How puf technology is securing iot @ONLINE. [Online]. Available: https://www.eetimes.com/how-puf-technology-is-securing-iot

[2] X. Liu, Y. Chen, K. Lu, D. Liu, B. Liu, and Q. Jiang, "Design and implementation of a low-cost aes coprocessor based on estt-mram ip," in 2020 IEEE 15th International Conference on Solid-State Integrated Circuit Technology (ICSICT), 2020, pp. 1–3.

[3] C. Duran, H. Gomez, and E. Roa, "Aes sbox acceleration schemes for low-cost socs," in 2021 IEEE International Symposium on Circuits and Systems (ISCAS). IEEE, 2021, pp. 1–5.

[4] E. N. Mui, R. Custom, and D. Engineer, "Practical implementation of rijndael s-box using combinational logic," Custom R&D Engineer Texco Enterprise Pvt. Ltd, 2007.

[5] K. Shahbazi and S.-B. Ko, "Area-efficient nano-aes implementation for internet-of-things devices," IEEE Transactions on Very Large Scale Integration (VLSI) Systems, vol. 29, no. 1, pp. 136–148, 2020.

[6] S. Banik, A. Bogdanov, and F. Regazzoni, "Compact circuits for combined aes encryption/decryption," Journal of Cryptographic Engineering, vol. 9, 04 2019.

[7] A. Gielata, P. Russek, and K. Wiatr, "Aes hardware implementation in fpga for algorithm acceleration purpose," in 2008 International Conference on Signals and Electronic Systems, 2008, pp. 137–140.

[8] K. Rahimunnisa, P. Karthigaikumar, S. Rasheed, J. Jayakumar, and S. SureshKumar, "Fpga implementation of aes algorithm for high throughput using folded parallel architecture," Security and Communication Networks, vol. 7, no. 11, pp. 2225–2236, 2014.

[9] S. Qu, G. Shou, Y. Hu, Z. Guo, and Z. Qian, "High throughput, pipelined implementation of aes on fpga," in 2009 International Symposium on Information Engineering and Electronic Commerce. IEEE, 2009, pp. 542–545.

[10] M. I. Soliman and G. Y. Abozaid, "Fpga implementation and performance evaluation of a high throughput crypto coprocessor," Journal of Parallel and Distributed Computing, vol. 71, no. 8, pp. 1075–1084, 2011.

[11] D.-H. Bui, D. Puschini, S. Bacles-Min, E. Beigné, and X.-T. Tran, "Aes datapath optimization strategies for low-power low-energy multisecurity-level internet-of-things applications," IEEE Transactions on Very Large Scale Integration (VLSI) Systems, vol. 25, no. 12, pp. 3281–3290, 2017.

[12] P.-K. Dong, H. K. Nguyen, and X.-T. Tran, "A 45nm high-throughput and low latency aes encryption for real-time applications," in 2019 19th International Symposium on Communications and Information Technologies (ISCIT). IEEE, 2019, pp. 196–200.

[13] A. Hodjat and I. Verbauwhede, "Area-throughput trade-offs for fully pipelined 30 to 70 gbits/s aes processors," IEEE Transactions on Computers, vol. 55, no. 4, pp. 366–372, 2006.

[14] P.-K. Dong, H. K. Nguyen, V.-P. Hoang, and X.-T. Tran, "Low-power implementation of a high-throughput multi-core aes encryption architecture," in 2020 IEEE Asia Pacific Conference on Circuits and Systems (APCCAS). IEEE, 2020, pp. 74–77.

[15] Y. Zhang, K. Yang, M. Saligane, D. Blaauw, and D. Sylvester, "A compact 446 gbps/w aes accelerator for mobile soc and iot in 40nm," in 2016 IEEE symposium on VLSI circuits (VLSI-circuits). IEEE, 2016, pp. 1–2.

[16] S. Morioka and A. Satoh, "An optimized s-box circuit architecture for low power aes design," in International Workshop on Cryptographic Hardware and Embedded Systems. Springer, 2002, pp. 172–186.

[17] A. Barrera, C.-W. Cheng, and S. Kumar, "A fast implementation of the rijndael substitution box for cryptographic aes," in 2020 3rd International Conference on Data Intelligence and Security (ICDIS), 2020, pp. 20–25.

[18] C.-C. Lu and S.-Y. Tseng, "Integrated design of aes (advanced encryption standard) encrypter and decrypter," in Proceedings IEEE International Conference on Application-Specific Systems, Architectures, and Processors. IEEE, 2002, pp. 277–285.

[19] M. Nabil, A. Khalaf, and S. M. Hassan, "Design and implementation of pipelined and parallel aes encryption systems using fpga," Indones. J. Electr. Eng. Comput. Sci, vol. 20, pp. 287–299, 2020.

[20] S. Taneja and M. Alioto, "Deep sub-pj/bit low-area energy-security scalable simon crypto-core in 40 nm," in 2020 IEEE International Symposium on Circuits and Systems (ISCAS). IEEE, 2020, pp. 1–5.

[21] M. Horowitz, "1.1 computing's energy problem (and what we can do about it)," in 2014 IEEE International Solid-State Circuits Conference Digest of Technical Papers (ISSCC). IEEE, 2014, pp. 10–14.

[22] W. I. El Sobky, A. A. Isamail, A. S. Mohra, and A. M. Hassan, "Implementation mini (advanced encryption standard) by substitution box in galois field $(2¡sup¿4¡/sup¿)$," in 2021 International Telecommunications Conference (ITC-Egypt), 2021, pp. 1–4.

[23] U. Banerjee, A. Pathak, and A. P. Chandrakasan, "2.3 an energy-efficient configurable lattice cryptography processor for the quantum-secure internet of things," in 2019 IEEE International Solid-State Circuits Conference-(ISSCC). IEEE, 2019, pp. 46–48.

A SDR Transmitter Baseband-to-IF IC with Digital Out-phasing Exponential Modulation in a 65nm CMOS LPE

Mihai Sanduleanu, Senior *Member, IEEE* , Ahmed Mahdy

Abstract— A digital pulsed modulation for high efficiency and linearity, switched-mode power amplifier (PA) is proposed. Based on a new representation of digital modulated signals at IF called Out-phasing Digital Exponential Modulation (ODEM), the paper presents the concept, simulations and measurements of an integrated solution in CMOS (baseband and IF) and GaN (PA) technologies. It exhibits a faster decaying spectrum, when compared to Out-phasing pulse-width modulation (OPWM), and a power spectrum that fulfills the spectral emission mask for 64QAM LTE standards. The implementation in a 65 nm CMOS LPE process from GF based on the baseband solution, shows that the SEM requirements for LTE standards are fulfilled. The circuit implementation of the baseband IC is presented is based on digital CMOS circuits.

Index Terms—Digital out-phasing, pulsed modulation, high efficiency, high linearity, switched-mode power amplifiers

I. INTRODUCTION

THE need to support multiple communication standards by mobile produced in the last years a plethora of papers [4-16] that try to approach the solution proposed by J. Mitola in 1995 paper [2] on the Software Defined Radio (SDR) transmitter concept with a D/A Converter controlled by DSP and sending signals directly to the antenna. However, we are still far away from that vision for various reasons related to different implementation that are not completely digital in their implementation. Apart from re-configurability and portability from one CMOS node to another, enabled by SDR, the increased demand for high data rate standards, calls for spectrally efficient modulation schemes such as high-order quadrature amplitude modulation (QAM) and orthogonal frequency-division multiplexing (OFDM). This comes at the expense of higher peak-to-average power ratio (PAPR) signals with stringent linearity requirements and additional power backoff (PBO) in order to control the transmitter's overall power [3]. For this reason, Class-A and Class-AB amplifiers, known for their linearity are heavily employed in different transmitter solutions. But this deviates from the Mitola's SDR Radio as this is an analog solution instead of a digital one. In consequence, it is difficult to achieve high efficiency as they operate more than 90% of their time in back-off in order to accommodate large PAPR. This is required by modern OFDM modulation formats with heavy amplitude modulation (e.g. 64 QAM). The efficiency in backoff drops to about 20%. On the other side, high efficiency power amplifiers using Class-E, D or D^{-1} are typically highly nonlinear [4]. Therefore, the transmitter efficiency is very high but the designs do not fulfill the spectral emission mask requirements and EVM requirements. Secondly, the growing wish of saving fabrication cost and achieving smaller form factor has necessitated the need for integrating radio frequency integrated circuit functionality along with digital baseband circuits in heavily scaled CMOS processes [5]. In order to attain high efficiency and high linearity at the same time, with a highly digitized concept we propose to change the type of representation of the digital signals, in order to improve the spectral characteristics of the signal before applying it to a classical, highly efficient power amplifier (PA) like Class D^{-1} operating at Watt level with high efficiency. To achieve this, a novel, digital pulsed modulation, called Digital Out-phasing Exponential Modulation (DOEM), for high efficiency and linearity, switched-mode PA is proposed. The paper shows the concept, simulations and measurement results based on a 65nm CMOS implementation. The paper is organized as follows: Section II introduces Digital Out-phasing Exponential Modulation, Section III presents the circuit design. Section IV presents measurement results and Section V concludes the paper.

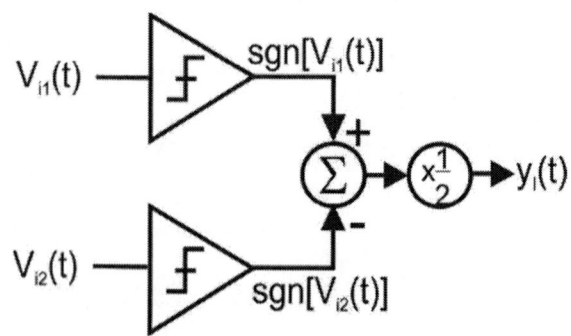

Fig. 1. Feeding $V_{i1}(t)$ and $V_{i2}(t)$ to two limiters

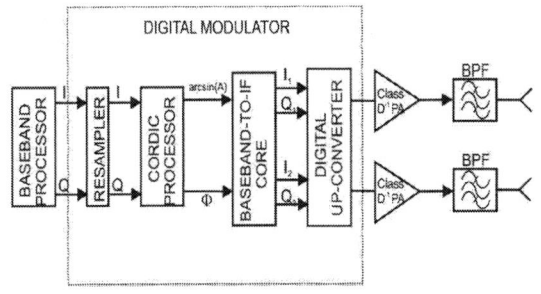

Fig. 2: Concept of proposed architecture

II. PROPOSED SYSTEM ARCHITECTURE

In [1] we introduced the Digital Exponential Modulation and the gist of it is presented below. An amplitude and phase-modulated signal can be represented as:

$$A(t)\cos[\omega t + \phi(t)]$$
$$= \sin[\arcsin A(t)]\cos[\omega t + \phi(t)] \qquad (1)$$
$$= \frac{1}{2}V_1(t) - \frac{1}{2}V_2(t)$$

where

$$V_{I1}(t) = \sin[\omega t + \phi(t) + \arcsin A(t)] \qquad (2)$$

$$V_{I2}(t) = \sin[\omega t + \phi(t) - \arcsin A(t)] \qquad (3)$$

Applying the two signals to two separate limiters (see Fig.1), yields:

$$y_I(t) = \frac{4}{\pi}A(t)\cos(\omega t + \phi(t)) \qquad (4)$$

$$+ \frac{4}{\pi}\frac{3A(t) - 4A^2(t)}{3}\cos[3(\omega t + \phi(t))]$$

$$+ \frac{4}{\pi}\frac{5A(t) - 16A^3(t) + 12A^5(t)}{5}\cos[5(\omega t + \phi(t))]$$
$$+ \cdots$$

Given that $|A(t)| \leq 1$ and if ω is sufficiently large, there is no aliasing and the signal can be recovered after low-pass filtering. This yields a form of modulation with constant amplitude that can be applied to two Class D^{-1} amplifiers in an

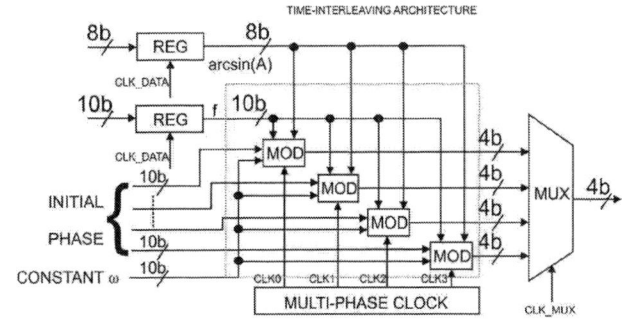

Fig. 4: Time interleaving Architecture to increase throughput

out-phasing manner, as shown in Fig.2.

Denote:

$$I_1(t) = \text{sgn}\big[\sin\big(\omega t + \phi(t) + \arcsin A(t)\big)\big], \qquad (5)$$

$$I_2(t) = \text{sgn}\big[\sin\big(\omega t + \phi(t) - \arcsin A(t)\big)\big]. \qquad (6)$$

To up-convert an IF signal to RF, the imaginary part of the IF signal is also needed. The imaginary part is:

$$\begin{aligned}&\text{Im}\big\{A(t)e^{j(\omega t + \phi(t))}\big\}\\ &= A(t)\sin[\omega t + \phi(t)] \qquad (7)\\ &= \sin\big[\arcsin\big(A(t)\big)\big]\sin[\omega t + \phi(t)]\\ &= -\left[\frac{1}{2}V_{Q1}(t) - \frac{1}{2}V_{Q2}(t)\right]\end{aligned}$$

where:

$$V_{Q1}(t) = \cos[\omega t + \phi(t) \\ + \arcsin A(t)] \qquad (8)$$

$$V_{Q2}(t) = \cos[\omega t + \phi(t) \\ - \arcsin A(t)] \qquad (9)$$

Applying the two signals to two limiters yields:

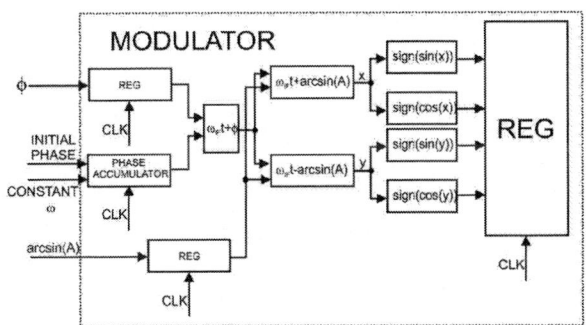

Fig. 3: Modulator built from a Phase Accumulator

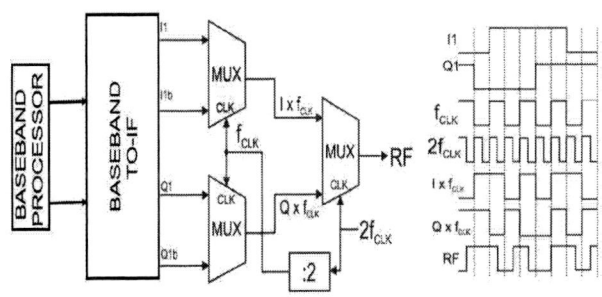

Fig. 5: Digital multiplexer based up-conversion to RF

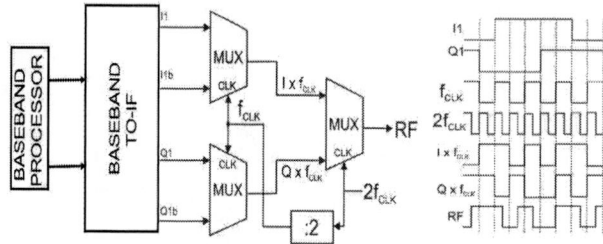

Fig. 6: Digital Multiplexer based up-conversion to RF

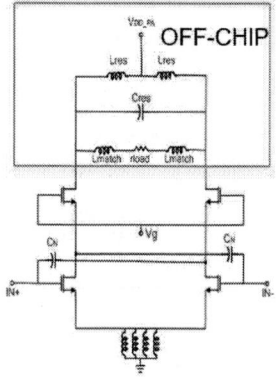

Fig. 8: GaN Class D^{-1} Output stage

$$y_Q(t) = -\left\{\frac{1}{2}\,\mathrm{sgn}[V_{Q1}(t)]\right.$$
$$\left.-\frac{1}{2}\,\mathrm{sgn}[V_{Q2}(t)]\right\}$$

$$y_Q(t)=$$
$$y_Q(t) == \frac{4}{\pi}\,A(t)\sin(\omega t + \phi(t)) \qquad (11)$$
$$-\frac{4}{\pi}\frac{3A(t)-4A^2(t)}{3}\sin[3(\omega t + \phi(t))]$$
$$+\frac{4}{\pi}\frac{5A(t)-16A^3(t)+12A^5(t)}{5}\sin[5(\omega t$$
$$+\phi(t))]+\cdots.$$

Equation (11) shows $y_Q(t)$ is only a scaled version of the imaginary part of the IF signal plus some harmonics, which can be low-pass filtered.

Let:

$$Q_1(t) = \mathrm{sgn}[\cos(\omega t + \phi(t) + \arcsin A(t))], \qquad (12)$$

$$Q_2(t) = \mathrm{sgn}[\cos(\omega t + \phi(t) - \arcsin A(t))] \qquad (13)$$

$I_1(t)$ and $Q_1(t)$, $I_2(t)$ and $Q_2(t)$ are fed to two up-converters separately, and the resultant signals can be applied to two Class-D^{-1} PAs in an out-phasing manner, as shown in Fig.2.

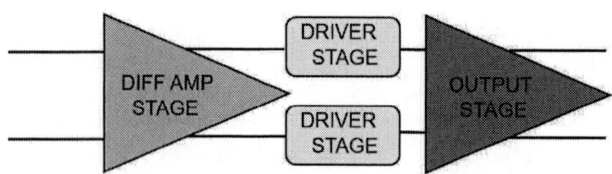

Fig. 7. GaN Class D^{-1} PA Architecture

III. CIRCUIT IMPLEMENTATION

In order to synthesize the instantaneous phase $\omega_F t + \phi$ we need a Phase Accumulator and a register that contains the phase ϕ. For implementing the architecture from Fig.2 we need to generate two signals $\omega_F t + \phi + arcsin(A)$ and the signal $\omega_F t + \phi - arcsin(A)$ as shown in Fig 3. The value arcsin(A) is provided by an FPGA. To increase throughput, we can use the interleaving Architecture from Fig. 4.

The next step is to up-convert the IF signal to RF. The digital based up-converter is presented in Fig.6. Its operation is presented on the right hand side of Fig.6. In order to measure the baseband to IF IC we need to apply the IF signal to a Class D^{-1}power amplifier built in GaN technology that we designed in-house. The architecture of GaN IC is shown in Fig.7. It consists of a differential amplifier stage followed by 2 single ended push-pull drivers and the final stage. The final stage is a differential cascode stage working in Class D^{-1} (see Fig.8). The capacitors C_N are neutrodyning capacitors to compensate for the parasitic gate drain capacitance. The differential load realized with inductors was implemented on PCB off-chip. The L_{LOAD} provides matching to the load.

IV. MEASUREMENT RESULTS

Figure 9 shows the chip photomicrograph of the baseband-to-IF circuitry and GaN PA. Baseband to IF CMOS ICI is realized in a 65nm CMOS LPE from GlobalFoundries. It measures 1.2mm x 2mm (2.4mm²). The GaN PA implemented

Fig. 9. Chip photomicrograph of the CMOS baseband/IF IC and GaN Class D^{-1} PA

979-8-3503-8083-5/23 $31.00 © 2023 IEEE

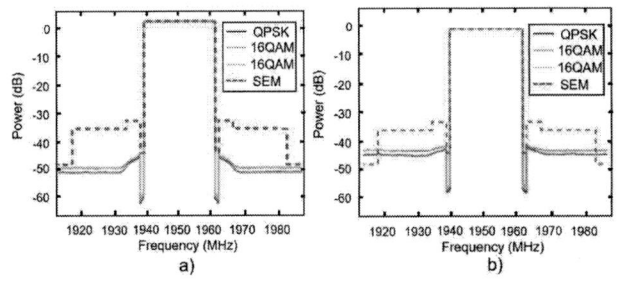

Fig. 8: a) Measured Power Spectrum of $S_{RF}(t)$ by EM in the RF channel and the frequency range specified by SEM. b) Power Spectrum of $S_{RF}(t)$ by PWM

in a GaN technology from MIT measures 3mm x 2mm (6mm^2). In order to prove the linearity of the proposed modulation and architecture we apply a 64- QAM LTE Signal at the input of the circuit (as baseband signals). The measured power spectrum for EM and PWM is shown in Fig.8. In both situations, three modulation methods, i.e. QPSK, 16QAM, 64QAM, have been tested.

As the spectrum of EM decays faster at higher frequencies, the PSD by EM is cleaner than that by PMW; in the frequencies close to the channel, the noise floor is also lower in the case of EM. In all cases, SEM is satisfied for EM.

Figure 9 a) shows the output spectrum for a 64QAM signal with PAPR of 7dB. The average output power is 10.4dBm. In Fig.9 b) the constellation diagram of the 64QAM signal at the output is presented. Measured EVM is -33.5dB.

V. CONCLUSIONS

The paper presents, a digital pulsed modulation, for high efficiency and linearity, switched-mode power amplifiers (PA) is proposed. It is based on a novel way of representing digital modulated signals at IF called Out-phasing Digital Exponential Modulation (ODEM). The paper shows the concept, simulations and measurements of a baseband-to-IF implementation. ODEM results on a faster decaying spectrum, when compared to pulse-width modulation (PWM) Out-phasing, and a power spectrum that fulfills the spectral emission mask for 64QAM LTE standards. An implementation in a 65 nm CMOS LPE process from GF, shows that the EVM and SEM requirements for LTE standards are fulfilled. The circuit

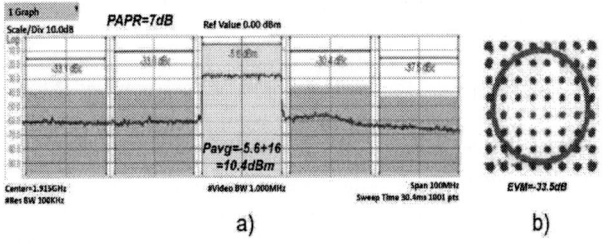

Fig. 9: Measured a) Spectrum by EM. b) Constellation for 64QAM and EVM

implementation of the baseband IC is based on digital circuits. The Inverse Class D amplifier was realized in GaN technology from MIT.

REFERENCES

[1] Y. Xu, M. A. T. Sanduleanu, "Digital outphasing exponential modulation: a constant amplitude pulsed modulati on for high efficency/linearity, switched-mode PA," *Microwave Conference (LAMC). IEEE MTT-S Latin America*, pp. 1-3, Dec 2018.

[2] J. Mitola, "The software radio architecture," *IEEE Commun. Mag.*, vol. 33, pp. 26–38, 1995

[3] S. H. Han and J. H. Lee, "An overview of peak-to-average power ratio reduction techniques for multicarrier transmission," *IEEE Wireless Commun.*, vol. 12, no. 2, pp. 56–65, Apr. 2005.

[4] M. Fulde et al., "A digital multimode polar transmitter supporting 40MHz LTE carrier aggregation in 28nm CMOS," in *Proc. IEEE Int. Solid-State Circuits Conf. (ISSCC)*, San Francisco, CA, USA, Feb. 2017, pp. 218–219.

[5] M. Hashemi, Y. Shen, M. Mehrpoo, M. S. Alavi, and L. C. N. de Vreede, "An intrinsically linear wideband polar digital power amplifier," *IEEE J. Solid-State Circuits*, vol. 52, no. 12, pp. 3312–3328, Dec. 2017.

[6] P. A. J. Nuyts, P. Reynaert, and W. Dehaene, Continuous-Time Digital Front-Ends for Multistandard Wireless Transmission. Cham, Switzerland: Springer-Verlag, 2014.

[7] P. A. J. Nuyts, P. Reynaert, and W. Dehaene, "A fully digital PWM-based 1 to 3 GHz multistandard transmitter in 40-nm CMOS," in *Proc. IEEE Radio Freq. Integr. Circuits Symp. (RFIC)*, pp. 419–422, , Jun. 2013.

[8] Mohsen Hashemi, Lei Zhou, Yiyu Shen, and Leo C. N. de Vreede, "A Highly Linear Wideband Polar Class-E CMOS Digital Doherty Power Amplifier", IEEE TMTT, Vol. 67, NO. 10, Oct, 2019

[9] Yun Yin, Baoyong Chi, Qian Yu, Bingqiao Liu and Zhihua Wang, "A 0.1-5GHz SDR Transmitter with Dual-Mode Power Amplifier and Digital-Assisted I/Q Imbalance Calibration in 65nm CMOS, In the Proceedings of ASSCC, pp.205-208, 2013

[10] A. Ravi et al., "A 2.4-GHz 20–40-MHz channel WLAN digital out-phasing transmitter utilizing a delay-based wideband phase modulator in 32-nm CMOS," IEEE J. Solid-State Circuits, vol. 47, no. 12, pp. 3184–3196, Dec. 2012.

[11] M. S. Mehrjoo, et al., "A 1.1-Gbit/S 10-Ghz Outphasing Modulator With 23-dBm Output Power And 60-dB Dynamic Range in 45-nm CMOS SOI," IEEE T-MTT, vol. 63, no. 7, pp. 2289-2300, July 2015.

[12] P. Godoy, S. Chung, T. Barton, D. Perreault, and J. Dawson, "A 2.4-GHz, 27-dBm asymmetric multilevel outphasing power amplifier in 65-nm CMOS," IEEE J. Solid-State Circuits, vol. 47, no. 10, pp. 2372–2384, Oct. 2012

[13] P. Madoglio, et al., "A 20dBm 2.4GHz Digital Outphasing Transmitter for WLAN Application in 32nm CMOS," ISSCC, pp. 168-169, Feb 2012.

[14] J. Lemberg et al., "A 1.5–1.9-GHz All-Digital Tri-Phasing Transmitter With an Integrated Multilevel Class-D Power Amplifier Achieving 100-MHz RF Bandwidth," in IEEE Journal of Solid-State Circuits, vol. 54, no. 6, pp. 1517-1527, June 2019.

[15] M. Kosunen et al., "13.5 A 0.35-to-2.6GHz multilevel outphasing transmitter with a digital interpolating phase modulator enabling up to 400MHz instantaneous bandwidth," 2017 IEEE International Solid-State Circuits Conference (ISSCC), San Francisco, CA, USA, 2017, pp. 224-225.

[16] J. Lemberg et al., "Digital Interpolating Phase Modulator for Wideband Outphasing Transmitters," in IEEE Transactions on Circuits and Systems I: Regular Papers, vol. 63, no. 5, pp. 705-715, May 2016.

A Review on Hyperdimensional Computing

Maram Abdulrahman[*], Sandy Wasif[†], Miran Wael[†], Eman Azab[†],
Maggie Mashaly[‡], Mohamed A. Abd El Ghany[†§]
[*]Media Engineering and Technology Department, German University in Cairo, Egypt
[†]Electronics Engineering Department, German University in Cairo, Egypt
[‡]Faculty of Information Engineering and Technology German University in Cairo
[§]TU Darmstadt, Darmstadt, Germany
[||]Emails: maram.abdulrahman@student.guc.edu.eg, sandy.abdelmalak@guc.edu.eg, miran.wael@guc.edu.eg,
eman.azab@guc.edu.eg, maggie.ezzat@guc.edu.eg, mohamed.abdel-ghany@guc.edu.eg

Abstract—Today, machine learning algorithms play a crucial role in analyzing data from interconnected Internet of Things (IoT) devices. With the rise of the Internet of things also known as IoT, numerous applications use machine learning algorithms to perform cognitive tasks. It has enabled a plethora of applications including personalized medical treatments, accurate face detection, self-driving cars, language detection, image recognition and even more applications. However, these algorithms often pose challenges when it comes to real-time learning on resource-constrained IoT devices. They require a lot of computational resources such as power and memory which makes them difficult to use on today's real-life embedded devices. To address these issues, a new computing paradigm called Hyperdimensional computing (HDC) has emerged. Utilizing HD vectors, known as "hypervectors," offers a brain-inspired alternative to conventional numerical computing. The basic HDC pipeline consists of encoding, training and inference stages. Nevertheless, the use of existing HDC solutions on today's real-life embedded devices imposes complex constraints on the overall design. In this paper, challenges faced when using HDC on low-power embedded devices are discussed. Contributions in this area are surveyed on encoding, training and inference methods along with benchmarking metrics and applications in HD classification. The surveyed work provides a comprehensive overview of existing real-time HDC architectures and gives insight into future promising research points in this field.

Index Terms—brain-inspired architectures, HDC, IoT, encoding, classification

I. INTRODUCTION

Advancements in deep learning (DL) algorithms have outperformed conventional machine learning (ML) approaches in many applications: image classification, voice recognition, activity recognition and language detection [1]. The rise of the Internet of Things (IoT) has led to the utilization of machine learning algorithms in cognitive tasks. However, the computational complexity and memory requirements of existing deep learning algorithms pose challenges for embedded IoT applications with limited resources [2]. To address this, alternative learning methods are needed that can achieve high classification accuracy while being suitable for less powerful IoT devices. Hyperdimensional computing (HDC) is a strategy inspired by human memory that operates on high-dimensional representations of data [3]. It utilizes hypervectors to represent memory and combines learning capabilities with memory

functions. These hypervectors represent various data types and have high dimensionality, often reaching thousands of dimensions, such as 10,000-d, where 'd' denotes the dimensionality [4]. As a computing paradigm that models human memory, Hyperdimensional computing exhibits robustness, scalability, energy efficiency, as well as faster training and inference times making it well-suited for IoT systems [5]. The HDC learning process involves encoding data points into high-dimensional space, also known as hypervectors, which is computationally expensive. During inference, similarity checks between the query hypervector and pre-trained class hypervectors contribute to computational costs and limit scalability, especially in applications with many classes [6]. In this paper, the benchmarking metrics of HDC are compared and discussed. The rest of this paper is organized as follows: Section II gives a brief background on HD Computing. Section III presents the general methodology of HD computing. Then three main encoding methods to form hypervectors from the input data are presented. Learning of HD Computing Benchmarking metrics, applications and some challenges are reviewed in Section IV. Then, in Section V, a comparative analysis is demonstrated comparing two advanced approaches known as lookHD and QuantHD Algorithms. Finally, the work is concluded and some suggestions for future work are discussed in section VI.

II. A BACKGROUND ON HYPERDIMENSIONAL COMPUTING

In the following section, HD computing is examined and it is compared with traditional computing. There is also an outline for an overview of the basic operations of HDC (Hyperdimensional Computing) [7]. Unlike traditional computing, which predominantly operates in low-dimensional binary states, HD computing ventures into multi-dimensional realms, creating a vast and diverse landscape for data representation and manipulation [8]. By employing hyperdimensional vectors, each comprising a multitude of dimensions, HD computing can encode and process information in a massively parallel and distributed manner. This departure from conventional binary computing facilitates more robust and efficient operations, making it particularly well-suited for complex problem-solving and pattern-recognition tasks [9].

979-8-3503-8083-5/23 $31.00 © 2023 IEEE

A. Classical Computing and HD Computing

There are several differences between classical computing and HD computing regarding several aspects including data representation, data transformation as well as data retrieval. Firstly, classical computing operates on bits, where each bit represents a value of either 0 or 1, while HD computing on the other hand operates on hypervectors. Classical computing uses data transformation operations of addition, multiplication, logic while the latter goes for add, multiply and permute operations [10]. Table I provides a summary comparing classical computing with HD computing paradigms.

TABLE I: Classifaction of Classical Computing vs HD Computing [4].

Computing method	Classical Computing	HD Computing
Data Type	Bit	Hypervector
Data Transformation	Addition, Multiplication, Logic	Addition, Multiplication, Permutation
Storage	Memory	Item Memory, Associative Memory
Training	Weights	Class Hypervectors
Model Complexity	High	Low
Accuracy	Very High	Acceptable
Feature Encoding	Easy	Difficult
Number of Features	Many	One

B. Data Representation

HD computing represents data in the form of hypervectors, which can be vectors of bits, integers, real, or complex numbers [11]. They are broadly classified into binary or non-binary hypervectors. Bipolar and integer hypervectors are commonly employed among the majority of non-binary hypervectors. In general, the use of non-binary HD algorithms tends to achieve higher accuracy, while using binary hypervectors is more suitable for hardware implementation, offering greater efficiency [12].

C. HD Basic Operations

The provided subsection discusses the fundamental operations in HyperDimensional Computing (HDC) and their application in representing human memory, perception, and cognitive abilities. In HD Computing, we utilize a set of operations known as (MAP) operations standing for multiplication, addition and permutation. The opposite operation of multiplication is commonly known as "release". The release operation is also used to represent the opposite of addition. In the examples provided, data transformations are demonstrated using binary hypervectors. Although the examples primarily demonstrate data transformations using binary hypervectors, it's crucial to know that similar principles apply when working with non-binary hypervectors. [13].

1) Multiplication (Binding)

The process of combining two High-Dimensional (HD) vectors is achieved by binding them together, often through an XOR bitwise operation, X = A⊕B. Eq. (1) shows the point-wise multiplication of two 10-bit binary vectors [14].

$$\begin{aligned} A &= 0000110011, \\ B &= 1011000101, \\ \hline A \oplus B &= 1011110110. \end{aligned} \tag{1}$$

2) Addition

Pointwise addition, also known as bundling, is used to combine different HD vectors into a single HD vector, it calculates a hypervector Z using the equation (2) with the input hypervectors X1, X2, ..., Xn. The resulting Z is designed to be highly similar to the n input hypervectors.

$$Z = [X1 + X2 + ... + Xn] \tag{2}$$

Equation (3) shows an example for the point-wise addition of three 10-bit binary vectors.

$$\begin{aligned} A &= 0000110011, \\ B &= 1011000101, \\ C &= 0010101101, \\ \hline [A + B + C] &= 0010100101. \end{aligned} \tag{3}$$

3) Permutation (Shifting)

An alternative method for combining HD vectors involves using a specific type of matrix known as a permutation matrix. This approach is particularly useful when dealing with data or sequences where the order is significant.

$$\begin{aligned} A &= 0000110011, \\ \hline permA &= 1000011001. \end{aligned} \tag{4}$$

III. LEARNING OF HD CLASSIFICATION METHODOLOGY

This section introduces the process and key steps involved in the encoding, training and inference phases of the High-Dimensional (HD) classification method. It aims to provide an in-depth understanding of how the HD classifier learns from the training data and makes predictions on new, unseen data.

A. HD Classification Methodology

In the context of high-dimensional (HD) computing, HD classification emerges as an innovative approach for pattern recognition and data classification tasks. Figure 1 demonstrates the flow of HDC and its different steps. Generally, the learning phase of the methodology involves the encoding step where hypervectors are generated (pre-stored in the item memory) to encode the training data into high-dimensional space. Multiple class hypervectors, denoted by k, are trained and stored in the associative memory. Moreover, during the inference phase, each test data point is encoded using the same encoding module transforming it into a query hypervector. The similarity between the query hypervector and each pre-trained class hypervector is checked in the associative memory. Finally, the label with the closest distance value is returned as

979-8-3503-8083-5/23 $31.00 © 2023 IEEE

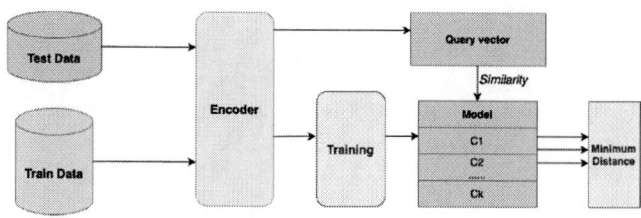

Fig. 1: The flow of HDC.

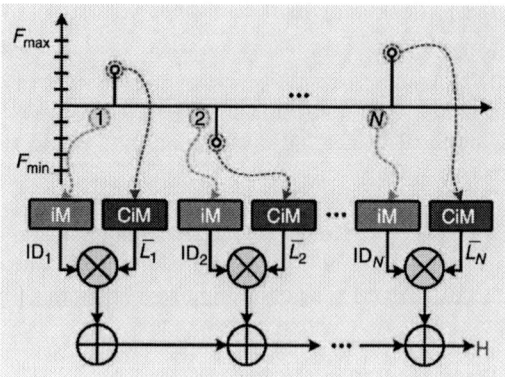

Fig. 2: Record-based encoding[18].

the classification result indicating highest similarity with that specific class hypervector[15].

B. Encoding Methods

High-Dimensional (HD) computing has the capability to handle diverse forms of input data, such as letters, signals, and images. Yet, it is essential to convert these input data into hypervectors, in the procedure previously defined as encoding. Few simple techniques have been used repeatedly for HD encoding. Within the realm of HD algorithms, three commonly utilized encoding methods are features superposition, record-based encoding and N-gram-based encoding.

1)Features Sueprposition
The HDC encoding module aims to map an n-dimensional feature vector (F) to a high-dimensional vector (H) with D dimensions, typically in the order of tens of thousands. HDC represents feature values as patterns of bitstreams in HDC space, combining them to maintain their positions. Alphabets generation involves assigning hypervectors (L1 to Lq) to quantized levels derived from the feature value range. These bipolar hypervectors have D dimensions and maintain similarity distance. The first hypervector is randomly generated, and subsequent ones are obtained by flipping random dimensions of the previous hypervector, ensuring orthogonality between the extreme hypervectors.

2)Record-based Encoding
The following encoding technique utilizes two types of hyper-vectors, which serve to represent the feature's position and feature value. In this encoding approach, the position hyper-vectors (IDi) are randomly generated to encode positional information within a feature vector, where $1 \leq i \leq N$. Simultaneously, feature value information is quantized and depicted using m-level hyper-vectors $L_1, L_2, \ldots, L_m$.
For a feature with N dimensions, a set of N-level hypervectors, denoted as rLi, is generated. These hypervectors are selected from m-level hypervectors L1, L2, ..., Lm according to their corresponding features [16]. It is important to know that the position hypervectors, IDi, are mutually orthogonal. This correlation is achieved through a continuous bit-flipping technique, where L1 represents the minimum feature value (Fmin), and subsequent levels are formed by randomly flipping d/m bits each time, with d denoting the dimensionality of

the hypervectors [17]. In the encoding process, each position hypervector is paired with its corresponding level hypervector. The final encoding hypervector, H, is obtained by summing up these results as shown in Eq.(5). A graphical representation of the dicussed process is provided as shown in Figure 2. Note iM refers to item memory, which stores the position hypervectors, and CiM refers to continuous item memory, which stores level hypervectors.

$$H = L_1 \oplus ID_1 + L_2 \oplus ID_2 + \ldots + L_N \oplus ID_N \quad (5)$$

$$Li \in \{L_1, L_2, \ldots, L_m\}, \text{ where } 1 \leq i \leq N \quad (6)$$

3)N-gram-based Encoding
N-gram-based encoding is a technique used in natural language processing and text analysis to represent words or sequences of characters in a text using n-grams. An n-gram refers to a consecutive sequence of n elements, where these elements can be characters, words, or larger units, contingent upon the specific context [19]. Initially, in this encoding technique, level hypervectors are generated randomly. Subsequently, the feature values are determined by the permutation of those level hyper-vectors. For instance, the level hypervector Li associated with the i-th feature position undergoes rotational permutation by (i-1) positions, where $1 \leq i \leq N$. The final encoded hypervector H is obtained using Equation (7). A visual representation of this encoding process is depicted in Figure 3. Note that CiM stores level hypervectors which are mutually orthogonal.

$$H = L_1 \oplus pL_2 \oplus \ldots \oplus p^{N-1} L_N \quad (7)$$

$$L_i \in \{L_1, L_2, \ldots, L_m\}, \text{ where } 1 \leq i \leq N \quad (8)$$

C. HD Training and Retraining

In HDC, the training process involves the summation of encoded hypervectors corresponding to each class. This operation yields k hypervectors, each possessing D dimensions, where k denotes the number of classes. For example, the hypervector of the ith class can be calculated as follows: $C_i = \sum_{j \in \text{class}_i} H_j$.

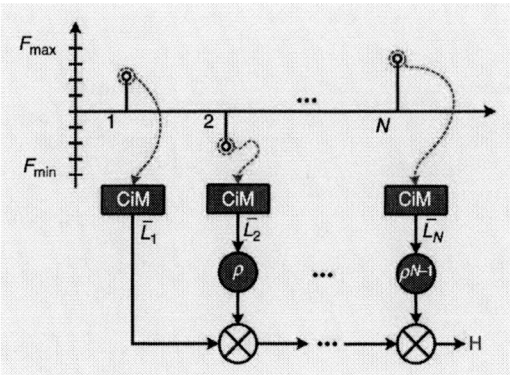

Fig. 3: N-gram-based encoding[20].

Measurement	Similarity	Orthogonality
Cosine Similarity	C1	0
Hamming Distance	0	0.5

TABLE II: A table with increased row height.

Once the initial training is completed, HDC employs a retraining phase, which involves iterating over the trained model. During each iteration of retraining, HDC examines the similarity between all training data points (represented by H) and the trained model. If a data point is misclassified, HDC updates the model by:

- Adding the hypervector of that data point to the correct class ($C_{correct} = C_{correct} + H$)
- Subtracting it from the class with which it was incorrectly matched ($C_{wrong} = C_{wrong} - H$)

The retraining process continues for several iterations until the HDC accuracy stabilizes over the validation data, which is a part of the training dataset.

D. Similarity Measurements

Two major similarity measurements shown in Table II are cosine similarity and Hamming distance.
Within the domain of non-binary hypervectors, their similarity is quantified using cosine similarity, as defined in Equation (9). This metric concentrates exclusively on the angle formed between the hypervectors, omitting consideration for their respective magnitudes denoted by the symbol ".". In contrast to the inner product operation, which factors in both magnitude and orientation, cosine similarity relies on the orientation of the hypervectors [21]. Consequently, in the majority of high-dimensional algorithms utilizing non-binary hypervectors, preference is given to cosine similarity over the inner product. A cosine similarity value that is almost to 1 indicates a very high similarity level between two hypervectors, signifying that they are nearly identical. Conversely, when the cosine similarity is 0, they are orthogonal and classified dissimilar [22].

$$\cos(A, B) = \frac{A \cdot B}{\|A\|\|B\|} \tag{9}$$

Concerning binary hypervectors characterized by a dimensionality denoted as d, whose components are values of either 0 or 1, their similarity is measured using the normalized Hamming distance, as expressed in Equation (10). As the Hamming distance between two hypervectors tends towards 0, they are

regarded as similar. For example, a Hamming distance of 0 signifies complete bit identity at each position, rendering hypervectors A and B entirely indistinguishable. In contrast, when the Hammin Distance is 0.5, hypervectors A and B are orthogonal or dissimilar. Finally, when Ham(A, B) = 1, it indicates that hypervectors A and B are diametrically opposed, presenting the outmost dissimilarity.

$$\text{Ham}(A, B) = \frac{1}{d} \sum_{i=1}^{d} \delta(A_i, B_i) \tag{10}$$

E. HD Inference

During the inference phase, HDC utilizes the identical encoding module used in training to map a test data point to a query hypervector. Within the HDC space, the classification process involves evaluating the similarity between the query hypervector and all class hypervectors. Based on this evaluation, each data point is assigned to the class that exhibits the highest similarity. Since HDC represents information as patterns of non-binary values, the cosine similarity metric is well-suited for measuring similarity in this context.

IV. BENCHMARKING METRICS, APPLICATIONS AND CHALLENGES

This section delves into benchmarking metrics, and the inherent challenges they present. Benchmarking metrics serve as crucial tools for objectively evaluating and comparing the performance of diverse systems and models. We explore their significance and the complexities involved in ensuring accurate and meaningful performance assessments.

A. The Benchmarking Metrics of HDC

Achieving accuracy and efficiency involves a compromise when it comes to HD computing. Figure 4 illustrates the considerable efforts made to enhance classification accuracy and energy efficiency, or strike a balance between the two aspects [23].

1)Accuracy
Choosing the encoding technique plays a crucial role in determining accuracy within HD computing. Opting for a suitable encoding method has the potential to improve accuracy, particularly with the efficient encoding techniques suggested earlier [24]. There are instances where combining various encoding methods may result in increased accuracy. Additionally, using binary hypervectors may lead to reduced accuracy, making non-binary models preferable when sufficient resources are available[25].
2)Efficiency

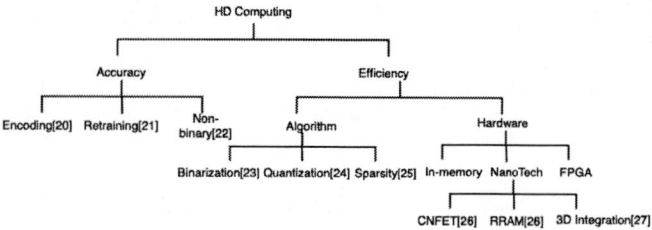

Fig. 4: Two benchmarking metrics and some possible ways to improve them.

Efforts to enhance efficiency in HD computing primarily focus on algorithmic and hardware characteristics [26]. From an algorithmic perspective, the reduction of dimensionality is a clear approach in order to enhance efficiency. Modeling indicates that reducing the dimensions of hypervectors slightly maintains acceptable classification accuracy while conserving hardware resources. Binarization, which is defined as involving the use of binary hypervectors over non-binary models, speeds up computational procedures and diminishes hardware requirements. However, precision may be compromised by the quantization of non-binary HD models. To address this concern, QuantHD was introduced, aiming for heightened efficiency with minimal impact on accuracy [27]. The concept of sparsity has also been introduced giving rise to SparseHD leading to the reduction of inference computations and enhancing efficiency[28].

V. COMPARATIVE ANALYSIS (LOOKHD VS QUANTHD)

In the realm of brain-inspired Hyperdimensional (HD) computing, the LookHD and QuantHD algorithms have emerged as significant advancements and architectural solutions to enhance the state-of-the-art HDC, each addressing distinct challenges in lightweight learning. In this section, we introduce both algorithms briefly and proceed to compare their key features, strengths, and limitations.

LookHD Algorithm: The LookHD algorithm, proposed as a solution for real-time HDC learning on low-power edge devices, presents an architecture tailored for immediate learning applications. By enabling HD computing in resource-constrained settings, LookHD sets the stage for efficient and prompt data processing on edge devices.

QuantHD Algorithm: The QuantHD algorithm, on the other hand, was developed to enhance classification accuracies in binarized and ternarized HD computing models. It aims to mitigate the limitations of existing approaches by leveraging innovative techniques, subsequently reducing the gap in classification accuracy and enhancing the overall stability and convergence rate.

Both LookHD and QuantHD algorithms address the classification accuracy gap in Hyperdimensional (HD) computing. LookHD focuses on real-time learning, while QuantHD specifically targets challenges in binarized and ternarized models. QuantHD's approach significantly reduces the classification

TABLE III: Comparison between LookHD and QuantHD Algorithms

Aspect	LookHD	QuantHD
Classification Accuracy Gap	Real-time focus Immediate apps	Enhances classification accuracies Addresses binarized/ternarized models
Convergence	Randomness for convergence	Improves convergence and stability
Similarity Checks	cosine similarity	Efficient Hamming distance checks
Energy Efficiency	Enables real-time on edge	Reduces energy consumption, enhances efficiency
Implementation	Tailored for real-time	Focus on accuracy, convergence, efficiency

accuracy gap, enhancing HD computing's reliability in various scenarios. Both algorithms emphasize improving convergence and stability, with LookHD introducing randomness and QuantHD attributing slower convergence to earlier lack of randomness. QubitHD employs Hamming distance for similarity checks, while LookHD prioritizes immediate learning and employs cosine similarity. While both enhance computational efficiency, they emphasize different aspects—QuantHD optimizes energy usage, while LookHD enables real-time learning on low-power edge devices [39]. The algorithms' distinct implementations reflect their adaptability to unique challenges. In summary, LookHD and QuantHD contribute to brain-inspired Hyperdimensional computing. LookHD suits real-time learning and edge compatibility, while QuantHD enhances accuracy, convergence, and energy efficiency. Researchers and practitioners can choose the algorithm aligning with their lightweight learning requirements [40]. Table III illustrates a comparison between both algorithms.

VI. CONCLUSION

In this paper, we have explored the emerging computing paradigm of Hyperdimensional computing (HDC) and its potential applications in resource-constrained Internet of Things (IoT) devices. The research on HDC shows promising results for real-time applications in IoT. It has the potential to achieve high classification accuracy while being energy-efficient and suitable for resource-constrained devices. Future research in HDC should focus on optimizing encoding and inference methods to reduce computational costs further. Additionally, exploring novel architectures that can handle large-scale HD computing tasks in depth such as the lookHD and QuantHD Algorithms against state-of-the-art machine learning algorithms and HDC will be essential for better understanding its full potential.

979-8-3503-8083-5/23 $31.00 © 2023 IEEE

REFERENCES

[1] Y. Xiang, A. Alahi, and S. Savarese, "Learning to track: Online multi-object tracking by decision making," in 2015 IEEE international conference on computer vision (ICCV), no. EPFL-CONF-230283, pp. 4705–4713, IEEE, 2015.

[2] K. He, X. Zhang, S. Ren, and J. Sun, "Deep residual learning for image recognition," in Proceedings of the IEEE conference on computer vision and pattern recognition, pp. 770–778, 2016.

[3] Imani, Mohsen, et al. "Revisiting hyperdimensional learning for fpga and low-power architectures." 2021 IEEE International Symposium on High-Performance Computer Architecture (HPCA). IEEE, 2021.

[4] Lulu Ge and Keshab K. Parhi. Classification using hyperdimensional computing: A review. IEEE Circuits and Systems Magazine, 20(2):30–47, 2020.

[5] Mohsen Imani, Yeseong Kim, Sadegh Riazi, John Messerly, Patric Liu, Farinaz Koushanfar, and Tajana Rosing. A framework for collaborative learning in secure high-dimensional space. In 2019 IEEE 12th International Conference on Cloud Com- puting (CLOUD), pages 435–446. IEEE, 2019.

[6] Abbas Rahimi, Pentti Kanerva, and Jan M Rabaey. A robust and energy-efficient classifier using brain-inspired hyperdimensional computing. In Proceedings of the 2016 international symposium on low power electronics and design, pages 64–69, 2016.

[7] Abbas Rahimi, Tony F Wu, Haitong Li, Jan M Rabaey, H-S Philip Wong, Max M Shulaker, and Subhasish Mitra. Hyperdimensional computing nanosystem. arXiv preprint arXiv:1811.09557, 2018.

[8] Faisal Rahutomo, Teruaki Kitasuka, and Masayoshi Aritsugi. Semantic cosine simi- larity. In The 7th international student conference on advanced science and technol- ogy ICAST, volume 4, page 1, 2012.

[9] Ajay Shrestha and Ausif Mahmood. Review of deep learning algorithms and archi- tectures. IEEE access, 7:53040–53065, 2019.

[10] Yonatan Vaizman, Katherine Ellis, and Gert Lanckriet. Recognizing detailed human context in the wild from smartphones and smartwatches. IEEE pervasive computing, 16(4):62–74, 2017.

[11] R. E. Bryant and D. R. O'Hallaron, "Computer systems: A programmer's perspective," 2015.

[12] A. Rahimi, P. Kanerva, and J. M. Rabaey, "A robust and energy-efficient classifier using brain-inspired hyperdimensional computing," in Proc. 2016 Int. Symp. Low Power Electronics and Design, pp. 64–69.

[13] M. Rastegari, V. Ordonez, J. Redmon, and A. Farhadi, "XNOR-net: Imagenet classification using binary con- volutional neural networks," pp. 525–542, 2016.

[14] R. W. Gayler, "Multiplicative binding, representation operators & analogy (workshop poster)," 1998.

[15] D. Widdows and T. Cohen, "Reasoning with vectors: A continuous model for fast robust inference," Logic Journal of the IGPL, vol. 23, no. 2, pp. 141–173, 2015.

[16] D. Kleyko and E. Osipov, "Brain-like classifier of temporal patterns," in Proc. 2014 Int. Conf. Computer and Information Sciences, pp. 1–6. doi: 10.1109/ICCOINS.2014.6868349.

[17] M. Imani et al., "QuantHD: A quantization framework for hyperdi- mensional computing," in Proc. IEEE Trans.Computer-Aided Design of Integrated Circuits and Systems, 2019. doi: 10.1109/TCAD.2019.2954472.

[18] A. Rahimi, P. Kanerva, J. R. Millán, and J. M. Rabaey, "Hyperdimensional computing for noninvasive brain-computer interfaces: Blind and one-shot classification of EEG error-related potentials," in Proc. 10th EAI Int. Conf. Bio-inspired Information and Communications Technologies, 2017. doi: 10.4108/eai.22-3-2017.152397.

[19] A. Joshi, J. T. Halseth, and P. Kanerva, "Language geometry using random indexing," in Proc. Int. Symp. Quantum Interaction, Springer, 2016, pp. 265–274. doi: 10.1007/978-3-319-52289-0_21.

[20] M. Imani, T. Nassar, A. Rahimi, and T. Rosing, "HDNA: Energy-efficient DNA sequencing using hyperdimensional computing," in Proc. 2018 IEEE EMBS International Conference on Biomedical & Health Informatics (BHI). IEEE, 2018, pp. 271–274. doi: 10.1109/BHI.2018.8333421.

[21] M. Hersche, J. d R. Millán, L. Benini, and A. Rahimi, Exploring embedding methods in binary hyperdimensional computing: A case study for motor-imagery based brain-computer interfaces. 2018. arXiv:1812.05705

[22] P. J. Olver and C. Shakiban, Applied Linear Algebra. Springer, 2018.

[23] D. Kleyko, A. Rahimi, D. A. Rachkovskij, E. Osipov, and J. M. Rabaey, "Classification and recall with binary hyperdimensional computing: Tradeoffs in choice of density and mapping characteristics," IEEE Trans. Neural Netw. Learn. Syst., vol. 29, no. 12, pp. 5880–5898, 2018. doi: 10.1109/TNNLS.2018.2814400.

[24] R. W. Gayler, Vector symbolic architectures answer jackendoff's challenges for cognitive neuroscience. 2004. [Online]. Available: arXiv:cs/0412059

[25] A. Rahimi, P. Kanerva, L. Benini, and J. M. Rabaey, "Efficient biosignal processing using hyperdimensional computing: Network templates for combined learning and classification of ExG signals," Proc. IEEE, vol. 107, no. 1, pp. 123–143, 2018. doi: 10.1109/JPROC.2018.2871163.

[26] M. Imani, J. Messerly, F. Wu, W. Pi, and T. Rosing, "A binary learning framework for hyperdimensional computing," in Proc. 2019 Design, Automation & Test in Europe Conference & Exhibition (DATE), pp. 126–131. doi: 10.23919/DATE.2019.8714821.

[27] D. A. Rachkovskij, "Representation and processing of structures with binary sparse distributed codes," IEEE Trans. Knowl. Data Eng., vol. 13, no. 2, pp. 261–276, 2001. doi: 10.1109/69.917565.

[28] M. Imani, S. Salamat, B. Khaleghi, M. Samragh, F. Koushanfar, and T. Rosing, "SparseHD: Algorithm-hardware co-optimization for efficient high-dimensional computing," in Proc. 2019 IEEE 27th Annu. Int. Symp. Field-Programmable Custom Computing Machines (FCCM), pp. 190–198. doi: 10.1109/FCCM.2019.00034.

[29] E.-J. Chang, A. Rahimi, L. Benini, and A.-Y. A. Wu, "Hyperdimensional computing-based multimodal- ity emotion recognition with physiological signals," in IEEE International Conference on Artificial Intelli- gence Circuits and Systems (AICAS). IEEE, 2019.

[30] D. Kleyko, "Vector Symbolic Architectures and their Applications Computing with Random Vectors in a Hy- perdimensional Space," Ph.D. dissertation, Luleå Uni- versity of Technology Luleå, Luleå, Sweden, 2018.

[31] A. Joshi, J. T. Halseth, and P. Kanerva, "Language ge- ometry using random indexing," in International Sym- posium on Quantum Interaction. Springer, 2016, pp. 265–274.

[32] D. Kleyko, S. Khan, E. Osipov, and S.-P. Yong, "Modal- ity classification of medical images with distributed representations based on cellular automata reservoir computing," in IEEE 14th International Symposium on Biomedical Imaging (ISBI). IEEE, 2017, pp. 1053– 1056.

[33] M. Imani, B. Rouhani, S. Bosch, M. Javaheripi, X. Wu, F. Koushanfar, and T. Rosing, "SemiHD: Semi- Supervised Learning Using Hyperdimensional Com- puting," 2019.

[34] D. Kleyko, E. Osipov, R. W. Gayler, A. I. Khan, and A. G. Dyer, "Imitation of honey bees' concept learning processes using vector symbolic architectures," Biolog- ically Inspired Cognitive Architectures, vol. 14, pp. 57– 72, 2015.

[35] G. Montone, J. K. O'Regan, and A. V. Terekhov, "Hyper-dimensional computing for a visual question- answering system that is trainable end-to-end," arXiv preprint arXiv:1711.10185, 2017.

[36] F. Sandin, B. Emruli, and M. Sahlgren, "Random in- dexing of multidimensional data," Knowledge and In- formation Systems, vol. 52, no. 1, pp. 267–290, 2017.

[37] G. Recchia, M. Sahlgren, P. Kanerva, and M. N. Jones, "Encoding sequential information in semantic space models: Comparing holographic reduced representation and random permutation," Computational intelligence and neuroscience, 2015.

[38] A. Joshi, J. T. Halseth, and P. Kanerva, "Language ge- ometry using random indexing," in International Sym- posium on Quantum Interaction. Springer, 2016, pp. 265–274.

[39] Hassan, Eman, et al. "Hyper-dimensional computing challenges and opportunities for AI applications." IEEE Access 10 (2021): 97651-97664.

[40] Bosch, Samuel, et al. "QubitHD: A stochastic acceleration method for HD computing-based machine learning." arXiv preprint arXiv:1911.12446 (2022).

Deep Neural Network Inference Processor

Ali Hegazy[1], Khaled Salah[2]
[1]Department of computer and systems, Faculty of Engineering, Ain Shams University, Cairo, Egypt.
Siemens, Fremont, USA.
Email: [2]Khaled.mohamed@siemens.com

Abstract— In recent years, neural networks' architecture has become complicated and it demands special computing requirements. Thus, new specialized microprocessors are proposed for this purpose. In this paper, a new scalable neural network inference processor architecture is proposed and implemented on FPGA. FPGA enables hardware customization to fit the specific computing requirements. Moreover, FPGA-based accelerators demonstrated high energy efficiency compared to GPUs and CPUs. In the proposed FPGA-based architecture, the size of the systolic matrix multiplication array, the size of data flowing inside the processor and all the primitive blocks of the processor are configurable and scalable. Also, in this paper, an illustrative example is introduced to show how a TensorFlow-Based neural network model can be inferred using the proposed architecture. Moreover, synthesis results in terms of area and power are very promising. The design is validated on Virtex UltraScale+ VCU118 Evaluation platform. It is tested using face recognition algorithm (160×160 face image at 25FPS) where it achieved an accuracy of 95%. Moreover, it shows performance and power efficiency of 55 GOPS and 10 GOPS/W at 100MHz, respectively.

Index Terms — Neural network processor – FPGA - Systolic array – TensorFlow.

I. INTRODUCTION

In recent years, neural networks have become complicated, often containing hundreds of layers. This led to increase in the computational requirements demands. Neural network (**NN**) processors have emerged as a promising solution to address these computational requirements. The elevated computational and storage complexity of neural networks presents hard challenges in its implementation [8]-[12]. CPUs are difficult to maintain enough computing capacity. GPUs are the first preference for a neural network processor due to their high computing capacity and simple development frameworks. However, FPGA-Based neural network inference processor is a next feasible option to exceed GPU in terms of performance and energy. (ASICs-based) processors even achieve higher efficiency but they need much longer development cycle and higher costs [1]-[4]. In this work, we propose an FPGA-based architecture for neural network inference processor. The proposed architecture is scalable. This paper

is organized as follows: In Section II, the proposed architecture is discussed with an illustrative example. In section III, validation and synthesis results are analyzed. In section IV, Conclusions are given.

II. THE PROPOSED ARCHITECTURE

Fig.1 (a) shows the proposed architecture for **NN inference soft processor.** This architecture is fully parameterized. **The matrix multiplication** unit plays a key part in the proposed architecture. The operation of each sub-block of the proposed architecture is explained in the following subsections.

A. Activation and Weight FIFOs

Activation and Weight FIFOs are responsible for fetching the activation and weight inputs respectively from the memory and providing them as an input to the matrix multiplication unit (MMU). FIFOs stores activation and weight inputs to the MMU to decrease the number of inputs on the FPGA. FIFO's implementation is generic can be reconfigured according to the data size and the MMU array's size. If write enable flag is high, the FIFO samples the input data and stores it, if read enable flag is high, the FIFO fetch the data outputs it to the MMU.

B. Matrix Multiplication Unit/Convolution Engine

MMU is the core of the neural network inference processor. Our whole design is focused around the systolic array architecture; a systolic array is a homogenous network of unit cells that focuses on increasing the performance by increasing number of computations that are performed per input/output access by propagating data within the array [1]. There are different systolic designs. In the proposed architecture, weight-stationary architecture is chosen where the weights are preloaded to the cells of the MMU array and stay there throughout the computation while the activation inputs move systolically. The reason behind this is that in deep learning operations weights are read much less frequently than the activations, so by preloading the weights once at the start of the operation, optimum bandwidth between the weights buffer and the MMU is achieved.

979-8-3503-8083-5/23 $31.00 © 2023 IEEE

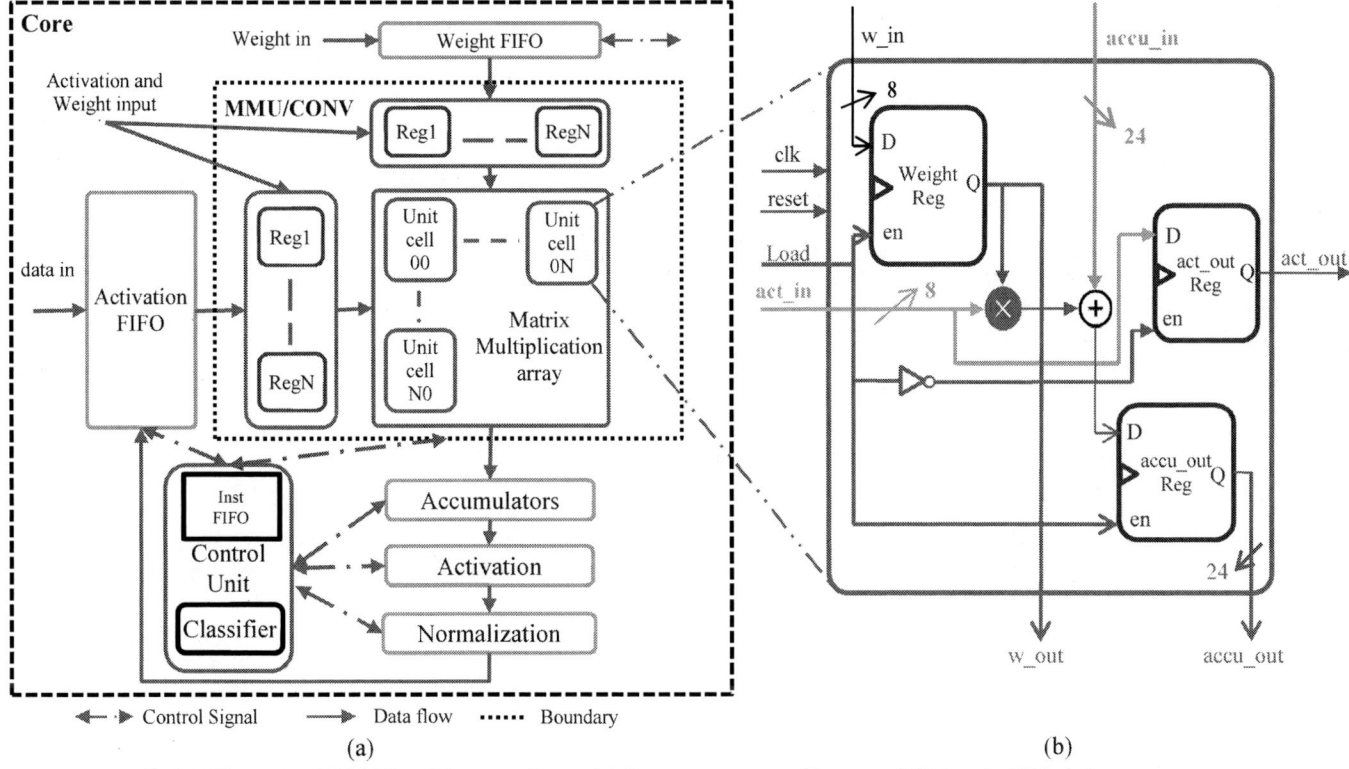

Fig.1 (a) The proposed FPGA-Based deep neural network inferenece processor architecture and (b) the unit cell block diagram.

The proposed processor can perform N x N 8-bit integer multiply and 24-bit integer accumulate per cycle, where N is the number of unit cells in a row/column inside the MMU array. MMU is responsible for carrying out the matrix multiplication operation of each layer inside the neural network. MMU is an array of unit cells (shown in Fig.1 (b)) and input registers that sample and store activation and weight inputs ejected from the activation and weight FIFOs. Each cell in the left most column is connected to an activation input register to load activation inputs from and each cell in the top row of the array is connected to a weight input register to load weight inputs from as shown in Fig.1 (a). Flow diagram of data input to data output process is shown in Fig. 2.MMU operations include: initialization, storing activation data and weights, load weights inside the unit cells, activation data systolic setup and multiply and accumulate.

1)   Initialization

If the MMU is not initialized, this implies that the weights and activation data required for calculations isn't loaded inside the input registers yet, so at the beginning of each neural network layer calculation we need to initialize the MMU with the required data from activation and weight FIFOs.

2)   Storing Activation Data and Weights

Fetching input data and weights by sampling the output of activation and weight FIFOs at each clock cycle and store it inside the input registers. This happens as follows: first output stored in the first register then output of the next clock cycle is stored in the second register and so on.

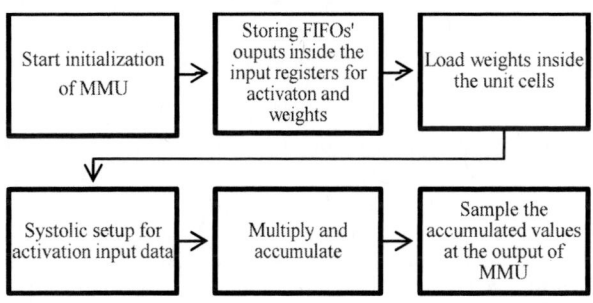

Fig. 2 Flow diagram of data input to data output process.

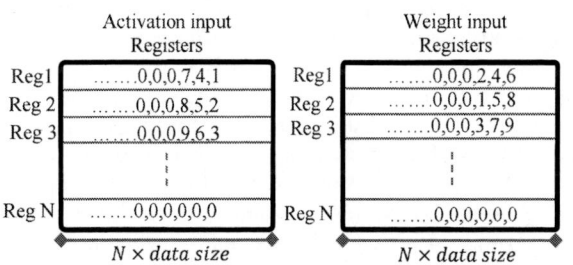

Fig.3 Activation and Weight input registers after MMU initialization.

As an example, if we have the following matrices:

$$Activation = \begin{bmatrix} A_{00} & A_{01} & A_{02} \\ A_{10} & A_{11} & A_{12} \\ A_{20} & A_{21} & A_{22} \end{bmatrix} = \begin{bmatrix} 1 & 2 & 3 \\ 4 & 5 & 6 \\ 7 & 8 & 9 \end{bmatrix} \quad (1)$$

$$Weight = \begin{bmatrix} W_{00} & W_{01} & W_{02} \\ W_{10} & W_{11} & W_{12} \\ W_{20} & W_{21} & W_{22} \end{bmatrix} = \begin{bmatrix} 2 & 1 & 3 \\ 4 & 5 & 7 \\ 6 & 8 & 9 \end{bmatrix} \quad (2)$$

979-8-3503-8083-5/23 $31.00 © 2023 IEEE

3) Activation data Systolic setup

Here, the activation inputs are driven systolically. Activation data is stored in the activation input registers and driven systolically as shown in Fig.5. Input data for each row is delayed by one clock cycle than the preceding row's data. After each clock cycle, activation data propagates to the right inside the array.

For the activation data FIFO, the first output will be {7,4,1} concatenated in the same order and will be stored in the input register of the first row of the MMU array, next clock cycle output will be {8,5,2} and stored in the input register of the second row and so on till last row. In the same manner, weight input FIFO first output will be {2, 4, 6} concatenated in the same order and this will be stored in the input register of the first column. In the next clock cycle, the output will be {1, 5, 8} and stored in the second input register of the second column and so on till the last column. At the end of initialization, input registers for activation and weights should be as depicted in Fig. 3.

4) Load weights inside the Unit cells

Unit cell shown in Fig.1 (b) is implemented as a finite state machine (FSM). This FSM has different types of signals: control signals as *load*, *clk* and *reset*. Input signals as *act_in*, w_in and *accu_in*. Output signals as *act_out*, *w_out* and *accu_out*. By using *load* control signal, we can toggle between two modes of operation inside the unit cell load mode and multiply & accumulate (**MAC**) mode. If load signal is high, unit cell will enter load mode, as we explained before we are using weight stationary systolic architecture so we need to preload the weights to the unit cells before executing the multiplication operation, at positive clock edge weight register samples the value of *w_in* and propagates the previously stored weight through *w_out*. If load signal is low, unit cell will enter MAC mode where at positive clock edge unit cell multiply *act_in* by the output of the weight register and add the result of the multiplication to *accu_in* to form new *accu_out*, and *act_in* is sampled by *act_out* register.

Unit cell has an active low asynchronous reset which put it in the MAC state and reset all its registers and outputs to zero. As shown in Fig.5, MMU is an $N \times N$ array of unit cells. After raising load signal to high, weights are loaded from the top side of the array through *w_in* and propagates from one cell to the cell below it at each clock cycle till it reaches the bottom of the array. After the first clock cycle, $W_{N0}..W_{NN}$ will be saved inside the weight registers of the unit cells of the top row and after each clock cycle weights will propagate to the next row until they reach the bottom of the array as depicted in the unit cells array in Fig.5.

5) Multiply and Accumulate

Once the weights are loaded and activation inputs are driven systolically to the array, multiply and accumulate operation starts. This requires *load* signal to go low in order to shift the mode of the unit cells to MAC mode and start executing the matrix multiplication operation. During one clock cycle, activation input moves one cell to the right, then multiplication of the activation input with the loaded weight inside the cell is performed at all cells simultaneously, then their results are added to the partial sum coming from the upper cell from the previous cycle and finally the result fanned-in to the cell below and so on until all the partial sums are accumulated at the output of the MMU.

C. Accumulators

Accumulators unit store and accumulate each accumulated output from each column. It is required when we need to perform calculations with neural network layers having more weight inputs than columns or layers having more activation inputs than rows. Accumulators unit has enabled signal that controls if it will store and wait to accumulate another value or store and pass directly to activation or normalization unit.

D. Activation unit

The purpose of activation function is to introduce non-linearity into the output of each neuron. A neural network without an activation function is a linear regression model, the activation function does the nonlinear transformation to the input making the neural network perform more complex tasks. In this unit, we apply rectified linear unit (**RELU**) activation function as computationally, it's less expensive than tanh and sigmoid as it involves simpler mathematical operations. RELU activation equation are given by equation (3):

$$F(x) = \max(0, x) \tag{3}$$

E. Normalization

The final value from the activation unit is then passed to the normalization unit. Normalization typically means rescales the values into a specific range. Since the output values from the accumulators is 24-bit wide, we need to rescale these values to 8-bit to ensure that these values remain within the expected domain for the inputs of the subsequent neural network layer.

Normalization function equation is as follows:

$$Normalized\ value(x) = (b - a)\frac{x - x_{min}}{x_{max} - x_{min}} + a \tag{4}$$

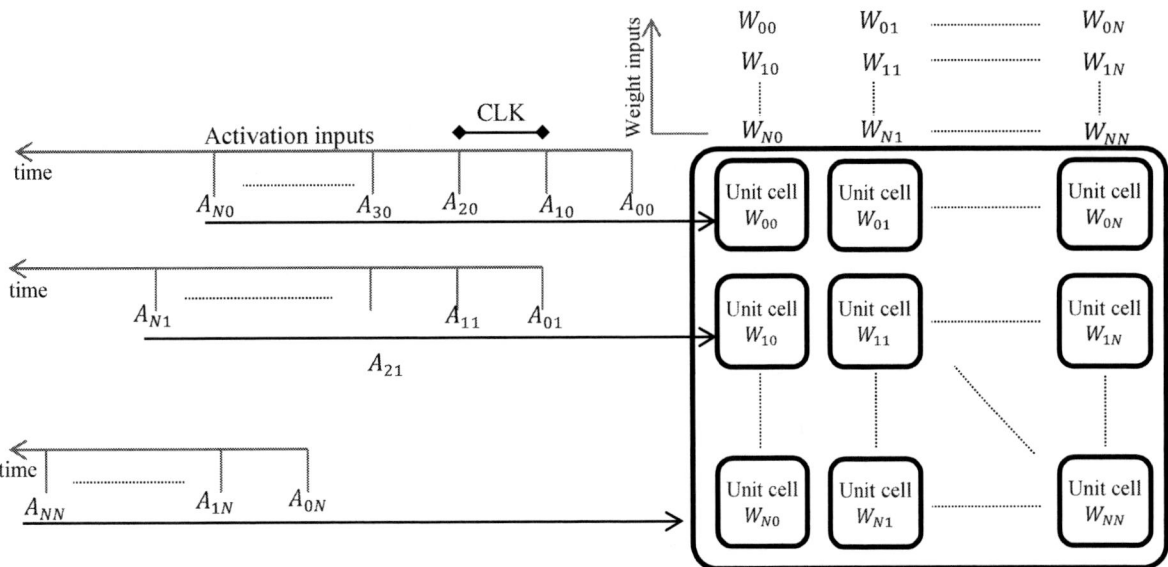

Fig.5 $N \times N$ matrix multiplication unit architecture with systolic data input and preloaded weights.

Where 'a' is the minimum value in the specified range, 'b' is the maximum value in the specified range, x_{min} is the minimum value of x, x_{max} is the maximum value of x. The output of the normalization unit is stored in the input FIFO as an input of the next neural network layer.

F. Control Unit

In order to synchronize between all the previous blocks a control unit is needed. Control unit is designed as an FSM that is responsible for controlling the control signals for each block and data flow throughout the neural network processor (NNP).

As shown in Fig. 6, there are 5 states for the control unit: idle, Load data, MAC, Activation, Normalize and Predict. High-level instructions are defined for the transition from one operation to another. At reset, NNP will go to idle state at which all blocks are disabled and waiting for "Start" instruction to go to load data state. The NNP starts loading weights and activation data from weight and activation FIFOs to weight and activation input registers unless the 1st layer output is already calculated in this case the activation inputs are read from the normalization output of the previous layer. After loading activation inputs and weights to the input registers MAC instruction is issued to start multiply and accumulate operation.

In MAC state, systolic data setup takes place along with the multiplication of the data with weights and accumulation of the results. Once the accumulation is done, activate command is issued to apply the activation function then a Normalize command is issued to rescale the output of the activation function and prepare it as an input for the next neural network layer. But, if it was the accumulation of the last layer in the neural network layers a predict instruction is issued and the result goes to the classifier to predict the inference result by choosing the index of the maximum number in the accumulated results. Table 1 summarizes all instructions and their functions.

Table 1 – Instructions and their functions

Instruction	Function
Start	• Shift from idle state to load data state.
Load data	• Read Weights from weights FIFO to weight input registers. • Read inputs to a layer from the Activation input FIFO or from the Normalization output of the previous layer to Activation input registers. • Shift from MAC and Normalization state to load data state
MAC	• Multiply the data with weights and accumulate the results. • Shift from load data state to MAC state.
Activate	• Apply RELU activation function. • Shift from MAC state to Activate state.
Normalization	• Normalize the output of the activation function.
Predict	• Predict the inference result from the output of the accumulated results of the MMU.

G. Illustrative Example: A Classification Problem

In this example, we take a deep neural network model and see how it will be computed and mapped to the proposed architecture. Fig.7 (a) shows the neural network model designed to classify images of clothing, like sneakers and shirts from 10 classes (0 → 9).

1) Model Training, Quantization and Weight Extraction

At first, we need to train the neural network in order to extract the weights for each layer in the neural network model that will be used later during inference. **TensorFlow** has been used to train this model using MNIST fashion dataset [5], 60,000 images used to train this model and 10,000 to evaluate how accurately the network learned, after 10 epochs the accuracy of the trained model became 90.1%. After training the model there is a mandatory step needed so we can use this model on our FPGA NNP architecture which is the quantization of the trained model.

Post training quantization reduce the model size while also improving hardware latency with little degradation in model accuracy. Post-training integer quantization technique is used, where it converts the entire model (weights and activations) to 8-bit during model conversion from TensorFlow to TensorFlow Lite, this technique matches the quantized model with the 8-bit calculations of the architecture proposed. Extraction of weights from the quantized model is performed using Netron. **Netron** is a viewer for neural network, deep learning and machine learning models that is used to visualize ".tflite" model and weights for each layer as shown in Fig.7.

2)   Neural Network Processor Initialization

In this step, we initialize of the processor with the extracted weights and the input image. The input is a 28x28 gray scale image which would be flattened to a vector of 784 elements after passing flatten layer, this vector is the input to the processor and the weights of the first layer is a 784x128 matrix already extracted from the quantized model. Each neuron in the neural network layer presented in one column inside the processor, having 128 neurons in the first layer and 10 neurons in the second layer means that at least 128 columns are needed inside the unit cells array. Each column of the activation input matrix requires a row, having 784 and 128 columns for the first layer and second layer respectively, means that 784 rows are needed inside the MMU which will increase the area and decrease performance drastically. Luckily, we have an accumulators unit that can help us decreasing number of rows inside the MMU by accumulating the sum of product value for each neuron at each iteration. In this example, we will consider an MMU array of size 256x256, all 784 pixels of the image would be driven after 4 accumulation iterations.

- *Weight Initialization*

Weight matrix size of the 1st layer is 784x128. Due to the array size of the MMU, it will be divided to 4 matrices, 3 matrices of size 256x128 and 1 matrix of size 16x128. At each iteration, the weight matrix is loaded to the unit cell array through the weight input FIFO as shown previously in Fig.5.

- *Activation initialization and driving*

Activation vector will be divided to four vectors one for each iteration 3 vectors of size 1x256 and 1 vector of size 1x16 as shown in Fig.8. Initialization starts by loading the vector to the activation input FIFO then store the value of each pixel inside the vector to the corresponding input register at each clock cycle till all input registers are initialized, then MMU starts fetching the stored pixels inside the input registers systollicaly to the unit cells array to start computation. At the end of each iteration accumulators unit will store the accumulated value from each column to accumulate it on the output of next iteration and so on. Note that weight and activation inputs initialization must be done at the start of each iteration.

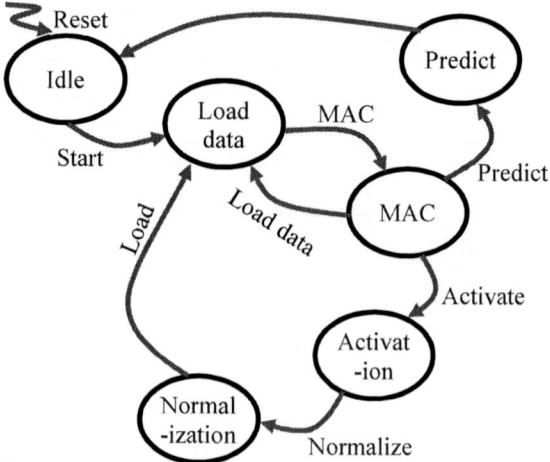

Fig.6 Control Unit FSM diagram.

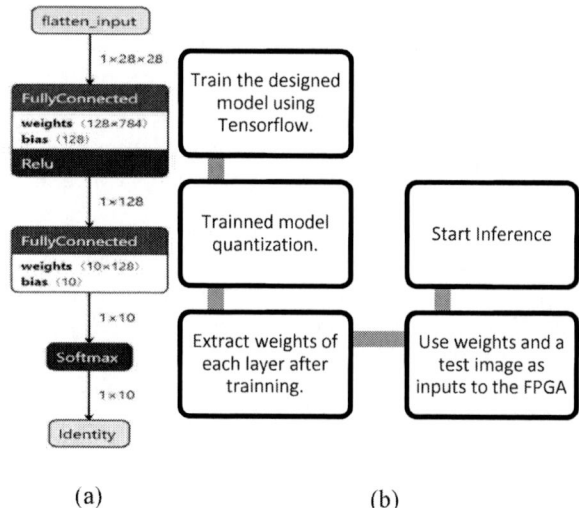

(a)   (b)

Fig.7 (a) Neural network architecture. (b) Flow diagram of the example.

- *Activation and Normalization*

After calculating the output for all iterations, accumulators unit has the output values for all 128 neurons of the first layer and ready to apply RELU activation function. Activation unit apply RELU function on each element in the output of the accumulators unit and pass the result to normalization unit to dynamically resize it to the input domain to be used as an input to the next neural network layer.

3)   Neural network Prediction

Once the processor finishes the calculation of all neural network layers, we have 10 values at the accumulators output corresponds to each class we have. As the value corresponds to each class increase, this implies that the neural network model is more confident that it's the correct prediction. So, we issue Predict command so that the classifier can begin calculating the maximum number out of these values to know the result of the inference.

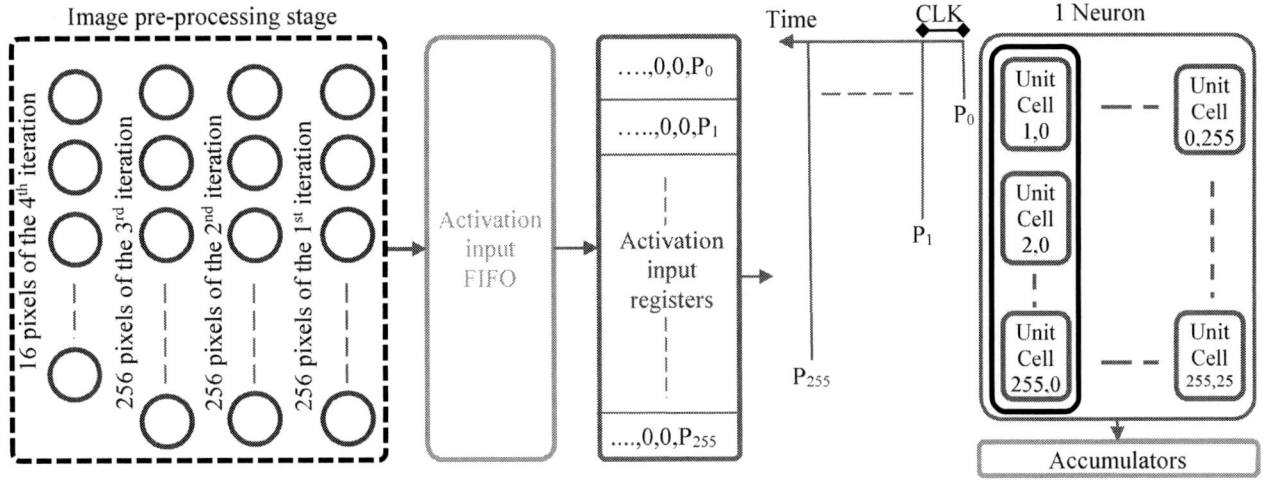

Fig. 8 Image input flow to the proposed neural network inference processor.

III. VALIDATION AND DISCUSSION

A. Validation

To implement the previously discussed architecture, Verilog is used as an HDL and SystemVerilog to stimulate the design. The output from each neuron in the last neural network layer is shown in Table 2; the class which has the largest output value is the class that the model believes the input image belongs to. Moreover, it is tested using AlexNet-based face recognition algorithm [6](160×160 face image at 25FPS) where it achieved an accuracy of 95%.

B. Synthesis Results

a NNP with MMU array size of 16x16 is implemented using Virtex UltraScale+ VCU118 Evaluation platform. Table 3 shows the area utilization of the synthesized design on the FPGA. The Dynamic power estimation from the synthesized netlist is 0.016 Watt.

C. Comparison with related work

Compared to related work [6]-[7], this work has better performance in terms of GOPS. The summary is shown in Table 4.

Table 2 – Inference output values

Class Neuron	Output Value	Class Neuron	Output Value
T-shirt	30943	Sandal	-29566
Trouser	-37996	Shirt	10354
Pullover	-6614	Sneaker	-101794
Dress	-4338	Bag	-31056
Coat	-27125	Ankle boot	-68754

Table 3 – Design utilization on FPGA

Resource	Utilization (%)
LUT	225 (0.02%)
LUTRAM	16 (0.00%)
FF	92 (0.00%)
IO	72 (8.65%)
BUFG	1 (0.06%)

Table 4 – Comparison with related work

	[6]	[7]	This Work
Frequency (MHz)	120	150	100
GOPS	117	17	55
GOPS/W	6	6	10

IV. CONCLUSIONS

In this work, an FPGA-based architecture for neural network inference processor is proposed. The design is coded with Verilog and successfully simulated in Questasim 10.0b simulator and implemented in Virtex UltraScale+ VCU118 Evaluation platform using Precision tool. Compared to related work, this design can achieve higher performance and lower power at 100MHz using less resources.

REFERENCES

[1] H. T. Kung, "Why systolic architectures?", IEEE Computer, vol. 15, Jan. 1982.

[2] N. P. Jouppi et al., "In-datacenter performance analysis of a tensor processing unit", International Symposium on Computer Architecture, pp. 1-12, 2017

[3] https://www.kaggle.com/zalando-research/fashionmnist.

[4] N. Suda, V. Chandra, G. Dasika, A. Mohanty, Y. Ma, S. Vrudhula, J. Seo, Y. Cao, "Throughput-optimized OPENCL-based FPGA accelerator for large-scale convolutional neural networks," FPGA - Proc. ACM/SIGDA Int. Symp. Field-Program. Gate Arrays, Monterey, CA, United States, February 2016, pp. 16-25.

[5] B. Zhang, J. Lai, "Design and Implementation of a FPGA-based Accelerator for Convolutional Neural Networks," Journal of Fudan University (Natural Science), vol.57, No. 2, pp. 236-242, 2018.

[6] Y. Taigman, M. Yang, M. Ranzato and L. Wolf, "DeepFace: Closing the Gap to Human-Level Performance in Face Verification," 2014 Proc IEEE Comput Soc Conf Comput Vision Pattern Recognit (CVPR), Columbus, OH, United States, June 2014, pp. 1701-1708.

[7] F. Schroff, D. Kalenichenko and J. Philbin, "FaceNet: A unified embedding for face recognition and clustering," IEEE Comput Soc Conf Comput Vision Pattern Recognit (CVPR, June 2015.

[8] K. S. Mohamed "Neuromorphic Computing and Beyond: Parallel, Approximation, Near Memory, and Quantum" Springer Nature, 2020.

[9] E. Qin et al., "SIGMA: A Sparse and Irregular GEMM Accelerator with Flexible Interconnects for DNN Training," 2020 IEEE International Symposium on High Performance Computer Architecture (HPCA), San Diego, CA, USA, 2020.

[10] G. Lacey and et al., "Deep learning on fpgas: Past, present, and future,"arXiv preprint arXiv:1602.04283, 2016.

[11] Zhang and et al., "Optimizing fpga-based accelerator design for deep convolutional neural networks," in Proc.s of the 2015 ACM/SIGDA Int. Symp. on Field-Programmable Gate Arrays. ACM, 2015, pp. 161–170.

[12] Chung, Eric, et al. "Serving dnns in real time at datacenter scale with project brainwave." IEEE Micro 38.2 (2018): 8-20.

A State-of-the-Art Design: Applying Forward Kinematics to Improve Patient Positioning in Radiosurgery

1st Alaa Saadah
Doctoral School of Informatics
University of Debrecen
Debrecen, Hungary
alaa.saadah@eng-unideb.hu

2nd Donald Medlin
Medical Beam Laboratories
Greenville, SC, USA
dmedlin@medbeamlabs.com

3rd Jad Saud
Dep of Robotics
Spinotron Kft
Debrecen, Hungary
jadsaud@spinotron.hu

4th Xiao Ran Zheng
Medical Beam Laboratories
Greenville, SC, USA
leonzheng@medbeamlabs.com

5th Géza Husi
Department of Air and Roads vehicles
University of Debrecen
Debrecen, Hungary
husigeza@eng.unideb.hu

Abstract—This paper presents a comprehensive study on the application of forward kinematics in the development and control of an innovative Patient Positioning System (PPS) for radiosurgery treatment. The proposed PPS, consisting of a linear rail system, a linkage system, and a tabletop, synergizes to enable precise human body positioning.

The standout feature of our PPS is the linkage system, equipped with four linkage arms—three active and one passive. Together, these arms facilitate diverse motions, allowing for rotation, translation, and inclination of the endpoint, thereby providing unmatched precision and versatility. Forward kinematics profoundly enhances the design and operational capabilities of the robotic linkage and the positioning system as a whole.

In addressing the positional challenges posed by the PPS's robotic components, remarkable accuracy has been achieved in the system's kinematic modeling. A crucial aspect of our findings revolves around the error analysis. By applying single and dual loop control methods, significant advancements have been made in minimizing positional errors, thereby enhancing the PPS's accuracy and reliability. Detailed insights into the error, including the Heat Matrix, will be discussed in-depth within the paper, following a robust experimental validation using MATLAB and a comparative assessment with CAD model measurements.

Index Terms—Forward Kinematics; Patient Positioning System (PPS) ; Robotic Medical Device ;DH Parameters; Kinematics Modeling ; Robotic Radiosurgery ; Precision Positioning.

I. INTRODUCTION

Robots have become indispensable in various fields, including healthcare, necessitating a strong emphasis on the safety and precision of their operations [1]. Different robotic applications bring forth unique challenges, particularly in ensuring operational safety and effectiveness [2]. Kinematics is essential in understanding and modeling the behavior of

This project was funded by the National Research Development and Innovation, Fund of the Ministry of Innovation and Technology.

robots, especially those with complex joint architectures [3] [4].

In this paper, we introduce a new Patient Positioning System (PPS) designed specifically for radiosurgery treatment. The PPS includes a parallel robot and a unique four-arm linkage system, which allows for precise and adaptable patient positioning.

We also discuss the practical aspects of implementing this system in real-world medical settings. This includes considerations of the technical requirements needed for the system to operate effectively, as well as any limitations or challenges in its application.

The paper is organized into sections that cover the system's design characteristics, development of its forward kinematic model, and our results and conclusions from testing its performance.

II. PATIENT POSITIONING SYSTEM PPS CHARACTERISTICS

As a 6-degree-of-freedom (DOF) robotic patient bed, it has been ingeniously engineered to position patients accurately and precisely. It does so by aligning the clinical targets with the Mechanical Isocenter (MIC), ensuring an accurate delivery of the prescribed radiation. The PPS operates as a comprehensive unit, comprised of three key elements: a Linear Rail System, a Linkage System, and a Tabletop.

A. The Linear Rail System

The linear rail system serves as the base of the Patient Positioning System (PPS), controlling movement along two axes. It shifts the main plate along a rail and uses a rotary table to rotate the Linkage System. This allows for flexible and convenient patient positioning within the operating area.

979-8-3503-8083-5/23 $31.00 © 2023 IEEE

B. The Linkage System

The linkage system is central to the PPS and comprises four linkage arms: three active and one passive. It's fixed to a stationary structure connected to the rotary table. Precision shafts join the arms in pairs, with three motors driving the active arms for 2D movement. A servo-driven gearbox moves three linkage arms. The system can adjust to different heights, accommodating patients with varying needs, and ensures precise adjustments for accurate patient positioning, using redundantly encoded motors for precision [5].

C. The Tabletop

The tabletop completes the system with its capacity to create pitching movements. This functionality is achieved through a helical cam-following system, driven by a servo motor.

III. PATIENT POSITIONING TABLE OVERVIEW

The PPS effectively controls patient position and orientation, meeting the high standards required for radiosurgery [6]. It marks a significant milestone in developing a reliable motion control system for patient positioning. With its 6-DOF in 3D space, covering both position and orientation, the PPS offers a complete solution for accurate patient alignment in radiosurgery.

Fig. 1. Patient Positioning Table.

A. Forward Kinematics

Using kinematic principles, we can understand the spatial layout and connections of a robotic system, effectively managing attributes like position, velocity, and acceleration through a kinematic chain that links each segment to its predecessor [7]. This chain determines the overall motion of the system. Forward kinematics allows us to calculate the precise location and orientation of the end-effector (or target point) in space, by transforming the joint-based description of the system into Cartesian coordinates [8].

B. Homogeneous Transformation Matrix

- Homogeneous Transformation Matrix. The homogeneous transformation matrix is key for capturing both translation and rotation within the robot's structure, condensing all essential information into a single matrix [9].

$$T_i = \begin{bmatrix} R & P \\ 0\,0\,0 & 1 \end{bmatrix}$$

- Denavit Hartenberg (DH),The Denavit-Hartenberg (DH) method is essential for kinematic modeling in robotics [10]. It simplifies the transition from the base frame to the end effector, capturing both rotational and translational motions [11]. This approach is widely used for robot control and is depicted in figure 2.

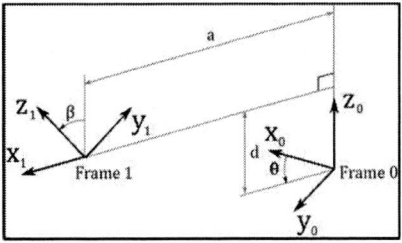

Fig. 2. Denavit Hartenberg (DH)

Starting from that we can generate the matrix of our robot after determining the assignment of the manipulator frames -which is defined in 2 as:

$$T_i = \mathrm{Rot}\,(z, \theta_i)\,\mathrm{Trans}\,(d_i)\,\mathrm{Trans}\,(a_i)\,\mathrm{Rot}\,(x, \alpha_i) \qquad (1)$$

Ti Final matrix will be the result of multiple of those matrices [12].

$$T_i = \begin{bmatrix} c\theta_i & -s\theta_i & 0 & 0 \\ s\theta_i & c\theta_i & 0 & 0 \\ 0 & 0 & 1 & d_i \\ 0 & 0 & 0 & 1 \end{bmatrix} \begin{bmatrix} 1 & 0 & 0 & a_i \\ 0 & c\alpha_i & -s\alpha_i & 0 \\ 0 & s\alpha_i & c\alpha_i & 0 \\ 0 & 0 & 0 & 1 \end{bmatrix} \qquad (2)$$

This equation can be translated into matrices and by multiplying them we get the final Ti as:

$$T_i = \begin{bmatrix} \cos\theta_i & -\cos\alpha_i\sin\theta_i & \sin\alpha_i\sin\theta_i & a_i\cos\theta_i \\ \sin\theta_i & \cos\alpha_i\cos\theta_i & -\sin\alpha_i\cos\theta_i & a_i\sin\theta_i \\ 0 & \sin\alpha_i & \cos\alpha_i & d_i \\ 0 & 0 & 0 & 1 \end{bmatrix}$$
$$(3)$$

where:

- a_i: The distance between the z_i and z_{i+1} axes along the x_i axis.
- α_i: The angle between the z_i and z_{i+1} axes along the x_i axis.
- d_i: The distance between the x_i and x_{i+1} axes along the z_i axis.
- θ_i: The angle between the x_i and x_{i+1} axes along the z_i axis.

C. PPS subsystems functionality description and coordinate frame assignment

This system is key for accurate patient alignment in diverse medical settings. It consists of three main components: the Linear Rail System, the Linkage System, and the Table Assembly. Each part brings unique features that collectively make the system efficient, versatile, and patient-friendly.

Fig. 3. PPS Frame Assignment

D. Linear Rail System from T0 to TRR

This subsystem controls the PPS's two-way movement: linear motion along a rail and rotational motion via a rotary table. It allows for safe patient placement and, in CT-equipped rooms, 180° patient rotation for imaging. Encoded motors ensure precise positioning.

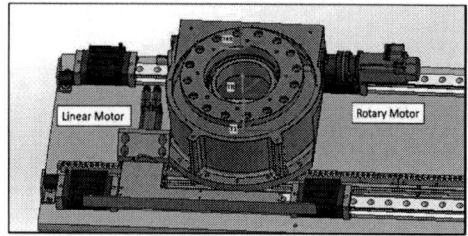

Fig. 4. Linear Rail Frames Assignment

E. Linkage System from TRLB To TLM:

The Linkage System, central to the PPS, consists of four two-arm linkage pairs connecting the PPS to a rotary plate. Two pairs are joined by precision shafts for stability. Three encoded motors manipulate three joints, allowing PPS adjustments in a 2D plane over a wide range. The design allows vertical table adjustments for various patient needs. The linkage kinematics was explored in detail in a separate paper[5].

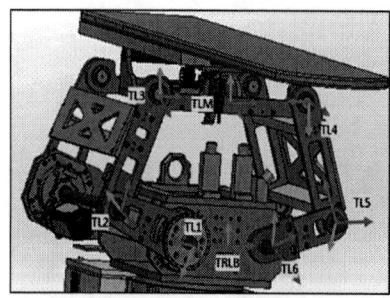

Fig. 5. Linkage system Frames Assignment

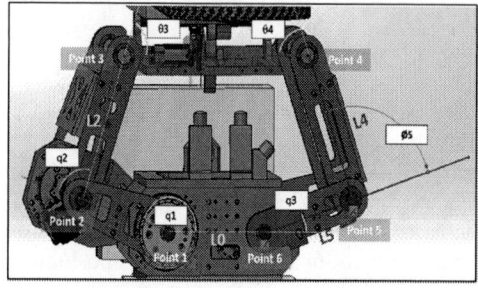

Fig. 6. Linkage Subsystems Joints, Lengths and Angles

F. Table Tob TLM to TE

The Table Assembly features a radio-friendly design and includes a pitch adjustment mechanism to maintain a level operating surface.

Fig. 7. Table Top Frames Assignment and Transitioning from TP to TE

IV. RESULTS

Now, after determining all needed parameters from figure 3 we can create the DH table as follows. Table 1 encompasses all the requisite translations and rotations for transitioning from one coordinate to the subsequent one, as stipulated by Equation 3. By substituting the appropriate values into the transformation matrices for each link and sequentially multiplying them, we obtain the final matrix. This matrix yields the comprehensive kinematic information, encapsulating both translational and rotational motions of the controlled point.

$$T_{E3}^{0} = \begin{bmatrix} 0 & 1 & 0 & 0 \\ -1 & 0 & 0 & 0 \\ 0 & 0 & 1 & 0 \\ 0 & 0 & 0 & 1 \end{bmatrix} \quad (4)$$

$$\begin{aligned} T_{0}^{E3} = {}& T_{1}^{R} \cdot T_{R}^{RR} \cdot T_{RR}^{RLB} \cdot T_{RLB}^{L1} \\ & \cdot T_{L1}^{L2} \cdot T_{L2}^{L3} \cdot T_{L3}^{LM} \cdot T_{LM}^{P} \\ & \cdot T_{P}^{E} \cdot T_{E}^{E1} \cdot T_{E1}^{E2} \cdot T_{E2}^{E3} \cdot T_{E3}^{0} \end{aligned} \quad (5)$$

$$= \begin{bmatrix} \mu_{x} & O_{x} & \alpha_{x} & p_{x} \\ \mu_{y} & O_{y} & \alpha_{y} & p_{y} \\ \mu_{z} & O_{z} & \alpha_{z} & p_{z} \\ 0 & 0 & 0 & 1 \end{bmatrix} \quad (6)$$

TABLE I
DH Parameters for the Linkage from Origin Frame to the End-Point Frame

Frame	a_i	α	d_i	θ
Linear Rail subsystem				
T_0	0	0	0	0
T_1	0	0	$-L + \text{Lin}$	0
T_R	0	$-\frac{\pi}{2}$	0	0
T_{RR}	0	0	L_R	Rot
Linkage subsystem				
T_{RLB}	0	$\frac{\pi}{2}$	0	0
T_{L1}	$L_0/2$	0	L_L	π
T_{L2}	L_1	0	0	$-\pi + q_1$
T_{L3}	L_2	0	0	q_2
T_{LM}	$L_3/2$	0	0	θ_3
Table Top subsystem				
T_P	L	$\frac{\pi}{2}$	L	$\frac{\pi}{2}$
T_E	L_{10}	0	0	$\frac{\pi}{2} + \text{Table Angle} - \theta_6$
Euler Rotation Matrices				
T_{E1}	0	0	0	$-\frac{\pi}{2} - \theta_6$
T_{E2}	0	$-\frac{\pi}{2}$	0	0
T_{E3}	0	0	0	$-\frac{\pi}{2}$

Utilizing MATLAB functions, we can compute the results of forward kinematics, which subsequently enables us to precisely determine the position and orientation of the target point [13]. Given the final transformation matrix T, the position of the target point can be directly extracted from its top-right column:

$$\begin{bmatrix} p_x \\ p_y \\ p_z \\ 1 \end{bmatrix} \quad (7)$$

This represents the position of the end-effector in the workspace [14]. Orientation, on the other hand, is commonly represented using Euler angles: Roll (ϕ), Pitch (θ), and Yaw (ψ) [15]. These angles can be derived from the rotation sub-matrix of T. There are various conventions for computing Euler angles, but for our chosen representation, the angles can be computed as:

$$\theta = \text{atan2}\left(-r_{31}, \sqrt{r_{11}^2 + r_{21}^2}\right)$$
$$\psi = \text{atan2}(r_{21}, r_{11})$$
$$\phi = \text{atan2}(r_{32}, r_{33})$$

where r_{ij} are the elements of the rotation sub-matrix of T. This approach provides a concise representation of the orientation of the target point, allowing for precise control and adjustments during system operations [16].

By applying single and dual loop control methods, significant advancements have been made in minimizing positional errors, thereby enhancing the PPS's accuracy and reliability. Detailed insights into the error, including the Heat Matrix, The subsequent figure exemplifies the precision of the Patient Positioning System (PPS), demonstrating that the system consistently achieves an accuracy level of less than 0.1 mm while applying dual loop control algorithm.

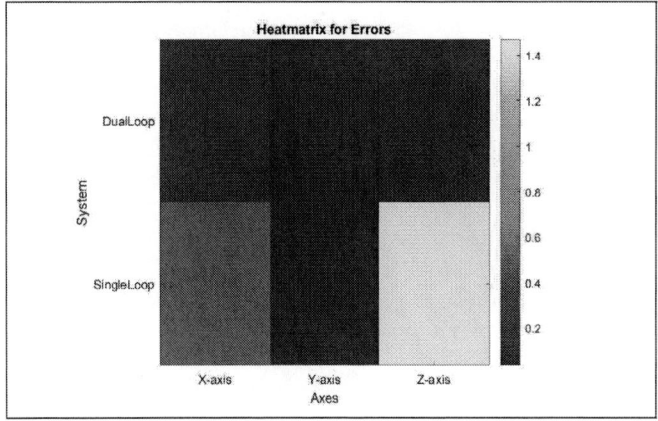

Fig. 8. Heat Map Matrix (PPS Accuracy Error).

V. Conclusion

In conclusion, this research elucidates the transformative power of forward kinematics in engineering a state-of-the-art Patient Positioning System (PPS) tailored for radiosurgery applications. Our novel PPS, distinguished by its advanced linkage system and linear rail, sets a new benchmark in achieving unparalleled precision and adaptability in patient positioning. In terms of real-world application, several considerations have been contemplated to ensure the PPS's operational integrity and effectiveness:

- **Technical Requirements:** The system necessitates specific hardware and software configurations to ensure robust performance, including high-precision sensors, durable actuators, and sophisticated control algorithms.
- **Operational Feasibility:** The PPS is designed to integrate seamlessly into existing medical workflows, embodying a user-centric design philosophy to facilitate ease of use and minimal disruption in the clinical environment.

MATLAB simulations, corroborated with CAD model measurements, authenticate the fidelity and accuracy of our kinematic model. Throughout the study, the PPS demonstrated remarkable precision and adaptability, essential attributes for seamless applicability in medical settings.

VI. Patent

The Patient Positioning System (PPS) is presently undergoing the patenting process, holding the application number PCT/US2019/048205.

Acknowledgment

Gratitude is extended to the contributors from Spinotron Kft, Hungary, and the Medical Beams Laboratory, USA. Their indispensable assistance was instrumental in the system's prototyping and examination phases conducted in Hungary.

REFERENCES

[1] John J Craig. *Introduction to Robotics: Mechanics and Control. 4th.* 2018.

[2] Vahid Nazari and Leila Notash. "Motion analysis of manipulators with uncertainty in kinematic parameters". In: *Journal of Mechanisms and Robotics* 8.2 (2016), p. 021014.

[3] Jochen Heinzmann and Alexander Zelinsky. "Quantitative safety guarantees for physical human-robot interaction". In: *The International Journal of Robotics Research* 22.7-8 (2003), pp. 479–504.

[4] Teun Koetsier. "Ludwig Burmester, Kinematics as Part of Geometry". In: *A History of Kinematics from Zeno to Einstein: On the Role of Motion in the Development of Mathematics.* Springer, 2023, pp. 265–285.

[5] Alaa Saadah et al. "Kinematics Study for Linkage System (Parallel Robotics System): linkage system of patient positioning system PPS to accurately position a human body for radiosurgery treatment". In: *2023 Advances in Science and Engineering Technology International Conferences (ASET).* IEEE. 2023, pp. 1–6.

[6] Robert J Webster III and Bryan A Jones. "Design and kinematic modeling of constant curvature continuum robots: A review". In: *The International Journal of Robotics Research* 29.13 (2010), pp. 1661–1683.

[7] Alaa Saadah. "Computing the kinematics study of a 6 DOF industrial manipulator prototype by matlab". In: *Recent Innovations in Mechatronics* 7.1. (2020), pp. 1–5.

[8] Josep M Porta, Lluís Ros, and Federico Thomas. "Inverse kinematics by distance matrix completion". In: (2005).

[9] Omar Martínez and Ricardo Campa. "Comparing methods using homogeneous transformation matrices for kinematics modeling of robot manipulators". In: *Multibody Mechatronic Systems: Papers from the MuSMe Conference in 2020 7.* Springer. 2021, pp. 110–118.

[10] Sébastien Briot et al. "Homogeneous transformation matrix". In: *Dynamics of Parallel Robots: From Rigid Bodies to Flexible Elements* (2015), pp. 19–32.

[11] Tao Zhang et al. "A novel method to identify DH parameters of the rigid serial-link robot based on a geometry model". In: *Industrial Robot: the international journal of robotics research and application* 48.1 (2020), pp. 157–167.

[12] Oscar Altuzarra et al. "Path analysis for hybrid rigid–flexible mechanisms". In: *Mathematics* 9.16 (2021), p. 1869.

[13] Peter Corke. *Robotics and control: fundamental algorithms in MATLAB®.* Vol. 141. Springer Nature, 2021.

[14] Pierre Larochelle. "Synthesis of planar mechanisms for pick and place tasks with guiding positions". In: *Journal of Mechanisms and Robotics* 7.3 (2015), p. 031009.

[15] Hao Ye et al. "Forward and inverse kinematics of a 5-DOF hybrid robot for composite material machining". In: *Robotics and Computer-Integrated Manufacturing* 65 (2020), p. 101961.

[16] James Diebel et al. "Representing attitude: Euler angles, unit quaternions, and rotation vectors". In: *Matrix* 58.15-16 (2006), pp. 1–35.

A Self-Aware Power Management Model for Epileptic Seizure Systems Based on Patient-Specific Daily Seizure Pattern

Shiva Maleki Varnosfaderani[1], Rihat Rahman[2], Nabil J. Sarhan[1], Mohammad Alhawari[1]

[1]Department of Electrical and Computer Engineering, Wayne State University, Detroit, USA

[2]Department of Computer Science, Wayne State University, Detroit, USA

Abstract—We analyze and compare various hardware-based epileptic seizure systems and discuss the challenges and opportunities for reducing power consumption and increasing the battery lifetime. Furthermore, we propose a power management model that employs patient-specific seizure patterns to manage the power consumption of the overall system. This model determines the patient-specific seizure pattern and switches the system to sleep mode when the likelihood of seizure occurrence is zero or very low. Our analysis shows that our proposed power management model could effectively reduce the power consumption by 49% compared to the complex model while the performance reduction is less than 1%.

Index Terms—Deep learning, energy efficiency, epileptic seizure monitoring, patient-specific daily seizure pattern, self-aware power management model.

I. INTRODUCTION

Epilepsy is a seizure disorder that is usually diagnosed after at least two seizures that are not caused by other medical disorders. Epileptic seizures may be caused by a brain injury or a genetic predisposition, while the reason is usually unclear. According to the World Health Organization (WHO), more than 65 million people worldwide suffer from epilepsy, with 3.4 million people in the United States [1]. Patients who had convulsive movements and loss of consciousness are at a greater risk of Sudden Unexpected Death in Epilepsy (SUDEP), which is a leading cause of death in patients with uncontrolled seizures [2].

Current commercial solutions monitor EEG in real-time to detect the seizure onset. For example, the *Responsive Neurostimulation System* (RNS) is an implantable device that detects the occurrence of certain types of seizures and provides stimulation to reduce the effect of a seizure [3]. Another device, called *Embrace Watch* detects possible convulsive seizures by measuring sympathetic nervous system activity (without using EEG signals), but it fails to detect all convulsive seizures. There is a need not only to detect seizures but also to predict them before they happen so that the patients can take precautions, stop certain activities (such as driving, swimming alone, or climbing ladders), and/or reduce their effects [4].

Deep Learning (DL) methods have become the cornerstone for modern artificial intelligence (AI) applications, especially for healthcare, due to the unprecedented achieved accuracy [5]. DL has been applied in a wide spectrum of healthcare applications, including predicting epileptic seizures [6]. predicting heart attacks [7], finding tumors in MRI images [8], and forecasting COVID-19 cases [9], [10]. Although DL methods can be designed to predict epilepsy with high accuracy [11], the high computational resources and memory bandwidth requirements are the key challenges in implementing DL models in low-power resource-constrained devices.

As illustrated in Figure 1, an automated EEG-based, battery-powered system for epilepsy prediction includes various hardware blocks: electrodes, data acquisition and digitization, hardware accelerator, wireless communication, and power management. Such a system is desired to achieve continuous monitoring and accurate prediction while maintaining low cost and a long battery lifetime [12]. The number of channels is

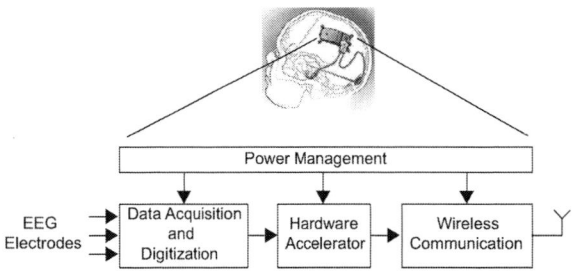

Fig. 1. A battery-powered system for predicting epileptic seizures.

determined by the number of used electrodes. These channels are filtered and digitized by the data acquisition and digitization block, and then processed by the hardware accelerator, which is a custom digital block that implements the DL model. The wireless communication block shares the results with the patient as well as the medical professionals, who can specify the patient-specific patterns and adjust antiseizure drugs accordingly. Finally, the power management module delivers power to all the blocks, ensuring low power consumption and a long battery lifetime.

Digital accelerators can be custom-made for DNNs to provide higher throughput, shorter latency, lower energy, and higher area efficiency [13]. Although digital accelerators provide better performance compared to GPUs, digital systems (including both GPUs and digital accelerators) are fundamentally limited in handling big data efficiently due to the separation of logic and memory (referred to as the Von Neumann bottleneck). Consequently, the system bandwidth is limited by the speed of accessing the data in the memory. Moreover, memory access requires at least 10x higher energy and longer delay compared to the actual computation in DNNs, specifically the multiply-accumulate (MAC) operation [14]. Recent advances in circuit design and memory technologies enable efficient computation through in-memory and near-memory computation, as well as more efficient computation in the analog domain and time domain. Further, it has been shown that computation can be done more efficiently in the analog domain [15] and time domain [16]. These technologies open the door for more innovation in hardware accelerators to achieve low-power and low-cost epilepsy prediction systems.

The use of a wearable accelerometer sensor in seizure detection systems is prevalent because it can detect sudden body movements, such as jerking motions, stumbling movements, and quick body falls, which are common phenomena for most seizure types, such as generalized tonic-clonic seizure (GTCS), tonic-clonic, myoclonic, and hyper motor. Accelerometers utilized in seizure detection systems are precise, three-dimensional, and energy efficient. The majority of accelerometer applications in seizure detection emerge as wearable devices, such as the SmartWatch Inspyre, E4 wristband, Embrace wristband, NightWatch, and a variety of

979-8-3503-8083-5/23 $31.00 © 2023 IEEE

other gadgets created by well-known technology corporations. An accelerometer can be worn in a variety of places on the body, including the wrist, chest, back, and ankle [17]. Channel selection is another approach, which selects only the channels that contain the most important information, thereby lowering the complexity, computation, and power consumption by decreasing the amount of data being processed. In [11], a selection algorithm is implemented to select 10 out of 23 channels on average based on the highest variance entropy product. The results showed that applying such an algorithm decreased 34% of parameters and 50% of training time while maintaining the same accuracy [11]. In [18], an SVM model is tested using 2 to 6 out of 22 channels. The results showed that applying channel selection decreased more than 93% of the computational costs. Channel selection can be an effective technique for reducing computational complexity, but the algorithm itself can be complex.

The power consumption of the hardware accelerator can be reduced by network pruning to remove unimportant connections [19], [20], reducing SRAM voltages based on the level of fault tolerance caused by bit masking, and optimizing the data types [20]. Feature extraction, bit-wise design, and estimation of the activation functions are some other solutions to simplify the design and reduce the design cost.

In this paper, we discuss various systems and hardware accelerators for epilepsy prediction and detection and compare them in terms of performance, size, and power consumption. Besides, According to the patient-specific daily seizure pattern, we propose a model that can switch to low-power mode when the chance of seizure incidence is minimal or zero. This work can help researchers to understand the design challenges for designing an energy-efficient, low-cost, and portable system for epileptic seizure prediction. The remainder of the paper is organized as follows. The prior work on hardware implementation for epilepsy monitoring system are discussed and different hardware structures for epilepsy prediction and detection systems are reviewed in section II. In Section III, we obtained the details of our proposed model. Finally, conclusions are drawn.

II. PRIOR WORK ON HARDWARE IMPLEMENTATIONS OF EPILEPSY MONITORING SYSTEMS

Table I summarizes the hardware implementations of various epilepsy monitoring systems. As depicted in the table, two commercial solutions are available for epileptic patients; RNS performs detection and requires surgery while Embrace is a non-invasive commercial device. RNS, an FDA-approved therapeutic option, is one of the interventional treatment options for people with refractory and focal epilepsy. The RNS system is the first commercially closed-loop responsive brain stimulation device [21]. RNS is implanted inside the brain to monitor EEG signals, detect seizures, and respond by applying electrical pulses to the patient's brain to reduce the effect of the seizure [3], [22]. Although RNS reduces the seizure frequency and severity, it has a high rate of false detections, raising the question of how much of this effect is attributable to closed-loop suppression of seizure-related ictal activity [23]. In addition, RNS is bulky, has a limited number of channels, and relies on basic hard thresholding with moderate seizure classification accuracy. Because of the power and size limits imposed by implanted devices, complex on-chip classification algorithms cannot be implemented [24].

Wearable sensors, such as the Embrace and Embrace 2 watches developed by MIT, are extremely valuable due to their precision in detecting epilepsy seizures. Embrace watch developed by Empathica is a wrist-worn wearable device that monitors various physiological signals, including heartbeat, temperature, and respiratory rate using EDA (electrodermal activity) sensor. The watch uses machine learning to detect patterns that may be associated with convulsive seizures [25], [26]. The reading is obtained from the watch and transmitted using a Bluetooth connection. The reading is then forwarded to a cloud server and database for storage. This is only possible if an internet connection is present. Users who have been

registered to the appropriate user can view the readings of the main user who is wearing the watch via the cloud server and database. When an epileptic seizure occurs, the watch notifies the other users who are registered to the appropriate user. Embrace2 is the second generation with improvements in battery life, weight, and connectivity [27]. Ictal Care365, Neuroon, Ricola, Vigil-Aide, Epi-Care free, and Zephyr are some examples of the possible potential solution for seizure monitoring using biomarker detection systems, which are available in the market [28].

A real-time seizure prediction system is presented in [29] based on an EEG dataset, which has 11 seizures extracted from 4 different focal epilepsy patients and obtained by the University of Texas Health Center using surface implanted electrodes. The system uses a one-class support vector machine (SVM) and is implemented on Zynq-7000 XC7Z045 FPGA which has the requisite FPGA fabric and DSP slices to host the computationally demanding method. Data are converted to Fixed-point to improve utilization of the FPGA fabric and DSP slices. The FPGA implementation consumes about 8 MB of the total 20 MB of block RAM (BRAM). The results showed that the system can predict seizures 3.6 minutes prior to clinical onset with a 3.9 mean false-positive rate (FPR) per hour. A hardware implementation of Bit-Serial Neural Network (BSNN) for epileptic seizure prediction is reported in [30]. The system utilizes a bit-serial data processing unit along with a finite state machine (FSM) to process the one-bit data at a time at each clock cycle to reduce power consumption and reduce area while running at low speed. The system is implemented on an ALTERA Cyclone V FPGA using 3931 ALMs which constitutes about 7% of the Cyclone V A7 capacity. The system achieved an accuracy of 90% when tested on the Bonn dataset [31]. For neural network computing, parallel hardware architecture is typically employed to boost performance. However, in this design, low power and low cost are more important than high performance, so a bit-serial architecture-based data processing unit (DPU) is presented for neural network computing to reduce power and cost. An SRAM is used to store the neural network's weights. For bit-serial processing, the ALU employs a proprietary multiplier. In [32], a deep neural network (DNN), implemented in a Xilinx Zynq-7000 Zybo-1 FPGA, is used to predict seizures. The system is tested on the American Epilepsy Society Seizure Prediction Challenge dataset [33]. The results show that the system predicts seizures with a 74% accuracy, 90% true-positive rate, and power consumption of 1.9 W. In [34], the RusBoosted classifier is applied to classify the data which is suitable for unbalanced data. One channel is used and chosen to reduce power usage and simplify the model. Four blocks from FPGA are used: BRAM, DSP48E, Flip Flop (FF), and Lookup Table with 19%, 52%, 20%, and 62% usage, respectively.

Due to its low power consumption and area requirements, epileptic seizure research has generally focused on ASIC implementation. [35] obtained accuracy higher than 92% at 2.8mW power with total area of 13.47 mm^2. Similarly, [36] and [37] demonstrated on-chip seizure detection with a sensitivity of 95.1% and 83.7% with a total area of $1mm^2$ at 2.73 J and 41.2 nJ per class. However, because these processors are not reprogrammable, ASIC-based wearable devices may face adaption and accuracy loss issues with a more diversified epileptic patient population.

Studies on FPGA-based seizure detection, on the other hand, have revealed partial reconfiguration, high-speed complicated feature extraction, and online training capabilities. [41] obtained 98.4% sensitivity on the Xilinx Zynq-7000 while utilizing 380mW of power at 100 MHz via FFT-based feature extraction. Similarly, [32] obtained 74% accuracy for seizure prediction on the Zynq-7000 with 1909 mW power. [43] extracted 256-point FFT-based features on an FPGA and classified them on-chip at 1.589 mW with a total area of 1409x1402 um^2 for the neural network chip. [45] demonstrated a hybrid approach in which online training is performed

979-8-3503-8083-5/23 $31.00 © 2023 IEEE

TABLE I
COMPARISON TABLE BETWEEN VARIOUS HARDWARE IMPLEMENTATIONS OF EPILEPSY MONITORING SYSTEMS

Ref	Goal	Implementation	Data (Num of Patient)	Sensitivity (%)	FPR (h^{-1})	ML Model	Power Consumption (Watt) & CLK(MHz)
[29]	Prediction	FPGA	sEEG (4) Private data	100	3.9	One-Class SVM	– & 50(MHz)
[30]	Prediction	FPGA	sEEG Bonn	90	–	BSNN	–
[32]	Prediction	FPGA	iEEG UPen	90	–	FC-DNN	1.9 & 25(MHz)
[34]	Prediction	FPGA	sEEG (10) European Epilepsy	77.3	0.04	RusBoosted	–
[38]	Prediction	FPGA	sEEG (8) CHB-MIT	–	–	–	–
[39]	Prediction	FPGA	sEEG (5) CHB-MIT	77.4	–	DL	0.013
[40]	Detection	FPGA	–	–	–	Generic Algorithm	–
[41]	Detection	FPGA	sEEG (24) CHB-MIT	98.4	0.356	SVM	0.38 & 100 (MHz)
[36]	Detection	–	sEEG (24) CHB-MIT	95.1	3.8	Dual LSVM	2.73 uJ/class
[42]	Detection	TSMC 180 nm	iEEG (14) CHB-MIT	95.7	0.02	Dual LSVM	–
[35]	Detection	TSMC 180 nm	iEEG (4) Long-Evans Rats	–	–	Linear Least Square(LLS)	0.028 & 3.125(MHz)
[43]	Detection	FPGA TSMC 180 nm	Data of Mice	–	–	FFNN	0.01589 & 8 (MHz)
[37]	Detection	TSMC 65 nm	iEEG (26)	83.7	–	Boosted Tree	41.2 nJ/class
RNS	Detection	Commercial Device	iEEG	–	–	–	–
Embrace– Embrace 2 [27], [28], [44]	Detection & Prediction	Commercial Device Apple-Android -Smartphone	Non-EEG Non-invasive	–	High-FPR	–	–

on the FPGA and seizure detection is performed on-chip. This system, however, has greater design costs, intermediate data transfer latency, and larger space needs. FPGA's high-performance capabilities come at the expense of increased power consumption and energy requirements. Therefore, designing a feasible and scalable, battery-powered, FPGA-based, seizure detection system is a major task.

III. PROPOSED MODEL BASED ON PATIENT-SPECIFIC SEIZURE PATTERNS

To reduce energy consumption, we propose a new generation of self-aware wearable or implantable systems based on the patient-specific daily seizure pattern. Epileptic seizures have been shown to have biases in distribution over time at various intervals that can be as long as 1 year or as short as 1 h [46]. In another word, the probability of seizure occurrence for each patient can follow a particular daily pattern; For each patient, there are some hours when the risk of having a seizure is high and certain hours when the risk is low or nearly zero. Therefore, we can find a daily seizure pattern for each patient. By analyzing the patient-specific daily seizure pattern, we can manage the power during the day for each patient. Therefore, during times of the day when seizures are not predicted or have a low chance of occurring, the system can switch to a low-power mode.

We determine and utilize the distribution of the number of seizures during the day, which could help to improve epilepsy systems to forecast patient-specific seizures through the addition of 24-hour-cycle information. The distribution of seizures over the 24-hour cycle for 30 patients from the European iEEG Epilepsy Dataset is shown in Figure 2 [47], [48]. The color in each block represents the fraction of seizures that occurred during a certain hour. For example, patient FR_916 had 52 seizures, and thus a block with a value of 0.2 at 2 a.m. depicts that one-fifth of the seizures occurred around 2 a.m. In this graph, dark blue blocks depict no-seizure intervals. Once the patient-specific daily seizure pattern is established, it can be incorporated into the prediction/detection system. The proposed model for power management based on the patient-specific pattern is represented in Figure 3. The patient's data are first read and digitalized. According to the daily seizure pattern for a patient, the power management unit reads the

daily pattern and assigns the low-power or high-power mode; low-power mode is given to ratio values greater than 0.025, and high-power mode is given to ratio values greater than 0.025. The key to a patient's daytime power management will be this pattern. Afterward, a simple or complex ML model is employed for the low-power or high-power mode, depending on the assigned mode, respectively.

In this research, we employed the simple and complex detection models in [49], [50] where the classification can be done based on a basic set of features and a more intricate set of features, respectively. Based on the computed threshold in the train section, the two-level classifier in [49] determines whether the simple model is confident to classify the incoming data or the complicated one. Our suggested two-level self-aware classifier, on the other hand, changes from a basic model to a complicated model or vice versa depending on the patient-specific seizure pattern. Making the choice to move between two states in this situation does not need any additional calculations. The influence of threshold (T) on the percentage of time that the system operates in the complex model per day ($P_{Complex}$) and the influence of threshold (T) on the model's performace are shown in Figure 5 and Figure 4. As it can be seen, increasing the threshold from zero to 225 reduces the time of being in the complex model. As a result, the basic model, which has a lower Gmean and power consumption, has a greater impact on system performance. The energy (E) and Gmean for our proposed power-management model are calculated based on the following equations:

$$E = P_{Complex} \times (E_{Complex}) + P_{Simple} \times (E_{Simple}) \quad (1)$$

$$Gmean - \frac{1}{225}(T \times (Gmean_S) + (225 - T) \times (Gmean_C)) \quad (2)$$

Where $P_{Complex}$ and P_{Simple} refer to the percentage of time per day using a complex and a simple model, respectively. $Gmean_C$ and $Gmean_{Simple}$ are the system's performance in the complex and simple models, respectively. T is the threshold which can be between 0 to 225.

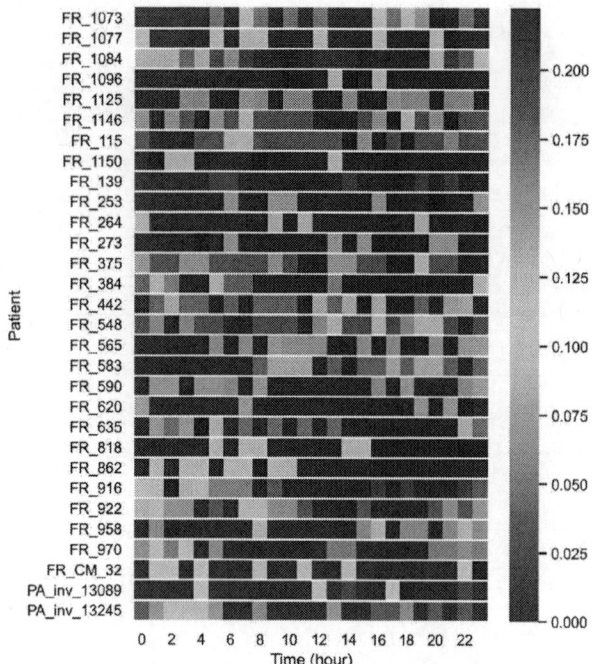

Fig. 2. Distribution seizures over the 24-hour period for the 30 patients.

The average results for 30 patients with $T = 25$ are obtained in Table II. The energy consumption and Gmean for our proposed sef-aware model are calculated and compared with the simple and complex models. As it can be seen, our proposed power-management model could effectively reduce the power consumption by 49% compared to the complex model however the performance reduction is less than 1 %.

The results indicate that our proposed power management model can greatly reduce power consumption for most patients while maintaining the performance.

IV. CONCLUSION

This paper performed a detailed analysis of the challenges and opportunities for designing low-power epilepsy prediction systems. Designing an efficient epilepsy prediction system requires various levels of software and hardware optimizations, including custom hardware accelerators, machine learning models, channel selection, and statistical analysis of patient-specific EEG data. We proposed a self-aware power-management model to reduce the power consumption for each patient based on the patient-specific pattern. This model is applicable for all types of epilepsy prediction and detection systems. The simulation results indicated that our proposed model can effectively reduce the power consumption by 49 percent compared to the complex model without losing the performance. It is worth to mention that the switch between basic and complicated models is done based on the daily

TABLE II
ESTIMATED RESULTS FOR EPILEPSY DETECTION BASED ON SIMPLE
MODEL, COMPLEX MODEL, AND OUR PROPOSED MODEL

Model	Low-power mode (Hours per Day)	Gmean (%)	Energy (uJ)
Simple Model	24	75.16	2.832
Complex Model	0	82.53	31.464
This Study	13	81.73	15.955

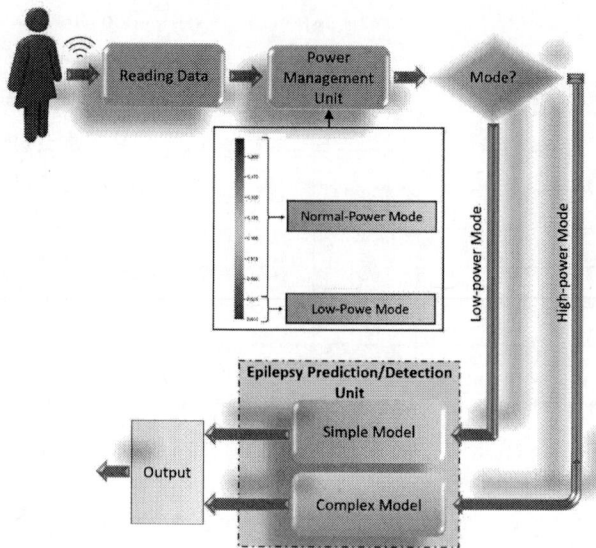

Fig. 3. Block diagram of the proposed self-aware model.

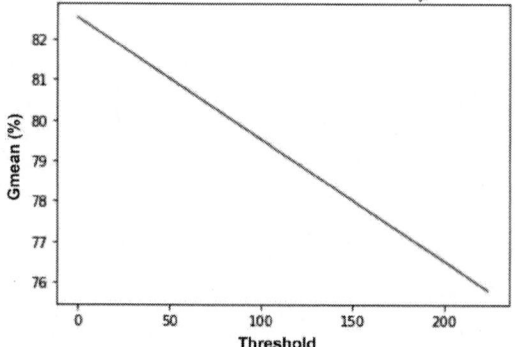

Fig. 4. The effect of threshold on the model's performance.

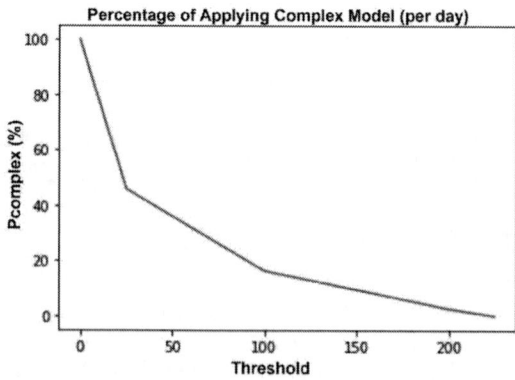

Fig. 5. The effect of threshold on the percentage of time of operating with the complex model per day.

seizure pattern that is unique for each patient and does not require any additional computations.

REFERENCES

[1] "World Health Organization, http://www.who.int/mediacentre/factsheets/fs999/en/, Accessed date: 2 April 2018.."

[2] T. Tomson, T. Walczak, M. Sillanpaa, and J. W. A. S. Sander, "Sudden unexpected death in epilepsy: A review of incidence and risk factors," *Epilepsia*, vol. 46, no. s11, pp. 54–61, 2005.

[3] https://www.epilepsy.com/learn/treating-seizures-and-epilepsy/devices/responsive-neurostimulation-rns.

[4] L. Iasemidis, "Seizure prediction and its applications," *The Neurodiagnostic journal*, vol. 22, pp. 489–506, 2011.

[5] R. U. Hagiwara, Oh S.L., "A deep learning approach for parkinson's disease diagnosis from eeg signals," *Neural Comput Applic 32*, vol. 22, pp. 10927–10933, 2020.

[6] R. Hussein, M. O. Ahmed, R. Ward, Z. J. Wang, L. Kuhlmann, and Y. Guo, "Human intracranial EEG quantitative analysis and automatic feature learning for epileptic seizure prediction," *CoRR*, vol. abs/1904.03603, 2019.

[7] F. Ali, S. El-Sappagh, S. R. Islam, D. Kwak, A. Ali, M. Imran, and K.-S. Kwak, "A smart healthcare monitoring system for heart disease prediction based on ensemble deep learning and feature fusion," *Information Fusion*, vol. 63, pp. 208 – 222, 2020.

[8] M. B. naceur, R. Saouli, M. Akil, and R. Kachouri, "Fully automatic brain tumor segmentation using end-to-end incremental deep neural networks in mri images," *Computer Methods and Programs in Biomedicine*, vol. 166, pp. 39 – 49, 2018.

[9] P. Arora, H. Kumar, and B. K. Panigrahi, "Prediction and analysis of covid-19 positive cases using deep learning models: A descriptive case study of india," *Chaos, Solitons Fractals*, vol. 139, p. 110017, 2020.

[10] Z. Mehdizadeh-Somarin, B. Salimi, R. Tavakkoli-Moghaddam, M. Hamid, and A. Zahertar, "Performance assessment and improvement of a care unit for covid-19 patients with resilience engineering and motivational factors: An artificial neural network method," *Computers in Biology and Medicine*, vol. 149, p. 106025, 2022.

[11] H. Daoud and M. A. Bayoumi, "Efficient epileptic seizure prediction based on deep learning," *IEEE Transactions on Biomedical Circuits and Systems*, vol. 13, no. 5, pp. 804–813, 2019.

[12] Z. Mei, X. Zhao, H. Chen, and W. Chen, "Bio-signal complexity analysis in epileptic seizure monitoring: A topic review," *Sensors*, vol. 18, no. 6, 2018.

[13] A. Shawahna, S. M. Sait, and A. El-Maleh, "Fpga-based accelerators of deep learning networks for learning and classification: A review," *IEEE Access*, vol. 7, pp. 7823–7859, 2019.

[14] J. Liu, H. Zhao, M. A. Ogleari, D. Li, and J. Zhao, "Processing-in-memory for energy-efficient neural network training: A heterogeneous approach," in *2018 51st Annual IEEE/ACM International Symposium on Microarchitecture (MICRO)*, pp. 655–668, 2018.

[15] Y. Chen, Y. Xie, L. Song, F. Chen, and T. Tang, "A survey of accelerator architectures for deep neural networks," *Engineering*, vol. 6, no. 3, pp. 264 – 274, 2020.

[16] H. Al Maharmeh, N. J. Sarhan, C. C. Hung, M. Ismail, and M. Alhawari, "Compute-in-time for deep neural network accelerators: Challenges and prospects," in *2020 IEEE 63rd International Midwest Symposium on Circuits and Systems (MWSCAS)*, pp. 990–993, 2020.

[17] S. E. M. F. Alzghoul *et al.*, "A scoring framework and apparatus for epilepsy seizure detection using a wearable belt," *Journal of Medical Signals and Sensors*, vol. 12, no. 4, pp. 326–333, 2023.

[18] N. Chang, T. Chen, C. Chiang, and L. Chen, "Channel selection for epilepsy seizure prediction method based on machine learning," in *2012 Annual International Conference of the IEEE Engineering in Medicine and Biology Society*, pp. 5162–5165, 2012.

[19] A. Parashar, M. Rhu, A. Mukkara, A. Puglielli, R. Venkatesan, B. Khailany, J. Emer, S. W. Keckler, and W. J. Dally, "Scnn: An accelerator for compressed-sparse convolutional neural networks," *SIGARCH Comput. Archit. News*, vol. 45, p. 27–40, June 2017.

[20] B. Reagen, P. Whatmough, R. Adolf, S. Rama, H. Lee, S. K. Lee, J. M. Hernández-Lobato, G. Wei, and D. Brooks, "Minerva: Enabling low-power, highly-accurate deep neural network accelerators," in *2016 ACM/IEEE 43rd Annual International Symposium on Computer Architecture (ISCA)*, pp. 267–278, 2016.

[21] F. T. Sun and M. J. Morrell, "The rns system: responsive cortical stimulation for the treatment of refractory partial epilepsy," *Expert review of medical devices*, vol. 11, no. 6, pp. 563–572, 2014.

[22] N. Sisterson, T. A. Wozny, and V. K. et al., "A rational approach to understanding and evaluating responsive neurostimulation," *Neuroinform 18*, pp. 365–375, 2020.

[23] https://www.verywellhealth.com/responsive-neurostimulation-for-epilepsy-4176500.

[24] M. Shoaran, M. Farivar, and A. Emami, "Hardware-friendly seizure detection with a boosted ensemble of shallow decision trees," in *2016 38th Annual International Conference of the IEEE Engineering in Medicine and Biology Society (EMBC)*, pp. 1826–1829, IEEE, 2016.

[25] A. Van de Vel, K. Cuppens, B. Bonroy, M. Milosevic, K. Jansen, S. Van Huffel, B. Vanrumste, P. Cras, L. Lagae, and B. Ceulemans, "Non-eeg seizure detection systems and potential sudep prevention: State of the art: Review and update," *Seizure*, vol. 41, pp. 141 – 153, 2016.

[26] M. Nasseri, E. Nurse, M. Glasstetter, S. Böttcher, N. M. Gregg, A. Laks Nandakumar, B. Joseph, T. Pal Attia, P. F. Viana, E. Bruno, A. Biondi, M. Cook, G. A. Worrell, A. Schulze-Bonhage, M. Dümpelmann, D. R. Freestone, M. P. Richardson, and B. H. Brinkmann, "Signal quality and patient experience with wearable devices for epilepsy management," *Epilepsia*, vol. n/a, no. n/a.

[27] P. Lentini, "Designing user experience in ehealth applications for young-age epilepsy," 2022.

[28] S. Tiwari, V. Sharma, M. Mujawar, Y. K. Mishra, A. Kaushik, and A. Ghosal, "Biosensors for epilepsy management: state-of-art and future aspects," *Sensors*, vol. 19, no. 7, p. 1525, 2019.

[29] S. Hooper, E. Biegert, M. Levy, J. Pensock, L. van der Spoel, X. Zhang, T. Zhang, N. Tandon, and B. Aazhang, "On developing an fpga based system for real time seizure prediction," in *2017 51st Asilomar Conference on Signals, Systems, and Computers*, pp. 103–107, 2017.

[30] S. M. Kueh and T. J. Kazmierski, "Low-power and low-cost dedicated bit-serial hardware neural network for epileptic seizure prediction system," *IEEE Journal of Translational Engineering in Health and Medicine*, vol. 6, pp. 1–9, 2018.

[31] R. G. Andrzejak, K. Lehnertz, F. Mormann, C. Rieke, P. David, and C. E. Elger, "Indications of nonlinear deterministic and finite-dimensional structures in time series of brain electrical activity: Dependence on recording region and brain state," *Phys. Rev. E*, vol. 64, p. 061907, Nov 2001.

[32] C. Tahar, D. I. Maulana, R. R. S. Pandia, and T. Adiono, "Fpga implementation of deep neural network for wearable pre-seizure detector on epileptic patient," in *2019 16th International Conference on Electrical Engineering/Electronics, Computer, Telecommunications and Information Technology (ECTI-CON)*, pp. 211–213, 2019.

[33] B. H. Brinkmann, J. Wagenaar, D. Abbot, P. Adkins, S. C. Bosshard, M. Chen, Q. M. Tieng, J. He, F. J. Muñoz-Almaraz, P. Botella-Rocamora, J. Pardo, F. Zamora-Martinez, M. Hills, W. Wu, I. Korshunova, W. Cukierski, C. Vite, E. E. Patterson, B. Litt, and G. A. Worrell, "Crowdsourcing reproducible seizure forecasting in human and canine epilepsy," *Brain*, vol. 139, pp. 1713–1722, 03 2016.

[34] E. Coşgun and A. Çelebi, "Fpga based real-time epileptic seizure prediction system," *Biocybernetics and Biomedical Engineering*, vol. 41, no. 1, pp. 278–292, 2021.

[35] W.-M. Chen, H. Chiueh, T.-J. Chen, C.-L. Ho, C. Jeng, M.-D. Ker, C.-Y. Lin, Y.-C. Huang, C.-W. Chou, T.-Y. Fan, *et al.*, "A fully integrated 8-channel closed-loop neural-prosthetic cmos soc for real-time epileptic seizure control," *IEEE Journal of Solid-State Circuits*, vol. 49, no. 1, pp. 232–247, 2013.

[36] C. Zhang, M. A. B. Altaf, and J. Yoo, "Design and implementation of an on-chip patient-specific closed-loop seizure onset and termination detection system," *IEEE journal of biomedical and health informatics*, vol. 20, no. 4, pp. 996–1007, 2016.

[37] M. Shoaran, B. A. Haghi, M. Taghavi, M. Farivar, and A. Emami-Neyestanak, "Energy-efficient classification for resource-constrained biomedical applications," *IEEE Journal on Emerging and Selected Topics in Circuits and Systems*, vol. 8, no. 4, pp. 693–707, 2018.

[38] Z. S. Zaghloul and M. Bayoumi, "Early prediction of epilepsy seizures vlsi bci system," *arXiv preprint arXiv:1906.02894*, 2019.

[39] C. Lammie, W. Xiang, and M. R. Azghadi, "Towards memristive deep learning systems for real-time mobile epileptic seizure prediction," in *2021 IEEE International Symposium on Circuits and Systems (ISCAS)*, pp. 1–5, IEEE, 2021.

[40] P. S. Thorbole, S. D. Kalbhor, V. K. Harpale, and V. Bairagi, "Hardware implementation of genetic algorithm for epileptic seizure detection and prediction," in *2017 International Conference on Computing, Communication, Control and Automation (ICCUBEA)*, pp. 1–5, IEEE, 2017.

[41] H. Wang, W. Shi, and C.-S. Choy, "Hardware design of real time epileptic seizure detection based on stft and svm," *IEEE Access*, vol. 6, pp. 67277–67290, 2018.

[42] M. A. B. Altaf, C. Zhang, and J. Yoo, "A 16-channel patient-specific seizure onset and termination detection soc with impedance-adaptive transcranial electrical stimulator," *IEEE Journal of Solid-State Circuits*, vol. 50, no. 11, pp. 2728–2740, 2015.

[43] C. Tsou, C.-C. Liao, and S.-Y. Lee, "Epilepsy identification system with neural network hardware implementation," in *2019 IEEE International Conference on Artificial Intelligence Circuits and Systems (AICAS)*, pp. 163–166, IEEE, 2019.

[44] T. Rukasha, S. I Woolley, T. Kyriacou, and T. Collins, "Evaluation of wearable electronics for epilepsy: A systematic review," *Electronics*, vol. 9, no. 6, p. 968, 2020.

[45] S.-A. Huang, K.-C. Chang, H.-H. Liou, and C.-H. Yang, "A 1.9-mw svm processor with on-chip active learning for epileptic seizure control," *IEEE Journal of Solid-State Circuits*, vol. 55, no. 2, pp. 452–464, 2019.

[46] N. D. Truong, A. D. Nguyen, L. Kuhlmann, M. R. Bonyadi, J. Yang, S. Ippolito, and O. Kavehei, "Convolutional neural networks for seizure prediction using intracranial and scalp electroencephalogram," *Neural Networks*, vol. 105, pp. 104–111, 2018.

[47] "European Epilepsy Dataset. EPILEPSIAE project, 2020, http://epilepsy-database.eu/,."

[48] R. Rahman, S. M. Varnosfaderani, O. Makke, N. J. Sarhan, E. Asano, A. Luat, and M. Alhawari, "Comprehensive analysis of eeg datasets for epileptic seizure prediction," in *2021 IEEE International Symposium on Circuits and Systems (ISCAS)*, pp. 1–5, 2021.

[49] F. Forooghifar, A. Aminifar, and D. Atienza, "Resource-aware distributed epilepsy monitoring using self-awareness from edge to cloud," *IEEE transactions on biomedical circuits and systems*, vol. 13, no. 6, pp. 1338–1350, 2019.

[50] F. Forooghifar, A. Aminifar, and D. A. Alonso, "Self-aware wearable systems in epileptic seizure detection," in *2018 21st Euromicro Conference on Digital System Design (DSD)*, pp. 426–432, IEEE, 2018.

Parkinson's Disease Detection Using Voice Features and Machine Learning Algorithms

Rumana Islam[1]
Dept. of Elect. and Comp. Engg.
University of Windsor,
Windsor, ON, Canada
islamq@uwindsor.ca

Esam Abdel-Raheem[2]
Dept. of Elect. and Comp. Engg.
University of Windsor,
Windsor, ON, Canada
eraheem@uwindsor.ca

Mohammed Tarique
Dept. of Elect. and Comp. Engg.
Univ. of Sci. and Tech. of Fujairah,
Fujairah, UAE
m.tarique@ustf.ac.ae

Abstract—This paper investigates the noninvasive screening for early signs of Parkinson's disease from voice signals using machine learning algorithms. It considers 752 audio features extracted from phonation of the sustained vowel '/a/' sound. Two machine learning algorithms, namely k-nearest neighbors (kNNs) and support vector machines (SVMs), are modeled to identify their effectiveness for the classification task. The results showed that cubic SVM and fine kNN algorithms could achieve 100% accuracy. However, the optimized kNN outperformed the optimized Gaussian SVM in terms of accuracy, detection speed, training time, and misclassification cost.

Keywords—Accuracy, classifier, machine learning, Parkinson's disease, voice features.

I. INTRODUCTION

Parkinson's disease (PD) is a progressive neurodegenerative disorder that affects the nervous system and the parts of the body controlled by the nerves. According to World Health Organization (WHO), global disability and death due to PDs are increasing faster than any other neurological disorder. The prevalence of PDs has doubled in the past 25 years.

PD was initially reported by Parkinson in his pioneer work [1]. In PD, specific brain nerve cells (neurons) gradually break down and fail to produce a chemical messenger called 'dopamine'. The consequence can be atypical brain activity, leading to impaired movement and other symptoms, including tremors, slowed movement, rigid muscles, and automatic, unconscious motion [2]. Not least, PD also causes changes in speech. The PD patients speak softly and quickly. They even slur and hesitate during speaking. Also, the speech becomes monotone rather than the usual speech. A combination of several symptoms can raise suspicion of PD. A clinical investigation of PD patients revealed that 89% of PD patients suffer from voice disorders [3]. Among them, 45% experience articulatory impairment, and 20% struggle with fluency. The authors also made a similar conclusion in another survey work [4]. They concluded that voice is the leading deficit, most frequently affected, and impaired to a greater extent at the initial stages of PDs. Hence, the generated speech samples contain useful information to detect PDs at their early stages.

Researchers have extensively used voice samples to detect certain diseases, including PDs, for a decade [5-8]. In [3], the authors have used 22 voice features and three machine learning algorithms, SVM, kNN, and discrimination function base (DBF), to identify PD patients. The highest accuracy achieved was 93.82% with the kNN algorithm. A new voice feature called pitch period entropy (PPE) has been used in [4] to detect PD. It was proved that this feature is robust to many uncontrollable confounding effects, including noisy acoustic environments. An artificial neural network (ANN) based PD detection system was designed in [8] from 44 audio features from voiced ('/a/') samples. Invasive audio signal processing technique could even detect specific vocal disease as demonstrated in [9]. A speech signal processing-based remote PD detection system has been introduced in [10]. Voice recordings of the PD patients using the Intel AHTD device could determine the progression level of PD in the patients. The authors achieved the best accuracy of 95.9% in [11] by applying a feature subset and four different classifiers.

A novel voice feature, the tunable Q-factor wavelet transform (TQWT) of voice signals, was introduced in [12]. The authors combined the Mel frequency cepstral coefficients (MFCCs) and TQWT features to detect PD. Four independent classification schemes, namely neural network, DMNeural, regression, and decision tree, were used in [13] to detect PD. They achieved the highest accuracy of 92.9% with the neural network. In [14], the authors selected the optimum voice features using a genetic algorithm. Once the optimum voice features were chosen, they employed a kNN for the classification. The highest accuracy achieved was 98.2%, with nine optimized voice features. In a similar work [15], the authors also used multiple classifiers, namely regression tree, Artificial neural network (ANN), and SVM, to detect PD. A decision support system for detecting PD was presented in [16], where the proposed method used ANN and SVM to aid specialists in diagnosing PD. Highly correlated speech features were integrated using principal component analysis (PCA) and linear discriminant analysis (LDA) in [17] to identify PD. The effect of selecting MFCC and TQWT on the efficiency of the voice-based PD detection algorithm was analyzed in [18]. The authors used several machine learning algorithms and achieved the highest accuracy with XGBoost. In [19], the authors proposed a novel multiple-feature evaluation approach (MFEA) to improve the accuracy of a PD detection algorithm. The improvement in the detection accuracy was 9.13% -15.22%.

This paper maneuvers multidimensional voice features and machine learning algorithms to investigate their performance in identifying PD patients. The main contributions of the proposed algorithms are as follows: (a) it employs multiple acoustic features in different domains to train the machine learning algorithms, (b) it investigates two machine learning

---

[1] Adjunct Faculty at the University of Science and Technology of Fujairah, Fujairah, UAE
[2] Visiting Scholar at the University of Sharjah, Sharjah, UAE from Sept. to Dec.2023

979-8-3503-8083-5/23 $31.00 © 2023 IEEE

algorithms to compare and correlate their contribution to PD identification, (c) furthermore it optimizes detection algorithms to improve the classification cost with Bayesian's rule, and (d) it provides a detailed performance analysis of the proposed system in terms of the statistical classification measures and model performance metrics. The rest of the paper is arranged as follows. Materials and methods are presented in section II. Section III explains simulation results with a discussion. And the paper is concluded with Section IV.

II. MATERIALS AND METHODS

The voice features used in this work are collected from the database provided by UCI Machine Learning Data Repository [20]. The features were collected from voice samples of 188 PD patients (107 males and 81 females) aged 33-87 years at the Department of Neurology, Faculty of Medicine, Istanbul University, Cerrahpasa. The control group comprises 64 healthy individuals (23 men and 41 women) in the age group of 41-82. During the data collection process, the microphone was set to 44.1 kHz. The sustained phonation of the vowel '/a/' was recorded from each subject with three (3) repetitions. So, seven hundred and fifty six (756) samples (192 healthy and 564 PDs) are considered for this investigation. The seven hundred fifty-two (752) voice features demonstrated in the database are classified as (a) baseline features, (b) intensity, (c) formant frequencies, (d) bandwidth, (e) vocal fold parameters, (f) MFCCs, (g) wavelet, and (h) TQWT. The distribution of these features is shown in Fig. 1. The design of machine learning algorithms, namely kNNs and SVMs, is conducted using MATLAB 2020. The system model is shown in Fig.2.

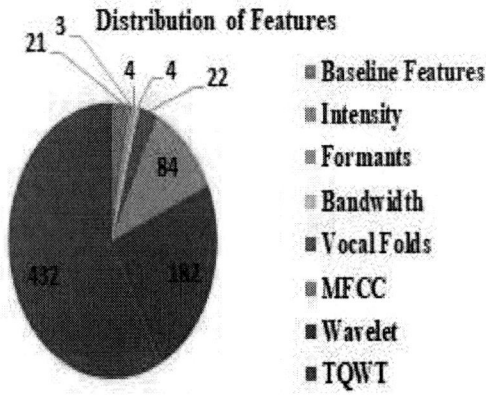

Fig. 1. The distribution of the voice features.

A. Feature matrix

The adopted database considered different signal processing and statistical based algorithms to extract clinically useful features from voice recordings of PD patients and healthy subjects. The baseline features considered are jitters, shimmers, fundamental frequencies, harmonicity, recurrence period density entropy (RPDE), detrended fluctuation analysis (DFA), and pitch period entropy (PPE). Minimum, maximum, and mean intensities are sub-grouped as intensity features. Four formants constitute the formants subgroup. Also, four bandwidths are considered for bandwidth features. The vocal fold parameters include GQs (glottal quotients), GNEs (glottal to noise excitation ratios), VFERs (vocal fold excitation ratios), and IMFs (intrinsic mode functions) of voice samples. Several spectral-based features, 13 MFCCs, and their delta, delta-delta coefficients are measured by processing voice signals with Mel filterbank. The wavelet coefficients were extracted from the voice signals' discrete wavelet transform (DWT). After wavelet decomposition, the energy, Shanon's entropy, log energy entropy, and Teager-Kaiser energy were computed. The TQWT features were obtained from the over-complete wavelet transforms. The Q factor of the bank of bandpass filters was tuned according to the oscillatory nature of the voice signal. Besides TQWT coefficients, each level's energy and entropy measures are also considered to form a significant part of the feature matrix.

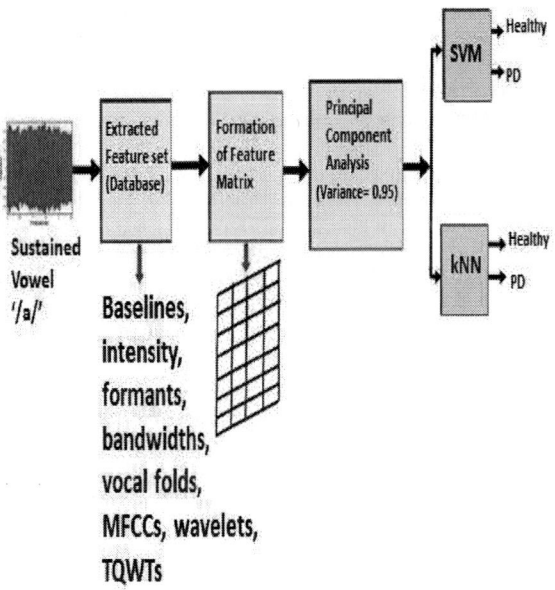

Fig. 2. The system model employing data reduction technique (PCA) and classification algorithms (SVM and kNN) to identify PD patients from the healthy ones with predominant voiced ('/a/') features.

B. PCA

High dimensional data are difficult to generalize and may lead to overfitting for the classification performance. PCA linearly transforms the feature vectors to remove redundancy in dimensions, thus generates a new set of principal components. To reduce the dimension of the feature matrix, PCA with 0.95 variances is used in this work. The correlated features are basically transformed to a set of uncorrelated features, well known as principal components. The generated first principal component explains the maximum variance of the feature set. The others are in descending order of variance. In this work, the principal components of individual category of features as described above are considered as the input for the designed classifiers.

C. SVM

SVM applies the statistical concept of support vector to classify data. It identifies a linear decision boundary that separates data into two categories. Let the linear decision boundary be defined by $W^T X + b = 0$, where X is the feature column vector of size $n \times 1$, W is the column vector of size $n \times 1$, b is a scalar constant. Considering two more parallel planes represented by $W^T X + b = 1$ and $W^T X + b = -1$, SVM obtains the optimal value for the W and b such that the following conditions are satisfied: (a) the region $W^T X + b > 1$ belongs to class I, (b) the region $W^T X + b < -1$ belongs to class II, and (c) the shortest Euclidian distance between two planes $W^T X + b = 1$ and $W^T X + b = -1$ is maximized. The shortest distance between two arbitrary points on $W^T X + b = 1$ and $W^T X + b = -1$ can be found as $d_{min} = \frac{2b}{W^T W}$ [21]. SVM optimizes the vectors W and the scalar b such that the distance, $d_{min} = \frac{2b}{W^T W}$, is maximum. However, it is difficult to identify the linear separation plane in the dimensional feature space formed by the investigated feature matrix mentioned above. Implementation of SVM with polynomial input transformation can result high generalization ability even in the case of non-separable training data [22]. Hence, 'kernel trick' is used in this work to map the feature matrix to higher dimension space (HDS) for better separation. This work uses several kernels: linear, quadratic, cubic, and gaussian for the classification tasks.

D. kNN

To avoid the implicit assumption of linearity in the SVM, the kNN algorithm has been introduced. Among the various methods of supervised statistical pattern recognition, the kNN achieves consistently high performance without a priori assumptions about the distributions of the training samples. It involves a training set of both diseased and healthy cases. A new sample is classified by calculating the distance to the nearest training case. If the Euclidean distance is used as the distance metric, this distance can be defined by

$$d(x, x') = \sqrt{(x_1 - x_1')^2 + \cdots + (x_n - x_n')^2} \qquad (1)$$

Finally, the data is assigned to a class with the largest probability defined by

$$\Pr(Y = j | X = x) = \frac{1}{k} \sum_{i \in A} I(y_i = j) \qquad (2)$$

where X = feature matrix, Y = class label, A = set of k-nearest observations, $I(y_i = j)$ evaluates 1 if an observation (x_i, y_i) in A is a member of class j and 0 otherwise. This algorithm determines the class of an unspecified point by counting the majority class votes from its kNN training points. The main advantage of the kNN is that it can generate a highly precise decision boundary compared to the SVM, as it is driven by the raw training data itself. However, for a given training data sets, the prediction may be less stable for the kNN compared to that of the SVM. The value of k is an important factor for the regularization of the kNN. The small value of k provides a more flexible boundary. On the other hand, the large value of k provides a more rigid boundary. It is common to select k small and odd to break ties (typically 1, 3, or 5). Larger k values help reduce the effects of noisy points within the training data set, and the choice of k is often determined through cross-validation.

III. THE RESULTS

The statistical performance of the proposed models is evaluated with the following parameters: (a) true positive (TP), (b) true negative (TN), (c) false negative (FN), and (d) false positive (FP). The performance measures are defined as follows:

$$Accuracy = \frac{TP + TN}{TP + FP + FN + TN} \qquad (3)$$

$$Precision = \frac{TP}{TP + FP} \qquad (4)$$

$$F1\ Score = \frac{2 * Recall * Precision}{Recall + Precision} \qquad (5)$$

$$Negative\ predictive\ value, NPV = \frac{TN}{TN + FN} \qquad (6)$$

$$Recall = \frac{TP}{TP + FN} \qquad (7)$$

$$Specificity = \frac{TN}{TN + FP} \qquad (8)$$

$$False\ negative\ rate, FNR = \frac{FN}{FN + TP} \qquad (9)$$

$$False\ detection\ rate, FDR = \frac{FP}{FP + TP} \qquad (10)$$

$$G-mean = \sqrt{Sensitivity * Specificity} \qquad (11)$$

The simulation results of the kNN algorithms are shown in Table I considering training, validation and testing of data samples. Eighty (80%) percent samples are used for training, the remaining 20% are equally divided between validation and testing purpose. Five (5) fold cross validation schemes are adopted to train the classifier. The performance measures are listed in Table II. An accuracy of 100% was achieved with the fine kNN (k =1). The other performance measures, precision, recall, F1 score, NPV, specificity, and G-mean, all equal 1. However, 100% accuracy is achieved with the optimized kNN as well, for $k's$ value of 2 with the distance metric Mahanalobis and distance weight squared inverse. The model's hyperparameters are tuned with the Bayesian's optimization rule. Though it is particularly suitable for low-dimensional data. The higher computational cost to process multidimensional PD speech samples is eliminated using this rule. The corresponding confusion matrix is shown in Fig.3. The model performance parameters: misclassification cost, prediction speed, and training time are listed in Table III.

The simulations are repeated for different SVM algorithms. The same number of samples are considered for training, validation, and testing purpose. Also, five (5) fold cross validation scheme is adopted to train the SVM model. The results are listed in Table IV. The cubic SVM provided

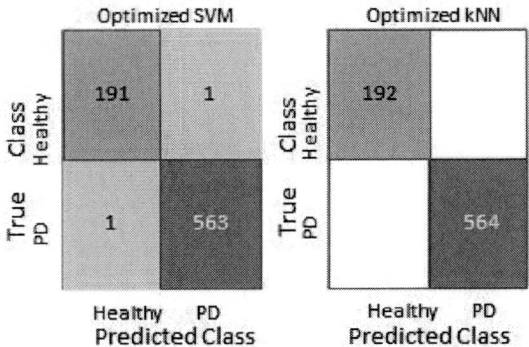

Fig. 3. The confusion matrices for optimized SVM and optimized kNN

separate classes, resulting in the highest prediction speed with the cost of significant misclassification. However, different kernels changed the shape of the decision manifold and provided better performances. The kNN algorithms performed better with smaller k values. Increasing the k value increases the number of neighbors, which leads to a decrease in performance because the chances of including a data point from a different class become higher with the increase of nearest neighbors for high dimensional data points. The optimized kNN resulted in a minimum classification error for $k = 2$. The estimated classification error for the optimized kNN model using the Bayesian's rule is shown in Fig. 4, demonstrating excellent performance. The best point for hyperparameters and the point for minimum error coincide with each other.

the best performance. It detected all PD patients and healthy subjects correctly. The classification performance measures are presented in Table V. The confusion matrix of the optimized SVM is shown in Fig. 3. The model performance of the investigated SVM algorithms is listed in Table VI. The optimized SVM provided 99.7% accuracy with the Gaussian kernel function.

TABLE I: The Simulation Results of the kNN Algorithms

kNN Classifier	TP	TN	FP	FN	k	Distance metric
Fine	564	192	0	0	1	Euclidean
Medium	550	122	70	14	10	Euclidean
Cosine	539	140	52	25	10	Cosine
Cubic	542	93	99	22	10	Cubic
kNN optimized	564	192	0	0	2	Mahalanobis

TABLE II: The Performance measures of the kNN Algorithms

Measures	Fine kNN	Medium kNN	Cosine kNN	Cubic kNN	kNN optimized
Accuracy (%)	100	88.9	89.8	84	100
Precision	1	0.88	0.91	0.85	1
Recall	1	0.97	0.96	0.96	1
F1 Score	1	0.93	0.93	0.90	1
NPV	1	0.89	0.85	0.81	1
Specificity	1	0.63	0.73	0.48	1
FNR	0	0.03	0.04	0.04	0
FDR	0	0.10	0.15	0.19	0
G-mean	1	0.79	0.83	0.68	1

TABLE III: The Performance parameters of the kNN Algorithms

kNN Classifier	Misclassification cost (nos)	prediction speed (obs/sec)	Training time (sec)
Fine	0	1000	2.9171
Medium	84	1000	3.4003
Coarse	164	1000	3.3002
Cosine	77	1000	3.1907
Cubic	121	60	41.378
kNN optimized	0	19000	44.260

Comparing the data presented in Table I-III and Table IV-VI, we can conclude that the optimized kNN algorithm outperformed the SVM algorithm in terms of all performance measures, including misclassification cost, prediction speed, and training time. The basic SVM uses linear hyperplanes to

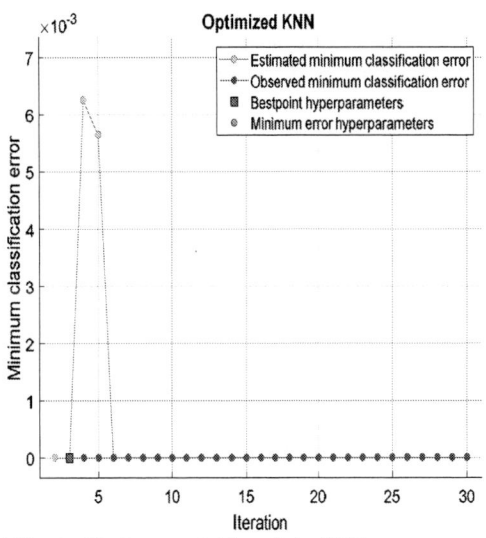

Fig. 4. The classification error plot for optimized kNN.

TABLE IV: The Simulation Results of the SVM Algorithms

SVM Classifier (Kernel Function)	TP	TN	FP	FN
Linear	558	128	64	6
Quadratic	564	184	8	0
Cubic	564	192	0	0
Medium Gaussian	560	133	59	4
Optimized (Gaussian)	563	191	1	1

TABLE V: The performance measures of the SVM Algorithms

Measures	Linear	Quadratic	Medium Gaussian	Cubic	Optimized (Gaussian)
Accuracy (%)	90.70	98.90	91.70	100	99.70
Precision	0.90	0.99	0.90	1	0.99
Recall	0.99	1	0.99	1	0.99
F1 Score	0.94	0.99	0.95	1	0.99
NPV	0.96	1	0.97	1	0.99
Specificity	0.67	0.96	0.69	1	0.99
FNR	0.01	0	0.01	0	0.001
FDR	0.04	0	0.03	0	0.001
G-mean	0.81	0.98	0.83	1	0.99

TABLE VI: THE PERFORMANCE PARAMETERS OF THE SVM ALGORITHMS

SVM (Kernel Functions)	Misclassification cost (nos)	prediction speed (obs/sec)	Training time (sec)
Linear	70	5700	2.7251
Quadratic	8	2300	5.6050
Cubic	0	2500	5.4518
Medium Gaussian	63	4300	5.2745
Optimized (Gaussian)	2	18000	71.205

IV. CONCLUSION

As a neurodegenerative disorder, PD is hard to diagnose at its early stage. PD patients sometimes have significant voice disorder that often happens earlier than dyskinesia. A wide variety of acoustic features in multiple domains from voice samples are considered to optimize both kNN and SVM classifiers to identify PD patients. It can be concluded that the features with high discriminative power and robustness from voice samples can identify PDs with excellent performance. Even the problem of imbalanced data samples for diseased and healthy classes (192 healthy and 564 PDs) is overcome with the selection of robust voice features and the design of a proper machine learning algorithm. This paper only focuses on the detection of PDs. The progression level of the PDs needs to consider in the future. The contribution of feature matrix in different domains and gender consideration for classification performance is left as future work.

REFERENCES

[1] J. Parkinson, "An Essay on Shaking Pulsy," *J. of Neuropsychiatry Clinical Neuroscience*, vol. 14, no. 2, pp. 223-236, May 2002.

[2] John Hopkins Medicine, "Parkinson's Symptoms," available at https://www.hopkinsmedicine.org/health/conditions-and-diseases/parkinsons-disease/parkinsons-symptoms accessed on July 31, 2022.

[3] J. A. Logemann, H. B. Fisher, B. Boshes, E. R. Blonsky "Frequency and Cooccurence of Vocal Tract Dysfunction in the Speech of a large samples of Parkinson's Patients," *J. of Speech and Hearing Disorders*, vol. 43, no. 1, pp. 47-57, February 1978.

[4] A. K. Ho, R. Iansek, C. Marigliani, J. L. Bradshaw, S. Gates, "Speech impairment in a large sample of patients with Parkinson's disease," *Behavioural Neurology*, vol. 11, no. 3, pp. 131-137, January 1999.

[5] R. Islam, M. Tarique and E. Abdel-Raheem,"A survey on signal processing based pathological voice detection techniques," *IEEE Access*, vol. 8, pp. 66749–66776, April 2020.

[6] R. Islam, E. Abdel-Raheem, and M. Tarique,"A study of using cough sounds and deep neural networks for the early detection of Covid-19," *Biomedical Engineering Advances*, vol. 3, 100025, June 2022.

[7] H. Karimi-Rouzbahani and M. Daliri, "Diagnosis of Parkinson's Disase in Human using Voice Signals," *Basic and Clinical Neuroscience*, vol. 2, no. 3, pp. 12-20, Februray 2011.

[8] R. Islam, E. Abdel-Raheem, and M. Tarique, "Voiced Features and Artificial Neural Network to Diagnose Parkinson's Disease Patients," *Proceedings of the International Conference on Electrical and Computing Technologies and Applications*, November 23-25, 2022, American University of Ras Al Khaimah, United Arab Emirates (UAE).

[9] R. Islam, E. Abdel-Raheem, and M. Tarique, "A Novel Pathological Voice Identification Technique through Simulated Cochlear Implant Processing Systems," *Applied Sciences*, vol. 12, no. 5, pp. 1-21, February 2022.

[10] A. Tsanas, M. A. Little, P. E. McSharry, L. O. Ramig, "Accurate Telemonitroing of Parkinson's Disease Progression by Noninvasive Speech Tests," *IEEE Transcation on Biomdical Engineering*, vol. 57, no. 4, pp. 884-893, April 2010.

[11] G. Solana-Lavalle and R. Rosa-Romero, "Analysis of voice as an assisting tool for detection of Parkinson's disease and its subsequent clinical interpretation," *Biomedical Signal Processing and Control*, vol. 66, 102415, April 2021.

[12] C. O. Sakar, G. Serbes, A. Gunduz et al. "A comparative analysis of speech signal processing algroithms for Parkinson's disease classification and the use of tunable Q-factor wavelet transform," *Applied Soft Computing*, vol. 74, pp. 255-263, January 2019.

[13] R. Das, "A comparison of multiple classifications methods for diagnosis of Parkinson's disease," *Expert Systems with Application*, vol. 37, no. 2, pp. 1568-1572, March 2010.

[14] R. A. Shirvan and E. Tahami, "Voice Analysis for Detecting Parkinson's Disease using Genetic Algorithm and KNN Classification methods," *Proceedings of the 18th Conference on Biomedical Engineering*, December 14-16, 2011, Tehran, pp. 278-284.

[15] Z. K. Senturk, " Early diagnosis of Parkinson's dusease using machine learning algorithms," *Medical Hypothesis*, vol. 138, 109603, May 2020.

[16] A. D. Gil and B. M. Johnson, "Diagnosing Parkinson by using Artificial Neural Networks and Support Vector Machine," *Global Journal of Computer Science and Technology*, vol. 9, no. 4, pp. 63-71.

[17] C. D. Anisha and N. Arulanand, "Early Detection of Parkinson's disease (PD) using Enseble Classifiers," *Proceedings of the International Conference on Innovative Trends in Information Technology*, Kottayam , India, February 13-14, 2020.

[18] I. Nissar, D. R. Rizvi, S. Masood, A. N. Mir, "Voice Based Detection of Parkinson's Disease through Ensemble Machine Learining Approach: A Performnace Study," *EAI Endorsed Transactions on Pervasive Health and Technology*, vol. 5, no. 19, pp. 1-8, August 2019.

[19] A. Salama, A. Mustapha, M. A. Mohammed et al., "Examining multiple feature evalutaion and classification methods for improving the diagnosis of Parkinson's Disease," *Congintive System Research*, vol. 54, pp. 90-99, May 2019.

[20] Parkinson's Data available at https://archive.ics.uci.edu/ml/datasets/Parkinsons, accessed on August 3,2022

[21] E. S. Gopi, Digital Speech Processing Using Matlab, Springer, New York, pp. 10-23.

[22] C. Cortes, and V. Vapnik, "Support-Vector Networks," *Machine Learning*, vol. 20, no. 3, pp. 273-297, 1995.

EEG experiment flow control using FPGA as an alternative to commercial devices

1st Cristian Y. Olivares, MSc.
Electrical and Electronics Engineering Department
Universidad Del Norte
Barranquilla, Colombia
ocristian@uninorte.edu.co

2nd Norelli Schettini, PhD.
Electrical and Electronics Engineering Department
Universidad Del Norte
Barranquilla, Colombia
nschettini@uninorte.edu.co

Abstract—Traditionally, Electroencephalography (EEG) experiments requiring precise timestamp analysis of a subject's exposure to stimuli involve the inclusion of timestamp marks within EEG records. This is typically achieved through the use of EEG amplifiers capable of receiving external triggers and synchronizing them with the EEG data. However, such equipment is often sourced as a compatible accessory from either third-party vendors or acquired separately from the original vendor after-sales, incurring in significant additional costs. This paper describes in detail the methodology and considerations taken into account for the design and cost-effective construction of an EEG signal-stimuli synchronization equipment, compatible with EEG devices equipped to record external triggers, making the development of such systems more accessible and affordable. Furthermore, it showcases a successful case study involving participants exposed to auditory stimuli, where stimuli-coherent timestamps, referred to as "onset triggers", were used. This data configuration is in high demand in neuropsychology research and our work demonstrates its attainability through the development and integration of custom-engineered components. These components are not only adaptable but are also open to enhancements and low-cost replication. Finally, a discussion about the significance of this alternative approach is presented, particularly for emerging neuropsychology and brain researchers exploring new paradigms through experimental setups that require flexibility and higher data manipulation capabilities.

Index Terms—EEG, Timestamp, Synchronization, Cost-effective, FPGA, Neuropsychology

I. INTRODUCTION

In EEG-based experiments, there are typically two types of analyses conducted on the recorded data: spectral analysis and temporal analysis. Both are essential for understanding brain activity patterns and their relationship to external events or stimuli. Accurate synchronization of EEG data with event timing is crucial for meaningful research, particularly in auditory experiments [1]. To minimize latency between brain activity and stimulus markers, EEG equipment often employs internal clocks. However, to ensure consistent synchronization across all devices in the EEG setup, it is advisable to use third-party clock-locking devices [2]. It's worth noting that the cost of such synchronization equipment can be substantial. These devices, which play a critical role in managing the flow of EEG experiments can be sourced as a compatible accessory from either third-party vendors or acquired separately from the original vendor after-sales, often incurring in additional

expenses that are frequently underestimated and overlooked during budget planning.

For instance, integrating audio signals, visual stimuli presentation, and response switch signals with EEG recordings requires the use of supplementary accessories and internal amplifier modules. These extra components represent a significant cost beyond the initial investment, particularly when considering additional expenses like taxes, shipping, handling fees from vendors.

The conventional approach for conducting EEG experiments with effective flow control involves the utilization of times-tamps that denote the precise moment when an event transpires. These experiments, often referred to as "phase-locking experiments" [3], primarily revolve around stimuli presented to participants. The objective is to establish a correlation between brain activity and the response to these stimuli, with a focus on quantifying the time it takes for participants to react in terms of electrical activity.

For example, in experiments related to steady-state visual evoked potentials (SSVEP) or P300 [4], [5], subjects are exposed to continuous flashes oscillating at specific frequencies [6]. These flashes are captured using the visual signal through a photo transistor positioned on the screen's target. In the case of SSVEP analysis, the goal is to extract phase-locking information from the brain signals in response to the presented stimuli. Meanwhile, P300 and other potential analyses follow a similar approach but focus on detecting desynchronization in the electrical brain activity triggered by anticipated and unanticipated stimuli-related prompts [7]. This analysis relies on time-locked assessments performed on the EEG signal.

Another case of use for stimuli timestamps is Auditory Brainstem Response (ABR) analysis. ABR signals encompass recordings of electrical activity near the site where the initial stages of the hearing process occur [8], [9], this is the pathway from the cochlea going through the brain till reaching the auditory cortex. To capture ABR signals effectively, the standard approach involves subjecting the individual to a continuous burst of auditory stimuli while concurrently recording both the electrical brain activity and the timestamps associated with the stimuli. Given the rapid nature of these responses, the latency between successive stimuli can be exceptionally short, sometimes as brief as 8 ms [9].

The easiest way to get the flow information during the EEG experiment is to use the tools provided by the EEG device manufacturer, in some cases the tools provided can be software-based, which is less precise than their hardware-based counterparts. In [10] a brief description of four commercially-oriented EEG systems shows that essentially these equipment were capable of recording external marks called "triggers" from external sources. However, it is noteworthy that the cost associated with procuring the proprietary hardware necessary for using these trigger inputs can be prohibitively high. This challenge becomes particularly pronounced in settings such as countries, hospitals, research centers, and universities, where research budgets are often constrained. Consequently, many researchers find themselves equipped with the essential EEG measurement equipment but lacking the accessories that would enable better flow control or the acquisition of reliable triggers during experiments.

Furthermore, even in cases where researchers have access to all these resources, the inherent third-party ownership and the fixed configurations of commercial EEG systems can pose challenges when attempting to tailor datasets to specific research requirements. In some instances, achieving this customization may be extremely challenging or virtually unattainable.

However, an examination of technical specifications for commercial EEG devices reveals that they predominantly rely on Field Programmable Gate Arrays (FPGAs) configured to manage flow control, often in conjunction with open-source communication protocols. The FPGA devices are based on semiconductor arrays organized in a matrix. These semiconductors can be interconnected using hardware description languages (HDL), with the most common being Verilog and Very High-Speed Integrated Circuit Hardware Description Language (VHDL). Each block in the matrix array is referred to as a logic block, and the quantity of these blocks is specified by the manufacturer. The programmable interconnection feature allows for high-speed switching and provides a high level of flexibility to perform various tasks simultaneously. These features also enable achieving low latency and a jitter close to 0 seconds when the FPGA is used to process signals, as in this work, where it emits triggers for EEG experiments. Having latency and jitter close to 0 seconds facilitates the proper approach to analysis, such as the common average over trials, as in the P300 paradigm, and increases the likelihood of obtaining relevant features in single-trial analysis.

This paper presents in detail the methodology and considerations taken into account for designing a custom-made synchronization device that can replace commercial proprietary equipment, utilizing an FPGA. This offers an opportunity for the construction of highly flexible data sets that confer higher data manipulation capabilities and challenges the notion of unaffordable costs associated with conducting neuro-scientific research.

II. USES OF TRIGGERS IN EEG EXPERIMENTS

The literature offers a variety of cases of use for FPGA-base devices in EEG experiments, from their uses as filters for EEG signals [11] to entire EEG signals processing units [12]. However, in most cases, the usage of FPGA devices is highlighted due to their capabilities to be used in real-time applications [13]. This shows that the use of FPGAs is not alien to EEG research, however, using them as a middleman to add additional information to the stream of data captured by specialized EEG equipment is not common, at least in DIY custom developments. Commercial devices are used to carry out these tasks in EEG experiments and their manufacturers agree that the principal source of error in timing analysis is the source of the stimuli [14], and the most straightforward way to correct it is by using a third-party device that monitors changes in the stimuli presented to the participant and their induced response. These third-party devices use Time-To-Live (TTL) lines and TTL signals, which are the standard signals on logical devices such as FPGAs.

The most common commercial devices for following EEG experiments are based on microprocessors, this is because the flexibility offered by the design of software-based tools. However, since the early 2000s, the development of FPGA boards as hardware-based counter tools that offer the same flexibility as software-based developments is more common, offering better performance in signal processing than the microprocessors-based solutions [15]. Nonetheless, either of the two alternatives used should be capable of monitoring the stimuli to which the participant is exposed to and their interfaces and ports must be compatible with the EEG equipment. The choice of stimuli for an EEG experiment depends on several factors. Typically, stimuli can be categorized as visual, auditory, or haptic. Thus, the primary considerations should include the average response time of the brain to the specific stimuli being used, the confidence level of the experimental results, and, in some cases, the sample rate of the EEG amplifier-recorder.

As indicated in [10], EEG equipment can vary in terms of sampling rates. Slower equipment may have a sampling rate starting at 128Hz, while faster equipment can go up to 2048Hz. This translates into the equipment's ability to detect changes in brain electrical activity as quickly as 7.8 ms down to 0.48 ms.

A. Visuals triggers

Frequencies under 30 Hz are typically the most common for Visual-based Brain-Computer Interfaces (BCI). However, it's worth noting that the signal-to-noise ratio within these frequency bands tends to be lower in comparison to higher-frequency signals, which can be around 50 Hz or higher [16]. When conducting experiments using this type of stimuli, it is crucial to maintain a latency between each trigger of no more than 1 ms, with jitter close to 0 ms, especially when using faster EEG devices.

To achieve this level of precision, the prevalent method for detecting changes in stimuli is by employing photo diodes or photo transistors as transducers for the stimuli to trigger conversion. These electronic devices exhibit rapid response times, typically in the order of nanoseconds. It is evident that the frequency of the stimuli is significantly lower than the sampling rate of the EEG equipment and even lower than the clock frequency of the FPGA.

B. Haptic triggers

Haptic triggers are employed to detect activities associated with the subject's movements. This application typically arises when, during the course of an experiment, the subject is instructed to press a button. In such experiments, the critical data revolves around measuring the time elapsed between the stimuli-related prompts and the subject's executed action. This action usually triggers a response switch, which, as reported by [17], can have a duration ranging from 150 to 220 ms.

C. Auditory triggers

The processing of auditory stimuli presents additional complexity due to the diverse nature of available stimuli. For instance, a single cue tone typically comprises a pure frequency component, which differs significantly from the complexity of natural speech. However, the analysis of the subject's elicited response begins with determining whether the EEG signal aligns with the presence of auditory stimuli, a phase known as the onset trigger.

In the case of rapid auditory brain responses, such as those associated with Auditory Brainstem Responses (ABRs), where the stimulus presentation rate can be as short as 4.35 ms [18], the equipment for detecting auditory stimuli must be equally fast. In the current landscape, commercially available and affordable Analog-to-Digital Converters (ADCs) with low response times and high-quality audio signal capabilities are commonplace. These ADCs can be effectively utilized for processing auditory stimuli and detecting sound onset through digital filtering, employing a threshold-based approach for sound detection.

As discussed earlier, each natural brain response, as reported in the state-of-the-art literature, is even slower than the more affordable FPGA. Therefore, considering the use of basic electronic circuits with suitable hardware descriptions for FPGA implementation, coupled with an appropriate communication protocol, presents a viable option for constructing a customized synchronization device capable of tracking the EEG experiment's flow.

In the subsequent section of this work, we present a proof of concept for the design and implementation of an electronic device based on FPGA. This device is tailored for monitoring an auditory experiment, focusing on achieving synchronization between auditory stimuli and the instances when the subject perceives the stimuli.

III. CASE OF USE: EEG SIGNAL DECODING OF AUDITORY STIMULI

As previously described, the most important factor in successfully searching for the relationship between the brain's electrical activity signal and the subject's behavior is knowing when the participant begins to perceive stimuli and/or the time relationship between both signals.

A successful case of use was developed in this work, using a BIOSEMI device as the EEG Equipment and a custom FPGA-based synching device. This new approach to exploring the use of FPGAs allowed building a tailor-made EEG dataset with 10 participants, and up to 540 repetitions per participant, with trigger marks for each event during the experiment, this is, at least 5400 marks that correspond to each stimulus presented to the participants. The experiment was based on decoding EEG signals recorded meanwhile the participants were exposed to auditory stimuli, as shown in Figure 1.

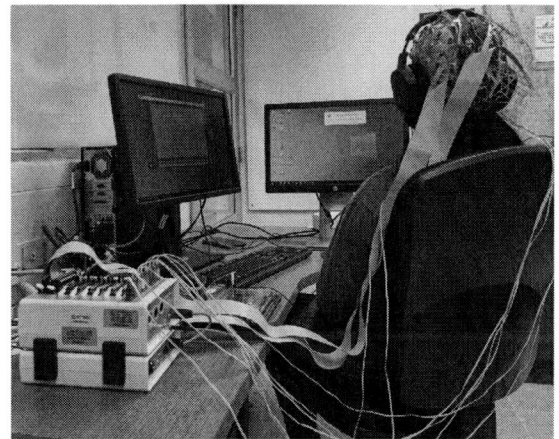

Fig. 1. Subject performing an experiment using the hardware design proposed.

In this proof of concept, the hardware was described using VHDL and tested using a general-purpose Altera DE2-115 FPGA. The device was used as a proxy between the subject and the PC that is conducting the experiment, as shown in Figure 2, and the described hardware was divided into two structures: The first structure received and sent the audio using the I2C Audio module converting the analog input delivered by the PC through their headphone output into a 24-bit digital signal to be compared with the reference value stored in a memory register of the FPGA; and the second structure, is a state machine designed to send the off code (0 in 8 bits) and set a rest time is activated to wait for the next experiment repetition (or the next control code).

The reference value was taken based on the value detected by the I2C module when a sound is played at half of the maximum volume recommended by the World Health Organization (WHO) for the continuous use of headphones. In the protocol, once the reference value is reached an internal flag is activated and a number between 1-255 in binary code (flow control code) is sent through 8 lines of the general-purpose input/output (GPIO) port to the BIOSEMI transceiver using their proprietary presentation cable (DB25 cable using a parallel protocol), as shown in Figure 3.

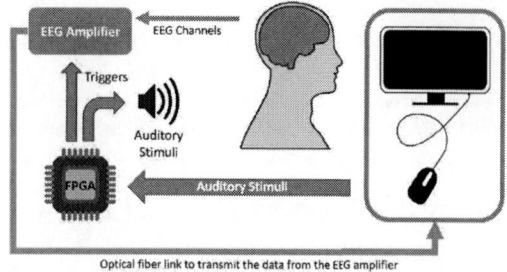

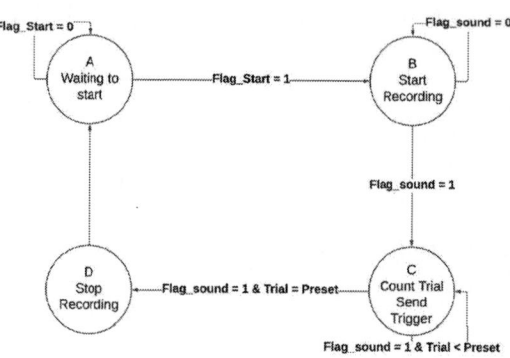

Fig. 2. Experiment setup based on the use of an FPGA as a proxy to figure out the flow control of the EEG experiment.

Fig. 4. State machine scheme.

Simultaneously, during the experiment, the digital signal perceived by the FPGA is sent out to the participant through the audio module headphone output, except, when the mute or reset button is pressed; these signals are asynchronous and were used to perform activities into the PC that produces the stimuli without interfering with the experiment.

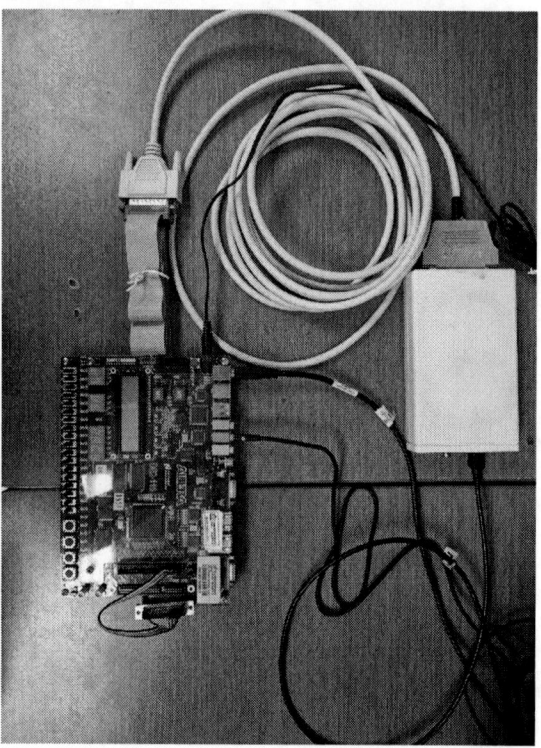

Fig. 3. Hardware connection using the DE2-115 FPGA as a proxy of the sound stimuli.

The hardware for the second structure follows a state machine design that uses the scheme presented in Figure 4, which operates as follows:

State A: Waiting for sound detection, this state has the only function of waiting for the first flag from the audio processing structure. During this state,

the number 0 is maintained in the 8 output lines, when the flag is received it sends the "start record" code to the BIOSEMI. The "start record" code must be defined in the configuration file of the EEG equipment.

State B: This state is in charge of monitoring the next flags from the audio processing structure, each time the audio processing structure detects a stimulus a "begin trial" code is sent to the BIOSEMI. After this occurs, the next state is activated.

State C: This state sends the "off" code and starts a rest timing clock for returning to state B, this state also stores the repetition number and compares it with the number of repetitions preset on the experiment setup, when this number matches with the preset repetition number then the next state must be the state D

State D: This state sends the "stop record" code to the BIOSEMI. The "stop record" code must be defined in the configuration file of the EEG equipment. Once the "stop record" code is sent, the state machine returns to state A.

Summarizing, the proposed FPGA-based device design requires a total of 613 logic elements, 31 I/O user ports mostly used for parallel communication with the EEG device used in the described experiment, along with a minimum of 12.3Kb of memory. Additionally, the FPGA must either adhere to the I2C or I2S standard, or an I2C or I2S logic block must be built if the chosen audio codec requires it. The selection between I2C and I2S is contingent upon the audio codec in use and the specific stimuli requirements.

In the tested design, a 24-bit embedded audio codec was used to receive and transmit the audio signals employed as stimuli for the participants. This audio codec established communication with the FPGA through the I2C protocol. The I2C interface was rigorously tested using auditory signals with distinct cue tones, maintaining consistent inter-tone intervals to detect any unpredictable latency. Although, it is worth noting that, due to the hardware's inherent characteristics, latency and jitter are not significant issues. This is attributed to the I2C VHDL controller implemented for communication, which possesses internal flags accessible to each component involved in the flow control of the EEG experiment. These flags allow the system to monitor and adjust for latency when an auditory stimulus is played.

In the proposed case of use, the BIOSEMI transceiver receives external triggers using parallel communication protocol; but due to the open Hardware/Software nature of this development, with some adjustments, it can be adapted to different interfaces, including EEG devices that rely on different communication protocols, such as serial or single TTL, as described in [10].

The EEG experiment conducted, using the hardware setup outlined above, proved to be highly successful. This innovative setup, built around non-commercial devices and utilizing a structure that can be easily replicated with low-cost equipment, enabled us to capture all triggers from each participant without missing a single event marker. In this context, the FPGA efficiently managed all aspects of the experiments, from initiating the recording to concluding each experimental session, encompassing both the start and stop record operations.

IV. DISCUSSION AND FUTURE WORK

In this work, the proof of concept for a DIY FPGA-based custom-made synching device used in an EEG decoding experiment was presented. The results obtained allowed us to propose a standardized way to build this device that allows any researcher to perform EEG experiments with timing analysis at low cost. In our case, we use a commercial FPGA device, which is commonly used for educational purposes, however, the description presented is susceptible to be implemented with cheaper devices like TinyFpga or TangNano devices which cost around 30 USD/unit and commercials ADC like the PCM1808 that allow up to 24bit of quality audio and their cost is around 3 USD/unit at the time of this work is presented.

Taking into consideration that each accessory for the recording of external triggers for EEG equipment can exceed one hundred dollars in most cases, it is easy to see how these alternatives can be used in each laboratory or research center. Additionally, given that this approach allows a highly detailed level of customization of each trigger sent according to the running EEG experiment, it allows for the research not to be limited to the traditional on/off stimuli or the inputs or outputs a commercial solution offers.

The greatest contribution to the state of the art of this approach is that it shows that a customizable-configurable solution offers a new way to brain exploration through a widely available technology such as EEG, despite it necessarily requires some electronics knowledge. Nonetheless, it is also clear that such developments open a wide range of possibilities to reduce the barriers imposed by the high cost of hardware when exploring new paradigms in EEG research,

that usually cannot conform to rigid configurations, which can result in high value and high impact research at lower cost.

V. CONCLUSION

To conclude, the uses of external devices that are not firstly designed for EEG experiments could be used in the same way as other devices if the communication protocol is open -which generally is-, as in the case of the use proposed in this paper. Due to the nature of the EEG amplifier that allows isolating the subject from the devices through an optical fiber link it was easier to build a proxy that emulates an auditory detector, and in the same way, the same concept could be used to perform experiments were the time response needs to be evaluated, and the stimuli latency is relevant for the analysis.

The performance acquired with this device is comparable with the commercial device in order that no missing marks were detected and according to the stimulus used and the trigger marks collected, the latency on each trigger was constant, allowing the use of onset timestamps as a relevant feature on the subsequent analysis over the EEG data.

ACKNOWLEDGMENT

This work is funded by Call 785 of the Colombian Minister of Science and Technology and the "scholarship for graduate students" by the academic development office supported by the Universidad del Norte, Barranquilla, Colombia

REFERENCES

[1] H. Jing, C. Chojnowska, S. Heim, J. Thomas, and A. A. Benasich, "Timing errors in auditory event-related potentials," *Journal of Neuroscience Methods*, vol. 138, pp. 1–6, 9 2004.

[2] G. Ionescu, A. Frey, N. Guyader, E. Kristensen, A. Andreev, and A. Guérin-Dugué, "Synchronization of acquisition devices in neuroimaging: An application using co-registration of eye movements and electroencephalography," *Behavior Research Methods*, vol. 54, pp. 2545–2564, 12 2021.

[3] V. Kolev and J. Yordanova, "Analysis of phase-locking is informative for studying event-related eeg activity," *Biological Cybernetics*, vol. 76, pp. 229–235, 4 1997.

[4] X. et al. Xiao, *Human Brain and Artificial Intelligence*, X. Ying, Ed. Springer Nature Singapore, 2023, vol. 1692. [Online]. Available: https://link.springer.com/10.1007/978-981-19-8222-4

[5] J. Zhang, K. Zhou, S. Mao, and Y. Chen, "A novel hybrid BCI system based on SSVEP and EOG," in *Fifth International Conference on Mechatronics and Computer Technology Engineering (MCTE 2022)*, D. Zhang, Ed., vol. 12500, International Society for Optics and Photonics. SPIE, 2022, p. 125002J. [Online]. Available: https://doi.org/10.1117/12.2660968

[6] D. Zhu, J. Bieger, G. G. Molina, and R. M. Aarts, "A survey of stimulation methods used in ssvep-based bcis," *Computational Intelligence and Neuroscience*, vol. 2010, pp. 1–12, 2010.

[7] E. Mussini, M. Berchicci, V. Bianco, R. L. Perri, F. Quinzi, and F. D. Russo, "The role of task complexity on frontal event-related potentials and evidence in favour of the epiphenomenal interpretation of the go/no-go n2 effect," *Neuroscience*, vol. 449, pp. 1–8, 11 2020.

[8] M. Don, C. W. Ponton, J. J. Eggermont, and A. Masuda, "Auditory brainstem response (abr) peak amplitude variability reflects individual differences in cochlear response times," *The Journal of the Acoustical Society of America*, vol. 96, pp. 3476–3491, 12 1994.

[9] J. J. Eggermont, *Chapter 30 - Auditory brainstem response*, ser. Handbook of Clinical Neurology. Elsevier, 2019, vol. 160, pp. 451–464. [Online]. Available: https://www.sciencedirect.com/science/article/pii/B9780444640321000308

[10] W. D. Hairston, K. W. Whitaker, A. J. Ries, J. M. Vettel, J. C. Bradford, S. E. Kerick, and K. McDowell, "Usability of four commercially-oriented eeg systems," *Journal of Neural Engineering*, vol. 11, p. 046018, 8 2014.

[11] K. Sundaram, Marichamy, and Pradeepa, "Fpga based filters for eeg pre-processing," in *2016 Second International Conference on Science Technology Engineering and Management (ICONSTEM)*, 2016, pp. 572–576.

[12] D. Liu, Q. Wang, Y. Zhang, X. Liu, J. Lu, and J. Sun, "Fpga-based real-time compressed sensing of multichannel eeg signals for wireless body area networks," *Biomedical Signal Processing and Control*, vol. 49, pp. 221–230, 3 2019.

[13] Y. T. Qassim, T. R. Cutmore, and D. D. Rowlands, "Fpga implementation of wavelet coherence for eeg and erp signals," *Microprocessors and Microsystems*, vol. 51, pp. 356–365, 6 2017.

[14] R. R. Plant, N. Hammond, and G. Turner, "Self-validating presentation and response timing in cognitive paradigms: How and why?" *Behavior Research Methods, Instruments, & Computers*, vol. 36, pp. 291–303, 5 2004.

[15] Z. Guo, W. Najjar, F. Vahid, and K. Vissers, "A quantitative analysis of the speedup factors of fpgas over processors," in *Proceedings of the 2004 ACM/SIGDA 12th International Symposium on Field Programmable Gate Arrays*, ser. FPGA '04. New York, NY, USA: Association for Computing Machinery, 2004, p. 162–170. [Online]. Available: https://doi.org/10.1145/968280.968304

[16] C.-C. Hsu, C.-L. Yeh, W.-K. Lee, H.-T. Hsu, K.-K. Shyu, L. P.-H. Li, T.-Y. Wu, and P.-L. Lee, "Extraction of high-frequency ssvep for bci control using iterative filtering based empirical mode decomposition," *Biomedical Signal Processing and Control*, vol. 61, p. 102022, 8 2020.

[17] "Reaction time ruler," Jun 2020. [Online]. Available: https://www.scienceworld.ca/resource/reaction-time-ruler/

[18] M. D. Sanfins, S. Hatzopoulos, C. Donadon, T. A. Diniz, L. R. Borges, P. H. Skarzynski, and M. F. Colella-Santos, "An analysis of the parameters used in speech abr assessment protocols," *The Journal of International Advanced Otology*, pp. 100–105, 5 2018.

An ECG-Based Blood Pressure Estimation Using U-Net auto-encoder and Random Forest Regressor

Elham Alaa Aldein
Electrical Engineering Department
Sohag University
Sohag, Egypt
elham.alaaaldein@eng.sohag.edu.eg

Mohamed Abdleraheem
Electrical Engineering Department
Assuit University
Assuit, Egypt
m.abdelraheem@aun.edu.eg

Usama Sayed Mohamed
Electrical Engineering Department
Assuit University, Sphinx University
Assuit, Egypt
usama@aun.edu.eg

Mohamed Atef*
Electrical Engineering Department
United Arab Emirates University
Al Ain, United Arab Emirates
moh_atef@uaeu.ac.ae

Abstract—Abstract— Measurements of Blood Pressure (BP) have become increasingly widespread in both clinical and private settings. In parallel, Electrocardiogram (ECG) monitors have also become increasingly prevalent. However, most ECG monitors currently available do not include the ability to estimate the value of BP. To address this gap, we have devised a novel BP estimation approach that relies solely on ECG signals. Our methodology involves a series of steps, including data filtering, and segmentation, and we thoroughly investigated the potential of using the auto-encoders of U-Net neural network, as an automatic feature extractor, followed by a regression algorithm in predicting the BP from the ECG. Using the MIMIC-II dataset, the model was trained. yielded mean absolute errors (MAE) of 6.0±4.49 mmHg (MAE ±STD) and 2.5 ±3.7 mmHg for Systolic Blood Pressure (SBP) and Diastolic Blood Pressure (DBP) respectively.

Index Terms—Electrocardiogram, Auto-encoder, Systolic blood pressure, Diastolic blood pressure Arterial blood pressure

I. INTRODUCTION

The World Health Organization (WHO) estimates that about 20.5% of women and 24% of men globally have hypertension, a condition often referred to as the "silent killer" in the medical field. This term is used because hypertension typically does not exhibit specific signs or symptoms, leading many individuals to be unaware of their condition [1].

The conventional mercury sphygmomanometer used to measure Blood Pressure (BP) requires an uncomfortable inflatable cuff to be worn around the arm [2]. However, in the past twenty years, extensive research has been carried out to develop noninvasive and cuffless methods for BP measurement. One such technique is the use of Pulse Transit Time (PTT), which includes determining the interval of time between the electrocardiogram's (ECG) R peak and the photoplethysmography (PPG) signal's initial peak [3].

In recent times, numerous researchers have employed engineered extraction features for measuring BP. Since numerous physiological and neurological factors have an impact on BP, adding feature extraction to a BP estimation model has the potential to significantly improve measurement accuracy [4]– [6].

However, most of this research depends on handcrafted, manually selected, and extracted features. This technique has

two significant drawbacks: the difficulty of simultaneously calculating multiple features due to individual-specific waveforms and motion artifacts, and the cost and time associated with extracting desired features for real-time monitoring.

Automatic ECG feature extraction, mainly using Convolution Neural Network (CNN) [7] [8] to extract has been investigated to overcome the above-mentioned challenges. A more efficient automatic feature extraction is done by using U-Net architecture as an auto-encoder to automatically select the important features from the ECG signal.

In this study, we aimed to enhance the feature extraction of the U-Net model process by utilizing the input for our model consisting of first- and second-order derivatives of the ECG signal with a small Number of Samples. To achieve this, we incorporated residual layers and augmented the model with Squeeze and Excitation (SE) modules for improved feature representation. The U-Net model was employed as a trainable feature extractor followed by various machine learning regression algorithms, namely SVR, AdaBoost, and Random Forest. These algorithms were employed as regression operators to determine the estimated value of the BP, which yielded the best results. By utilizing this approach, we sought to optimize the feature extraction process and identify the most effective regression operator for our specific task.

Paper Organization –The rest of the paper is organized as follows: Section II presents a brief review of related work. Section III describes the main stages of the proposed system. Section IV presents the experiments setup and the proposed system performance evaluation. Finally, the paper is concluded in Section V.

II. RELATED WORKS

In this section, we present a review of recent research work that utilizes ECG signals for blood pressure estimation. Simjanoska *et al.* [6], presented a novel filtering and segmentation algorithm, and feature extraction of ECG signals for BP estimation. The algorithm's mean absolute error (MAE) for SBP and DBP was 7.72 mmHg and 9.45 mmHg, respectively. the technique used complexity analysis and machine learning techniques.

Mousavi *et al.* [4], suggested a new algorithm for feature vector extraction. The algorithm transfers all ECG signal

* Corresponding Author

979-8-3503-8083-5/23 $31.00 © 2023 IEEE

values at a given interval between two R peaks to the frequency domain using the Fast Fourier Transform (FFT). The algorithm's MAE for SBP and DBP was 12.75 mmHg and 6.04 mmHg, respectively.

Monika *et al.* [5], proposed a strategy for estimating blood pressure using only ECG signals that is based on complexity analysis and machine learning techniques. The algorithm's MAE for SBP and DBP was 7.1 mmHg and 6.3 mmHg, respectively.

Slapnivcar *et al.* [9], developed a technique based on a brand-new residually connected spectro-temporal deep neural network. Using a PPG to estimate blood pressure. The algorithm produced MAE for SBP and DBP of 9.43 mmHg and 6.88 mmHg, respectively.

Qin *et al.* [10], developed a technique to predict the arterial blood pressure waveform from the PPG signal based on a deep generative model with domain adversarial training. The algorithm produced MAE for SBP and DBP of 7.945 mmHg and 4.114 mmHg, respectively.

Baek *et al.* [11], proposed An innovative cuffless method for forecasting BP that makes use of a deep CNN that accepts input from both the ECG and the PPG. The approach produced MAE for SBP and DBP of 9.30 mmHg and 5.12 mmHg, respectively.

Zhang *et al.* [12], proposed An algorithm utilizing a Convolutional auto-encoder (CAE) that calculates continuous BP without manual feature extraction or calibration. With an MAE of 9.61 mmHg for SBP and 6.73 mmHg for DBP, the algorithm produced remarkable results.

Mahmud *et al.* [13] proposed a U-Net Model using a fully connected dense layer for feature extraction and MLP algorithm to Predict BP Using ECG and PPG Signals. 147,116 samples were used in this algorithm's training. The method produced accurate findings, with MAE for SBP and DBP of 2.728 mmHg and 1.166 mmHg, respectively.

In this research, we enhanced the performance of the U-Net model used in [13] as we utilized the U-Net model with residual layers and SE mechanism using the ECG signal only. This method provides some advantages such as improved feature representation, enhanced information flow, and effective channel-wise feature recalibration. Moreover, it reduces model complexity with a lightweight computational module, enabling efficient training and inference on resource-constrained devices. The combination improves generalization, robustness, and segmentation accuracy, making it a powerful approach for data segmentation tasks as we use only ECG signals with 3736 samples for training.

III. SYSTEM ARCHITECTURE AND DESIGN

In the following Section, we will present the architecture and detailed design of the proposed system. The system consists of three main parts namely; the Pre-processing stage, Feature Extraction Stage, and regression stage. Below, we will provide the detailed design of each stage.

A. ECG Database

This study utilized a portion of the online waveform database for multi-parameter intelligent monitoring in critical care MIMIC-III), specifically version 3 from 2015, as the data source [14]. The selected portion of the database contains simultaneously sampled ECG and Aortic Blood Pressure (ABP) signals from individuals. The sampling frequency for both signals is 125 Hz. The ABP signals were recorded invasively from the aorta, while the ECG signals correspond to channel II.

B. Pre-Pocessing Stage

The objective of this stage is to perform essential pre-processing procedures on the ECG signal to prepare it for feature extraction and regression stages. The ECG signals obtained from the MIMIC dataset [14] exhibit significant baseline drift in numerous cases. As a result, baseline wandering was eliminated before signal normalization. Once the baseline drifts were corrected and the signals were appropriately normalized, the first two derivatives of the ECG were computed and saved alongside their corresponding ECG signals to be employed as predictors. Additionally, severely distorted signals were excluded. For this stage, we follow the steps suggested by [13].

- **Signal segmentation:** At first, The ECG signals extracted from the MIMIC dataset were divided into segments of 4000 samples, corresponding to a duration of 32 seconds, while ensuring that the original sampling rate of 125 Hz was preserved.

- **Baseline drift correction:** Baseline drift correction is used to Remove Baseline Wandering. This step is performed using a Sliding Window of the specified size (peak_dist_median) that moves across the ECG signal from left to right. The window starts at the beginning of the signal and slides one sample at a time until it reaches the end of the signal. determines the minimum value within the window. It scans all the samples within the window. then A polynomial curve is fitted to the baseline. The resulting baseline fit is subtracted from the original signal.

- **Signal Normalization:** ECG signals were normalized to standardize the distribution of the ECG signal by dividing by the standard deviation after subtracting the mean. the standard deviation is 1 and the mean is 0 for the normalized ECG signal. This normalization process helps in comparing and analyzing ECG signals that may have different scales or baseline variations. The normalized value of the ECG (ECG_{norm}) is calculated as follows:

$$ECG_{norm} = range \left(\frac{(ECG - \mu_{ECG})}{\sigma_{ECG}} \right) \quad (1)$$

where ECG is the original sampled data of ECG signal μ_{ECG} is mean of ECG signal, and σ_{ECG} is standard deviation for ECG signal.

Global min-max normalization was applied to the complete dataset of the ABP waveforms. By ensuring that the

ABP waveform values are scaled within a predetermined range, often between 0 and 1, this normalizing technique makes the waveform values similar and facilitates further analysis.

$$ABP_{norm} = \frac{ABP_i - MIN_{ABP}}{ABP_{AMP}}, \qquad (2)$$

where ABP_{norm} is the normalised ABP $ABP_i - ABP$ is The Sampled ABP and ABP_{AMP} is the difference between maximum arterial blood pressure and minimum arterial blood pressure.

- **ECG Derivatives Calculation:** The first and second derivatives of ECG can also be used to extract features that are useful for blood pressure prediction. the first derivative of ECG can be used to identify the P wave, the QRS complex, and the T wave. the second derivative of ECG can be used to identify the ST-segment elevation. the first and second derivatives of ECG can be used to track changes in the heart rate and the QT interval [15] [16]. However, The derived signals exhibited distortions as a result of applying derivatives, with higher-order derivatives leading to increased distortion levels. To address these high-frequency distortions, a Digital Filter was utilized to filter the signals and mitigate the undesired artifacts [17]. The selection of cutoff frequencies for the bandpass filter was done meticulously to allow the passage of crucial frequency components present in the ECG derivatives, while simultaneously reducing the impact of low- and high-frequency distortions. This careful consideration ensured that the filtered signals retained the essential information while minimizing unwanted frequency artifacts. Nevertheless, the application of filters to the signals introduces a certain amount of delay, which tends to increase with the increase of the order of the derivative. To address this, the average filter delay was computed. Subsequently, the signals were shifted to the left by an amount equal to their respective calculated delays. To ensure consistency in the length of the first and second derivative signals after the delay, the original length of the ECG signals was adjusted accordingly, so that each signal had a length of 4000 samples.

- **Outlier Removal:** To ensure optimal performance, the deep learning model addressed highly distorted signals in the MIMIC dataset. Certain samples were eliminated based on specific criteria: ABP signals with extreme SBP and DBP values were removed. SBP values below 80 or above 190, and DBP values above 120 or below 50 were excluded. ABP signals with a BP range (SBP-DBP) less than 20 or exceeding 120 were also discarded. This decision was made as highly distorted signals often showed such extreme BP ranges, with few exceptions. Blank samples resulting from extreme distortion after signal processing and derivative calculations were removed from the dataset.

C. Feature Extraction

The purpose of this stage is to automatically extract the useful feature from the raw signal using the auto-encoder of the U-Net network. Figure 1 depicts the different stages of the feature extraction and regression processes.

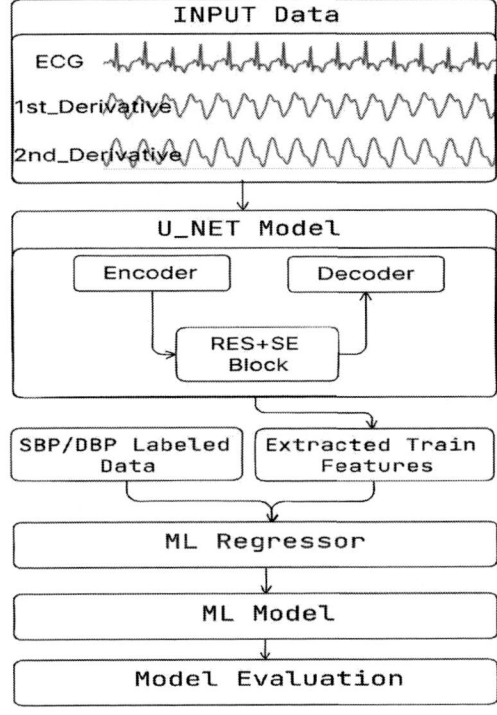

Fig. 1. The two-stage deep features extraction and regression process used to estimate the BP value.

An auto-encoder based on the U-Net architecture is utilized to extract features from input data. Basically, the U-Net architecture consists of an encoder and decoder connected by a "bridge" layer. The encoder is a typical CNN that downsamples the input data, while the segmentation mask is created by the decoder upsampling the feature maps. The bridge layer connects the encoder and decoder and provides high-resolution features to the decode. The U-Net architecture has a "U" shape, with the encoder and decoder parts separated by a middle layer that has a high number of feature maps. This architecture enables the network to collect contextual data from both the local and global levels, which is important for accurate segmentation [18].

In this research, we utilized the U-Net model with residual layers, supplemented with Squeeze and Excitation (SE). A residual block is a type of building block for deep neural networks that allows the model to efficiently learn from very deep architectures by creating a shortcut connection that bypasses one or more layers [19].

A squeeze-and-excitation (SE) [20] module is a component used in CNNs to enhance the representation and attention of important features within the network. The SE module is typically applied to a feature map or activation map from a CNN layer. It consists of two main steps: squeeze and excitation.

- **Squeeze:** In this step, the SE module performs a global spatial pooling operation on the input feature map. This pooling operation reduces the spatial dimensions (width and height) of the feature map to a single value per channel. The purpose of this step is to capture channel-wise statistics of the feature map.
- **Excitation:** After the squeeze step, the SE module applies a set of fully connected layers to model the channel dependencies and relationships. These layers learn channel-wise weights or coefficients that represent the importance of each channel. The aim is to emphasize informative channels and suppress less relevant ones.

The excitation step is typically implemented using a Multi-Layer Perceptron (MLP) or a series of convolutional layers followed by global average pooling and a sigmoid or softmax activation function. The output of the excitation step is a set of channel-wise weights that are multiplied element-wise with the original feature map. This weight multiplication operation allows the network to give more attention to important features and suppress less informative ones.

The introduction of the SE module helps the network to adaptively recalibrate the channel-wise feature responses, enabling it to focus on more discriminative and informative channels. This can lead to improved performance in BP prediction using ECG signals.

The feature extraction is performed as follows: The U-Net network was trained using the ECG, its first and second derivative as inputs, while the corresponding ABP signal is set as a target output of the network. After the training was performed, the useful feature set was extracted from the U-Net encoder, which will be used later for the regression stage.

To establish the optimal settings for the training process of the feature extractor (auto-encoder) based on U-Net architecture, a standard configuration was adopted, including a batch size of 5, 200 epochs, Adam optimizer using Mean Squared Error (MSE) as the chosen loss function and Mean Absolute Error (MAE) as the monitored metric. Multiple experiments were conducted on Jupyter notebook platform coded Python language using Tensorflow machine learning libraries, modifying the parameters, to determine the ideal settings for the batch size and training epoch count.

D. ECG Regression Stage

Regression was performed on the collected features using traditional machine learning (ML) techniques, including Random Forest, Support Vector Regression (SVR), and adaptive boosting. A range of parameters were adjusted and optimized for these ML algorithms to achieve the best possible outcome.

IV. PERFORMANCE EVALUATION

Within this section, we will describe the series of experiments used to evaluate the effectiveness of the suggested method. Also, we will provide a performance comparison with related work.

A. Experiment Setup

The MIMIC dataset was divided into 60%,20%, and 20% of the dataset and a training set, a test set, and a validation set, respectively. The experiment involved three input signals, namely ECG, 1st Derivative represents the First derivative of the ECG signal, and 2nd Derivative, represents the second derivative of the ECG signal, while the target signal for prediction was ABP as Shown in Table I.

TABLE I
DESCRIPTION OF THE EXPERIMENT'S TRAIN AND TEST SETS.

Input	Target	Number of Samples in the Train Set	Number of Samples in the Test Set
ECG 1st_Derivative 2nd_Derivative	ABP	3736	1168

In order to ascertain the most suitable U-Net architecture for utilization as an auto-encoder, multiple sub-experiments were carried out empirically. The MAE was measured for each architecture, and the following parameters were considered:

- **Encoder depth:** The depth was systematically changed from 1 to 5 to examine the effect of the encoder's depth (number of levels) on the latent characteristics that were recovered from the U-Net auto-encoder. The goal of this investigation was to ascertain whether the extracted features were affected by the depth of the architecture.
- **kernel size:** To assess the influence of kernel size on performance, a range of kernel sizes from 1 to 9 were tested.

Regression experiments were conducted using the extracted features to train various traditional Machine Learning regression techniques, including SVR, AdaBoost, and random forest. The objective was to predict blood pressure (BP) using these regression models.

B. Evaluation Metrics and Benchmarks

British Hypertension Society (BHS) Standard: The BHS created a standardized process [21] that is often used in the literature for evaluating the accuracy of BP monitoring devices and methods. This method, often known as the BHS standard, grades results according to the absolute inaccuracy of the predictions made to assess performance, allocating grades A, B, or C. The grades are based on the proportion of prediction absolute errors that lie within, or equal to, 5 mmHg, 10 mmHg, and 15 mmHg, respectively. Furthermore, research that doesn't fit the requirements for a B is assigned a D [21].

C. Experimental Results:

Based on the preceding information, our experimental approach involved a comprehensive set of distinct studies focused on identifying the network architecture that best suited our objectives and determining the optimized parameters for optimal performance. In the subsequent sections, we will present and discuss the significant findings derived from these studies.

1) Effect of U-Net parameter: Encoder depth: There is a decrease in mean absolute error (MAE) for predicting BP as the encoder depth increases. This direct correlation suggests that deeper encoders are more effective in capturing signal features, including peripheral ones such as SBP and DBP. Consequently, the deeper version of U-Net, functioning as an auto-encoder model, exhibited superior performance in BP prediction.

kernel size: When the kernel size was adjusted from 1 to 9, the kernel size with k = 7 performed best.

Results for 200 epochs as an indication for kernel size effect.

TABLE II
MAE of BP PREDICTION GIVEN THE KERNEL OR FILTER SIZE.

| Parameters | Kernel Size | MAE | |
		SBP	DBP
Encoder Type:U-Net+Resnet	1	10.31	4.44
Encoder Width: 66	3	7.09	2.97
No.of Extracted Feature:4000	5	6.48	2.71
Regressor: RF	7	6.0	2.5
Encoder depth:5	9	6.8	2.94

We utilized the U-Net model with residual layers, supplemented with squeeze and excitation(SE).

The actual and predicted ABP values from the U-Net model are displayed in Fig. 2. It is worth mentioning that we do not benefit from the predicted ABP signal but we use the features set resulting from the U-Net stage for the regression model which includes much lower data compared to the original ECG signal and its derivatives.

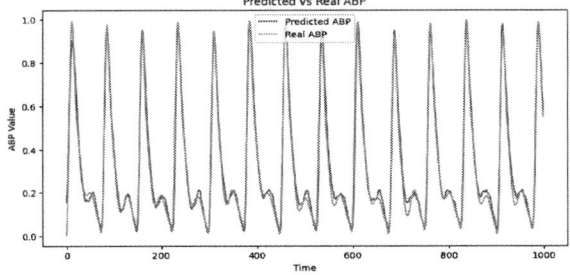

Fig. 2. U-Net Predicted ABP signal VS actual ABP signal.

D. Regression methods comparison

The features that were taken from the U-Net model were utilized for training various traditional machine learning regression techniques to predict the BP values. As listed in Table III Random forest (RF) demonstrated superior performance compared to other classical machine learning methods.

Figure 3 displays the response plots for the SBP and DBP regression results using the Random forest model. A strong association between the target values and the ground truths can be seen in the charts.

TABLE III
MAE of BP PREDICTION FOR DIFFERENT ML REGRESSION METHODS.

| Parameters | Regressor Algorithm | MAE | |
		SBP	DBP
Encoder Type:U-Net+Resnet	Random forest	6.0	2.5
Encoder Width: 66,Encoder depth:5	Ada Boost	9.7	3.5
No.of Samples:4000	SVR	6.8	3.0

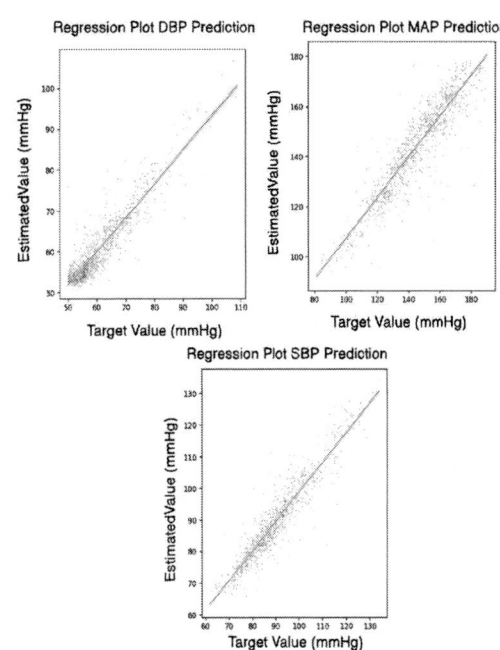

Fig. 3. Regression Plot for DBP, MAP, and SBP .

E. Benchmarking using standard methods

To validate the practical feasibility of our study, we present an evaluation of the proposed system against the standard benchmark which is Standard BHS. Table IV lists the standards for the three grades as well as the study's model performance. As can be inferred from the Table, our proposed technique meets grade A for DBP and grade B for SBP.

TABLE IV
EVALUATION OF BP PREDICTION IN EXPERIMENT IN TERMS OF BRITISH HYPERTENSION SOCIETY (BHS) STANDARD.

| | | Cumulative Error Percentage | | |
		≤ 5 mmHg	≤ 10 mmHg	≤ 15 mmHg
Our results	DBP	86.64%	97.52%	99.4%
	MAP	82.79%	96.4%	98.97%
	SBP	54.71%	82.36%	93.41%
BHS Metric	Grade A	60%	85%	95%
	Grade B	50%	75%	90%
	Grade C	40%	65%	85%

F. Comparison with related work

Here, we provide a numerical comparison with related work that utilizes ECG alone or is associated with other physiological signals like (PPG). The MAE of the estimated value of the BP is used to compare the different techniques for both SBP and DBP. Table V lists the related work included in the comparison, the techniques used, the physiological signal utilized for the estimation process, and the MAE for both SBP and DBP. As can be inferred from the table, In our study we utilize only ECG signal with a small number of samples compared to [13] which used ECG and PPG signal with a greater number of samples, However, we could achieve grade A for DBP, and grade B for SBP.

TABLE V

Study	Model	Input Signals	MAE SBP (mmHg)	MAE DBP (mmHg)
Simjanoska *et al.* [6]	Complexity analysis + ML	ECG	7.72	9.45
Mousavi *et al.* [4]	Random Forest	ECG	12.75	6.04
Simjanoska *et al.* [5]	Machine learning	ECG	7.1	6.3
slapnivcar *et al.* [9]	DNN with residual	PPG	9.43	6.88
Qin *et al.* [10]	Deep generative	PPG	7.945	4.114
Rastegar *et al.* [15]	Hybrid CNN-SVR	ECG,PPG	1.23	3.08
Mahmud *et al.* [13]	CNN + ANN	ECG, PPG	2.728	1.166
Ertu *et al.* [22]	ELM	ECG, PPG	4.25	3.95
Zhang *et al.* [12]	CNN	ECG, PPG	9.61	6.73
Kachuee *et al.* [23]	Random Forest	ECG, PPG	9.87	5.71
Ibtehaz *et al.* [24]	CNN	PPG	5.73	3.45
Choi *et al.* [25]	equation	ECG,PPG	2.72	2.29
Baek *et al.* [11]	CNN	ECG,PPG	9.30	5.12
Our study	U-Net+RF	ECG	6.0	2.5

V. Conclusion

This study focused on the development of a unique approach to predicting BP using an ECG signal. The research includes extracting important features from the ECG signal using the U-Net architecture as an auto-encoder. The extracted features were used in machine learning approaches to regress and predict SBP, and DBP values, as opposed to using the raw signals directly. RF regressor achieved the best result which achieves grade A (2.5 mmHg) for DBP and grade B (6.0 mmHg) for SBP according to BHS. In our model we utilize only ECG signal with a smaller number of samples compared to state-of-the-art, however, we could achieve accurate BP prediction. In future work, we will deploy the model on a hardware platform with complete data acquisition and processing stages.

References

[1] "World health organization (who).the top 10 causes of death. available online: https://www.who.int/news-room/fact-sheets/detail/the-top-10-causes-of-death (accessed on september 2022)."

[2] G. Ogedegbe and T. Pickering, "Principles and techniques of blood pressure measurement," *Cardiology clinics*, vol. 28, no. 4, pp. 571–586, 2010.

[3] L. Peter, N. Noury, and M. Cerny, "A review of methods for non-invasive and continuous blood pressure monitoring: Pulse transit time method is promising?" *Irbm*, vol. 35, no. 5, pp. 271–282, 2014.

[4] S. S. Mousavi, M. Hemmati, M. Charmi, M. Moghadam, M. Firouzmand, and Y. Ghorbani, "Cuff-less blood pressure estimation using only the ecg signal in frequency domain," in *2018 8th International Conference on Computer and Knowledge Engineering (ICCKE)*. IEEE, 2018, pp. 147–152.

[5] M. Simjanoska, M. Gjoreski, M. Gams, and A. M. Bogdanova, "Novel data processing approach for deriving blood pressure from ecg only," in *ICT Innovations 2018. Engineering and Life Sciences: 10th International Conference, ICT Innovations 2018, Ohrid, Macedonia, September 17–19, 2018, Proceedings 10.* Springer, 2018, pp. 273–285.

[6] M. Simjanoska, M. Gjoreski, M. Gams, and A. Madevska Bogdanova, "Non-invasive blood pressure estimation from ecg using machine learning techniques," *Sensors*, vol. 18, no. 4, p. 1160, 2018.

[7] F. Miao, B. Wen, Z. Hu, G. Fortino, X.-P. Wang, Z.-D. Liu, M. Tang, and Y. Li, "Continuous blood pressure measurement from one-channel electrocardiogram signal using deep-learning techniques," *Artificial Intelligence in Medicine*, vol. 108, p. 101919, 2020.

[8] J.-H. Kim and J.-H. Chang, "Attention wave-u-net for acoustic echo cancellation." in *Interspeech*, 2020, pp. 3969–3973.

[9] G. Slapničar, N. Mlakar, and M. Luštrek, "Blood pressure estimation from photoplethysmogram using a spectro-temporal deep neural network," *Sensors*, vol. 19, no. 15, p. 3420, 2019.

[10] K. Qin, W. Huang, and T. Zhang, "Deep generative model with domain adversarial training for predicting arterial blood pressure waveform from photoplethysmogram signal," *Biomedical Signal Processing and Control*, vol. 70, p. 102972, 2021.

[11] S. Baek, J. Jang, and S. Yoon, "End-to-end blood pressure prediction via fully convolutional networks," *Ieee Access*, vol. 7, pp. 185 458–185 468, 2019.

[12] J. Zhang, D. Wu, and Y. Li, "Cuff-less and calibration-free blood pressure estimation using convolutional autoencoder with unsupervised feature extraction," in *2019 41st Annual International Conference of the IEEE Engineering in Medicine and Biology Society (EMBC)*. IEEE, 2019, pp. 3323–3326.

[13] S. Mahmud, N. Ibtehaz, A. Khandakar, A. M. Tahir, T. Rahman, K. R. Islam, M. S. Hossain, M. S. Rahman, F. Musharavati, M. A. Ayari *et al.*, "A shallow u-net architecture for reliably predicting blood pressure (bp) from photoplethysmogram (ppg) and electrocardiogram (ecg) signals," *Sensors*, vol. 22, no. 3, p. 919, 2022.

[14] M. Kachuee, M. M. Kiani, H. Mohammadzade, and M. Shabany, "Cuffless high-accuracy calibration-free blood pressure estimation using pulse transit time," in *2015 IEEE international symposium on circuits and systems (ISCAS)*. IEEE, 2015, pp. 1006–1009.

[15] J. Arteaga-Falconi, H. Al Osman, and A. El Saddik, "R-peak detection algorithm based on differentiation," in *2015 IEEE 9th international symposium on intelligent signal processing (WISP) proceedings*. IEEE, 2015, pp. 1–4.

[16] P. H. Langner Jr and D. B. Geselowitz, "First derivative of the electrocardiogram," *Circulation research*, vol. 10, no. 2, pp. 220–226, 1962.

[17] M. C. Saxena, M. V. Upadhyaya, H. K. Gupta, and A. Sharma, "Denoising of ecg signals using fir & iir filter: A performance analysis," in *Proceedings on International Conference on Emerg*, vol. 2, 2018, pp. 51–58.

[18] O. Ronneberger, P. Fischer, and T. Brox, "U-net: Convolutional networks for biomedical image segmentation," in *Medical Image Computing and Computer-Assisted Intervention–MICCAI 2015: 18th International Conference, Munich, Germany, October 5-9, 2015, Proceedings, Part III 18.* Springer, 2015, pp. 234–241.

[19] K. He, X. Zhang, S. Ren, and J. Sun, "Deep residual learning for image recognition," in *Proceedings of the IEEE conference on computer vision and pattern recognition*, 2016, pp. 770–778.

[20] J. Hu, L. Shen, and G. Sun, "Squeeze-and-excitation networks," in *Proceedings of the IEEE conference on computer vision and pattern recognition*, 2018, pp. 7132–7141.

[21] "O'brien, e.; petrie, j.; littler, w.; de swiet, m.; padfield, p.l.; altman, d.; bland, m.; coats, a.; atkins, n. the british hypertension society protocol for the evaluation of blood pressure measuring devices. j hyper. tens. 1993, 11, s43–s62."

[22] Ö. F. Ertuğrul and N. Sezgin, "A noninvasive time-frequency-based approach to estimate cuffless arterial blood pressure," *Turkish Journal of Electrical Engineering and Computer Sciences*, vol. 26, no. 5, pp. 2260–2274, 2018.

[23] M. Kachuee, M. M. Kiani, H. Mohammadzade, and M. Shabany, "Cuffless blood pressure estimation algorithms for continuous health-care monitoring," *IEEE Transactions on Biomedical Engineering*, vol. 64, no. 4, pp. 859–869, 2016.

[24] N. Ibtehaz, S. Mahmud, M. E. Chowdhury, A. Khandakar, M. A. Ayari, A. Tahir, and M. S. Rahman, "Ppg2abp: Translating photoplethysmogram (ppg) signals to arterial blood pressure (abp) waveforms using fully convolutional neural networks," *arXiv preprint arXiv:2005.01669*, 2020.

[25] J. Choi, Y. Kang, J. Park, Y. Joung, and C. Koo, "Development of real-time cuffless blood pressure measurement systems with ecg electrodes and a microphone using pulse transit time (ptt)," *Sensors*, vol. 23, no. 3, p. 1684, 2023.

Examining the Performance of Melanoma Classification using Superpixel Segmentation: A Comparative Analysis

Faezeh Mohammadi Aydoghmishi[1], Sudipta Modak[1], Esam Abdel-Raheem[1], and Luis Rueda[2]

[1] *Department of Electrical and Computer Engineering*
[2] *School of Computer Science*
University of Windsor
Windsor, Ontario, Canada
Email: {mohamm1g, modak, eraheem, lrueda}@uwindsor.ca

Abstract—Skin cancer, characterized by the abnormal growth of skin cells, is a severe and prevalent condition. Despite the advancement in digital diagnosis techniques, existing skin cancer detection methods often fail to achieve satisfactory performance in melanoma identification in dermoscopy imaging. This study presents a new melanoma classification algorithm that utilizes a superpixel-based segmentation technique, simple linear iterative clustering, and transfer learning-based convolutional neural networks to achieve the objective. A key innovation of our methodology is the introduction of a novel approach for efficiently merging superpixels, that enhances the quality of segmentation resulting in better classification performance. Following the segmentation phase, several convolutional neural networks have been utilized for feature extraction from segmented images and classification. The proposed method shows an enhanced performance in melanoma classification which is 91.67 %, 95.33 %, 91.23 %, and 90.48 % in terms of accuracy, precision, recall and $F1$-score, respectively. Our results indicate that the superpixel segmentation technique considerably enhances the classification models' accuracy compared to k-means segmentation methods. A comparative analysis between the proposed method and several state-of-the-art methods in the field has also been presented in the context.

Index Terms—Skin cancer, Lesion segmentation, Convolutional neural networks, Superpixels

I. INTRODUCTION

Skin cancer, characterized by the abnormal growth of skin cells, is a common type of cancer among individuals. The primary causative factor of this type of cancer is prolonged exposure of the skin to sunlight [1]. Typically, skin lesions are categorized into two principal types, which are malignant and benign [2]. Benign lesions seldom metastasize and are typically treatable by medicine. Conversely, malignant lesions are highly aggressive and often present themselves without apparent symptoms. Melanoma is the most lethal of all skin cancers and accounts for approximately 75 % of all skin cancer-related deaths [3]. Delayed diagnosis allows it to penetrate deeper and proliferate rapidly in the skin, complicating treatment efforts and threatening survival chances [4]. As reported by the World Health Organization, the usage of sunbeds escalates the risk of skin cancer development. In the United States, skin cancer is estimated to affect 1.1 million

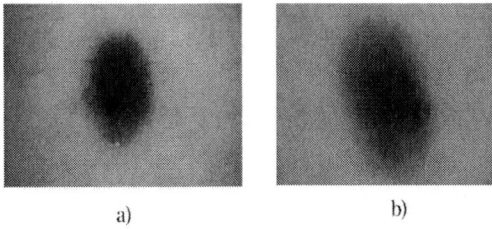

Fig. 1. Examples of skin lesions a) Malignant b) Benign

people annually [5]. An example of malignant and benign skin lesions is presented in Figure 1.

In the last decade, many different approaches to cancer diagnosis have emerged. The traditional approach involves a physical inspection and biopsy. Lately, advancements in non-invasive diagnostic tools, particularly macroscopic and dermoscopic imaging, have offered substantial assistance to dermatologists [6]. These methods can be laborious and time-consuming, especially for less experienced dermatologists, due to the complex structure of skin lesions [7]. To quickly diagnose melanoma at its earliest stage and improve the accuracy of overall skin cancer diagnosis, significant research has been carried out to create powerful automatic algorithms within the realm of computer-aided image analysis. Yet, the main difficulties in classifying skin lesions arise due to similar texture, color, or positioning of the lesion on the body [8]. These variations make it complex to distinguish between harmless and harmful lesions.

To counter this obstacle, recent studies have built several deep-learning frameworks to aid in the classification of melanoma. Araújo *et al.* [9], has used segmentation of lesions combined with transfer learning to achieve high performance in melanoma diagnosis. Similarly, Murugan *et al.* [10], has used watershed segmentation along with k-NN (k-Nearest Neighbors), Random Forest, and Support Vector Machine (SVM) for classification of melanoma. Thapar *et al.* [11] has presented an advanced method for the diagnosis of skin cancer

979-8-3503-8083-5/23 $31.00 © 2023 IEEE

aided by swarm intelligence (SI) algorithms for segmenting skin lesion regions of interest (RoI) and speeded-up robust features (SURF) for feature extraction. CNN are then used to classify the lesions into their respective categories. Afza et al. [12] has proposed a hierarchical framework that utilizes a combination of two-dimensional superpixels and deep learning for melanoma detection. Using an improved grasshopper optimization algorithm, the framework employs a ResNet-50 to extract optimized features for classification.

Annaby et al. [13], in their work, has presented a superpixel generation method from skin lesion images by constructing a region adjacency graph where each node is a superpixel, and edge weight is a function of the distance between superpixel descriptors. Features are then extracted from this graph model and the color, geometry, and texture of the skin lesion images and fed into an ensemble classifier for classification. Anand et al. in their work in [14], has developed a classification pipeline that combines the U-Net++ segmentation and traditional image processing techniques for melanoma classification. The method puts focus on irregular pigment networks in dermoscopic images. Gilani et al. in their work in [15], using the ISIC 2019 [16] dataset, has utilized a deep spiking neural network for classifying melanoma and non-melanoma skin lesions. Highlighting the inefficiencies of traditional visual inspections, they showcased their spiking VGG-13 model's capability, achieving 89.57 % accuracy and a 90.07 % $F1$ score. Notably, the model outperformed standard models such as VGG-13 and AlexNet, all while being more power-efficient and utilizing fewer parameters. El-Khatib et al. [17], has developed a skin lesion diagnosis system combining deep learning and feature-based methods utilizing neural networks, classifiers such as GoogleNet, ResNet-101, and NasNet-Large, and a support vector machine. The study has introduced a fusion-based decision system that integrates all methods to optimize accuracy and has been validated on two different databases.

Segmentation is a crucial task when it comes to finding the boundary of the lesions. As displayed by Modak et at. in their work presented in [18] for lung nodule segmentation, superpixel-based unsupervised segmentation algorithms perform better than their supervised counterparts both in terms of time efficiency and performance. Therefore, it is evident that unsupervised segmentation techniques play a crucial role in the reduction of the number of learnable parameters and the training time of a deep learning-based classification model. In this study, we aim to improve melanoma classification in dermoscopy images by utilizing various image preprocessing techniques such as the Black Hat Operation, the Inpainting method, and Otsu thresholding similar to the works presented in [19], [20], and [21] respectively, a superpixel generation method for segmentation, namely, simple linear iterative clustering (SLIC) that is presented in [22] and [23] and various CNN architectures that have been trained on ImageNet [24]. We present a novel technique of superpixels merging to achieve better segmentation and classification accuracy. The process involves the construction of a graph where each superpixel acts as a node. The relationships between these nodes are determined by two key factors: spatial centroids and intensity differences of superpixels. By employing a radius-based technique, we efficiently identify neighboring superpixels, ensuring the spatial integrity of the image. We then use an adjacency matrix to map the relationships between superpixels. This matrix is further refined by considering both intensity differences and centroid distances, and specific thresholds are applied to these factors to determine adjacency. Subsequently, a binary mask is produced to represent the most significant connected component. We compare the performance of the proposed method with the several state-of-the-art methods in the field based on the evaluation metrics as specified by Hossin et al. [25]. After segmenting the images using our proposed method, we trained six transfer learning models, including VGG16, ResNet50, Inceptionv3, Mobilenet, Xception, and Densenet121 [26], and evaluated their performance outcomes.

II. METHODOLOGY

This section uses various preprocessing techniques to enlarge the dataset and remove artifacts from the images. This is followed by the application of superpixel generation and merging of superpixels to get the segmentation masks of the lesions. Consequently, several pre-trained CNN models are utilized to train the segmented masks to classify melanoma in dermoscopy images.

A. Preprocessing

Dermoscopic images often contain irrelevant elements like hair, which can interfere with accurate image analysis. Eliminating such distractions improves the efficacy of diagnostic models. To enhance the accuracy of our models, we go through several preprocessing steps aimed at removing hair from the images. We first change the images to grayscale to enhance feature extraction accuracy. By turning color images into grayscale, we highlight patterns and simplify the data. We achieve this by showing the image in shades of gray, from black to white. The Black Hat Operation [19] is then used to emphasize small dark areas against brighter backgrounds. This method contrasts the original image with its morphologically closed version, revealing dark details that might otherwise be hidden. The following step is thresholding which is dividing the image into two segments by setting a specific intensity value. Pixels with intensities above this value turn white, while those below turn black. In the context of hair removal, this step creates a mask, showing the exact locations of hair. Lastly, we apply Inpainting using the Telea Method to blend these changes. This method uses information from nearby pixels to fill areas marked as hair, ensuring a consistent and natural look in the processed image [20].

After these steps, we apply data augmentation strategies to increase the dataset size. In this research, through the utility of Keras' ImageDataGenerator, various augmentation techniques were employed. We use both horizontal and vertical flips, catering to different object orientations. Additionally, the

979-8-3503-8083-5/23 $31.00 © 2023 IEEE

dataset underwent rotations within 45°. The images also experience zoom levels of up and shifts across both axes—width, and height—up to a magnitude of 10 %. Overall, these augmentation techniques played a crucial role in enriching our dataset and enhancing the model's adaptability to varied real-world conditions.

B. Superpixel Segmentation

The proposed method utilizes two distinct methods to segment skin lesions, namely superpixel generation and merging.

1) Superpixel Generation: Superpixel generation is a crucial step often used in image processing and computer vision for segmentation. Essentially, it groups individual pixels with similar features into larger blocks called superpixels. Using these superpixels makes it easier to work with images since there is less data to process than dealing with each pixel separately. In the proposed framework, the SLIC algorithm has been utilized in this research for the fast generation of superpixels. SLIC groups pixels based on their color and position in the image, resulting in well-defined superpixel blocks.

2) Superpixel Merging: This research explores innovative methods for merging these superpixels for more precise segmentation. Merging superpixels starts by constructing a graph, where each superpixel represents a node. The distances between the centroids of these superpixels, as well as their intensity differences, are calculated. The centroid provides a representative point for each superpixel. Equations (1) and (2) provide these centroids' x and y coordinates. The average intensity for each superpixel, I_{avg}, is also computed, giving an idea about the brightness or darkness of a superpixel. In these equations, N represents the total number of pixels within a specific superpixel and i represents an index used to sum over all the pixels in a specific superpixel.

$$C_x = \frac{1}{N}\sum_{i=1}^{N} x_i, \tag{1}$$

$$C_y = \frac{1}{N}\sum_{i=1}^{N} y_i, \tag{2}$$

$$I_{avg} = \frac{1}{N}\sum_{i=1}^{N} I_i. \tag{3}$$

A vital aspect of this analysis is the radius-based search, which identifies neighboring superpixels based on their spatial relationships. This approach ensures that the spatial consistency of the image remains intact. A notable benefit of the radius-based approach is its computational efficiency. Constraining the search within a specific radius limits the number of potential neighbors to be evaluated. The measure of spatial separation between two superpixels is captured using the distance D, which is derived using the Euclidean distance formula between their centroids and is given by,

$$D = \sqrt{(C_{1x} - C_{2x})^2 + (C_{1y} - C_{2y})^2}. \tag{4}$$

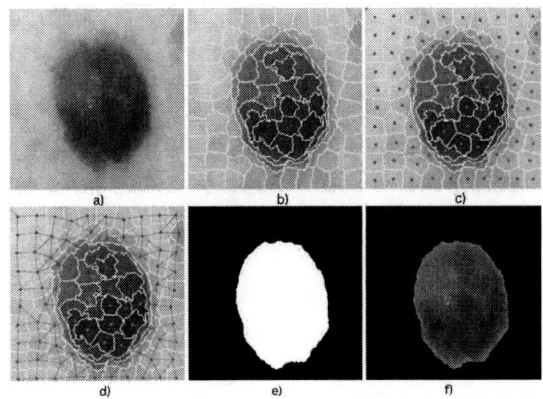

Fig. 2. Stages of segmentation dermoscopic melanoma image. (a) Original skin lesion image. (b) Superpixel clustering using the SLIC technique. (c) the centroid of each superpixel. (d) Graph construction. (e) Binary mask. (f) Final segmented image.

C_{1x} and C_{1y} are the coordinates of the centroid of the first superpixel, and C_{2x} and C_{2y} are the coordinates of the centroid of the second superpixel. The next significant component in this analysis is the adjacency matrix. This matrix acts as a structure that captures the relationships between superpixels. Every entry in this matrix signifies the relationship or adjacency between two superpixels. Specifically, if two superpixels are deemed neighbors based on predetermined criteria, their respective entry in the matrix is marked as 1; otherwise, it remains 0. Intensity differences and centroid distance between potential neighboring superpixels are assessed to refine this relationship further. Thresholds derived from the 95th percentile of the centroid distances and Otsu's method [21] on the intensity differences are applied. Only when the centroid distance and intensity difference fall within these thresholds are two superpixels considered neighbors, and an edge is established between them in the graph. Figure 2 shows these processes.

Upon analyzing the adjacency matrix, connected components are detected. These components determine clusters of interconnected superpixels following specific criteria. Essentially, they group superpixels that are both near each other and share similar intensity values. Each of these clusters or components is given a unique label, effectively combining the individual superpixels they contain. In the subsequent step, a binary mask is formulated to emphasize the largest connected component. This entails initializing a mask identical in size to the image and marking every pixel as zero. Then, pixels linked to the largest component in the mask are set to one. The resulting mask distinctively highlights the principal connected region, where the associated pixels are marked as one, and all others remain zero. While various methods are available within morphological operations in this research, the Closing technique is specifically employed. The rationale behind choosing the Closing operation is its ability to fill small holes within the binary mask effectively. This ensures a more unified and accurate representation, which is crucial for the integrity of subsequent analyses.

C. Pre-Trained Models

Upon completing the image segmentation process, two distinct segmentation methods are employed on the images: k-means clustering and our novel superpixel segmentation. Each set of segmented images is then used to train a range of deep learning models, aiming to analyze and compare the efficacy of the segmentation techniques.

The deep-learning models chosen for this task were VGG16, ResNet50, DenseNet, InceptionV3, Mobilenet, and Xception. These models have previously been trained on ImageNet [24], a vast dataset with millions of labeled images across thousands of categories, making it a benchmark for many state-of-the-art models in computer vision. The convolutional layers in these models are powerful feature extractors due to their extensive training on ImageNet [24]. For our specific tasks and datasets that differ from ImageNet, adjusting or replacing the fully connected layers, especially the final classification layer, became necessary. Given that ImageNet encompasses 1,000 classes and our dataset might differ in class count, the convolutional base from these pre-trained models combines with our custom fully connected layers. This strategy ensured that the models retained their vast knowledge from ImageNet while becoming specialized for our dataset. Of all the models, only in ResNet50 were the first six layers frozen, preserving their acquired features due to the model's complexity and depth. For the others, all layers remained trainable, optimizing the features for our dataset further. A batch size of 64 is maintained for 50 epochs, with other significant hyperparameters being a learning rate of 0.00001, weight decay of 0.2, and regularization parameter of 0.01. Figure 3 depicts these models' training and validation results, segmented with novel superpixel segmentation. All chosen methods for this research are implemented using Python, using the capabilities of Keras and the Scikit-Learn library. Google Colab Pro, a premier cloud service for deep learning research, was the platform for all research, training, and testing phases.

III. EXPERIMENT AND RESULTS

A. Evaluation Metrics

We have evaluated the efficiency of our models by using various metrics: accuracy (Acc), the $F1$-score, precision ($Prec$), and recall (Rec). To assess these metrics, we grouped predicted images into four categories: True Positives (TP), False Positives (FP), True Negatives (TN), and False Negatives (FN). In this context, TP denotes the quantity of correctly identified positive instances, while TN signifies the count of accurately classified negative instances. FP refers to the number of cases inaccurately identified as positive, and FN indicates the number of instances mistakenly labeled as negative. Performance metrics are determined according to the quantities of True Positives, False Positives, True Negatives,

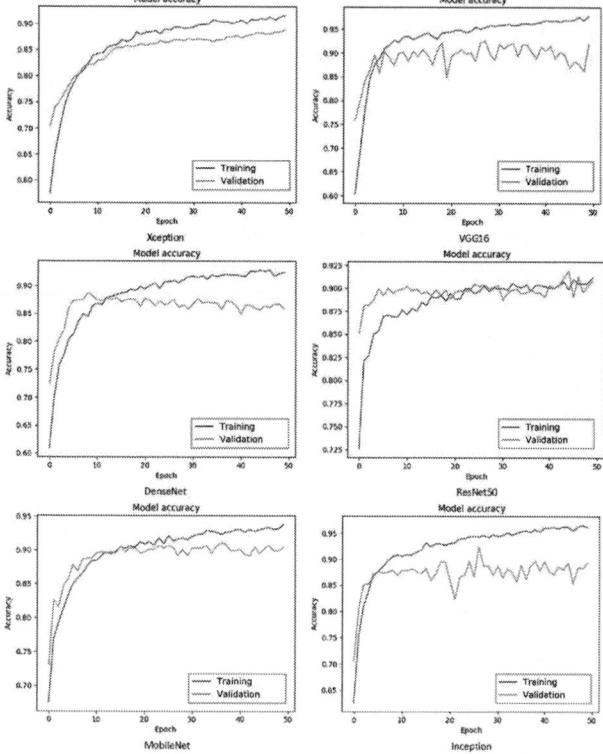

Fig. 3. Training and validation accuracy of various pre-trained models.

and False Negatives. The following equations demonstrate how these metrics are calculated,

$$Acc = \frac{TP + TN}{TP + FP + FN + TN}, \tag{5}$$

$$Rec = \frac{TP}{TP + FN}, \tag{6}$$

$$Prec = \frac{TP}{TP + FP}, \tag{7}$$

$$F1\text{-score} = \frac{2 \times P \times R}{P + R}. \tag{8}$$

B. Dataset Description

For our study we have used the ISIC 2019 [16] dataset, which consists of 4,523 melanoma images. We have utilized all 4,523 melanoma images and chosen 4,250 benign images to have a balanced dataset for training and testing. Considering the diverse resolutions, we standardized all images to 224 × 224 pixels to streamline the computational workload. Importantly, we split the dataset, allocating 80 % of the images for training and the remaining 20 % for testing purposes.

C. Results and Discussion

Table I shows the classification results based on images segmented by the proposed method and the traditional k-means approach. The k-means approach fails on several

TABLE I
OVERALL PERFORMANCE OF THE PROPOSED METHOD COMPARED TO k-MEANS

Model	*Acc (%)*		*F1-score (%)*		*Prec (%)*		*Rec (%)*	
	Superpixel	k-mean	Superpixel	k-mean	Superpixel	k-mean	Superpixel	k-mean
VGG16	91.67	88.14	90.48	87.45	95.33	83.71	91.23	86.11
Resnet50	90.68	84.58	85.87	79.83	77.90	69.45	92.18	91.08
Inceptionv3	87.06	86.80	85.07	84.50	90.65	87.68	81.71	80.14
Xception	86.17	86.00	82.74	82.28	97.15	84.90	81.25	72.05
Mobilenet	90.78	86.90	89.44	85.06	94.52	82.00	84.88	84.00
Densenet121	89.32	83.86	87.59	83.75	94.14	78.56	85.20	81.89

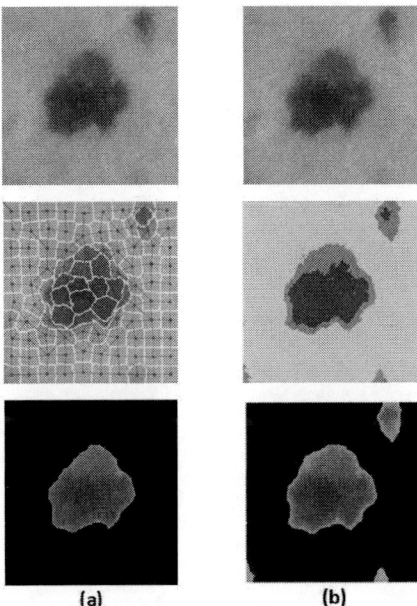

(a) (b)

Fig. 4. Comparison of skin lesion segmentation (a) SLIC superpixel-based segmentation (b) k-means clustering-based segmentation.

images and this results in low accuracy for the preceding classification task. Figure 4 presents a comparison between the segmentation performed using the proposed method and the k-means clustering technique. Furthermore, several performance metrics, including accuracy, $F1$-score, precision, and recall, are compared. The evaluation spans six deep learning models, including VGG16, Resnet50, Inceptionv3, Xception, Mobilenet, and Densenet121. Notably, the VGG16 model, when utilizing the superpixel method, excelled with an accuracy of 91.67 %, significantly higher than the 88.14 % achieved with k-means. Regarding $F1$ score, precision, and recall, VGG16 again outperformed with respective values of 90.48 %, 95.33 %, and 91.23 %, as opposed to the k-mean values of 87.45 %, 83.71 %, and 86.11 %.

The Resnet50 model also demonstrated an improved performance with the superpixel approach. The accuracy, $F1$ score, and precision were recorded at 90.68 %, 85.87 %, and 77.90 %, respectively, compared to the k-means values of 84.58 %, 79.83 %, and 69.45 %. Inceptionv3's performance was relatively consistent between the two methods. However, the superpixel technique still marginally outperformed the

k-means in accuracy and precision. Similarly, the Xception model performed noticeably better with the proposed segmentation method with an even higher precision. Both Mobilenet and Densenet121 also displayed improved results with the superpixel method. For instance, Densenet121's accuracy, $F1$ score, and precision with the superpixel approach were 89.32%, 87.59 %, and 94.14 %, respectively, showcasing its superiority over the corresponding k-means values. This proves that for the classification task, it is very important to input essential features for training the model and if the model is trained on a high amount of nonessential features then the classification accuracy will decrease significantly.

D. Comparison with the State-of-the-Art Methods

Based on the results presented in Table II, our study highlights the superior performance of the chosen VGG16 model. When compared with other models evaluated on the same dataset, the effectiveness of the VGG16 model is clear. Anand *et al.*'s model [14], which uses Resnet50 and Random Forest, had an accuracy of 85.00, an $F1$-score of 85.80 %, a precision of 82.50 %, and a recall of 89.30 %. In comparison, our VGG16 model got a higher accuracy of 91.60 %, which is better than this model. Gilani *et al.*'s Spiking VGG-13 model [15] had an accuracy of 89.57 % and an $F1$ score of 90.07 %. While these are good numbers, they're still about 2.03 % less accurate than the proposed model. Finally, El-Khatib *et al.*'s method [17] had an accuracy of 88.33 %, which is less compared to our proposed architecture, signifying an overall enhancement of performance over these state-of-the-art methods.

IV. CONCLUSION

In this research, we have developed a superpixel-based segmentation technique to segment skin lesions and have reduced the number of trainable parameters for several classification models. We have introduced a novel approach to merging superpixels based on intensity and distance. This process has significantly enhanced the performance of melanoma detection using the proposed segmentation method. The proposed framework has also been compared to many state-of-the-art methods in the field of melanoma classification and it has shown overall better performance compared to these methods. Additionally, a comparative examination of six distinct CNN models has also been conducted in this study, highlighting their classification performance on images segmented via the proposed superpixel segmentation technique in contrast to those segmented

979-8-3503-8083-5/23 $31.00 © 2023 IEEE

TABLE II
COMPARISON WITH STATE-OF-THE-ART METHODS ON THE ISIC 2019 DATASET

Author	Model	Acc (%)	$F1$-score (%)	$Prec$ (%)	Rec (%)
Anand *et al.* [14]	Resnet50 + Random Forest	85.00	85.80	82.50	0.893
Gilani *et al.* [15]	Spiking VGG-13	89.57	90.07	90.68	89.46
El-Khatib *et al.* [17]	GoogleNet+ResNet101+NasNet	88.33	-	-	88.46
The proposed method-VGG16	**VGG16**	**91.67**	**90.48**	**95.33**	**91.23**

using k-means. The results conclusively pointed toward the superior efficacy of the superpixel-based segmentation. For future studies, we intend to investigate several superpixel generation methods and analyze each of these methods in terms of evaluation metrics for segmentation. The goal is to develop a fast superpixel generation method that will produce segmented lesions with better boundary adherence than SLIC and therefore reduce the number of unwanted features to train the classification model. Our research introduces an innovative method for melanoma detection and contributes to the development of more accurate and efficient diagnostic tools. By advancing segmentation techniques, we aim to enhance diagnostic accuracy, leading to improved patient outcomes and reduced healthcare burdens associated with melanoma

ACKNOWLEDGMENT

This work has been partially funded by the Council of Canada (NSERC) Discovery Grant (RGPIN−2018−05523).

REFERENCES

[1] D. Didona, G. Paolino, U. Bottoni, and C. Cantisani, "Non melanoma skin cancer pathogenesis overview," *Biomedicines*, vol. 6, no. 1, p. 6, 2018.

[2] P. Autier and J.-F. Doré, "Ultraviolet radiation and cutaneous melanoma: a historical perspective," *Melanoma Research*, vol. 30, no. 2, pp. 113–125, 2020.

[3] S. Pathan, K. G. Prabhu, and P. Siddalingaswamy, "Techniques and algorithms for computer aided diagnosis of pigmented skin lesions—a review," *Biomedical Signal Processing and Control*, vol. 39, pp. 237–262, 2018.

[4] S. Jain, N. Pise, *et al.*, "Computer aided melanoma skin cancer detection using image processing," *Procedia Computer Science*, vol. 48, pp. 735–740, 2015.

[5] S.-P. Sinikumpu, J. Jokelainen, S. Keinänen-Kiukaanniemi, and L. Huilaja, "Skin cancers and their risk factors in older persons: a population-based study," *BMC geriatrics*, vol. 22, no. 1, p. 269, 2022.

[6] M. Efimenko, A. Ignatev, and K. Koshechkin, "Review of medical image recognition technologies to detect melanomas using neural networks," *BMC bioinformatics*, vol. 21, no. 11, pp. 1–7, 2020.

[7] J. Ferlay, M. Colombet, I. Soerjomataram, D. M. Parkin, M. Piñeros, A. Znaor, and F. Bray, "Cancer statistics for the year 2020: An overview," *International journal of cancer*, vol. 149, no. 4, pp. 778–789, 2021.

[8] A. G. Goodson and D. Grossman, "Strategies for early melanoma detection: Approaches to the patient with nevi," *Journal of the American Academy of Dermatology*, vol. 60, no. 5, pp. 719–735, 2009.

[9] R. L. Araújo, F. H. d. Araújo, and R. R. e. Silva, "Automatic segmentation of melanoma skin cancer using transfer learning and fine-tuning," *Multimedia Systems*, vol. 28, no. 4, pp. 1239–1250, 2022.

[10] A. Murugan, S. A. H. Nair, and K. S. Kumar, "Detection of skin cancer using svm, random forest and knn classifiers," *Journal of medical systems*, vol. 43, pp. 1–9, 2019.

[11] P. Thapar, M. Rakhra, G. Cazzato, and M. S. Hossain, "A novel hybrid deep learning approach for skin lesion segmentation and classification," *Journal of Healthcare Engineering*, vol. 2022, 2022.

[12] F. Afza, M. Sharif, M. Mittal, M. A. Khan, and D. J. Hemanth, "A hierarchical three-step superpixels and deep learning framework for skin lesion classification," *Methods*, vol. 202, pp. 88–102, 2022.

[13] M. H. Annaby, A. M. Elwer, M. A. Rushdi, and M. E. Rasmy, "Melanoma detection using spatial and spectral analysis on superpixel graphs," *Journal of digital imaging*, vol. 34, no. 1, pp. 162–181, 2021.

[14] A. K. Nambisan, A. Maurya, N. Lama, T. Phan, G. Patel, K. Miller, B. Lama, J. Hagerty, R. Stanley, and W. V. Stoecker, "Improving automatic melanoma diagnosis using deep learning-based segmentation of irregular networks," *Cancers*, vol. 15, no. 4, p. 1259, 2023.

[15] S. Qasim Gilani, T. Syed, M. Umair, and O. Marques, "Skin cancer classification using deep spiking neural network," *Journal of Digital Imaging*, pp. 1–11, 2023.

[16] P. Tschandl, C. Rosendahl, and H. Kittler, "The ham10000 dataset, a large collection of multi-source dermatoscopic images of common pigmented skin lesions," *Scientific data*, vol. 5, no. 1, pp. 1–9, 2018.

[17] H. El-Khatib, D. Popescu, and L. Ichim, "Deep learning–based methods for automatic diagnosis of skin lesions," *Sensors*, vol. 20, no. 6, p. 1753, 2020.

[18] S. Modak, E. Abdel-Raheem, and L. Rueda, "Lung nodule segmentation on ct scan images using patchwise iterative graph clustering," in *2023 IEEE International Symposium on Circuits and Systems (ISCAS)*, pp. 1–5, IEEE, 2023.

[19] M. Ahammed, M. Al Mamun, and M. S. Uddin, "A machine learning approach for skin disease detection and classification using image segmentation," *Healthcare Analytics*, vol. 2, p. 100122, 2022.

[20] P. Chatterjee, S. Jana, and S. Ghosh, "Comparative study of opencv inpainting algorithms," *Global Journal of Computer Science and Technology*, vol. 21, no. 2, pp. 27–37, 2021.

[21] D. Liu and J. Yu, "Otsu method and k-means," in *2009 Ninth International conference on hybrid intelligent systems*, vol. 1, pp. 344–349, IEEE, 2009.

[22] R. Achanta, A. Shaji, K. Smith, A. Lucchi, P. Fua, and S. Süsstrunk, "Slic superpixels compared to state-of-the-art superpixel methods," *IEEE transactions on pattern analysis and machine intelligence*, vol. 34, no. 11, pp. 2274–2282, 2012.

[23] C. Y. Ren and I. Reid, "gslic: a real-time implementation of slic superpixel segmentation," *University of Oxford, Department of Engineering, Technical Report*, pp. 1–6, 2011.

[24] J. Deng, W. Dong, R. Socher, L.-J. Li, K. Li, and L. Fei-Fei, "Imagenet: A large-scale hierarchical image database," in *2009 IEEE conference on computer vision and pattern recognition*, pp. 248–255, Ieee, 2009.

[25] M. Hossin and M. N. Sulaiman, "A review on evaluation metrics for data classification evaluations," *International journal of data mining & knowledge management process*, vol. 5, no. 2, p. 1, 2015.

[26] A. Adegun and S. Viriri, "Deep learning techniques for skin lesion analysis and melanoma cancer detection: a survey of state-of-the-art," *Artificial Intelligence Review*, vol. 54, pp. 811–841, 2021.

Spatial Multiplexing MIMO for Remote Areas employing MMSE Parallel Interference Cancellation for Non-Orthogonal GFDM

Danilo Gaspar[*], Vanessa Mendes Rennó[†], Luciano Leonel Mendes[†], Tales Pimenta[‡] and Shahab Ehsanfar[§]

[*]Hitachi Kokusai Linear, Santa Rita do Sapucaí - MG, Brazil
Email: dgaspar@linear.com.br
[†]Instituto Nacional de Telecomunições - Inatel, Santa Rita do Sapucaí - MG, Brazil
Email: vanessarenno@inatel.br and lucianol@inatel.br
[‡]Universidade Federal de Itajubá - Unifei, Itajubá - MG, Brazil
Email: tales@unifei.edu.br
[§]Professorship of Communications Engineering, Technische Universität Chemnitz, Germany
Email: shahab.ehsan-far@etit.tu-chemnitz.de

Abstract—Remote areas are regions isolated from urban areas, mainly characterized by a low population density, such as small settlements and medium-size rural properties. Grant broadband Internet access in remote areas figures as a technical and economic challenge to any terrestrial wireless network. The high cost of spectrum license and low revenue associated with remote areas restrains this sort of investment. To keep the evolutionary trend initiated by the Remote Area Access Network for the 5th Generation (5G-RANGE), next generation of mobile communications must provide reliable long-range and cost-effective connection in these regions by exploring unused TV channels in TV White Space (TVWS) regime with high spectral efficiency. In this sense, we propose to make use of spatial multiplexing (SM)-multiple-input-multiple-output (MIMO) and a non-orthogonal (NO) waveform, compromised to meet low out-of-band (OOB) emission while harvesting multiplexing and extra diversity gain at the same time when employing a low complexity minimum mean squared error (MMSE)-parallel interference cancellation (PIC) detector. We checked out the performance of this maximum likelihood (ML) approaching method, reproducing the frame error rate (FER) from the base reference and analyze the proposed frame structure face the requirements of remote areas targeted for smart farm scenario in the agribusiness vertical.

Index Terms—Spatial Multiplexing, MIMO, MMSE-PIC, GFDM, Non-orthogonal

I. INTRODUCTION

Fifth Generation (5G) of mobile network has recently started to be deployed in capitals and high populate centres. It was discussed and designed in the last years mostly to address three main scenarios: i) the enhanced mobile broadband (eMBB) communications scenario, to provide high data rate; ii) the ultra-reliable and low-latency communications (URLLC), allowing several new applications to run over mobile networks and; iii) massive machine type communications (mMTC), where a large number of power-limited devices are expected to be connected to the network. Nonetheless, in face of the scope of 5G, there is still many challenges to bring broadband connectivity to the rural and remote regions. Particularly in Brazil, a big digital gap between urban and rural

areas remains. Urban areas have a typical coverage around 65% while the rural penetration is just 34% [1]. In order to meet this demand, the Remote Area Access Network for the 5th Generation (5G-RANGE) project, a Brazil-Europe bilateral cooperation project, were discussed to provide reliable long-range and cost-effective connection in these regions. It also incorporates a cognitive engine allowing local and rural operators to exploit unused TV channels, also known as TV White Space (TVWS), as secondary network in a opportunistic usage of vacancy channels [2]. As a result, a practical transceiver, encompassing frame structure and diversity multiple-input-multiple-output (MIMO) support were designed and evaluated [3], opening opportunities for improvements [4]. Aiming to continue the evolutionary trajectory started by 5G-RANGE and keep the track of the Sixth Generation (6G) of mobile network, under discussion by the academic community, this paper proposes to make use of spatial multiplexing (SM)-MIMO and non-orthogonal (NO) waveform, more specifically the Generalized Frequency Division Multiplexing (GFDM), in order to achieve a higher spectral efficiency with low out-of-band (OOB) emission, besides harvesting multiplexing and extra diversity gain at the same time when employing a low complexity minimum mean squared error (MMSE)-parallel interference cancellation (PIC) detector [5]. These achievements are essential for a mobile network designed for remote areas since low OOB emissions are mandatory for coexistence with legacy technologies, higher data rates are needed to cover the requirements for imaging and real-time positioning in agribusiness, and robustness is a requirement for increasing coverage and reliability. Also, a comprehensive analysis of estimation and detection algorithms for future mobile communication generations is crucial for applying and advancing innovative technologies, including reconfigurable intelligent surfaces (RIS) [6], and novel waveforms [7].

To this end, this work is organized as follows: Section II discusses the requirements for remote areas envisioning

979-8-3503-8083-5/23 $31.00 © 2023 IEEE

the agribusiness and smart farm scenario. In Section III, we present the necessary theoretical background about GFDM and SM-MIMO linear model, brief reviewing the intricate working flow involving the MMSE-PIC detection. Section IV details a suitable frame structure considering a NO waveform and reproduces the frame error rate (FER) simulation results in [5], taking it as a reference, demonstrating the correctness of the algorithm interpretation, a necessary step prior to a future hardware implementation. Finally, in Section V, we conclude this work listing the open topics and future research opportunities.

II. REQUIREMENTS FOR REMOTE AREAS

Remote areas are regions isolated from urban areas, mainly characterized by a low population density and lack of basic services, including under served rural areas, such as countryside, villages and farms, which normally experience limited internet access and few essential facilities, from basic schools to health centers and small supermarkets. These areas are prone to innovation, not only for the agribusiness segment, but also for small to medium-sized producers, leading to improvements in practically all agriculture production process. For example, in the smart farm scenario, this use case attempts to provide capabilities for automated machinery, production traceability, transportation of goods, Internet of Things (IoT) gateways, remote maintenance, cattle monitoring and other services. According to [2], these applications corresponds to numerology number 3, where the radio access network shall operate mainly on 700 MHz band or below it, in order to achieve long distances of at least 30 km of cell radius. To explore the channel vacancy in TVWS regime, typically with bandwidth between 6 to 8 MHz, where the aggregation of several channels is not always possible, high spectral efficiency is a key factor, once high order modulation might be accompanied by high transmission power to keep long distance coverage. In this sense, SM-MIMO in VHF and UHF bands with a moderate number of antennas at the transmitter and receiver is feasible. Moreover, low OOB emission is mandatory to grant the coexistence with incumbents. The channel typically has long-range and doubly-dispersive profile [8]. Mobility is loose assuming that in-the-field vehicles, including aerial vehicles like drones, usually travel at 60 km/h or less. Thus, maximum Doppler deviation at 800 MHz is 44.5 Hz for a 200 kHz coherence bandwidth, the 50% coherence time is 4 ms and the maximum channel delay spread is 1253 ns. With respect to user experience key performance indicators (KPIs), the end-to-end latency is 50 ms and the throughput is typically 1 Mbps, thus the network must provide at least 100 Mbps at the edge of the cell while assuring at least 30 kbps per connected device during busy hours.

III. THEORETICAL BACKGROUND

In a broad context, the digital communication physical layer (PHY) has two primary functions: converting digital information into a transmission waveform and recovering information from the distorted received signal. At the transmitter, data bits undergo processing techniques such as randomization, forward error correction (FEC), and interleaving to enhance the system's resilience against the challenges posed by the mobile channel. These processed bits are then modulated into waveforms that are finely tuned for mobile channels using methods like symbol modulation and multicarrier techniques. In modern mobile systems, the PHY must adeptly handle complex double-dispersive MIMO channels characterized by time-variant impulse responses and the presence of additive white Gaussian noise (AWGN). On the receiver side, waveform demodulation encompasses time and frequency synchronization, channel equalization, waveform demodulation, antenna decoupling, and symbol-to-bit recovery. The subsequent bit decoding step corrects errors introduced by the channel. Figure 1 depicts the main PHY blocks.

One of the main characteristic of the 5G-RANGE transceiver is the use of a flexible waveform, named GFDM [9]. This section will provide an overview of the fundamental theoretical concepts of the GFDM modulation technique, delve into the details of the MIMO communication channel [10], and revisit the MMSE-PIC [5], an iterative detection technique. These concepts are essential for a comprehensive analysis of the numerical simulation results, which will be discussed in Section IV.

A. Fundamentals on GFDM

GFDM is a flexible multicarrier modulation scheme where each subcarrier conveys data symbols within sequential time segments, known as subsymbols [11]. A GFDM block can hold a maximum of N data symbols, where $N = M \times K$. Here, M represents the count of subsymbols, and K denotes the number of subcarriers [9]. Each individual subcarrier undergoes pulse shaping to control the waveform behaviour, tailoring it for a given application. This process involves filtering by a prototype pulse represented as $\vec{g}_{k,m}$, which is shifted in both time and frequency domains. These shifted versions can be methodically organized within a comprehensive modulation matrix G. Depending on the choice of the transmission pulse, there can be overlap between adjacent subcarriers and subsymbols, resulting in a non-orthogonal (NO) system. Then, the GFDM transmitter and receiver processing can be described as matrix operations [12]. The modulated signal vector $\vec{x}$ can be expressed as

$$\vec{x} = G\vec{d}, \tag{1}$$

where $\vec{d}$ is the N-element data vector, and G is the $N \times N$ transmit matrix. This matrix is composed of all possible combinations of the K frequency-shifted and the M time-shifted versions of the transmitter pulse.

To protect the M subsymbols from inter-block interference (IBI) introduced by the dispersive communication channel, a cyclic prefix (CP) is added to the GFDM block. This CP is designed to be longer than the channel delay profile, ensuring effective protection against IBI [9].

979-8-3503-8083-5/23 $31.00 © 2023 IEEE

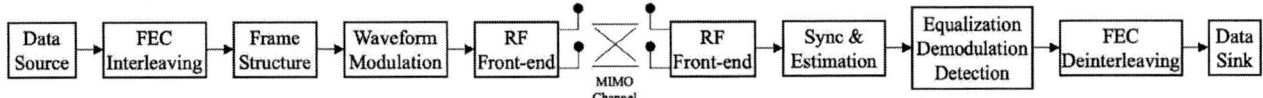

Fig. 1: Simplified block diagram of the radio PHY.

On the receiver side, assuming perfect synchronization and after removing the CP, the received signal can be characterized as

$$\vec{y} = \mathring{H}\vec{x} + \vec{w}. \tag{2}$$

Here, $\mathring{H}$ represents the channel's circulant matrix derived from channel impulse response (CIR), denoted as $\vec{h}$. The matrix $\mathring{H}$ represents the circular convolution matrix between the vector $\vec{x}$ and the impulse response $\vec{h}$. Also, $\vec{w}$ represents the AWGN vector with zero mean and variance $\sigma_{\vec{w}}^2$.

After equalization, the transmitted symbols can be retrieved using a receiving prototype pulse, denoted as $\vec{v}$, suitably adjusted both in the time and frequency domains [13]. The receiving filters can also be organized in a receiving matrix V, leading to

$$\hat{\vec{d}} = V\hat{\vec{y}}. \tag{3}$$

Distinct demodulation matrices can be employed to recover the transmitted data symbols.

B. SM-MIMO Linear Model

Multi-antenna techniques are crucial for improving communication performance, either by mitigating or capitalizing on the multipath scattering within the communication channel [14]. Among these techniques, SM-MIMO is notable for its ability to increase spectral efficiency. It achieves this by transmitting multiple data streams simultaneously over the same time and frequency resources, resulting in higher data rates without the need for extra bandwidth. This concept can be extended to a system with N_T transmitting antennas and N_R receiving antennas, representing a broader mobile communication setup. When combined with GFDM, N_T parallel streams, each carrying N complex data symbols per block are simultaneously transmitted [10].

Each of the N_T transmitting antennas sends a modulated signal $\vec{x}_j \in \mathbb{C}^{N \times 1}$ in the time domain ($j = 1, 2, \ldots, N_T$). This signal is generated through a linear transformation of the jth data stream, represented in (1) with the subscript j. To describe the modulated signal for the entire system, we stack these complex data streams into $\vec{d} \in \mathbb{C}^{N_T N \times 1}$. Additionally, we introduce an extended generic modulation matrix $\bar{G} \in \mathbb{C}^{N_T N \times N_T N}$, which contains N_T GFDM modulation matrices G along its main diagonal. This allows us to define the modulated signal for the entire system as $\vec{x} \in \mathbb{C}^{N_T N \times 1}$.

It is considered the equivalent baseband CIR between the jth transmitting antenna and the ith receiving antenna as a finite discrete sequence of L taps, denoted by $\vec{h}_{i,j} \in \mathbb{C}^{L \times 1}$, where $i = 1, 2, \ldots, N_R$. Assuming perfect synchronization, and following CP removal, the received block symbol at the ith receiving antenna is represented by $\vec{y}_i \in \mathbb{C}^{N \times 1}$, while

$\vec{y} \in \mathbb{C}^{N_R N \times 1}$ encompasses all received signals vertically stacked, as depicted in (2). Here, $\bar{H} \in \mathbb{C}^{N_R N \times N_T N}$ stands as a structured linear transformation matrix. Its elements, labeled as $\mathring{H}_{i,j} \in \mathbb{C}^{N \times N}$, are equivalent circulant matrices derived from the CIR between the ith and jth receiving-transmitting antennas, correspondingly. The vector $\vec{w} \in \mathbb{C}^{N_R N \times 1}$ denotes the AWGN for all receiving antennas, arranged vertically.

To recover the data at the receiver, it is feasible to incorporate (1) into (2), based on the aforementioned definitions. In this manner, a new matrix is formulated, denoted as $\tilde{H} \in \mathbb{C}^{N_R N \times N_T N}$, which encompasses both transformation matrices, obtained by the product $\bar{H}\bar{G}$. However, the direct solution in time domain involves an undesired and even prohibitive $N_R N \times N_T N$ matrix inversion for the entire system transformation matrix $\tilde{H}$. Hence, it is advantageous to represent the system in the frequency domain and factor it appropriately, ensuring that the majority of the energy is concentrated at the corners of the main diagonal of $\tilde{H}$.

In [5], the concepts of system factorization for (2), in the context of GFDM and SM-MIMO, were introduced. Permutation matrices reorganize the system's structure in a more practical manner, focusing on storage efficiency and facilitating parallel computation. Figure 2a illustrates an example of a factorized and permuted orthogonal SM-MIMO system with $M = 1$ and $K = 12$, while Figure 2b represents a factorized and permuted NO SM-MIMO system with $M = 3$ and $K = 4$. In both scenarios, $N_T = N_R = 2$ and $\vec{y}_s = H_s \vec{d}_s$ represents a noiseless factorized subsystem.

The orthogonal factorization case is extensively discussed in the literature [15], where the elements of the corresponding matrix $\bar{H}$ in the frequency domain represent the equivalent flat channel between each pair of antennas on the subcarrier k. Thus, the orthogonal multicarrier detection problem can be divided into K subsystems, each with dimensions $N_T \times N_R$. In the NO case, the permutation operation proposed in [5] results in the reorganization of the system in frequency domain into M distinct linear subsystems. Each of these subsystems carry $N_T K$ data symbols, s.t. each mth subsystem, for $m = 0 \ldots M - 1$ subsymbols, has dimensions of $N_R K \times N_T K$, mainly defined by a band diagonal matrix. For K_{on} active subcarriers, the dimensions in the transmitter side can be replaced by $N_{on} = MK_{on}$ active resources.

The factorization procedure proposed in [5] indeed yields a novel system representation, better suited for storage and parallel computation, carrying significant potential for achieving considerable complexity reduction when utilizing a NO waveform based on GFDM. In [16], an algorithm is introduced to estimate the diagonal elements of an inverse matrix without explicitly calculating the inverse, capitalizing on its band-

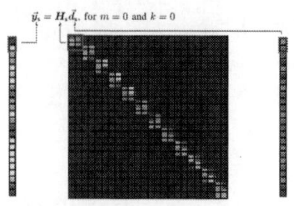 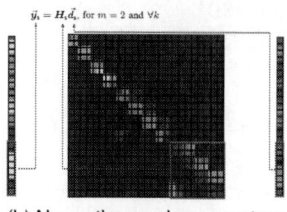

(a) Orthogonal case for $M = 1$ and $K = 12$.   (b) Non-orthogonal case system for $M = 3$ and $K = 4$.

Fig. 2: Factorized and permuted MIMO systems with $N_\mathrm{T} = N_\mathrm{R} = 2$.

diagonal structure [17]. This approach turns the NO algorithm complexity comparable to orthogonal case.

With these established definitions, the iterative MMSE-PIC detection technique can be revisited [5]. This technique excels in data recovery, endeavoring to harness both diversity and multiplexing advantages from the SM-MIMO approach, particularly when employing the NO GFDM waveform.

C. Iterative MMSE-PIC Detector

In [5], Matthé *et al.* proposed an iterative, low-complexity, and low-latency MMSE-PIC method. This technique is well-suited for, though not confined to, NO GFDM detection within SM-MIMO applications. Its primary goal is to simultaneously leverage both multiplexing and diversity gains. The complexity reduction relies on the equivalence between a PIC process and a single linear minimum mean square error (LMMSE) estimation [18], [19], facilitated by a-priori knowledge about system parameters [17]. Additionally, provided the presumption that both inputs and outputs are continuous Gaussian random variables, linear estimators become a compelling facilitator for iterative detection. They furnish estimates and uncertainties for soft-input-soft-output (SISO) demapping and decoding operations. This assumption plays a key role in enabling the use of factorized systems such as described in Section III-B, thus allowing the exploration of low-complexity estimation techniques based on specific band-structured matrices [20].

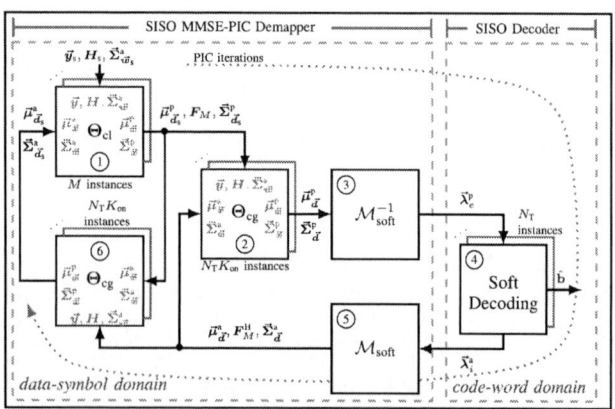

Fig. 3: MMSE-PIC detector.

Figure 3 outlines the schematic of the proposed iterative SISO MMSE-PIC detector. All blocks are enumerated according to respective execution order. The variable superscripts (a) and (p) within the block diagram refer, respectively, to a-priori and posteriori parameters. The starting point encompasses the received signal vector $\vec{y}$, assuming perfect synchronization and channel state information at the receiver (CSIR). Post CP removal, the equivalent MIMO channel $\tilde{H}$ and the variance of the corresponding AWGN $\sigma_{\tilde{w}}^2$ are considered. On sequence, the linear system, as presented in (2), undergoes a factorization into M subsystems in the frequency domain. Each resultant subsystem is the entry point to estimate the expectation and variance of $\vec{d}_\mathrm{s}$. To achieve this, a linear estimation function, called Θ_cl, comes into play, taking into account available a-priori information as the expected value $\vec{\mu}_{\vec{d}_\mathrm{s}}^\mathrm{a}$ and the variance $\vec{\Sigma}_{\vec{d}_\mathrm{s}}^\mathrm{a}$ from a given subsystem. The posteriori parameters, jointly to the M-point discrete Fourier transform (DFT) matrix, F_M, forms another linear system, used by Θ_cg to estimate the expected value $\vec{\mu}_{\vec{d}}^\mathrm{p}$ and the variance $\vec{\Sigma}_{\vec{d}}^\mathrm{p}$ in time domain for the soft demodulator $\mathcal{M}^{-1}$ and SISO decoder. In a similar manner, the feedback data symbol parameters from the soft modulator $\mathcal{M}$ are further used with F_M^H as a third linear system, but now to estimate, through Θ_cg, the subsystem expectancy and variance in frequency domain, resulting in a-priori information for a new iteration. In the absence of any initial a-priori information, these parameters are initialized as null and unitary vectors, respectively.

Blocks 1, 2 and 6 employ the component-wise conditionally unbiased (CWCU)-LMMSE estimator [21], with the difference that Θ_cl explores the band-diagonal structure of H_s, in order to reduce the computational cost involved in matrix inversion, while the estimators, identified by Θ_cg, use the conjugate gradient (CG) to estimate only the term $\hat{\vec{x}} = A^{-1}\vec{b}$ for a generic linear system in the form $A = \vec{x}\vec{b}$, without explicitly inverting A [17], [22]. These techniques, besides frequency domain factorization, are responsible for complexity reduction of the NO MMSE-PIC detector close to the orthogonal case. Inter-block processing, s.t. permutation, concatenation and indexing were intentionally omitted for simplicity of illustration in Figure 3, but are detailed in aforementioned references [5].

The function $\mathcal{M}_\mathrm{soft}^{-1}$ performs soft demapping and it is responsible for estimating individual extrinsic bit probabilities $\vec{\lambda}_e^\mathrm{p}$ for each data symbol, constrained by the constellation set $\mathcal{C}$. Assuming uncorrelated noise and without requiring prior knowledge of the bit sequence, the approximated log-likelihood ratios (LLRs) can be efficiently derived, yielding minimal impact on the overall detection performance [22]. It is worth to highlight that various LLRs sequences are necessary in the context of SM-MIMO systems that utilize frame structures accommodating integer multiples of a code-word. These code-words are simultaneously transmitted by N_T antennas across N_s block symbols per frame. Consequently, after appropriately aggregating and arranging each code-word, the process of soft decoding constitutes the final step in each PIC iteration.

979-8-3503-8083-5/23 $31.00 © 2023 IEEE

Soft channel decoding is responsible for retrieving the transmitted bit information from demapper estimates, adhering to the code constraints. Generally, this decoding process employs algorithms capable of accurately calculating or approximating the posteriori probability of the information bits or a more general measure of reliability for each information bit. In order to enhance the robustness of channel code and the accuracy of its decisions, bit interleaving operations among different streams are performed internally in block 4, according to transmitter interleave rules. As a result of soft decoding, decoded LLRs are produced. These LLRs are needed either for generating re-encoded $\vec{\lambda}_i^a$ for the subsequent iteration or for making a definitive hard bit decision $\vec{b}$.

This encloses the iterative outer loop of the MMSE-PIC process, enabling the commencement of a new iteration while incorporating the a-priori information acquired from soft decoding. Throughout successive iterations, both the demapper and decoder procedures mutually benefit from the exchange of enhanced information between them.

Next, we present the results obtained by a Monte Carlo simulation focused solely on the aforementioned MMSE-PIC approach within the context of GFDM modulation. This method, as highlighted in [20], stands out as the preeminent detection technique for SM-MIMO applications in terms of both performance and complexity.

IV. NUMERICAL SIMULATION RESULTS

In this section, we present the bit error rate (BER) and the FER obtained by a Monte Carlo simulation of the MMSE-PIC detector suitable for the smart farm scenario in agribusiness vertical. We identified that the proposed frame structure is quite similar to numerology number 3 in [2]. The simulation considers a 4×4 MIMO system under practical extended typical urban (ETU) channel model, with a maximum delay spread of $5\,\mu s$. Table I shows the power delay profile (PDP) and its equivalent base band nearest discrete time indexes, for a sampling frequency of $f_s = 23.04$ MHz. Coincident indexes have their average power σ_l^2 summed. Every simulated CIR between any transmitting and receiving antenna is normalized by its squared root energy.

TABLE I: Power delay profile of 3GPP ETU channel.

τ_l [ns]	0	50	120	200	230	500	1600	2300	5000
round($f_s\tau_l$)	0	1	3	5	5	12	37	53	115
σ_l^2 [dB]	-1.0	-1.0	-1.0	0.0	0.0	0.0	-3.0	-5.0	-7.0

Although we consider perfect synchronization and CSIR, the frame structure aims to employ N_T preambles, simultaneously transmitted, where each antenna carries an orthogonal sequence to jointly achieve synchronization and channel estimation [3], [23]. This is a first immediate countermeasure to avoid using pilot subcarriers and, consequently, a more sophisticated mechanism to overcome inter-symbol interference (ISI), inter-carrier interference (ICI) and inter-antenna interference (IAI) by considering pilot-aided and NO waveforms to recover synchronization and CSIR [24], [25]. Following the preamble, $N_s = 7$ data blocks are transmitted, carrying

TABLE II: Simulation parameters.

Description	Symbol/Parameter	Value
Number of antennas	$[N_T, N_R]$	[4, 4]
Bits per symbol and coding	$[\mu, r]$	[4, 1/2]
Channel coding	{Convolutional; BCJR}	{7, [171, 133]; Log-MAP}
Code-word length	N_c	1008
Channel model	{PDP, CSIR}	{ETU, Perfect}
Blocks per frame	[#Preambles, #Data]	[1, 7]
Allocated subcarriers	$[K_{on}, K]$	[3, 128]
Allocated subsymbols	$[M_{on}, M]$	[12, 12]
Prototype filter and roll-off	$\{g, \alpha\}$	{RC, 1}
CP and time-window samples	$[N_{cp}, N_w]$	[384, 16]
Block and frame duration [µs]	$[T_B, T_F]$	[84.7, 931.4]
Block and frame efficiency [dB]	$[\eta_B, \eta_F]$	[-1.04, -0.58]
SNR	$1/\sigma_{\tilde{w}}^2$	$\mu r E_b/N_0$

N_T code-words of 1008 bits after rate 1/2 convolutional encoding. Data symbol modulation uses quadrature amplitude modulation (QAM) s.t. $\mu = 4$ bits per symbol is transmitted by each resource element. The channel decoder employs SISO log-MAP Bahl-Cocke-Jelinek-Raviv (BCJR) algorithm [26]. In order to meet the low OOB emission, a ramp-up and ramp-down raised cosine (RC) time-window is used. To mitigate IBI, the CP duration is $16.7\,\mu s$. Table II summarizes the simulated numerology, where the bandwidth of a single GFDM subcarrier equals a physical resource block (PRB) of a Long-Term Evolution (LTE) system.

Figure 4 illustrates the BER and the FER of the simulated system. The BER curves represents a 1x1 system used to demonstrate the exactness of noise calibration by comparing the orthogonal case, when $\alpha = 0$, to theoretical curves of AWGN and flat Rayleigh channel. In the same figure, the FER curve is plotted against the reference curve from [5], showing the preciseness of our interpretation about the intricate details involving the MMSE-PIC algorithm for NO waveforms. It simulates the reception of 3 allocated PRB, simultaneously transmitted by $N_T = 4$ antennas in ETU channel model, assuming constant CIR during the frame period.

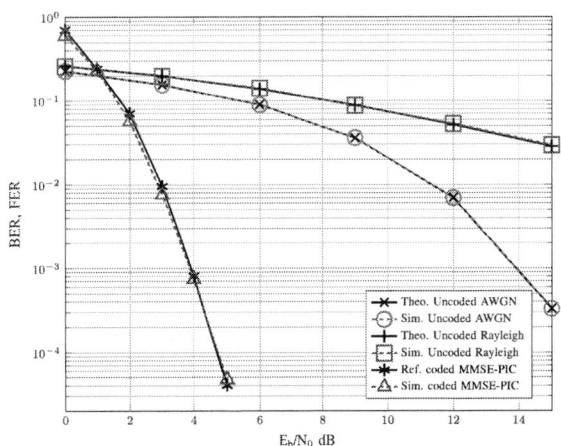

Fig. 4: Uncoded BER of a 1x1 system in AWGN and flat Rayleigh channel besides coded FER of a 4x4 SM-MIMO-NO-GFDM employing the MMSE-PIC detector.

In the context of smart farm use case, allocating all 83 available PRB, results in an occupied bandwidth (BW) of

15 MHz and a gross bit rate of 120 Mbps, equivalent to a total throughput of approximately 60 Mbps considering the channel coding rate $r = 1/2$, insufficient in face of the required 100 Mbps. In terms of coverage, the CP protection ensures a 30 km, which is considered the minimum coverage for a remote are mobile network. As the results presented in this paper shows, the MMSE-PIC detection method is indeed a prominent tool capable to achieve the maximum likelihood (ML) lower bound with reasonable complexity and low latency for remote and rural areas applications. Further improvements might be achieved by improving the efficient of the coding technique, while a new numerology for the frame structure can be proposed to improve the coverage.

V. CONCLUSIONS

In this work, we present a review regarding on the remote areas requirements, especially for the smart farm scenario in the agribusiness vertical. We also present theoretical background encompassing the GFDM waveform, the basic on SM-MIMO and an explanation on the MMSE-PIC detector aiming NO waveforms as an advance technique to harvest both multiplexing and diversity gains at the same time. We analyzed the frame structure and detection method performance w.r.t. energy efficiency, throughput and cell coverage against the 5G-RANGE parameters for remote areas. The MIMO detection technique can indeed provide close to optimum performance with a more affordable complexity when considering the parameters for the remote area scenario. Nevertheless, there are new challenges and research opportunities in this subject. A new numerology for 5G-RANGE can provide better coverage, while research in the coding scheme can allow for better spectrum efficiency and robustness. Also, exploiting precoded pilots for the channels estimation in the NO GFDM frame can lead to a more efficient frame structure.

ACKNOWLEDGMENT

The authors would like to thank Fapemig and CNPq-Brasil (Grant No. 303282/2021-5) for their financial support. This work was partially supported by Brasil 6G Project (RNP/MCTIC, Grant No. 01245.010604/2020-14), SAMURAI project (FAPESP, Grant No. 20/05127-2) and Advanced Academic Education in Telecommunications Networks and Systems project (Huawei, contract No PPA6001BRA23032110257684).

REFERENCES

[1] A. M. Cavalcante, M. V. Marquezini, L. L. Mendes, and C. S. Moreno, "5G for Remote Areas: Challenges, Opportunities and Business Modeling for Brazil," *IEEE Access*, vol. 9, pp. 10 829–10 843, 2021.

[2] L. L. Mendes *et al.*, "Enhanced Remote Areas Communications: The Missing Scenario for 5G and Beyond 5G Networks," *IEEE Access*, vol. 8, pp. 219 859–219 880, 2020.

[3] A. Ferreira *et al.*, "5G-RANGE Project Field Trial," in *European Conference on Networks and Communications (EuCNC)*, 2019, pp. 490–494.

[4] W. Dias, A. Ferreira, R. Kagami, J. S. Ferreira, D. Silva, and L. L. Mendes, "5G-RANGE: A transceiver for remote areas based on software-defined radio," in *European Conference on Networks and Communications (EuCNC)*, 2020, pp. 100–104.

[5] M. Matthé, D. Zhang, and G. Fettweis, "Low-Complexity Iterative MMSE-PIC Detection for MIMO-GFDM," *IEEE Trans. Commun.*, vol. 66, no. 4, pp. 1467–1480, 2018.

[6] V. Tapio, I. Hemadeh, A. Mourad, A. Shojaeifard, and M. Juntti, "Survey on reconfigurable intelligent surfaces below 10 ghz," *EURASIP Journal on Wireless Communications and Networking*, vol. 2021, pp. 1–18, 2021.

[7] D. Zhang, M. Matthé, L. L. Mendes, and G. Fettweis, "A study on the link level performance of advanced multicarrier waveforms under mimo wireless communication channels," *IEEE Transactions on Wireless Communications*, vol. 16, no. 4, pp. 2350–2365, 2017.

[8] A. M. Pessoa *et al.*, "CDL-based Channel Model for 5G MIMO Systems in Remote Rural Areas," in *16th International Symposium on Wireless Communication Systems (ISWCS)*, 2019, pp. 21–26.

[9] G. Fettweis, M. Krondorf, and S. S. Bittner, "GFDM - Generalized Frequency Division Multiplexing," in *IEEE 69th Vehicular Technology Conference*, 2009.

[10] E. Öztürk, E. Basar, and H. A. Çırpan, "Spatial modulation GFDM: A low complexity MIMO-GFDM system for 5G wireless networks," in *IEEE International Black Sea Conference on Communications and Networking (BlackSeaCom)*, 2016.

[11] N. Michailow, M. Matthé, I. S. Gaspar, A. N. Caldevilla, L. L. Mendes, A. Festag, and G. Fettweis, "Generalized Frequency Division Multiplexing for 5th Generation Cellular Networks," *IEEE Trans. Commun.*, vol. 62, no. 9, pp. 3045–3061, 2014.

[12] L. L. Mendes, N. Michailow, M. Matthé, I. Gaspar, D. Zhang, and G. Fettweis, "GFDM: providing flexibility for the 5G physical layer," *Opportunities in 5G Networks: A Research and Development Perspective*, p. 325, 2016.

[13] J. S. Ferreira, H. D. Rodrigues, A. A. Gonzalez, A. Nimr, M. Matthe, D. Zhang, L. L. Mendes, and G. Fettweis, "GFDM frame design for 5G application scenarios," *Journal of Communication and Information Systems*, vol. 32, no. 1, 2017.

[14] J. R. Hampton, *Introduction to MIMO communications*. Cambridge university press, 2013.

[15] N. Suga, R.Sasaki, and T. Furukawa, "Channel Estimation Using Matrix Factorization Based Interpolation for OFDM Systems," in *IEEE 90th Vehicular Technology Conference*, 2019.

[16] C. Bekas, E. Kokiopoulou, and Y. Saad, "An estimator for the diagonal of a matrix," *Applied numerical mathematics*, vol. 57, no. 11-12, pp. 1214–1229, 2007.

[17] M. Matthe, D. Zhang, and G. Fettweis, "Iterative Detection using MMSE-PIC Demapping for MIMO-GFDM Systems," in *22th European Wireless Conference*, 2016.

[18] M. Huemer, O. Lang, and C. Hofbauer, "Component-wise conditionally unbiased widely linear MMSE estimation," *Signal Processing*, vol. 133, pp. 227–239, 2017.

[19] M. A. Albreem, W. Salah, A. Kumar, M. H. Alsharif, A. H. Rambe, M. Jusoh, and A. N. Uwaechia, "Low Complexity Linear Detectors for Massive MIMO: A Comparative Study," *IEEE Access*, vol. 9, pp. 45 740–45 753, 2021.

[20] D. Gaspar, L. L. Mendes, and T. C. Pimenta, "A review on principles, performance and complexity of linear estimation and detection techniques for MIMO systems," *Frontiers in Communications and Networks*, vol. 4, p. 968370, 2023.

[21] M. Huemer, O. Lang, and C. Hofbauer, "Component-wise conditionally unbiased widely linear MMSE estimation," *Signal Processing*, vol. 133, pp. 227–239, 2017.

[22] C. Studer, S. Fateh, and D. Seethaler, "ASIC Implementation of Soft-Input Soft-Output MIMO Detection Using MMSE Parallel Interference Cancellation," *IEEE Journal of Solid-State Circuits*, vol. 46, no. 7, pp. 1754–1765, 2011.

[23] J. S. Ferreira, H. Rodrigues, A. Gonzalez, A. Nimr, M. Matthé, D. Zhang, L. Mendes, and G. Fettweis, "GFDM Frame Design for 5G Application Scenarios," *Journal of Communication and Information Systems*, vol. 32, no. 1, Jul. 2017.

[24] S. Ehsanfar, M. Matthe, D. Zhang, and G. Fettweis, "Interference-free pilots insertion for MIMO-GFDM channel estimation," in *IEEE Wireless Communications and Networking Conference (WCNC)*, 2017, pp. 1–6.

[25] S. Ehsanfar, M. Matthé, M. Chafii, and G. P. Fettweis, "Pilot- and CP-Aided Channel Estimation in MIMO Non-Orthogonal Multi-Carriers," *IEEE Trans. Wireless Commun.*, vol. 18, no. 1, pp. 650–664, 2019.

[26] L. Bahl, J. Cocke, F. Jelinek, and J. Raviv, "Optimal decoding of linear codes for minimizing symbol error rate (corresp.)," *IEEE Trans. Inf. Theory*, vol. 20, no. 2, pp. 284–287, 1974.

Comparison of ANFIS and ANN for Small-Signal Modelling of GaN HEMT up to 40 GHz

Bagylan Kadirbay
Electrical and Computer Engineering
Nazarbayev University
Astana, Kazakhstan
bagylan.kadirbay@nu.edu.kz

Saddam Husain
Electrical and Computer Engineering
Nazarbayev University
Astana, Kazakhstan
saddam.husain@nu.edu.kz

Anwar Jarndal
Electrical Engineering
University of Sharjah
Sharjah, UAE
ajarndal@sharjah.ac.ae

Mohammad Hashmi
Electrical and Computer Engineering
Nazarbayev University
Astana, Kazakhstan
mohammad.hashmi@nu.edu.kz

Abstract—This paper compares the performance of the Adaptive Neuro-Fuzzy Interface System (ANFIS) and Artificial Neural Network (ANN) by developing small-signal models for Gallium Nitrite (GaN) High electron Mobility Transistor (HEMT). The theoretical background of ANFIS modeling is considered, and implemented in the context of the GaN HEMT. The studied GaN HEMT device has a wide frequency range of 40 GHz. Moreover, the full dataset comprises of 36400 data samples, which enables a through comparison of ANN and ANFIS based models. Through the integration of ANFIS, we unveil a comprehensive approach that captures intricate relationships in small signal responses, allowing for more accurate predictions and an improved understanding of the GaN HEMT behavior. The proposed algorithm has achieved high accuracy in terms of mean square error, mean absolute error, and coefficient of regression (R^2) for both training and testing data sets. To demonstrate the significance of this paper, the popular feed-forward ANN architecture is utilized with the same internal layer configuration as of ANFIS based models. The results show that ANN based models are more accurate and require less training time as compared to ANFIS based models.

Index Terms—ANN, ANFIS, Fuzzy Logic, GaN-on-Diamond HEMT, Small-Signal Modeling

I. INTRODUCTION

Gallium Nitride High Electron Mobility Transistors (GaN HEMTs) have exceptional material properties including high electron mobility, high breakdown field, high saturation velocity, outstanding thermal behavior, which enable these devices to operate at higher voltages, higher frequencies, high power density compared to conventional Silicon (Si) based transistors [1]–[4]. As a result, GaN HEMTs have found their way into Radio Frequency (RF) power electronic applications, wireless communication systems and radar to satellite communications [5]–[7]. Central to harnessing the full potential of GaN HEMTs are two indispensable processes: Small-Signal Modeling (SSM) and Large-Signal Modeling (LSM) [8]–[15]. These modeling techniques serve as the compass, guiding researchers and engineers in their pursuit of developing cutting-edge RF systems. For GaN HEMTs, SSM serves as a starting point for designing and optimizing RF amplifiers, oscillators, and other

RF components [15], [16]. It communicates how the transistor responds to small input signals, offering critical information about gain, phase shift, and stability.

Among the several SSM approaches, Machine Learning (ML) based techniques are getting popular due to their high accuracy and relatively low turn-around time compared to other methods [17]–[22]. Adaptive Neuro-Fuzzy Interface System (ANFIS) is a hybrid ML modeling technique that combines the strengths of neural networks and fuzzy logic, enabling the capture of intricate relationships within the data. Specifically, Neural networks excel at pattern recognition and nonlinearity, while fuzzy logic provides a framework for dealing with imprecise or uncertain information [23], [24]. The synergy of these approaches empowers ANFIS to create models that can approximate complex functions, making it a promising candidate for tackling the challenges posed by the SSM of Device Under Test (DUT). For example, researchers have successfully implemented the ANFIS technique in predicting terrestrial trunked radio (TETRA) RF signal [25]. The other paper showed that the ANFIS outperformed the popular ML techniques such as the Artificial Neural Network (ANN), and the Support Vector Machine (SVM) in classifying the heating value of municipal solid waste [26]. However, the implementation of ANFIS in the development of SSM for GaN HEMTs and comparison of corresponding models with other ML algorithms in terms of accuracy and efficiency are still missing.

This paper develops ANFIS based small-signal model for a GaN HEMT device grown on a diamond substrate. The developed models are assessed using various regression metrics such as Mean Squared Error (MSE), Mean Absolute Error (MAE), and Coefficient of determination (R^2). The computational efficiency of the technique for different sample sizes is also computed. Moreover, the performance on the dataset up to 40 GHz is compared with the results obtained using the ANN based models.

This paper initially discusses the DUT's physics, and then

979-8-3503-8083-5/23 $31.00 © 2023 IEEE

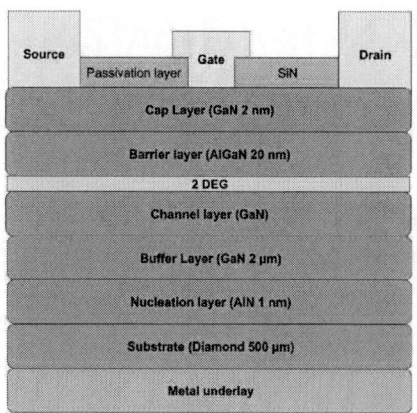

Fig. 1. The generic epitaxial structure of the DUT [27].

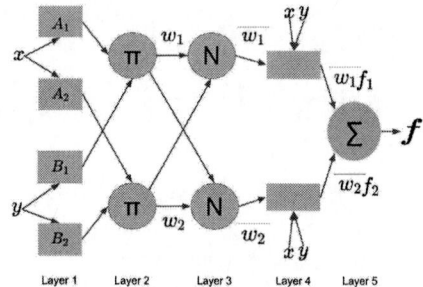

Fig. 2. ANFIS structure [23].

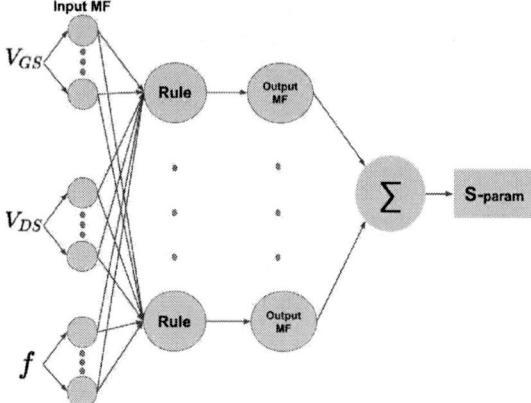

Fig. 3. Structure of the ANFIS based models.

the basic configuration of ANFIS and ANN in the Section II and Section III, respectively. Thereafter, Section IV shows data processing and modeling parts. Section V presents the results and discussion derived from that, and finally the Section VI sums up the main points.

II. PHYSICS OF THE DUT

The basic configuration of the DUT (GaN-on-Diamond HEMT) is presented in Fig. 1. The DUT has 4 gate-fingers and each has a width and length of 125 μm and 0.25 μm, respectively. Whereas, the E-beam lithography-designed gate has a length and width of 0.25 μm and 100 μm, respectively. The width between the gate-to-drain and gate-to-source are 2 μm and 1 μm, respectively. Under the cap (2 nm) and barrier (20 nm) layers is the two-dimensional electron gas, which is formed as the result of the difference in energy band structures between the doped ALGaN (barrier layer) and undoped GaN (channel layer) and serves as a highly conductive channel that can be controlled by a gate voltage. AlN-based nucleation layer and Fe-doped GaN buffer layer are located under the channel layer with 1 nm and 2 μm dimensionality, respectively. The subsequent substrate layer is composed of the diamond and has a dimension of 500 μm. Finally, the ground plane is constructed from metal underlay. The researcher papers [7], [27] and [28] have more information about the device configuration.

III. BASIC COMPOSITION OF ANFIS AND ANN

A. Fuzzy Logic

Fuzzy logic recognizes that many real-world situations are not simply false or true, but rather exist in several possible outcomes just like the way human reasoning handles uncertainty and imprecision [23]. Therefore, instead of using precise numerical values, fuzzy logic employs linguistic variables such as "less, more, high, very" to describe input and output characteristics at its core.

B. ANFIS

ANFIS generally integrates fuzzy logic concepts into a neural network framework, which results in a system that can capture both the linguistic interpretability of fuzzy logic and the adaptive learning capability of neural networks [24]. The structure of the ANFIS algorithm is illustrated in Fig. 2. Similar to traditional fuzzy logic systems, ANFIS begins with the formulation of fuzzy rules. These rules consist of linguistic variables, membership functions, and associated parameters that define the relationships between input and output variables. Then ANFIS uses a structure called a "fuzzifier" to calculate the degree of membership of input data to different linguistic terms. ANFIS employs the feed-forward neural network architecture to combine multiple layers, including

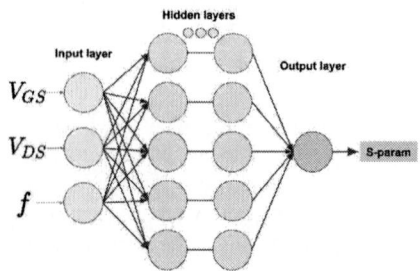

Fig. 4. ANN structure [21].

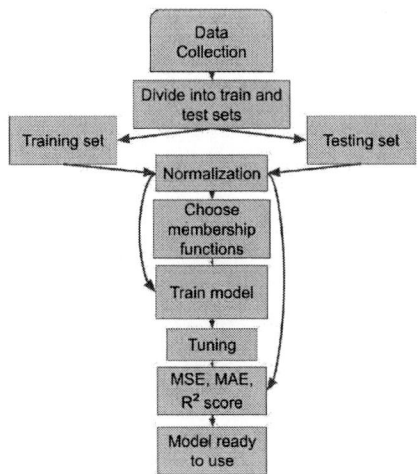

Fig. 5. Model building process.

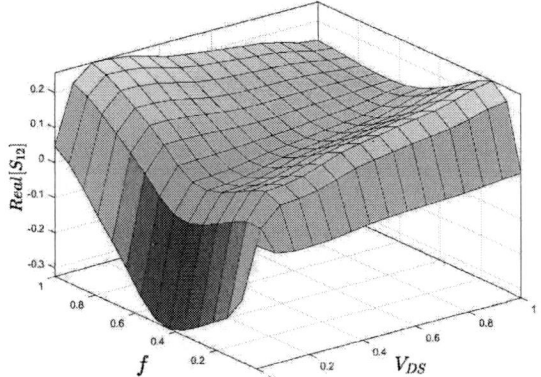

Fig. 6. 3D representation of the ANFIS based model for Real $[S_{12}]$.

input, membership function parameters, and output layers. One thing to note is that the ANFIS implements a hybrid learning algorithm, which combines the least squares method and gradient descent optimization. This algorithm adjusts the parameters of both the membership functions and the neural network connections to minimize the error between the predicted output and the actual output from the data. The ANFIS learning process also involves a forward pass to compute the output of the network and a backward pass to update the parameters based on the error signal. This process is iterated over the training data until the model converges to an optimal configuration [23].

C. ANN

The background theory for ANN is well-known and the references [7], [19], [21] have detailed information about the ANN configuration used in this research. In general, we exploited the feed-forward ANN as shown in Fig. 4. It is also known as Multilayer Perceptron (MLP) which connects the nodes without any cycle. In other words, the data flows in

one direction, from the input layer to the output layer, in the absence of any feedback loops or recurrent connections.

IV. DATA PROCESSING AND MODEL DEVELOPMENT

The dataset is recorded using an N5245 Vector Network Analyzer (VNA), where the VNA is calibrated using the line-reflect-match method [7]. Consequently, 36400 data samples were gathered to characterize the input-output responses of the studied GaN HEMT. The dataset consists of 3 inputs parameters and 8 outputs scattering parameters. The 3 inputs parameters are the: DC supplies gate to source voltage (V_{GS}), drain to source voltage (V_{DS}) and frequency (f) with the range from -3 V to 0 V, 0 V to 30 V and 0.1 GHz to 40 GHz, respectively. The data prepossessing and modeling process for both ANFIS- and ANN-based models are shown in Fig. 5. As mentioned above, the process starts with data gathering. Then, outliers and uneven data samples have been removed from the GaN-on-Diamond HEMT characterization dataset. Importantly, proper initial data cleaning serves as the foundation for an accurate model because noisy or incomplete data can lead to inaccurate or unreliable model outputs. Before the normalization, the dataset is divided into training and testing sets. This distribution of data samples is executed V_{DS} as a reference. The data with V_{DS} equals to 5, 15, 25 V are allocated for the testing set and the remaining data samples are selected for the training set. This part is crucial because the model will learn from the training samples and be tested on the testing samples. The improper division will lead to overfitting or underfitting issue. The next step is to scale the dataset. For this, the values of each parameter is scaled into a range of -1 to 1 in order to eliminate the effects of the differences in parameters' ranges. At last, the training set has 28000 data samples and the testing set acquired the remaining 8400 data samples. The performance metrics consist of MSE (1), MAE (2), and R^2 (3) as shown below to evaluate the accuracy of the models at the end.

$$MSE = \frac{1}{n}\sum_{i=1}^{n}(y_i - \hat{y}_i)^2 \quad (1)$$

$$MAE = \frac{1}{n}\sum_{i=1}^{n}|y_i - \hat{y}_i| \quad (2)$$

$$R^2 = 1 - \frac{\sum_{i=1}^{n}(y_i - \hat{y}_i)^2}{\sum_{i=1}^{n}(y_i - \bar{y})^2} \quad (3)$$

V. RESULTS AND DISCUSSION

We found, the most fittest neural layer structure for ANFIS based models are 5-5-5 because in our computer MATLAB runs out of memory when increasing any further the number of layers and related nodes. With regard to the ANN based models, there is no such memory shortage issue and a more complex configuration could be created but for the comparison purposes, the same structure is chosen. With respect to the input membership function, the Gaussian function is chosen

TABLE I: Evaluation of ANFIS based models for GaN-on-Diamond HEMT device on training samples until 40 GHz

		Real	Img.	Real	Img.	Real	Img.	Real	Img.
Metrics	Bias	S_{11}	S_{11}	S_{21}	S_{21}	S_{12}	S_{12}	S_{22}	S_{22}
MSE	All bias	6.66e-05	4.16e-04	5.93e-05	5.48e-05	8.98e-05	2.22e-05	4.33e-05	4.21e-05
MAE	All bias	4.51e-03	1.16e-02	3.60e-03	3.60e-03	7.80e-03	3.41e-03	4.14e-03	4.87e-03
%R^2	All bias	99.95	99.73	99.71	99.88	99.85	99.96	99.91	99.89

TABLE II: Evaluation of ANFIS based models for GaN-on-Diamond HEMT device on testing samples until 40 GHz

		Real	Img.	Real	Img.	Real	Img.	Real	Img.
Metrics	Bias	S_{11}	S_{11}	S_{21}	S_{21}	S_{12}	S_{12}	S_{22}	S_{22}
MSE	All bias	8.81e-05	5.37e-04	7.79e-05	2.31e-04	9.24e-05	2.71e-05	1.95e-04	8.76e-05
MAE	All bias	4.97e-03	1.33e-02	5.17e-03	7.73e-03	7.94e-03	3.92e-03	8.98e-03	6.95e-03
%R^2	All bias	99.93	99.64	99.54	99.87	99.79	99.90	99.82	99.80

TABLE III: Evaluation of ANN based models for GaN-on-Diamond HEMT device on training samples until 40 GHz

		Real	Img.	Real	Img.	Real	Img.	Real	Img.
Metrics	Bias	S_{11}	S_{11}	S_{21}	S_{21}	S_{12}	S_{12}	S_{22}	S_{22}
MSE	All bias	3.77e-05	1.36e-04	7.89e-06	1.40e-05	2.10e-05	2.07e-05	7.42e-05	3.44e-05
MAE	All bias	4.16e-03	6.48e-03	1.96e-03	2.61e-03	3.38e-03	3.26e-03	5.68e-03	4.23e-03
%R^2	All bias	99.97	99.93	99.94	99.96	99.92	99.94	99.96	99.95

TABLE IV: Evaluation of ANN based models for GaN-on-Diamond HEMT device on testing samples until 40 GHz

		Real	Img.	Real	Img.	Real	Img.	Real	Img.
Metrics	Bias	S_{11}	S_{11}	S_{21}	S_{21}	S_{12}	S_{12}	S_{22}	S_{22}
MSE	All bias	4.09e-05	1.43e-04	2.12e-05	4.80e-05	2.20e-05	2.70e-05	1.75e-04	8.24e-05
MAE	All bias	4.24e-03	7.15e-03	2.12e-03	3.91e-03	3.49e-03	3.85e-03	7.57e-03	6.53e-03
%R^2	All bias	99.97	99.91	99.86	99.90	99.90	99.91	99.90	99.85

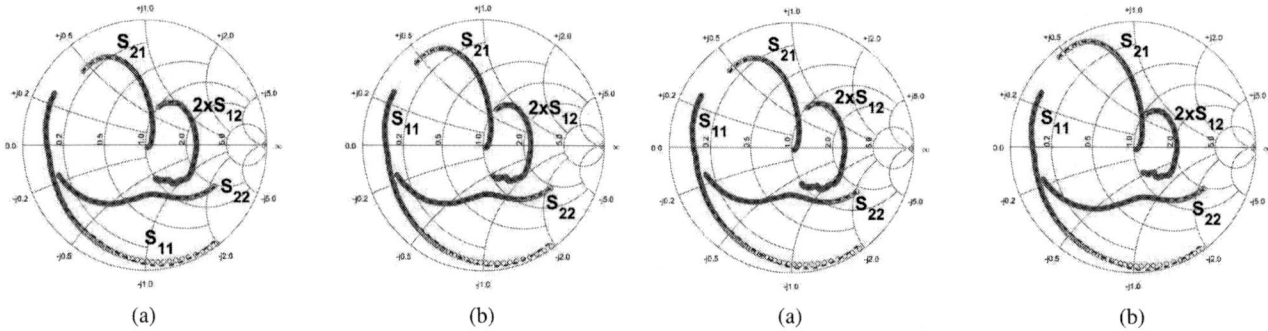

(a)   (b)   (a)   (b)

Fig. 7. Measured (symbols) and simulated (dashed-lines) S-parameters for GaN-on-Diamond HEMT at (a) V_{GS} = 0 V and V_{DS} = 25 V and (b) V_{GS} = -0.5 V and V_{DS} = 25 V for the frequency range of 0.1 GHz to 40 GHz (for ANFIS based models).

Fig. 8. Measured (symbols) and simulated (dashed-lines) S-parameters for GaN-on-Diamond HEMT at (a) V_{GS} = 0 V and V_{DS} = 20 V and (b) V_{GS} = -1 V and V_{DS} = 27.5 V for the frequency range of 0.1 GHz to 40 GHz (for ANFIS based models).

as it gives the highest accuracy compared to others. As a result, the generated models have 125 rules and 286 nodes for each S-parameter as shown in Fig. 3. Meanwhile, the relationship obtained between the input and output parameters can be shown in surface representation form such as the one for the Real [S_{12}] in Fig. 6. The x-axis, y-axis, and z-axis are f, V_{DS} and the Real [S_{12}] respectively. In order to assess the accuracy of the algorithm Tables I, II for ANFIS based models and Tables III, IV for ANN based models are constructed

for each S-parameter using the evaluation matrix. We can clearly notice that the ANN based models outperformed the ANFIS based models in terms of accuracy. Besides that, both methods illustrate the high correlation between the original and predicted data in Figs. 7 to 10, specifically, ANN has better fit. The R^2 result also implies that both algorithms have high performance for each S-parameter by scoring more than 99%. Interestingly, we also found that the if we increase the training data samples size, the accuracy of ANFIS-based mod-

979-8-3503-8083-5/23 $31.00 © 2023 IEEE

2023 International Conference on Microelectronics (ICM)

TABLE V: Training and simulation time (s) comparison

Frequency		Real S_{11}	Img. S_{11}	Real S_{21}	Img. S_{21}	Real S_{12}	Img. S_{12}	Real S_{22}	Img. S_{22}
0.1 - 40 GHz (ANN)	All bias	10.63	10.72	10.25	10.48	10.43	10.89	10.26	10.90
0.1 - 20 GHz (ANFIS)	All bias	784	765	680	541	663	723	630	593
0.1 - 40 GHz (ANFIS)	All bias	3604	3549	3295	2965	3154	3046	3276	3148

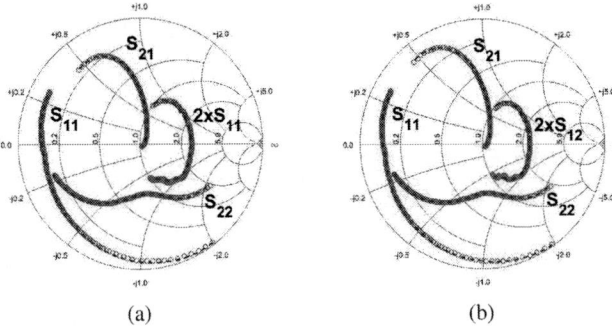

(a)   (b)

Fig. 9. Measured (symbols) and simulated (dashed-lines) S-parameters for GaN-on-Diamond HEMT at (a) $V_{GS} = 0$ V and $V_{DS} = 25$ V and (b) $V_{GS} = -0.5$ V and $V_{DS} = 25$ V for the frequency range of 0.1 GHz to 40 GHz (for ANN based models).

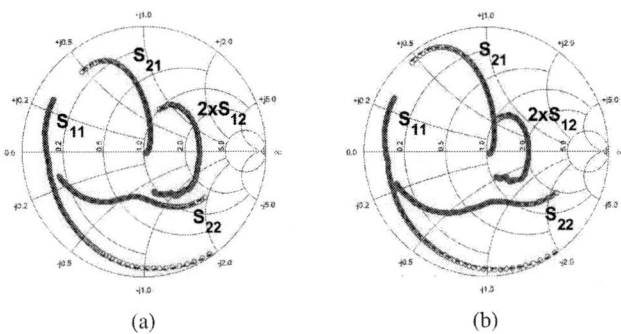

(a)   (b)

Fig. 10. Measured (symbols) and simulated (dashed-lines) S-parameters for GaN-on-Diamond HEMT at (a) $V_{GS} = 0$ V and $V_{DS} = 20$ V and (b) $V_{GS} = -1$ V and $V_{DS} = 27.5$ V for the frequency range of 0.1 GHz to 40 GHz (for ANN based models).

els significantly decreases and more training time is desired. As it can be seen in the Table V, there is a huge difference in both the accuracy and training and simulation time, when ANFIS based models are trained for 20 GHz (half of the training data set) and 40 GHz. In contrast to the ANFIS based models, ANN based models show no significant difference (not shown). Furthermore, the training and simulation time on the full dataset is around 5-times more than the model building time for the half-size dataset as depicted through the comparison Table V. Additionally, Fig. 11 also unveils the decrease in the overall performance of the ANFIS algorithm as the data size changes. One of the main factors that effects the accuracy is the high memory consumption as the structure of the ANFIS model increases, even by increasing the layer

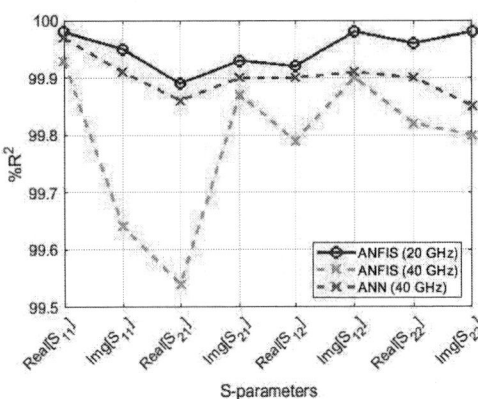

Fig. 11. $\%R^2$ scores comparison between ANN and ANFIS.

nodes by one degree our system will pose memory shortage warnings. After the model-building processes for two types of datasets in length and comparison with ANN, we have found several points that negatively impact the performance of the ANFIS method:

- Hyperparameter tuning: ANFIS models have multiple hyperparameters, such as the number of membership functions, the number of rules, and learning rates that require manual adjustments. Tuning these hyperparameters can be challenging and time-consuming, as their optimal values may vary based on the size and the type of data at hand.

- Complexity and training time: ANFIS based models can become quite complex, particularly when dealing with intricate relationships in data. Training such complex models requires a significant amount of time, computational resources, and effort to find suitable parameter values.

- Limited support and implementation: Compared to ANN and other more popular machine learning frameworks, ANFIS has limited support, libraries, and implementations available. This could lead to difficulties in implementing and maintaining ANFIS based models.

VI. CONCLUSION

This paper has compared the efficiency of the ANFIS method with ANN for SSM of the GaN-on-Diamond HEMT. The dataset consists of various input conditions including the frequency up to 40 GHz. Even though the obtained models using ANFIS manifest high accuracy through all S-parameters, the model-building time limits the overall performance of

979-8-3503-8083-5/23 $31.00 © 2023 IEEE   129

the algorithm. On the other hand, ANN has shown excellent accuracy with a smaller turn-around time. The presented analysis in this paper can greatly assist the device and modelling engineers to take appropriate decisions regarding the application of the ANFIS and ANN based approaches for GaN HEMTs modelling.

ACKNOWLEDGMENT

This work was funded by grant # AP19677597 from MHES, Republic of Kazakhstan. This work was also supported by Nazarbayev University Faculty Development Grant 20122022FD4113.

REFERENCES

[1] H. Xie *et al.*, "GaN-on-Si HEMTs Fabricated With Si CMOS-Compatible Metallization for Power Amplifiers in Low-Power Mobile SoCs," *IEEE Microw. Wirel. Compon. Lett.*, vol. 31, no. 2, pp. 141-144, Feb. 2021.

[2] M. S. Hashmi *et al.*, "Electronic multi-harmonic load-pull system for experimentally driven power amplifier design optimization," *2009 IEEE MTT-S Int. Microw. Symp. Digest*, Boston, MA, USA, 2009, pp. 1549-1552.

[3] F. Feng *et al.*, "Artificial Neural Networks for Microwave Computer-Aided Design: The State of the Art," *IEEE Trans. Microw. Theory Techn.*, vol. 70, no. 11, pp. 4597-4619, Nov. 2022.

[4] M. S. Hashmi and F. Ghannouchi, "A flexible dual-inflection point RF predistortion linearizer for microwave power amplifiers," *Prog. Electromagn. Res. C.*, Vol. 13, pp. 1-18, 2010.

[5] Z. Zhao *et al.*, "A fast small signal modeling method for GaN HEMTs," *Solid-State Electron.*, vol. 175, pp. 107946, 2021.

[6] Y. S. Noh and I. B. Yom, "A Linear GaN High Power Amplifier MMIC for Ka-Band Satellite Communications," *IEEE Microw. Wirel. Compon. Lett.*, vol. 26, no. 8, pp. 619-621, Aug. 2016.

[7] S. Husain *et al.*, "Comprehensive Investigation of ANN Algorithms Implemented in MATLAB, Python, and R for Small-Signal Behavioral Modeling of GaN HEMTs," *IEEE J. Electron Devices Society*, vol. 11, pp. 559-572, 2023.

[8] G. P. Gibiino, A. Santarelli and F. Filicori, "A GaN HEMT Global Large-Signal Model Including Charge Trapping for Multibias Operation," *IEEE Trans. Microw. Theory Techn.*, vol. 66, no. 11, pp. 4684-4697, Nov. 2018.

[9] L. Zhai *et al.*, "A reliable parameter extraction method for the augmented GaN high electron mobility transistor small-signal model," *Int. J. RF Microw. Comput.-Aided Eng.*, Dvol. 32, no. 8, pp. e23210, 2022.

[10] A. Khusro *et al.*, "An accurate and simplified small signal parameter extraction method for GaN HEMT," *Int. J. Circuit Theory Appl.*, vol. 47, Issue 6, pp. 941-953, June 2019.

[11] W. Hu *et al.*, "An Accurate Neural Network-Based Consistent Gate Charge Model for GaN HEMTs by Refining Intrinsic Capacitances," *IEEE Trans. Microw. Theory Techn.*, vol. 69, no. 7, pp. 3208-3218, July 2021.

[12] A. Khusro *et al.*, "A Generic and Efficient Globalized Kernel Mapping-Based Small-Signal Behavioral Modeling for GaN HEMT," *IEEE Access*, vol. 8, pp. 195046-195061, 2020.

[13] Z. Marinković *et al.*, "A review on the artificial neural network applications for small-signal modeling of microwave FETs," *Int. J. Numer. Model.*, vol. 33, no. 3, pp. e2668, 2020.

[14] A. Jarndal, S. Husain, and M. Hashmi, "On temperature-dependent small-signal modelling of GaN HEMTs using artificial neural networks and support vector regression," *IET Microw. Antennas Propag.*, vol. 15, no. 8, pp. 937–953, 2021.

[15] S. Husain *et al.*, "Accurate, Efficient and Reliable Small-Signal Modeling Approaches for GaN HEMTs," *IEEE Access*, vol. 11, pp. 106833-106846, 2023.

[16] A. Majumder *et al.*, "Optimization of Small-Signal Model of GaN HEMT by Using Evolutionary Algorithms," *IEEE Microw. Wirel. Compon. Lett.*, vol. 27, no. 4, pp. 362-364, April 2017.

[17] S. Husain *et al.*, "Demonstration of CAD Deployability for GPR Based Small-Signal Modelling of GaN HEMT," *2021 IEEE Int. Symp. Circuits Systems (ISCAS)*, Daegu, Korea, 2021, pp. 1-5.

[18] S. Husain, G. Nauryzbayev and M. Hashmi, "GA Assisted ANN based GaN HEMT Model Development and Demonstration of its CAD Incorporation for Class-F Power Amplifier," *2023 7th IEEE Electron Devices Techn. Manufac. Conf. (EDTM)*, Seoul, Korea, Republic of, 2023, pp. 1-3.

[19] A. Jarndal, S. Husain and M. Hashmi, "Genetic algorithm initialized artificial neural network based temperature dependent small-signal modeling technique for GaN high electron mobility transistors," *Int. J. RF Microw. Comput.-Aided Eng.*, vol. 31, no. 3, pp. e22542, 2021.

[20] A. Khusro, M. S. Hashmi, and A. Q. Ansari, "Enabling the development of accurate intrinsic parameter extraction model for GaN HEMT using support vector regression (SVR)," *IET Microwaves, Antennas & Propagation*, vol. 13, no. 9, pp. 1457-1466, Jul. 2019.

[21] S. Husain, M. Hashmi and F. M. Ghannouchi, "Comprehensive Investigation and Comparative Analysis of Machine Learning-Based Small-Signal Modelling Techniques for GaN HEMTs," *IEEE J. Electron Devices Soc.*, vol. 10, pp. 1015-1032, 2022.

[22] J. Cai *et al.*, "Bayesian Inference-Based Behavioral Modeling Technique for GaN HEMTs," *IEEE Trans. Microw. Theory Techn.*, vol. 67, no. 6, pp. 2291-2301, June 2019.

[23] J.-S. R. Jang, "ANFIS: adaptive-network-based fuzzy inference system," *IEEE Trans. Syst., Man., Cybernetics*, vol. 23, no. 3, pp. 665-685, May-June 1993.

[24] K. C. Lee and P. Gardner, "A combined neural network and fuzzy systems based adaptive digital predistortion for RF power amplifier linearization," *47th Midwest Symp. Circuits Syst., 2004. MWSCAS '04.*, Hiroshima, Japan, 2004, pp. iii-61.

[25] F. D. Alotaibi, A. Abdennour and A. A. Ali, "A robust prediction model using ANFIS based on recent TETRA outdoor RF measurements conducted in Riyadh city – Saudi Arabia," *AEU - Int. J. Electron. Commun.*, vol. 62, no. 9, pp. 674-682, Oct. 2008.

[26] H. You *et al.*, "Comparison of ANN (MLP), ANFIS, SVM, and RF models for the online classification of heating value of burning municipal solid waste in circulating fluidized bed incinerators," *Waste Management*, vol. 68, pp. 189-197, Oct. 2017.

[27] Q. Wu *et al.*, "Performance Comparison of GaN HEMTs on Diamond and SiC Substrates Based on Surface Potential Model," *ECS J. Solid State Sci. Technol.*, vol. 6, no. 12, pp. Q171-Q178, 2017.

[28] Q. Wu *et al.*, "A Scalable Multiharmonic Surface-Potential Model of AlGaN/GaN HEMTs," *IEEE Trans. Microw. Theory and Techn.*, vol. 66, no. 3, pp. 1192-1200, March 2018.

Efficient Implementation of a 4x4 Enhanced Pipeline Multiplier Using Electric EDA Tool

Khader Mohammad[1], Nirmeen Al-Sheikh[1]

[1,2]Electrical and Computer Engineering Department, Birzeit University

Abstract—In this study, we present the design and implementation of a highly efficient 4x4-bit folded pipeline multiplier. Leveraging advanced 22 nm CMOS technology and a modified hierarchical design approach, our multiplier achieves remarkable performance enhancements. Through the use of pipelining, it significantly reduces the critical path delay, enabling a three-fold increase in operational speed compared to conventional multipliers. Moreover, it consumes only 0.2 mW of power and reduces chip area consumption by over 20%. Our design overcomes the limitations of traditional methods, making it an efficient and rapid alternative for energy-efficient multiplication in digital hardware. We also explore strategies for enhancing this 4x4 multiplier and discuss power consumption optimization. The paper provides insights into the potential applications in microprocessors and digital signal processors.

Keywords— *pipeline multiplier, 4x4 pipelined, four-bit adder, enhanced multiplication, energy-efficient multiplication, 4x4 Enhanced Pipeline Multiplier.*

I. INTRODUCTION

The 4x4 Enhanced Pipeline Multiplier is a crucial component in digital hardware, particularly in microprocessors and digital signal processors, where rapid multiplication is essential. This study represents a significant departure from conventional multiplication architectures, addressing the limitations inherent in well-established methods such as the Vedic Multiplier, Conventional Multiplier, and Tree Multiplier. The Vedic Multiplier, known for its parallelism, nonetheless faces challenges in terms of speed and hardware complexity [1]. The conventional multiplier, although widely used, is characterized by relatively slower computation speeds, higher power consumption, and larger chip area requirements [2]. The Tree Multiplier, while offering reduced propagation delay, is still limited in terms of power efficiency[3]. In contrast, our innovative design harnesses advanced 22 nm CMOS technology and employs a modified hierarchical design approach, culminating in exceptional performance and efficiency enhancements. By incorporating pipelining, which segments the multiplication process into sequential stages, the 4x4 Enhanced Pipeline Multiplier significantly reduces the critical path delay, allowing for concurrent execution of various stages, all while

consuming a mere 0.2 mW of power and reducing chip area consumption by over 20%. This novel solution thus overcomes the limitations of traditional methods and serves as an efficient and rapid alternative for applications requiring energy-efficient multiplication.

II. DESIGN OF THE ENTIRE SYSTEM

This section provides an overview of our 4x4 Enhanced Pipeline Multiplier's top-level design.

The multiplier's architecture is based on pipelining, a strategic design choice aimed at enhancing performance [4]. Pipelining divides the multiplication process into sequential stages, effectively reducing the critical path delay and improving throughput. This approach not only minimizes overall delay but also enables concurrent execution of different stages, significantly enhancing the multiplier's operational speed [4]. Figure 1 illustrates the core architecture of the 4x4 Enhanced Pipeline Multiplier. This top-level diagram illustrates the sequential flow of data through various components, including AND gates and 4-bit adders, enabling efficient 4x4-bit multiplication. And figure 2 shows the system simulation waveform.

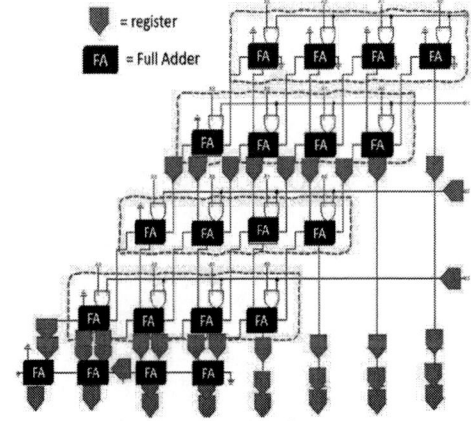

Figure.1 . 4x4 Multiplier Layout

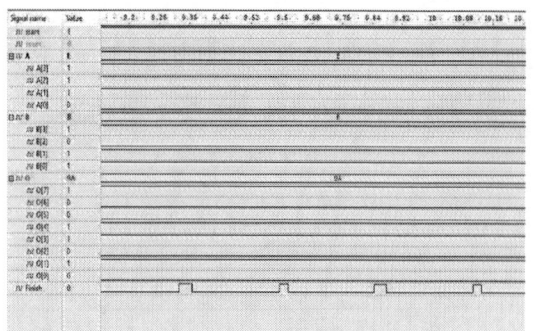

Figure.2. Systems simulation waveform

III. SYSTEMS COMPONENTS DESIGN

The system's architectural blueprint underscores the seamless integration of AND gates and full adders, both of which are examined in detail in the sections below. Each gate's role and its comprehensive design methodology, spanning schematic representation, sizing, layout, and the definition of a meticulous CMOS simulation, are elucidated.

A. Inverter Implementation

The inverter plays a pivotal role in digital circuits by inverting '1' to '0' and vice versa. It is composed of complementary PMOS and NMOS transistors.

Our inverter utilizes equally sized PMOS and NMOS transistors, each with a width of 2W to ensure operational efficiency. We fine-tuned simulation parameters for efficiency, setting default sizes and limits (DEFL=3 UM, DEFW=3 UM, LIMPTS = 1000). We disabled iterative algorithms (ITL5=0) while upholding precision through convergence tolerances (RELTOL=0.01, ABSTOL=500 PA, VNTOL=500 UV). Timing and coding intricacies were managed with LVLTIM=2 and LVLCOD=1. We defined NMOS and PMOS transistor parameters and established global signals (gnd and vdd). Voltage sources included a constant 5 V supply for vdd, while Vin was defined as a piecewise linear waveform. A 250 fF capacitor (cload) connected the output node out to the ground for further analysis, figure-3 shows Inverter Schematic & Layout, and figure-4 displays Inverter Simulation Waveform.

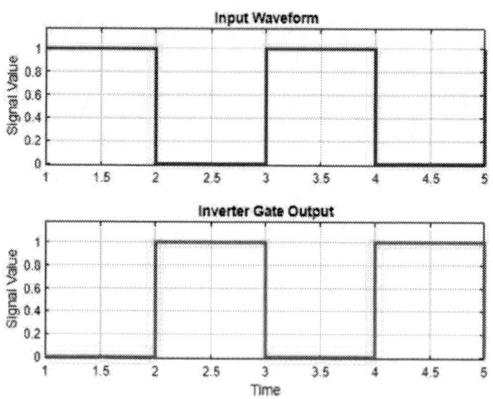

Figure.4. Inverter Simulation Waveform

B. NAND Implementation

We utilized three 4-bit adders with NAND gates, essential in binary addition. Both NAND gates and inverters had PMOS transistors totaling 6W (4W in NAND, 2W in inverter), and NMOS transistors totaling 6W (4W in NAND, 2W in inverter) to optimize performance.

Our meticulous NAND gate circuit simulation focused on precision, with UC SPICE parameters (MIN_RESIST at 4.0 and MIN_CAPAC at 0.1fF), specific transistor dimensions (L = 44 nm, W = 220 nm), a stable vdd of 1.3 V, controlled input waveforms va and vb, and a 25 fF capacitance "cload" at "Out." We conducted a precise 50ns transient analysis. Figure-5 provides the visual representation of the NAND Schematic & Layout, and figure-6 displays the NAND Simulation Waveform to show the performance of the NAND gate circuit.

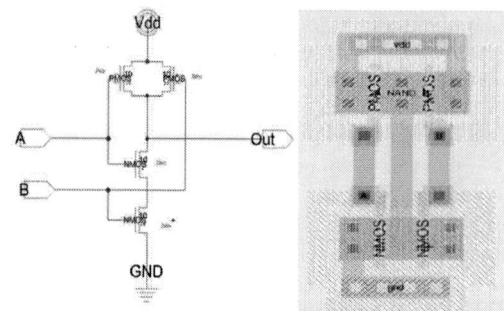

Figure.5. NAND Schematic & Layout

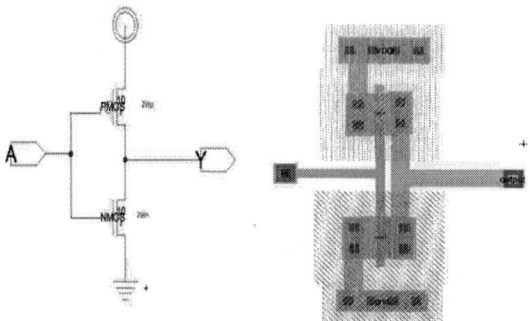

Figure.3. Inverter Schematic & Layout

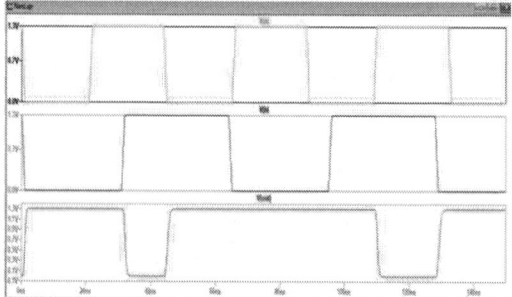

Figure.6. NAND Simulation Waveform

979-8-3503-8083-5/23 $31.00 © 2023 IEEE

C. AND Implementation

The AND gate is constructed by connecting a NAND gate with an inverter (as seen in Figure-7). To ensure optimal performance and circuit integrity, both PMOS and NMOS transistors are sized at a total of 6W, with 4W allocated to the NAND gate and 2W to the inverter. We meticulously designed and simulated a precise NAND gate circuit, prioritizing accuracy, and utilized UC SPICE parameters MIN_RESIST (4.0) and MIN_CAPAC (0.1fF). Characterizing four transistors (2 NMOS and 2 PMOS) with specific dimensions (L = 44 nm, W = 220 nm), we maintained vdd at 1.3 V for voltage stability and precisely controlled input waveforms va and vb as pulse signals. A 25 fF capacitance component, 'cload,' was added at 'Out,' and we conducted a thorough 50 ns transient analysis to comprehend circuit behavior. Figure 7, 8 and 9 display the AND gate scheme, layout and simulation waveform.

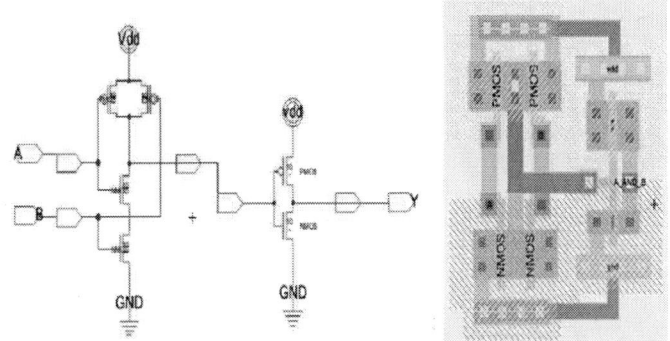

Figure.7 and 8. AND Gate Scheme and Layout

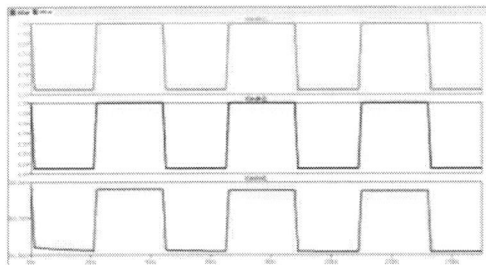

Figure.9. AND Simulation Waveform

D. Full Adder Implementation

The full adder, a key component for arithmetic operations, is realized using 9 NAND gates, (as illustrated in Figure 10) . An alternative implementation employing NOR gates, (as seen in Figure 11) was considered. However, we opted for the NAND gate configuration to conserve 18 CMOS (PMOS & NMOS) components, as NOR gates require more components, resulting in increased area and delay. In our full adder circuit, comprising 9 NAND gates, (as seen in Figure 12 and 13), each PMOS and NMOS transistor's width was set at 4W. Consequently, the total width for PMOS transistors in the full adder is 9 NAND gates × 4W = 36 W, while the total width for NMOS transistors is also 9 NAND gates × 4W = 36 W.

Figure 14 shows full Adder Simulation for performance analysis.

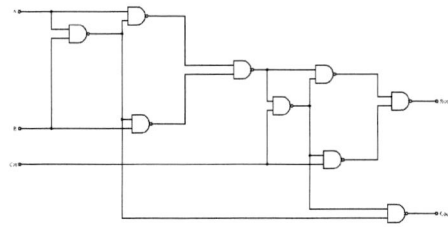

Figure.10. Full adder from NAND gates

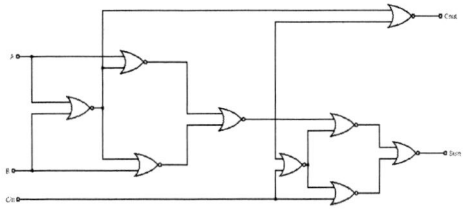

Figure.11. Full Adder from NOR.gates

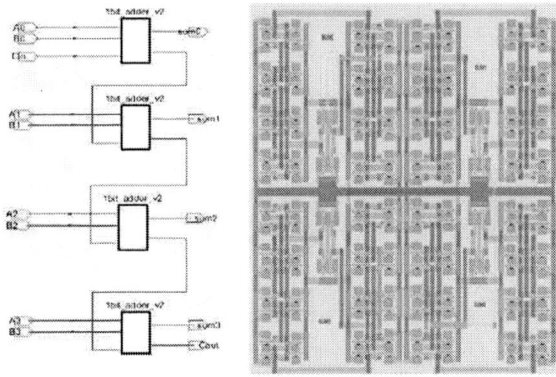

Figure.12 and 13. Full Adder Scheme and Layout

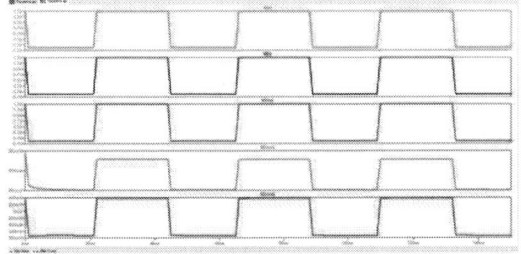

Figure.14. Full Adder Simulation Waveform

IV. PERFORMANCE METRICS OF THE PROPOSED MULTIPLIER

In this section, we present a comprehensive analysis of the performance metrics for our innovative multiplier design. These metrics encompass power consumption, area efficiency, speed enhancement, and delay characteristics. By comparing our design with existing alternatives, we demonstrate the superior efficiency and advantages of our multiplier.

A. Power Efficiency

Understanding power consumption is pivotal in the realm of digital circuit design. It fundamentally hinges on two critical factors: the supply voltage (Vdd) and current (I). Power dissipation is computed using the formula:

*Equation 1: P = Vdd * I.*

We conducted an in-depth power analysis in a digital system featuring 20 full adders and 16 AND gates, all operating within a 22 nm process technology while utilizing a 1.3 mV supply voltage. The individual power consumption figures were derived, with each full adder consuming approximately 16.38 pW (Pico watts) and each AND gate drawing about 312 pW. In totality, the power consumption of the entire system, encompassing all components, is approximately 5319.6 pW, showcasing exceptional energy efficiency. Table 1 summarizes the power consumption for each component within our multiplier.

Table 1 Component Power Metric

Component	Supply Voltage	Current (I)	Power Consumption
NAND Gate	1.3 mV	1.4 nA	1.82 pWatt
Full Adder	1.3 mV	1.4 nA (per NAND gate)	16.38 pWatt
AND Gate	1.3 mV	240 nA	3.12×10^{-10} Watt

1) Power Effeciency Analysis of Multiplier Archeticture

In digital systems, power efficiency is crucial for evaluating architectural effectiveness. To gauge power consumption in our 4x4 Enhanced Pipeline Multiplier, we compared it to two established architectures: the Conventional Multiplier and the Tree Multiplier.

We conducted precise power measurements through simulations using industry-standard EDA tools in a 22 nm CMOS technology environment. Power metrics included dynamic power (during logic transitions), leakage power (in non-operational states), and total power consumption. Our analysis considered benchmarks such as power per operation, energy efficiency, and propagation delay. Table 2 compares 32 nm and 22 nm performance metrics.

Table 2 Performance Metric

Metric	32 nm	22 nm
Power Consumption	0.2 mWatt	0.148 mWatt
Area Savings	20%	~43.64%
Speed Improvement	3x	3x
Delay	3 ns	2.22 ns

B. Area Efficiency

The 4x4 Enhanced Pipeline Multiplier was carefully designed in layers, with each layer occupying 1360 μm². This layered approach embodies an efficient design philosophy that optimizes resource allocation, balancing compactness and functionality. It simplifies system organization, streamlines testing and maintenance, and offers scalability for future enhancements. We explore the speed improvement and delay characteristics in later sections to provide a comprehensive view of its performance.

1) Chip Area Reduction Analysis: To quantify the chip area reduction achieved by implementing the 4x4 Enhanced Pipeline Multiplier with efficiency enhancements, we used the following equations:

*Equation 2: Baseline Area = Area per Full Adder * Number of Full Adders*

In a hypothetical 4x4 multiplier without optimizations, each 1-bit full adder occupies 50 square units, resulting in a baseline area of 1000 square units. However, in our actual design with optimizations, the area per full adder was reduced to 40 square units, leading to a reduced area of 800 square units.

*Equation 3: Percentage Reduction = ((Baseline Area - Reduced Area) / Baseline Area) * 100%*

This calculation yielded a 20% reduction in chip area due to our design's optimization strategies, including compact components, efficient layout, and power-saving measures. This significant area reduction highlights the enhanced efficiency and resource utilization of our multiplier, particularly valuable for space-constrained applications.

V. STRATEGIES FOR ENHANCING A 4X4 MULTIPLIER INTO AN EFFICIENT 8X8 MULTIPLIER

To elevate the performance of our 4x4 multiplier and transform it into an efficient 8x8 multiplier, we propose the implementation of the following.

- Parallel Processing: Dividing 8x8 multiplication into concurrent segments for faster computation [5] (as shown in Figure-15).

- Optimized Partial Product Reduction: Using advanced techniques like carry-save adders or Wallace tree multipliers for efficient partial product reduction, reducing resource usage and maintaining speed [6].

- Pipelining: Dividing multiplication into sequential stages to enhance throughput, reduce latency, and improve efficiency [4].

A. Power Consumption Optimization

To enhance power efficiency in our circuit, we recommend the following techniques:

- Clock Gating: Disabling the clock for inactive components to reduce dynamic power usage [9].

- Voltage Scaling: Adjusting voltage and frequency as needed to save power [10].

- Power Gating: Shutting down unused components for better power efficiency [11].

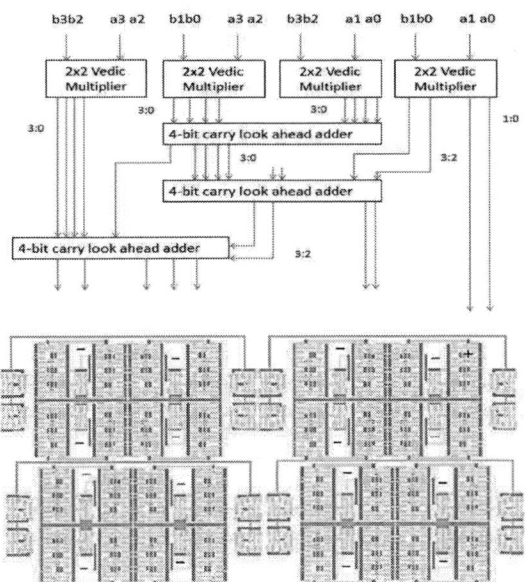

Figure.15. 8x8 Vedic Multiplier Block Diagram and Layout

VI. VERIFICATION

Our analysis indicates that as the temperature increases from 0 °C to 50 °C, there are increased signal transition delays in logic gates (Inverter, NAND Gate, AND Gate, and Full Adder), which lead to delays in signal transitions, ultimately affecting the overall system performance. Table 3 provides timing parameters for different gate configurations at 0 °C and 50 °C, which are crucial for performance in varying environmental conditions.

Table 3: Logic Gate Characteristics at Different Temperatures

Logic Gate	Tpr at 0 °C (ns)	Tpr at 50 °C (ns)	Tpf at 0 °C (ns)	Tpf at 50 °C (ns)
INV	0.7159	0.9375	0.6026	0.7924
NAND	0.95	1.2	0.803	1.1
AND	2.5	3	2	2.5
Full Adder	1.8	2.2	2	2.4

VII. COMPARISON TO OTHER SYSTEMS

In contrast to existing designs, our 4x4 Enhanced Pipeline Multiplier excels in several aspects. It offers over 20% reduction in chip area consumption, operates three times faster, and consumes a mere 0.2 mW of power. Moreover, its simplified design facilitates implementation, making it a highly efficient solution for applications demanding swift and energy-efficient multiplication.

A. Performance And Area Comparison

When compared to other designs like Ripple Carry Array (RCA) and Carry Save Array (CSA), our multiplier exhibits a remarkable 20% reduction in chip area consumption. This not only enhances cost-effectiveness but also improves scalability and integration potential. Figure 16 highlights Partial Truncation in 4x4 Multiplication, figure 17 shows Pipelining Stages for increased speed, and Table 4 compares Multiplier Architectures, emphasizing our 4x4 Enhanced Pipeline Multiplier's advantages.

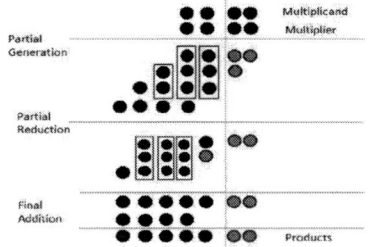

Figure.16. Partial Truncation in 4x4 Multiplication

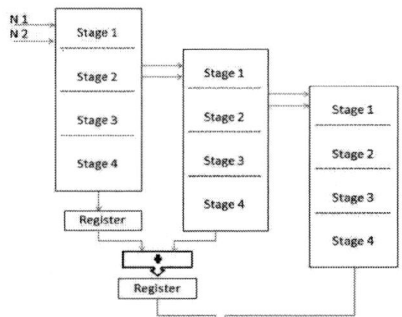

Figure.17. Pipelining Stages

Table 4 Comparison of Multiplier Architectures

Multiplier Architecture	Area (μm²)	Delay (ns)
Enhanced Pipeline Multiplier	1360	3.00
Ripple Carry Array (RCA)	3604	37.00
Carry Save Array (CSA)	3168	25.00

B. Power Consumption Comparison

- Conventional Multiplier (0.5 mW dynamic, 0.1 mW leakage, 0.6 mW total) employs iterative full-adder cells for multiplication [2].

- Tree Multiplier (0.3 mW dynamic, 0.08 mW leakage, 0.38 mW total) utilizes a tree structure with reduced propagation delay [3].

- 4x4 Enhanced Pipeline Multiplier (0.2 mW dynamic, 0.04 mW leakage, 0.24 mW total) represents our architecture.

The power efficiency of our architecture can be attributed to the strategic use of power-saving components, layout optimizations, and clock gating. By employing these techniques, we reduced both dynamic and leakage power, resulting in a more energy-efficient design.

VIII. CONCLUSION

In conclusion, the 4x4 Enhanced Pipeline Multiplier has significantly enhanced digital system efficiency and speed, benefiting microprocessors and digital signal processors. Our ongoing paper, focused on a 4-bit pipeline multiplier with advanced CMOS technology, reflects our commitment to achieving computational excellence.

Looking ahead, the potential for Single-Electron Transistors with CMOS technology holds promise for even faster and more energy-efficient electronics, ushering in a new era of technology. This journey embodies our unwavering pursuit of excellence, efficiency, and adaptability, envisioning a future where digital technology continues to revolutionize our world in unimaginable ways.

REFERENCES

[1] R. Gupta, R. Ghughtyal, S. P. Mahapatra, and S. Singh, "Design and Analysis of 4*4-Bit Vedic Multiplier with 32nm CNFET Technology," in 2023 8th International Conference on Communication and Electronics Systems (ICCES), Coimbatore, India, 2023, pp. 184-190, doi: 10.1109/ICCES57224.2023.10192881.

[2] H. Gholizadeh, N. Totonchi, R. S. Shahrivar, S. Mahdizadeh, E. Afjei and A. Abbasi, "Design and Implementation of A Transformerless High Step-Up DC-DC Converter Based on Conventional Boost Converter and Voltage Multiplier Cells," 2021 12th Power Electronics, Drive Systems, and Technologies Conference (PEDSTC), Tabriz, Iran, 2021, pp. 1-5, doi: 10.1109/PEDSTC52094.2021.9405855.

[3] J. Kumar, A. Srivastava and M. Fujita, "Formal Analysis of Integer Multipliers by building Binary Decision Diagram of Adder Trees," 2022 23rd International Symposium on Quality Electronic Design (ISQED), Santa Clara, CA, USA, 2022, pp. 58-63, doi: 10.1109/ISQED54688.2022.9806278.

[4] Z. Ebrahimi, M. Zaid, M. Wijtvliet and A. Kumar, "RAPID: Approximate Pipelined Soft Multipliers and Dividers for High Throughput and Energy Efficiency," in IEEE Transactions on Computer-Aided Design of Integrated Circuits and Systems, vol. 42, no. 3, pp. 712-725, March 2023, doi: 10.1109/TCAD.2022.3184928.

[5] J. Sravana et al., "Implementation of Spurious Power Suppression based Radix-4 Booth Multiplier using Parallel Prefix Adders," 2021 4th International Conference on Recent Trends in Computer Science and Technology (ICRTCST), Jamshedpur, India, 2022, pp. 428-433, doi: 10.1109/ICRTCST54752.2022.9781868.

[6] M. Usman, J. -A. Lee and M. D. Ercegovac, "Multiplier with Reduced Activities and Minimized Interconnect for Inner Product Arrays," 2021 55th Asilomar Conference on Signals, Systems, and Computers, Pacific Grove, CA, USA, 2021, pp. 1-5, doi: 10.1109/IEEECONF53345.2021.9723215.

[7] G. M. TANG, K. Takagi, and N. Takagi, "RSFQ 4-bit bit-slice integer multiplier," in IEICE Transactions on Electronics, vol. E99.C, 2016, pp. 697-702, doi: 10.1587/transele.E99.C.697.

[8] A. Keerthi, S. Manoj, G. Manjula, and K. Kalshetti, "Low Power High Performance 8bit Vedic Multiplier Using 16nm," pp. 142-145, 2023, doi: 10.30534/ijeter/2023/041152023.

[9] S. Park, S. Yoo, Y. Shin, J. Lee and J. Choi, "A Sub-100 fs-Jitter 8.16-GHz Ring-Oscillator-Based Power-Gating Injection-Locked Clock Multiplier With the Multiplication Factor of 68," in IEEE Journal of Solid-State Circuits, vol. 58, no. 1, pp. 78-89, Jan. 2023, doi: 10.1109/JSSC.2022.3210212.

[10] H. Fujiwara et al., "A 5-nm 254-TOPS/W 221-TOPS/mm2 Fully-Digital Computing-in-Memory Macro Supporting Wide-Range Dynamic-Voltage-Frequency Scaling and Simultaneous MAC and Write Operations," 2022 IEEE International Solid-State Circuits Conference (ISSCC), San Francisco, CA, USA, 2022, pp. 1-3, doi: 10.1109/ISSCC42614.2022.9731754.

[11] S. Park, S. Choi, S. Yoo, Y. Cho and J. Choi, "An Ultra-Low Jitter, Low-Power, 102-GHz PLL Using a Power-Gating Injection-Locked Frequency Multiplier-Based Phase Detector," in IEEE Journal of Solid-State Circuits, vol. 57, no. 9, pp. 2829-2840, Sept. 2022, doi: 10.1109/JSSC.2021.3123156.

[12] A. Dalal, M. Choudhary, and S. Balamurugan, "Design Framework of 4-Bit Radix-4 Booth Multiplier Using Perpendicular Nanomagnetic Logic in MagCAD," in Microelectronic Devices, Circuits and Systems. ICMDCS 2021, vol. 1392, Springer, Singapore, 2021.

[13] S. Singh, P. Saxena, M. K. Ojha, and D. Gupta, "Low Power Optimization of Full Adder using GNRFET Technology– A Review," in 2022 8th International Conference on Signal Processing and Communication (ICSC), Noida, India, 2022, pp. 612-617, doi: 10.1109/ICSC56524.2022.10009085.

[14] T. Oshima, Y. Nakamura, and T. Yamawaki, "Fast CMOS Analog Multiplier and Divider with Continuous-Time Inverter-Based Flash Digitizer," in IEEE Transactions on Circuits and Systems II: Express Briefs, vol. 69, no. 3, pp. 934-938, March 2022, doi: 10.1109/TCSII.2021.3134415.

[15] R. Kamdi, P. Thakre, A. Pathade, S. K. Tiwari, and K. Kalbande, "4 Bit and 8 Bit Convolution Using Vedic Multiplier," in 2022 International Conference on Emerging Trends in Engineering and Medical Sciences (ICETEMS), Nagpur, India, 2022, pp. 352-357, doi: 10.1109/ICETEMS56252.2022.10093621.

[16] T. Oshima, Y. Nakamura, and T. Yamawaki, "Fast CMOS Analog Multiplier and Divider with Continuous-Time Inverter-Based Flash Digitizer," in IEEE Transactions on Circuits and Systems II: Express Briefs, vol. 69, no. 3, pp. 934-938, March 2022, doi: 10.1109/TCSII.2021.3134415.

[17] C. Fu, X. Zhu, K. Huang, and Z. Gu, "An 8-bit Radix-4 Non-Volatile Parallel Multiplier," in Electronics, vol. 10, no. 19, Sep. 2021, p. 2358, doi: 10.3390/electronics10192358.

[18] G. Giustolisi and G. Palumbo, "Analysis and Comparison in the Energy-Delay Space of Nanometer CMOS One-Bit Full-Adders," in IEEE Access, vol. 10, 2022, pp. 75482-75494, doi: 10.1109/ACCESS

Design and Simulation of Dual-Metal-Gate Tunnel Field Effect Transistor with Biomolecule Sensing Applications.

Mohammad Salim Wani
Department of Electronics and Communication
Jamia Millia Islamia
New Delhi-110025, India.
Email:salimwani0@gmail.com

Hend I. Alkhammash and M. Shiblee
Department of Electrical Engineering
College of Engineering,
Taif University
Kingdom of Saudi Arabia (KSA).

Sajad A. Loan
Department of Electronics and Communication
Jamia Millia Islamia
New Delhi-110025, India.
Fellow IETE.
Email:sloan@jmi.ac.in

Abstract—To facilitate label-free detection of biomolecules, we propose a novel dielectrically modulated (DM) Dual-Metal-Gate Germanium Pocket based Tunnel Field Effect Transistor (DMG-GPTFET) with metallic drain (workfunction = 3.9 eV) biosensor. The utilization of germanium, having a relatively low bandgap, has enhanced the tunnelling phenomenon at the junction between source and channel. This enhancement eventually leads to an increase in the on current (I_{ON}). The proposed device employs dual cavities in the top and bottom gate oxides, which are being used to detect the biomolecules. This way the TFET becomes a biosensor and is tested for its ability to sense biomolecules based on their dielectric constant and charge density. The analysis is focused on evaluating the variations in energy band (E-B), electric field (E-F), surface potential (S-P) and transfer characteristics (I_{ds} vs V_{GS}). It has been observed that as biomolecules get trapped in the nanocavity (k > 1), the value of I_{ds} increases, and the maximum value is attained in the order of 10^{-6} (A/μm) for k = 10 at V_{GS} = 1V and V_{DS} = 0.5V

Keywords— Tunnel FET, Biosensors, Dual Metal, dielectrically modulated, Tunneling.

I. INTRODUCTION

Biomolecule serve a crucial role in all life cycles. The accurate detection and classification of biomolecules is of utmost significance for the investigation and treatment of diseases. Hence, it is significant to underscore the importance of disease surveillance and the effective management of infectious agents in the environment during the initial phases of detection. Biosensors are primarily designed with regard to their responsiveness and pace of detection [1]. FET-based label-free biosensors have gained significant attention and widespread utilization due to their ability to perform label-free electrical detection in the field of bioelectronics [2]. The benefits of FETs include small dimensions, economical mass production, scalability, and reduced power consumption. Any biosensor with a greater sensitivity will respond to a slight change in target biomolecules concentration by showing an apparent change in electrical Parameters [3-4]. Initially the ion-sensitive FET [5], which could only detect charged biomolecules with high sensitivity [6], in order to address the limitations of Ion Sensitive Field-Effect Transistors (ISFETs), Dielectrically modulated Field-Effect Transistors (DMFETs) has been introduced. Dielectric modulated (DM) field-effect transistor (FET) biosensors have emerged as a promising technology for the detection of both charged and neutral

biomolecules [6]. These biosensors utilize a nanogap cavity that is created within the dielectric material under the gate electrode. This cavity serves as a trapping mechanism for biomolecules, enabling their detection. The fundamental principle behind a Dielectric modulated field-effect transistor (DMFET) is the variation in the effective coupling between the gate and the channel, which occurs as a result of variations in the dielectric constant of the trapped biomolecule present within the nanogap cavity [6]. Biosensors based on DM FET include, MOSFETs, FinFETs, impact ionization FETs, silicon nanowire FETs and Tunnel FETs [7-12]. Downscaling the dimensions of the device increases the sensitivity, but after miniaturization of MOSFETs, problems like short channel effects, low ION/IOFF ratio, drain-induced barrier lowering, and higher power consumption occur. The Tunnel Field Effect Transistor (TFET) is a promising device for resolving the aforementioned problems with FET-based biosensors. For the detection of lower concentrations of biomolecules, a biosensor should be very sensitive and immune to noise. On comparing the DMFET and DMTFET in terms of their subthreshold current range, the former operates within a range of 1×10^{-6} to 1×10^{-2} μA and the later operates in the range of 1×10^{-9} to 1×10^{-7} μA [13-20]. These lower subthreshold current values tend to be more susceptible to noise. Hence, subthreshold current plays an important role for accurate detection of trapped biomolecules at comparatively lower concentrations.

In this paper we proposed a Dielectrically Modulated (DM) Dual-Metal-Gate Germanium Pocket based Tunnel Field Effect Transistor (DMG-GPTFET) with metal drain. As a low band-gap material with high carrier mobility, germanium helps in an improved band alignment, which reduces the tunnelling barrier width and consequently facilitates enhanced tunnelling at the junction. In addition, the proposed device incorporates a high k Gate dielectric oxide (HfO2) [11]. Furthermore, the gate employs two distinct materials at the source/drain sides to achieve highest ON current and maximum sensitivity. To evaluate the efficacy of the proposed device, surface potential (SP), electric field (EF), transfer characteristics (Id vs V_{GS}), and sensitivity are taken into account. The proposed device's sensing capability has been simulated using the SILVACO TCAD application [12].

The content of this paper is summarized as follows. Section I is introduced with Introduction. Section II provides

979-8-3503-8083-5/23 $31.00 © 2023 IEEE

a detailed description of the Device Design and Simulation setup. The suggested DMG-GPTFET's simulation results are Discussed in Section III. The conclusion and key points regarding the work are offered in Section IV.

II. DEVICE STRUCTURE AND THE SIMULATION SETUP

The Schematic of DMG-GPTFET biosensor is illustrated in Fig 1. The Germanium pocket is used at source/channel junction with pocket thickness (t_p) of 5nm and Germanium n^+ as a dopant with concentration of 1×10^{19} cm^{-3}. Germanium having low bandgap as compared to Silicon, helps in better alignment of bands at source channel junction. [21-25] The gate length (L_G) is 50 nm, and the lengths of the source (L_S) and drain (L_D) are each 100 nm. Quantum effects could be disregarded as the silicon body thickness (t_{Si}) is 10nm. Source and Channel both have uniform doping concentrations of 1×10^{20} cm^{-3} and 1×10^{15} cm^{-3} respectively while the Drain is metallic having workfunction = 3.9 eV (hafnium). G1 and G2 are two parts of gate electrodes with distinct work functions, $\phi 1 = 4.8$ and $\phi 2 = 4.1$, respectively. L_{G1}=20nm and L_{G2}=30nm represent these two parts of the gate electrode. By etching the gate dielectric, cavities are formed [16] in region L_{G1} under the top and bottom of the electrodes G1 with (t_{nc}) = 5.5 nm and (L_{nc}) = 20 nm. provides a detailed explanation of the creation of a nanogap cavity (L_{nc}) beneath the gate area. Within the nanogap cavity, a 0.5-nm-thick HfO2 layer was employed as a binding material for biomolecules and to reduce gate leakage current and consequent sensitivity degradation [15]. Furthermore, to depict the presence or absence of any biomolecule that has been trapped into the nanogap cavity (where k = 1 represents the absence of biomolecules), the material's charge density/dielectric constant has been used, and it is presumed that the biomolecules completely fill the nanogap cavity. The simulation of DMG-GPTFET was performed using the 2D Atlas Silvaco TCAD Version: Atlas 5.26.1.R. To achieve accurate sensitivity levels, the meshing at the quantum tunnelling zone is kept dense (0.0001) in comparison to the rest of the device. The calculation of tunnelling rate at source/channel junction is carried out by the Nonlocal band to band tunnelling (BTBT) model.

In addition to that, essential models include Shockley-Read-Hall (SRH), parallel electric field dependence (FLDMOB), Auger, concentration-dependent mobility (CONMOB), Universal Schottky Tunneling (UST), band-gap narrowing (BGN) for heavily doped regions, and numerical tunneling probability using the Wentzel-Kramers-Brillouin (WKB) method. The proposed structure can be fabricated by following the process flows as proposed in our earlier work [21-24]. The various steps involved can be the realization of p and n regions in an intrinsic substrate. This can be followed by the selective deposition of hafnium oxide (HfO2) and then dual material gates deposition. A cavity can be created in the HfO2. Another similar structure can be created, flipped and bonded with the first structure. This way the proposed structure can be fabricated.

TABLE I

Biomolecules with their dielectric constants and nature of charge [16].

Biomolecules	Dielectric constant (K)	Charge
DNA	1-64	Charged
Uricase	1.54	Neutral
Streptavidin	2.1	Neutral
Biotin	2.63	Neutral
APTES	3.57	Neutral

I. RESULTS AND DISCUSSION

The biomolecules that are found in all living organisms may be neutral, positively, or negatively charged. We examined the impact of different biomolecules on the device performance by substituting dielectric values (k > 1) for air (k = 1) and charge densities (ρ = - 10^{11} C cm^{-2} to -10^{12} C cm^{-2}) in the nanogap cavities. Energy band diagrams of DMG-GPTFET are shown in Fig. 2(a) for ON (V_{DS} = 1 V, V_{DS} = 0.5 V) and OFF (V_{DS} = 0 V, V_{DS} = 0.5 V) state. When the device is in OFF state the barrier between the source and the channel is wide, more energy is needed to tunnel electrons from the valence band (VB) to the conduction band (CB). Contrary to the OFF state, the ON state requires a lesser amount of energy for electron tunneling from VB to the CB since these energy bands are much closer in proximity. Fig. 2(b) depicts the E-B diagram of DMG-GPTFET in the ON-state (V_{GS} = 1.0 V and V_{DS} = 0.5 V) for different values of k (ρ = 0). It demonstrates that as the value of k increases, band bending increases, causing the barrier width to decrease. However, Fig. 2(c) depicts the effect of charge density (ρ) on the DMG-GPTFET's E-B diagram. When negative charge density (-ρ) increases, the formation of holes in the channel also increases, and band bending decreases, causing the barrier width to decrease.

The corresponding variations in the electric field (E-F) of the DMG-GPTFET (all along the device's length) for various values of k (at ρ = 0) and ρ (at k = 5) have shown that as the value of k increases (ρ = 0), it consequently increases the ability of the gate to control the channel. Therefore, the tunnelling barrier width is reduced at the tunnel junction, and as a result, the E-F increases. The maximum value of E-F is obtained at tunnelling junction for

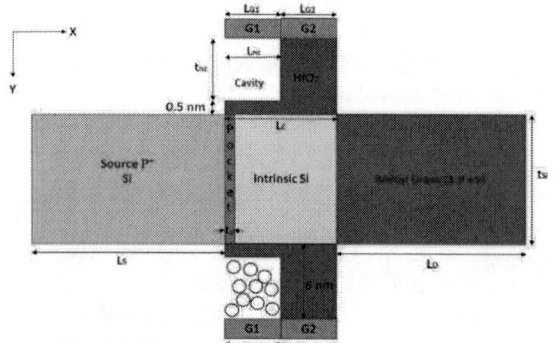

Fig.1 Schematic diagram of DMG-GPTFET biosensor

k = 10 is 3.65×10^6 V/cm. Also, when the value of -ρ is increased (k = 5), the E-F is seen to go downwards the maximum value of E-F for varying ρ is 3.54×10^6 at ρ = 0.

Further, it has been observed that the surface potential (S-P) exhibits a lower value when the air is present (k = 1). As the dielectric constant increases (k > 1), the S-P subsequently increases due to the improved coupling within the channel and the gate electrode. Flat band voltage given by **qρ/C_eff** is increased in the presence of negatively charged biomolecules, where ρ is charge density, and C_{eff} is the effective capacitance between the gate and the channel. Thereby causing the S-P (in the channel area on the source side) within the L_{GI} region to decrease. Besides, it has been observed that the barrier width decreases when there is presence of biomolecules (k > 1), which causes an increase in the e⁻ tunnelling rate. As a result, the overall drain current (Ids) in the region increases, which in turn affects the device's ON current. The effects of geometrical factors, namely the nanocavity thickness (tnc), and the nanocavity length (Lnc), on device transfer characteristics at k = 5 and ρ = 0 are depicted in Figs. 3(a) and (b), respectively. As we increase the thickness of nanocavity (tnc) in DMG-GPTFET there is increment in the tunnelling barrier at the tunnel junction. As a result, the device's ON current and overall drain current decrease. Contrarily, when the nanocavity length (Lnc) in the DMG-GPTFET varies, there doesn't seem to be any significant difference in Ids.

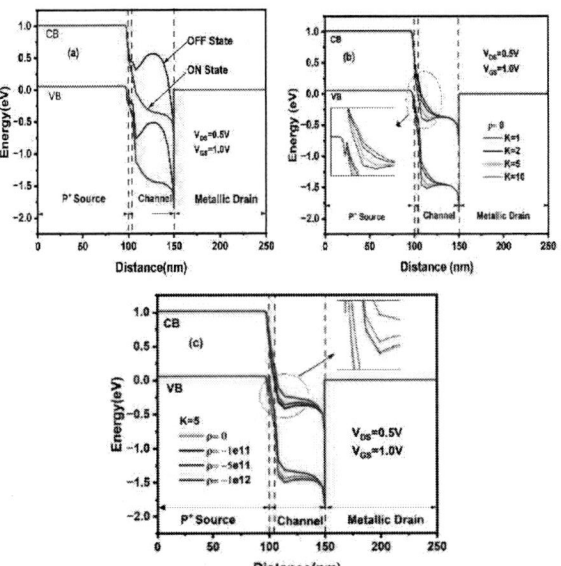

Fig. 2(a): Energy band diagram of DMG-GPTFET for ON and OFF state Fig. 2(b) Energy band diagram of DMG-GPTFET with different dielectric constants. Fig 2(c) Energy band diagram for different charge densities.

Since the DMG-GPTFET's working mechanism is based on tunnelling phenomena rather than bulk as in the case of conventional FET [16]. As reported in [16], ON current changes with the change in Lnc only when the nanocavity is filled with air (k = 1) but in the case of biomolecules (k > 1) Ion does not change with the variation in Lnc. The corresponding Ids – V_GS characteristics of the DMG-GPTFET for various dielectric constants (k) is depicted in Fig. 4. When air is present in the nano cavities (k = 1), the

minimum value of the drain current (Ids) is attained in the order of 10^{-10} (A/μm) with V_{GS} = 1V and V_{DS} = 0.5V. Figure 5(a) and (b) are illustrations of the drain current Ids sensitivity with gate bias of the DMG-GPTFET for various dielectric constants, k's (ρ = 0), and charge densities, ρ's (k = 5,7 and 10). At ρ = 0, we see that the sensitivity increases as k increases, reaching its maximum value at k = 10. The significant change in Ids sensitivity with gate bias demonstrates the DMG-GPTFET's capacity to detect the charge in its nanocavity [46]. However, when the negative charge density (ρ) increases, the Ids sensitivity decreases, which can be seen in Fig. 5(b).

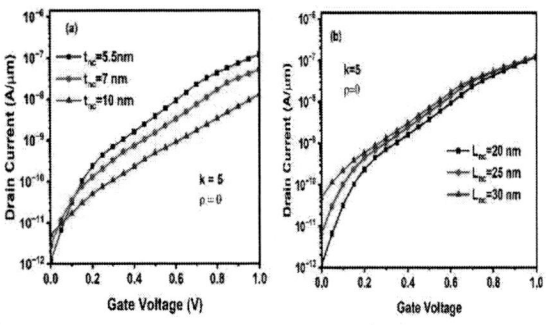

Fig. 3. I_d vs V_{GS} plots on (a) varying cavity thickness (tnc) (b) cavity length (Lnc) at k = 5 and ρ = 0

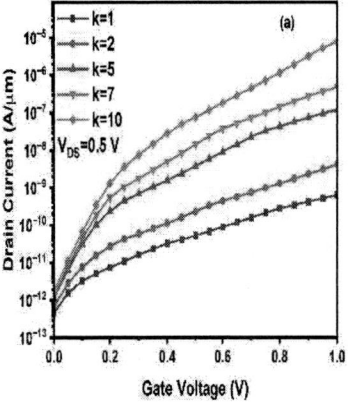

Fig. 4. Transfer characteristics for different values of k

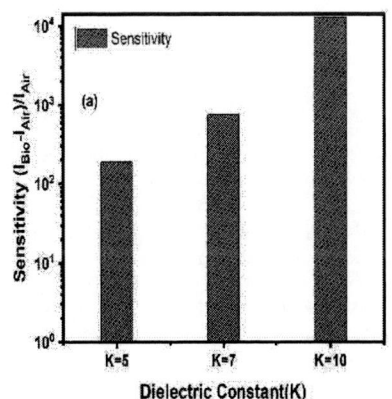

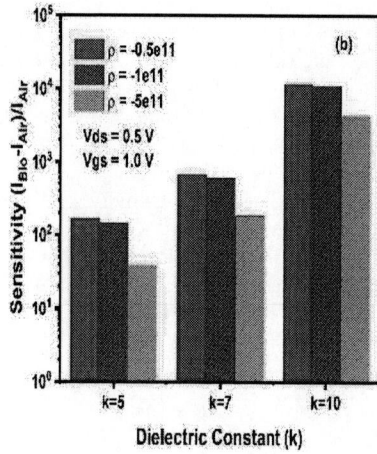

Fig. 5 Sensitivity analysis of the DMG_GPTFET biosensor at (a) different dielectric constants, (k's) (at $\rho = 0$) (b) different negative charge densities (r) (at k = 5, 7 and 10)

II. CONCLUSION

This paper presents the design and simulation of a DMC-GPTFET for the purpose of biomolecule sensing. The device incorporates the creation of two cavities through the process of etching out the oxide material beneath the gate electrode of the source side that can trap and sense both neutral and charged biomolecules. The coupling between gate and channel region beneath nanocavity is influenced by the dielectric constant and charge density associated with the biomolecules. Neutral and charged biomolecules have been used to study electrical properties such as energy band, transfer characteristics, surface potential, and electric field. Also, the special case of Uricase biomolecule has been studied for the better understanding of sensitivity. It has been seen that the position of biomolecules in the nanocavity can affect the value of drain current sensitivity. Also, the cavity dimensions play an important role in a higher drain current. Thus, with the right choice of cavity dimensions and placement of probes near the tunnelling junction inside the nanocavity, better sensitivity can be achieved with a small amount of detection samples.

V. REFERENCES

[1] Barbaro M, Bonfiglio A, RaffoL, "A charge-modulated FET for detection of biomolecular processes: conception, modeling, and simulation," IEEE Trans. Electron Devices, vol. 53, no. 1, pp. 158–166, Feb. 2006.

[2] Kim C.-H, Jung C et al., "Novel dielectric modulated field-effect transistor for label-free DNA detection," Biochip J., vol. 2, no. 2, pp. 127–134, June 2008.

[3] Seabaugh A.C, Zhang Q, "Low-voltage tunnel transistors for beyond CMOS logic," Proc. IEEE., vol. 98, no. 12, pp. 2095–2110, Dec. 2010.

[4] Choi W.Y et al., "Tunneling field-effect transistors (TFETs) with subthreshold swing (SS) less than 60 mV/dec," IEEE Electron Device Lett., vol. 28, no. 8, pp. 743–745, July 2007.

[5] P. Bergveld, "Development, operation, and application of the ion sensitive field-effect transistor as a tool for electrophysiology," IEEE Trans. Biomed. Eng., vol. BME-19, no. 5, pp. 342–351, Sep. 1972.

[6] H. Im, X.-J. Huang, B. Gu, and Y.-K. Choi, "A dielectric-modulated field-effect transistor for biosensing," Nature Nanotechnol., vol. 2, no. 7, pp. 430–434, 2007.

[7] Rajani Gautam et al., "Numerical model of Gate-All-Around MOSFET with vacuum gate dielectric for biomolecule detection,"

IEEE Electron Device Lett., vol. 33, no. 12, pp. 1756-1758, Dec. 2012.

[8] N. Kannan and M. J. Kumar, "Dielectric-modulated impact-ionization MOS transistor as a label-free biosensor," IEEE Electron Device Lett., vol. 34, no. 12, pp. 1575–1577, Dec. 2013.

[9] C.-H. Kim, J.-H. Ahn, K.-B. Lee, C. Jung, H. G. Park, and Y.-K. Choi, "A new sensing metric to reduce data fluctuations in a nanogap-embedded field-effect transistor biosensor," IEEE Trans. Electron Devices, vol. 59, no. 10, pp. 2825–2831, Oct. 2012.

[10] J.-M. Choi, J.-W. Han, S.-J. Choi, and Y.-K. Choi, "Analytical modelling of a nanogap-embedded FET for application as a biosensor," IEEE Trans. Electron Devices, vol. 57, no. 12, pp. 3477–3484, Dec. 2010.

[11] R. Pandey, B. Rajamohanan, H. Liu, V. Narayanan, and S. Datta, "Electrical noise in heterojunction interband tunnel FETs," IEEE Trans. Electron Devices, vol. 61, no. 2, pp. 552–560, Feb. 2014.

[12] M.-L. Fan, V. P.-H. Hu, Y.-N. Chen, P. Su, and C.-T. Chuang, "Analysis of single-trap-induced random telegraph noise and its interaction with work function variation for tunnel FET," IEEE Trans. Electron Devices, vol. 60, no. 6, pp. 2038–2044, Jun. 2013.

[13] J. Wan, C. L. Royer, A. Zaslavsky, and S. Cristoloveanu, "Tunneling FETs on SOI: Suppression of ambipolar leakage, low-frequency noise behavior, and modeling," Solid-State Electron., vols. 65–66, pp. 226–233, Nov./Dec. 2011.

[14] R. Narang, M. Saxena, R. S. Gupta, and M. Gupta, "Dielectric modulated tunnel field-effect transistor—A biomolecule sensor," IEEE Electron Device Lett., vol. 33, no. 2, pp. 266–268, Feb. 2012.

[15] Y. Khatami and K. Banerjee, "Steep subthreshold slope n- and p-type tunnel-FET devices for low-power and energy-efficient digital circuits," IEEE Trans. Electron Devices, vol. 56, no. 11, pp. 2752–2761, Nov. 2009.

[16] R. Asra, M. Shrivastava, K. V. R. M. Murali, R. K. Pandey, H. Gossner, and V. R. Rao, "A tunnel FET for VDD scaling below 0.6 V with a CMOS-comparable performance," IEEE Trans. Electron Devices, vol. 58, no. 7, pp. 1855–1863, Jul. 2011.

[17] K. K. Bhuwalka, J. Schulze, and I. Eisele, "Scaling the vertical tunnel FET with tunnel bandgap modulation and gate workfunction engineering," IEEE Trans. Electron Devices, vol. 52, no. 5, pp. 909–917, May 2005. pp. 103–114, May 2015.

[18] 11 Khan, A., & Loan, S. A. (2021). Double gate TFET with germanium pocket and metal drain using dual oxide. International Conference on Microelectronics (ICM), 2021, 170–173. https:// doi. org/ 10. 1109/ ICM52 667. 2021. 96649 49.

[19] 12 ATLAS Device Simulation Software, Silvaco, Santa Clara, CA, USA, 2016.

[20] Boucart, K., & Ionescu, A. M. (2007). Double gate tunnel FET with high-k gate dielectric. IEEE Transactions on Electron Devices, 54(7), 1725–1733.

[21] Anam Khan and Sajad A Loan, Pocketed Dual Metal Gate TFET: Design and Simulation, accepted in Materials Today Communications,

[22] Anam Khan A., and Loan, Sajad A.,(2021), Double Gate TFET with Germanium Pocket and Metal Drain Using Dual Oxide in proceedings of IEEE ICM 2021 Egypt.

[23] Anam Khan. A and Loan, Sajad A., (2021), Metal Drain Double-Gate Tunnel Field Effect Transistor with Underlap : Design and Simulation, Silicon Journal, vol. 13, no. 5, pp.1421-1431.

[24] Anam Khan, A., Alharbi, A.G. and Sajad A Loan; Germanium pocket based tunnel FET with underlap: design and simulation. Analog Integr Circ Sig Process (2022)

Design and Fabrication of Nanofibrous membrane and Microelectrodes for highly robust biocompatible humidity sensing

Afaque Manzoor Soomro
Department of Electrical Engineering,
Sukkur IBA University
Sukkur, Pakistan, 65200
afaque.manzoor@iba-suk.edu.pk
0000-0002-0352-9319

Faheem Ahmed
Department of Mechatronics
Engineering, Jeju National University
Jeju, South Korea, 63243
ahmedfaheem@stu.jejunu.ac.kr

Muhammad Waqas
New Materials Department, Yangtze
Delta Research Institute of University
of Electronic Science &Technology
of China (Huzhou)
mwaqas@csj.uestc.edu.cn

Abstract—**This study reports the fabrication of a self-sensing humidity sensor based on a nanofibrous membrane and microelectrodes (comb-shaped). Unlike the conventional approach, where separate substrates are used for mechanical support of the electrodes, we have proposed a membrane based on polylactic glycolic acid (PLGA) that serves both objectives; sensing humidity with full range (0%-100% RH) and also providing mechanical support to the electrodes with minimum compromise over the flexibility of the sensor. Various electrical and surface characterizations were also performed. The as-developed sensor showed excellent response and recovery time (1.8 s and 4.3 s, respectively), sensitivity, and ability to distinguish between different breathing styles when practically used with volunteer subjects.**

Keywords—Nanofibrous membrane, microelectrodes, humidity sensor, wearable monitoring

I. INTRODUCTION

It's important to monitor and control a variety of environmental variables such as temperature, humidity, light, pressure, etc. Controlling these variables is intended for various applications, such as those in the fields of medicine [1], [2], agriculture[3]–[5], food[6], [7], and manufacturing [8], [9]. Generally, the humidity of the environment is measured through the relative humidity [10], [11]. There may be variations in resistive [12], capacitive [13], impeditive [14], acoustic [15], or optical [16] characteristics depending on the amount of water vapour present. Recently, personnel health monitoring has become the main focus of humidity sensing [17], which uses noninvasive humidity measurement to track several physiological features of the human body, including stress and breathing patterns. Furthermore, it is quite important to achieve intuitive monitoring in various fields of life[18].

Furthermore, humidity sensors are known for their high sensitivity, quick reaction and recovery times, repeatability (stability/cyclic performance), fabrication procedures, and cost. Various approaches are employed, such as Field Effect Transistor, IDTs, Quartz Crystal-based sensors, and Surface Acoustic Waves, have an impact on the efficiency of humidity sensors as well . However, it can be difficult to reach all of the merit numbers. Practically, extremely sensitive sensors, like metal dichalcogenides (MoS2, WS2, CdS, and copper sulphide)[4], [14], [19]–[23], may respond slowly, whereas fast-reacting sensors may have lower sensitivity. Numerous novel materials with humidity sensing capabilities are being researched in order to achieve the best outcomes. These characteristics include high surface area to volume ratio, crystalline structure, and hydrophilicity. Moreover, different functional materials, including Polymers, ceramics and oxides have been used [8], [24]–[26], graphene and ZnO, etc.), as well as 2D materials like graphene, graphene oxide, reduced graphene oxide, and ML-MOS2 can all be used to create the sensing However, an extensive amount of literature's lack of essential features eventually limits its applications in the monitoring of personal health [11], [23], [27], [28]. To achieve such bio-related features, research into biopolymer materials with humidity sensing properties is essential.

This study proposes a biocompatible substrate-less humidity sensor based on polylactic glycolic acid (PLGA; lactic: glycolic = 50:50). The device is fabricated using electrospinning and screen printing techniques. Silver ink was used to fabricate Inter Digited Electrodes (IDTs) while the sensing membrane was developed using an in-house electrospinning setup. The as-developed sensor has excellent response to the whole range of relative humidity (0%RH-100% RH) with excellent response and recovery time. The sensor was also utilized in the human breathing application which could significantly distinguish various breathing patterns.

II. EXPERIMENTAL

A. Materials and methods

The PLGA was supplied by Sigma-Aldrich. From Silverjet, silver conductive ink was ordered with the following specifications: 1.5 cps viscosity, 24.4 mN/m surface tension, and octane-based dispersion matrix. In contrast, silver epoxy was purchased from 4science to work with IDTs. Glycolic acid (HOCH2COOH) is the smallest -hydroxy acid. To make the solution (10 mL), 100 mg of PLGA were dissolved in TFE (2, 2, 2-Trifluoroethanol). The mixture was continuously stirred for 30 minutes at room temperature (25 0C) to produce a homogeneous solution. Then, until it was utilized, it was kept at room temperature.

B. Sensor fabrication

The electrospinning solution was first thoroughly mixed until it was a homogeneous solution. High viscosity ink was electrospun to create a 25 um thick membrane of nanofibers, which was then sliced into 15mm*25mm pieces (Applied Potential: 8.5 kV, Collector distance: 20 cm, Flow rate: 0.7 ml/hr). To modulate the ES process, different concentrations of the substance were utilised. The electrospun fibers were collected on aluminum foil. Moreover, the ES process was performed in temperature and humidity controlled chamber (~20 °C and 40%RH). The Interdigital electrodes (IDEs) on the ES-PLGA membrane were initially created using EAGLE version 10 in Dxf format in order to produce electrodes. The file was next transformed into a BMP file, and Dimatix Drop Manager changed the BMP file's format to open. The inkjet printing system received the ptn file and began printing. A total of 3 ml of Ag ink was put into the 1 pL 16 nozzle cartridge. For reliable printing, the 25 V was applied to cartridge nozzles. The printing platform's temperature was set at 50 °C. Later, the suggested IDEs were cured for two hours at 60 °C. The electrode has a total of 40 fingers, each measuring 50 um thick, 40 um apart, and 8 mm long. Sensor-2 has 10 fingers (thickness of 200um, spacing of 250um and

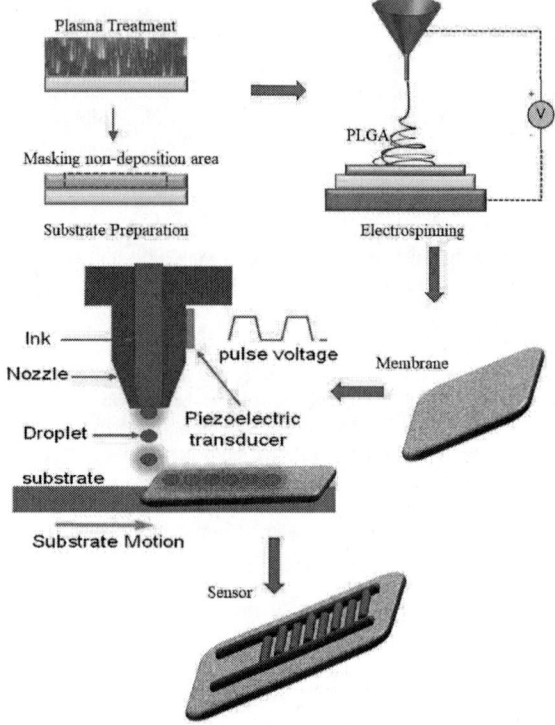

Figure 1 Fabrication of nanofibrous membrane using electrospinning process, followed by microelectrode (IDE based) fabrication.

length of 5mm). The step by step fabrication process is shown in Fig. 1.

C. Characterization

Using a custom-built controlled environment box, the electrical response of the sensor to changes in relative humidity was determined. The setup has the capacity to use a proportional feedback controller (PID) to regulate its humidity between 0% and 100% RH. Dry nitrogen gas was used to reduce humidity to a minimum value of zero or to adjust it to any other value of choosing, whereas compressed dry air was favored to achieve humidity levels up to 10% RH.

To increase the RH in the chamber, water vapour was introduced using a desktop humidifier. Except for dry nitrogen gas, which was manually controlled, all inlets were electronically controlled using electronic mass flow controllers (MFC). Using a USB connection and a smart device, the sensor's answer was continuously logged to a desktop computer. Moreover, the sensor's data was immediately logged in the personnel computer through a serial connection. Total of 50 samples were taken in each experiment as there was gradual change of 2% during each step.

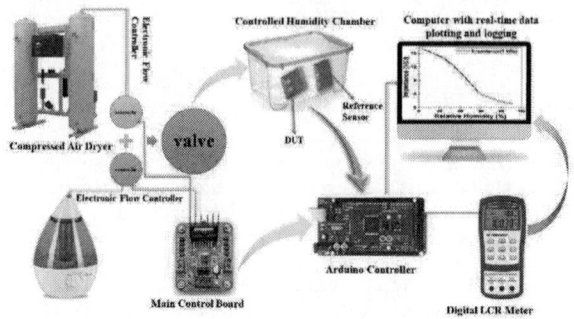

Figure 2 Experimental setup used to characterize the proposed sensing device.

III. RESULTS AND DISCUSSION

The target of developing the proposed sensing device was to utilize such sensors not only for environmental monitoring for also for human health monitoring, specially breathe activity monitoring. For such applications, biocompatibility plays a pivotal role. Herein, we selected PLGA, highly biocompatible material, which is commonly used for drug delivery in various medical applications. Moreover, in our previous study, the toxicity test was successfully performed, which proved that PLGA can be reliably used in the wearable applications[1]. Before the testing of the device, the membrane was physically examined using SEM analysis. As shown in Fig. 3, the nanofibers are formed, though they are random in nature, but still show excellent formation.

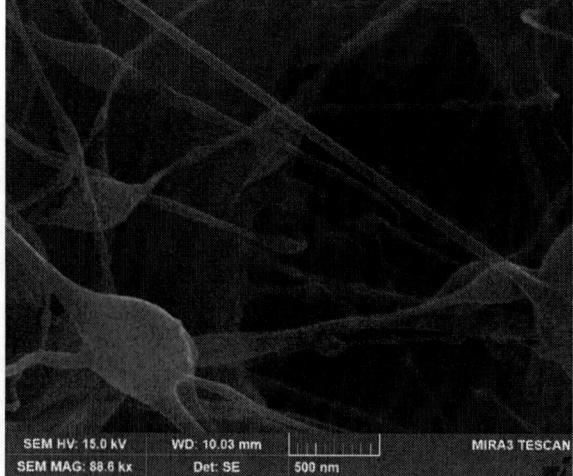

Figure 3 SEM image of humidity sensing nanofibrous membrane.

Moreover, once the nanofibrous membrane based on PLGA was ready, the screen printing was used to design comb type electrodes using silver ink. It is also worth mentioning that there is no need for a separate substrate for mechanically holding the electrodes, in fact the same nanofibrous membrane serves both the purposes, humidity sensing, and mechanical support for the electrodes. By comparing the output impedance to variations in relative humidity of an electrically controlled chamber, the electrical performance of the sensor was evaluated. The obtained results show a behaviour of decreasing impedance as water vapour concentration rises. Differently developed sensing layers with different thicknesses presented various intrinsic impedances, with the main behaviour being an inverse relationship between thickness and impedance. Fig. 3 displays the sensor's impedance response to changes in relative humidity at two frequencies of 1 kHz.

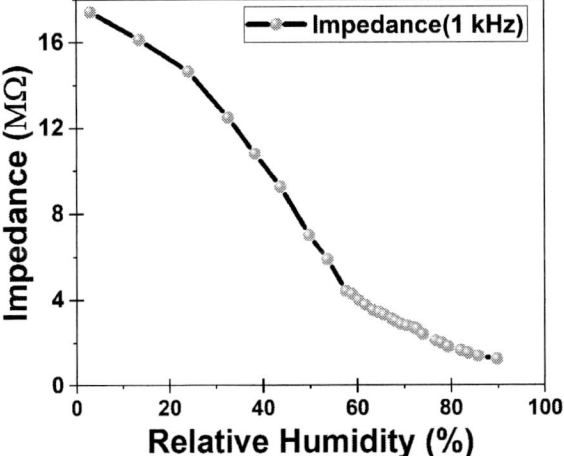

Figure 5 Sensor's response against changing relative humidity at 1 kHz.

Moreover, the proposed humidity sensor operates on the theory that when water molecules are captured by the detecting layer, impedance changes. The behaviour of the sensor decreases impedance as the controlled chamber's relative humidity rises. The magnitude of the impedance is determined by the interaction between the capacitance of the manufactured film and the electrodes' resistance. Therefore, an increase in humidity value either results in a higher dielectric constant or a higher conductivity, which lowers resistance. In Fig. 4, the mechanism is shown. In the beginning, the primary behavior of the sensor is caused by chemisorption, which results in the generation of positive ions (H+).

The results demonstrate an inverse relationship between the sensor's impedance and the relative humidity of the testing chamber. The response of both sensors is extremely sensitive and linear over the whole range of humidity. Each sensor's overall response can be divided into two unequal parts: the first part of each sensor is extremely sensitive and experiences a rapid change in impedance (which contributes to more than 70% of the response), while the second part experiences a decreased change in impedance because of sufficient saturation brought on by adsorption. The device's response, however, demonstrates an excellent match.

Due to the sensor's eco-friendliness and biocompatibility, respiration monitoring is accomplished in order to apply the actual application of designed and built sensor. The sensor was positioned under the volunteer's nose to measure their ability to breathe. Responses were separated into four main groups: normal breathing, deep breathing, quick breathing, and deliberate pauses between two breathing patterns known as apneas in order to stabilise the RH value. In figure 5, the respiratory response is shown. For normal, deep, and fast breathing, the sensor showed rapid impedance fluctuations.

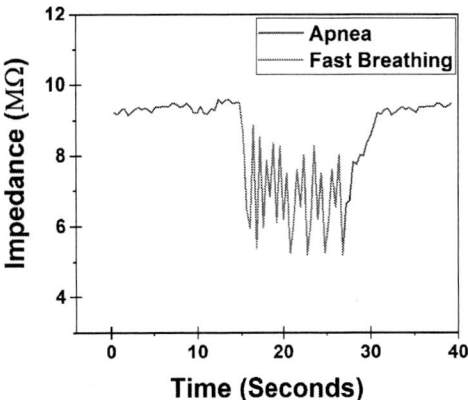

Figure 4 Sensor's response to two different breathing styles, fast breathing and apnea (stoppage of breathing).

IV. CONCLUSION

In summary, a self-sensing humidity sensor that uses comb-shaped microelectrodes and a nanofibrous membrane is created. We have proposed a membrane that accomplishes both of our goals—sensing humidity with a full range (0%–100% RH) and also providing mechanical support to the electrodes with the least amount of compromise to the flexibility of the sensor—in contrast to the conventional approach, which uses separate substrates for this purpose. Additionally, numerous electrical and surface characterizations were carried out. When employed in practise with volunteer individuals, the sensor's as-developed performance displayed good reaction and recovery times (1.8 s and 4.3 s, respectively), sensitivity, and the capacity to discern between various breathing techniques.

ACKNOWLEDGMENT

We are thankful to Office of Research and Innovation, Sukkur IBA University, for supporting this research work.

REFERENCES

[1] A. M. Soomro, F. Jabbar, M. Ali, J. W. Lee, S. W. Mun, and K. H. Choi, "All-range flexible and biocompatible humidity sensor based on poly lactic glycolic acid (PLGA) and its application in human breathing for wearable health monitoring," *J. Mater. Sci. Mater. Electron.*, vol. 30, no. 10, pp. 9455–9465, 2019, doi: 10.1007/s10854-019-01277-1.

[2] M. A. U. Khalid *et al.*, "A highly sensitive biodegradable pressure sensor based on nanofibrous dielectric," *Sensors Actuators, A Phys.*, vol. 294, pp. 140–147, 2019, doi: 10.1016/j.sna.2019.05.021.

[3] F. Ahmed *et al.*, "Multi-material Bio-inspired Soft Octopus Robot for Underwater Synchronous Swimming," *J. Bionic Eng.*, no.

0123456789, 2022, doi: 10.1007/s42235-022-00208-x.

[4] A. M. Soomro, M. A. U. Khalid, I. Shah, S. W. Kim, Y. S. Kim, and K. H. Choi, "Highly stable soft strain sensor based on Gly-KCl filled sinusoidal fluidic channel for wearable and water-proof robotic applications," *Smart Mater. Struct.*, vol. 29, no. 2, 2020, doi: 10.1088/1361-665X/ab540b.

[5] H. Ashraf, B. Shah, A. M. Soomro, Q. ul A. Safdar, Z. Halim, and S. K. Shah, "Ambient-noise Free Generation of Clean Underwater Ship Engine Audios from Hydrophones using Generative Adversarial Networks," *Comput. Electr. Eng.*, vol. 100, no. March, p. 107970, 2022, doi: 10.1016/j.compeleceng.2022.107970.

[6] M. J. C. Santbergen, M. van der Zande, A. Gerssen, H. Bouwmeester, and M. W. F. Nielen, "Dynamic in vitro intestinal barrier model coupled to chip-based liquid chromatography mass spectrometry for oral bioavailability studies," *Anal. Bioanal. Chem.*, vol. 412, no. 5, pp. 1111–1122, 2020, doi: 10.1007/s00216-019-02336-6.

[7] Andreescu, "Trends and challenges in biochemical sensors for clinical and environmental monitoring*," *Pure Appl. Chem*, vol. 76, no. 4, pp. 861–878, 2004, doi: 10.1351/pac200476040861.

[8] F. Jabbar *et al.*, "Robust Fluidic Biocompatible Strain Sensor Based on PEDOT : PSS / CNT Composite for Human-wearable and High-end Robotic Applications," vol. 32, no. 12, pp. 1–17, 2020.

[9] A. M. Soomro *et al.*, "Fully 3D printed multi-material soft bio-inspired frog for underwater synchronous swimming," *Int. J. Mech. Sci.*, vol. 210, no. July, p. 106725, 2021, doi: 10.1016/j.ijmecsci.2021.106725.

[10] K. Khorsand Kazemi *et al.*, "Low-Profile Planar Antenna Sensor Based on Ti3C2Tx MXene Membrane for VOC and Humidity Monitoring," *Adv. Mater. Interfaces*, vol. 9, no. 13, pp. 1–11, 2022, doi: 10.1002/admi.202102411.

[11] T. Li *et al.*, "Porous Ionic Membrane Based Flexible Humidity Sensor and its Multifunctional Applications," *Adv. Sci.*, vol. 4, no. 5, 2017, doi: 10.1002/advs.201600404.

[12] H. M. Zeeshan Yousaf, S. W. Kim, G. Hassan, K. Karimov, K. H. Choi, and M. Sajid, "Highly sensitive wide range linear integrated temperature compensated humidity sensors fabricated using Electrohydrodynamic printing and electrospray deposition," *Sensors Actuators, B Chem.*, vol. 308, no. January, p. 127680, 2020, doi: 10.1016/j.snb.2020.127680.

[13] Q. Gao, H. Li, J. Zhang, Z. Xie, J. Zhang, and L. Wang, "Microchannel Structural Design For a Room-Temperature Liquid Metal Based Super-stretchable Sensor," *Sci. Rep.*, vol. 9, no. 1, pp. 1–8, 2019, doi: 10.1038/s41598-019-42457-7.

[14] A. M. Soomro *et al.*, "Flexible Fluidic-Type Strain Sensors for Wearable and Robotic Applications Fabricated with Novel Conductive Liquids: A Review," *Electron.*, vol. 11, no. 18, 2022, doi: 10.3390/electronics11182903.

[15] A. I. Buvailo, Y. Xing, J. Hines, N. Dollahon, and E. Borguet, "TiO2/LiCl-based nanostructured thin film for humidity sensor applications," *ACS Appl. Mater. Interfaces*, vol. 3, no. 2, pp. 528–533, 2011, doi: 10.1021/am1011035.

[16] H. Khan *et al.*, "Highly sensitive mechano-optical strain sensors based on 2D materials for human wearable monitoring and high-end robotic applications," *J. Mater. Chem. C*, vol. 10, no. 3, pp.

932–940, 2022, doi: 10.1039/d1tc03519c.

[17] Zeeshan, A. M. Soomro, and S. Cho, "Design and Fabrication of a Robust Chitosan/Polyvinyl Alcohol-Based Humidity Sensor energized by a Piezoelectric Generator," *Energies*, vol. 15, no. 20, 2022, doi: 10.3390/en15207609.

[18] A. M. Soomro *et al.*, "Textile Based Flexible Temperature Sensors for Wearable and Sports Applications," doi: 10.1002/pssa.202300523.

[19] V. J. Babu *et al.*, "Intelligent Nanomaterials for Wearable and Stretchable Strain Sensor Applications: The Science behind Diverse Mechanisms, Fabrication Methods, and Real-Time Healthcare," *Polymers (Basel).*, vol. 14, no. 11, p. 2219, 2022, doi: 10.3390/polym14112219.

[20] J. W. Lee, A. M. Soomro, M. Waqas, M. A. U. Khalid, and K. H. Choi, "A highly efficient surface modified separator fabricated with atmospheric atomic layer deposition for high temperature lithium ion batteries," *Int. J. Energy Res.*, vol. 44, no. 8, pp. 7035–7046, 2020, doi: 10.1002/er.5371.

[21] Y. Zheng, H. Li, W. Shen, and J. Jian, "Wearable electronic nose for human skin odor identification: a preliminary study," *Sensors Actuators, A Phys.*, doi: S0924424718313827.

[22] N. Benmoussa, A. Benichou, N. Medjahdi, and K. Rahmoun, "Modeling and optimization of a capacitive humidity sensor response in MEMS technology," *2014 North African Work. Dielectr. Mater. Photovolt. Syst. NAWDMPV 2014*, pp. 8–11, 2014, doi: 10.1109/NAWDMPV.2014.6997606.

[23] H. Y. Li, C. S. Lee, D. H. Kim, and J. H. Lee, "Flexible Room-Temperature NH3 Sensor for Ultrasensitive, Selective, and Humidity-Independent Gas Detection," *ACS Appl. Mater. Interfaces*, vol. 10, no. 33, pp. 27858–27867, 2018, doi: 10.1021/acsami.8b09169.

[24] K. H. Choi *et al.*, "Hybrid Surface Acoustic Wave-Electrohydrodynamic Atomization (SAW-EHDA) for the Development of Functional Thin Films," *Sci. Rep.*, vol. 5, no. March, pp. 1–14, 2015, doi: 10.1038/srep15178.

[25] K. H. Choi, M. Sajid, S. Aziz, and B. S. Yang, "Wide range high speed relative humidity sensor based on PEDOT:PSS-PVA composite on an IDT printed on piezoelectric substrate," *Sensors Actuators, A Phys.*, vol. 228, pp. 40–49, 2015, doi: 10.1016/j.sna.2015.03.003.

[26] G. U. Siddiqui, M. Sajid, J. Ali, S. W. Kim, Y. H. Doh, and K. H. Choi, "Wide range highly sensitive relative humidity sensor based on series combination of MoS2and PEDOT:PSS sensors array," *Sensors Actuators, B Chem.*, vol. 266, pp. 354–363, 2018, doi: 10.1016/j.snb.2018.03.134.

[27] A. D. Smith *et al.*, "Resistive graphene humidity sensors with rapid and direct electrical readout," *Nanoscale*, vol. 7, no. 45, pp. 19099–19109, 2015, doi: 10.1039/c5nr06038a.

[28] K. H. Choi, M. Sajid, S. Aziz, and B.-S. Yang, "Wide range high speed relative humidity sensor based on PEDOT:PSS–PVA composite on an IDT printed on piezoelectric substrate," *Sensors Actuators A Phys.*, vol. 228, pp. 40–49, Jun. 2015, doi: 10.1016/j.sna.2015.03.003.

InGaAs Self-Switching Diode With Suppressed Harmonics For High Frequency Applications

B. Sharma, S. Garg, P. Singh, Supriya
Garg, G. Das, D. K. Sharma, N. Gupta,
and
A. K. Singh*
Semiconductor Research Centre,
Department of Electronics and
communication Engineering,
Punjab Engineering College,
Sector-12, Chandigarh, India
*arun@pec.edu.in

S. Kumar
Department of Physics,
Punjab Engineering College,
Sector-12, Chandigarh, India

S. R. Kasjoo
School of Microelectronics
Engineering, University Malaysia
Perlis, 02600 Arau, Perlis, Malaysia

Abstract— In this study, a novel InGaAs-based nano- rectifier known as self-switching diode is presented to exhibit the suppressed harmonics for high frequency applications. The self-switching diode device exhibits current-voltage characteristics analogous to the conventional diodes, eliminating the need for a p-n junction and/or Schottky barrier. The direct and alternating current characteristics of the proposed device are investigated by filling its trenches with different dielectric materials. Further, the total harmonic distortion is quantified by implementing Fast Fourier Transform to estimate the corresponding harmonic components. The results suggest the introduction of dielectric materials with permittivity ranging from 1.0 to 9.3 into the trenches results in the significant reduction in total harmonic distortion from 69% to 60.4% at high frequencies.

Keywords— *Self-switching diode, harmonics, THD, dielectric constant, trench capacitance.*

I. INTRODUCTION

In recent years, there has been a growing demand for high frequency rectifiers, driven by their wide-range applications in fields such as biology, medicine, telecommunications, and energy harvesting [1- 3]. While advancements in rectifier and converter technologies have primarily been directed at high power devices, relatively less attention has been paid to high frequency rectifiers. One of the significant challenges associated with these devices is the presence of harmonics in the output waveform, which can adversely affect the performance and efficiency of the electronic devices and often require additional filter circuitry to reduce the harmonics. Generally, rectifiers are developed using various technologies like Schottky barrier diodes (SBD), Silicon and GaN MOSFETS [4] with different topologies such as half bridge [5], full bridge [6] and buck-boost converters [7]. Therefore, the development of effective techniques for harmonic reduction has become a significant area of research. Self-switching diodes have emerged as promising devices for high frequency rectification due to their high switching speed, low conduction losses and easy integration to form bridge rectifiers [8-10]. However, their inherent characteristics can lead to the generation of harmonic content in the output waveform.

The main objective of this research is to address the issue of harmonics reduction, we propose the filling of

dielectrics in the trenches of self-switching diode-based rectifiers. Dielectrics offer unique electrical properties that can modify the electric field distribution within the device, thereby influencing the generation and propagation of harmonics. By strategically placing dielectrics in the trenches, it is possible to alter the current flow and mitigate the unwanted harmonic components.

Despite previous studies of SSD filled with dielectrics on temperature [11], enhanced performance by nonmonotonic frequency dependence [12], rectification performance [13], and electrical behavior [14], the harmonic reduction in self-switching diode-based rectifiers have not been studied yet. Hence in this work, we investigate the reduction of harmonics through the utilization of dielectrics in the trenches of self-switching diode-based rectifiers. The rectifying characteristic of the proposed devices are examined using Silvaco TCAD software, up to the tens of GHz frequencies, and total harmonic distortion (THD) is evaluated for various dielectric materials through comprehensive simulations and theoretical validations and compared with previously reported devices and circuits. This work will further lead to the advancement of high frequency rectifiers and facilitates the deployment of more efficient and reliable high power electronic devices.

II. DEVICE STRUCTURE AND WORKING

The SSDs have planar structure which leads to a very low parasitic capacitance and are easily integrable with other circuits on chip. They can be fabricated through a straightforward process involving single step lithography and

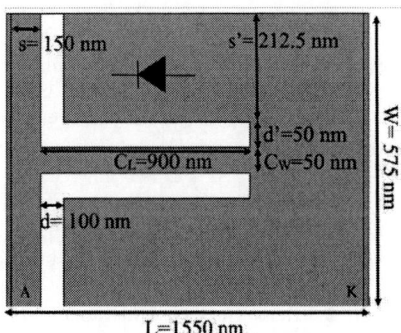

Fig. 1. The schematic of InGaAs self-switching diode (SSD).

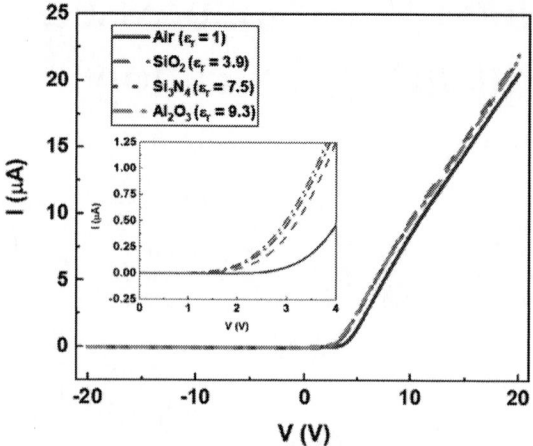

Fig. 2. *I-V* characteristics of SSD by varying dielectric materials, the inset shows the magnified version of the *I-V* characteristics depicting the change in the threshold voltage (V_t) of the SSD.

then wet-chemical/dry etching [15]. SSDs have two L-shaped trenches that are formed by removing semiconductor material to create the necessary channel and break the device symmetry between the anode and cathode terminals. These trenches are wide enough to minimize leakage current, and the vertical trenches are extended to the device boundary to ensure that current conducts exclusively through the channel. The asymmetrical edges of trenches along the channel, leads to the formation of surface states eliminating need for any external bias [8]. As a result, an etched interface inside the channel naturally gives rise to a depletion region, even in the absence of any applied bias. This depletion region, spanning across the trenches, governs the movement of charge carriers within the channel in response to the bias condition established between the anode and cathode terminals. In the case of a positive bias applied to the anode terminal, the device operates under a forward bias condition, permitting the passage of charge carriers through the channel. Conversely, a negative bias applied to the anode induces a reverse bias condition, thereby impeding the transportation of charge carriers across the channel [16].

In this study, the two-dimensional structure of the InGaAs SSD is simulated using Silvaco Atlas software [17]. The structure of self-switching diode is shown in Fig. 1. To introduce non-linearity in the *I-V* characteristics, L-shaped trenches are etched. The dimensions of SSD are 0.575 x 1.55 μm with channel length (C_L) and channel width (C_W) as 0.9 μm and 0.05 μm respectively. Since, the dimensions of SSD are greater than 10 nm, semi-classical physics-based equations along with SRH recombination, field dependent mobility models have been utilized for 2D simulations. In the simulation of the device, a background doping of 2×10^{16} cm^{-3} and an electron mobility of 8000 cm²/V-s are considered [18]. Since the device is simulated in 2D, a fixed surface charge density of -0.2×10^{12} cm^{-2} is defined in account for the influence of other substrate layers [19]. The structure of the SSD is adapted from our previous work [20], and all the simulation models have undergone calibration with experimental results previously. The etched trenches can be easily filled with different dielectric materials through spin coating and other CVD techniques. The addition of dielectric results in change of vertical and horizontal trench capacitance

of the device leading to change in the device performance [21].

III. RESULTS AND DISCUSSION

A. *I-V characteristic analysis*

We conducted simulations of the Self Switching diode (SSD) to examine their *I-V* characteristics and rectification performance. In these simulations, we modified the surface states located at the interface between the insulating trenches and the conducting channel by varying the dielectric material of the SSD trenches. The *I-V* characteristics of the SSD for different dielectric materials are shown in Fig. 2. The inset of Fig. 2 shows the magnified version of the electrical characteristics depicting the change in the threshold voltage (V_t) of the SSD. It is observed through previous theoretical studies on SSD that V_t is inversely proportional to Capacitance (C) leading to decrease in threshold voltage as the relative permittivity (ε_r) of the dielectric material filled in the trenches increases. There is an increase in the channel current due to decrease in the resistance (R). This is attributed to the increased capacitance in the horizontal trenches that decrease the leakage of the charge carriers through the trenches, facilitating a greater movement of charge carriers through the channel. However, the difference in the channel current in not very large due to already low amount of leakage of charge carriers through channel due to planar structure of SSD.

B. *AC analysis*

The AC analysis has been performed at range of frequencies to measure the highest frequency of rectification. The SSD has an *I-V* behavior analogous to a diode leading to half-wave rectification as observed in Fig. 3. However, the planar structure allows for a higher frequency of operation when compared to a conventional diode. The Fig. 3(a) – (d) depict the half waveform rectified output current (red) for various dielectric materials, while applying an input sinusoidal waveform (black) with a peak voltage of 10 V at 100 MHz at anode keeping cathode at ground.

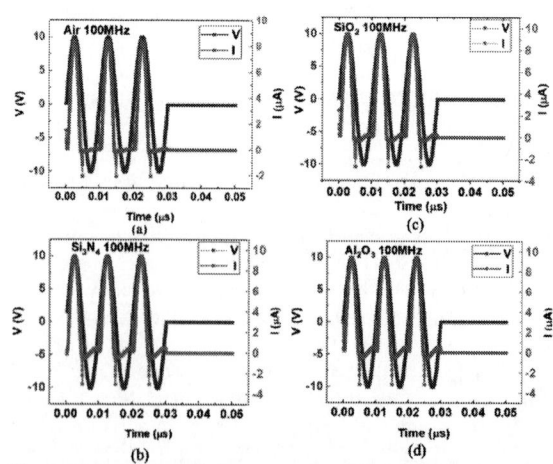

Fig. 3. Half-wave rectified output current (red) while applying the sinusoidal input having a peak voltage 10V at 100 MHz at anode of SSD having trenched filled with (a) Air, (b) SiO₂, (c) Si₃N₄, (d) Al₂O₃.

Table 1: Parameter comparison for different dielectric materials

Material & permittivity (ε)	Average Relative permittivity (ε_r)	Vertical Capacitance (C_V) 1×10^{-17} F	Horizontal Capacitance (C_H) 1×10^{-17} F	Capacitance (C) 1×10^{-17} F	Resistance (R) 1×10^3 ohm	Cut-off frequency (f_c) GHz
Air (1)	7.6	10.6	30.23	7.84	1374	1.48
SiO$_2$ (3.9)	9.05	12.62	35.99	9.34	1160	1.47
Si$_3$N$_4$ (7.5)	10.85	15.13	43.15	11.20	1120	1.31
Al$_2$O$_3$ (9.3)	11.75	16.38	46.73	12.12	1000	1.31

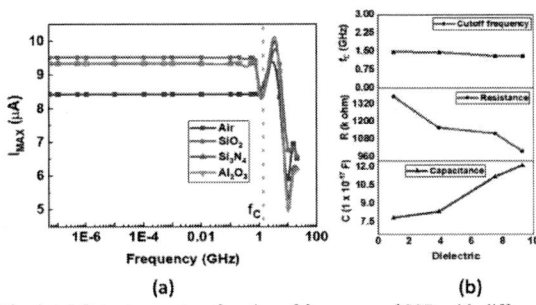

(a)   **(b)**

Fig. 4. (a) Output current as function of frequency of SSD with different dielectric materials demonstrating cut-off frequency ~2 GHz. (b) capacitance (C), resistance (R) and cut-off Frequency (f_C) based on variation of dielectric constant.

C. Frequency analysis

The frequency response of rectified current is shown in Fig. 4(a). It demonstrates a simulated cut-off frequency of ~2 GHz. The same has been theoretically verified by calculating the cut-off frequency as $f = 1/2\pi RC$ and can be visualized from Fig. 4(b), which implicates variation of capacitance (C), resistance (R) and cut-off frequency w.r.t dielectric constants of insulating materials. The conformal mapping technique introduced in [22] and in our previous studies [23] is used for theoretically measuring the device capacitance.

The capacitance arises because of a single vertical trench (C'_V) is given as follows

$$C'_V = \varepsilon_0 \varepsilon_r \left(\frac{K'(k)}{K(k)} \right) = \varepsilon_0 \varepsilon_r \frac{1}{\pi} \left(2 \frac{1 + \sqrt{k'}}{1 - \sqrt{k'}} \right) \quad (1)$$

The complementary modulus of k is k' and its value is $k' = \sqrt{1 - k^2}$. The total vertical trench capacitance is equal to $C_V = 2C'_V$ because the capacitances are in parallel to each other. The value of k is taken as $k = d/(2s + d)$, it can be seen that modulus k is related to vertical trench width (d) and InGaAs layer width (s). The values of $s = 0.15$ μm and $d=$

0.1 μm are as shown respectively in Fig. 1. For $s \gg d$, logarithmic approximation results in $k \approx 4 \exp\left(-\frac{\pi K'(k)}{K(k)}\right)$. By putting the value of K (k) / K'(k) in (1) yields

$$C'_V = \varepsilon_0 \varepsilon_r \frac{1}{\pi} \ln\left(\frac{4}{k}\right) \approx \varepsilon_0 \varepsilon_r \frac{1}{\pi} \ln\left(8\frac{s}{d}\right) \quad (2)$$

Likewise, the capacitance arises due to horizontal trench (C_H), is calculated using the previous mentioned technique. To derive C_H as shown below only one quarter plane is used because SSD is a symmetric device in horizontal plane.

$$C_H = 4\varepsilon_0 \varepsilon_r \left(\frac{K(k_1)}{K'(k_1)}\right) \quad (3)$$

Where $k_1 = \frac{a}{b}\sqrt{\frac{c^2 - b^2}{c^2 - a^2}}$, the values of $c = (C_W/2 + d' + s')$, $b = (C_W/2 + d')$ and $a = C_W/2$ are as shown in Fig. 1. Here C_W, d' and s' denotes channel width, horizontal trench width and In$_{0.53}$Ga$_{0.47}$As layer width, respectively. By solving (3), we get

$$C_H = 4\varepsilon_0 \varepsilon_r \left(\frac{\pi}{\ln\left(\frac{4}{k_1}\right)}\right) \quad (4)$$

ε_r is the average permittivity of dielectric constant of In$_{0.53}$Ga$_{0.47}$As ($\varepsilon_s = 14.2$) and insulating materials (ε) as listed in Table 1. The substitution of relative permittivity, device parameters and $\varepsilon_0 = 8.85 \times 10^{-18}$ F/μm as shown in Fig. 1, results in values of vertical capacitance (C_V), horizontal capacitance (C_H) and equivalent capacitance (C), by taking the derivative of I-V characteristics and resistance (R) of the device can be calculated for different dielectric materials. The cut-off frequency is calculated as $f = 1/2\pi RC$ and the resulted values are given in Table 1.

D. Total Harmonic Distortion (THD) analysis

The total harmonic distortion (THD) was evaluated by performing Fast Fourier Transform (FFT) on the simulated rectified current waveform as it is an important performance parameter for any AC to DC converter. Fig. 5(a) demonstrates the FFT response for different dielectric materials at 100 MHz. It is observed that the magnitude of output current at fundamental frequency (100 MHz for this case) is higher for higher permittivity materials and lower for 2nd and 3rd harmonics. It leads to a decrease in the THD as the permittivity increases as shown in Fig. 5(b). The THD is computed by taking into consideration, the currents of 1st, 2nd, and 3rd harmonics for the dielectric materials, given as:

$$THD \% = \frac{\sqrt{I_2^2 + I_3^2}}{I_1} \times 100 \quad (5)$$

The results for SSD with dielectric-filled trenches were obtained at a cutoff frequency of 1.47 GHz. In contrast to the references listed in Table 2. typically report their findings at much lower frequencies ranging from 60 Hz to 7 MHz.

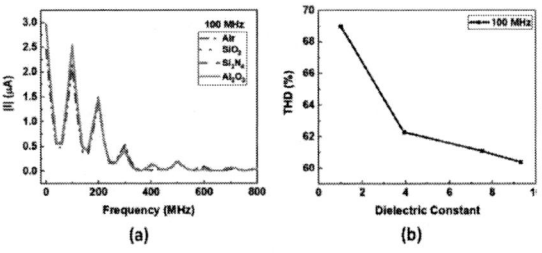

(a)   **(b)**

Fig. 5. (a) Fast Fourier transform (FFT) of device with various dielectric materials. (b) Variation of THD as a function of dielectric constant.

Table 2: Comparison with other state-of-the-art rectifiers

Ref.	Device	Topology	Cut-off Frequency	THD %
[4]	MOSFET	Dual half wave rectifier	71 kHz	15.6
[5]	Diode	Single phase half bridge	60 Hz	4.45
[6]	GaN MOSFET	full bridge rectifier	6.87 MHz	25.04
[7]	Buck-boost	Diode bridge rectifier	100 kHz	14.38
This work	InGaAs SSD	Diode half-wave	1.47 GHz	60.4

Furthermore, the outcomes presented in the previous studies in Table 2. are based on different circuit topologies of rectifiers integrated with discrete components and additional filtering circuits. While our work leverages a novel half-wave rectifier known as the Self-Switching Device, integrated on the same substrate. What sets our approach apart is the absence of the additional requirement for filtering circuits to achieve rectification.

IV. CONCLUSION

To summarize, we have successfully conducted two-dimensional TCAD simulations of $In_{0.53}Ga_{0.47}As$ based self-switching diode (SSD) for various dielectric materials. Our work showed that the SSD with higher dielectric constant results in the significant reduction in total harmonic distortions from 69% to 60.4% at high frequencies. Also, the cutoff frequency is simulated at the approximate value of 2GHz and is theoretically calculated at 1.47 GHz. The change in cut-off frequency is insignificant while there is significant reduction in harmonics. This phenomenon can be attributed to the presence of a dielectric material. In the context of the cutoff frequency, the increase in the capacitance is negated by the decrease in the resistance of the device. While for the harmonic suppression, the increase in capacitance plays a pivotal role in mitigating ripple effects, thereby leading to a significant reduction in harmonics. Therefore, this technique can readily be integrated into existing devices, as the fundamental device structure remains unchanged, with the only modification being the filling of the trenches with dielectric materials.

ACKNOWLEDGMENT

B. Sharma and S. Garg have equal contribution in this work. This work was supported by Science and Engineering Research Board, Department of Science and Technology, Government of India under CRG (CRG/2022/003235).

REFERENCES

[1] W. Zhu, K. Zhou, M. Cheng and F. Peng, "A High-Frequency-Link Single-Phase PWM Rectifier," *IEEE Transactions on Industrial Electronics*, vol. 62, no. 1, pp. 289-298, 2015.

[2] M. Ito, K. Hosodani, K. Itoh, S. Betsudan, S. Makino, T. Hirota, K. Noguchi and E. Taniguchi, "High efficient bridge rectifiers in 100 MHz and 2.45 GHz bands", *IEEE Wireless Power Transfer Conference*, pp. 64–7, 2014.

[3] B. Singh, S. Singh, A. Chandra and K. Al-Haddad, "Comprehensive Study of Single-Phase AC-DC Power Factor Corrected Converters With High-Frequency Isolation," *IEEE Transactions on Industrial Informatics*, vol. 7, no. 4, pp. 540-556, 2011.

[4] NA. Zawawi, S. Iqbal, MK. Jamil. "A single-stage power factor corrected LED driver with dual half-wave rectifier." *IEEE Industrial Electronics and Applications Conference*, pp. 12-19, 2016.

[5] R. Ghosh, G.Narayanan. "A simple analog controller for single-phase half-bridge rectifier." *IEEE transactions on power electronics*, vol. 22, no. 1,1pp. 86-98, 2007.

[6] S. Cochran, F. Quaiyum, A. Fathy, D. Costinett, and S. Yang, "A GaN-based synchronous rectifier for WPT receivers with reduced THD." IEEE PELS Workshop on Emerging Technologies: Wireless Power Transfer (WoW), pp. 81-87, 2016.

[7] C. K. Lee, S. Kiratipongvoot, S. C. Tan. "High-frequency-fed unity power-factor AC–DC power converter with one switching per cycle." *IEEE Transactions on Power Electronics*, vol. 30, no. 4, pp. 2148-2156, 2014.

[8] A. M. Song, M. Missous, P. Omling, A. R. Peaker, L. Samuelson, W. Seifert; "Unidirectional electron flow in a nanometer-scale semiconductor channel: A self-switching device". *Appl. Phys. Lett.*, vol 83, no. 9, pp. 1881–1883, 2003.

[9] A.M Song, I. Maximov, M. Missous, W. Seifert, "Diode-like characteristics of nanometer-scale semiconductor channels with a broken symmetry", *Physical E: Low-dimensional Systems and Nanostructures*, vol. 21, pp. 2–4, 2004.

[10] S. Garg, B. Kaushal, S. Kasjoo, S. Kumar, N. Gupta, N. Song, A.K. Singh, "InGaAs self-switching diode-based THz bridge rectifier". *Semiconductor Science and Technology*, vol. 36 no. 7, p. 075017, 2021.

[11] N.F. Zakaria , Z. Zailan, M. M. Isa, S. Taking, M. K. M. Arshad, and S. R. Kasjoo. "Permittivity and temperature effects to rectification performance of self-switching device using two-dimensional simulation." *5th International Symposium on Next-Generation Electronics (ISNE)*, pp. 1-2, 2016.

[12] X. F. Lu, K. Y. Xu, G. Wang, and A. M. Song. "Material and process considerations for terahertz planar nanodevices." *Materials science in semiconductor processing*, vol 11, no. 5-6 , pp.407-410, 2008.

[13] S. Garg, A. Garg, S. Bansal, S. Chaudhary, A.K. Singh ,and S.R. Kasjoo, "Effect of filling dielectric in etched trenches of novel unipolar nanodiode," *IEEE International Conference Microelectron, Comput. Commun.*, Jan. 23–25, 2016.

[14] G. Farhi, D. Morris, S. A. Charlebois, and J.P. Raskin. "The impact of etched trenches geometry and dielectric material on the electrical behaviour of silicon-on-insulator self-switching diodes." *Nanotechnology*, vol 22, no. 43, p. 435203, 2011.

[15] K. Y. Xu, X. F. Lu, A. M. Song and G. Wang, "Enhanced terahertz detection by localized surface plasma oscillations in a nanoscale unipolar diode," Journal of App. Phys., vol. 103, p.113708, 2008.

[16] A. K. Singh, G. Auton, E. Hill, and A. Song. "Estimation of intrinsic and extrinsic capacitances of graphene self-switching diode using conformal mapping technique." *2D Materials*, vol. 5 no. 3, p.035023, 2018.

[17] B. Kaushal , S. Garg, K. Prakash, S.R. Kasjoo, S. Kumar, N. Gupta, and A.K. Singh, "Graphene self-switching diode-based thermoelectric rectifier". *Electron. Lett.*, vol. 56, pp.1069-1072, 2020.

[18] ATLAS User's Manual, SILVACO Int., Santa Clara, CA, USA, 2010.

[19] H. Temkin, Y.K. Chen, P. Garbinski, T. Tanbun Ek, and R.A. Logan, "Insulating gate InGaAs/InP field effect transistors," *App. Phy. Lett.*, vol. 53, pp. 2534–2536, 1988.

[20] T. González, I. Iñiguez-de-la Torre, D. Pardo, J. Mateos, and A. M. Song, "Monte Carlo analysis of Gunn oscillations in narrow and wide bandgap asymmetric nanodiodes," *J. Phys., Conf. Series*, vol. 193, no. 01, 2018.

[21] S. Garg, B.Kaushal, S.Kumar, A.K. Singh, and S. Mahapatra, "Parametric optimization of self switching diode," *IEEE Nanomater. Devices Conf.*, Oct.14–17, 2018.

[22] P. Cattaneo, "Capacitance calculation in a microstrip detector and its applications to signal processing," *Nucl. Instrum. Methods Phys. Res.*, vol. 295, pp. 207–218, 1990.

[23] S. Garg, B. Kaushal, S. Kumar, S. R. Kasjoo, S. Mahapatra and A. K. Singh, "Extraction of Trench Capacitance and Reverse Recovery Time of InGaAs Self-Switching Diode", in *IEEE Transactions on Nanotechnology*, vol. 18, pp. 925-931, 2019

Secured and Optimized Hardware Accelerators using Key-Controlled Encoded Hash Slices and Firefly Algorithm based Exploration

Anirban Sengupta
Computer Science and Engg.
Indian Institute of Technology Indore,
Indian Statistical Institute Kolkata,
India
asengupt@iiti.ac.in

Aditya Anshul
Computer Science and Engg.
Indian Institute of Technology Indore,
India
phd2101101007@iiti.ac.in

Chirag Kothari, Sumer Thakur
Indian Institute of Technology Indore,
India
chiragkothari2503@gmail.com,
ce190004039@iiti.ac.in

Abstract—The design process of application-specific integrated circuits (ASICs) as hardware accelerators or reusable hardware intellectual property (IP) cores require considering various design goals like area, latency, and security against attacks such as IP piracy and false claims of IP ownership during architectural synthesis. The problem of IP piracy and false IP ownership claim poses a significant threat to hardware IP cores, jeopardizing innovation, fair competition, and economic growth. The proposed work introduces a novel security framework to design secured and optimized hardware accelerators using key-controlled encoded hash slices integrated with firefly algorithm (FF) based design space exploration (DSE) during high level synthesis (HLS). The proposed methodology also demonstrates the embedding of covert security constraints into an optimal 8-point DCT hardware IP design, obtained by performing a tradeoff between area and latency through the FF-based DSE. The proposed methodology achieves stronger tamper tolerance and entropy than recent approaches. The experimental results also provide design cost assessment and optimality analysis.

Keywords— Hardware security, Firefly based design space exploration, IP cores, Hash-slices, HLS

I. INTRODUCTION

The growing need for application-specific computing is becoming increasingly apparent in the modern digital era. As traditional general-purpose processors struggle to keep up with the demands of specialized tasks, application-specific computing offers tailored solutions that significantly enhance performance, efficiency, and scalability. The explosion of data-intensive applications, artificial intelligence, machine learning, and other emerging technologies requires high-speed processing and optimized computational power. Application-specific computing, such as application-specific integrated circuits (ASICs), provides the ability to execute specific tasks with incredible speed and energy efficiency. These ASICs are designed as a reusable hardware intellectual property(IP) core using a high level synthesis framework. Some examples of ASIC hardware IP cores include discrete cosine transform (DCT), finite impulse response filter (FIR), discrete wavelet transform (DWT), etc. However, security threats to hardware IP cores pose significant risks to innovation, financial stability, safety, reliability and fair competition in the technology industry. These threats include IP piracy, unauthorized access, and false claims of IP ownership. Malicious actors (present in the system-on-chip (SoC) integration house of design supply chain process) may illegally pirate, distribute, or use hardware IP cores without permission, leading to financial losses for the original creators and potential safety risks for end consumers due to compromised product integrity. Moreover, pirated (illegitimate) copies of IPs can lead to unpredictable consequences, such as the leakage of sensitive or critical information, excessive heat dissipation, and incorrect computation, among other potential issues. In addition to security, designing an optimal secure architecture is challenging, as it involves considering multiple objective parameters, including area, latency, energy efficiency, and more. One essential aspect of HLS is DSE, which helps to generate optimal architectural solutions corresponding to the target hardware IP by considering multiple conflicting parameters. [1]

Several works on hardware security have been explored. One of them proposes a palmprint biometric-based watermarking technique for securing hardware IP cores [2]. However, this approach has limitations, including weaker security due to lower hardware security constraints generation and incapable of generating optimal architecture solutions. Another approach focuses on using the DNA impression of the IP vendor to secure DSP IP cores [3], offering more robust security than [4] but lacking an optimal design solution and generating sufficient security constraints. Further, authors in [4] have discussed binary variable watermarking for generating security constraints. However, [4] show vulnerabilities when the signature size or encoding mechanisms are compromised, besides providing weaker security due to the generation of limited security constraints. Another approach [5] utilizes MD5 and SHA1 [5] to enhance security. Additionally, a hardware description language (HDL) design-level IP watermarking approach using SHA1 and RSA has also been presented in [6]. While [6] offer robustness, it becomes vulnerable when the RSA key value is compromised. Additionally, [9] utilizes mathematical relations for hardware watermarking by establishing links between input data, internal computation initial values, and output data to create the final watermark. However, all the discussed approaches fail to produce a low-overhead secured hardware IP datapath while providing weaker security. The proposed approach presents a hardware security framework capable of generating an optimal architecture solution corresponding to secure hardware IP using key-based encoded hash slices and firefly algorithm-based design space exploration with more robust security.

II. PROPOSED METHODOLOGY

A. Threat model

Threat model: An adversary in the untrustworthy SoC integrator design house may pirate and fraudulently claim

---

This work is technically and financially supported by Council of Scientific and Industrial Research (CSIR) grant no. 22/0856/23/EMR-II.

979-8-3503-8083-5/23 $31.00 © 2023 IEEE

the original IP design without the consent of original IP vendor. Therefore, securing the hardware IP against hardware security threats is essential.

B. Details of the proposed methodology

The details of proposed approach are highlighted in Fig. 1.

B.1 Firefly-algorithm based design space exploration

The proposed firefly-algorithm based design space exploration (FF-DSE) algorithm utilizes the firefly algorithm for finding an optimal architectural solution corresponding to secure hardware IP design. Compared to other techniques like genetic algorithm and bacterial foraging based optimization, FF-DSE offers several advantages: (a) FF-DSE incorporates essential hyperparameters, such as step-size control and absorption coefficient, to control randomness during the design search, (b) FF-DSE employs a divide-and-conquer approach based on the attraction parameter. Fireflies with higher attractiveness gather around local optimums in separate subgroups, eventually leading to the discovery of the final optimal solution, (c) The linearly decreasing step size control and absorption coefficient in FF-DSE strikes a balance between exploration and exploitation, ensuring faster convergence to the optimal solution.

The FF-DSE algorithm takes several parameters as input, including the initialization parameter, step size control parameter ('α'), attractiveness parameter ('β'), absorption coefficient ('γ'), design constraints, dimensions (resource types: #adders, #multipliers), and terminating criterion. Firefly positions are initialized based on resource types, with the first position set to the maximum value and the second and third positions set to the minimum and average of the first two, respectively. After initialization,

fitness (design cost) is calculated using equation (1), and the initial fitness values are considered local best, while the firefly with the minimum fitness is declared global best. The new firefly positions are then determined by adding a drift factor (computed using local best firefly, global best firefly, α, β, and γ) to the original positions. If any new positions violate the design boundary limits, they are clipped within those limits. The fitness of the new positions is computed, and local and global best are updated accordingly, if necessary. The algorithm continues until the terminating criterion is met, which can be either no improvement in fitness for ten consecutive iterations or reaching fifty iterations. The final global best solution is considered the optimal solution. The FF-DSE block's initial, intermediate, and optimal resource configurations are used to generate scheduled data flow graphs (SDFGs) and corresponding encoded templates. These templates are later used to generate secret hardware security constraints, which are embedded in the hardware IP design during the register allocation phase of the HLS framework. This results in covert digital evidence embedded secure hardware IP design. *Note:* As discussed above, fitness values of fireflies are calculated post-embedding of security constraints.

B.2 Key-controlled encoded hash slice based security

As shown in Fig. 1, this block accepts the encoded template as its primary input. The encoded template is derived from the SDFG (obtained using the outputs of FF-DSE and scheduling block) and the IP vendor's chosen encoding mechanism. The encoding mechanism used in our proposed approach is as follows: when both the operation and control step number in the SDFG is even, the output bit of the encoded template is '0'; otherwise, it's '1'.

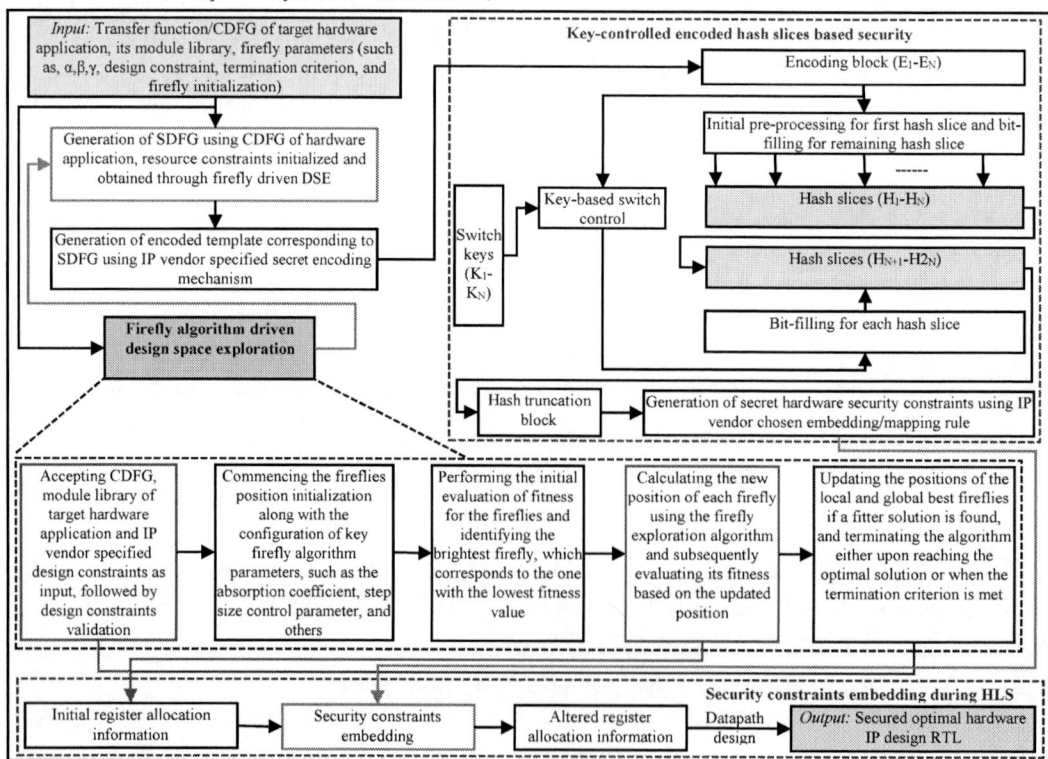

Fig. 1. Proposed hardware security framework

Additionally, eight other (total of nine) IP vendor-selected mechanisms (E_1-E_N, where the value of 'N' is nine) are utilized in the proposed approach. However, this selection may vary depending on the IP vendor's choice. Once generated, the encoded templates are fed into primary hash slices (H_1-H_N) after undergoing initial preprocessing and bit-filling. For instance, the first encoded template (template 1) is expanded to 896 bits by appending bit '1' followed by bit '0's. Furthermore, the expanded template-1 is appended with a 128-bit representation of its original length (# of operations in the SDFG, denoted as z). The resulting 1024-bit expanded *template-1* is then fed into SHA-512 *hash slice-1* to produce an encrypted hash digest of 512-bit. The subsequent encoded templates go through a different process of expansion. For example, the second template is prefixed with ($380-z$) zeroes to convert into a 380-bit template. Further, the 512-bit output of the first hash slice is appended with "1000," followed by an expanded template-2 of 308 bits and a 128-bit representation of the previous output's length (*i.e.*, 512). This expanded template-2 is then fed into *hash slice-2* to generate its corresponding 512-bit encrypted hash digest. This process continues until 'H_N'.

Following this, the final 512-bit encrypted output from the hash slice (H_1-H_N, *i.e.,* output of hash slice-H_N) serves as the input for the first block, *i.e.,* H_N+1, of hash slice (H_{N+1}-H_{2N}) block. The remaining 'N' hash slices have their encoded template inputs controlled by a key-based switch control unit composed of multiplexers (Muxes). These Muxes select the encoded templates from the encoding block using IP vendor-selected switch keys (K_1-K_N). These selected encoded templates for the remaining hash slices are then subjected to bit-filling to expand them to 308 bits (similar to *template-2* expansion as discussed above) before being fed into the corresponding hash slices. The process for generating encrypted hash digests is repeated until the hash slice (H_{2N}), ultimately resulting in a final output of a 512-bit encrypted hash digest. This final output is then passed through a hash truncation block, where it is truncated to the length chosen by the IP vendor. Subsequently, the truncated encrypted output is converted into secret security constraints based on the IP vendor's chosen embedding or mapping rule (explained in the next subsection). The generated security constraints are then forwarded to embedding block to produce a secure hardware IP design.

C. Demonstration of proposed approach on 8-point DCT

Fig. 2 shows the SDFG of the 8-point DCT application. It comprises 15 (z) operations scheduled in eight control steps (CS), using the optimal configuration of 1(+) and 2(*) obtained through FF-DSE. Table I shows the register allocation information in the register allocation table (RAT) before embedding security constraints corresponding to the 8-point DCT (shown in black color) generated from SDFG [2]. Initially, a z- bit encoded template is determined based on the IP vendor selected encoding-1 (*E1*), represented as "111011111011111". Subsequently, all remaining encoded templates are generated. These templates are then used as input for the key-controlled encoded hash slice-based security module to produce a final 512-bit encrypted output and its corresponding security constraints. The embedding rule employed in the proposed approach for generating security constraints involves implanting an additional

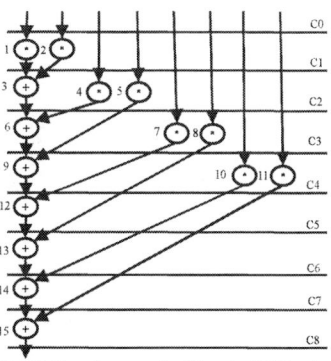

Fig. 2. Data flow graph of 8-point DCT using 1(+) and 2(*)

(artificial) edge between (even, even) storage variable pairs of RAT if the bit is '0' and embedding an edge between (odd, odd) storage variable pairs of the RAT otherwise. The final obtained security constraints are: *(D0,D2),* ----- *(D28,D30), (D1,D3),* ----- *,(D25,D29).* Encrypted security constraints are implanted in the 8-point DCT design using the RAT framework in high-level synthesis. Two storage variable pairs of the incoming artificial edge must be allocated to different registers. This allocation is achieved by either swapping register colors or assigning new registers [2]. Table I shows the register allocation information before and after embedding hardware secret security constraints (Red colored registers indicate modified locations).

D. Detection of IP piracy

In order to validate authenticity of the IP, the proposed key-controlled encoded hash slice based security constraints are regenerated and matched with the embedded security constraints. Only an authentic IP vendor would be able to do so. However, an adversary would fail to do that. This also protects against false IP ownership claim by adversary.

III. RESULTS AND ANALYSIS

The experimental analysis of the proposed security approach was conducted on a system with a 2.30 GHz workstation and 4GB main memory (implemented in Python). The proposed approach employed the following parametric values: firefly population size is 3, u_1 and u_2 set to 0.5; β_0 set to 1; γ gradually decreasing from 0.5 to 0.1, and α_x and α_y decreasing linearly from the maximum value of the first and second dimensions, respectively. The evaluation of area and latency for the 8-point DCT is performed using a 15nm scale with the NanGate library [7].

Entropy analysis (X_E) and tamper tolerance (T_Z) are the two parameters used for performing security analysis of the proposed approach. Entropy refers to the difficulty an adversary faces and the uncertainty encountered when decoding hidden information within an IP design. To estimate the proposed approach's entropy, we use the formula [8] $X_E = ((1/2^y)(1/E_N)(1/P)*(1/2^{64}))$, where '$y$' represents the generated signature's strength (magnitude), 'E_N' is the IP vendor's specified encoding rules to generate encoded templates, 'P' is the maximum value of round computations, and $(1/2^{64})$ represents the probability of finding the exact key hash buffer initialized value in the SHA-512 security module. Furthermore, the tamper tolerance (T_Z) is formulated as [4, 8] $T_Z = q^t$, where 'q'

Table I
Register allocation table pre and post embedding generated signature

CS	R	G	I	B	Y	Bl	V	P	L	O	A	T	G	M	S	K
0	D0	D1	D2	D3	D4	D5	D6	D7	D8	D9	D10	D11	D12	D13	D14	D15
1	D16/D17	D17/D16	D2	D3	D4	D5	D6	D7	-	-	-	-	-	-	-	-
2	D18	D19	D24/D19	D18	D4	D5	D6	D7	-	D24	-	-	-	-	-	-
3	D20	D19	D21/D19	D25	D21	D20	D6	D7	D25	-	-	-	-	-	-	-
4	D20	D22	D21	D23	D26/D21	D20	D23	D22	-	-	D27	-	-	-	-	-
5	D27	D22	D21	D23	-	-	D23	D22	-	-	-	-	-	D28	-	-
6	D28	D22	-	D23	-	-	-	-	-	-	-	-	-	-	-	-
7	D29	-	-	D23	-	-	D23	-	-	D29	-	-	-	-	-	-
8	D30	-	-	-	-	-	-	-	-	-	-	-	-	-	-	D30

Table II
Comparison of entropy and tamper tolerance between the proposed approach, [2], [3], [4], [5], and [6] corresponding to 8-point DCT

Security approach	Security parameters		
	Embedded security constraints	Entropy	Tamper tolerance
Proposed approach	496	3.68E-172	2.04E+149
Palmprint biometric [2]	125	1.93E-55	4.25E+37
Encrypted signature [5]	160	2.01E-87	1.46E+48
Watermarking [4]	240	1.66E-111	1.76E+72
DNA biometric [3]	128	2.9E-39	3.40E+38
HDL watermarking [6]	256	5.85E-99	1.15E+77

Table III
Optimality analysis and design cost of proposed technique for 8-point DCT

Parameters	Values
Spacing (S_A)	0.48
Generational distance (G_D)	0
Weighted metric (W_M)	0.17
Spread (S_D)	0.34
Design cost	-0.132
Area	182.45 um^2
Latency	1324.85 ps

optimality metrics: spacing, generational distance, weighted

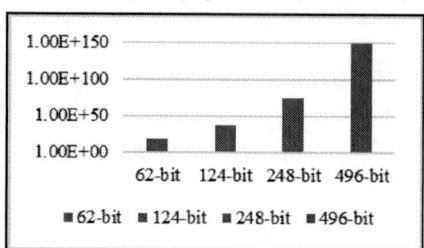

Fig. 3. Security analysis of the proposed approach in terms of tamper tolerance with varying signature sizes

sum, and spreading. Table III displays the computed values for each optimality metric, representing the results of the methodology. A zero or near zero value of all metrics indicates that list of solutions obtained using this approach aligns with true Pareto front with even scatting on curve.

IV. Conclusion

This paper presents a novel low-cost security framework using firefly driven design space exploration and key-controlled encoded hash slices to secure hardware IPs from IP piracy and false IP ownership claims.

denotes the total embedded constraints, and $'t'$ represents distinctive encoding variables utilized in the proposed security approach. A larger value of T_Z results in a bigger signature space, making it challenging for attackers to decode the precise signature combination and extract the covert constraints, indicating more robust security. Consequently, a higher value of T_Z enhances the security technique's resilience against tampering attacks. Table II reports the comparison of entropy and tamper tolerance between the proposed approach, [2], [3], [4], [5], and [6] corresponding to 8-point DCT. The proposed approach outperforms all the previously mentioned methods, primarily due to its higher tamper tolerance (T_Z) value, as seen in Table II. Moreover, the proposed approach provides a stronger entropy (lower probability value) than all above mentioned related approaches. Additionally, Fig. 3 reports the security analysis of the proposed approach in terms of variation in T_Z with varying signature sizes. The value of T_Z increases with an increase in signature size.

Further, Table III highlights the optimality analysis of the proposed approach, design cost, area and latency corresponding to secured 8-point DCT hardware IP design. The design cost metric used is as follows [8]:

$$Fitness\ (design\ cost) = u1*((AR-AR_{cons})/AR_{max}) + u2*((LT-LT_{cons})/LT_{max}) \quad (1)$$

where $'u1'$ and $'u2'$ represent weighing factors specified by the IP vendor. AR and LT refer to the design area and latency, respectively. $'AR_{cons}'$ and $'LT_{cons}'$ are design constraints set by the IP vendor or designer, while $'AR_{max}'$ and $'LT_{max}'$ denote maximum area and latency. Moreover, the presented approach's optimality analysis for determining the explored architectural solution is evaluated using several

References

[1] C. Pilato, S. Garg, K. Wu, R. Karri and F. Regazzoni, "Securing Hardware Accelerators: A New Challenge for High-Level Synthesis," *IEEE Embedded Systems Letters*, vol. 10, no. 3, (2018), 77-80, Sept.

[2] A.Sengupta, R. Chaurasia, T. Reddy, "Contact-Less Palmprint Biometric for Securing DSP Coprocessors Used in CE Systems," *IEEE Transactions on Consumer Electronics*, vol. 67, no. 3, (2021), 202-213.

[3] A. Sengupta, R. Chaurasia, "Securing IP Cores for DSP Applications Using Structural Obfuscation and Chromosomal DNA Impression," *IEEE Access*, vol. 10, 2022, 50903-50913.

[4] F. Koushanfar, I. Hong, M. Potkonjak, Behavioral synthesis techniques for intellectual property protection, *ACM Trans. Des. Autom. Electron. Syst.* 10, 3 (2005), 523–545, July.

[5] E. Castillo, L. Parrilla, A. Garcia, U. Meyer-Baese, G. Botella, A. Lloris, "Automated Signature Insertion in Combinational Logic Patterns for HDL IP Core Protection," *2008 4th Southern Conference on Programmable Logic, Bariloche, Argentina*, (2008), 183-186.

[6] T. Yu and Y. Zhu, "A new watermarking method for soft IP protection," 2011 International Conference on Consumer Electronics, *Communications and Networks (CECNet)*, China, 2011, 3839-3842.

[7] Open Cell NanGate Library, 15 nm open cell library, Available: https://si2.org/open-cell-library/, last accessed on March 2023.

[8] A. Sengupta, R. Chaurasia and A. Anshul, "Robust Security of Hardware Accelerators Using Protein Molecular Biometric Signature and Facial Biometric Encryption Key," *IEEE Transactions on Very Large Scale Integration (VLSI) Systems*, vol. 31, no. 6, 826-839, 2023.

[9] B. L. Gal and L. Bossuet, Automatic low-cost IP watermarking technique based on output mark insertions. *Des. Autom. Embedded Syst.* 16, 71–92, 2012.

On Structure Design Optimization of GaN Based Semiconductor Device for Reduced Trapping

Arivazhagan L and Anwar Jarndal

Department of Electrical Engineering, University of Sharjah, United Arab Emirates, ajarndal@sharjah.ac.ae

Abstract – **The material and structure have a significant impact on trapping, which is the most relevant degradation mechanism that limit the performance of GaN based semiconductor devices. In this paper, the impact of three different field-plate (FP) structures on buffer trapping in GaN High Electron Mobility Transistor (HEMT) is investigated by means of TCAD simulation. Gate, source and gate-source connected field plates are considered in the investigation. The impact of the FP structure on the trapping effect is analyzed in terms of trapped electron concentration, potential contour, electric field and current collapse. Pulsed I-V simulations are carried out to characterize the trapping induced current collapse and evaluate the effectiveness of the three FP structures on reducing the buffer trapping. It is found that the gate-connected field plate provides optimal performance with respect to the other structures.**

Keywords- CAD, GaN HEMT, Physical Modeling, Advanced Structural Design, Field plate, Trapping.

I. INTRODUCTION

GaN is a promising material for high power and microwave semiconductor devices such GaN power diode and GaN high electron mobility transistor (HEMT). It is due to its intrinsic properties such as wide band gap, higher electron mobility, higher thermal conductivity and higher electron velocity [1]. However, GaN material based devices suffer from trapping effects, which induces current collapse [2], [3]. This traps originates from intrinsic material defect, dislocation and lattice mismatch between two materials [4]. The trap is also depends on the material used in the device [5]. The GaN HEMT consists of passivation layer, barrier layer, buffer layer and substrate. In general, the traps are located in SiO_2 passivation/AlGaN barrier interface (or surface), AlGaN barrier layer or GaN buffer layer [6-7]. In the case of the SiO2/AlGaN interface, the surface trap induced by the lattice mismatch between these two materials. In the case of AlGaN barrier layer and GaN buffer layer, the trap is induced by both intrinsic material defect and lattice mismatch. These traps could be reduced by using material and structural optimization [5]-[6], [8]. In material optimization, the passivation and substrate material plays important role on the traps. SiN material is an alternative to SiO_2 material for the passivation as it reduces the surface traps. As a substrate material, the Silicon Carbide is a better than Silicon substrate as it reduces buffer traps [1]. Further,

thin Aluminium Nitride layer is used between AlGaN barrier layer and GaN channel to reduce the channel traps and interface.

In structural optimization, field plate (FP) technique is used to reduce both surface and buffer traps. However, the FP also induces defects or cracks in the surface of the layer [9] and this needs more investigation on the field plate design in the future. Furthermore, most of the work in field plate structure is used for improving breakdown voltage (V_{BR}). FP is also used for reducing surface traps. However, there is limited research on the impact of the FP the buffer traps and this needs more investigation. Hence, our work is focused on the FP structure and its influence on the buffer trapping and the associated current collapse. The investigation is carried out using Technology Computer Aided Design (TCAD) physical simulator.

II. DEVICE SCHEMATIC AND SIMULATION MODELS

The schematic of AlGaN/GaN HEMT analyzed in this paper is depicted in Fig. 1. The device consists of AlGaN barrier, GaN-buffer, transition layer and Silicon substrate with the thickness of 30 nm, 2 µm, 0.5 µm and 98 µm, respectively. The device is passivated with a 0.8-µm-thick of SiN layer. Aluminum composition of 30% is used in AlGaN barrier. The gate-source spacing (L_{SG}) gate-length (L_G), gate-drain spacing (L_{GD}) are 1µm, 0.2 µm and 4 µm, respectively. The gate width (W_G) of the device is 400 µm. The structural details are the same of the reported GaN HEMT in [10]. Silvaco Atlas TCAD simulator is used in this paper, which is a software solution to develop new semiconductor devices and processes [11]. In the simulation, four different HEMT devices are considered including devices with the gate field plate (G-FP), source field plate (S-FP), gate-source field plate (G&S-FP) and without field plate. These cases are presented in Fig. 1. The length of the G-FP and S-FP are 1 µm and 2.5 µm, respectively. Thickness of Silicon Nitride (t_{SiN}) below the gate FP and source FP are 50 nm and 75 nm, respectively. Drift-diffusion model is used for carrier transport. Shockley–Read–Hall recombination model is used to include the trap capture and emission process. High field mobility model is used to incorporate mobility degradation at higher drain voltage [12]. Polarization model is used to induce 2-Dimentional Electron Gas (2DEG). Initially, the model has been tested by simulating the DC I-V characteristics of the

979-8-3503-8083-5/23 $31.00 © 2023 IEEE

device with No-FP as shown in Fig. 2. As can be seen, the model reproduces the reported measured data in [13] and shows the expected output (I_{DS}-V_{GS}) characteristics.

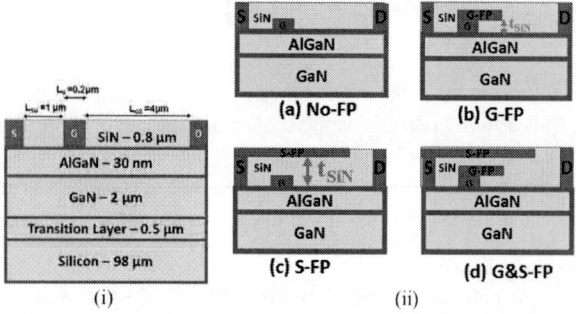

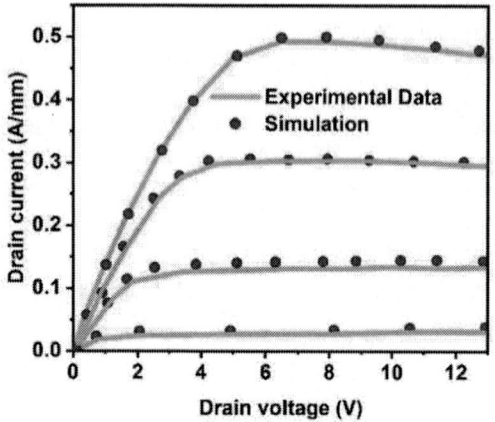

Fig. 1. (i) Schematic of AlGaN/GaN HEMT. (ii) AlGaN/GaN HEMT of four different structures: (a) No-field plate (No-FP); (b) Gate field plate (G-FP); (c) Source field plate (S-FP); (d) Gate & source field plate (G&S-FP).

Fig. 2. Output characteristics of GaN-HEMT with No-FP.

III. RESULTS AND DISCUSSION

In order to study the effect of buffer trapping, the electron concentration and potential distribution in GaN region are analyzed. The drain and gate voltages are initiated from V_{DS} = 4 V and V_{GS} = 0 V (pre-stress). Then the device is stressed for short time by high voltages of V_{DS} = 25 V and V_{GS} = -8 V and then returned back to V_{DS} = 4 V and V_{GS} = 0 V (post-stress) see Fig. 3. The simulated electron concentration in the channel at pre-stress and post-stress conditions is shown in Fig. 4. It can be observed that the electron concentration along AlGaN/GaN interface at post-stress is lower than the pre-stress condition. This is attributed to the injected electrons from the channel into buffer trap sites during the high stress voltage of V_{DS}= 25 V and V_{GS}= 0 V. Furthermore, the injected electrons into the GaN-buffer region are verified by the simulated potential distribution shown in Fig. 5. As can be seen, the potentials in the GaN buffer under the gate at pre-stress and post-stress are 1.95 V and -4.4 V, respectively. The negative potential of -4.4 V at post-stress condition is due to the trapped electrons.

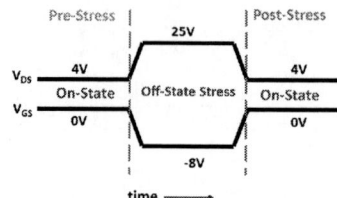

Fig. 3. Gate and drain voltages of pre-stress, off-state stress and post stress conditions.

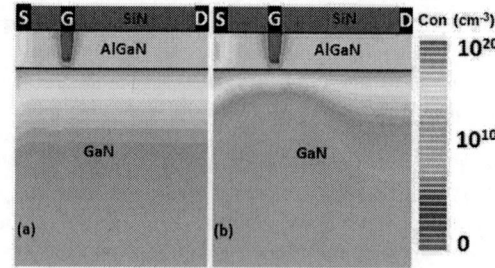

Fig. 4. Electron concentration in the GaN buffer: (a) before and (b) after the stress voltages of V_{DS} = 25 V, V_{GS} = -8 V. The quiescent voltages are V_{DS} = 4 V and V_{GS} = 0 V.

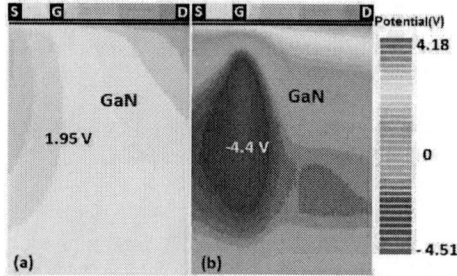

Fig. 5. Potential in the GaN buffer: (a) before and (b) after the stress voltages of V_{DS} = 25 V, V_{GS} = -8 V. The quiescent voltages are V_{DS} = 4 V and V_{GS} = 0 V.

The DC and pulsed I-V characteristics of the device for various field plates and without field plate are depicted in Fig. 6 at different quiescent voltage. In the pulsed I-V simulation, a narrow pulse width of 1 μs over 1 ms time period is used. It is well known that the trapped electrons take longer time in the order of 1 ms to de-trap [14]. Hence, a narrow pulse width (1μs) is used to characterize the trapped electrons induced current collapse. The drain and gate quiescent voltages are V_{DSO} = 25 V and V_{GSO} = -8 V. The DC I-V does not show the collapse because the device is biased for longer time enough for the electrons to de-trap and contribute in the channel conduction [15]. For that reason, all four devices show the same I-V characteristics. As can be noted also in Fig. 6, the simulated pulsed I-Vs show the expected current collapse with respect to the I-V. It is interesting to note that the current collapse for the device with G-FP is lower than the S-FP, G&S-FP, and No-FP devices. This could be attributed to the reduced peak electric field at the gate edge of the G-FP device as illustrated in Fig. 7.

979-8-3503-8083-5/23 $31.00 © 2023 IEEE

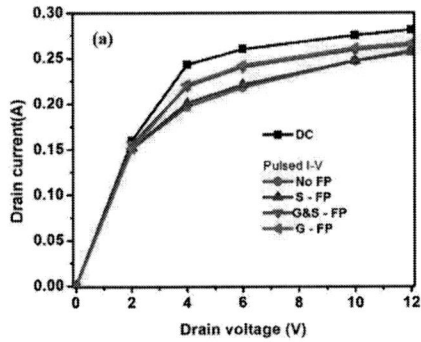

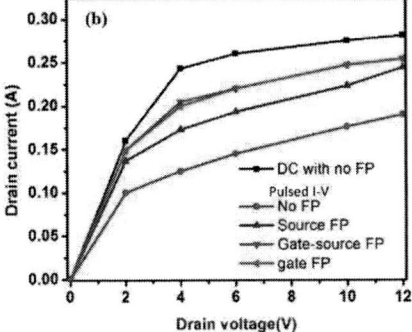

Fig. 6. DC drain current and pulse drain current for various field plate techniques with quiescent voltages of: (a) $V_{DSO} = 25$ V and $V_{GSO} = -8$ V and (b) $V_{DSO} = 50$ V and $V_{GSO} = -8$ V.

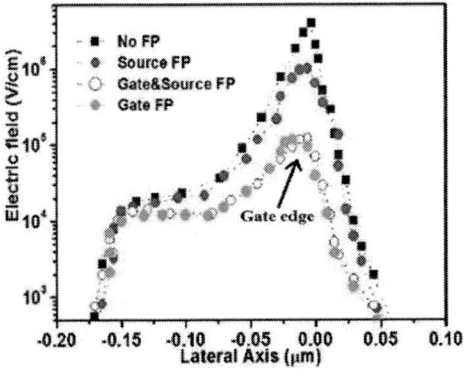

Fig. 7. Electric field along the channel axis (lateral axis) for various field plate techniques at quiescent voltage stress ($V_{DSO} = 25$ V, $V_{GSO} = -8$ V).

The trapping effect and the related current collapse depends on the quiescent bias voltages [15]. Hence, the current collapse with quiescent voltages of $V_{DSO} = 50$V and $V_{GSO} = -8$V are also analyzed (See Fig. 6(b)). The current collapse is more pronounced for $V_{DSO} = 50$V than $V_{DSO} = 25$V. Even at higher quiescent voltage ($V_{DSO} = 50$V), the current collapse for the G-FP and G&S-FP is still lower than the S-FP and No-FP (See Fig. 6(b)). The electric field is also analyzed for $V_{DSO} = 50$ V, which is shown in Fig. 8. For both $V_{DSO} = 25$ V, $V_{DSO} = 50$ V and lower drain-gate spacing, the peak electric field for the G-FP is lower than the G&S-FP, S-

FP and No-FP (See Fig. 6 and Fig. 8). The lower electric field is also desirable to enhance the breakdown voltage. In addition, the S-FP induces extra-parasitic capacitances, which is not desirable. Thus based on these results, the G-FP is more effective than the S-FP, especially for short gate-drain spacing ($\leq 4\mu m$).

To show the injected electron into the buffer for various FP techniques, simulated potential profile is analyzed under the same stress voltages of $V_{DS} = 25$ V and $V_{GS} = -8$ V (see Fig. 9). As can be observed in Fig. 9, the potential in the GaN-buffer of the G-FP, S-FP, S&G-FP and No-FP devices are -2.11 V, -2.11 V, -4.05 V and -4.4 V, respectively. It is interesting to note that the S-FP and No-FP cases exhibit more negative potential in the GaN-buffer than the G-FP and G&S-FP. This higher stress voltage stimulates more electrons to be injected into buffer traps and thus enhances the current collapse.

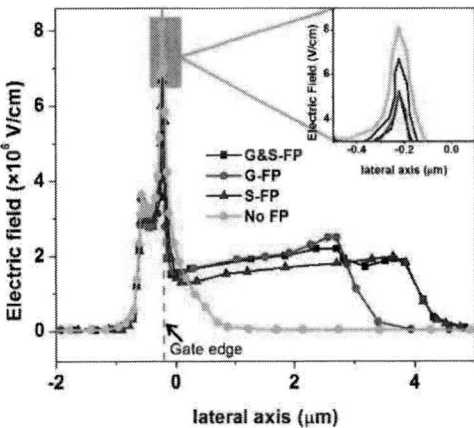

Fig. 8. Electric field along the channel axis (lateral axis) for various field plate techniques at quiescent voltages of $V_{DSO} = 50$ V and $V_{GSO} = -8$ V.

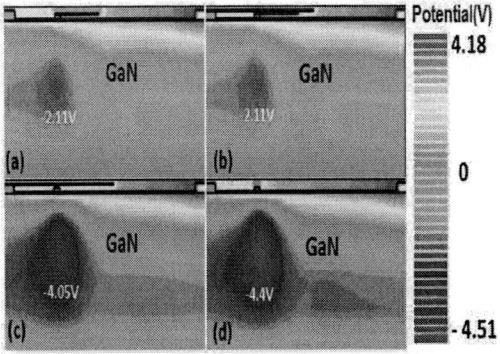

Fig. 9. Potential distribution in GaN-buffer at post stress for various field plate techniques: (a) Gate field plate; (b) Gate & source field plate; (c) Source field plate and (d) No-field plate.

For further clarification, the electrons along the channel are extracted to show the effectiveness of FP techniques on reducing the trapped electrons. These electrons are extracted at $V_{DS} = 4$ V and $V_{GS} = 0$ V after high voltage stress ($V_{DS} = $

979-8-3503-8083-5/23 $31.00 © 2023 IEEE

25 V and $V_{GS} = 0$ V), which is shown in Fig. 10. As can be seen, the electron concentration is reduced more in high field region (gate edge of the drain side). This is due to the trapped electrons during high voltage stress. Furthermore, it is observed that electron concentration under the gate for G-FP is higher than the G&S-FP, No-FP, and S-FP techniques. This indicates that the trapped electrons for the G-FP are lower than other techniques.

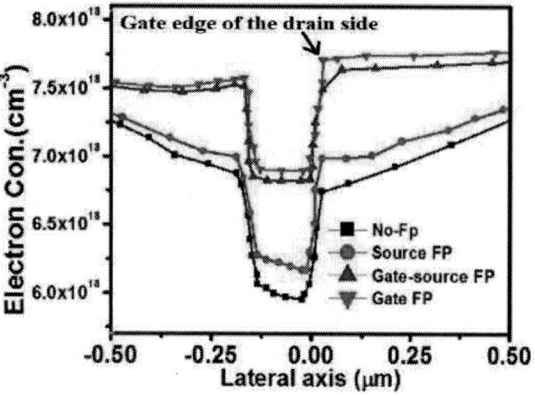

Fig. 10. Electron concertation along the channel axis (lateral axis) for various field plate techniques at $V_{DS} = 4$ V and $V_{GS} = 0$ V after high voltage stress ($V_{DS} = 25$ V & $V_{GS} = -8$ V).

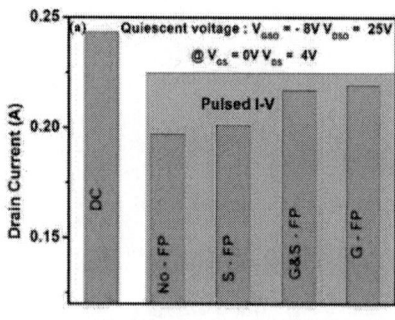

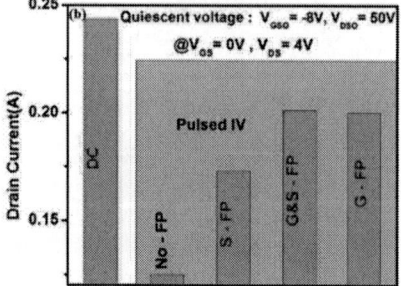

Fig. 11. DC drain current versus Pulsed I-V drain current at $V_{DS} = 4$ V and $V_{GS} = 0$ V after the high voltage stress of: (a) $V_{DSO} = 25$ V and $V_{GSO} = -8$ V and (b) $V_{DSO} = 50$ V and $V_{GSO} = -8$ V.

The DC and pulsed I-V drain current is also analyzed for the G-FP HEMT with respect to other devices. Fig. 12 shows the DC and pulsed I-V characteristics for the G-FP and S-FP. As expected, the current collapse increases, for both devices, by increasing the drain current and its associated buffer

trapping. In addition, the characteristics show the typical buffer-trapping induced knee-voltage walkout, which results in reducing the RF output power of the GaN device [16].

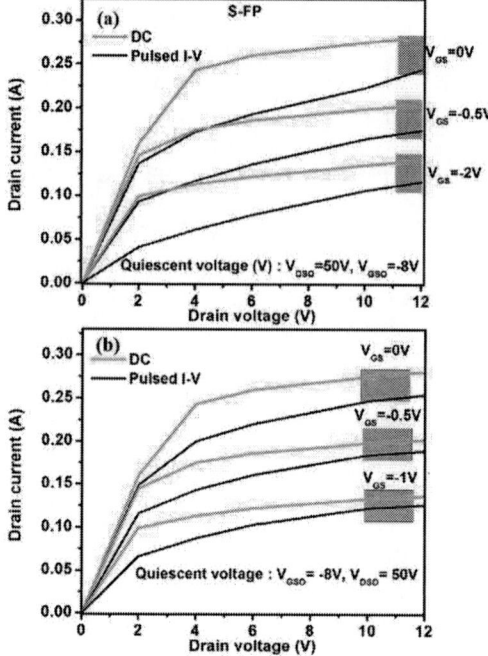

Fig. 12. DC and pulsed I-V drain current of GaN-HEMT with: (a) S-FP and (b) G-FP.

As it has been mentioned, the quiescent voltage (V_{DSO}) plays a major role in the trapping and current collapse since the peak electric field at gate edge depends on V_{DSO}. Therefore, the impact of FP techniques on the current collapse are analyzed and compared at different quiescent voltages ($V_{DSO} = 25$ V and $V_{DSO} = 50$ V), as can be seen in Fig. 11. At $V_{DSO} = 25$ V, the S-FP is not reducing the current collapse much with respect to the No-FP. However, at $V_{DSO} = 50$ V, the S-FP exhibits lower current collapse than No-FP. In both quiescent conditions ($V_{DSO} = 25$ V and $V_{DSO} = 50$ V), current collapse is almost the same for both G-FP and G&S-FP. Overall, the G-FP and G&S-FP exhibit lower current collapse than S-FP. However, the additional source connection in G&S-FP structure induces an extra parasitic capacitance. Hence, this support our previous observation of better performance of the G-FP in terms of higher breakdown voltage and lower buffer trapping.

In order to show the impact of the design space parameter on the current collapse, the drain current is simulated under different thicknesses (t_{SiN}) of the dielectric under the FP. This thickness represents the spacing between the FP and the top surface of AlGaN (See Fig. 1). It has been varied from 50 nm to 500 nm for both S-FP and G-FP. From the pulsed I-V simulation at $V_{DSO} = 50$ V and $V_{GSO} = -8$ V, the drain current is observed at $V_{DS} = 12$ V and $V_{GS} = 0$ V. The current collapse is defined as the difference between the pulsed and DC drain current and as follow:

$$\Delta I_{DS} = (I_{DS,DC} - I_{DS,Pulsed}) \qquad (1)$$

Fig. 13(a), shows ΔI_{DS} for the G-FP and S-FP devices. As can be seen, ΔI_{DS} decreases with decreasing t_{SiN} thickness. For both cases, it is significantly reduced for t_{SiN} below 200 nm. For t_{SiN} above 200 nm, the current collapse saturates and exhibits a weak dependency on t_{SiN}. This could be attributed to the higher effectiveness of the FP to influence and reduce the electric field at the gate edge for lower t_{SiN}. In other words, if the FP is far away from the channel (larger t_{SiN}), the FP has less influence on the electric field along the channel and the buffer traps. Furthermore, the impact of FP length on the current collapse (I_{DS}) is also analyzed (See Fig. 13(b)). The FP length used in the simulation is relative to the gate edge of the drain side and it has been varied from 0 μm to 1.75 μm under the same t_{SiN} of 50 nm (for both G-FP and S-FP). It is observed that the current collapse (ΔI_{DS}) reduces with increasing the length of the field plate length for both G-FP and S-FP. This could be explained by the results in Fig. 8, which show the effect of FP on reducing the peak electric field. Thus, increasing the FP length improves its effectiveness in reducing the electric field and the associated current collapse.

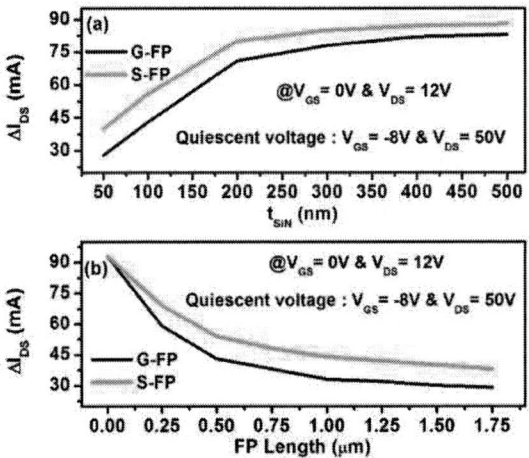

Fig. 13. Current collapse (ΔI_{DS}) for various t_{SiN} and FP length at V_{DS}=12 V and V_{GS}=0 V.

IV. Conclusion

GaN HEMT with various field plate (FP) structure are investigated and their impact on the buffer trapping is quantified. The effect of field plate on the buffer trapping induced current collapse is characterized using TCAD simulation. In the simulation, electric field, potential distribution, and electron concentrations are observed, which is helpful to evaluate the effect of the FP structure on buffer trapping. The peak of the electric field at the gate edge of the drain side is considered to increase the electron trapping in

the buffer. More reduction in current collapse and the electric field is observed for GaN-HEMT with gate field plate with respect to the other field plate techniques.

Acknowledgment

The authors gratefully acknowledge the support from the University of Sharjah, Sharjah, United Arab Emirates.

Reference

[1] Anwar Jarndal, L Arivazhagan, D Nirmal, "On the performance of GaN-on-Silicon, Silicon-Carbide, and Diamond substrates," *International Journal of RF and Microwave Computer-Aided Engineering*, vol. 30, no. 6, Jun. 2020.

[2] T. Oishi et al., "Bias Dependence Model of Peak Frequency of GaN Trap in GaN HEMTs Using Low-Frequency Y₂₂ Parameters," *IEEE Transactions on Electron Devices*, vol. 68, no. 11, pp. 5565-5571, Nov. 2021

[3] O. Axelsson et al., "Application Relevant Evaluation of Trapping Effects in AlGaN/GaN HEMTs With Fe-Doped Buffer," *IEEE Transactions on Electron Devices*, vol. 63, no. 1, pp. 326-332, Jan. 2016

[4] A. Prasad et al., "Accurate Modeling of GaN HEMT RF Behavior Using an Effective Trapping Potential," *IEEE Transactions on Microwave Theory and Techniques*, vol. 66, no. 2, pp. 845-857, Feb. 2018.

[5] T. J. Anderson et al., "Effect of Reduced Extended Defect Density in MOCVD Grown AlGaN/GaN HEMTs on Native GaN Substrates," *IEEE Electron Device Letters*, vol. 37, no. 1, pp. 28-30, Jan. 2016.

[6] R. Ye et al., "An Overview on Analyses and Suppression Methods of Trapping Effects in AlGaN/GaN HEMTs," *IEEE Access*, vol. 10, pp. 21759-21773, Dec. 2022.

[7] Elangovan S, Cheng S, Chang EY, "Reliability Characterization of Gallium Nitride MIS-HEMT Based Cascode Devices for Power Electronic Applications," *Energies*, vol. 13, no. 10, pp. 2628, May 2020.

[8] P. -Å. Nilsson, et al., "Influence of Field Plates and SurfaceTraps on Microwave Silicon Carbide MESFETs," *IEEE Transactions on Electron Devices*, vol. 55, no. 8, pp. 1875-1879, Aug. 2008.

[9] Bie Y-N et al., "Effect of Source Field Plate Cracks on the Electrical Performance of AlGaN/GaN HEMT Devices," *Crystals*, vol. 12, no. 9, pp. 1195, Aug. 2022.

[10] A. Jarndal, R. Essaadali and A. B. Kouki "A Reliable Model Parameter Extraction Method Applied to AlGaN/GaN HEMTs," *IEEE Transactions on Computer-Aided Design of Integrated Circuits and Systems*, vol. 35, no. 2, pp. 211-219, Feb. 2016.

[11] *Device Simulator Atlas Ver. 5.10.0.R. Atlas User's Manual*, Silvaco Int., Santa Clara, CA, Jul. 2005.

[12] M. Farahmand, et. al., "Monte Carlo Simulation of Electron Transport in the III-Nitride Wurtzite Phase Materials System: Binaries and Terniaries," *IEEE Trans. Electron Devices*, vol. 48, no. 3, pp. 535-542, Mar. 2001.

[13] A. -J. Tzou, et al., "An Investigation of Carbon-Doping-Induced Current Collapse in GaN-on-Si High Electron Mobility Transistors," *Electronics*, vol. 5, no. 28, pp. 1-11, Jun. 2016.

[14] B. Syamal, et al., "A Comprehensive Compact Model for GaN HEMTs, Including Quasi-Steady-State and Transient Trap-Charge Effects," *IEEE Transactions On Electron Devices*, vol. 63, no. 4, pp. 1478-1485, Apr. 2016.

[15] A. Y. Polyakov and I. -H. Lee, "Deep traps in GaN-based structures as affecting the performance of GaN devices," *Materials Science and Engineering*, vol. 94, pp.1-56, Aug. 2015.

[16] S. J. Doo, et al., "Effective Suppression of IV Knee Walk-Out in AlGaN/GaN HEMTs for Pulsed-IV Pulsed-RF With a Large Signal Network Analyzer," *IEEE Microwave and Wireless Components Letters*, vol. 16, no. 12, pp. 681-683, Dec. 2006.

Fault simulation Framework using PyUVM

Mina Hanna Fayez
Faculty of Engineering Electronics
and Communications Department
Cairo University
Cairo, Egypt
mina.hannaone@gmail.com

Mohamed Ahmed ElAdawy
Faculty of Engineering Electronics
and Communications Department
Cairo University
Cairo, Egypt
mahmededawy2@gmail.com

Micheal Safwat Sahyon
Faculty of Engineering Electronics
and Communications Department
Cairo University
Cairo, Egypt
michealsafwat310@gmail.com

Islam Osama Ahmed
Faculty of Engineering Electronics
and Communications Department
Cairo University
Cairo, Egypt
islam.osama.ahmed@gmail.com

Omar Hossam El-Din
Faculty of Engineering Electronics
and Communications Department
Cairo University
Cairo, Egypt
omar21hossam@gmail.com

Mohamed Ahmed ElShafie
Faculty of Engineering Electronics
and Communications Department
Cairo University
Cairo, Egypt
mohamedelshafie818@gmail.com

Mohamed Ayman Taha
Faculty of Engineering Electronics
and Communications Department
Cairo University
Cairo, Egypt
mayman7795@icloud.com

Mohamed Gamal Talaat
Faculty of Engineering Computer
Department
Ain Shams University
Cairo, Egypt
mohgamaltaleat@gmail.com

Abstract—This paper presents a fault simulation environment developed using PyUVM to address challenges in simulating faults during the manufacturing process of integrated circuits (ICs). The two main challenges tackled are simulating pre-manufacturing fault behavior and overcoming performance issues. The contributions of this work include expanding the fault simulation capacity beyond existing fault simulation methodologies, improving simulation time through efficient algorithms, and enabling compatibility with Universal Verification Methodology (UVM) and SystemVerilog testbenches. The environment allows for injecting various types of faults into a design, reporting the resulting behavior, and facilitating comprehensive analysis under different fault scenarios. The proposed solution enhances fault simulation capabilities, reduces simulation time, and provides compatibility for seamless integration with existing designs and verification methodologies.

Keywords— *ICs-FuSa-E/E-UVM-ASILs-FMEDA-CoCotb-EDA- DUT-SV – GUI.*

I. INTRODUCTION

The growing demand for sophisticated technology has pushed the manufacturing industry into the fast lane. As technology is evolving rapidly, manufacturers face the challenge of developing components and systems quickly and at low cost. They must also ensure that their products meet all necessary quality requirements, such as ISO26262 or other safety standards. ISO 26262 is an international functional safety standard for the development of electrical and electronic systems in road vehicles. It defines guidelines to minimize the risk of accidents and ensure that automotive components perform their intended functions correctly and at the right time. It also provides an automotive-specific approach for determining risk classes known as Automotive Safety Integrity Levels (ASILs). According to ISO26262 the term functional safety (FuSa) is defined as the absence of unacceptable risk due to hazards caused by malfunctioning behavior of electrical and/or electronic (E/E) systems.[4]

The safety levels which are governed by ASIL determine the safety metrics. Safety metrics determine the injected safety mechanisms and the overall area of the design. The functional safety flow starts with Safety Analysis phase where calculations of Failure Modes, Effects, and Diagnostic Analysis (FMEDA) metrics take place to predict electronic design failure rates and provide early safety architectural guidance, then we move on to Design for Safety phase where we insert and verify safety mechanisms to create safe designs that can mitigate the effect of random hardware faults, to the final phase which is our concern in this paper which is Safety Verification where fault simulation takes place to check whether the design achieved the target ASIL level through fault campaigns or not. Safety Verification is based on validating the effectiveness of the inserted Safety Mechanisms by observing the deviation of the design behavior due to the injected fault and how the Safety Mechanism will react. Faults are injected on critical nodes then a Fault Summary Report is generated to report back the behavior of each fault, in this report, we see the effect of these faults on other nodes called Alarm nodes where an alarm is triggered when the fault propagates and reaches that node.[5]

In this paper, a fault simulation environment solution is introduced to determine how the design will behave under the effect of hazards after manufacture. The objective is to create this environment using PyUVM to inject faults, monitor critical nodes, and print a safety report summary of the design under test(DUT) due to these injected faults. This environment is built on Universal Verification Methodology (UVM) concepts but using Python, not SystemVerilog. In Python, we used a library called CoCotb which stands for Coroutine-based Cosimulation testbench, and from it also comes the name " CocoSim". To use the hierarchical class-based testbench concepts of UVM we used another library built on CoCotb which is called PyUVM. A user-friendly Graphical User Interface(GUI) is also available with the framework to take input design files and data then use them to create the environment and generates the reports. This environment using CoCotb packages can interact with the simulator which in our case is QuestaSim and can reach any node in the design at hand. The main advantage of CocoSim compared to previous ideas or implementations is that it can run multiple faults in one run/simulation using one single core, unlike the typical method of injecting faults which can only simulate one fault on a single core.

II. LITERATURE REVIEW

Functional safety is an important aspect of electronic system design, particularly in safety-critical domains such as automotive, aerospace, and medical devices. The goal of functional safety is to ensure that a system behaves safely even in the presence of faults or errors.[4]

To achieve functional safety, designers use a combination of design techniques and verification tools. An Electronic Design Automation (EDA) tool aids in the functional safety verification of digital circuits. There are more advanced tools for Fault Simulation provides a comprehensive set of safety-related metrics, coverage analysis and capabilities to help designers quantify the level of safety achieved by their designs.[2]

All Fault Simulation Simulators can perform fault simulation but with constraint as it can simulate "N" faults in parallel such that N is the number of cores of the machine, so it's constrained by the number of cores.[7]

However, there is still a need for additional frameworks that can improve the efficiency of functional safety verification. One such framework is CocoSim (Fault Simulation Framework using PyUVM), a new framework presented in this paper that aims to primarily mimic fault simulation in a PyUVM environment. CocoSim is designed to work in tandem with other EDA tools used to provide a new set of safety metrics that can improve the overall effectiveness of functional safety verification by simulating multiple "N" faults independent of the number of machine cores.

III. UNDERSTANDING FAULTS IN INTEGRATED CIRCUITS

The presence of imperfections in a circuit, known as faults, can have a significant impact on the functional behavior of a system, either permanently or for a limited period. The focus of this paper, however, is solely on the stuck-at fault and Time delay fault model and this can be categorized into the following faults.

A. Stuck-at fault

Stuck-at Faults have two categories[3]

- **Stuck-at 0 fault**

This kind of fault occurs when a gate or output of a transistor is shorted to the ground and becomes irresponsive regardless of any input applied for all the time.

- **Stuck-at 1 fault**

This kind of faults occurs when a gate or output of a transistor is shorted to VDD and becomes irresponsive regardless of any input applied for all the time.

B. Time delay fault

This kind of faults occurs when a gate or output of a transistor doesn't follow the supposed transition from 1 to 0 or 0 to 1 immediately but after some delay time due to transistor capacitor doesn't charge or empty fast enough and there are two types of it.

- **Time delay from 0 to 1**

This kind of delay happens when there is a slow transition from 0 to VDD causing a Delay in response or fault in the whole system.

- **Time delay from 1 to 0**

This kind of delay happens when there is a slow transition from VDD to 0 causing a Delay in response or fault in the whole system.

IV. FAULTS CLASSIFICATION

A Fault has different categories based on what kind of outcome happened in the circuit, we will discuss some of those outcomes:

A. Detected Fault

As shown in Figure 1, we injected a fault and triggered its alarm (the Alarm node can be any net in the design). In the

next Figure, we see an example where B is stuck-at 0. After injecting the fault, the expected output on the flip flop is 1, but due to the fault it is 0. Since the alarm state differs from the golden reference, the fault is classified as detected

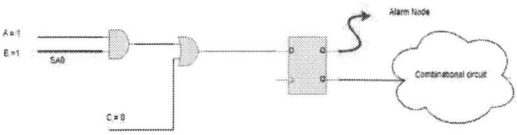

Fig. 1. Detected Fault example 1

B. Safe Fault

When injecting a fault and the input applied to the design doesn't trigger the alarm nor change the state from the golden reference and it is masked by the values applied.

C. Unsafe Faults

An unsafe fault is the kind of fault triggered and any of the Endpoints *(An Endpoint is considered as any flip flop in the design)* changed its state from the golden reference and the alarm is not triggered.

V. INPUTS FOR CREATING PyUVM ENVIRONMENT

There are some things that need to be done to help construct our PyUVM Environment.

A. Data entry

To build the environment we need the design file(s), testbench file(s), fault list, and alarm list, choose the top module file between the design file(s) and the name of the clock signal in the design then files pass through the Data processing phase to extract All needed information for simulation.

B. Data Processing

We wrote some Python scripts to help us get the necessary information from the inputs mentioned above and also prevent the user from entering multiple inputs. Data processing consists of two processes Reading Process Which mainly uses regular expressions also the Generating Process which uses Text processing to generate multiple files each process consists of some APIs

- Reading Process APIs
  - Extract input Signals from the Top Module
  - Extract module name from Top Module
  - Extract fault Node, fault Value, trigger time, and release time from fault list
  - Extract alarm nodes from alarm list
  - Extract End points (flip-flops or latches or primary output ports) from Design Files.

After this process the extracted data is dumped in a file then it is passed to the PyUVM Environment

- Generating Process APIs
  - Generate tcl, Makefile, and SV files
  - Generate a new top module

We have generated a tcl file which contains commands to execute Makefile through PowerShell which contains some commands designed to compile and simulate a design in Verilog/SystemVerilog.It sets up a work library, compiles the design files and a testbench, runs the simulation of PyUVM

979-8-3503-8083-5/23 $31.00 © 2023 IEEE

We have mentioned we are compiling and running the testbench mentioned in the Data entry, so we have generated an SV file to trace the inputs from the running testbench simulation and dump it in data_file as shown in Figure 2, so this idea gave us advantage to run the testbench in parallel with PyUVM simulation

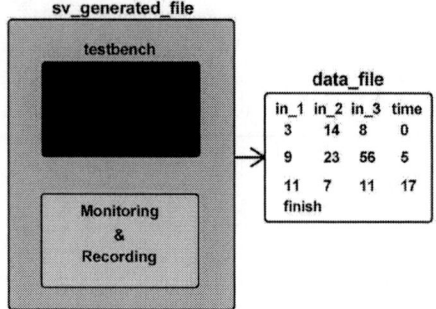

Fig. 2. Stimuli recording and dumping

We also construct a new top module from the top module file included in the data entry files. In which we create a golden instance where we don't inject any faults from the fault list, and N faulty instances where N is the total number of injected faults as shown in Figure 3.

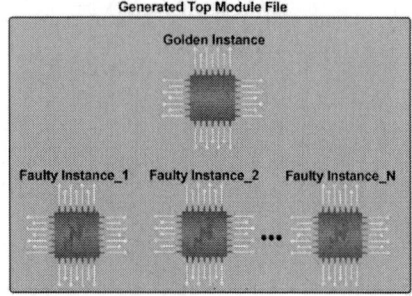

Fig. 3. Top Module Generated

VI. PyUVM Environment components

This is the main part of the fault simulation, where we drive the input stimuli, inject faults, observe alarms and generate final safety report. PyUVM gives us a lot of features to create our environment as shown in Figure 4.[5]

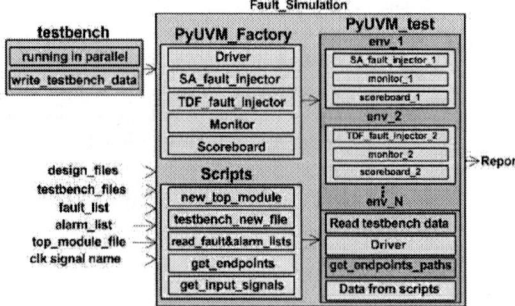

Fig. 4. PyUVM Fault simulation Environment 1

A. Driver

This is the first component. We drive the DUT using input stimuli we have got from the dumped file into the input signals. With the help of CoCotb, which PyUVM is based on, we can drive the inputs very easily using a handle to the DUT. In our driver, we depend on using the given testbench file(s) to drive the DUT. We can use either directed testing with simple input stimuli at specified times or randomized testing Like using a UVM testbench, because we treat with the test bench as a black box.

B. Fault injector

This is the second component and the most important one. Also, with the help of the handle to the DUT, we can inject the fault at any part of the DUT even if it is a net buried deep in the hierarchy of the DUT. We have two fault injectors in our environment, one for the stuck-at (SA) faults, and the other for the Time-Delay Faults (TDF). In our fault injector, we first wait for the injection time before we inject the fault on the fault node, after we either force a value on the fault node in case of SA fault (for example if we are injecting SA0, we force a zero on the fault node), or freeze the current value of fault node in case of TDF (for example if we are injecting TDF01, we freeze the fault node if it is at zero, and if it is at one we wait for it to go to zero then we freeze the node). Then we wait for the release time to release the fault whether it is SA fault or TDF.

To release the fault, we inject it twice in the same SystemVerilog (SV) time slot: once at the beginning, and once more if the golden reference changes value. This makes the faulty node follow the golden reference in any case. We also ensure that the fault is released by the next time slot, to avoid creating another fault.

C. Monitor

This is the third component and the most complicated one. It senses the values of the fault node in both golden and faulty instances. The monitor continuously checks for any deviations in the fault node, alarm node, and endpoint node between golden and faulty instances. If deviation occurs at a certain time, this time is recorded as trigger time or alarm time, or fail time for this particular fault. After recording all this data, it will be sent to the scoreboard.

D. Scoreboard

We collect all the information from the monitor and organize it to print the final Safety report. The final report contains each fault node and its corresponding error type, error value, trigger time, alarm time, fail time, and fail node.

E. Environment

This is the last component, and it is constructed with the PyUVM Factory Concept as we create an individual environment by name for each fault and every one contains a Fault injector, monitor, and scoreboard and this idea helped us to simulate all faults in only one test that contains multiple environments that run in pseudo-parallel. [1]

VII. PyUVM Environment Execution

After we constructed the environment and registered its components in PyUVM factory, we created our PyUVM test where we first take the output of our scripts which is dumped in a temporary file by CocoSim to pass it to the environment. Secondly, we use an important feature provided by CoCotb, it gets all the paths of all signals in the DUT, we use this feature to get the paths of the endpoints that we extracted using a Python script. Thirdly, we have constructed a Friendly Graphical User Interface (GUI) as shown in Figure 5 using PyQt5 technology [6]   to help the user to interact with the fault simulation framework and this GUI Consists of

- Main Window (Frontend of the framework)
- Fault Configuration Window (Configure simulating one fault or multiple faults)
- CSS Style Sheet (Themes to change appearance)
- Code editor (Preview user files)

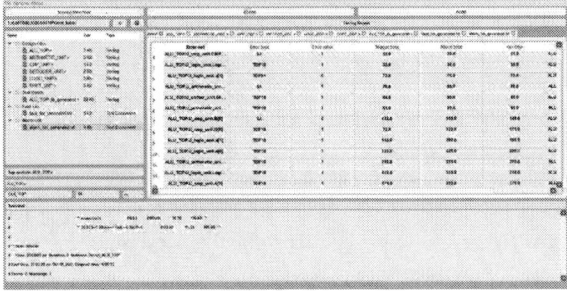

Fig. 5. Fault Simulation for ALU Design on CocoSim.

VIII. RESULTS AND DISCUSSION

The result that we achieved so far is a framework that takes the design files, test-bench either it is a simple or sophisticated UVM test-bench, fault list, and alarm list then it generates a Python file that contains our fault injection and monitoring classes in PyUVM and invokes QuestaSim which runs the modified simulation with the multiple instances (one instance for each fault) to validate whether the design can sense the faults that may happen during the fabrication or the design needs more safety mechanism implementation.

Our framework's output is a table as shown in Table I which contains the fault name, type, value, and trigger time is the time when the deviation between the fault node in the faulty instance and fault node in the golden instance happens, alarm time is the time which the deviation between the alarm node in the faulty instance and the alarm node in the golden instance, fail time is the time which a flip-flop sense that deviation and the fail-node is the node that sensed the deviation.

TABLE I.   FAULT RESULT EXAMPLE FROM CocoSim

Error Net	Error Type	Error Value	Trigger Time	Alarm Time	Fail Time	Fail Node
ALU_TOP.U_logic_unit.a[1]	TDF01	0	75.0	80.0	80.0	ALU_TOP.U_shifter_unit.Shift_OUT[0]

We conducted a comparison between CocoSim and a legacy fault simulation method and we found that we have prevailed in many cases and we are going to mention some of them.

A. Complete Mismatch

TABLE II.   COMPARISON BETWEEN A LEGACY FAULT SIMULATION METHOD AND CocoSim

	Error Net	Error Type	Error Value	Trigger Time	Alarm Time	Fail Time	Fail Node
CocoSim	ALU_TOP.U_logic_unit._Logic_D[2]	TDF 10	1	743	750	750	ALU_TOP.U_logic_unit.Logic_OUT[1]
	ALU_TOP.U_logic_unit._Logic_D[2]	TDF 10	1	803	810	810	ALU_TOP.U_logic_unit.OUT_TOP[2]

In Table II, we can see that although legacy fault simulation method captured a deviation it accepted only the late one and missed the first deviation but CocoSim captured it.

B. Fail Time Mismatch

TABLE III.   COMPARISON BETWEEN A LEGACY FAULT SIMULATION METHOD AND CocoSim

	Error Net	Error Type	Error Value	Trigger Time	Alarm Time	Fail Time	Fail Node
CocoSim	ALU_TOP.U_decoder_unit._ALU_FUN[1]	SA	0	763	770	770	ALU_TOP.U_cmp_unit.CMP_OUT[1]
	ALU_TOP.U_decoder_unit._ALU_FUN[1]	SA	0	763	770	763	ALU_TOP.U_arithmetic_unit.Arith_OUT[1]

In TABLE III, there is no deviation at the fail-time given by a legacy fault simulation method and even that given time doesn't make sense as the fail time happens at flip flops and the clock here is 5 and 763 is not a multiple of 5.

C. Performance Results

Impressively, our framework has revolutionized the fault injection process, surpassing the limitations of traditional approaches that could only run a limited number of faults simultaneously and restricted to the number of available cores. By harnessing the power of a single core, our framework now enables the execution of a significantly higher number of faults. Figure 9 depicts that we have achieved speed improvements, ranging from 20% up to 67% faster execution when dealing with a number of faults, ranging from 50 to 100. The results, graphically depicted below, exemplify the unparalleled efficiency and effectiveness of our optimized fault injection methodology.

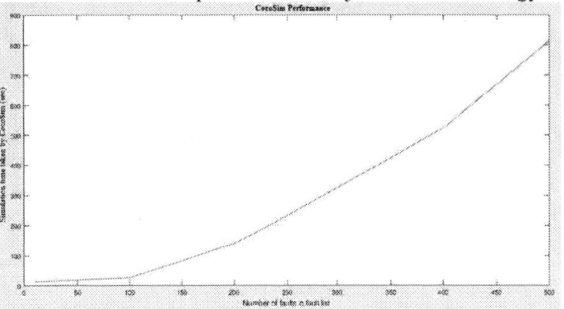

Fig. 9. CocoSim Performance with increasing number of faults

IX. CONCLUSION

In this paper, we have presented CocoSim, a new methodology for the safety verification of Verilog designs. CocoSim is built using PyUVM, a powerful verification framework. By injecting stuck-at faults and time-delay faults into a design and monitoring for alarms, CocoSim is able to identify potential safety hazards and provide users with a comprehensive safety metric.

CocoSim exhibits an average simulation time of 1 second per fault in terms of performance. This observation was derived by conducting simulations ranging from 10 to 500 faults. Notably, as the number of faults within a simulation increases, we observed a corresponding increase in the time required to simulate each fault so in comparison with the traditional methods of injecting faults, CocoSim made an improvement of 67%-20% in the number of faults ranging from 50 to 100.

REFERENCES

[1] R. Salemi, "Python for RTL Verification: A Complete Course in Python, CoCotb, and Pyuvm," 2021.

[2] Functional safety and fault simulation," Siemens Digital Industries Software. Available at: https://eda.sw.siemens.com/en-US/ic/verification-and-validation/functional-safety/.

[3] K. Osama, "Handbook of Digital CMOS Technology, Circuits, and Systems"

[4] ISO 26262-1:2018, "Road vehicles - Functional safety - Part 1: Vocabulary," in International Organization for Standardization, 2018

[5] Academy courses Academy Courses | Verification Academy. Available at: https://verificationacademy.com/academy-courses.

[6] Qt for python# Qt for Python. Available at: https://doc.qt.io/qtforpython-6/.

[7] B. Skaggs and P. Khondkar, "Questa Advanced Verification", https://resources.sw.siemens.com/en-US/white-paper-primary

AUTG: An Automatic UVM-based TestBench Generator for VLSI Chip Design Verification

Mohammad Ismael, Ayman Hroub, Abdellatif Abu-Issa

Department of Electrical and Computer Engineering, Birzeit University, Birzeit, Ramallah, Palestine

Abstract—VLSI (Very-Large-Scale Integration) chip design is a complex process, especially with the increasing systems complexity. This process comprises many steps. Each design step has a corresponding verification step. Chip design verification is the process of ensuring that the RTL (Register Transfer Level) design of a chip captures exactly the chip specifications. This process is time and resource-consuming. Almost 70% of the chip design effort is spent on verification. Thus, improving the verification productivity will accelerate the overall chip design process, and make it more efficient. Automation is an efficient technique for accelerating the verification process. There are many attempts in the literature for TestBench auto-generation. Although these techniques contributed to design verification productivity improvement, these tools are not easy to use, too expensive, and have limited features. UVM (Universal Verification Methodology) is a standardized and popular methodology for VLSI chip design verification. In this paper, we propose AUTG (Automatic UVM TestBench Generator). AUTG receives the GRef (Golden Reference) and the port list of the DUT (Design under Test). Then, it transforms the GRef into a complete reference model with checkers. Moreover, it generates the components of the UVM-based TestBench of the DUT. AUTG has been implemented and tested with multiple benchmarks. The experimental results showed that AUTG can successfully generate functional verification code. The auto-generated code was tested, and it successfully hunted all the planted bugs in the benchmarks' RTL designs.

Index Terms—Automatic Design Verification, UVM TestBench Generation, SystemVerilog, RTL Verification, Automatic Verification

I. INTRODUCTION

The increasing complexity of VLSI chip design poses challenges on manual chip functional verification, which consumes significant time and resources. The verification process can consume around 70% of the design effort [1] [2]. Moreover, manual verification can be biased, error-prone, miss corner cases, etc. Thus, there is a crucial need for TestBench auto-generation to improve the efficiency and reliability.

Simulation-based verification is the dominant verification method in VLSI chip design verification. In this method, the design verification engineers write a testbench around the DUT. The DUT is instantiated within the TestBench. Then, it is driven by the generated stimuli. The simulation environment simulates how the DUT will respond to the stimuli. Finally, the actual output obtained from the simulation is compared to the expected output obtained from the reference model. If they match, then the DUT passes in this test. Otherwise, the DUT has a bug. The reference model can be thought as executable specifications. In other words, it is a high level

functional implementation of the DUT specifications that is expected to be always correct.

UVM (Universal Verification Methodology) [3] [4] is an open-source IEEE-standard verification methodology for SystemVerilog. It has been specifically designed for complex digital integrated circuit designs.

UVM provides built-in components and APIs (Application Programming Interfaces) to simplify the verification process. Moreover, it also provides a structured framework for creating and organizing verification tests. It standardizes the verification process, leading to increased efficiency, quality, and reusability.

There are many attempts in the literature to improve the design verification productivity via automation. Although these techniques succeeded to recover some productivity, they still suffer from the lack of usability and the lack of features.

In this paper, we introduce a flexible and scalable tool called AUTG, an Automatic UVM-based TestBench Generator. AUTG takes a golden reference (GRef) file and the port list of the Design under Test (DUT) to auto-generate a UVM-based TestBench. AUTG saves time and resources. Moreover, it is easy to use, and it has better accuracy and productivity.

AUTG has been implemented in Python and tested on various benchmarks. The experimental results showed that AUTG is working correctly, and it is able to generate UVM based verification code. The auto-generated code was used to verify defective RTL codes of the DUTs, and it succeeded to hunt the bugs.

The rest of this paper is organized as follows. Section 2 surveys the related work. Section 3 provides details about AUTG architecture and workflow. Section 4 provides details about AUTG implementation and the TestBench autogeneration mechanism. Section 5 evaluates AUTG and presents the experimental results. Finally, we conclude in section 6.

II. RELATED WORK

There are numerous efforts in the literature to automate TestBench generation. Murtza et al. [5] introduced VerTGen, an open-source tool for Verilog TestBench auto-generation. VerTGen, with its GUI (Graphical User Interface), facilitates TestBench generation for diverse designs, and supports random test vector generation for thorough testing.

Xian et al. [6] developed an Automatic VHDL TestBench Generation tool with versatility, capable of generating TestBench for various VHDL design types, such as, synchronous, asynchronous, and finite state machine designs.

979-8-3503-8083-5/23 $31.00 © 2023 IEEE

TABLE I
THE SURVEYED TESTBENCH AUTOMATION FRAMEWORKS, TECHNIQUES AND TOOLS

Article / Tool	Test Vectors/DUT Stimuli Generation	TestBench Generation	Coverage / Assertions Generation
VerTGen [5]	Random test vector generation	Generate RTL TestBench	No Assertions support
VHDL TestBench [6]	User-defined	Generate RTL TestBench	No Assertions support
AVERT [7]	User-defined / Automated	Generate RTL TestBench	No Assertions support
eTBc [8]	User-defined	Semi-Automatic RTL TestBench	No Assertions support
Guiding Intelligent [9]	User-defined	Intelligent TestBench generation	Automatic Coverage Closure Framework
VeriSC [10]	User-defined	SystemC TestBench	No Assertions support
UVM Environment [11]	Generated by the Sequencer	SystemC TestBench	User-defined
GoldMine [12]	User-defined	/ Generator module	Generate assertions
AAG [13]	Uses VerTGen	Uses VerTGen	Uses GoldMine
Python/Sonar [14] [15]	User-defined	Uses Sonar library	Used for monitor changes across an entire command

Maia et al. [8] proposed a Semi-Automatic TestBench Generation Tool to assist in RTL TestBench creation. This tool saves time, particularly for large and complex designs. It enables engineers to focus on stimuli and assertion constraints instead of writing low-level code. However, it requires learning eTBc (A Semi-Automatic TestBench Generation Tool) Design Language (eDL) and eTBc Template Language (eTL).

Moreover, there are methods that utilize data mining for improved hardware design verification. For example, Murtza et al. [13] proposed an Automatic Assertion Generation framework that combines data mining, static analysis, and dynamic analysis to create high-quality assertions, later validated using a formal verifier. In addition to that, Mandouh et al. [9] proposed an approach that fuses simulation, data mining, and formal methods. Moreover, it employs coverage-directed test generation (CDTG) and data mining algorithms for intelligent TestBench automation. Vasudevan et al. [12] developed GoldMine, a tool that combines data mining algorithms and static analysis techniques to enhance functional verification efficiency. These methods collectively demonstrate the potential of data mining to optimize hardware design verification and boost design coverage.

Mohanty et al. [16] proposed an automation flow for creating a UVM-based TestBench for verifying interconnect buses in system-on-chip (SoC) designs. This approach reduces the time and effort needed for these critical and time-consuming tasks.

Da Silva et al. [10] introduced VeriSC, an automatic TestBench generator that creates SystemC-based object-oriented TestBench for functional verification.

Mefenza et al. [11] presented an automatic UVM based environment to enhance coverage and assertion-based verification of SystemC designs, reducing the effort required for TestBench implementation.

McEllin et al. [7] developed AVERT, a tool that utilizes grammatical evolution (GE) to automate TestBench generation for digital hardware designs. AVERT is customizable for specific constraints and design goals, potentially handling large and complex designs. However, it relies on GE, which can be complex and computationally intensive.

Saraswati et al. [14] used sonar for simulating and test-benching hardware [15]. Sonar is a python library that is used for creating testbenches. The authors used this library to generate a **.sv** and **.data** files, which contain the user input test vectors for testbench creation and design verification.

SN Nag et al. [17] also discussed using Python SystemVerilog in the simulation-based verification and how it leverages the power of Python and SystemVerilog to write less code, minimize errors, and reduce the verification time.

H Liang et al. [18] proposed a Python-based testbench for coverage driven verification. A serial peripheral interface (SPI) and an advanced peripheral bus (APB) controller are used as case studies to demonstrate the effectiveness of the proposed method. The results show that the proposed method can effectively verify the correctness of the APB-SPI controller functions.

Table I summarises the surveyed existing tools for Test-Bench autogeneration.

III. AUTG ARCHITECTURE AND WORKFLOW

Fig. 1 shows the architecture of the AUTG generated TestBench.

AUTG framework needs information about the DUT to be able to generate a fully functioning UVM TestBench. AUTG does not require the complete RTL design to be ready. It only needs the port list (input/output signals) of the DUT in addition to the reference model. The reference model is provided to the tool in a specific way that tells how the input and output signals are related to each other.

The workflow of AUTG tool comprises the following steps :

- Step 1: The user provides the tool with the DUT's port list and reference model as two SystemVerilog files, namely, design.sv and GRef.sv.
- Step 2: This step is optional. AUTG can generate checkers for the scoreboard, if the user provides the Golden reference (GRef) model in AUTG-specific format. Fig. 2 shows the GRef of a full adder. For each assignment(=), AUTG will create a checker.
  - The provided reference model will be automatically modified via the tool by creating checkers.
  - The user should make sure that the provided SystemVerilog files are syntax error-free, because AUTG will not fix such errors.

979-8-3503-8083-5/23 $31.00 © 2023 IEEE

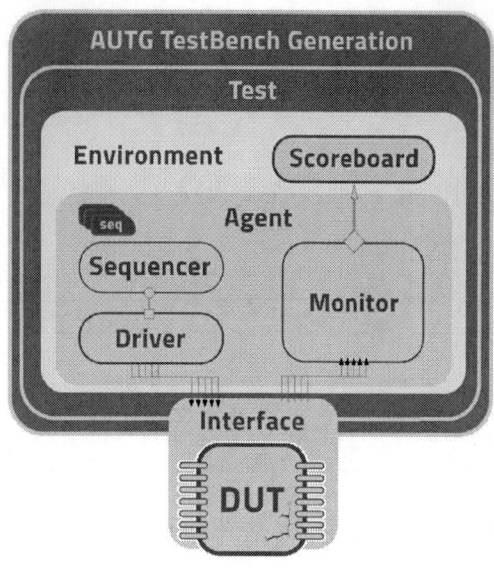

Fig. 1. AUTG Generated TestBench Architecture

```
task GRef(sequence_item in_tran, sequence_item exp_tran);
  `uvm_info(get_type_name(), "-----------------------------", UVM_NONE)
  `uvm_info(get_type_name(), "--- Inside Reference Model  ---", UVM_NONE)
  `uvm_info(get_type_name(), "-----------------------------", UVM_NONE)

  if (in_tran.A == 0 && in_tran.B == 0 && in_tran.Cin == 0) begin
    exp_tran.Sum  = 1'b0;
    exp_tran.Cout = 0;
  end else if (in_tran.A == 0 && in_tran.B == 0 && in_tran.Cin == 1) begin
    exp_tran.Sum  = 1'b1;
    exp_tran.Cout = 0;
  end else if (in_tran.A == 0 && in_tran.B == 1 && in_tran.Cin == 0) begin
    exp_tran.Sum  = 1'b1;
    exp_tran.Cout = 0;
  end else if (in_tran.A == 0 && in_tran.B == 1 && in_tran.Cin == 1) begin
    exp_tran.Sum  = 1'b0;
    exp_tran.Cout = 1;
  end else if (in_tran.A == 1 && in_tran.B == 0 && in_tran.Cin == 0) begin
    exp_tran.Sum  = 1'b1;
    exp_tran.Cout = 0;
  end else if (in_tran.A == 1 && in_tran.B == 0 && in_tran.Cin == 1) begin
    exp_tran.Sum  = 1'b0;
    exp_tran.Cout = 1;
  end else if (in_tran.A == 1 && in_tran.B == 1 && in_tran.Cin == 0) begin
    exp_tran.Sum  = 1'b0;
    exp_tran.Cout = 1;
  end else if (in_tran.A == 1 && in_tran.B == 1 && in_tran.Cin == 1) begin
    exp_tran.Sum  = 1'b1;
    exp_tran.Cout = 1;
  end

endtask : GRef
```

Fig. 2. Full Adder Reference Model

- Step 3: After generating the TestBench, the user should check the **TestBench_Top.sv** for the DUT instance to verify whether all signals are connected properly.

After creating the required directories, the design.sv file is placed in the Design directory, and GRef in the Golden_Reference directory. AUTG will execute any **.py** file inside the directory **Tool Codes**, which contains sequencer.py,

sequence_item.py, interface.py, Test_bench_top.py, agent.py, monitor.py, driver.py, random_test.py, generic_test.py, base_test.py, environment.py, and scoreboard.py scripts. The generation of the **.sv** files will be explained in the next section.

If the user only wants to create a specific UVM file, for example, driver.sv, he or she can place only the driver.py inside the **Tool Codes** directory and the main.py script will only generate the driver.sv, because each script is independent from each other.

IV. AUTG Tool Implementation and Generation Mechanism

As shown in Fig. 3, the **Tool Codes** directory contains the required Python script files to create a fully functioning UVM TestBench. Each Python script in this directory is independent from others. Which means the driver.sv, scoreboard.sv, or any other UVM classes will be generated based on the provided RTL only. This will give the script stability and accuracy in creating the classes. Thus, if there is a mistake in any class, it will not affect other classes. After executing main.py, the SystemVerilog files will be generated in this order: Test_bench_top.sv, interface.sv, sequence_item.sv, sequencer.sv, random_test.sv, driver.sv, monitor.sv, agent.sv, scoreboard.sv, environment.sv, base_test.sv, and generic_test.sv. This is the normal order of creating a UVM TestBench components. Excluding the scoreboard, each Python script will retrieve the design.sv file from the design directory and analyze its input/output signals. The procedure is as follows:

- The interface class will take the input and output signals and declare them as logic. It will also create clocking blocks for the driver and monitor. The driver clocking block will take the input signals and define them as outputs. On the other hand, the monitor clocking block will take all the signals and declare them as inputs.
- Regarding the uvm_sequence_item generated class, it will take the input signals, and identify them as **rand logic** to support randomization tests. On the other hand, the output signals will be defined as logic only. Defining the signals as logic ensures a clear representation of their logical behavior. Moreover, all input/output signals will be registered in the UVM macros in the uvm_sequence_item class. Thus, there is no need to modify or add anything to the sequence item after the auto-generation finishes.
- The uvm_driver generated class on the other hand will take the input signals of the DUT, and send them to the interface object by the drive task that is called when the run phase of the driver class is activated. After sending the data to the DUT, "Data Sent to DUT" will be printed on the terminal to indicate the success of driving the test. Fig. 4 shows a code snippet of the generated driver of a Full Adder.
- The uvm_moniter generated class will take all the signals back from the interface and send them via the UVM port **item_collected_port.write(mon_req)** to the scoreboard.

979-8-3503-8083-5/23 $31.00 © 2023 IEEE

```
virtual task run_phase(uvm_phase phase);
  super.run_phase(phase);

  forever begin
      @(posedge vif.monitor.clk)

    // Capture input signals
    mon_req.A = vif.A;
    mon_req.B = vif.B;
    mon_req.Cin = vif.Cin;

    // Capture output signals
    mon_req.Sum = vif.Sum;
    mon_req.Cout = vif.Cout;

    `uvm_info(get_type_name(), ("Sampling data by
    // Send the captured signals to the analysis
    item_collected_port.write(mon_req);
  end
endtask
```

Fig. 5. Sampling Input/Output Signals by the Generated Monitor

to generate the scoreboard.sv. First, it will take the interface class name and object from the design itself and generate the reference model and its checkers from the given GRef. The generated reference model will include the checkers. If the GRef is not given, the generated scoreboard will not include the reference model and its checkers.

Fig. 6 shows a code snippet of the modified reference model imported by the user in Fig. 2.

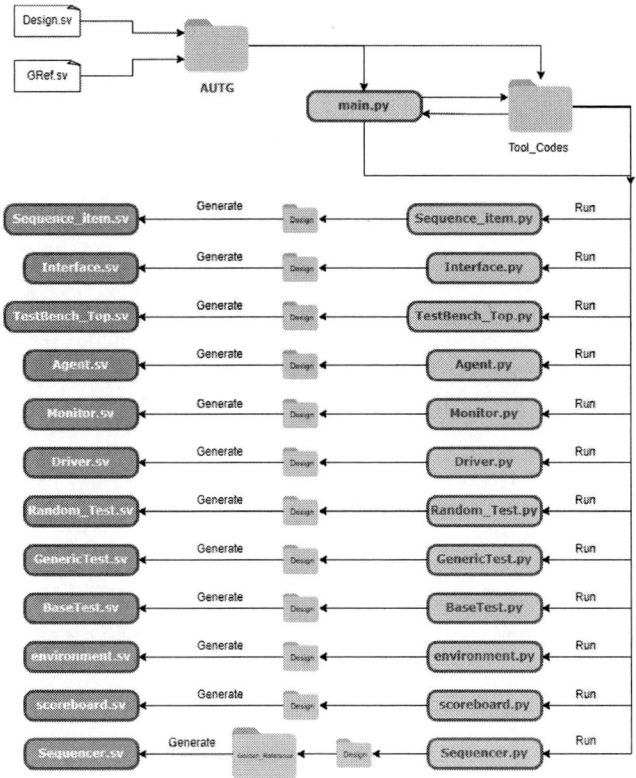

Fig. 3. AUTG Workflow

```
// run phase
virtual task run_phase(uvm_phase phase);
  super.run_phase(phase);

  forever begin
    seq_item_port.get_next_item(req);
    drive(req);
    @(posedge vif.driver.clk) seq_item_port.item_done();
  end
endtask : run_phase

// drive task
virtual task drive(FullAdder_sequence_item req);
  vif.A <= req.A;
  vif.B <= req.B;
  vif.Cin <= req.Cin;

  `uvm_info(get_type_name(), $sformatf("Data Sent to DUT:
endtask : drive
```

Fig. 4. Driving Input Signals to The DUT by The Generated Driver

When this step finishes, the terminal prints "Sampling data by the monitor is done" to indicate the success of sampling the values of the input/output signals of the DUT for this test. Fig. 5 shows a code snippet of sampling input/output signals from a full adder circuit.
- The rest of the generated UVM classes except the scoreboard are generated in the same way for any design.

scoreboard.py will take the design and the imported GRef

```
task GRef(FullAdder_sequence_item in_tran, FullAdder_sequence_item exp_tran);
  `uvm_info(get_type_name(), "-----------------------------------------------
  `uvm_info(get_type_name(), "----        Inside Reference Model
  `uvm_info(get_type_name(), "-----------------------------------------------

  if (in_tran.A == 0 && in_tran.B == 0 && in_tran.Cin == 0) begin
    exp_tran.Sum = 1'b0;
    checker_Sum(in_tran, exp_tran);
    exp_tran.Cout = 0;
    checker_Cout(in_tran, exp_tran);
  end else if (in_tran.A == 0 && in_tran.B == 0 && in_tran.Cin == 1) begin
    exp_tran.Sum = 1'b1;
    checker_Sum(in_tran, exp_tran);
    exp_tran.Cout = 0;
    checker_Cout(in_tran, exp_tran);
  end else if (in_tran.A == 0 && in_tran.B == 1 && in_tran.Cin == 0) begin
    exp_tran.Sum = 1'b1;
    checker_Sum(in_tran, exp_tran);
    exp_tran.Cout = 0;
    checker_Cout(in_tran, exp_tran);
  end else if (in_tran.A == 0 && in_tran.B == 1 && in_tran.Cin == 1) begin
    exp_tran.Sum = 1'b0;
    checker_Sum(in_tran, exp_tran);
    exp_tran.Cout = 1;
    checker_Cout(in_tran, exp_tran);
  end else if (in_tran.A == 1 && in_tran.B == 0 && in_tran.Cin == 0) begin
    exp_tran.Sum = 1'b1;
    checker_Sum(in_tran, exp_tran);
    exp_tran.Cout = 0;
    checker_Cout(in_tran, exp_tran);
```

Fig. 6. Generated Modified Reference Model

A checker is added for each assignment with the name of the output signal in the reference model. Moreover, each of these output checkers has its own checker as shown in Fig. 7. Since the Full Adder has only two output signals, two checkers are generated for each one of them. Each checker checks for failing scenarios and will print a UVM error message if the

979-8-3503-8083-5/23 $31.00 © 2023 IEEE

test fails.

```
task checker_Sum(FullAdder_sequence_item in_tran, FullAdder_sequence_item exp_tran);
    // Checker for exp_tran.Sum = in_tran.Sum
    `uvm_info(get_type_name(), "-------------------------------------------", UVM_NONE)
    `uvm_info(get_type_name(), "----          Inside Checkers Section       ----", UVM_NONE)
    `uvm_info(get_type_name(), "-------------------------------------------", UVM_NONE)
    `uvm_info(get_type_name(), "-------------------------------------------", UVM_NONE)
    `uvm_info(get_type_name(), "----     Checker_Sum: Sum Testing...    ----", UVM_NONE)
    `uvm_info(get_type_name(), "-------------------------------------------", UVM_NONE)

    if (exp_tran.Sum == in_tran.Sum) begin
        `uvm_info(get_type_name(), "----          in_tran.print        ----", UVM_NONE)
        in_tran.print();
        `uvm_info(get_type_name(), "----          exp_tran.print       ----", UVM_NONE)
        exp_tran.print();
        `uvm_info(get_type_name(), "Sum TEST PASSED", UVM_NONE)
    end else begin

        `uvm_info(get_type_name(), "----          in_tran.print        ----", UVM_NONE)
        in_tran.print();
        `uvm_info(get_type_name(), "----          exp_tran.print       ----", UVM_NONE)
        exp_tran.print();
        `uvm_error(get_type_name(), "Sum TEST FAILD")
    end
endtask

task checker_Cout(FullAdder_sequence_item in_tran, FullAdder_sequence_item exp_tran);
    // Checker for exp_tran.Cout = in_tran.Cout
    `uvm_info(get_type_name(), "-------------------------------------------", UVM_NONE)
    `uvm_info(get_type_name(), "----          Inside Checkers Section       ----", UVM_NONE)
    `uvm_info(get_type_name(), "-------------------------------------------", UVM_NONE)
    `uvm_info(get_type_name(), "-------------------------------------------", UVM_NONE)
    `uvm_info(get_type_name(), "----     Checker_Cout: Cout Testing...    ----", UVM_NONE)
    `uvm_info(get_type_name(), "-------------------------------------------", UVM_NONE)

    if (exp_tran.Cout == in_tran.Cout) begin
        `uvm_info(get_type_name(), "----          in_tran.print        ----", UVM_NONE)
        in_tran.print();
        `uvm_info(get_type_name(), "----          exp_tran.print       ----", UVM_NONE)
        exp_tran.print();
        `uvm_info(get_type_name(), "Cout TEST PASSED", UVM_NONE)
    end else begin

        `uvm_info(get_type_name(), "----          in_tran.print        ----", UVM_NONE)
        in_tran.print();
        `uvm_info(get_type_name(), "----          exp_tran.print       ----", UVM_NONE)
        exp_tran.print();
        `uvm_error(get_type_name(), "Cout TEST FAILD")
    end
```

Fig. 7. Generated Checkers

V. EXPERIMENTAL RESULTS

AUTG tool has been implemented and tested on a bunch of benchmarks listed in Table II. Moreover, this table contains the EDA Playground links to both the bug-free and the defective RTL designs of these benchmarks in addition to the generated TestBenches.

The experimental results proved the functional correctness and the efficiency of the AUTG tool. The generated verification code succeeded to hunt the implanted bugs in the RTL design.

VI. CONCLUSIONS

In this paper, we presented AUTG (Automatic UVM Test-Bench Generator) tool to auto generate UVM-based Test-Benches for VLSI chip design verification. The motivation behind AUTG was to improve the design verification productivity that suffers from the time and resource-consuming manual processes. Moreover, the auto-generated TestBenches do not suffer from biasing, missing corner cases, and human errors. AUTG has been implemented and tested on various benchmarks. The experimental results proved the functional correctness and the efficiency of AUTG. In the future, AUTG can be extended by supporting functional coverage, Assertion based Verification (ABV), constraints, sequences, and multiple interfaces.

TABLE II
THE LIST OF BENCHMARKS

Benchmark	Gref	Bug Free Design	Defective Design	Reference Model	Generat TestBench
Arithmetic Logic Unit (ALU)	YES	ALU TestBench	ALU TestBench	YES	Success
Three-bit Up Down Counter	YES	Counter TestBench	Counter TestBench	YES	Success
4-bit parity checker	YES	P.Checker TestBench	P.Checker TestBench	YES	Success
BCD to Excess3	YES	Model TestBench	BCD TestBench	YES	Success
One-bit Full Adder	YES	Full Adder TestBench	Full Adder TestBench	YES	Success
2-Bit Magnitude-Comparator	YES	Mag.C TestBench	M.Comparator TestBench	YES	Success
Shift Register	YES	Shift Register TestBench	Shift Register TestBench	YES	Success

Note. N/A = Not Applicable. Links provide access to simulation results.

REFERENCES

[1] P. Rashinkar, P. Paterson, and L. Singh, *System-on-a-chip Verification: Methodology and Techniques.* Springer Science & Business Media, 2007.

[2] C. Pixley, A. Chittor, F. Meyer, S. McMaster, and D. Benua, "Functional verification 2003: technology, tools and methodology," in *ASIC, 2003. Proceedings. 5th International Conference on*, vol. 1, pp. 1–5, IEEE, 2003.

[3] N. Kim, Y.-N. Yun, Y.-R. Cho, J. B. Kim, and B. Min, "How to automate millions lines of top-level uvm testbench and handle huge register classes," in *2012 International SoC Design Conference (ISOCC)*, pp. 405–407, IEEE, 2012.

[4] E. M. Hamed, K. Salah, A. H. Madian, and A. G. Radwan, "An automated lightweight uvm tool," in *2018 30th International Conference on Microelectronics (ICM)*, pp. 136–139, IEEE, 2018.

[5] S. A. Murtza, O. Hasan, and K. Saghar, "Vertgen: An automatic verilog testbench generator for generic circuits," in *2016 International Conference on Emerging Technologies (ICET)*, pp. 1–5, IEEE, 2016.

[6] K. T. K. Xian and N. K. Thulasiraman, "An automatic vhdl testbench generator for medium complexity design," in *2021 IEEE 19th Student Conference on Research and Development (SCOReD)*, pp. 113–118, IEEE, 2021.

[7] J. McEllin, R. Conway, and C. Ryan, "Avert: An automatic verilog testbench generation tool for grammatical evolution," in *2022 33rd Irish Signals and Systems Conference (ISSC)*, pp. 1–8, IEEE, 2022.

[8] I. Maia, K. R. da Silva, L. Max, R. Camara, and E. U. Melcher, "etbc: A semi-automatic testbench generation tool," in *IP Based SoC Design Conference & Exhibition*, pp. 1–5, 2007.

[9] E. El Mandouh and A. G. Wassal, "Guiding intelligent testbench automation using data mining and formal methods," in *2015 10th International Design & Test Symposium (IDT)*, pp. 60–65, IEEE, 2015.

[10] K. R. Da Silva, E. U. Melcher, G. Araujo, and V. A. Pimenta, "An automatic testbench generation tool for a systemc functional verification methodology," in *Proceedings of the 17th symposium on Integrated circuits and system design*, pp. 66–70, 2004.

[11] M. Mefenza, F. Yonga, and C. Bobda, "Automatic uvm environment generation for assertion-based and functional verification of systemc designs," in *2014 15th international microprocessor test and verification workshop*, pp. 16–21, IEEE, 2014.

[12] S. Vasudevan, D. Sheridan, S. Patel, D. Tcheng, B. Tuohy, and D. Johnson, "Goldmine: Automatic assertion generation using data mining and static analysis," in *2010 Design, Automation & Test in Europe Conference & Exhibition (DATE 2010)*, pp. 626–629, IEEE, 2010.

[13] S. A. Murtza, O. Hasan, and K. Saghar, "Aag: An automatic assertion generation framework for rtl designs," in *2018 International Conference on Computing, Mathematics and Engineering Technologies (iCoMET)*, pp. 1–6, IEEE, 2018.

[14] S. Huggi and S. Jamuna, "Design and verification of memory elements using python," in *2020 IEEE International Conference on Electronics, Computing and Communication Technologies (CONECCT)*, pp. 1–4, IEEE, 2020.

[15] V. Sharma, N. Tarafdar, and P. Chow, "Sonar: Writing testbenches through python," in *2019 IEEE 27th Annual International Symposium on Field-Programmable Custom Computing Machines (FCCM)*, pp. 311–311, IEEE, 2019.

[16] S. K. Mohanty, S. Sengupta, and S. Mohapatra, "Test bench automation to overcome verification challenge of soc interconnect," in *2015 International Conference on Man and Machine Interfacing (MAMI)*, pp. 1–4, IEEE, 2015.

[17] S. N. Nag, "Python systemverilog (python sv)," *World Journal of Engineering and Technology*, vol. 11, no. 3, pp. 409–416, 2023.

[18] H. Liang, N. Tan, Y. Ren, W. Hu, J. He, and J. Xia, "Python based testbench for coverage driven functional verification," in *2022 7th International Conference on Integrated Circuits and Microsystems (ICICM)*, pp. 361–365, IEEE, 2022.

2023 International Conference on Microelectronics (ICM)

A Sub-1dB Noise Figure Ku Band GaN Low Noise Amplifier for Space Applications

Husna Hamza

Electrical Engineering Department
University of Sharjah
Sharjah, United Arab Emirates
hhamza@sharjah.ac.ae

Anwar Jarndal

Electrical Engineering Department
University of Sharjah
Sharjah, United Arab Emirates
ajarndal@sharjah.ac.ae

Abstract— **AlGaN/GaN HEMT technology have shown excellent results for the construction of low-noise, high-dynamic-range, and very robust amplifiers in addition to being extensively used for microwave high-power applications. In this work, we report the design and simulation of a Ku band two stage cascaded low-noise amplifier (LNA) using 250 nm gate length AlGaN/GaN HEMT on silicon carbide (SiC) for space application with sub 1 dB noise figure. The device structure was improved by using AlInN back barrier (BB) and its performance was compared with the conventional AlGaN/GaN HEMT. The AlGaN/GaN HEMT LNA has a noise figure (NF) of about 2.5 dB and over 11.5 dB linear gain at 13 GHz. The LNA designed using AlGaN/GaN HEMT with AlInN BB exhibited a similar gain with a noise figure of 0.98 dB at 13 GHz. Thus, AlGaN/GaN HEMT LNA performance can be enhanced by incorporating AlInN BB.**

Keywords—*Cascade, Noise Figure, Low Noise Amplifier, Ku band, GaN HEMT, back barrier.*

I. INTRODUCTION

Receivers with low noise, high dynamic range, and excellent reliability are essential parts of commercial navigation, communication, and defense systems. The need for reliable and secure satellite communication systems is growing as a result of the present trends in world security. One of the most crucial, yet delicate, links in the satellite communication chain is a receiver. These receivers must continue to work even under intense jamming, and no deterioration brought on by high input energies from adversarial electromagnetic attacks is allowed. Recent developments in AlGaN/GaN HEMT technologies have amply illustrated the merits of this technology for both transmitter and reception applications [1], [2]. AlGaN/GaN HEMTs are particularly promising for applications demanding low noise, excellent linearity, and durability in addition to their appealing power attributes [3]. Furthermore, they have exceptional input power handling capabilities due to their high breakdown voltage characteristics [4], which eliminates the need for extra RF limiting circuitry in the receiver frontend. As a result, the size, weight, and noise of the system are all reduced. Thus GaN HEMT ensures very high robustness for the low noise amplifier (LNA) and offers good noise performance.

Ku band GaN LNA are reported in the literature with different design configuration. In [5] a GaN LNA for the Ku-band with a lower noise figure (greater than 4.4 dB) and associated gain (19.8 dB) at 14 GHz was reported. Also, [6] reports a balanced GaN HEMT amplifier with a noise figure below 7.5 dB and associated gain (20 dB) over the 4-16 GHz band. This paper describes the design of 0.25μm AlGaN/GaN HEMT two stage cascaded LNA for satellite telecommunication applications at Ku-band frequencies in the band 12-14 GHz. This work aims at designing a Ku-band two stage cascaded GaN low noise amplifier based on improved structure GaN HEMT with AlInN back barrier (BB). Section II describes the GaN HEMT device structure used for this design and the simulation approach. Section III describes the circuit design of the LNA. Section IV illustrates the outcomes with respect to gain, input/output matching, noise figure, and stability at the targeted frequency of 13 GHz.

II. GaN HEMT DEVICE AND SIMULATION APPROACH

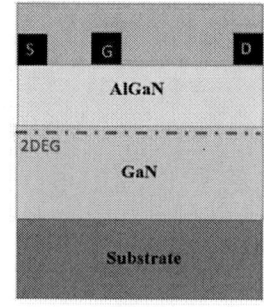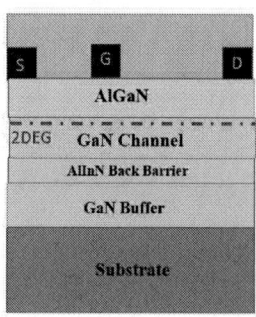

(a)   (b)

Fig 1. Schematic diagram of the GaN HEMT on SiC substrate

The design approach is to initially build the physical model of the GaN HEMT in Silvaco Atlas TCAD and calibrate to the measurement results. Next, S-parameter and noise parameters are extracted to create a touchstone file at an optimum bias condition for LNA design. This touchstone file is then imported to ADS to design LNA at the desired frequency of operation. Fig. 1(a) depicts the schematic diagram of the AlGaN/GaN HEMT under investigation. Silvaco Atlas TCAD's physical simulator is used to build this device. The device is made up of a 3 μm GaN epi-layer and a 22 nm AlGaN barrier layer. Silicon Nitride (SiN) is utilized as a passivation layer for protecting the ambient contamination. The substrate used is Silicon carbide (SiC). The device has 8 fingers (8×50 μm) with a unit-gate width of 50 μm. Gate length (LG), gate source spacing (LGS), and gate drain spacing (LGD) are, respectively, 0.25 μm, 1 μm, and 2.7 μm [7]. The physical parameters are listed in Table.I. Fig. 2 displays the GaN HEMT's measured and simulated transfer characteristics and output characteristics. The results in Fig. 2 demonstrate a pinche-off gate voltage of -3 V for the device. As can be seen, the simulation output matches the measured data fairly closely. Also the simulated output characteristics similarly display a very excellent fit between the simulated results and the measured data. This guarantees the accuracy of TCAD DC simulation. As shown in Fig. 3, the measured data and the simulated S_{11} and S_{22} correlate extremely well. This proves that the physical model employed in this investigation was accurate. The S_{21} and S_{12} plots are also shown in Fig. 3. Fig. 4 shows the simulation and measurement of NFmin at $V_{DS} = $ 10V and $V_{GS} = $ -2.7 V. Both the simulated and measured noise

979-8-3503-8083-5/23 $31.00 © 2023 IEEE

values have a good correlation and increase linearly with the frequency.

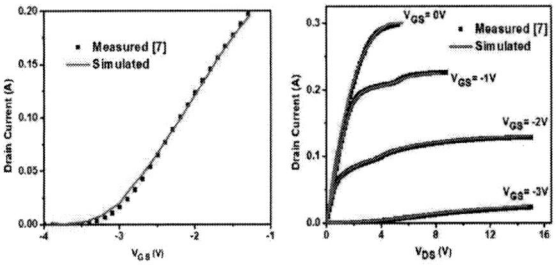

Fig 2. Simulated and measured (a) transfer characteristics and (b) output characteristics of 8×50-μm GaN-on SiC HEMT.

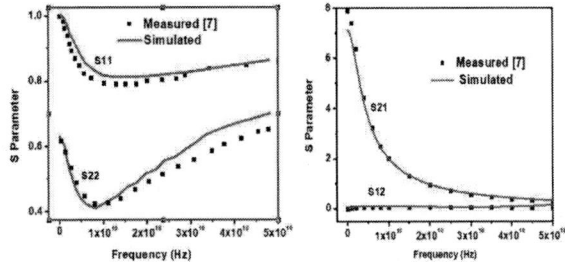

Fig 3. Simulated and measured (a) S11, S22 and (b) S21, S12 of 8×50-μm GaN-on SiC HEMT at V_{GS}=-2.7V and V_{DS} = 10V.

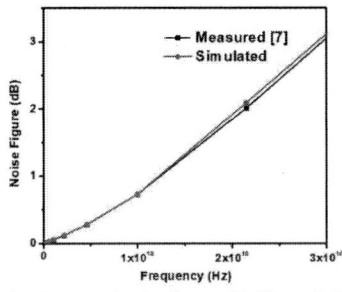

Fig 4. Simulated and measured noise figure of 8×50-μm GaN-on SiC HEMT at V_{GS}=-2.7V and V_{DS} = 10V.

TABLE I.

Parameter	Dimension
Gate Length	250 nm
Gate drain distance	2.7 μm
Gate Source distance	1 μm
AlGaN Thickness	22 nm

To design the HEMT with BB, the GaN buffer of the AlGaN/GaN HEMT is divided into two regions, as GaN channel and GaN buffer and a 100 nm thick layer is incorporated in between the channel and buffer as shown in Fig.1 (b). The drain current and transconductance characteristics are depicted in Fig.5 and Fig.6 respectively. A small increase in transconductance peak can be observed for the AlGaN/GaN HEMT with AlInN BB, which can be attributed to the steeper 2DEG conduction band edge due to the presence of the back barrier. The noise analysis of the AlGaN/GaN HEMT with and without BB are compared in Fig. 7. It can be seen better noise performance is achieved for the device with AlInN BB, which can be attributed to the

better confinement of electrons in the 2DEG channel due to the raised conduction band edge [8].

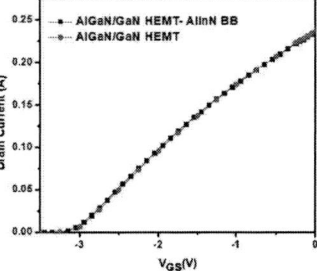

Fig 5. Simulated drain current of GaN HEMT with and without BB at V_{DS} = 10V.

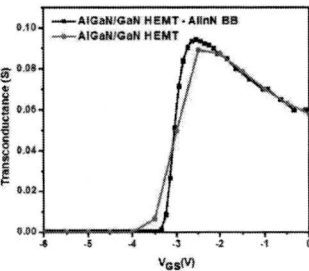

Fig 6. Simulated transconductance of GaN HEMT with and without BB at V_{DS} = 10V.

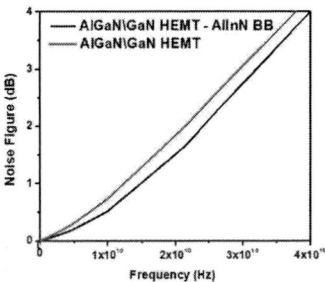

Fig 7. Simulated noise figure of GaN HEMT with and without BB at V_{GS}=-2.7V and V_{DS} = 10V.

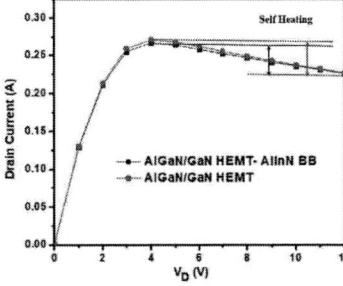

Fig 8. Simulated drain current versus drain voltage at V_{GS}=0V for temperature 300K. The self-heating effect of the devices are also depicted.

Also the self-heating effect of both the HEMT are analyzed. Fig.8 depicts the simulated I_{DS}-V_{DS} characteristics at V_{GS}=0V for temperature 300K. It is observed that, the drain current decreases with the self-heating effect above V_{DS}=4V. It can be seen in Fig.8 that the impact of the self-heating on the drain current is more on the AlGaN/GaN HEMT. The impact of different back barrier materials on the noise figure for GaN HEMT LNA design can be explored. Research on

979-8-3503-8083-5/23 $31.00 © 2023 IEEE

the impact of different back barrier materials on the GaN HEMT is reported in [9].

III. LOW NOISE AMPLIFIER DESIGN

In an LNA, the maximum small signal gain can be obtained when biasing the HEMT within the linear region. Once the calibration is done, the S-parameters and noise parameters of the GaN HEMT was extracted at V_{GS}= -2.7 V and V_{DS}=10 V (at maximum transconductance). These data were used to create the touchstone file and was imported to Keysight ADS.

The input impedance and output impedance were calculated using S-parameter simulation in ADS. Table II depicts the input impedance (Z_{in}) and output impedance (Z_{out}) of both the GaN HEMT at 13 GHz. The circuit schematic of the designed two stage cascade LNA is depicted in Fig.9 and Fig.10. The components SnP1 and SnP2 represent the touchstone files of the GaN HEMT at V_{GS} =-2.7V and V_{DS}=10V, representing the first stage and second stage respectively. The values in Table.1 were used to design the input matching network, inter stage matching network and output matching network.

TABLE II.

Parameter	AlGaN/GaN HEMT	AlGaN/GaN HEMT with AlInN BB
Z_{in}	17.2-j3.9	0.26-j12.7
Z_{out}	3.7-j20.6	40.55-j44.15

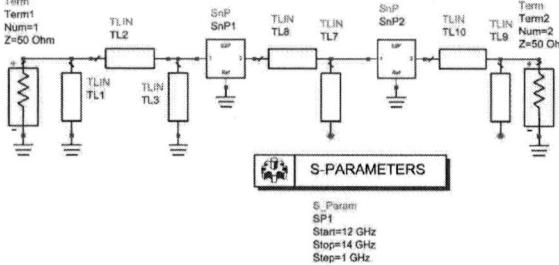

Fig. 9. ADS simulation schematic of two stage GaN HEMT Ku band LNA.

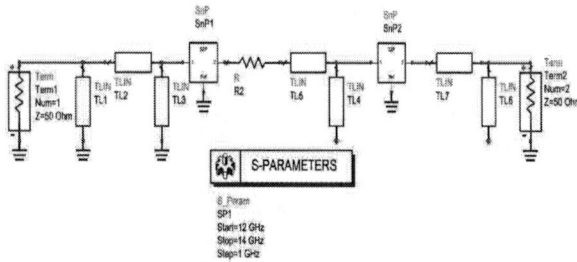

Fig.10. ADS simulation schematic of two stage GaN HEMT with BB Ku band LNA.

For the AlGaN/GaN HEMT LNA schematic in Fig. 9, the components SnP1 and SnP2 represent the two stages of GaN HEMT used. The input matching network comprises of TL1, TL2 and TL3. The inter stage matching of the stage SnP1 and SnP2 were done with TL7 and T8. The output matching network comprises of TL9 and TL10. For the AlGaN/GaN HEMT with AlInN BB LNA schematic in Fig.10, the components SnP1 and SnP2 represent the two stages of GaN

HEMT with AlInN BB used. The input matching network comprises of TL1, TL2 and TL3. The inter stage matching of the stage SnP1 and SnP2 were done with TL4 and T5. For ensuring stability the resistor R_2 is added at the output of first stage. The output matching network comprises of TL7 and TL8. A fine tuning of the electrical length and impedance of the transmission lines were done to obtain the best results in terms of noise figure, gain and reflection coefficients while maintaining the stability of the circuit. The S parameter simulation of the circuit was done with noise analysis in ADS.

IV. RESULTS AND DISCUSSIONS

The noise performance is represented by the LNA's noise figure, and a higher noise figure denotes a worsening of the signal. Therefore, when designing the LNA within the operational frequency range, low noise figure is the goal. The LNA's noise figure is calculated as [10],

$$NF = NF_{min} + {R_n}/{G_s} (Y_s - Y_{opt})^2 \qquad (1)$$

where R_n is the noise resistance, G_s is the source conductance, and NF is the noise figure. NFmin is the lowest noise figure that the LNA can achieve when the noise match is perfect. It is clear that the source termination Y_s, whose source impedance must match the ideal noise impedance Y_{opt} to achieve NF=NF_{min}, determines the noise figure of the LNA. The noise figure and gain of the designed LNA is depicted in Fig. 11(a). As can be seen, at 13 GHz, the noise figure is 2.1 dB and the gain is 11.5 dB.

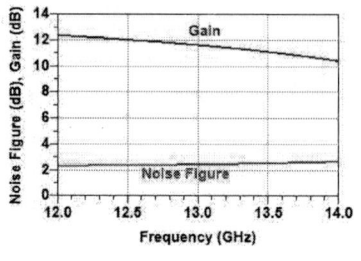

(a)

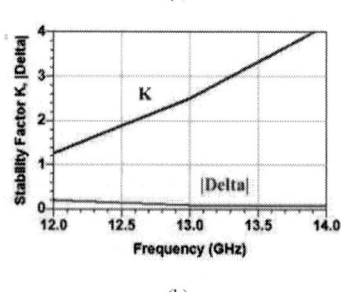

(b)

Fig.11 (a) The gain and noise figure of the two stage GaN HEMT LNA from 12 GHz to 14 GHz (b) The stability analysis of two-stage cascaded GaN HEMT LNA.

The stability can be investigated by using Rollet's Condition given below [11]:

$$Stability\ Factor\ K = \frac{(1-(|S11|^2-|S22|^2+|\Delta^2|))}{(2|S21||S12|)} \geq 1 \quad (2)$$

$$Delta\ |\Delta| = |S_{11}S_{22} - S_{12}S_{21}| < 1 \qquad (3)$$

To achieve unconditional stability in the LNA, the stability factor (K) must be more than 1 and delta ($|\Delta|$) should be less than 1. The stability analysis of the designed LNA is depicted in the Fig. 11(b). As illustrated in the figure, the stability factor is greater than1 and delta is less than 1 in the frequency range 12 GHz to 14 GHz thus ensuring the stability of the LNA.

The simulation results of AlGaN/GaN HEMT with AlInN BB is depicted in Fig. 12(a) and Fig. 12(b). The noise figure and gain are depicted in Fig. 12(a). The LNA designed with AlGaN/GaN HEMT with AlInN BB exhibited again of 12 dB and a noise figure of 0.98 dB at a frequency of 13 GHz. The stability analysis is done in Fig. 12(b). The stability factor is greater than 1 and magnitude of delta is less than 1 for the frequency range of 12 GHz to 14 GHz. Thus, the LNA is stable.

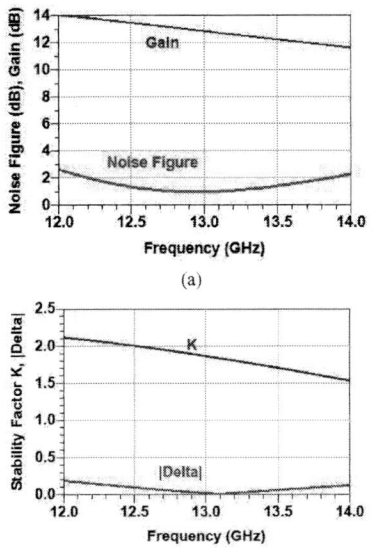

Fig. 12. (a) The gain and noise figure of the two stage AlGaN/GaN HEMT with AlInN BB LNA from 12 GHz to 14 GHz. (b) The stability analysis of two-stage cascaded AlGaN/GaN HEMT with AlInN BB LNA.

V. CONCLUSION

Ku Band AlGaN/GaN HEMT based two stage LNA was designed for space applications. The design of LNA with HEMT with and without back barrier was done. The designed LNA using the HEMT with back barrier attained a gain of 12 dB and a noise figure of 0.98 dB at 13 GHz frequency. This work emphasizes the use of the physical model of the device to study how to optimize the LNA performance by varying the physical parameters of the HEMT like thickness and material of barrier layer, back barrier layer, and substrate. Thus, further optimization of the designed LNA in terms of gain is possible in future.

ACKNOWLEDGMENT

The authors gratefully acknowledge the support from the University of Sharjah, Sharjah, United Arab Emirates.

REFERENCES

[1] Shi C, Yang L, Zhang M, Wu M, Hou B, Lu H, Jia F, Guo F, Liu W, Yu Q, Ma X. High-Efficiency AlGaN/GaN/Graded-AlGaN/GaN Double-Channel HEMTs for Sub-6G Power Amplifier Applications. IEEE Transactions on Electron Devices. 2023 Apr 3;70(5):2241-6.

[2] H. B. Ahn, H. -G. Ji, Y. Choi, S. Lee, D. M. Kang and J. Han, "25–31 GHz GaN-Based LNA MMIC Employing Hybrid-Matching Topology for 5G Base Station Applications," in *IEEE Microwave and Wireless Technology Letters*, vol. 33, no. 1, pp. 47-50, Jan. 2023, doi: 10.1109/LMWC.2022.3201075.

[3] Moon, J.S., Wong, J., Grabar, B., Antcliffe, M., Chen, P., Arkun, E., Khalaf, I., Corrion, A., Chappell, J., Venkatesan, N. and Fay, P., 2020. 360 GHz f MAX graded-channel AlGaN/GaN HEMTs for mmW low-noise applications. *IEEE Electron Device Letters*, 41(8), pp.1173-1176.

[4] S. Gao, X. Liu, J. Chen, Z. Xie, Q. Zhou and H. Wang, "High Breakdown-Voltage GaN-Based HEMTs on Silicon With Ti/Al/Ni/Ti Ohmic Contacts," in IEEE Electron Device Letters, vol. 42, no. 4, pp. 481-484, April 2021, doi: 10.1109/LED.2021.3058659.

[5] Suijker EM, Rodenburg M, Hoogland JA, van Heijningen M, Seelmann-Eggebert M, Quay R, Brückner P, van Vliet FE. Robust AlGaN/GaN Low Noise Amplifier MMICs for C-, Ku- and Kaband Space Applications. IEEE Compound Semiconductor Integrated Circuit Symposium 2009; pp. 1-4, DOI: 10.1109/csics.2009.5315640.

[6] Seo S, Pavlidis D, Moon JS. A wideband Balanced AlGaN/GaN HEMT MMIC Low Noise Amplifier for Transceiver Front-ends. IEEE Gallium Arsenide and Other Semiconductor Application Symposium 2005; pp. 225-228.

[7] D. Floriot et al., "GH25-10: New qualified power GaN HEMT process from technology to product overview," in Proc. 9th European Microwave Integrated Circuit Conference, Rome, Italy, 2014, pp. 225-228.

[8] Sanabria, Christopher, et al. "Influence of epitaxial structure in the noise figure of AlGaN/GaN HEMTs." *IEEE transactions on microwave theory and techniques* 53.2 (2005): 762-769.

[9] Hamza KH, Nirmal D, Fletcher AA, Ajayan J, Natarajan R. Enhanced drain current and cut off frequency in AlGaN/GaN HEMT with BGaN back barrier. Materials Science and Engineering: B. 2022 Oct 1;284:115863.

[10] J -W. Lee, A. Kuliev, V. Kumar, R. Schwindt and I. Adesida, "Microwave noise characteristics of AlGaN/GaN HEMTs on SiC substrates for broad-band low-noise amplifiers," in IEEE Microwave and Wireless Components Letters, vol. 14, no. 6, pp. 259-261, June 2004, doi: 10.1109/LMWC.2004.828026.

[11] David M. Pozar (2012). Microwave Engineering 4th Edition. In John Wiley & Sons Inc.

A Low-Power Analog Integrated Gaussian-based Neural Network Classifier with Application to Hepatitis Disease Recognition

Vassilis Alimisis, Nikolaos P. Eleftheriou and Paul P. Sotiriadis

Department of Electrical and Computer Engineering
National Technical University of Athens, Greece
E-mail: alimisisv@gmail.com, eleftheriou_nikos@hotmail.com , pps@ieee.org

Abstract—**Hepatitis is a medical condition characterized by inflammation of the liver. It can be triggered by various factors, including viral infections, excessive alcohol consumption, specific medications, or autoimmune disorders. Recognizing hepatitis early is crucial in reducing its symptoms, as it allows for timely treatment. In this study, a low-power ($4.31 \mu W$) and low-voltage (0.6V) analog artificial neural network classifier is introduced, utilizing a Gaussian-based activation function. The architecture comprises a hidden layer: with a Gaussian activation function circuit and tanh approximation, an output layer with a softmax function circuit, and an argmax operator. A comparative analysis is performed to evaluate the performance of this methodology against commonly used analog classifiers. To conduct this assessment, a real-world hepatitis dataset is employed. The models are trained and results processed using the Python programming language. The hardware design and result processing are executed using Cadence IC Suite, utilizing the TSMC 90nm CMOS process technology, demonstrating the practical applicability and effectiveness of this methodology.**

Index Terms—**Artificial Neural Network, analog VLSI implementation, hepatitis classification, low-power design**

I. INTRODUCTION

Machine Learning (ML) and Artificial Intelligence (AI) have revolutionized bioengineering and healthcare, ushering in a new era of personalized, data-driven medical practices [1]. These technologies enable in-depth analysis of vast sets of biological and clinical data, offering unprecedented insights into disease mechanisms, treatment responses, and patient outlooks [1]. In bioengineering, ML algorithms are utilized to craft and enhance tailored medical devices, prosthetics, and implants [2]. Additionally, AI-powered image analysis has greatly enhanced diagnostic precision in medical imaging, swiftly detecting abnormalities and tumors.

Regarding healthcare, prognostic models utilize patient data to predict the trajectory of diseases, enabling proactive interventions [2]. Additionally, natural language processing facilitates the extraction of invaluable information from clinical records and academic literature, expediting medical research and knowledge acquisition [3]. The integration of ML and AI in bioengineering and healthcare not only elevates the standard of patient care but also lays the foundation for pioneering

innovations with the potential to revolutionize the future of medicine [4].

The hardware implementation of ML and AI in bioengineering is a pivotal frontier in modern healthcare tech [1]. Specialized systems process vast datasets, enabling real-time decision-making in medical applications. Customized circuits and accelerators handle complex ML algorithms, optimizing tasks like image recognition and predictive modeling. Neuromorphic computing mimics the human brain's neural networks, offering parallel processing for tasks like pattern recognition [5]. These hardware advances boost the speed and efficiency of ML and AI in bioengineering, with potential to revolutionize patient care and medical research.

Analog computing is rapidly advancing in the field of bioengineering for ML and AI applications [6]. This approach utilizes continuous signals and physical phenomena to process complex biological data, mirroring the continuous nature of biological systems. This is particularly beneficial for real-time responses in bioengineering, where intricate physiological processes are common. Specialized analog circuits mimic neuron behavior, efficiently processing neural network models. Analog computing excels in tasks like signal processing and pattern recognition, proving invaluable in medical imaging and biosignal analysis [7]. Its precision and speed are poised to drive significant progress in personalized medicine, prosthetic design, and bioinformatics, leading to more effective healthcare solutions and improved patient outcomes.

Motivated by the low-power and area efficiency requirements of smart sensors for biomedical applications [8], [9], this work proposes a low-power ($4.31 \mu W$) and low-voltage (0.6V) analog artificial neural network (ANN) classifier, utilizing a Gaussian-based activation function. The implemented classifier is a promising approach, appropriate for battery dependent smart sensor classification systems, since it achieves 96.42% accuracy. It is designed and verified on a real-world hepatitis disease recognition dataset [10]. The post-layout simulation results, performed in a TSMC 90nm CMOS process and simulated using Cadence IC Suite, confirm the accuracy of the proposed implementation through comparison with a software-based approach and analog related classifiers.

The remainder of this paper is organized as follows. Section II refers to the background; explains the ANN approach and the hepatitis disease. In Section III the proposed architecture and the basic building blocks of the proposed classifier are thoroughly explained. In Section IV, the proposed classifier's performance is validated using a real-world dataset for hepatitis disease recognition and is contrasted with the software-based implementation. Section V conducts a comparative study with analog classifiers in the field. Concluding remarks are presented in Section VI.

II. BACKGROUND

A. Artificial Neural Network Model

An ANN, inspired by the human brain's neural network system, is a potent computational model [11]. It comprises layers of interconnected neurons. Each neuron processes and transmits information, with the input layer receiving data. This data is weighted and processed through hidden layers for complex computations. The output layer produces the final prediction or classification. Through backpropagation during training, the network adjusts its internal parameters, minimizing the error between predicted and actual outcomes. This capability allows ANNs to discern intricate patterns in data, making them invaluable in tasks such as image recognition, speech processing, and predictive modeling. ANNs have transformed various fields, showcasing their adaptability and potential in solving complex problems.

ANNs can be mathematically represented at different levels of abstraction. Here are some of the fundamental equations [11] that describe the workings of a simple feedforward neural network; neuron input (weighted sum), activation function and feedforward process. The input to a neuron in a feedforward network is calculated as the weighted sum of its inputs, followed by the addition of a bias term and the application of an activation function. This can be represented as:

$$z_j = \sum_{i=1}^{n} w_{ij} x_i + b_j. \tag{1}$$

Where:

- z_j is the weighted sum for neuron j.
- w_{ij} represents the weight connecting neuron i to neuron j.
- x_i is the output of neuron i in the previous layer.
- b_j is the bias term for neuron j.

The output of a neuron is obtained by applying an activation function to the weighted sum. Common activation functions include *sigmoid*, Rectified Linear Unit (ReLU) , tanh, and softmax. For instance, the *sigmoid* activation function is defined as:

$$a_j = \sigma(z_j) = \frac{1}{1 + e^{-z_j}}. \tag{2}$$

Here, a_j is the output (activation) of neuron and σ represents the *sigmoid* function. In a feedforward network, the outputs of

one layer serve as the inputs to the next. This process continues until the final layer provides the network's output.

$$x_j^{(l+1)} = \sigma(\sum_{i=1}^{n} w_{ij}^{(l)} x_i^{(l)} + b_j^{(l)}). \tag{3}$$

Here, $x_j^{(l)}$ is the output of neuron j in layer l, $w_{ij}^{(l)}$ are the weights connecting neurons between layers l and $l + 1$ and $b_j^{(l)}$ is the bias term for neuron j in layer l. These equations form the basis for understanding the computations that occur within a feedforward ANN.

B. Hepatitis Disease

Hepatitis is a widespread medical condition characterized by inflammation of the liver [12]. It can be caused by various factors, with viral infections being the most common culprits. Hepatitis viruses are categorized from A to E, each with distinct transmission methods and potential outcomes. For instance, Hepatitis A and E are typically contracted through contaminated food or water, while Hepatitis B, C, and D are primarily transmitted through contact with infected blood or other bodily fluids. Chronic forms of Hepatitis, particularly B and C, can lead to serious liver complications over time, such as cirrhosis and even liver cancer if left untreated. It's imperative to prioritize prevention through vaccination, safe hygiene practices, and public health education to mitigate the spread of hepatitis.

III. PROPOSED ARCHITECTURE

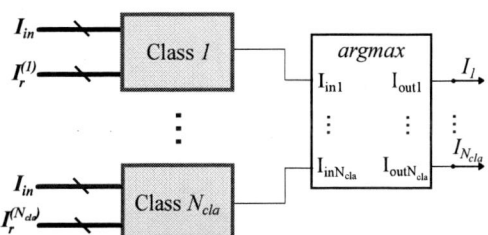

Fig. 1: The high level architecture of the proposed ANN classifier which combines N_{cla}, N_d input features. It consists of N_{cla} class cells and $argmax$ operator with N_{cla} inputs.

In this section, the proposed analog implementation of the ANN is analysed. The introduced architecture is adaptable, capable of handling different quantities of classes and input dimensions. The structure of the suggested classifier, depicted in Fig. 1, is tailored for a classification task with N_{cla} classes and N_d input dimensions. The number of layers in each class is a hyper-parameter, typically determined through exploratory data analysis. For simplicity, it consists of 1 hidden layer in each class. The proposed architecture's k-th ($k \in \{1, ..., N_{cla}\}$) class comprises a hidden layer: with a Gaussian activation function circuit (GHL) and tanh approximation (THL), an output layer with a softmax function circuit, and a Voltage-to-Current (V/I) converter, as shown in Fig. 2.

Firstly, the inputs are sent to a hidden layer which performs a nonlinear Gaussian function as activation function, on the

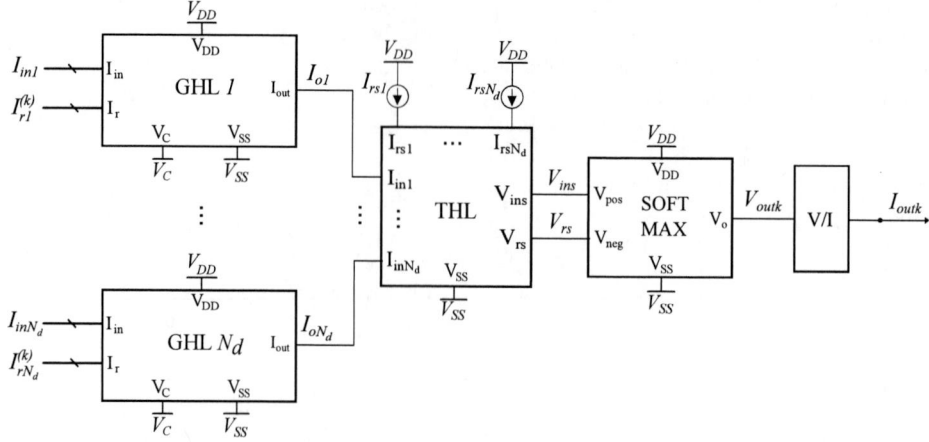

Fig. 2: The proposed architecture's k-th ($k \in \{1, ..., N_{cla}\}$) class comprises a hidden layer: with a Gaussian activation function circuit (GHL) and tanh approximation (THL), an output layer with a softmax function circuit, and a Voltage-to-Current (V/I) converter.

weighted sum of all inputs. The input currents are adjusted to restrict the input range for the hidden layer between $-0.3V$ and $0.1V$, ensuring that the input transistors in the hidden layer never enter the linear region. For the implementation of the Gaussian function as activation function, a current-mode Gaussian function circuit [13] is employed. It is shown in Fig. 3 and the dimensions of the transistors are summarized in Table I. In particular, the I_{in} is the input current and the current parameter I_r, the voltage parameter V_c, and the bias current I_{bias} regulate the mean value, variance, and amplitude, respectively.

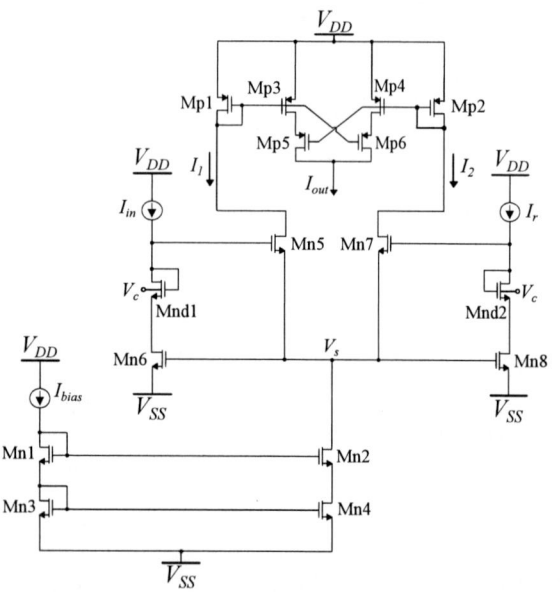

Fig. 3: The current-mode Gaussian activation function circuit.

The second component in the hidden layer is a straightforward NMOS cascode current mirror with a PMOS diode load, which is shown in Fig. 4. Since all transistors operate in the

TABLE I: MOS Transistors' Dimensions (Fig. 3).

Block	W/L (μm/μm)	Current Correlator	W/L (μm/μm)
M_{n1}-M_{n6}	0.8/1.6	M_{p1}-M_{p2}	0.8/1.6
M_{n7}, M_{n9}	0.4/1.6	M_{p3}-M_{p6}	0.4/1.6
M_{n8}, M_{n10}	1.6/1.6	-	-

sub-threshold region, this unit imparts an approximately tanh behavior to the output current of the Gaussian function circuit. This method significantly trims down hardware expenses when compared to conventional circuit implementations of these activation functions. In the proposed ANN classifier, weight vectors are realized by adjusting the bias current I_{bias} of the preceding Gaussian function circuit block. For each class, a set of N_d basic NMOS cascode current mirrors (CCMs) with a PMOS diode load is employed. These circuits are linked to the output of the respective Gaussian function circuits, receiving an input current I_{ok} ($k \in \{1, ..., N_{cla}\}$). A second set of N_d basic NMOS CCMs with a PMOS diode load is put into operation and biased with a current I_{rsk} ($k \in \{1, ..., N_{cla}\}$). The current I_{rsk} serves as a parameter current, supplied during the classifier's training.

Both sets of NMOS cascode current mirrors generate output voltages, namely V_{ins} and V_{rs}, as depicted in Fig. 4. These voltages are subsequently fed into the output layer. The output neuron is implemented as a pseudo-differential current mirror, carrying out the softmax operation on the weighted sum of the outputs from the hidden layer, which is shown in Fig. 4 (middle block). Following this, the output voltage (V_o) from each softmax block is directed into a V/I converter [14], shown in Fig. 5, which produce the appropriate output current. Then, all the output currents are fed into the argamx operator circuit (determining the winning class).

Moving forward, the ANN is established using an argmax operator circuit, specifically employing a Winner-Take-All (WTA) circuit [15]. In a classification problem featuring N_{cla}

2023 International Conference on Microelectronics (ICM)

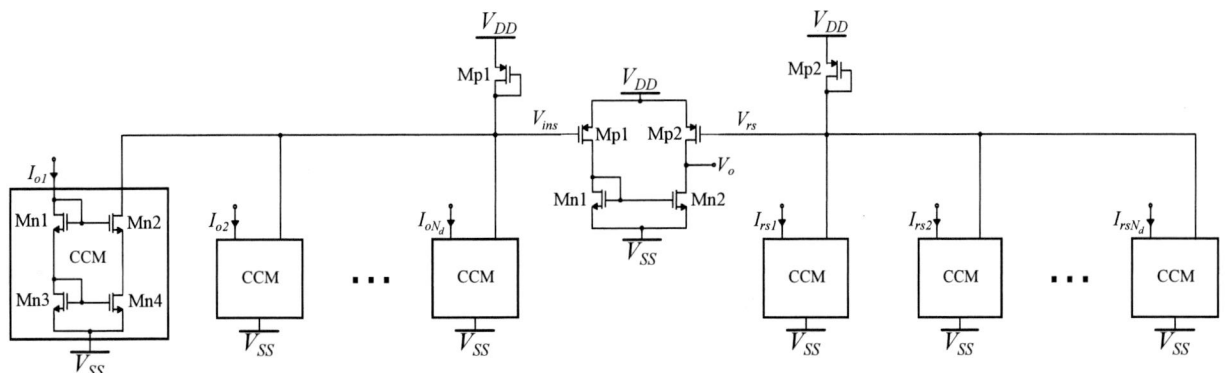

Fig. 4: A set of N_d basic NMOS CCMs with a PMOS diode load linked to the output of the respective Gaussian function circuits (left). A second set of N_d basic NMOS CCMs with a PMOS diode load is put into operation and biased with a current I_{rsk} (right). A pseudo-differential current mirror, carrying out the softmax operation (middle)

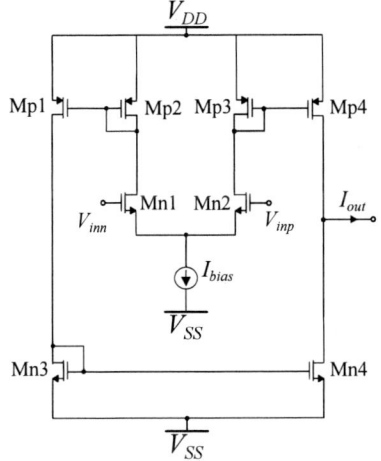

Fig. 5: The implementation of V/I converter.

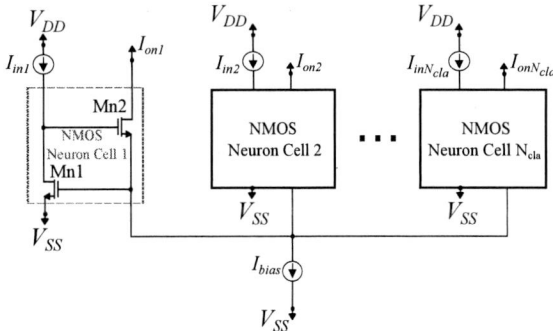

Fig. 6: A N_{cla}-neuron Standard Lazzaro NMOS Winner-Take-All (WTA) circuit.

IV. HEPATITIS DATASET AND SIMULATION RESULTS

In this Section, the proper operation of the proposed classifier is confirmed via a real-world hepatitis disease recognition dataset [10]. It is provided by University of California, Irvine (UCI) Machine Learning Repository. It contains valuable information related to the liver disease, hepatitis, which is a significant global health concern. It encompasses various laboratory test results, and presence of hepatitis B and C viruses. By leveraging it, experts can gain deeper insights into

classes, the standard Lazzaro WTA circuit encompasses N_{cla} neurons. These neurons collectively share a common bias current, as illustrated in Fig. 6. Each neuron within the WTA circuit corresponds to an individual class. The WTA circuit efficiently discerns the class with the highest input current and allocates a non-zero output current to the respective neuron. Concurrently, the remaining neurons receive an output current of zero. Notably, all transistors in the mentioned designs operate in the sub-threshold region, with power supply rails set as $V_{DD} = -V_{SS} = 0.3$V.

the factors influencing the onset and progression of hepatitis, ultimately contributing to improved diagnostic and treatment strategies for individuals affected by this condition.

The proposed architecture is implemented using the TSMC $90nm$ CMOS process with the aid of the Cadence IC suite. The entire classifier is powered with $V_{DD} = -V_{SS} = 0.3V$. All the simulation results are provided from the implemented layout, which is depicted in Fig. 7 (post-layout simulations). This study addresses a hepatitis disease recognition problem, involving $N_{cla} = 2$ classes and $N_d = 19$ inputs. It's a binary classification task wherein the classifier determines whether a patient is healthy or afflicted with hepatitis, making it a binary classifier. The relevant metrics (attributes) mentioned earlier are directly inputted into the classifier. The system's essential parameters are derived by calculating the mean value, variance, and prior probability for each class.

To comprehensively evaluate the proposed classifier in terms of classification specificity and its performance under varying Process, Voltage, and Temperature (PVT) conditions, two distinct tests are conducted on the layout. To account for experimental variability, the results from 20 different training-test iterations are shown in Fig. 8. The circuit's sensitivity is further confirmed through a Monte Carlo analysis. Specifically,

979-8-3503-8083-5/23 $31.00 © 2023 IEEE

Fig. 9 illustrates the Monte Carlo Histogram based on N = 100 points, providing further insights into the circuit's robustness and performance characteristics It has a mean value of $\mu_M = 95.98\%$ and a standard deviation of $\sigma_M = 1.37\%$.

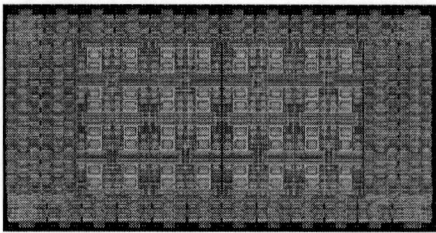

Fig. 7: The layout of the proposed architecture.

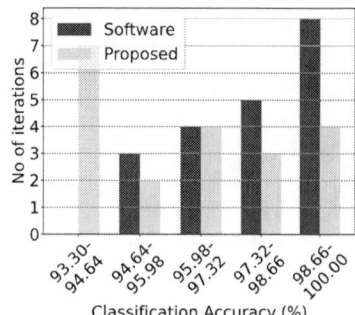

Fig. 8: Classification results of the proposed architecture and the equivalent software model on the hepatitis disease recognition dataset over 20 iterations.

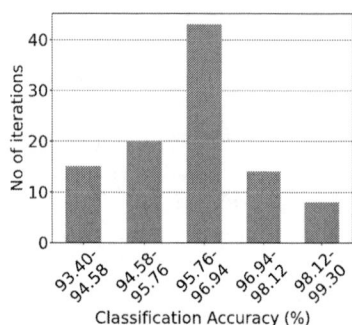

Fig. 9: Post-layout Monte-Carlo simulation results of the proposed architecture on the hepatitis disease recognition dataset with $\mu_M = 95.98\%$ and a standard deviation of $\sigma_M = 1.37\%$

V. COMPARISON AND DISCUSSION

In the realm of current literature, it is evident that a significant portion of analog classifiers is customarily designed to cater to particular applications. This circumstance poses a challenge when attempting to execute an unbiased comparison across varied implementations. Nevertheless, this challenge offers us the opportunity to modify analog classifiers to suit a common application, thereby streamlining a performance evaluation encompassing both machine learning models and alternative methodologies Notably, Table III provides an overview

of the performance metrics of our study in conjunction with comparable classifiers like ANN [16], Radial Basis function [17], Multilayer Perceptron (MLP) [18], Long Short-Term Memory (LSTM) [19], K-means [20], Bayesian [13], Gaussian Mixture Model (GMM) [22], Fuzzy [21], Threshold [23], Support Vector Machine (SVM) [24] and centroid-based [25] all within the context of hepatitis disease recognition.

The proposed work presents an intriguing solution as it offers a trade-off between accuracy, power, and energy consumption per classification when compared to related analog classifiers. It's crucial to emphasize that, in this specific application, the design is handling a high input dimensionality. The proposed topology provides a notable advantage by eliminating the need for Principal Component Analysis (PCA), allowing for the utilization of all 19 input dimensions without any loss of information. To attain optimal accuracy, several other topologies need to reduce the dimensions to 12, which represents a notable limitation in previous related works, except from the more complex models [16], [18]–[20]. While the proposed classifier showcases the ability to accurately classify more classes, we select a binary classification scenario for fair comparison. This adjustment enables a more meaningful comparison with binary analog classifiers [21], [23], [24].

In terms of classification accuracy, the proposed architecture surpasses all related classifiers except for MLP [18] and LSTM [19]. These models, while achieving higher accuracy, are more complex and require more power and hardware area (due to having more components). The Threshold classifier achieves the lowest power consumption, albeit at the expense of accuracy and processing speed, owing to its straightforward model design [23]. It's worth emphasizing that in biomedical applications of this nature, rapid processing speed is not a critical requirement, primarily due to their low occurrence frequency. Consequently, in the proposed approach, processing speed is intentionally reduced to enhance accuracy and optimize power consumption performance. Furthermore, it boasts lower energy consumption per classification compared to all classifiers, with the exception of ANN [16], which achieves lower classification accuracy.

VI. CONCLUSION

In this work, a low power ($4.31\mu W$), low voltage (0.6V) architecture of an analog Gaussian-based ANN classifier for hepatitis disease recognition was proposed. The presented architecture consists of a hidden layer with a Gaussian activation function circuit, an output layer with a softmax function circuit, and an argmax operator. The post-layout simulation results were conducted utilizing the TSMC 90nm CMOS process and were compared with both a software-based implementation and a range of analog classifiers. The realized architecture attains a classification accuracy of 96.42% along with satisfactory sensitivity characteristics. It can act as a fundamental component in diverse wearable biomedical devices, particularly those with stringent power consumption requirements.

TABLE II: Analog classifiers' comparison on the Hepatitis Disease Recognition

	Classifier	Worst accuracy	Mean accuracy	Best accuracy	Power consumption	Processing speed	Energy per classification	No. of Dimensions
This work	ANN	93.30%	96.42%	99.40%	$4.31\mu W$	$1.2M\frac{\text{classifications}}{s}$	$\frac{3.59 \text{ pJ}}{\text{classification}}$	19
[16]	ANN	91.70%	94.78%	97.90%	$3.12\mu W$	$14M\frac{\text{classifications}}{s}$	$\frac{0.22 \text{ pJ}}{\text{classification}}$	19
[17]	RBF	86.90%	90.12%	92.90%	$29.43\mu W$	$200k\frac{\text{classifications}}{s}$	$\frac{147.15 \text{ pJ}}{\text{classification}}$	12
[18]	MLP	94.70%	97.48%	99.70%	$434.32\mu W$	$930k\frac{\text{classifications}}{s}$	$\frac{466.97 \text{ pJ}}{\text{classification}}$	19
[19]	LSTM	97.30%	99.12%	100.00%	$31.21mW$	$870M\frac{\text{classifications}}{s}$	$\frac{35.87 \text{ pJ}}{\text{classification}}$	19
[20]	K-means	91.90%	95.87%	97.70%	$138.42\mu W$	$5M\frac{\text{classifications}}{s}$	$\frac{27.68 \text{ pJ}}{\text{classification}}$	19
[21]	Fuzzy	88.30%	93.55%	97.30%	$2.49\mu W$	$4.55K\frac{\text{classifications}}{s}$	$\frac{547.2 \text{ pJ}}{\text{classification}}$	12
[22]	GMM	87.30%	90.44%	93.40%	$3.14\mu W$	$100K\frac{\text{classifications}}{s}$	$\frac{31.4 \text{ pJ}}{\text{classification}}$	12
[13]	Bayes	83.90%	88.33%	91.90%	$1.94\mu W$	$100K\frac{\text{classifications}}{s}$	$\frac{19.4 \text{ pJ}}{\text{classification}}$	12
[23]	Threshold	88.30%	89.21%	93.10%	$1.09\mu W$	$100K\frac{\text{classifications}}{s}$	$\frac{10.9 \text{ pJ}}{\text{classification}}$	12
[24]	SVM	89.90%	90.21%	91.20%	$73.28\mu W$	$140K\frac{\text{classifications}}{s}$	$\frac{523.43 \text{ pJ}}{\text{classification}}$	12
[25]	Centroid	91.20%	93.28%	96.70%	$3.18\mu W$	$100K\frac{\text{classifications}}{s}$	$\frac{31.8 \text{ pJ}}{\text{classification}}$	12

REFERENCES

[1] G. Rong, A. Mendez, E. B. Assi, B. Zhao, and M. Sawan, "Artificial intelligence in healthcare: review and prediction case studies," *Engineering*, vol. 6, no. 3, pp. 291–301, 2020.

[2] H. B. Mamo, M. Adamiak, and A. Kunwar, "3d printed biomedical devices and their applications: A review on state-of-the-art technologies, existing challenges, and future perspectives," *Journal of the Mechanical Behavior of Biomedical Materials*, p. 105930, 2023.

[3] E. M. Davidson, M. T. Poon, A. Casey, A. Grivas, D. Duma, H. Dong, V. Suárez-Paniagua, C. Grover, R. Tobin, H. Whalley *et al.*, "The reporting quality of natural language processing studies: systematic review of studies of radiology reports," *BMC medical imaging*, vol. 21, no. 1, pp. 1–13, 2021.

[4] A. Abernethy, L. Adams, M. Barrett, C. Bechtel, P. Brennan, A. Butte, J. Faulkner, E. Fontaine, S. Friedhoff, J. Halamka *et al.*, "The promise of digital health: Then, now, and the future," *NAM perspectives*, vol. 2022, 2022.

[5] K. Aboumerhi, A. Güemes, H. Liu, F. Tenore, and R. Etienne-Cummings, "Neuromorphic applications in medicine," *Journal of Neural Engineering*, vol. 20, no. 4, p. 041004, 2023.

[6] W. Haensch, T. Gokmen, and R. Puri, "The next generation of deep learning hardware: Analog computing," *Proceedings of the IEEE*, vol. 107, no. 1, pp. 108–122, 2018.

[7] B. J. MacLennan, "A review of analog computing," *Department of Electrical Engineering & Computer Science, University of Tennessee, Technical Report UT-CS-07-601 (September)*, 2007.

[8] C. Bachmann, M. Ashouei, V. Pop, M. Vidojkovic, H. De Groot, and B. Gyselinckx, "Low-power wireless sensor nodes for ubiquitous long-term biomedical signal monitoring," *IEEE Communications Magazine*, vol. 50, no. 1, pp. 20–27, 2012.

[9] M. U. Anjum, A. Fida, I. Ahmad, and A. Iftikhar, "A broadband electromagnetic type energy harvester for smart sensor devices in biomedical applications," *Sensors and Actuators A: Physical*, vol. 277, pp. 52–59, 2018.

[10] [Online]. Available: https://archive.ics.uci.edu/dataset/46/hepatitis

[11] C. M. Bishop and N. M. Nasrabadi, *Pattern recognition and machine learning*. Springer, 2006, vol. 4, no. 4.

[12] D. Castaneda, A. J. Gonzalez, M. Alomari, K. Tandon, and X. B. Zervos, "From hepatitis a to e: A critical review of viral hepatitis," *World journal of gastroenterology*, vol. 27, no. 16, p. 1691, 2021.

[13] V. Alimisis, G. Gennis, C. Dimas, and P. P. Sotiriadis, "An analog bayesian classifier implementation, for thyroid disease detection, based on a low-power, current-mode gaussian function circuit," in *2021 International conference on microelectronics (ICM)*. IEEE, 2021, pp. 153–156.

[14] P. Bertsias, C. Psychalinos, A. G. Radwan, and A. S. Elwakil, "High-frequency capacitorless fractional-order cpe and fi emulator," *Circuits, Systems, and Signal Processing*, vol. 37, no. 7, pp. 2694–2713, 2018.

[15] J. Lazzaro, S. Ryckebusch, M. A. Mahowald, and C. A. Mead, "Winner-take-all networks of o (n) complexity," *Advances in neural information processing systems*, vol. 1, 1988.

[16] S. T. Chandrasekaran, R. Hua, I. Banerjee, and A. Sanyal, "A fully-integrated analog machine learning classifier for breast cancer classification," *Electronics*, vol. 9, no. 3, p. 515, 2020.

[17] S.-Y. Peng, P. E. Hasler, and D. V. Anderson, "An analog programmable multidimensional radial basis function based classifier," *IEEE Transactions on Circuits and Systems I: Regular Papers*, vol. 54, no. 10, pp. 2148–2158, 2007.

[18] K. Lee, J. Park, and H.-J. Yoo, "A low-power, mixed-mode neural network classifier for robust scene classification," *Journal of Semiconductor Technology and Science*, vol. 19, no. 1, pp. 129–136, 2019.

[19] Z. Zhao, A. Srivastava, L. Peng, and Q. Chen, "Long short-term memory network design for analog computing," *ACM Journal on Emerging Technologies in Computing Systems (JETC)*, vol. 15, no. 1, pp. 1–27, 2019.

[20] R. Zhang and T. Shibata, "An analog on-line-learning k-means processor employing fully parallel self-converging circuitry," *Analog Integrated Circuits and Signal Processing*, vol. 75, pp. 267–277, 2013.

[21] E. Georgakilas, V. Alimisis, G. Gennis, C. Aletraris, C. Dimas, and P. P. Sotiriadis, "An ultra-low power fully-programmable analog general purpose type-2 fuzzy inference system," *AEU-International Journal of Electronics and Communications*, vol. 170, p. 154824, 2023.

[22] V. Alimisis, G. Gennis, K. Touloupas, C. Dimas, M. Gourdouparis, and P. P. Sotiriadis, "Gaussian mixture model classifier analog integrated low-power implementation with applications in fault management detection," *Microelectronics Journal*, vol. 126, p. 105510, 2022.

[23] V. Alimisis, G. Gennis, E. Tsouvalas, C. Dimas, and P. P. Sotiriadis, "An analog, low-power threshold classifier tested on a bank note authentication dataset," in *2022 International Conference on Microelectronics (ICM)*. IEEE, 2022, pp. 66–69.

[24] V. Alimisis, G. Gennis, M. Gourdouparis, C. Dimas, and P. P. Sotiriadis, "A low-power analog integrated implementation of the support vector machine algorithm with on-chip learning tested on a bearing fault application," *Sensors*, vol. 23, no. 8, p. 3978, 2023.

[25] V. Alimisis, V. Mouzakis, G. Gennis, E. Tsouvalas, C. Dimas, and P. P. Sotiriadis, "A hand gesture recognition circuit utilizing an analog voting classifier," *Electronics*, vol. 11, no. 23, p. 3915, 2022.

An Adaptive Analytical FPGA Placement flow based on Reinforcement Learning

C. Barn and S. Vermeulen and S. Areibi and G. Grewal
University of Guelph, School of Computer Science/Engineering
Guelph, Canada N1G 2W1
Email: {cbarne07, sverme01, sareibi, ggrewal}@uoguelph.ca

Abstract—Maximizing solution quality and minimizing CPU time are two competing objectives in developing Field Programmable Gate Array (FPGA) placement tools. Placement tools utilize a number of optimizations that may benefit one objective while sacrificing the other. In this paper, we explore a reinforcement-learning *placement agent* for use in established FPGA placements tools to efficiently and dynamically determine which costly optimizations may be avoided and which are necessary for a given placement problem. Placement flows guided by the proposed agent achieve 8.63% average improvement in runtime and result in an average reduction in wirelength of 1.5% when deployed in a state-of-the-art academic placement tool.

Index Terms—Machine Learning; Reinforcement Learning; Sequential Decision Making; FPGA Placement

I. Introduction

Due to their versatility, high performance per watt, and fast turnaround time, Field Programmable Gate Arrays (FPGAs) are being used to implement a wide range of digital circuit designs. However, as these designs grow in size and complexity, the software tools used to map a digital circuit onto an FPGA device are experiencing increased pressure to produce high-quality solutions in reasonable amounts of time. Mapping a circuit onto an FPGA involves performing a sequence of tasks each modeled as an optimization problem. Since these problems are NP-hard, heuristic solution methods must be used that forego optimality for reduced CPU runtime. One key example is placement, which is responsible for finding the optimal location of the circuit components on the FPGA device. Academic FPGA placement tools, like those in [1] [2] [3], all perform a fixed set of optimization tasks, regardless of the features of the original design or the target FPGA device. This leads to excessive runtimes, when unnecessary (or possibly detrimental) optimizations are performed.

For example, [1] and [3] mitigate congestion by adjusting the logical size of circuit components to reduce placement density, nevertheless, the methods used to do this are unique to each tool. [1] begins placement by uniformly inflating all Look-up Tables (LUTs) and Flip-Flops (FFs) then partitions and spreads the components. The congestion of the placement is re-estimated at various points later in the flow and circuit components are re-sized accordingly. [3] does not repeatedly re-size the circuit components nor partition the netlist. Instead, the logical size of circuit components are adjusted according to a congestion estimation made after an initial placement is generated that minimizes Half-Perimeter Wire Length (HPWL).

Despite the differences in strategy, each of the tools mentioned above can produce high quality placement solutions indicating the potential benefit of optionally selecting optimizations in response to circuit characteristics and placement progress.

Machine Learning (ML), a powerful tool in pattern detection, has been proposed as a conduit for dynamic decision making problems. Reinforcement Learning (RL) is a class of ML in which an agent interacts with an environment to maximize rewards received. Focusing our work on two key optimizations, we compare RL frameworks tasked with modifying the placement flow in [1] and observing the outcome to decide mid-placement which optimizations are necessary in terms of solution quality and which can be skipped to reduce runtime. The main contributions of this paper can be summarized in the following points:

1) We formulate the process of flow selection as an RL problem to utilize the framework in state-of-the-art RippleFPGA as a method of dynamic flow selection [1].
2) We investigate three reinforcement learning algorithms using realistic benchmarks from which static features of the benchmark are retrieved and dynamic features of partial placement solutions are generated.

The remainder of the paper is organized as follows. Necessary background is introduced in Sec. II and related work in Sec. III. Section IV describes our methodology and the results are presented in Sec. V. Conclusions are summarized in Sec. VI.

II. Background

In this section, we briefly describe the architecture of the modern heterogeneous FPGA device that we target, the benchmark circuits used to train and evaluate the model, the placement problem, and the state-of-the-art RippleFPGA placement tool [1].

A. Modern Heterogeneous FPGA Architecture

The Xilinx UltraScale FPGA [4] targeted in this work is an example of a modern, heterogeneous programmable chip. Depicted in Fig. 1, this architecture features columns of configurable LUTs and FFs interleaved with columns of specialized hardware such as IO blocks, Random Access Memory (RAM), and Digital Signal Processing (DSP) blocks. LUTs and FFs are organized into *slices*, which are composed of 8 sets of one LUT and two FFs. Components of the FPGA

979-8-3503-8083-5/23 $31.00 © 2023 IEEE

are wired together using a preset network of long and short routing segments connected via re-configurable switches.

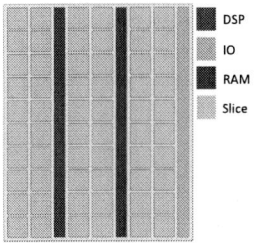

Fig. 1: Example of a Heterogeneous FPGA Architecture

B. Benchmark Circuits

In this work we use realistically sized, complex circuits that target the Xilinx Virtex UltraScale device. These benchmark circuits are sourced from the 12 provided for the 2016 ISPD contest [5] and 360 additional circuits provided by Xilinx Inc. available at [6].

C. FPGA Placement

During placement, circuit components are placed at legal locations on an FPGA such that the wirelength is minimal, timing constraints are satisfied, and/or routability is maximized. Analytic placers produce placement solutions by solving a system of equations that model the placement in terms of the chosen objective (wirelength, timing, routability, etc.). Since the objective function must be differentiable, certain constraints imposed by the architecture of the target FPGA cannot be accounted for. For example, each component must be placed in a site at the device dedicated to the component type (LUT, FF, DSP, etc.). A separate routine is invoked to relocate illegally placed components in a way that does not violate architectural constraints and minimizes degradation of results in terms of the objective function.

D. An Overview of RippleFPGA

Figure 2 details the RippleFPGA [1] placement flow that we seek to guide using RL. This tool uses an analytic solver

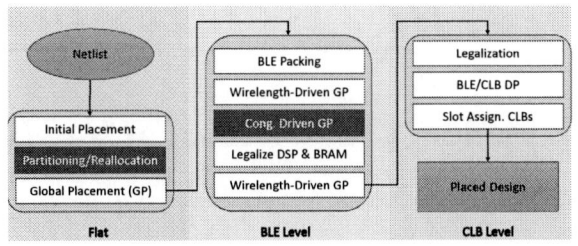

Fig. 2: The RippleFPGA Placement Flow

to minimize B2B [7] wirelength and a rough legalization technique that ensures the solution is within a range of feasibility. Global Placement (GP) iterations are combined with optimization procedures to target objectives such as minimizing wirelength and maximizing routability. To investigate

the viability of using an RL placement agent to dynamically select a placement flow, we constrain the problem to two key optimizations in the RippleFPGA [1] placement flow detailed below.

1) Partitioning and Reallocation: Xilinx UltraScale FPGAs have fewer routing segments per switch in vertical channels compared to horizontal channels. To avoid solutions that over utilize routing resources, the netlist is partitioned into clusters that maximize internal connectivity and then reallocated to distribute demand from vertical routing segments to horizontal routing segments. The netlist is modeled as a graph, $G(V, E)$, each block is an element of V and there is an edge between two elements on V if a net connects them. Recursive bi-partitioning is performed on this graph until the maximum number of elements of V in a partition is 25% of the netlist or the minimum number of edges between any two partitions is greater than 5% of the edges in the graph. Clusters are stacked vertically and spread across the width of the FPGA whilst maintaining the relative ordering of clusters and blocks within a cluster.

2) Congestion-Driven GP: The routing resource requirements in highly populated regions of the placement can easily exceed the fixed specifications of the FPGA fabric and cause a placement to be unroutable. Congestion is estimated as the relative wirelength per switch with adjustments made to account for net fan-out. The size of blocks in highly congested regions are inflated to cause blocks to be placed further apart, while cells in regions of lower congestion are deflated to improve wirelength.

At present, [1] applies both previous optimizations to all placement problems irrespective of the specific circuit features. This strategy may benefit some placement problems while negatively impacting the quality-of-result and runtime for others. The potential to improve both runtime and quality-of-result serves as a driving force for incorporating RL within RippleFPGA. This would enable the determination of whether to perform none, one, or both optimizations based on the circuit's features.

III. PREVIOUS WORK

Work appearing in the literature has demonstrated advantages gained by guiding FPGA placement tools via integrated ML frameworks. [8] presents an efficient ML model used to predict effective placement tool parameter settings based on features derived from the circuit netlist. In [9], an adaptive decision making ML model is trained to select the most appropriate wire-length model to use during placement stages. Results obtained demonstrate that the wire-length models predicted improve HPWL and decrease runtime when compared to the default wire-model. An efficient tabular RL approach is presented in [10] that does not require offline training yet achieves a significant reduction in runtime while preserving solution quality. Work in [11] presents an ML placement adviser that uses static netlist features to determine a placement flow in RippleFPGA. The results presented in [11] show that tailoring a placement flow to individual circuits can improve HPWL and reduce runtime. Our proposed work in this paper differs from the previous works in the respect that it is the first work to explore the use of RL as a smart decision making tool that leverages both static netlist features and characteristics of intermediary placement solutions to dynamically derive the most effective optimization strategy.

IV. METHODOLOGY

A. Decision-Making Problem

RL is a type of machine learning that involves an agent learning to make decisions by interacting with an environment. In the context of the problem described in Sec. II-D, RL is used to train an agent to determine the best optimization strategy for a given placement problem based on static features of the circuit and features of the current placement. The agent then receives feedback in the form of rewards or penalties based on the quality-of-result achieved by applying different optimization strategies. Over time, the agent learns to make better decisions, leading to improved runtime and quality-of-result for the placement problem.

In this work we explore the decision to perform *both*, *none*, and *one* of Partition/Reallocation and Congestion-Driven GP. Before placement begins, it is not yet clear if the early placement steps sufficiently address congestion as the only information available to the agent comes from the static netlist features. Rather than relying on this information alone, we decouple and delay the decision to perform each optimization to allow the agent to leverage the most recent placement characteristics in addition to netlist features. To accomplish this, we interrupt placement at two points as shown in Fig. 3, once before Partition/Reallocation and once before Congestion-Driven GP. At each interruption, dynamic features of the current placement are surveyed and combined with the static netlist features; the RL agent is invoked to decide whether to perform or skip the next optimization based on these features and previous rewards with the goal of maximizing future rewards.

Formally, an RL task is modeled as a Markov Decision Process (MDP). An MDP is a 3-tuple (S, A, R) where S is a set of states defined by the environment, A is a set of actions that dictate the interaction between the agent and the environment, and R is a reward function. At time $t \in 0..T$ where T is the time a terminal state is reached, s_t is the current state, a_t is the action taken and r_t is the reward received.

We model this decision-making problem in terms of an RL task, (S, A, R), as follows:

- $S \subseteq \mathbb{R}$: States that describe the environment derived from 11 static circuit features, extracted from the circuits netlist; 5 dynamic features, generated during the placement flow; and the current stage of placement.
- $A = \{0, 1\}$: Actions, *1 (perform optimization)*, and *0 (do not perform optimization)*.
- R: $S \to \mathbb{R}$, Reward function.

In the subsequent subsections, we present three potential RL algorithms that progress from the most straightforward to the most advanced.

B. Q-Learning

To explore the efficacy of simple RL in the context of this problem, we implement the Q-Learning algorithm. Q-Learning is an iterative approach to estimating the value of state-action pairs, denoted $Q(s, a)$, in terms of the sum of future rewards that an agent can expect to receive if action a is executed in the state s. The agent uses these values to determine the next action to take in the current state s_t. We select the next action according to a Softmax function, $softmax(a_t) = \frac{e^{Q(a_t)}}{\sum_{\forall a \in A} e^{Q(a)}}$, as a result, the probability of an action being selected is proportional to the state-action value. Q-Learning algorithms typically store and update $Q(s, a)$ values in an $|S| \times |A|$ table, therefore, the state-action space must be finite and discrete. Since the features in this work (discussed in Sec. II) are continuous, we first reduce and discretize the state space by assigning each feature within a range of values to a bin represented by an integer $[0...n]$ where n is the number of bins. The features used to derive state and the bin boundaries are detailed in Tab. I.

TABLE I: Q-Learning Features

Dynamic Features	Range	Step size
HPWL	1×10^5 - 1×10^9	powers of 10
90th Percentile Congestion	0.2 - 1	0.2
Static Features	**Range**	**Step size**
SUM(FDRE,LUT2-6)	1×10^5 - 1×10^6	1×10^5

The value of each state-action pair is initialized to 0, and updated as shown in Alg. 1 according to the function

$$Q(s_t, a_t) = Q(s_t, a_t) + \alpha(r_{t+1} + \gamma max_a Q(s_{t+1}, a) - Q(s_t, a_t)) \quad (1)$$

where s_t is the current state, a_t is the action taken, r_{t+1} is the reward received, and s_{t+1} is the next state given s_t and a_t. $max_a Q(s_{t+1}, a)$ is the maximum expected future

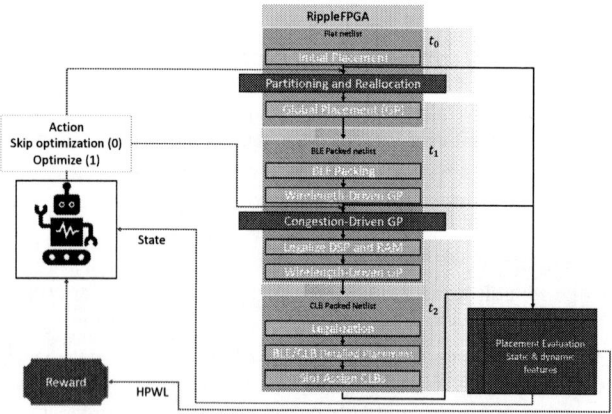

Fig. 3: RL Sequential Decision Making in RippleFPGA [1]

reward. α is a parameter representing the learning rate and γ discounts future rewards. Small values of γ emphasize short terms rewards.

Algorithm 1 Q-Learning Algorithm

Require: Initial q-table, Q
Require: Learning rate α, discount factor γ
1: **for** $t = 0, 1, 2, \ldots$ **do**
2:    **for** $s \in S$ **do**
3:      $a_t = softmax(Q(s_t))$
4:      take action a_t, observe r_{t+1}, s_{t+1}
5:      $Q(s_t, a_t)$ updated as shown in Equation 1
6:    **end for**
7: **end for**

C. Policy Gradient RL

In the context of RL, a *policy* is a function $\pi : A \to [0, 1]$ that returns the probability with which an action is selected. In this section, the policy is approximated using a neural network to accommodate the continuous state space. The policy used in our Q-Learning algorithm, the softmax function, was indirectly improved as the agent gained experience and updated state-action pair values. In contrast, this class of RL iteratively improves by making adjustments based on the gradient of the policy function.

The policy gradient is determined from a sample of trajectories of the form $\tau_i = (s_t, a_t, r_{t-1}, s_{t-1}, a_{t-1}, r_T, s_T)$ where s_T is a terminal state indicating the end of placement and a_t, a_{t+1} are selected under the current policy. Updates to the policy network are performed using stochastic gradient descent. We explore two policy gradient methods: REINFORCE as described in [12] and PPO, a constrained policy gradient method introduced by Schulman et al. in [13].

1) REINFORCE: As shown by Alg. 2, REINFORCE estimates the gradient of the policy network as $G_t \nabla \log \pi(a_t|s_t)$, where G_t is the future return at time t.

Algorithm 2 REINFORCE with Baseline Algorithm

Require: Initial policy function parameters, θ_k
Require: Learning rate α
1: **for** $k = 0, 1, 2, \ldots$ **do**
2:    Generate a trajectory $s_0, a_0, r_{-1}, s_{-1}, a_{-1}, r_T$ following policy π_{θ_k}
3:    **for** $t = 0, 1, \ldots, T - 1$ **do**
4:      $G_t = \sum_{k=t+1}^{T} R_k$
5:      $\theta_{k+1} = \theta_k + \alpha G_t \nabla \log \pi(a_t|s_t, \theta_k)$
6:    **end for**
7: **end for**

---

2) PPO: As shown by Alg. 3, PPO reduces the variance by preventing large policy updates when performing stochastic gradient descent. This is done by using the following clipped objective function

$$L(\theta) = \hat{\mathbb{E}}_t \left[\min(r_t(\theta)\hat{A}_t, \text{clip}(r_t(\theta), 1 - \epsilon, 1 + \epsilon)\hat{A}_t) \right], \quad (2)$$

Algorithm 3 Clip Proximal Policy Optimization (PPO-Clip) Algorithm

---

Require: Initial policy function parameters, θ_k
Require: Initial value function parameters, ϕ_k
1: **for** $k = 0, 1, 2, \ldots$ **do**
2:    Collect trajectories $\mathcal{D}_k = \{\tau_i\}$ from environment following policy π_{θ_k}
3:    Compute future reward, $\hat{R}_t = \sum_{t'=t}^{T} r_{t'}$
4:    Compute advantage estimates,
    $\hat{A}_t \leftarrow -V_{\phi_k}(s_t) + r_t + \gamma V_{\phi_k}(s_{t+1}) + \ldots + \gamma^T V_{\phi_k}(s_T)$
5:    Update policy function by maximizing the clipped objective function (2) using the Adam optimizer
6:    Fit value function using Adam optimizer on mean-squared error:

$$\phi_{k+1} = \arg\min_{\phi} \frac{1}{|\mathcal{D}|} \sum_{\tau \in \mathcal{D}_k} \sum_{t=0}^{T} \left(V_{\phi}(s_t) - \hat{R}_t \right)^2$$

7: **end for**

---

where $r_t(\theta) = \frac{\pi_\theta(a_t|s_t)}{\pi_{\theta_{\text{old}}}(a_t|s_t)}$, ϵ is a hyperparamter that controls how conservative the clipping function is, and $\hat{A}_t$ is an estimate of the advantage. The advantage of a state-action pair is the difference between the value of the state-action pair, and the estimated value of the state. If $r_t(\theta) < 1 - \epsilon$ and $\hat{A}_t < 0$ then the min function returns $(1 - \epsilon)\hat{A}_t$ to prevent extreme decreases in the probability of an action. Similarly if $r_t(\theta) > 1 + \epsilon$ and $\hat{A}_t > 0$ then the clip function returns $(1 + \epsilon)\hat{A}_t$ to prevent extreme increases in the probability of an action.

D. Features

Table II summarizes the 11 static features derived from the benchmark circuit netlist and five dynamic features gathered during the placement flow used to develop the models. Features are individually normalized to a range of [0,1]. The static features include the total number of FFs, LUTs (2-6 pins), DSPs, RAMs, input buffers, output buffers and pins. These features denote the size and composition of the netlist.

To obtain the dynamic features, 369 benchmark circuits are placed with four placement flows derived from placement flow in [1]. Placement flows differ in the optimizations executed, that is *both*, *none*, or *one* of Partition/Reallocation and Congestion-Driven GP. The dynamic features change as the placement flow progresses and denote the utilization of the FPGA, the congestion, and the estimated wirelength. The utilization is measured using the number of occupied sites on the FPGA. Minimum, maximum, and 90th percentile congestion values describe the congestion of the placement. 90th percentile congestion is the congestion value at which the congestion level of 90% of sites on the FPGA are less than or equal to. Estimated wirelength is recorded as the sum of the HPWL of each net in the netlist. The HPWL of a net is half of the perimeter of the smallest rectangle that encloses it.

979-8-3503-8083-5/23 $31.00 © 2023 IEEE

TABLE II: List of Features

#	Dynamic Features	Calculation
1	HPWL	$\sum_{net \in E} HPWL_{net}$
2	Sites	$|SI_d|$
3	Min Congestion	$min(Cong_{sw}|sw \in SW_d)$
4	Max Congestion	$max(Cong_{sw}|sw \in SW_d)$
5	90th Percentile Congestion	$Cong_{sw}|Cong_{sw} = P_{90} \wedge sw \in SW_d$
#	Static Features	Calculation
6	FDRE	$|SI_d \cap SI_{FDRE}|$
7-11	LUT2-6	$|SI_d \cap SI_{LUTi}|, i = 2..6$
12	DSP48E2	$|SI_d \cap SI_{DSP48E2}|$
13	RAMB32E3	$|SI_d \cap SI_{RAMB32E3}|$
14	IBUF	$|SI_d \cap SI_{IBUF}|$
15	OBUF	$|SI_d \cap SI_{OBUF}|$
16	# pins	$\sum_{net \in E} fanout_{net}$

Note: SW_d and SI_d represent the set of switches and sites used by the design respcetively, SI_i represents the sites of type i

E. Reward Signal

In FPGA placement, HPWL is a common metric to target in optimization and to evaluate placement solutions relative to those produced by other tools as it offers a lower bound of the final routed wirelength. Since the distribution of HPWL across benchmarks is very large, using negative HPWL alone is insufficient as larger netlists will have a larger effect on model. To combat this bias, we choose the reward to be relative to change in HPWL. HPWL is noted at 3 points in the flow: immediately before Partitioning/Reallocation ($HPWL_0$), immediate before Congestion-driven GP ($HPWL_1$) and at then end of placement ($HPWL_2$). The reward at time $t + 1$ is calculated as $1 - \frac{HPWL_{t+1}}{HPWL_t}$.

F. Hyper-parameter Tuning

Hyper-parameter tuning is an important step in ML that is used to improve accuracy by reducing over/under fitting. We conducted an exhaustive search of the parameter space defined by the range of feasible values for each hyper-parameter as determined by the problem at hand and the algorithm specification. Hyper-parameters were evaluated on 15% of the data held out from training and testing.

Hyper-parameters common to Q-Learning, REINFORCE, and PPO include γ, a value used to discount future rewards, and α which is the learning rate of the RL agent. In addition, we tuned the learning rate of the advantage network (λ), ϵ discussed in Sec. IV-C2, and $\epsilon_{entropy}$ which is used to increase the rate of exploitation in the PPO implementation. The tuned values of these hyper-parameters are show in Tab. III

TABLE III: Hyper-Parameters

Algorithm	Q-Learning	REINFORCE	PPO
γ	1	1	1
α	0.001	0.001	0.95
ϵ	-	-	0.01
$\epsilon_{entropy}$	-	-	0.001
λ	-	-	0.0001

V. RESULTS

We evaluate the proposed RL application in terms of the model performance, reduction in runtime, and HPWL.

A. Model Performance

Table IV shows the results for the model trained using Q-Learning, REINFORCE and PPO. A reasonably high accuracy, between 0.79 and 0.82, is achieved by each model in choosing to perform Partition/Reallocation. The accuracy of REINFORCE drops significantly to 0.3750 when selecting the second action, whether or not to perform Congestion-Driven GP. The accuracy of the Q-Learning and PPO algorithms in selecting the second action is within ± 0.03 of the accuracy in selecting the first action. Q-Learning does not require

TABLE IV: RL Overall Performance

Algorithm	Accuracy	Matthew's Coef	Precision	Recall	F1-Score
Partition and Reallocation					
Q-Learning	0.7917	0.5612	0.7714	0.9310	0.8438
REINFORCE	0.8125	0.6149	0.7778	0.9655	0.8615
PPO	0.8125	0.6149	0.7778	0.9655	0.8615
Congestion-Driven Global Placement					
Q-Learning	0.7708	0.5475	0.7200	0.8182	0.7660
REINFORCE	0.3750	-0.2414	0.3889	0.6364	0.48285
PPO	0.8333	0.6690	0.7917	0.8636	0.8261
Overall Flow					
Q-Learning	0.7812	0.5544	0.7457	0.8746	0.8049
REINFORCE	0.5938	0.1867	0.5833	0.8009	0.6721
PPO	0.8229	0.6419	0.7847	0.9146	0.8438

approximations of key functions but constrains the problem to a discrete space, resulting in unavoidable loss of information. REINFORCE, unlike PPO, does not restrict the change in policy between training steps. Therefore, REINFORCE will often overfit to policies that result in extreme rewards. To demonstrate the difference between REINFORCE and PPO, we performed 2500 updates to each policy (see loss functions used to update in Alg. 2 and Alg. 3) and plot the associated loss values in Fig. 4. The large fluctuations in the REIN-FORCE loss values represent large changes in the policy. Average inference times for each model are shown in Tab. V.

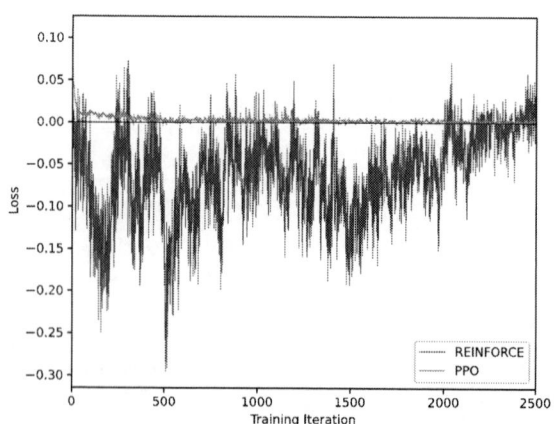

Fig. 4: Loss values over 2500 training iterations

TABLE V: Model Inference Time

Algorithm	Q-Learning	REINFORCE	PPO
Inference Time (s)	0.0015	0.0008	0.0014

B. Placement Improvement

To assess the performance of the model in terms of the placement problem, we deploy PPO in the modern placement tool RippleFPGA [1]. As shown in Tab. VI, by avoiding costly optimizations the runtime of the placement tool is reduced as much as 30% when guided by the PPO RL model. On average, the runtime of the placement tool is reduced by $> 9\%$ when selecting the optimizations that result in a minimum HPWL and $> 8\%$ when performing the optimizations dictated by the PPO model. Table VI shows that the reduction in runtime

TABLE VI: Runtime and Wirelength Improvement

Alg	Runtime				Wirelength			
	Min	Mean	Std Dev.	Max	Min	Mean	Std Dev.	Max
Best	-1.81%	9.27%	7.44%	23.05%	0.00%	2.39%	2.80%	12.15%
Q-L	-2.40%	8.68%	8.79%	32.60%	-7.24%	1.23%	3.29%	12.15%
RE.	0.00%	4.90%	8.73%	30.43%	-6.87%	1.01%	2.88%	12.15%
PPO	0.00%	8.63%	8.01%	30.43%	-6.87%	1.50%	2.99%	12.15%

achieved by the PPO model does not result in a significant decrease in solution quality, and on average the HPWL is improved by 1.5%.

C. Performance by Benchmarks

Twelve categories of the benchmark circuits (FP01 - FP12) can be formed by grouping the circuits based on similarity in the number of blocks of each type, control sets, and the Rent exponent. The percentage of successful predictions in each benchmark category is shown in Tab. VII. In all clusters other than FP08, PPO performs equally as well or better than Q-Learning and REINFORCE. HPWL is successfully reduced by the PPO agent in greater than 75% of benchmarks in 7 of 12 categories. The average percent improvement by benchmark

TABLE VII: Benchmarks: Successful Predictions

Alg.	FP01	FP02	FP03	FP04	FP05	FP06	FP07	FP08	FP09	FP10	FP11	FP12
Q-L	75%	75%	50%	100%	50%	75%	50%	50%	50%	50%	75%	25%
RE.	0%	25%	50%	0%	50%	75%	0%	0%	0%	75%	50%	25%
PPO	100%	75%	50%	100%	50%	75%	75%	25%	50%	75%	75%	25%

category is shown in Tab. VIII. The solution quality of 11/12 benchmark categories was either improved or unchanged. In the case of FPGA06, all but one benchmark circuit in the category achieve the best HPWL when both optimizations are performed and therefore the possible improvement for those circuits is 0.0%.

TABLE VIII: Wirelength Improvement By Benchmark

Alg.	FP01	FP02	FP03	FP04	FP05	FP06	FP07	FP08	FP09	FP10	FP11	FP12
Best	6.21%	3.90%	0.21%	1.36%	5.83%	1.16%	2.24%	2.28%	2.08%	0.10%	1.04%	2.32%
Q-L	6.21%	3.87%	-1.81%	1.36%	3.59%	0.00%	0.72%	-0.09%	0.65%	-0.64%	0.88%	0.00%
RE.	5.33%	3.17%	0.00%	0.00%	3.59%	0.00%	0.00%	0.00%	0.00%	0.00%	0.00%	0.00%
PPO	6.21%	3.87%	0.00%	1.36%	3.59%	0.00%	1.53%	-0.09%	0.65%	0.00%	0.88%	0.00%

D. Summary of Results

The main points regarding RL algorithm performance are:

1) PPO, an algorithm that avoids the loss of information that can occur as a result of a continuous state space and limits large policy updates, has less variability in training than REINFORCE and achieves the greatest accuracy compared to Q-Learning and REINFORCE,

2) PPO successfully selects the best overall optimization strategy in terms of HPWL for at least 75% of circuits in 7/12 benchmark categories which is equal to or greater than the results of REINFORCE and Q-Learning,

3) The optimization strategies selected by PPO results in an average reduction in HPWL for 11/12 benchmark categories.

VI. CONCLUSIONS

This paper presented an RL model used to advise an analytic FPGA placer in selecting between different possible optimization strategies. We conclude that, when deployed in a modern placement flow, the RL model accurately suggests optimizations for individual benchmarks that result in an average reduction in runtime of 8.63% without degrading solution quality.

REFERENCES

[1] G. Chen et al., "RippleFPGA: Routability-Driven Simultaneous Packing and Placement for Modern FPGAs," *IEEE Transactions on CAD and Systems*, vol. 37, no. 10, pp. 2022–2035, October 2018.

[2] W. Li, S. Dhar, and D. Pan, "UTPlaceF: A Routability-driven FPGA Placer with Physical and Congestion Aware Packing," *IEEE Transactions on CAD and Systems*, vol. 37, no. 4, pp. 869–882, 2018.

[3] Z. Abuowaimer et al., "GPlace3.0: Routability-Driven Analytic Placer for UltraScale FPGA Architectures," *ACM Trans. on Design Automation of Electronic Systems*, vol. 23, no. 5, pp. 1–33, Sept. 2018.

[4] Xilinx, ""UltraScale Architecture Configurable Logic Block User Guide"," http://www.xilinx.com/support/documentation/user_guides/ug574-ultrascale-clb.pdf, 2017, online; Accessed 23 October 2019.

[5] ——, "ISPD 2016 Routability-Driven FPGA Placement Contest," 2016, [accessed 2017-03-17].

[6] G. Grewal and S. Areibi, "Guelph FPGA CAD Group," https://fpga.socs.uoguelph.ca/.

[7] M.-C. Kim, D.-J. Lee, and I. L. Markov, "SimPL: An Algorithm for Placing VLSI Circuits," *Communications of the ACM*, vol. 56, no. 6, pp. 105–113, 2013.

[8] G. Grewal et al., "Automatic Flow Selection and Quality-of-Result Estimation for FPGA Placement," in *IEEE Reconfigurable Architectures Workshop (RAW)*, Orlando, Florida, USA, May 2017, pp. 115–123.

[9] T. Martin et al., "An Adaptive Sequential Decision Making Flow for FPGAs using Machine Learning," in *IEEE Int'l Conference on Microelectronics*, Morocco, December 2022, pp. 1–4.

[10] M. A. Elgammal, K. E. Murray, and V. Betz, "Rlplace: Using reinforcement learning and smart perturbations to optimize fpga placement," *IEEE Transactions on Computer-Aided Design of Integrated Circuits and Systems*, vol. 41, no. 8, pp. 2532–2545, 2022.

[11] T. Martin et al., "FPGA Placement: Dynamic Decision Making Via Machine Learning," in *36th Symposium on Integrated Cicuits and Systems Design (SBCCI)*, Brazil, August 2023, pp. 1–6.

[12] R. S. Sutton, F. Bach, and A. G. Barto, *Reinforcement learning: An introduction.* MIT Press Ltd, 2018.

[13] J. Schulman, F. Wolski, P. Dhariwal, A. Radford, and O. Klimov, "Proximal policy optimization algorithms," *CoRR*, vol. abs/1707.06347, 2017. [Online]. Available: http://arxiv.org/abs/1707.06347

Supporting Dynamic Control-Flow Execution for Runtime Reconfigurable Processors

Hassan Nassar, Rafik Youssef, Lars Bauer, Jörg Henkel

Chair for Embedded Systems, Karlsruhe Institute of Technology, Germany
hassan.nassar@kit.edu

Abstract—As the need for more computing power grows, traditional methods are hitting limits. To boost performance, we're expanding Central Processing Unit (CPU) capabilities and using specialized hardware accelerators. For example, mobile devices usually have cameras, video encoding, and audio accelerators. To perform the different tasks, these accelerators execute microcode programs. These accelerators, however, take up space and often sit idle. Reconfigurable processors offer a solution. They have a normal core connected to several accelerator slots. These accelerator slots can be filled during runtime to accommodate the application running. Once one application finishes and another application is running, the accelerators can be switched. For example, playing music after using the camera.

In this work, we introduce dynamic control-flow execution for the microcode of runtime reconfigurable processors, i.e., support for loops, conditional jumps, and exception handling. We benchmark using four different applications from four domains (object detection, ocean movement simulation, artificial intelligence and security) that all are compute-intensive and would require the dynamic control-flow when executed on reconfigurable processors. We show that the dynamic control-flow allows different applications to be executed with significant speedup in comparison with execution on general-purpose processors.

I. INTRODUCTION

The conventional methods for enhancing processors have hit their technological limits, necessitating fresh approaches to meet the increasing demands of applications. One such approach is the Application-Specific Instruction Set Processor (ASIP), which enables a higher level of parallelism. These processors are fairly basic, featuring a reduced instruction set architecture (RISC). There is also the option to introduce instruction set extensions or incorporate special instructions (SIs) to enhance their functionality. These SIs, much like Application-Specific Integrated Circuits (ASICs), boast unique hardware implementations that enable them to achieve greater parallelism [1].

Field Programmable Gate Arrays (FPGAs) offer a very attractive solution for hardware acceleration. FPGAs consist basically of Look Up Tables (LUTs) which are able to implement basic logic gates. By interconnecting the LUTs and usage of registers they can implement anything from simple logic circuits up to processors and even more complex hardware. Therefore, FPGAs are used in a wide range of applications,

This work was partially funded by the "Helmholtz Pilot Program for Core Informatics (kikit)" at Karlsruhe Institute of Technology. The authors would like to thank Niklas Lorenz, Sascha Herring, and Andreas Bühner for their help in implementing the accelerators.

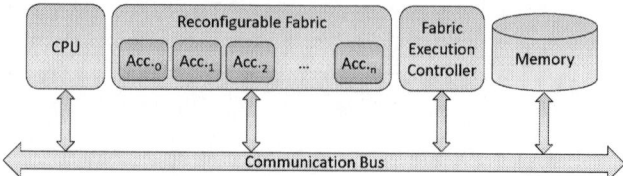

Fig. 1: Target Reconfigurable Processor Architecture. A main CPU core is connected to a reconfigurable Fabric. The execution of the accelerators residing on the fabric is controlled by a controller executing microcode.

e.g., digital signal processing, cloud computing, and machine learning acceleration [2]–[5].

By combining CPUs performing SIs with FPGAs and employing reconfiguration or alternative configuration streams, the hardware implementation can be modified, potentially even during runtime [6]. This capability paves the way for the exploration of novel strategies and facilitates the development of reconfigurable systems. This is especially relevant for resource-constraint devices, e.g., mobile devices where the resource allocation can be challenging. Runtime reconfiguration gives the chance to load the accelerators only when needed and to switch from one accelerator to another whenever needed.

A typical reconfigurable processor would look as shown in Figure 1. A CPU is connected to a Reconfigurable Fabric which in itself is divided into several accelerator slots. Each accelerator slot can be used to load the different accelerators at runtime. The accelerators execute the SIs which consist of microcode performed by the accelerator. A fabric execution controller controls the execution of the SIs [1], [6].

Applications accelerated on such systems can be rather complex. Therefore, the microcode executed needs to be able to fulfill all needed control steps of the application. Such applications might need jumps, conditional loops that cannot be unrolled, etc. Therefore, in this work, we tackle the issue of having a dynamic control flow for reconfigurable processors. Our novel contributions are as follows:

- We introduce dynamic control flow execution support to accelerators executing SIs of reconfigurable Processors
- We analyze several applications on how to make them benefit from the dynamic control flow execution
- We show the timing improvement in executing these applications on hardware utilizing the dynamic control flow

The rest of the paper continues as follows. The background needed is given in Section II. We present our implementation

979-8-3503-8083-5/23 $31.00 © 2023 IEEE

for the dynamic execution in Section III. The benchmark applications are detailed in Section IV. We evaluate our results in Section V and draw conclusions in Section VI.

II. BACKGROUND

Reconfigurable Processors are continuously gaining attraction. From academia, several architectures have been proposed, e.g., RISPP [1], *i*-Core [6], Molen [7], and KHARISMA [8]. From industry, the Zynq MPSOC from Xilinx-AMD combines embedded ARM processing system (PS) with a reconfigurable programmable logic (PL) [9]. Similarly, Intel provides the Intel Stratix SoCs with similar architecture [10].

The reconfigurable fabric of reconfigurable processors is usually divided into several accelerator slots as Figure 1 shows. The reconfigurable fabric can then be used by the applications running on the GPP to accelerate their hotspots and guarantee that they meet their timing requirements [11]. This is advantageous as for out-of-order processors which can also increase the performance, guarantees for meeting timing requirements are not provided [11].

The accelerators residing on the reconfigurable fabric execute SIs consisting of a microcode. It consists of the successive steps that have to be executed for the accelerator to complete its intended task. The microcode of the SI is usually given to the reconfigurable fabric in VLIWs [1], [6], [12]. Each VLIW consists of several sub-instructions with each sub-instruction controlling the execution on one of the accelerators. The reconfigurable fabric can even be shared between multiple GPPs [13]. If no conflict occurs, e.g., two applications requiring different sets of accelerators, the VLIWs for two or more SIs can be merged together and the performance on several cores will be enhanced.

When reconfigurable processors execute Special Instructions (SIs), the control flow is defined in a static manner. The VLIWs are processed sequentially, one after another, with no opportunity for the loaded accelerators to influence the program flow. At maximum, loops that can be unrolled will be unrolled and executed [12]. To address this constraint, we develop a dynamic execution control system for these special instructions. *To the best of our knowledge, our work is the first to tackle the need and the implementation of dynamic execution of reconfigurable processors.*

III. DYNAMIC CONTROL-FLOW SUPPORT

Several features are needed to support the acceleration of more complex algorithms. Firstly, there is a need to introduce support for jumps within the SIs. These jumps can function as conditional statements, enabling the creation of loops within the execution flow. Secondly, the inclusion of exception-handling mechanisms is essential. This is required not only to identify errors in the program flow but also to handle errors originating from the accelerators.

To incorporate the necessary functions, it is imperative to introduce new commands into the existing VLIWs. Each VLIW serves as a map for commands related to all components within the fabric, including the control logic. All the parameters

related to jumps, along with their predefined structure, are established in advance using distinct commands. To enable the implementation of concepts like nested loops within the SI, there are four separate sets of parameters that are internally reserved by the controller. Consequently, a jump command encapsulates the entire parameter set, requiring only 2 bits to represent it. Each of these parameter sets includes a counter spanning 12 bits in length. This choice balances command length and the range of values it can accommodate.

Our implementation offers three distinct categories of jump commands. The first group comprises static commands that are executed unconditionally. Within this group, we have "No Jump," which signifies that no jump should occur. Notably, this command doesn't require a parameter set. Additionally, the static commands include "Always Jump," an instruction that unconditionally redirects execution to its specified destination.

The second category encompasses jumps whose conditions are tied to one of the four counters. These jumps, in addition to the parameter set, require an operand that is compared with the counter to determine whether the jump occurs. The final category comprises conditional jumps, where the conditions are associated with a 2-bit control signal from one or more accelerators. These jumps will only occur if the control signals of all selected accelerators meet the specified criteria. In addition to the parameter set and the operand, the selection of the accelerators is also required for these jumps. It's important to note that all conditional jumps, whether they rely on the counter or control signals, are available in four distinct variants. Table I shows all the newly added jump instructions.

We assume that the underlying CPU comes equipped with exception-handling capabilities. Therefore, we use this to create our support for exceptions. The first group of traps is logic related. An invalid jump target, that falls outside its permissible boundaries triggers a trap event. If one of the accelerators runs into an error, e.g., returns a NaN, it triggers a trap event. In cases where a VLIW requires 512 clocks or more due to a set stall signal, an abort and trap event will be triggered. It's worth noting that this threshold for clock cycles can be adjusted if needed. Figure 2 shows how the stall signal support is implemented. As long as the data is not available, the stall is held and the same VLIW stalls. Once the data is there, the VLIWs continue execution on the reconfigurable fabric. If the stall limit is passed, the trap event will be raised to signal the exception.

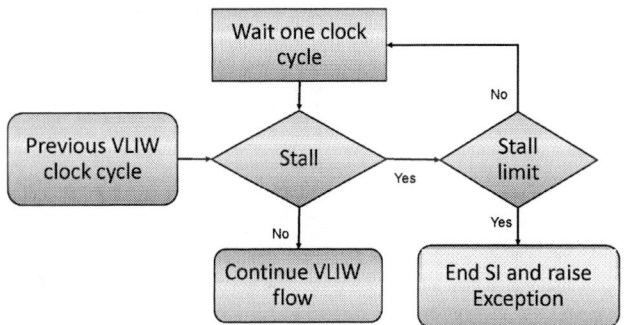

Fig. 2: Exception Support for the Dynamic Control-Flow

TABLE I: Dynamic Execution jump sub-instructions of the SI VLIW. The different jumps allow it to go dynamically within the microcode based on accelerator output or user-defined counter values. Hence, more complicated algorithms can be accelerated on the reconfigurable fabric.

Instruction	Arguments	Explanation
NO_JMP	none	don't jump
ALW_JMP	destination	jumps always to destination
JMP_IF_CNT_EQ	destination, value	jump to destination if counter equals value
JMP_IF_CNT_NEQ	destination, value	jump to destination if counter does not equal value
JMP_IF_CNT_LT	destination, value	jump to destination if counter is less than value
JMP_IF_CNT_GT	destination, value	jump to destination if counter is greater than value
JMP_IF_ACC_EQ	destination, value, accelerator slot	jump to destination if output from accelerator equals value
JMP_IF_ACC_NEQ	destination, value, accelerator slot	jump to destination if output from accelerator does not equal value
JMP_IF_ACC_LT	destination, value, accelerator slot	jump to destination if output from accelerator is less than value
JMP_IF_ACC_GT	destination, value, accelerator slot	jump to destination if output from accelerator is greater than value

The second group of traps is user-specific. Users can initiate these traps via a command within the VLIW based on the logic of the algorithm accelerated, e.g., the output is negative where output should be positive only. To achieve this, 3 bits are included in the VLIW which determines the specific trap value.

IV. BENCHMARKING APPLICATIONS

To show the benefit of the dynamic execution support for reconfigurable processors, we develop accelerators and SIs for four different applications. Then we highlight the need for dynamic execution within each SI and how does it benefit from it.

A. SIFT

David G. Lowe introduced the Scale-Invariant Feature Transform (SIFT) algorithm for image recognition, aiming to enhance various applications such as tracking, object detection, 3D image matching, and mobile robots [14]. The SIFT algorithm has different steps. One of these steps is the SIFT-match step. In this step, the features extracted from the object are compared to a reference features value. Based on the Euclidean distance between the extracted and reference values, the decision on whether objects match or not is taken. This comparison follows Eq. (1). a is the reference object, b is the detected object, $d(a, b)$ is the euclidean distance between both objects, a_i is the i-th feature of object a, b_i is the i-th feature of object b, and n is the total number of features.

$$d(a,b) = \sum_i^n (a_i - b_i)^2 \quad (1)$$

We develop the SIFT-match-SI, a Special Instruction specifically designed to perform the object recognition matching kernel for the SIFT algorithm. In this process, the typical operation involves subtracting two single-precision values and then squaring the result. This operation is performed for each of the values within the feature vectors. After executing the Siftmatch-SI, the sum of all squared results is presented to the Integer Pipeline as a single numerical value. To expedite the computation, it is divided into four accelerators to parallelize the execution as shown. Each accelerator first performs the subtraction of the two values for the features and then squares the output to match the equation.

Figure 3 shows the design of the accelerator. It contains two floating point units. One is dedicated for addition and the other is dedicated for multiplication. The control units and the muxes control which unit is enabled an which input is given either from the accelerator input or from the data saved. The registers are used to provide the correct inputs, as well to save the accumulated value till the calculation within the accelerator is finalized.

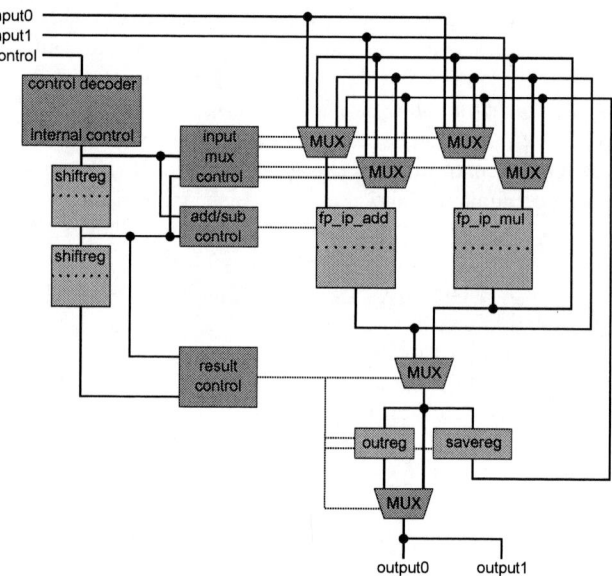

Fig. 3: Accelerators design executing the SIFT-match SI

Benefit to SIFT from Dynamic Execution: The features vector can have arbitrary length, moreover, based on different constraints on the system a reduced version with a lower number of features might be executed. Having dynamic execution, where the number of features can be set dynamically at runtime enables this. Otherwise, the SI will only be able to support a fixed number of features.

B. SWE

The Shallow Water Equations are a system of hyperbolic differential equations that describe the behavior of incompressible fluids, such as water, in scenarios where the horizontal dimensions are significantly smaller than the depth. These equations are derived from the Navier-Stokes equations and are typically applied to model fluid flow in 2D domains divided into cells using a consistent rectangular grid. To simulate the flow over discrete time intervals, Riemann problems are solved at the cell edges. The state of the simulation grid in the subsequent time step is then calculated using the solutions obtained from these edge-local Riemann problems [15].

The Riemann equation involved in this process is quite lengthy and complex, making it an excellent candidate for testing the dynamic control. The initial solver used for this purpose was the FWave solver [15], which provides a straightforward Riemann solution with reasonable performance. However, it has a notable limitation—it can only model bodies of water and cannot represent features like shorelines because it lacks the capability to simulate the wetting or drying of cells. Therefore, the HLLE solver which is more complex but also more general is also needed [15].

We build the SWE-SI which is able of performing the calculations from both solvers. For this, we develop four different accelerators. One accelerator performs the addition, multiplication, and subtraction of floating points (very similar to the accelerator from Figure 3), another performing floating point division, a third performs square root calculation, and one does utility services like calculating max and min, and so on. For the final design at the reconfigurable fabric, we use two FMAV accelerators to deal with the several additions, multiplications, and substractions. Then we have one accelerator from each type. to do the different calculations.

Figure 4 shows the design of the utility accelerator as an example of the accelerators developed. We have one floating point comparator unit and one unit that calculate absolute value of floating point numbers. Inputs are given to the units directly from the inputs of the accelerators. The control units and muxes decide which output from both units.

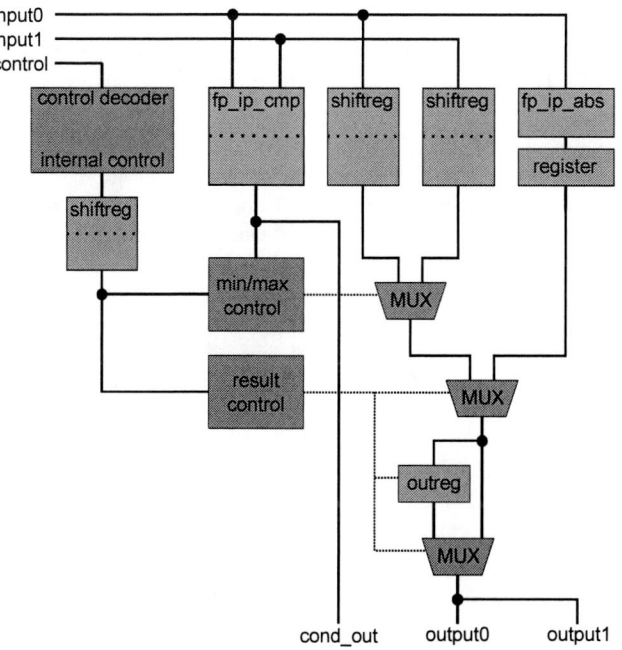

Fig. 4: Accelerator designed for utility acceleration for the SWE-SI.

Benefit to SWE from Dynamic Execution

For SWE as we have two solvers, we need to dynamically switch between them based on the surrounding environment. If we have wetting of dry surroundings, we can use the HLLE microcode, otherwise, we use the FWAVE microcode. Therefore, dynamic execution helps us to jump between both microcodes at runtime.

C. CNN

The utilization of Convolutional Neural Networks (CNNs) presents a significant challenge due to their demanding computational and storage requirements [16]–[18]. In order to accelerate CNNs with reconfigurable processors we design two accelerators. The first accelerator is the CNN-MAC accelerator. This is a multiply accumulate unit capable of accelerating 3×3 matrices needed for CNN calculation. CNNs suffer from a memory bottleneck. For this purpose, each CNN-MAC accelerator has 3 line buffers implemented in block RAMs as Figure 5 shows. Data is continuously streamed to the buffers so they are ready once needed.

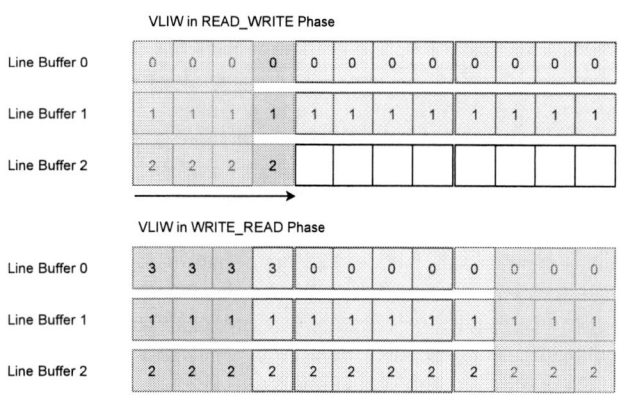

Fig. 5: Line buffers used for the CNN-MAC accelerator. Streaming data to the buffers significantly reduces the bottleneck for both writing and reading the data.

The second accelerator we designed is CNN-SUM. This accelerator performs the activation, pooling and quantization steps for the CNN. Similar to the case with CNN-MAC, CNN-SUM also has line buffers where data are streamed to resolve the memory bottleneck. The final setup of the reconfigurable fabric to accelerate CNNs contains two CNN-SUM and two CNN-MAC atoms are used to have as much parallel execution as possible.

Benefit to CNN from Dynamic Execution: To build a generic accelerator for CNNs, it needs to support an arbitrary number of channels. Without the dynamic execution, our reconfigurable processor would not have the capability to execute an arbitrary number of channels. It would have needed instead to design several SI microcodes to support each and every channel size. Figure 6 shows the steps of the developed CNN SI. It contains three nested loops wich show the need for the dynamic execution, specially to deal with the arbitrary unumber of channels.

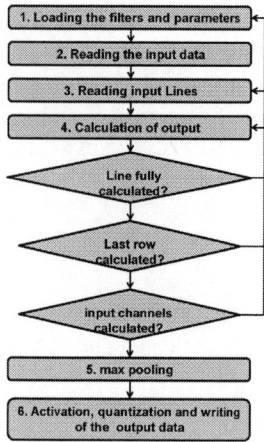

Fig. 6: Steps performed for the CNN-acceleration. The SI contains three nested loops which require our dynamic execution. Other SIs are not shown for brevity.

D. SHA-3

SHA-3 stands for Secure Hash Algorithm 3, and it's a family of cryptographic hash functions that have been defined by the U.S. National Institute of Standards and Technology (NIST) and are proven to be post-quantum secure [19]. To process queries of varying lengths, hash functions typically comprise three components. First, a padding function is employed to expand the input data to an integer multiple of the block length required by the hash function. This enables the input to be divided into several blocks, each of equal size.

Following this, a compression function (or, in the case of SHA-3, a permutation) is constructed. This function sequentially combines the blocks with the output of the permutation and further processes them. While many widely used hash functions follow the Merkle-Damgård construction, SHA-3 adopts the sponge construction. In this process, the input data is divided into multiple equally sized blocks through the use of padding. These blocks are then successively combined using the sponge construction, resulting in a 1600-bit bit vector. The final hash is extracted from this wide-bit vector [19].

To accelerate SHA-3 we design two accelerators. The first accelerator is SHA-Buff. It is used to fetch data from memory. As for SHA, the data can be quite large and of any arbitrary length it can easily become a bottleneck. Therefore, we need to stream data to the BRAMs to resolve the bottleneck. The second accelerator is SHA-Comp. It does all the computation needed for SHA. Figure 7 shows the internal design of the accelerator. It contains two RAM components the GAM-memory which stores the output of the gamma step of the application. The second RAM component is the Res-memory which stores the output of the whole algorithm, it also stores the intermediate outputs till the final output is reached. It has four calculation units, Gamma for the gamma part of the application, read and result units which get the input and give out the output and the Rho buffer which shifts the data to perform the permutation step.

The architecture of the rho-buffer is shown in Figure 8. It gets four inputs, they get splitted using the splitter and then

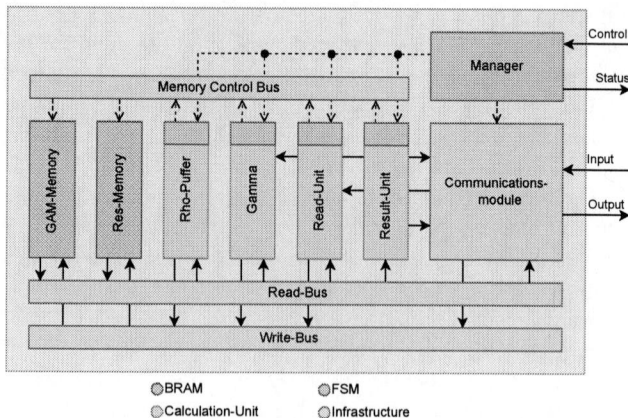

Fig. 7: The internal structure of the SHA-Comp accelerator

fed to seven shift registers. After shifting the data, using a selector, the data is selected and then combined to get the outputs.

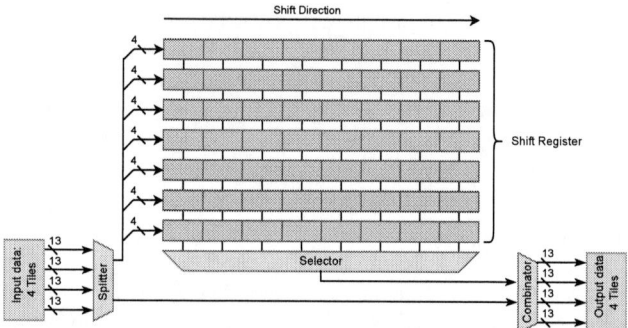

Fig. 8: Structure of the Rho buffer

The final structure of the reconfigurable fabric uses two SHA-Buff accelerators and two SHA-Comp accelerators. The data streamed from each SHA-Buff accelerator is then fed directly to one of the SHA-Comp accelerators. Then each of the SHA-Comp accelerators performs the post-quantum-secure algorithm on part of the data in parallel to have maximum acceleration.

Benefit to SHA from Dynamic Execution: Hash algorithms have to be able to support several input data sizes. To achieve this, it needs a dynamic loop that can adapt to the size of the data at runtime. This is not simple, as for example, the amount of padding needed will change based on the exact size of the data. Therefore, the dynamic execution allows the SHA-3 SI to be able to work with data of any arbitrary size instead of fixing on one size.

V. Evaluation

A. Experimental Setup

All tests were conducted on a Xilinx VC707 development board, which features a Virtex-7 XC7VX485T-2FFG1761C FPGA. The CPU core used is a Leon3 core [20]. The reconfigurable fabric is implemented to contain 5 accelerator slots. Each slot has 1600 LUTs, 3200 Flip-Flops, 10 Brams, and 20 DSPs. The benchmark applications were loaded onto the board and executed as "bare-metal" applications and were compiled using the Gaisler BCC version 4.4.2 compiler [20], which we modified to support using the SIs.

To configure the FPGA, bitstreams were generated using Vivado. To assess FPGA utilization and timing performance, data was collected through the "Report Timing Summary" and "Report Utilization" features provided by Vivado.

B. Resource Usage

To introduce the dynamic execution, a hardware extension is developed. Moreover, while all accelerators have an upper bound of resources, each accelerator designed for the different SIs has its own resource utilization. The resource utilization of the different components of our system are shown in Table II. It can be seen that in comparison to the Leon3 core, the dynamic execution hardware is $7\times$ smaller which is an acceptable overhead.

As for the accelerators, the CNN accelerators are the most resource-intensive with CNN-MAC hitting the upper limit of available LUTs. The SHA-Comp accelerator has also a similar resource utilization. Moreover, for both CNN and SHA accelerators having BRAM within the accelerator is very crucial as they use several of them. This is especially relevant for the SHA-Buff accelerator which is basically a RAM buffering needed data. In contrast, for SIFT and SWE, the accelerators are less resource hungry. However, as both basically execute mathematical equations, the usage of DSP is required.

TABLE II: Resources needed for each component

Component	DSP	LUT	BRAM	Flip-Flop
Leon3	0	14013	18	6924
Dynamic Execution	0	2100	0	682
SIFT-FMAV	4	627	0	277
SWE-SQRT	0	526	0	'85
SWE-UTIL	0	179	0	206
SWE-FMAV	4	627	0	277
SWE-DIV	0	857	0	258
SHA-Comp	0	1205	2	628
SHA-Buff	0	0	4	0
CNN-SUM	5	1333	8	249
CNN-MAC	9	1600	10	3200

C. Timing Improvement

The goal of developing the dynamic execution for the SIs executed on the reconfigurable fabric is to improve the performance. Figure 9 shows the timing improvement for each of the accelerated algorithms. The numbers are taken by running the same application on the same data-set once using the hardware acceleration and another using pure software. The same experiment is run in a loop then for 1000 times then averaged to make sure that any noise in the performance gets eliminated. CNN and SHA-3 get the highest speedups. This is expected as both applications are memory bound. Using the buffering of the data to the accelerators to use them right away solves the problem. Having the dynamic execution eases this by allowing to process a dynamic number of channels for CNN or data with arbitrary size for for SHA-3. SIFT and SWE have speedups of $14\times$ and $7\times$ respectively which is still significant.

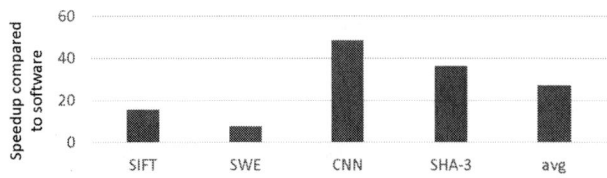

Fig. 9: Execution time improvement of the benchmarking applications. The memory intensive applications (SHA-3 and CNN have remarkably higher timing improvement. On average we have $27\times$ improvement of execution time.

VI. CONCLUSIONS

In conclusion, this work introduces dynamic execution for reconfigurable processors. We add support for different jump sub-instructions, stall the processor, and several exception classes. We show that the dynamic execution is very useful in accelerating a spectrum of applications ranging from object detection to security applications. We create special instructions for four different applications and develop the accelerators needed for each of the applications. Our results show that using our SIs and accelerators we have a relatively low overhead of 6% and we are able to reach an execution time improvement up to $42\times$ and on average to $27\times$.

REFERENCES

[1] L. Bauer et al., "RISPP: A run-time adaptive reconfigurable embedded processor", in *IEEE FPL*, 2009.
[2] H. Nassar et al., "LoopBreaker: Disabling interconnects to mitigate voltage-based attacks in multi-tenant FPGAs", in *ICCAD*, 2021.
[3] K. Elsaid et al., "Optimized FPGA Architecture for Machine Learning Applications using Posit Multipliers", in *ICM*, 2022.
[4] M. G. Jordan et al., "MUTECO: A Framework for Collaborative Allocation in CPU-FPGA Multi-tenant Environments", in *SBCCI*, 2021.
[5] A. Hassan et al., "Exploiting the dynamic partial reconfiguration on NoC-based FPGA", in *NGCAS*, 2017.
[6] L. Bauer et al., "Adaptive application-specific invasive micro-architectures", in *Invasive Computing*. FAU University Pres, 2022.
[7] G. Kuzmanov et al., "The Virtex II ProTM MOLEN Processor", in *Computer Systems: Architectures, Modeling, and Simulation*, 2004.
[8] R. Koenig et al., "KAHRISMA: A Novel Hypermorphic Reconfigurable-Instruction-Set Multi-Grained-Array Architecture", in *DATE*, 2010.
[9] *Zynq UltraScale+ Device Technical Reference Manual*, AMD, 2023.
[10] *Intel Stratix 10 SoC FPGA Boot User Guide*, Intel, 2023.
[11] M. Damschen et al., "WCET Guarantees for Opportunistic Runtime Reconfiguration", in *ICCAD*, 2019.
[12] T. Harbaum et al., "Auto-SI: An adaptive reconfigurable processor with run-time loop detection and acceleration", in *IEEE SOCC*, 2017.
[13] A. Grudnitsky et al., "COREFAB: Concurrent Reconfigurable Fabric Utilization in Heterogeneous Multi-Core Systems", in *CASES*, 2014.
[14] D. G. Lowe, "Distinctive image features from scale-invariant keypoints", *International journal of computer vision*, 2004.
[15] A. Breuer et al., "Teaching parallel programming models on a shallow-water code", in *IEEE ISPDC*, 2012.
[16] Z. Dai et al., "Coatnet: Marrying convolution and attention for all data sizes", in *Advances in Neural Information Processing Systems*, M. Ranzato et al., Eds., 2021.
[17] A. Krizhevsky et al., "Imagenet classification with deep convolutional neural networks", in *Advances in Neural Information Processing Systems*, F. Pereira et al., Eds., 2012.
[18] O. Russakovsky et al., "Imagenet large scale visual recognition challenge", 2015.
[19] M. Dworkin, "Sha-3 standard: Permutation-based hash and extendable-output functions", 2015.
[20] *GRLIB VHDL IP Core Library: Configuration and Development Guide*, Cobham Gaisler AB, 2023.

2023 International Conference on Microelectronics (ICM)

Fast Parallel Multiple Access Distributed Arithmetic (FPMA-DA) Reconfigurable FIR Filter

Ahmed H.A. Bayoumi
Electronics and Communication
Department
Ain Shams University
Si-Vision Company
Cairo, Egypt
ahmed.h.a.bayoumi@gmail.com

Sameh A. Ibrahim
Electronics and Communication
Department
Ain Shams University
Cairo, Egypt
sameh.ibrahim@eng.asu.edu.eg

Hossam A.H. Fahmy
Electronics and Communication
Department
Cairo University
Cairo, Egypt
hossam.a.h.fahmy@gmail.com

Abstract—A finite impulse response (FIR) filter is an influential block in various signal processing applications. The complexity in VLSI implementation of FIR filters is dominated by the number of multiply and accumulate (MAC) operations. Distributed Arithmetic (DA) is an alternative technique where the MAC operations can be replaced by a series of memory elements and addition operations. FIR filters based on DA are computationally efficient because of the high degree of mechanization involved in the implementation of MAC operations using DA. Many reconfigurable and non reconfigurable FIR filter architectures can be developed using DA. This paper reviews the direct FIR filter implementation and compares it to the existing FIR filter architectures based on DA and the proposed fast parallel multiple access FPMA DA FIR filter different architectures. The FPMA DA FIR filter can achieve 26% better than direct implementation in terms of power consumption numbers and 36% better in terms of area numbers at high computation complexity, and can operate at a maximum frequency 120MHz higher than the direct FIR filter implementation.

Keywords—DSP, FIR, Distributed Arithmetic, registers bank, OBC, Reconfigurable FIR Filter, SerDes, ADC, ASIC, FPGA.

I. INTRODUCTION

A finite impulse response (FIR) digital filter is one of the most fundamental blocks in digital signal processing (DSP) systems [1]. As it offers numerous advantages that make them attractive as stability and linear phase properties. So, it has been involved in various applications like digital loudspeakers [2], image processing [3], multichannel filters [4] and digital communication. That's why the design and implementation of effective FIR filters have gained significant attention.

In the context of serial links, 112-Gb/s ADC-based SerDes receiver [5] has gained more attention especially for long reach, due to higher flexibility in signal processing and equalization, and improving signal integrity and robustness. The ADC is a time-interleaved design [6] of many sub-ADCs, each operating at 875MHz or 437.5MHz. Each sub-ADC generates an output which is time-delayed by the sampling frequency compared to the preceding sub-ADC. So, we can use these outputs directly to the multipliers of the direct FIR filter implementation.

The direct FIR filter implementation as in Fig. 1 can be characterized by the number of the MAC operations required for each input sample. An FIR filter of order K requires K number of MAC operations for each input sample. Therefore, the overall complexity of the direct implementation is significantly increased by increasing the order of the filter (K), the number of bits of each input sample (N) and the number of bits of each tap-coefficient (W). The dotted rectangles between multipliers and adders are registers, a

pipeline stage is implemented to meet timing at higher operating frequencies.

Croisier et al. [7] introduced the concept of distributed arithmetic in 1973. White [8] proposed using DA as an efficient implementation of MAC operations in digital filters. DA technique provides an alternative approach to implement FIR filters that overcome some of the mentioned challenges. DA is a multiply-less FIR filter implementation where all the combinations of filter coefficients are precomputed and stored in a memory element. The input samples are then used as indices to access the corresponding precomputed combination from the memory element.

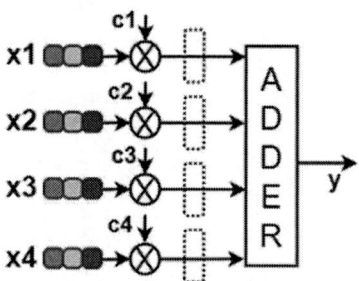

Fig. 1 Direct FIR Filter

The organization of this paper is as follows. Section II gives an overview of DA. Section III describes FIR filter's different architectures based on DA. Section IV contains the proposed FIR filter. Section V contains simulation results and comparative analysis between different FIR filter architectures. Section VI concludes the paper.

II. OVERVIEW OF DA

A. Binary DA

DA technique [8] breaks down vector to vector multiplication into a combination of smaller operations without using multipliers.

The Formulation of the DA algorithm involves considering the inner product of two vectors c and x as shown

$$y = \sum_{k=1}^{K} x_k c_k \qquad (1)$$

where c_k is a tap-coefficient, x_k is the input signal and K is the filter order. x_k is an N-bit scaled 2's complement binary number and can be expressed as::

$$x_k = -b_{k0} + \sum_{n=1}^{N-1} b_{kn} 2^{-n} \qquad (2)$$

979-8-3503-8083-5/23 $31.00 © 2023 IEEE

$$x_k = \{b_{k0}, \ldots, b_{k(N-1)}\}$$

where b_{kn} represents N_{th} bit of x_k.

By substituting Eq.(2) in (1), the expanded from of y is

$$y = \sum_{k=1}^{K} c_k \left[-b_{k0} + \sum_{n=1}^{N-1} b_{kn} 2^{-n} \right] \qquad (3)$$

The conventional form of inner product expression is given by Eq.(3).

By interchanging Eq. (3) in Eq. (2), we get the following

$$y = \sum_{n=1}^{N-1} \left[\sum_{k=1}^{K} c_k b_{kn} \right] 2^{-n} + \sum_{k=1}^{K} c_k (-b_{k0}) \qquad (4)$$

Computation of DA is described by Eq. (4). For the first K-summation, the value of b_{kn} can only be either 1 or 0, resulting in a limited number of possibilities, specifically 2^k possible combinations which can be pre-computed and stored in a memory element.

To calculate the second K-summation, the memory element will store the negative values of all the 2^k possible combinations, where b_{k0} represents the sign bit of the input word. Consequently. the total size of the memory element will be 2^{k+1}. The algorithm's flowchart in Fig. 2.

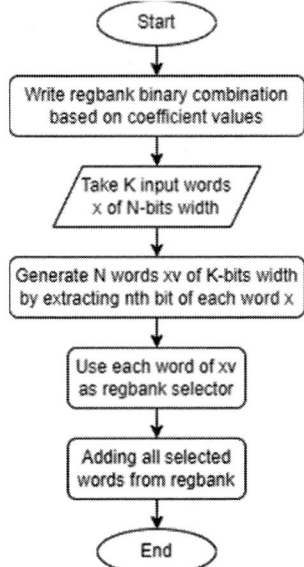

Fig 2. Binary DA Algorithm

The serial binary DA FIR filter of 4 taps and 3-bit input sample (K=4 and N=3) is in Fig. 3, The K-summations in Eq.(4) are calculated by taking all the N_{th} bit of each input word to be the word selector of the memory element to produce an output, where the first K-summation selects from the first half of the memory element, while the second K-summation selects from the second half. Then adding all the outputs using the accumulator to implement the N-summation in Eq.(4). The accumulator completes its operation after N clock cycles and produces the result y.

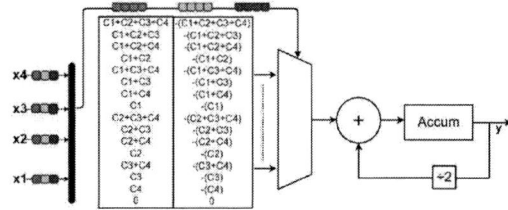

Fig 3. Serial Binary DA FIR Filter

The combination in the left half side of the memory element of binary DA is a mirrored version of the combination in the right half side differing only in having a negative sign. Modified binary DA (MBDA) algorithm [9],[12],[13] can reduce the memory element to its half to be 2^k combinations only, by adding a control signal S_0 to distinct the sign bit. So, the accumulator can add or subtract the memory element output words to produce Eq. (5) as shown in Fig. 5. The algorithm's flowchart in Fig. 4.

$$y = \sum_{n=0}^{N-1} \left\{ \left[\sum_{k=1}^{K} c_k b_{kn} \right] . (-1 * S_0) 2^{-n} \right\} \qquad (5)$$

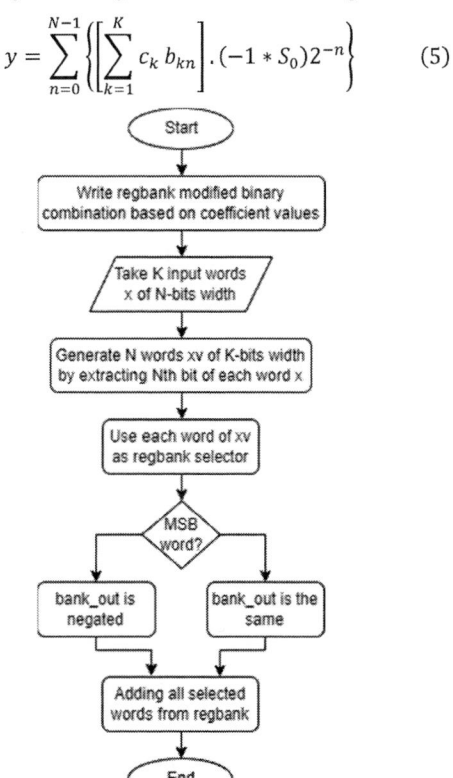

Fig 4. Modified Binary DA Algorithm

B. OBC DA Derivation

The gradual increase in filter order leads to a corresponding increase in the size of the memory element, resulting in escalated power and area requirements and performance degradation. The offset-binary code (OBC) [10],[12] can be employed where it reduces the memory element size in DA. However, this reduction comes at the cost of additional XOR gate usage. In OBC, the input data for signed values is represented as {-1, 1} instead of the conventional binary {0, 1}. The architecture of the M-OBC DA FIR filter is shown in Fig. 7.

979-8-3503-8083-5/23 $31.00 © 2023 IEEE

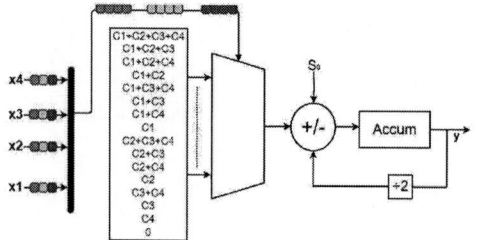

Fig 5. Serial MBDA FIR Filter

Let us write x_k as

$$x_k = \frac{[x_k - (-x_k)]}{2} \quad (6)$$

The two's complement of Eq. (2) can be represented as

$$-x_k = -b_{k0} + \sum_{n=1}^{N-1} b_{kn} 2^{-n} + 2^{-(N-1)} \quad (7)$$

On substituting x_k and $-x_k$ in Eq. (6), we get

$$x_k = \frac{1}{2}\left[-(b_{k0} - \overline{b_{k0}}) + \sum_{n=1}^{N-1} (b_{kn} - \overline{b_{kn}})2^{-n} - 2^{-(N-1)} \right] (8)$$

For the sake of simplicity, let us assume

$$s_{kn} = \left\{ \begin{array}{l} b_{kn} - \overline{b_{kn}}, n \neq 0 \\ -(b_{kn} - \overline{b_{kn}}), n = 0 \end{array} \right\} \; Where \; s_{kn} \, \epsilon \, \{-1, 1\} \quad (9)$$

On substituting these values we get

$$y = \frac{1}{2} \sum_{k=1}^{K} c_k \left(\sum_{n=0}^{N-1} s_{kn} 2^{-n} - 2^{-(N-1)} \right)$$

$$= \frac{1}{2} \sum_{n=0}^{N-1} \left(\sum_{k=1}^{K} c_k s_{kn} \right) 2^{-n} - \frac{1}{2} 2^{-(N-1)} \left(\sum_{k=1}^{K} c_k \right) \quad (10)$$

$$y = \sum_{n=0}^{N-1} Q(b_n) 2^{-n} + 2^{-(N-1)} Q(0) \quad (11)$$

where

$$Q(b_n) = \frac{1}{2} \sum_{k=1}^{K} c_k s_{kn} \; \& \; Q(b_n) = -\frac{1}{2} \sum_{k=1}^{K} c_k \quad (12)$$

$Q(b_n)$ has 2^{k-1} possible combination and $Q(0)$ represents the initial condition register or the offset. Hence the memory size in M-OBC DA is halved, where it becomes 2^{k-1} other than 2^k possible combination.

Upon observing the contents or the combination of the memory element in the OBC DA FIR filter after the addition of the offset $Q(0)$, the lower half of the combination values are a mirrored version of the upper half but with a negative sign. In order to access the required combination, the address to the memory element is calculated by XORing b_{k0} with the other input sample bits. Additionally, b_{k0} and S_0 are XORed too to generate the appropriate control signal to the add/subtract unit to implement M-OBC other than normal OBC DA FIR filter. Initially when $S_1 = 1$, the accumulator will have the value of $Q(0)$. At sign bit, the control signal $S_0 = 1$, the output of memory element which is $Q(b_{N-1})$, is added

to $Q(0)$ for the first clock cycle, for next clock cycles, the accumulator will generate the outputs as $Q(b_{N-1}) + [Q(0) + Q(b_{N-2})]2^{-1}$ After N clock cycles. The accumulator achieves an output as in Eq.(11). This architecture eventually reduces the size of the memory element but at the expense of adding some overhead XOR gates and a single mux at the input of the add/subtract unit. The algorithm's flowchart in Fig. 6.

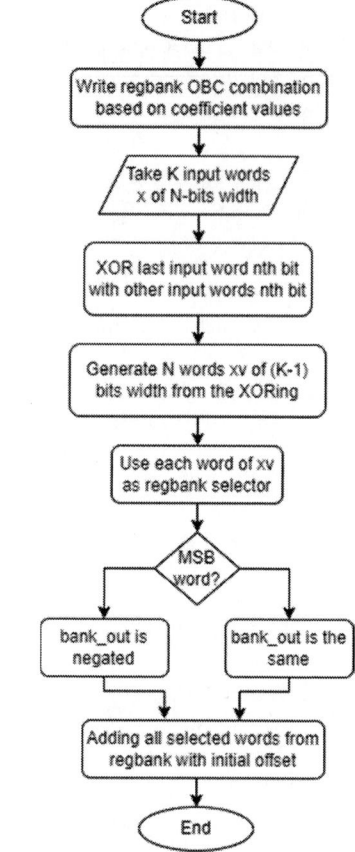

Fig 6. Modified OBC DA Algorithm

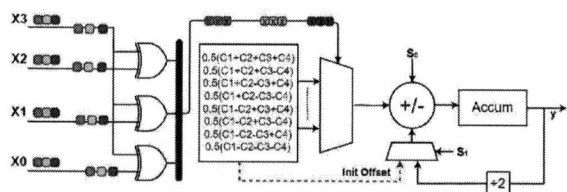

Fig 7. Serial Unpartitioned-Bank M-OBC DA FIR Filter

III. FIR FILTER IMPLEMENTATION BASED ON DA

DA FIR filters can be implemented using lookup tables (LUTs) as memory elements, but we will focus here on using register banks in order to implement reconfigurable FIR filters for more configurability and adaptability. Any DA FIR can be implemented in serial or in parallel. In this section we will start with the serial DA FIR filter and then we will focus on the traditional implementation of parallel DA FIR Filter

A. Serial DA FIR Filter

Serial DA approach [11],[12],[13] as in Fig. 3,5,7 is a bit serial operation used to perform inner product generation, its

979-8-3503-8083-5/23 $31.00 © 2023 IEEE

disadvantage is its slowness as it limits the processing to only one sample at a time. This slowness decreases the latency and throughput as it needs N cycles to produce one output which decreases the throughput to $\frac{1}{N}$ of the direct FIR filter throughput.

It will be unfair to compare power, area and speed of serial DA FIR filters to direct FIR filters implementation due to throughput difference. So, we will focus in this paper on parallel DA FIR filters as their throughput is the same as direct FIR filters.

B. Traditional Parallel DA FIR Filter

Traditional Parallel DA approach [14],[15],[16],[17] has N register banks, N muxes, and only one read at a time from each register bank. We can say that N expresses the order of parallelism in a parallel DA FIR filter.

The selector of each mux is K-width word concatenated from the same bit index from all input words, where there is a mux for each bit index, then adding all the N-outputs from all the N muxes, weighted depending on the bit weight to generate the FIR filter output y.

The traditional parallel MBDA approach is in Fig. 8, where the register bank is reduced to half the register bank size of the traditional parallel binary approach by adding by adding a 2's complement block which negates the output to implement the add/subtract block function without using of a control signal as there is a mux for each bit index word.

The dotted rectangles before the weighted adder are registers. A pipeline stage is implemented to meet timing at higher operating frequencies. The traditional parallel OBC DA approach is shown in Fig. 9 where the same parallelism is implemented but with half the register bank size with adding extra XOR gates.

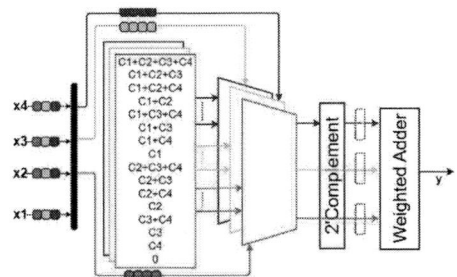

Fig 8. Traditional Parallel Unpartitioned-Bank MBDA FIR Filter

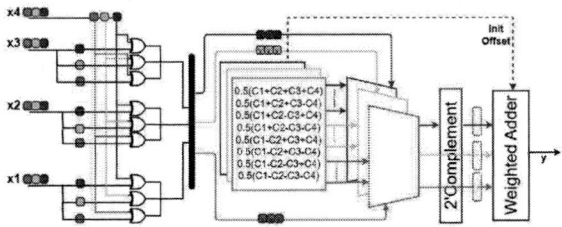

Fig 9. Traditional Parallel Unpartitioned-Bank M-OBC FIR Filter

IV.   FPMA-DA FIR FILTER

The proposed FPMA-DA FIR filter is implemented by replacing all the banks in the traditional parallel DA FIR filter by only one bank, So, there is only one bank that is shared by

all the muxes and this is the multiple access from the N-muxes to the one shared register bank.

A. FPMA MBDA FIR Filter

The proposed FPMA-DA FIR filter can be implemented as MBDA FIR filter where there are 2^k pre-computed combinations in the register bank, and adding the 2's complement block which we discussed before, the implementation is shown in Fig. 10.

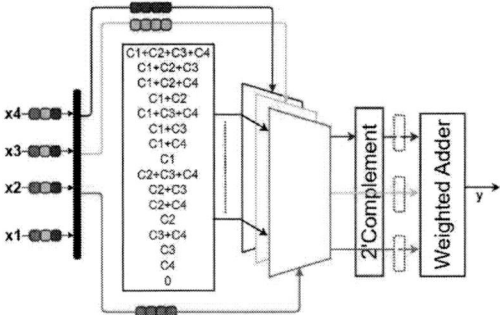

Fig 10. FPMA Unpartitioned-Bank MBDA FIR Filter

The implementation of the FPMA-DA FIR filter can be elaborated by bank partitioning as the traditional parallel DA FIR filter as shown in Fig. 11. DA partitioning means that the large bank of registers is divided into number of banks (L) of fewer number of registers, the number of registers in each bank will be $2^{\frac{K}{L}}$ to optimize bank utilization and access efficiency than using one large bank but at the expense of adding extra adders to the output of each small bank, to calculate the output which corresponds to the old output which was calculated before from the one large bank.

The input words are divided as shown in Fig. 11, where half the input words are used to select from the upper bank that consists of pre-computed combination of half of the coefficients, where the second half of input words are used as selectors for the lower bank to choose from the pre-computed combinations of second half of coefficients.

Further partitioning till K=L will lead to only 1 coefficient of each bank that has only 2 combinations that will lead to bank-less DA approach which is shown in Fig. 12. Bank-partitioning architecture is explained in Eq.(13) and we can relate Eq.(13) to Eq.(4)

$$y = \sum_{n=1}^{N-1}\sum_{l=0}^{L-1}\sum_{k=l*\left(\frac{K}{L}\right)+1}^{(l+1)*\left(\frac{K}{L}\right)} c_{kl}b_{knl}2^{-n} + \sum_{l=0}^{L-1}\sum_{k=l*\left(\frac{K}{L}\right)+1}^{(l+1)*\left(\frac{K}{L}\right)} c_{kl}(-b_{k0l}) \quad (13)$$

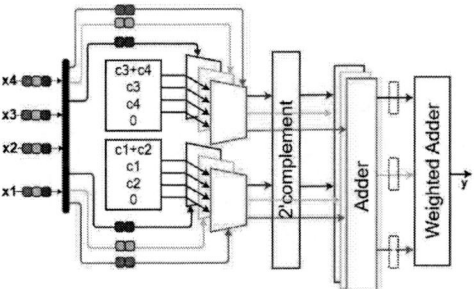

Fig 11. FPMA Partitioned-Bank MBDA FIR Filter

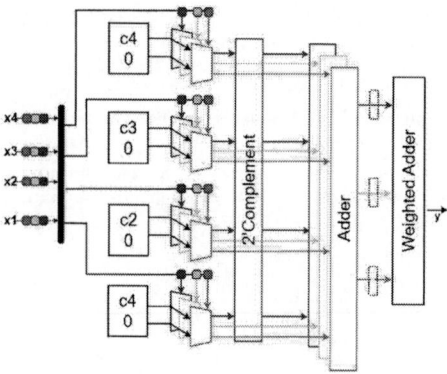

Fig 12. FPMA Bank-Less MBDA FIR Filter

B. FPMA M-OBC DA FIR Filter

FPMA M-OBC DA FIR filter is similar to parallel traditional M-OBC FIR filter implementation but also using one register bank instead of N banks, and multiple read operation are implemented simultaneously, these multiple access read are addressed by (K-1) width word concatenated from the same bit index from all input words, with the xoring by b_{k0} as shown in Fig. 13.

FPMA partitioning-bank M-OBC FIR filter can be implemented by subdividing the register bank into number of banks, each bank has registers of $2^{\left(\frac{K}{L}\right)-1}$, where half of the input words are used to address the upper bank which contains the pre-computed combinations of half of the filter coefficients, and other half of input words address the lower bank of the second half of the filter coefficients as shown in

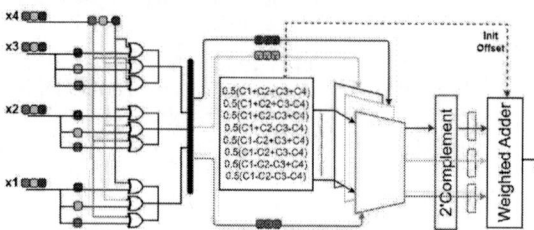

Fig. 13. FPMA Unpartitioned-Bank M-OBC DA FIR Filter

Fig. 14 and can be expressed in Eq.(14), we can relate Eq.(14) to Eq.(10)

$$y = \frac{1}{2}\sum_{n=0}^{N-1}\left(\sum_{l=0}^{L-1}\sum_{k=l*\left(\frac{K}{L}\right)+1}^{(l+1)*\left(\frac{K}{L}\right)} c_{kl}s_{knl}\right)2^{-n} - $$
$$\frac{1}{2}\,2^{-(N-1)}\left(\sum_{l=0}^{L-1}\sum_{k=l*\left(\frac{K}{L}\right)+1}^{(l+1)*\left(\frac{K}{L}\right)} c_{kl}\right) \quad (14)$$

Also can be expressed as in Eq.(11) where

$$Q(b_n) = \frac{1}{2}\sum_{l=0}^{L-1}\sum_{k=l*\left(\frac{K}{L}\right)+1}^{(l+1)*\left(\frac{K}{L}\right)} c_{kl}s_{knl}$$

$$Q(0) = -\frac{1}{2}\sum_{l=0}^{L-1}\sum_{k=l*\left(\frac{K}{L}\right)+1}^{(l+1)*\left(\frac{K}{L}\right)} c_{kl} \quad (15)$$

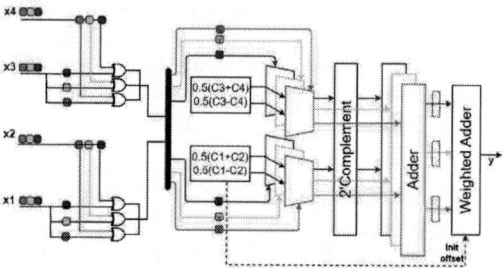

Fig 14. FPMA Bank-Less M-OBC DA FIR Filter

V. Results and comparisons

The RTL is parametrized in different parameters which are input sample width (N), the coefficient width (W), and order of the filter (K), and an automated synthesis environment run at 437.5MHz using SAED (Synopsys educational design kit) 32nm for education purposes.

A. Comparing DA FIR filter architectures

Comparing unpartitioned, partitioned and bankless architectures can be done by sweeping the number of banks in the FIR filter of the same order, starting by L=1 as unpartitioened architecture, ending by L=K as bank-less MBDA architecture, or till L=K/2 as bank-less M-OBC architecture. When the number of banks is small, there will be many registers in the bank which consumes much power and occupies much area. By increasing the number of banks, the number of registers decreases, but at the expense of the extra adders being implemented.

Based on experimental results done on many filter orders starting from order of 2 till order of 16, we find a curve that has a minimum optimum point, which corresponds to minimum power and area and this optimum point is found at L=K/2, which means that the number of banks is half the order of the filter, Fig. 15 and Fig. 16 shows the power and area numbers on different number of banks at K=6 and 12 respectively.

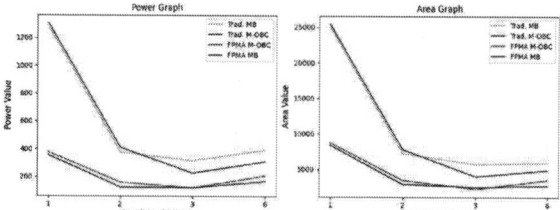

Fig 15. 6-Tap Bank Partitioning Effect On DA FIR Filters at N=6 &W=6

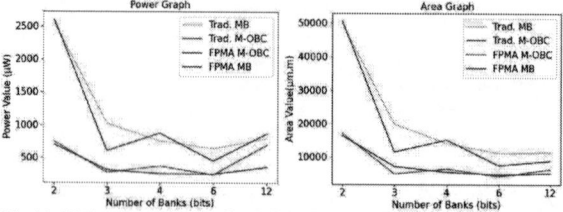

Fig 16. 12-Tap Bank Partitioning Effect On DA FIR Filters at N=6 &W=6

B. Comparing DA architectures Vs direct FIR filter

Comparing different FIR filters architectures can be done by sweeping the influential parameters that affect the power and area numbers, and see their effect on the numbers of different architectures. So, we can sweep the input sample width N which expresses the order of parallelism in parallel

DA FIR filters as in Fig. 17, we will see that FPMA MBDA achieves 21% better than direct implementation in power and 36% better in area numbers at high computational complexity. Also and we can sweep tap coefficient width as in Fig. 18, we will see FPMA MBDA architecture is 26% better than direct implementation and FPMA M-OBC in terms of power consumption and both FPMA architectures are 12% better than direct implementation in terms of area numbers especially when the computational complexity increases, and also we can sweep the order of the filter as in Fig. 19, FPMA implementations are 7% better than direct implementation in terms of power numbers and FPMA M-OBC achieves 30% better in area comparing to the direct implementation numbers. Regarding operating speed, FPMA MBDA implementation operates at 120MHz maximum frequency higher than direct implementation and 70MHz higher than FPMA M-OBC implementation as shown in Table 1.

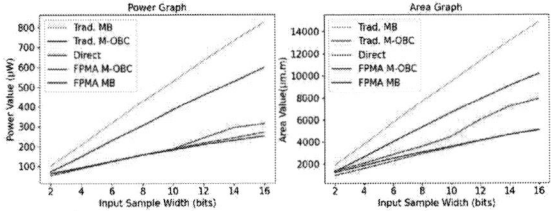

Fig 17. Power and Area Graph By Sweeping N at K=6 & W=6

VI. Conclusion

A survey of various DA FIR filter implementations has been presented in this paper targeting ASIC starting from serial binary and MBDA architectures, serial OBC and M-OBC architectures, then moving into parallel architectures as parallel traditional binary and MBDA architectures, parallel traditional OBC and M-OBC architectures and then the proposed FPMA binary, MBDA, OBC and M-OBC architectures. Each architecture of the previous DA architectures can be an unpartitioned bank, partitioned bank or bank-less based implementation.

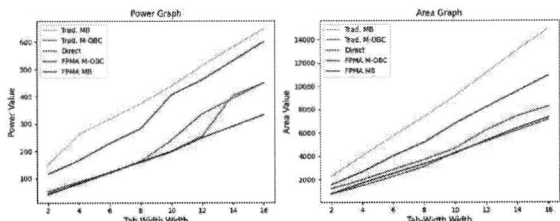

Fig. 18. Power and Area graph By Sweeping W at K=6 & N=6

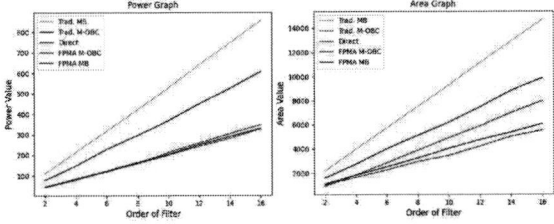

Fig 19. Power and Area Graph By Sweeping K at N=6 & W=6

Direct	FPMA MBDA	FPMA M-OBC	Trad MBDA	Trad M-OBC
550	675	605	675	605

Table 1. Maximum Operating Frequency In MHz

It is obvious from the synthesis results that the FPMA MBDA and FPMA M-OBC FIR filter architectures have much less area and less power than the parallel traditional implementations, and also have better performance than direct FIR implementation as they consume less power and have lower area numbers, especially when computation complexity increases, depending on the defined values of the parameters and depending on the power and area spec, we can choose the appropriate implementation from FPMA MBDA and FPMA M-OBC FIR filter architectures.

REFERENCES

[1] K. Parhi, VLSI digital signal processing systems: design and implementation. New Delhi, India: Wiley India Pvt Ltd., 2007.

[2] P. Valousek, "A digital loudspeaker: experimental construction," Acta Polytechnica, vol. 46, no. 4, 2006.

[3] S. Gross, T. Stehle, "Real time frame-real time processing framework for medical video sequences," Acta Polytechnica, vol. 48, no. 3, 2008.

[4] P. Vaidyanathan, Multirate systems and filter banks. Englewood Cliffs, NJ: Prentice Hall, 1993.

[5] Ting et al., "A blind ADC-based CDR with digital data interpolation and adaptive CTLE and DFE," in Custom Integrated Circuits Conference (CICC), 2014 IEEE Proceedings of the , vol., No., pp. 1-4, Sep. 15-17, 2014.

[6] Frans, et.al., "A 56Gb/s PAM4 wireline transceiver using a 32-way time-interleaved SAR ADC in 16nm FinFET," 2016 IEEE Symposium on VLSI Circuits (VLSI-Circuits), 2016

[7] A. Croisier, D. J. Esteban, M. E. Lecilion, and V. Rizo, "Digital filter for PCM encoded signals." U.S. Patent 3777130, 1973

[8] S. A. White, "Applications of distributed arithmetic to digital signal processing: a tutorial review," IEEE ASSP Magazine, vol. 6, no. 3, pp. 4–19, 1989.

[9] S. F. Ghamkhari and M. B. Ghaznavi-Ghoushchi, "Low-power low-area architecture design for distributed arithmetic (DA) unit," in Electrical Engineering (ICEE), 2012 20th Iranian Conference on. IEEE, May 15–17, 2012

[10] Bo Hong, Haibin Yin, Xiumin Wang, and Ying Xiao, Implementation of FIR filter on FPGA using DAOBC algorithm, IEEE 2010.

[11] G. NagaJyothi and S. SriDevi, "Distributed arithmetic architectures for FIR filters–A comparative review," in Proc. Int. Conf. Wireless Commun., Signal Process. Netw. (WiSPNET), Mar. 2017, pp. 2684–2690.

[12] Sanzhar Yergaliyev and Muhammad Tahir Akhtar, "A Systematic Review on Distributed Arithmetic-Based Hardware Implementation of Adaptive Digital Filters", IEEE 2023

[13] Ratnala Vinay, T.S.V.S. Vijayakumar, L.M. Saini, and Brahmjit Singh, "Power efficient FIR filter Architecture using Distributed Arithmetic Algorithm", IEEE ICMICA, 2020.

[14] W. Sen, T. Bin and Z. Jun, "Distributed Arithmetic for FIR Filter Design on FPGA," Communications, Circuits and Systems, 2007. ICCCAS 2007. International Conference on, Kokura, 2007, pp. 620-623.

[15] Shaheen Khan and Zainul Abdin Jaffery, "Low power FIR filter implementation on FPGA using parallel Distributed Arithmetic", IEEE INDICON, 2015.

[16] Mengxue Lei and Zhongsong Ma, "Design of high-speed FIR filter with distributed parallel structure", IEEE Information Technology, Networking, Electronic and Automation Control Conference, 2016

[17] Martin Kumm, Konrad Möller, and Peter Zipf, "Reconfigurable FIR filter using distributed arithmetic on FPGAs", IEEE ISCAS, 2013

Hardware Acceleration For Deep Learning Model

Shereen Afifi
Media Engineering and Technology
German University in Cairo
Cairo, Egypt
shereen.moataz@guc.edu.eg

Abdelrahman Amgad Abdallah
Media Engineering and Technology
German University in Cairo
Cairo, Egypt
abdelrhmanamggad14@gmail.com

Radwa Taha
Media Engineering and Technology
German University in Cairo
Cairo, Egypt
radwa.hussein@student.guc.edu.eg

Abstract—In an era defined by fast technological growth, deep learning has sparked a revolution in artificial intelligence. With this incredible breakthrough, the possibility to unlock previously unimaginable possibilities lies within the depths of innovative hardware acceleration approaches. Deep learning models are computationally intensive and require a significant amount of processing power. FPGAs have emerged as a popular choice for accelerating the training and inference of deep learning models due to their highly parallel architecture, customization, and flexibility, while achieving high parallelism, low power consumption, and better programmability.

In this study, we propose an approach that combines FPGA-based hardware acceleration with software-based optimization techniques, using the latest design tools to achieve high performance and efficiency in deep learning applications. Through trials on a facial recognition dataset, we show that our technique outperforms existing software-based approaches in terms of speed and accuracy. Our methodology includes Int-8 bit quantization as a critical step in preparing the trained facial recognition network for hardware deployment. This quantization reduces the numerical precision of the weights and activations in a deep learning model to 8 bits, significantly reducing memory usage and computational complexity while maintaining acceptable accuracy. Our approach showed a significant speedup and better utilization of hardware resources, obtaining a 3.52x increase in FPS, a 2.83x decrease in latency, a 1.5x reduction in LUT utilization, and a 1.25x reduction in BRAM utilization. In conclusion, this study tackles the challenges of enhancing efficiency and performance in deep learning models, with a particular focus on the application of face recognition.

Index Terms—Hardware Acceleration, FPGA, CNN, Deep Learning, Quantization, Optimization

I. INTRODUCTION

Deep learning models have revolutionized the area of computer vision in recent years, allowing considerable breakthroughs in many technologies. Traditional processors, on the other hand, encounter substantial difficulties in satisfying the processing requirements for building deep learning applications. Deep learning models, such as those used in facial recognition, are becoming increasingly complicated, requiring significant computing and memory resources. They frequently struggle to provide the required performance and efficiency, stifling the development of real-time and scalable systems. Current computing techniques frequently lack the required parallelism, use too much power, and have restricted programmability, limiting their efficacy in real-world circumstances. These constraints interrupt the general acceptance and practical application of facial recognition technologies in

crucial applications. As a result, there is an urgent need to solve these issues and create unique approaches to improve efficiency and performance. By achieving high parallelism, minimizing power consumption, and improving programmability, we can unleash the true capabilities of hardware acceleration, allowing its smooth integration into multiple domains.

Convolutional Neural Networks (CNN) is a famous deep learning model that is inspired by the way our visual system works. It has layers of artificial neurons that mimic the cells in our brains, allowing it to process and understand visual information. These networks excel at identifying shapes, edges, textures, and patterns in images.

Field Programmable Gate Arrays (FPGAs) are programmable integrated circuits that offer great flexibility and customization for specific applications. Unlike GPUs or CPUs, which have traditional architectures, FPGAs are featured with high programmability and flexibility for realizing efficient design of various embedded applications. This flexibility allows developers to optimize FPGAs for machine learning and deep learning workloads by tailoring the hardware architecture to match the specific requirements of the models and meet challenging constraints of embedded systems [1], [2], [3], [4]. The main objective of this study is to investigate FPGA-based hardware acceleration and optimization techniques, using the recent design tools aiming to achieve high performance computing for deep learning applications.

This paper is organized as follows. Section II reviews related work on the existing strategies and studies discussing their approaches, implementation plans, and results. Section III outlines the methodology and the proposed workflow. Subsequently, analyzing the results obtained from testing the proposed hardware accelerator is discussed in Section IV. The conclusion is given in Section V.

II. LITERATURE REVIEW

Research and development activities in the area of hardware acceleration for convolutional neural networks (CNNs) have increased significantly. The demand for effective and high-performance computing in a variety of fields, such as object identification and picture recognition, has increased to the point where it is essential to have specialized hardware designed for CNNs. In this study, we present recent and key studies on hardware acceleration for CNNs, so we can

979-8-3503-8083-5/23 $31.00 © 2023 IEEE

investigate the most important trends, problems, and possible research areas in this rapidly developing topic.

Zhang et al. [5] introduced a pipelined architecture approach using OpenCL-based FPGA accelerators, achieving a speedup of up to 1.7x and an energy efficiency improvement of up to 5.5x compared to previous FPGA-based accelerators, while maintaining similar accuracy. However, implementing this design methodology may pose challenges for those with limited FPGA development experience. Other researchers focused on reducing latency and improving memory bandwidth. Another approach [6] used the pin-pong method to reduce latency, while another one [7] combined column-based pipeline techniques with ping-pong architecture-based filter storage to reduce filter loading latency. However, these methods did not allow for local data reuse, resulting in increased memory bandwidth. In contrast, another study [8] proposed hardware that optimized data fetching from Local Memory and Control Unit (LMCU), enhancing throughput and energy efficiency. This hardware stored weights inside the processor element (PE) using a dual-port memory for simultaneous read and write capabilities. Additionally, it employed the Random Access Line Buffer (RALB) method for storing input feature maps, enabling simultaneous read-write operations and eliminating the need for multiple shift registers for local data reuse.

The demand for deep learning applications on resource-limited mobile and embedded systems has led to the development of hardware acceleration techniques. An approach by Kechiche L. et al. [9] proposes FPGA-based hardware acceleration for image classification, achieving high performance within the constraints of small devices. The authors plan to optimize power consumption in future work. Another study [10] investigates weight pruning on FPGA platforms for Transformer-based language models, achieving high pruning ratios with minimal accuracy decay. The proposed framework demonstrates significant speed-ups compared to CPU and GPU platforms. In the realm of deep learning algorithms, an FPGA-based hardware accelerator [11] outperforms existing systems in terms of accuracy, processing speed, and power consumption. Similarly, an OpenCL-based FPGA accelerator [12] improves the real-time performance of the Faster R-CNN object identification method, demonstrating faster processing speeds without sacrificing accuracy. These studies highlight the potential of FPGA-based accelerators for efficient and high-performance execution of deep learning algorithms.

Mao N. et al. [13] presents a parameterized design strategy for putting a CNN on an FPGA, achieving improved performance and resource utilization compared to earlier work. Their approach reduces time, logic resources, and latency, outperforming previous efforts in terms of latency, FPS, and resource consumption. Jiang C. et al. [14] introduces a DL accelerator systolic array design optimization for sparse CNN models on FPGA platforms. Using a sparse matrix packing approach, the proposed accelerator achieves significant performance and energy efficiency advantages compared to a server-class CPU and an NVidia GPU. Wu D. et al. [15] focuses on optimizing MobileNet for mobile devices, introducing specialized com-

putational engines and a specific architecture called Channel Augmentation. The proposed accelerator achieves significant speedups compared to CPUs for picture classification and object identification tasks. Luo L. et al. [16] utilizes a lightweight CNN called "LiteCNN" to speed up the identification of plant diseases. LiteCNN achieves high accuracy while having a simpler design and fewer parameters, making it suitable for implementation on resource-constrained FPGAs. Gilan A. et al. [17] focuses on optimizing hardware for each layer of the CNN to meet embedded system constraints, achieving high performance and computational efficiency compared to previous accelerators. The accelerator outperforms modern accelerators on average, with convolution layer throughput of 153.8 frames per second and a total throughput of 82 frames per second for object identification. Bouafia M. et al. [18] introduces IP HW/SW kernels for CONV and FC layers, demonstrating processing speeds of up to 150 MHz with low hardware costs. The suggested technique offers high throughput, low power consumption, and flexibility, making it a viable alternative for CNN accelerator systems. Wang J. et al. [19] presents a unique strategy for efficient model parallelism in distributed CNN inference. By decoupling the network topology, employing group convolution and channel shuffle procedures, and designing a parallel FPGA accelerator, the method achieves lower inference time. However, the speed-up ratio faces limitations with increased device synchronization time. The channel optimization strategy reduces data transfers between FPGA and CPU, enabling neural network inference on devices with limited resources.

III. METHODOLOGY

Fig. 1 illustrates the proposed system design to achieve the main objective of this study. Face recognition is chosen as a case study for our system. This will provide insights into the challenges, techniques, and potential benefits of employing FPGA-based hardware acceleration for face recognition using CNNs. This section provides a comprehensive overview of the research design, data collection methods, hardware and software setup, and experimental procedures.

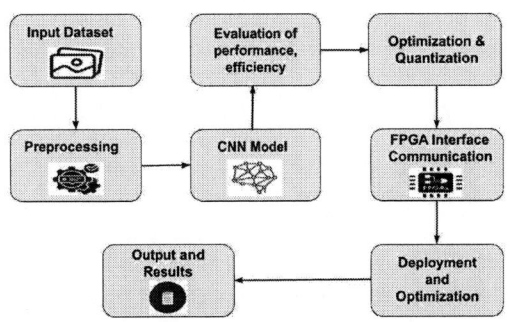

Fig. 1. Proposed System Design

A. Dataset

The face recognition dataset used in this study was obtained from the Celebrity Face Image dataset [20] combined with the LFW dataset [21]. The dataset includes a total of 5780 unique individuals and more than 13,000 images. The dataset offers an extensive collection of face photos for the face recognition application by playing changes in lighting conditions, positions, facial emotions, and ethnicity as seen in Fig. 2.

Fig. 2. Dataset Sample [20]

B. Pre-processing Methods

1) Image Resizing: To maintain standardization and speed up further processing, the captured facial images have been resized to an appropriate resolution. The photos can be resized to a certain size, as [224 224 3], which enables consistent input across the facial recognition system. The "augmentedImage-Datastore" function of MATLAB is used to resize the photos while maintaining their aspect ratio. This function is often used during training; instead of saving the transformed pictures, it modifies the training data for each epoch.

2) Data Augmentation: An imageDataAugmenter object is generated using a predefined MATLAB function in the data augmentation section. By creating a set of augmentation techniques that may be used on the pictures during training to broaden the dataset's variety and enhance the deep learning model's generalization capabilities.

C. Investigated Models

In the preliminary stages, various projects were conducted using popular deep learning frameworks such as TensorFlow and Keras on platforms like MATLAB and Google Colab. Different architectures and models suitable for face recognition were explored to determine the best fit for the research.

State-of-the-art deep learning architectures like ResNet, GoogLeNet, and SqueezeNet were utilized in these models. These pre-trained models, developed by experts, have shown success in various tasks and serve as a solid starting point for many deep learning applications. MATLAB's Deep Learning Toolbox offers high-level abstractions that simplify the development process, providing convenient methods and objects for constructing and configuring deep learning networks.

Using pre-trained models in MATLAB reduces the time and computing resources required to build deep learning models from scratch. These models can be fine-tuned or used as feature extractors since they are already optimized, accelerating

development and minimizing the need for extensive data and computing infrastructure.

D. SqueezeNet

The SqueezeNet convolutional neural network (CNN) model is employed as the core architecture for face recognition. SqueezeNet is chosen due to its lightweight nature and high efficiency, making it suitable for deployment on resource-constrained devices such as FPGAs. This model leverages deep learning techniques to learn discriminative facial features from the preprocessed images and provides an effective representation for face recognition tasks. The model's small architecture makes it suitable for deployment on platforms with limited resources, including embedded systems or mobile devices, where memory and computing capacity are restricted. The face recognition system may execute in real-time and with efficient inference on such platforms by utilizing Squeezenet. It also recorded the highest accuracy while training on the collected datasets compared to other pre-trained models provided by MATLAB as illustrated in Table I.

TABLE I
PERFORMANCE RESULTS

Network	Accuracy	Learning Rate	Input Size
GoogleNet	69.23%	0.0004	13248
ResNet-50	76.92%	0.0003	13248
SqueezeNet	**92.23%**	0.0004	13248

Although, Squeezenet is a strong and efficient deep learning architecture, however it was initially intended for generic image classification problems. Face recognition requires specialised model modifications to recognise unique face traits and discriminate between people. For this reason, the MATLAB's pre-trained model was modified in order to allow us leveraging the benefits of transfer learning.

However, the network architecture's adaptability makes it easier for the model to be adjusted to varied face recognition datasets, assuring efficacy in a range of scenarios and domains. The network can manage the complexity and variability of face photos with more ease thanks to the addition of new layers, which eventually improves the face recognition system's overall performance.

E. Quantization

To prepare the trained facial recognition network for hardware deployment, quantization is a critical step. The quantization procedure, which involves lowering the accuracy of network weights and activations to lower memory needs and increase efficiency, is implemented by MATLAB's new tool, Deep Learning Toolbox Model Quantization Library that was released this year [22]. The network is set up for FPGA execution and simulation using the dlQuantizer object. By mapping the input range to the range provided by the destination hardware, the calibration procedure makes sure that the quantized network operates as efficiently as possible.

The network is quantized, which reduces the size of the model and enables more rapid inference and more effective storage. The code demonstrates the significance of quantization in producing deep learning models that are hardware-friendly, enabling deployment on systems with limited resources, such as FPGAs. Quantization minimized the network's memory footprint by expressing model parameters and activations with 8-bit integer values rather than higher precision floating-point values. This decrease in memory needs enables more efficient storage and memory utilization, which is especially essential when deploying the model on resource-constrained devices or hardware accelerators, that is FPGA in our case.

Additionally, certain layers in the redesigned deep learning network were classified as software (SW) layers and hardware (HW) layers. The allocation of layers to distinct domains denotes the use of a heterogeneous computing paradigm, in which calculation workloads are spread across software and hardware components.

Software (SW) layers are executed on general-purpose processors or CPUs, leveraging their flexibility for tasks like complex computations, data processing, and control flow. On the other hand, Hardware (HW) layers are offloaded to specialized components such as FPGAs. HW layers handle computationally intensive tasks that can be parallelized and accelerated in hardware, utilizing FPGAs' high parallelism, low latency, and dedicated resources.

By applying a calibration approach and using FPGA-based hardware accelerators, the quantized network achieves efficient calculations, enabling high-speed processing and real-time face recognition. While hardware layers enhance performance, software layers provide flexibility when hardware acceleration is not feasible. The quantized network can be deployed and executed on general-purpose processors, making the face recognition program compatible with a wide range of devices and platforms. This accessibility and adaptability make face recognition more widely applicable.

F. Optimization

In MATLAB, the dlhdl.processorConfig class allowed us to set up the processor for deploying deep learning models on our hardware target (FPGA). This class allows the configuring of many features and settings to optimize the quantized model's deployment and operation. It has been successful in achieving customized parallel processing, memory management, optimization algorithms, and other hardware-specific options.

Moreover, it also allows full utilization of the FPGA's capabilities and optimization of the performance and efficiency of the deployed model. It may adjust memory parameters for the deployed model, such as on-chip memory size and organization. Efficient memory management can improve execution speed and resource utilization. In conclusion, using the dlhdl.processorconfig class and its configuration parameters, optimized the deployment of our quantized model using the DLQuantizer object on the FPGA. This guaranteed that our network uses resources efficiently performs optimally, and executes at a faster overall pace.

G. FPGA Implementation

Configuring the FPGA interface and establishing communication between the host device and the FPGA accelerator using a custom-generated bitstream was required for the deployment of the quantized network for face recognition. The dlhdl.Target MATLAB function was used with the supplied arguments to help with this. In this example, the 'Xilinx' target platform was chosen, and the communication interface was set to 'Ethernet'.

DlquantizationOptions were established for the deployment to indicate the bitstream and target settings. The 'Bitstream' argument was set to 'zcu102 int8', representing the Xilinx ZYNQ ZCU102 FPGA development board's particular pre-built bitstream file. The Xilinx ZYNQ board (Fig. 3) [2] is featured with a hybrid structure combining Processing System (PS) and Programmable Logic (PL) on the same board, which simplifies embedded systems realization.

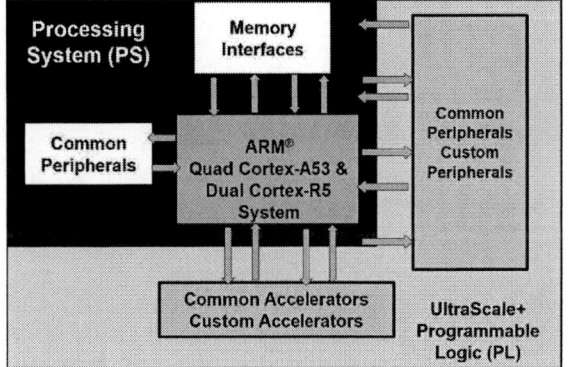

Fig. 3. ZYNQ board [2]

Validation was performed using the simulation and FPGA deployment options. This enabled the network's accuracy and performance to be evaluated in simulation and FPGA execution environments, respectively. It guaranteed that the quantized network's behavior and correctness were thoroughly tested by undertaking validation on both simulation and FPGA platforms. This phase was necessary to confirm the network's appropriateness and dependability for face recognition tasks. The validation predictions offered insights into the network's performance and allowed comparison between the two execution contexts, both for simulation (prediction simulation) and FPGA deployment (prediction FPGA).

IV. RESULTS

A. Software Implementation Results

The trained SqueezeNet model had achieved an impressive 92.23% accuracy on the validation set. Several adjustments were made to the training options, such as increasing epochs, changing the optimization algorithm, and adjusting the initial learning rate, to achieve that high accuracy. These results demonstrate the effectiveness of the SqueezeNet model in learning and classifying facial features in our dataset.

Using the trained network, we applied it to predict results for new unknown data. By utilizing the previously trained SqueezeNet model on the dataset, we fed the test data into the network and used the predict function to generate predicted

979-8-3503-8083-5/23 $31.00 © 2023 IEEE

labels. Fig. 4 shows some of the prediction results. The overall prediction results highlight the model's capability to accurately categorize facial characteristics in previously unseen photos.

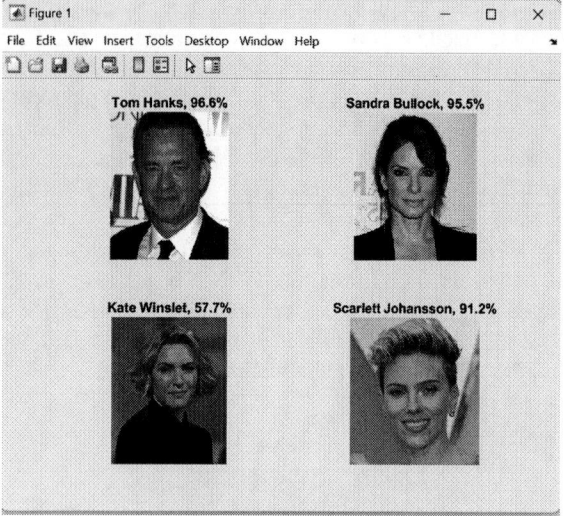

Fig. 4. Prediction Results

B. Hardware Implementation Results

The findings show that a deep-learning processor outperforms a general-purpose processor in handling the computing needs of deep-learning models. The higher FPS and lower latency suggest that the CPU can effectively perform neural network computations, allowing for quicker and more responsive inference for face recognition applications. It is important to highlight that these performance enhancements not only improve the user experience by giving faster results, but also enable real-time face recognition applications in a variety of sectors such as surveillance, biometrics, and human-computer interaction. Table II shows the overall hardware results before and after applying the quantization and optimization methods. When compared to the original network, the quantized network displayed substantial acceleration (Table II), with a 3.52x increase in frames per second (FPS). This increase in FPS implies a quicker inference speed, allowing for real-time processing of facial recognition applications. Furthermore, the quantized network had decreased latency, with a 2.83x reduction when compared to the original network. This lower latency translates to faster response times in face recognition activities, which improves the overall user experience.

The use of hardware resources improved as well. The quantized network reduced Look-Up Table (LUT) utilization by 1.5x, showing improved resource efficiency and the possibility of integrating new functionality or future developments. Furthermore, the use of Block RAM (BRAM) decreased by a factor of 1.25x as seen in (Table II), meaning that these specialized hardware components are being used more effectively for faster mathematical computations in deep learning algorithms.

Table III shows a high-level comparison of our results to some related work in the literature, which highlights great

optimization in our results that promises hardware acceleration of deep learning models on FPGA.

TABLE II
OVERALL RESULTS

Quantization&Optimization	LUTs	BRAM	DSPs	FPS	Latency (ms)
Without	79%	54%	16%	21	38.2
With	52%	43%	32%	74	13.5

TABLE III
COMPARISON WITH EXISTING WORK

	[4]	[11]	[15]	Proposed Work
Model	YOLO-V2	LeNet-5	AlexNet	SqueezeNet
Accuracy	77.04%	-	90%	92.2%
Dataset	KITTI	-	-	LFW
Quantization	8-bit	16-bit	16-bit	8-bit
LUTs (Utilization%)	38%	11.5%	32%	52%
BRAM (Utilization%)	67%	56.43%	72%	43%
DSPs (Utilization%)	46%	54.55%	64%	32%
Frames Per Second	43.7	71	52	74
Latency (ms)	27.7	24.8	153.5	13.5

V. CONCLUSION

In conclusion, this study demonstrates quantization and optimization of a pre-trained deep-learning model (SqueezeNet) on FPGA (ZYNQ board) for a face recognition application using the latest design and optimization tools. Our proposed hardware-accelerated design achieved significant hardware acceleration with a 3.52x increase in frames per second and a 2.83x reduction in latency. This enabled real-time face recognition capabilities. The hardware implementation results achieved high performance and low resource utilization due to the applied quantization and optimization techniques, meeting the challenging embedded system's constraints.

The findings advance the area of hardware-accelerated deep learning by emphasizing the potential of FPGA-based solutions for optimizing and deploying deep learning models in resource-constrained contexts.

For future work, The developed model can serve as a valuable reference due to its utilization of novel tools and state-of-the-art technologies. Our proposed hardware-accelerated design could be used for other applications not only face recognition. Also, critical parameters like power consumption, which are critical under embedded system restrictions should be examined for FPGA deployment.

REFERENCES

[1] A. HajiRassouliha, A. J. Taberner, M. P. Nash, and P. M. Nielsen, "Suitability of recent hardware accelerators (dsps, fpgas, and gpus) for computer vision and image processing algorithms," Signal Processing: Image Communication, vol. 68, pp. 101–119, 2018.

[2] S. Afifi, H. GholamHosseini, and R. Sinha, "Dynamic hardware system for cascade svm classification of melanoma," Neural Computing and Applications, vol. 32, no. 6, pp. 1777–1788, 2020.

[3] S. Afifi, H. GholamHosseini, and R. Sinha, "Fpga implementations of svm classifiers: A review," SN Computer Science, vol. 1, pp. 1–17, 2020.

[4] S. Afifi, H. GholamHosseini, and R. Sinha, "A system on chip for melanoma detection using fpga-based svm classifier," Microprocessors and Microsystems, vol. 65, pp. 57–68, 2019.

[5] J. Zhang and J. Li, "Improving the performance of opencl-based fpga accelerator for convolutional neural network," in Proceedings of the 2017 ACM/SIGDA Inter-national Symposium on Field-Programmable Gate Arrays, pp. 25–34, 2017.

[6] J. Zhang, L. Cheng, C. Li, Y. Li, G. He, N. Xu, and Y. Lian, "A low-latency fpga implementation for real-time object detection," in 2021 IEEE International Symposium on Circuits and Systems (ISCAS), pp. 1–5, IEEE, 2021.

[7] C. Wu, J. Zhuang, K. Wang, and L. He, "Mp-opu: A mixed precision fpga-based overlay processor for convolutional neural networks," in 2021 31st International Conference on Field-Programmable Logic and Applications (FPL), pp. 33–37, IEEE, 2021.

[8] M. N. Islam, R. Shrestha, and S. R. Chowdhury, "An uninterrupted processing technique-based high-throughput and energy-efficient hardware accelerator for convolutional neural networks," IEEE Transactions on Very Large Scale Integration(VLSI) Systems, vol. 30, no. 12, pp. 1891–1901, 2022.

[9] L. Kechiche, "Hardware acceleration for deep learning of image classification," in 2021 International Conference of Women in Data Science at Taif University (WiDSTaif), pp. 1–5, IEEE, 2021.

[10] H. Peng, S. Huang, T. Geng, A. Li, W. Jiang, H. Liu, S. Wang, and C. Ding, "Accelerating transformer-based deep learning models on fpgas using column balanced block pruning," in 2021 22nd International Symposium on Quality Electronic Design(ISQED), pp. 142–148, IEEE, 2021.

[11] Y. Chi, Z. Zheng, R. Liu, and W. Cui, "Design of hardware acceleration system based on fpga and deep learning algorithm," in 2020 IEEE International Conference on Artificial Intelligence and Computer Applications (ICAICA), pp. 1332–1337, IEEE, 2020

[12] J. An, D. Zhang, K. Xu, and D. Wang, "An opencl-based fpga accelerator for fasterr-cnn," Entropy, vol. 24, no. 10, p. 1346, 2022

[13] N. Mao, H. Yang, and Z. Huang, "A parameterized parallel design approach to efficient mapping of cnns onto fpga," Electronics, vol. 12, no. 5, p. 1106, 2023.

[14] C. Jiang, D. Ojika, B. Patel, and H. Lam, "Optimized fpga-based deep learning accelerator for sparse cnn using high bandwidth memory," in 2021 IEEE 29th Annual International Symposium on Field-Programmable Custom Computing Machines(FCCM), pp. 157–164, IEEE, 2021.

[15] D. Wu, Y. Zhang, X. Jia, L. Tian, T. Li, L. Sui, D. Xie, and Y. Shan, "A high performance cnn processor based on fpga for mobilenets," in 2019 29th International Conference on Field Programmable Logic and Applications (FPL), pp. 136–143, IEEE, 2019.

[16] Y. Luo, X. Cai, J. Qi, D. Guo, and W. Che, "Fpga–accelerated cnn for real-time plant disease identification," Computers and Electronics in Agriculture, vol. 207, p. 107715, 2023.

[17] A. A. Gilan, M. Emad, and B. Alizadeh, "Fpga-based implementation of a real-time object recognition system using convolutional neural network," IEEE Transactions on Circuits and Systems II: Express Briefs, vol. 67, no. 4, pp. 755–759, 2019.

[18] S. Bouaafia, S. Messaoud, R. Khemiri, and F. E. Sayadi, "An fpga-soc based hardware acceleration of convolutional neural networks," in 2022 IEEE 9th International Conference on Sciences of Electronics, Technologies of Information and Telecommunications (SETIT), pp. 537–542, IEEE, 2022.

[19] J. Wang, W. Tong, and X. Zhi, "Model parallelism optimization for cnn fpga accelerator," Algorithms, vol. 16, no. 2, p. 110, 2023.

[20] K. Dataset, "Available online: https://www. kaggle. com/datasets," Accessed on Apr, 2022.

[21] A. Anand, Celebrity face image dataset. https://www.kaggle.com/datasets/atulanandjha/lfwpeople?resource=download, 2019.

[22] Deep learning toolbox model quantization https://www.mathworks.com/matlabcentral/fileexchange/74614-deep-learning-toolbox-model-quantization-library, 2023.

EFFICIENT MUX-BASED MULTIPLIER FOR MAC UNIT

Huruy Tekle Tesfai, Hani Saleh, Mahmoud Meribout, Mahmoud AL-Qutayri, Thanos Stouraitis

Department of Electrical and Computer Engineering, Khalifa University, Abu Dhabi, United Arab Emirates

Multiply-Accumulate Units (MACs) are crucial for compute-intensive algorithms that are used in many application fields, such as machine learning, real-time embedded systems, etc. Therefore, the choice of the multiplier for MAC units has significant impact in the efficiency of these compute-intensive operations, particularly those executed on edge devices. The cost of fast full-tree or parallel-tree multipliers are high and hence limits applications to systems whose speed requirement is critical. The slow serial implementations are also usually mixed with the full tree to lower the cost to an acceptable range resulting in partial tree. In this paper, we propose a full tree of lower height with pre-computation of multiplicand multiples using multiplexer networks. It is demonstrated that for a radix-16 version of the multiplier, the partial product reduction network shrinks by a factor of two. The proposed design is more efficient consuming 35% less power and requires 13% less area than a standard Booth-recoded multiplier with comparable speed. Hence, the proposed multiplier design is suitable for integration in multiply-intensive applications such as DNNs (Deep Neural Networks) and other algorithms.

Index Terms—**MAC, Multiplier, Arithmetic, AI, DNN.**

I. INTRODUCTION

The continuing advancements in artificial intelligence and post quantum applications make meeting the computing demands of such applications on edge devices using traditional arithmetic circuits increasingly challenging. The need for carry propagation in operations using operands represented in positional number systems is the main reason for the slow computer arithmetic operations. As a result, much of the research effort has been devoted to the investigation of fast multiplier units and various non-conventional numbering systems to explore potential parallelism opportunities for improved efficient arithmetic circuit implementations [1] [2]. To achieve significant gains in speed, various multiplication schemes to reduce the number of cycles are used. The application of recoding to the operands and carry-save network for higher-radix multiplication is common in literature [1]. The use of fast full-tree or parallel tree multipliers in extreme cases improves the speed of computation down to one clock cycle. The choice of the numbering system has also significant impact on determining the complexity and throughput of a design [2] or its accuracy, and power consumption [1] [3] [4].

Multi-operand addition and shift operations are required for partial products of a multiplication process. Addition, which is the primary element of all arithmetic operations, however, requires carry propagation from least significant bit to most significant bit in worst case. This is the major source of delay in arithmetic circuits causing increased complexity and cost of its hardware implementation. Among the techniques used

to deal with this carry network problem is to either limit the carry propagation to smaller lengths or eliminate the inter-digit propagation by applying a number representation.

This paper proposes a multiplexer-based full tree multiplier to improve the latency by using higher radix. Its efficiency is compared to the Booth-recoded multiplier, which is common in the industry. The functionality of both designs has been verified with simulation and synthesized using Synopsys Design compiler. Moreover, it was compared to Synopsys DesignWare built-in multipliers to shed more light on it's performance.

The rest of this paper is organized as follows: Section II presents some existing multiplier architectures, Section III introduces the proposed radix-4 and radix-16 MUX multiplier designs, Section IV presents verification, performance and synthesis results of this work, and finally, the paper is concluded in Section V.

II. EXISTING MULTIPLIER HARDWARE IMPLEMENTATIONS

Due to its complexity, the design of a multiplier dictates the performance and implementation cost of computation elements to a large extent [5]. Fast multipliers have large chip area requirement in ASIC design flow or high logic and interconnect usage in FPGA design. Existing solutions provide either speed at a lower cost or high throughput with parallel independent multipliers. The trade-off is between cost-effectiveness with serial-parallel and high throughput with array multipliers [1, 6]. In serial multiplication schemes, the number of cycles needed in multiplying two k-bit operands can be realized in k clock cycles of shift-add operations. While in parallel full tree multiplier, the multiply operation can be done in one clock cycle. Recursive multiplication applies a divide-and-conquer strategy, where multiplication operation on chunks of two operands produces partial products. This is then followed by a multi-operand addition.

III. PROPOSED MULTIPLEXER-BASED MULTIPLIERS

Higher-radix operation results in a smaller number of partial products in a multiplication process. Although radix-4 is more common, some industrial units of radix-8 [7], [8] and radix-16 multipliers [9] also exist in microprocessor implementations. In this section, we introduce a new multiplexer-based multiplier architecture. The multiplication process is performed in four stages demonstrated for signed and unsigned operands. Single CPA and simple shift operation first generates multiples of the input multiplicand. While power-of-two multiples are easily obtained with shift operations, the odd multiples increase the complexity for higher-radix implementations. Hence, the odd multiples were limited to 3x

979-8-3503-8083-5/23 $31.00 © 2023 IEEE

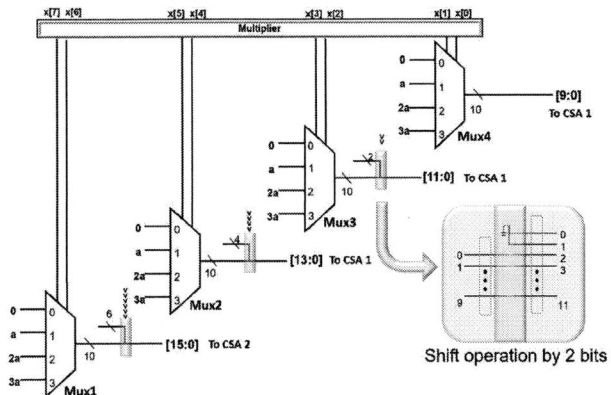

Fig. 1: Generation of partial products for an 8-bit by 8-bit multiplication.

only in this work, adding just one more partial product. This holds also for radix-8 and radix-16 implementations of our multiplier. In the second stage, a set of multiplexers are used to route one of the multiples based on the multiplier bits. As shown in Fig 1, this multiplication scheme uses four radix-4 multipliers. The output of the first partial product is passed into the reduction network as is, while the partial products from higher pairs of multiplier bits are shifted to the left by r-bits incrementally, based on its position for a radix-2^r multiplier (e.g. 2 for radix-4, Fig 1). This reduces the number of input operands to the reduction network by a factor of two, resulting in a smaller number of CSA networks. It is evident from Fig. 1 that the design can be easily pipelined with paths with small delay differences between registers, as MUXs are more compact compared to the imbalanced CSA trees. Reducing the CSA network in the reduction circuit minimizes the glitches.

A Radix-8 version of the multiplier can be implemented in two different ways as shown in Fig. 2. In Fig. 2(a), the $6a$ multiple can be obtained by shifting the $3a$ one and passing a through the 2:1 MUX whenever $5a$ or $7a$ multiples are needed resulting in simpler encoding block compared to Fig. 2(b). In the second implementation, a 4:1 MUX is required to pass all a, $2a$, and $3a$ multiples whenever $5a$, $6a$ or $7a$ multiples are needed.

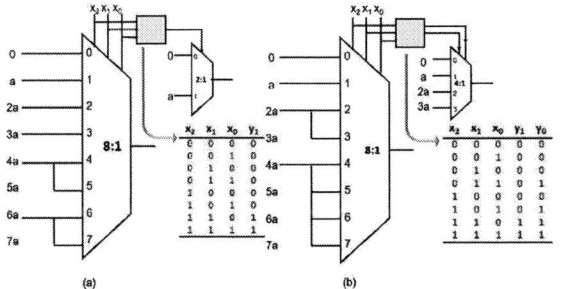

Fig. 2: Two ways of generating the partial products for radix-8 implementation.

This PPG can be extended to multipliers of larger bit-width. The reduction factor increases, if higher radix is used, but at the cost of an extra or more complex combinational circuit. A PPG circuit for unsigned Radix-16 multiplier is presented in Fig. 3, where the odd multiples of the multiplicand, $5a$, $6a$, $7a$, are connected to $4a$ and the combinational circuit in the control path of the 4:1 multiplexer passes 0, a, $2a$, or $3a$, to add the required offset based on the 4-bit multiplier. In other words, we realize $5a$ as $4a + a$, $6a$ as $4a + 2a$ and $7a$ as $4a + 3a$. Similar logic works for the other multiples, which are not power-of-two multiples like $9a$, $10a$, $11a$, connected to $8a$, and $13a$, $14a$, $15a$, connected to $12a$, which is a shifted version of $3a$. In this PPG, we have two partial products of varying bit width that need to be added with a full-adder network and avoid the need for a CSA network tree. So, for the radix-16 implementation of multiplication with a multiplicand of arbitrary bit width by 4-bit multiplier using this setup, no CSA network is required for PPG reduction, as we have only two partial products from the two multiplexers. The largest possible partial product from the 16:1 MUX is $15a$, while the largest possible partial product from the 4:1 MUX is $3a$. Therefore, it can be directly connected to a CPA. The gray combinational block shown in Fig. 3 provides the control signal to the 4:1 MUX, so the corresponding multiples of a are passed to the carry ripple addition stage. The implementation of this combinational logic is shown in the truth table provided in Fig. 3 where input $x3$, $x2$, $x1$, $x0$ are to encoded to $y1$, $y0$.

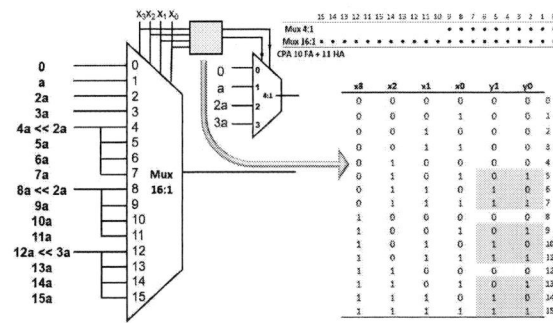

Fig. 3: Higher-radix (16) alternative of the proposed multiplier with 4-bit multiplier. For 32-bit implementation, eight radix-16 blocks are used to generate all partial products.

A 32x32-bit multiplier was designed using the above techniques to compare the proposed design with a multiplier that uses the Booth recoding technique [10]. The radix-16 multiplier blocks shown in Fig. 3 are cascaded to perform 32x32-bit multiplication. Modifying the radix-16 block in Fig. 3 for negative multiplicand and positive multiplier is straight forward. The partial products are usually sign-extended. In general, to minimize the overhead introduced from this sign extension, various techniques such as compensation vector approach, Baugh–Wooley method and its modified version can be used. However, since the partial products are precomputed in our case, the Baugh–Wooley method can not be used. The

979-8-3503-8083-5/23 $31.00 © 2023 IEEE

compensation vector best suits the partial products generated with our approach. The compensation vector exploits the fact that a 2's complement number can be expressed as the sum of a number represented with sign extension and a constant vector [11], as shown in equation 3. Hence, a similar scheme is used in our work.

$$\underbrace{x_{n-1}....x_{n-1}}_{k} x_{n-1}...x_0 = \underbrace{1...1}_{k} \bar{x}_{n-1}...x_0 + \underbrace{0...0}_{k} 1 \underbrace{0....0}_{n-1} \quad (3)$$

If the multiplier is negative, the radix-16 block in Fig 3 can be modified to pass 2's complement multiples of the multiplicand when the MSB of the multiplier is one (Fig 4), effectively subtracting 8x of the multiplicand from the partial sums.

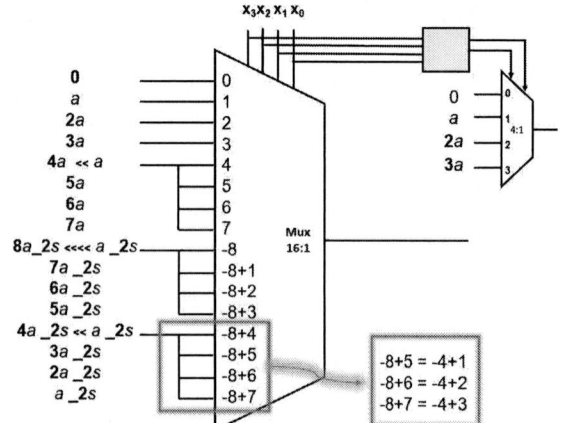

Fig. 4: Radix-16 block for signed multiplication (The leftmost unit from the series of radix-16 blocks). a_2s is a 2's complement version of multiplicand a.

The complexity of wiring and glitches due to unbalanced signal routes of the CSA reduction network, introduces a serious challenge. Booth recoding applies a high-radix multiplication scheme to reduce partial product stages. However, with growing radices, the reduction factor in partial products is dominated by the need to precompute odd multiples of a multiplicand. So we came up with a MUX-based partial product generation to minimize the complexity of encoding for high-radix implementation, while restricting the odd multiples to 3x only.

IV. VERIFICATION AND SYNTHESIS

The proposed multiplexer-based multiplier design was implemented in Verilog and verified for correct functionality. To prove the efficiency of this multiplier, an HDL design was synthesized with Synopsys DC (Design Compiler) tool using GF 22nm FDSOI standard cell library. Moreover, the same synthesis setting and technology library were used to ensure fairness of the comparison of the proposed multiplier with the full-tree implementation of Wallace multiplier and Booth recoding-based multiplier. Both our design and DesignWare multipliers were implemented for 4x4, 8x8, and 16x16-bit multiplications. The reports generated from the synthesis tool

	Critical path (ps)	Area (μm^2)	Power (mW)
signed 32x32 bit radix-16 multipliers			
Proposed design	1533.46	2990.60	0.5980
Booth multiplier [10]	1533.69	3413.86	0.9130
Unsigned Nx4 bit radix-4 multipliers			
4x4-BIT (Proposed design)	532.43	66.76	1.36E-02
4x4-BIT (DW_mult_dx)	541.46	72.68	1.39E-02
8x8-BIT (Proposed design)	910.10	239.01	2.06E-02
8x8-BIT (DW_mult_dx)	910.51	281.09	2.04E-02
16x16-BIT (Proposed design)	1288.93	871.93	3.99E-02
16x16-BIT (DW_mult_dx)	1279.99	626.53	4.37E-02

TABLE I: Comparing various multipliers. DW_mult_dx is a Booth-recoded Wallace-tree synthesis model from Synopsys DesignWare library.

are summarized in Fig. 5 and Table I. The radix-4 version of our multipliers is energy and area-efficient for smaller bit sizes, but requires larger area for higher-bit size operands, with better power consumption. The area, power, and timing of the multiplier presented in [10] that uses Booth recoding was also compared with our design using the same technology. Fig. 6 provides comparison of a 32x32-bit, radix-16 implementation of our design with the Booth multiplier reported in [10]. The result in Table I shows that our design consumes 35% less power and requires 13% less area compared to the design in [10].

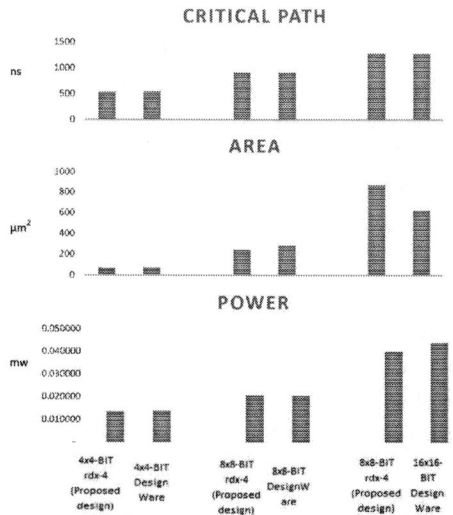

Fig. 5: Synthesis report GF FDSOI.

The critical path length in our work is shorter than the single-pipeline implementation of Wallace multiplier and the multiplier obtained from Synopsys DesignWare library. Our multiplier is also more area-efficient, as shown in Table II. From the synthesis report obtained, the modified version of the 16-bit multiplier architecture shown in Fig. 3 provides better performance and efficiency, when compared to the multiplier realized by the DC synthesis tool. The ICC2 Synopsys PNR tool was used to generate the Floorplan layout shown in Fig.

7.

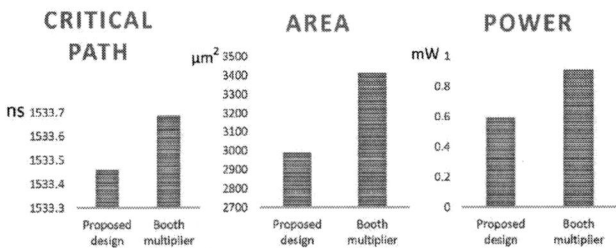

Fig. 6: Performance of the 32x32-bit proposed design vs Booth multiplier.

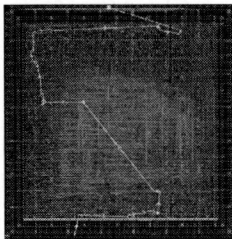

Fig. 7: Layout of the proposed 32x32-bit multiplier (Critical path shown in yellow).

V. Conclusion

In this paper, a MUX-based multiplier was designed to improve efficiency of multiplications for speed-critical applications. With comparable speed and the number of the generated partial products, the design is more area and power-efficient compared to Booth multiplier and a multiplier design synthesized from Synopsys Designware with the DC tool. It is hard to directly compare performance of the various multiplier architectures in the literature, as these were implemented with different technology libraries. In this work, multiplexer-based multipliers were implemented in Verilog-HDL with and without pipeline to be compared against a multiplier synthesized by DC tool, full tree implementation of Wallace and Booth multiplier using GF 22nm FDSOI technology.

To cope with the dynamics of DNN models and increased feature-map bandwidth requirements of these networks, a programmable platform needs to be used to explore the benefits of using non-conventional numbering systems and estimate efficiency of these implementations.

As future work, applying Booth recoding to the multiplier may reduce the height of the partial products generated by a factor of 4.

References

[1] Behrooz Parhami. *Computer arithmetic: Algorithms and hardware designs*. Vol. 39. 7-8. 2000, p. 266. ISBN: 9780195328486. DOI: 10.1016/s0898-1221(00)90295-5.

[2] Kamyar Givaki, Ahmad Khonsari, MH Gholamrezaei, Saeid Gorgin, and M Hassan Najafi. "A generalized residue number system design approach for ultra-low power arithmetic circuits based on deterministic bit-streams". In: *IEEE Transactions on Computer-Aided Design of Integrated Circuits and Systems* (2023).

[3] T Stouraitis and V Paliouras. "Considering the alternatives in low-power design". In: *IEEE Circuits and Devices Magazine* 17.4 (2001), pp. 22–29.

[4] Mohammad Alhawari, Temesghen Tekeste, Baker Mohammad, Hani Saleh, and Mohammed Ismail. "Power management unit for multi-source energy harvesting in wearable electronics". In: *2016 IEEE 59th International Midwest Symposium on Circuits and Systems (MWS-CAS)*. IEEE. 2016, pp. 1–4.

[5] Behrooz Parhami. "A theoretical analysis of square versus rectangular component multipliers in recursive multiplication". In: *2016 50th Asilomar Conference on Signals, Systems and Computers*. IEEE. 2016, pp. 157–161.

[6] Divyansh Jangalwa, M Nagabushanam, and MC Parameshwara. "Design and Analysis of 8-Bit Multiplier for Low Power VLSI Applications". In: *2022 IEEE 2nd Mysore Sub Section International Conference (MysuruCon)*. IEEE. 2022, pp. 1–5.

[7] Daniel W Dobberpuhl, Richard T Witek, Randy Allmon, Robert Anglin, David Bertucci, Sharon Britton, Linda Chao, Robert A Conrad, Daniel E Dever, Bruce Gieseke, et al. "A 200-MHz 64-b dual-issue CMOS microprocessor". In: *IEEE Journal of Solid-State Circuits* 27.11 (1992), pp. 1555–1567.

[8] Glenn Colón-Bonet and Paul Winterrowd Jr. "Multiplier evolution: A family of multiplier VLSI implementations". In: *The Computer Journal* 51.5 (2008), pp. 585–594.

[9] Reid Riedlinger, Ron Arnold, Larry Biro, Bill Bowhill, Jason Crop, Kevin Duda, Eric S Fetzer, Olivier Franza, Tom Grutkowski, Casey Little, et al. "A 32 nm, 3.1 billion transistor, 12 wide issue Itanium® processor for mission-critical servers". In: *IEEE Journal of solid-state circuits* 47.1 (2011), pp. 177–193.

[10] Hani H. Saleh, Baker S. Mohammad, and Earl E. Swartzlander. "The optimum Booth radix for low power integer multipliers". In: *2013 8th IEEE Design and Test Symposium*. 2013, pp. 1–4. DOI: 10.1109/IDT.2013.6727119.

[11] Xin Lou, Ya Jun Yu, and P.K. Meher. "New Approach to the Reduction of Sign-Extension Overhead for Efficient Implementation of Multiple Constant Multiplications". In: *IEEE Transactions on Circuits and Systems I: Regular Papers* 62 (Nov. 2015), pp. 1–11. DOI: 10.1109/TCSI.2015.2476319.

2023 International Conference on Microelectronics (ICM)

Development and optimization of a planar wideband ultrathin absorber based on equivalent circuit model analysis

Yasmine Abdalla Zaghloul
Electrical Engineering Departement
German International University
Cairo, Egypt
yasmine.zaghloul@giu-uni.de

Hany Hammad
Information Engineering and Technology
German University in Cairo
Cairo, Egypt
hany.hammad@guc.edu.eg

Abstract— **Grounded ultrathin single layer absorbers are studied in detail in this paper under normal incidence. The analysis proposed aims to model the theory of near unity absorption using non-rigorous equations. In addition to implementing the near perfect design perspectives for ultrathin absorbers using non-complicated structures. The analysis results show that planar patch absorber structure can be adapted to generate multiple adjacent resonances providing a wider bandwidth. Furthermore, fabricated prototypes have been measured for model validation and calculations verification. The agreement between the analytical calculations, simulations results, and measurements has demonstrated the efficiency of the proposed analysis technique and design methodology. Different absorbers' configurations available in literature are summarized in comparison to the work presented in this paper.**

Keywords— *absorber, equivalent circuit model, frequency selective surface, planar, oblique incidence, wideband.*

I. INTRODUCTION

LATELY, there has been a growing interest in the area of electromagnetic wave absorbers for their wide variety of applications [1-6]. An ideal electromagnetic wave absorber should realize near unity absorption for a broad frequency band independent of the polarization angle (ϕ) and the incidence angle (θ) [7-9]. However, it is challenging to design a planar single layer non-complicated absorber structure and yet achieve the near perfect characteristics. Up to date, no simplified guidelines are available in literature to model, design and scale ultrathin absorber structures with stable near unity performance [10]. Hence, the main objectives of this paper are; firstly, define a methodology to analyze the theory of near unity absorption using non-rigorous equations for ultrathin absorbers and then implement the near perfect design perspectives for ultrathin absorbers using non-complicated planar structures. The work presented followed a simple design analysis technique (equivalent circuit model) to understand the basic theory of absorption and give an initial physical insight about the absorption mechanism. This analysis provides a realization for the absorption response (in terms of resonance frequency and absorption ratio) for non-complicated rectangular ultrathin patch absorbers using simple equations (substrate thickness < $\lambda/10$). The proposed equations conclude an initial estimation for the design structure dimensions (patch and substrate

geometrical length and width) for a given dielectric material (substrate thickness and permittivity). The analytical calculations proposed is then integrated to implement the design of wide band ultrathin planar configurations. This paper is organized as follows: Section II explains the equivalent circuit derivation process of the analytical calculations and their physical interpretations. The proposed guidelines for the circuit model calculations is then extended in Section III for multiple *N*-sub cell pattern to generate multiple resonances instead of single resonance cell analysis. Moreover, various reported techniques [11-13] have been studied and a comparison to the proposed design is also summarized. Finally, a conclusion is presented in Section V.

II. EQUIVALENT CIRCUIT MODEL FORMULATION FOR RECTANGULAR PATCH ABSORBERS

Generally, the patch absorber unit cell structure consists of three layers as shown in Figure. 1; the first layer is a patch metallic pattern; the second layer is a lossy dielectric substrate and the third layer is a full ground plane. The absorption mechanism of ultrathin absorbers has been studied in detail under normal incidence and an equivalent circuit model has been formulated accordingly for the ultrathin grounded patch elements [14]. As demonstrated in [14], in z-direction of TM mode (E_x&H_y field components), the patch absorber can be represented by two anticipated absorbing slots of width (W_p) and height (h), separated by the distance (L_p), these two slots can give an approximate account for the structure absorption.

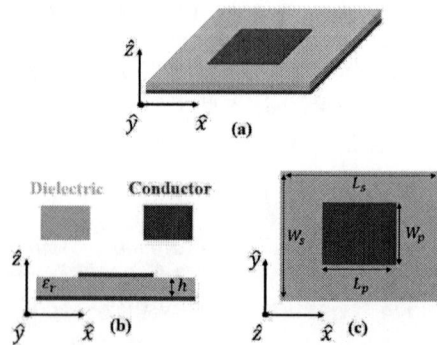

Figure. 1: Single layer grounded patch absorber unit cell (a) Isometric view, (b) Side view, (c) Top view

979-8-3503-8083-5/23 $31.00 © 2023 IEEE

The proposed model shown in Figure.2 can be implemented following the below simple guidelines [18]:

- The stimulated non-transverse E-field component modeled as the capacitance C_{RLC}.
- The surface currents at the patch edges modeled as the inductance L_{RLC}.
- The losses of the dielectric substrate modelled as the resistance R_{RLC}.

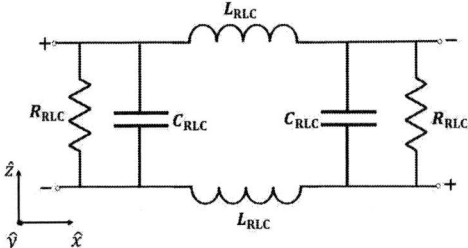

Figure. 2: patch absorber unit cell equivalent circuit model

The total input impedance Z_{in} under normal incidence is verified in [15] as a series combination of two identical parallel RLC circuit model as shown in Figure. 2. and is given by (1)

$$Z_{in}(\omega) = \frac{2}{\left(\frac{1}{R_{RLC}} + \frac{1}{j\omega L_{RLC}} + j\omega C_{RLC}\right)} \qquad (1)$$

where ω is the angular frequency, R_{RLC}, L_{RLC} and C_{RLC} are the circuit model resistance, inductance and capacitance. Note that a square patch absorber configuration is initially used for a symmetrical response under normal incident for both TE and TM polarizations [16]. Following the basic concepts of electromagnetic theory [17] considering a lossless metallic layer, an approximation for the RLC lumped parameters in terms of the absorber physical dimensions can be deduced as given in (2-4) where λ is the material space wavelength, k is the wave number, η is the characteristic impedance for the dielectric material, c is the velocity in free space, σ and ε_{eff} is the conductivity the effective substrate permittivity of the substrate calculated using the formulas presented in [17]. L_p and W_p are the patch length and width. L_s and W_s are the substrate periodic cell length and width.

$$C_{RLC} = \frac{W_p}{\eta \lambda \omega_o}[1 - 0.636 \ln(kh)] \qquad (2)$$

$$L_{RLC} = \frac{L_p}{W_p} \frac{\eta\sqrt{\varepsilon_{eff}}}{c\omega_o} \qquad (3)$$

$$R_{RLC} = \frac{h}{\sigma L_s W_s} \qquad (4)$$

The input impedance can then be analytically calculated for minimal reflection at ($Z_{in}(\omega_o) = Z_0 = 377\Omega$) matching condition and max absorption achieved at the desired frequency accordingly. A periodic arrangement of a grounded metallic square patch on lossy $FR4$ substrate ($h=0.8mm$, $\varepsilon_r=4.3$ & $tan\delta=0.025$) is then designed at resonance $f_0=8.5$ GHz. Initial optimization for the absorber dimensions with minimal reliance on full wave simulation during the design process is established using incorporated equations in this

section. The optimized physical dimensions ($L_p=W_p=8mm$, $L_s=W_s=20mm$) are extracted from (1-4) to be simulated on CST using unit cell boundary conditions. A fabricated prototype of the periodic arrangement of 14×14 cells ($28 \times 28\ cm^2$) is also measured for verification. As illustrated in Fig. 4, the measurement test set up is held where two horn antennas were used to measure the reflections from the absorber. The distance from horn antennas is considered as far as possible (approximately $2m$) to realize normal incident plane wave.

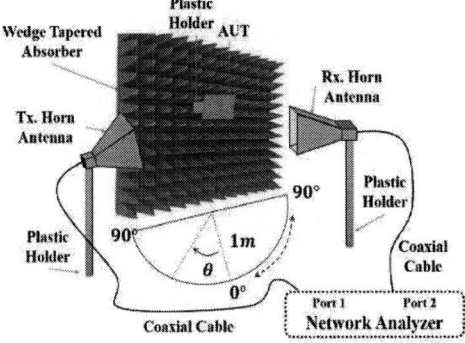

Figure. 3: Measurement setup diagram for electromagnetic wave absorbers

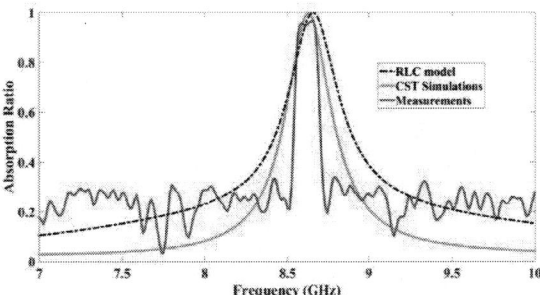

Figure. 4: Comparison between the RLC model calculations, CST simulations and measurements for the designed patch absorber at 8.5 GHz

As demonstrated in Fig. 4, the RLC model calculations are validated using CST simulations and measurements for the single resonance patch absorber. The slight deviation simulations and calculations may go back to the assumptions ignoring the mutual coupling between the periodic cells of the absorber structure and the conductor losses in the analytic equations. Yet, the developed analysis (see (1-4)) has concluded that, for a given substrate parameters (ε_r, $tan(\delta)$ and h), the absorption resonance frequency is controlled through the metallic patch dimensional variations (L_p and W_p) while the absorption amplitude ratio is controlled through substrate dimensional variations (L_s and W_s) with no significant effect on the resonance frequency. Based on the aforementioned guidelines, the analysis is extended for multiple N-sub cells instead of single cell to generate multiple resonances serving a broader bandwidth for grounded planar absorber structures as discussed in the next section.

III. PROPOSED WIDE-BAND DESIGN IMPLEMENTATION

A planar periodic arrangement of N patch cells' pattern is then employed to generate multiple adjacent resonances under normal incidence. The wideband implementation is based on introducing slight variation (δ_n) in the patch size ($L_p \& W_p$) for each n-subcell so that a variation in the lumped parameters ($L_{RLC-n} \& C_{RLC-n}$) occurs and the resonance frequency accordingly ($f_0 = 1/(2\pi\sqrt{L_{RLC}C_{RLC}})$). Keeping in mind a fixed cell size ($L_s \& W_s$) for which the real impedance parameter (R_{RLC}) is adjusted for free space matching impedance, one can then calculate the total effective input impedance as an average impedance ($Z_{in-average}$) for the N-subcell pattern using (5)

$$Z_{in-average}(\omega) = \frac{1}{N}\sum_{n=1}^{N} Z_{in-n}(\omega) \qquad (5)$$

where Z_{in-n} is the n sub-cell input impedance calculated under normal incidence for each ($L_{RLC-n} \& C_{RLC-n}$) separately using (1-3). To proceed, square symmetrical design is chosen for the rectangular patch and the dielectric substrate to ensure the four-fold symmetry of the structure keeping a polarization insensitive configuration. The unit cell is assumed to be divided into N number of sub-cells as shown in Fig. 5. Each sub-cell has its own resonating dimensions ($L_1 \neq L_2 \neq .... L_N$) with fixed square substrate dimensions ($L_{s1} = L_{s2} = ...P$) as presented in Table I.

TABLE I
TABLE OF DIMENSIONS FOR THE PERIODIC ABSORBER STRUCTURE ESTIMATED FROM THE RLC MODEL CALCULATIONS, $f_0 = 8.5$ GHz

	L_1	L_2	L_3	L_4	L_5	L_6	P	FBW
Value (mm)	N=1 Cell Pattern ($\delta_n = 0$)							
	8	8	8	8	8	8	20	4%
	N=3 Cell Pattern ($\delta_n = 0.25$)							
	8.5	8.25	8	8.5	8.25	8	16	12.5%
	N=6 Cell Pattern ($\delta_n = 0.25$)							
	8.5	8.25	8	7.75	7.5	7.25	12	25%

Using equal patch size with $\delta_n = 0$ for the $N=1$ sub-cell pattern, which is basically similar to the single cell design response ($L_1 = L_2 = L_3 = L_4 = L_5 = L_6$). The dimensions calculated at resonance 8.5 GHz following (1-5) are as reported in Table I. The simulated impedance view of the s-parameters presented in Fig. 5 shows a single resonance at the center frequency (simulated $f_{min} = 7$ GHz & $f_{max} = 10$ GHz) giving a fractional bandwidth of 4%. For $N=3$ and $N=6$, a slight variation $\delta_n = 0.25$ is introduced as shown in Table I. The impedance view of the S-parameters demonstrates multiple peak resonances during simulation giving a fractional bandwidth of 12.5% and 25% for $N=3$ and $N=6$ respectively. A decrease can be noticed in the periodic substrate cell size (P) from 20 to 16 mm and to 12 mm following the inverse relation between the substrate dimensions and real impedance R_{RLC} as given in (4). This happens to achieve higher n sub-cell impedance in order to be able compensate the average impedance calculation reported in (5) fulfilling free space matching condition. To be stated, a greater number of sub-cells can achieve a wider bandwidth response however a limitation on the minimal cell size shall be considered. A prototype of the N-sub cell pattern structures ($N=1$, $N=3$ and $N=6$) is then fabricated for verification. The increase in bandwidth from 4% to 25% at absorption higher than 85% is noticed in Fig. 9.

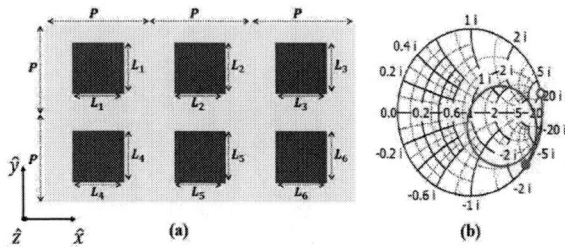

Figure. 5: Proposed N sub-cell pattern (a)$N=1$ (b)S-parameters (impedance view)

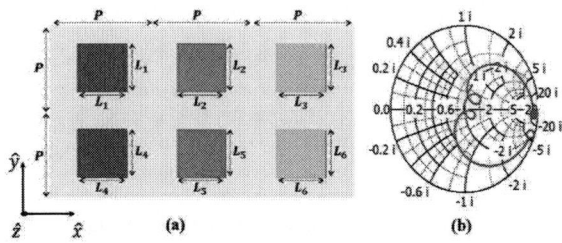

Figure. 6: Proposed N sub-cell pattern (a)$N=3$ (b)S-parameters (impedance view)

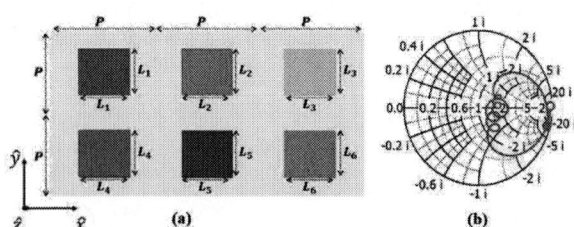

Figure. 7: Proposed N sub-cell pattern (a)N=6 (b)S-parameters (impedance view)

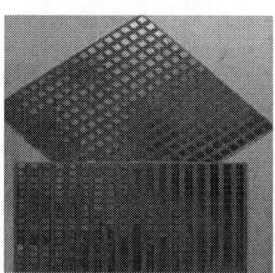

Figure. 8: Fabricated Prototypes of the proposed planar N sub-cell Pattern

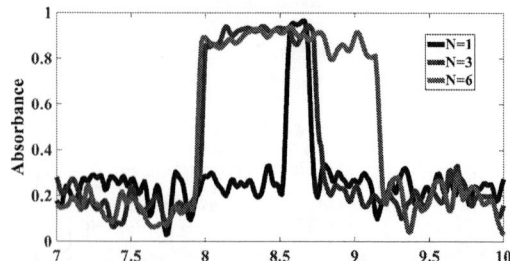

Figure. 9: The measured absorbance for the proposed N sub-cell pattern

979-8-3503-8083-5/23 $31.00 © 2023 IEEE

TABLE II
COMPARISON SUMMARY FOR VARIOUS STRUCTURES ACHIEVING WIDEBAND RESPONSE WITH ABSORPTION RATIO ABOVE 85%

Design Structure	FBW (%) (f_c)	COMPLEXITY	CIRCUIT MODEL
Thick substrate [11]	80% (10GHz)	Simple (planar)	✓
Thick substrate [12]	100% (6GHz)	Complex (multilayer structure)	✓
Thin substrate [13]	55% (6GHz)	Complex (loaded with lumped elements)	☒
Proposed pattern Thin substrate	25% (8.5GHz)	Simple (planar)	✓

As presented in Table II, in comparison to selected structures presented in literature [11-13], the developed single layer patch design achieves a wide band response keeping the ultrathin planar feature. Moreover, the analysis formulated guidelines is based on equivalent circuit model parameters' optimization following non-rigorous equations.

IV. CONCLUSION

The work in this paper presents a methodology to understand the absorption mechanism of grounded ultrathin patch elements under normal and oblique incidence. An equivalent circuit model is formulated that provides a realization for the absorption response (in terms of resonance frequency and absorption ratio) for non-complicated rectangular patch absorbers using simple equations. The analytical calculations proposed is then integrated to implement the design of wide-band ultrathin absorber configurations. The design methodology proposed is based on using multiple sub-cells (N) of simple single layer patches to add another degree of freedom for the design equations keeping the ultrathin planar feature. A planar single layer wideband ultrathin absorber structure is developed with an increase in bandwidth from 5% to 25% at absorption higher than 85%. Different prototypes were fabricated & measured providing a good agreement with the RLC model calculations and design simulations. Finally, a Comparison summary with previously reported techniques is presented to validate the competence of the proposed work

REFERENCES

[1] C. M. Watts, X. Liu, and W. J. Padilla, "Metamaterial Electromagnetic Wave Absorbers (Adv. Mater. 23/2012)," *Advanced Materials*, vol. 24, no. 23, Dec. 2012.

[2] N. Liu, M. Mesch, T. Weiss, M. Hentschel, and H. Giessen, "Infrared perfect absorber and its application as plasmonic sensor," Nano Lett. 10, 2342–2348, 2010.

[3] O. T. Gunduz and C. Sabah, "Energy harvesting through lumped elements located on metamaterial absorber particles," *2015 IEEE 5th International Conference on Consumer Electronics - Berlin (ICCE-Berlin)*, Berlin, pp. 314-317, 2015.

[4] M. Diem, T. Koschny, and C. M. Soukoulis, "Wide-angle perfect absorber/thermal emitter in the terahertz regime," *Phys. Rev. B 79*, 033101, 2009.

[5] D. Jaggard, N. Engheta and J. Liu, "Chiroshield: a Salisbury/Dallenbach shield alternative", *Electronics Letters*, vol. 26, no. 17, p. 1332, 1990.

[6] L. D. C. Folgueras and M. C. Rezende, "Multilayer radar absorbing material processing by using polymeric nonwoven and conducting polymer," *Materials Research*, vol. 11, no. 3, pp. 245–249, 2008.

[7] S. Kalraiya, R. K. Chaudhary, R. K. Gangwar, and M. A. Abdalla, "Compact quad-band polarization independent metamaterial absorber using circular/square metallic ring resonator," *Materials Research Express*, vol. 6, no. 5, p. 055812, 2019.

[8] A. Sellier, T. Teperik and A. de Lustrac, "Resonant circuit model for efficient metamaterial absorber", *Optics Express*, vol. 21, no. 6, 2013, p. A997, 2013.

[9] M. Zhang, T. Jiang, and Y. Feng, "Design and Measurement of Microwave Absorbers Comprising Resistive Frequency Selective Surfaces," *Journal of Electromagnetic Analysis and Applications*, vol. 06, no. 08, pp. 203–208, 2014

[10] A. Kazemzadeh and A. Karlsson, "On the Absorption Mechanism of Ultrathin Absorbers", *IEEE Transactions on Antennas and Propagation*, vol. 58, no. 10, pp. 3310-3315, 2010.

[11] D.-U. Sim, J.-H. Kwon, S.-O. Park, and Y.-J. Chong, "Design of electromagnetic wave absorber using periodic structure and method to broaden its bandwidth based on equivalent circuit-based analysis," *IET Microwaves, Antennas & Propagation*, vol. 9, no. 2, pp. 142–150, 2015.

[12] Z. Zhou, K. Chen, B. Zhu, J. Zhao, Y. Feng, and Y. Li, "Ultra-Wideband Microwave Absorption by Design and Optimization of Metasurface Salisbury Screen," IEEE Access, vol. 6, pp. 26843–26853, 2018.

[13] X. Lin, P. Mei, P. Zhang, Z. Chen and Y. Fan, "Development of a resistor-loaded ultra-wideband absorber with antenna reciprocity," IEEE Transactions on Antennas and Propagation, Vol. 64, noon the. 11, pp.4910-4913, 2016.

[14] Y. Abdalla and H. Hammad, "Simplified formulation and realization for ultrathin near perfect absorbers with rectangular microstrip elements using cavity and transmission line models," *The Loughborough Antennas & Propagation Conference (LAPC 2018)*, Loughborough, pp.1-5, 2018.

[15] B. Munk., "On Designing Absorbers for an Oblique Angle of Incidence," in *Metamaterials: Critique and Alternatives*, New York: John Wiley & Sons, ch. 4, pp. 71-91, 2009.

[16] D. Kundu, A. Mohan and A. Chakrabarty, "Comment on 'Wide-angle broadband microwave metamaterial absorber with octave bandwidth'," in *IET Microwaves, Antennas & Propagation*, vol. 11, no. 3, 2017, pp. 442-443, 2017.

[17] A. B. Constantine, Antenna theory: analysis and design, *John wiley & sons*, 2005

[18] M. William, Y. Abdalla and H. Hammad, "Optimizing angular independency and polarization insensitivity of FSS absorbers by adopting far-field analysis of patch antennas," *The Loughborough Antennas & Propagation Conference (LAPC 2018)*, Loughborough, 2018, pp. 1-5.

Augmented Reality as an Educational Enrichment Tool: Integrating the Virtual with the Real

Eder Costa Maciel*, Miller Henrique Lúcio Fernandes*, Tales Cleber Pimenta†,
Jaqueline Corrêa Silva de Carvalho† and Marcos Alberto de Carvalho†
*José do Rosario Vellano University, BRA
Faculty of Computer Science, Alfenas, Minas Gerais 37132-440
†Federal University of Itajubá, BRA
Institute of Systems Engineering and Information Technology (IESTI), Itajubá, Minas Gerais, 37500-903

Abstract—In recent decades, we have witnessed significant advancements in the application of Augmented Reality (AR) in education. This article explores the integration of AR into the educational system and examines its concrete impacts on the learning process. Through the development of a mobile AR application, initially focused on the field of Science, we aim to provide a clearer and more tangible perspective on the improvements that this technology can bring to education. The results of this study reveal that AR has stimulated increased classroom engagement, interaction, and content retention. Experiments were conducted in a local school classroom with the participation of 18 elementary school students. It was observed that the average accuracy rate in identifying planets reached 92.5%. The findings suggest that AR has the potential to significantly enhance the learning process, serving as an incentive for future research and the expansion of this approach to other disciplines, promoting a more interactive and engaging education.

Index Terms—Educational Enhancement, Augmented Reality, Technology in Education.

I. INTRODUCTION

Augmented Reality (AR) is a technological feature that overlays real-world objects with virtual resources generated through computers, typically represented in three dimensions (3D), in a real-time and real-world environment. In the field of education, AR can be employed in various ways. For example, it can encourage students to analyze the real-world environment by providing supplementary information (virtual objects) linked to the real world. Moreover, AR has the capacity to extend the real world by incorporating digital resources, like representing objects that can't be physically present due to size, shape, or composition constraints, such as geometric objects and natural elements. This provides students with a more dynamic experience of the subject matter being studied [1] [2].

However, AR alone cannot solve the problems of education, but it can offer significant benefits. Therefore, this technology should act as support for teachers and students, enhancing the learning experience in current educational settings. As such, AR should not be viewed solely as a technological resource, but rather as a complement to the teaching concept, as the focus is on the learning outcome of the entire process [3] [4].

A major challenge of augmented reality is to make virtual elements appear as part of the real environment and seamlessly integrate with it. Thus, it is believed that AR can offer a new way to present content compared to conventional methods, enabling greater student interaction with subjects taught in the classroom [5]. This technology is capable of representing objects and phenomena that were previously impossible to observe with such reality and proximity. Consequently, students can observe what was once only viewable through images, textual descriptions, or imagination.

Augmented reality and virtual reality (VR) are related concepts. In a VR environment, the observer or user is completely immersed in a synthetic environment developed for the application, which can optionally mimic the characteristics of a real-world environment, such as physics and materials. On the other hand, AR enhances the strictly real-world environment, maintaining compliance with the laws of physics. AR incorporates a mechanism that blends the real and virtual worlds, keeping the user in the real world and emphasizing quality advantages and interaction [6].

Using AR not only stimulates but also facilitates students acquisition of knowledge and assists teachers in their educational practices, expanding options for more dynamic teaching methods [7]. This methodology is well-suited for subjects where student abstraction becomes more complex, such as studying the planets in the solar system.

The main objective of this work was to construct, through AR analysis and foundations, a mobile application that delivers significant results for education, particularly in the subjects of science and geometry. Through the application, the aim was to provide an enhanced learning experience for elementary school students, dynamically revitalizing the teaching method and making it less monotonous.

As a result of this work, it is expected to reduce the complexity of student learning through their interaction with the elements created by the application and to enable the integration of AR into the conventional education system, allowing it to act as a complement to the learning method.

II. MATERIAL AND METHODS

A. Vuforia

Vuforia is a cross-platform software development kit (SDK) used to create applications on iOS, Android, and Windows platforms [8]. This tool can be integrated with various development environments such as Android Studio, Unity, or

979-8-3503-8083-5/23 $31.00 © 2023 IEEE

Microsoft Visual Studio, allowing programming in languages such as Java, C#, and C++. The SDK provides a range of AR features including visualization, 3D object rendering, marker (trigger) tracking, and 3D mapping of real-world environments, enabling surface detection and other functionalities. When combined with development platforms, it offers a considerable range of resources that simplify the creation of complex AR applications [9].

B. Markers

Marker-based AR uses a camera along with some form of visual marker, such as a QR/2D code (Fig. 1), to generate a result only when the marker is recognized by a reader. Applications utilizing markers utilize the device's camera to distinguish a marker from other objects in the real environment. QR codes are often used as markers due to their easy recognition and low processing power demand for reading. The position and orientation of the marker are also calculated, allowing some type of content or information to be superimposed onto the marker. Markers, also known as triggers, play a crucial role in AR technology [6].

Fig. 1. Example of a Marker (trigger).

Markers are used to capture and register virtual objects as follows: the marker's symbol is identified, indicating which object should be incorporated into the scene to obtain the geometric transformation matrix that defines the relative positioning between the marker and the video camera. With this information available, the software can insert interactive 3D objects into the real scene, seamlessly integrating them with the surrounding elements.

AR relies heavily on capturing images in real environments, and for this purpose, there are techniques to perform tracking and recognition of these images, known as computer vision techniques.

C. Unity

With its first version released in 2005, Unity is a complete game engine equipped with a powerful rendering engine and an intuitive workflow tool. Originating as a proprietary game engine, Unity was developed and is maintained by Unity Technologies. The Unity development environment is highly suitable for implementing AR software. It is a high-level environment for applications and games [9].

This platform is available for various operating systems such as Windows, Mac, and Linux. Additionally, it offers free versions for individual developers.

Unity features an advanced 3D engine that enables the rendering of 3D models and animations in various formats. In this environment, elements can be created, positioned, animated, and incorporated into a 3D scene. It is also possible to develop scripts in the C# language, which can be connected or not to scene objects, leveraging all the resources available on the platform.

III. RESULTS

Using the Unity 3D tool in conjunction with the C# language, scenes and objects were created and displayed. Through these tools, it became possible to manipulate all the objects. Object capture was performed through markers. As for the storage of used markers, Vuforia was employed.

With the aid of these tools, the development of the first version of the mobile application for the Android operating system was achieved. The application is capable of recognizing markers present on the student's school materials and projecting them in 3D on the mobile device screen. In this way, the student can manipulate and view the object dynamically.

After creating the first version of the application, experiments were conducted in a private school located in Alfenas–MG, Brazil, involving a class of 18 elementary school students. The application was initially made available to the teacher for prior familiarization and later to the students to complete the proposed activity.

To evaluate the performance of students during the application's use, a booklet containing recognition markers was provided (Fig. 2). For each marker, the student identified the planet projected on the mobile device screen.

Usability and ease of navigation aspects within the application were considered to facilitate its usage. Fig. 3 illustrates some screens of the application.

Fig. 2. Student using the application.

Through experiments conducted throughout the study, an increase in student engagement was observed, resulting in satisfactory outcomes. The data presented in Table I shows that the average accuracy rate for planet identification was 92.5%, demonstrating good performance.

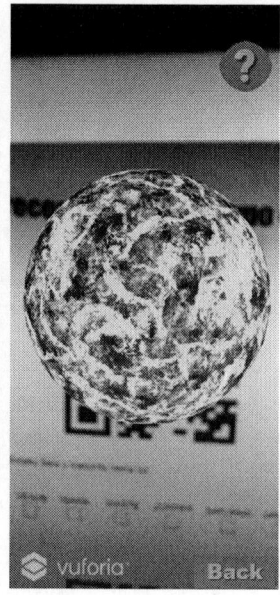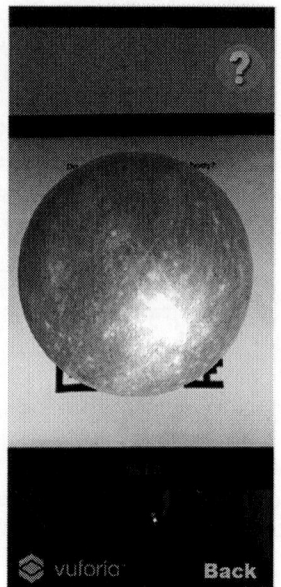

Fig. 3. Screens of the mobile application.

TABLE I
RESULTS OBTAINED FROM THE QUESTIONNAIRE

Planets	Percentage of Correct Answers
Sun	100%
Mercury	88.8%
Venus	88.8%
Earth	100%
Mars	100%
Jupiter	94.4%
Saturn	100%
Uranus	77.7%
Neptune	83.3%
Average	92.5%

The utilization of the application in the school environment facilitated greater interaction by students during classes. This encouraged them to engage more with their classmates and teachers, promoting targeted and pertinent questions related to the content presented. Additionally, the application stimulated students to learn collaboratively, dynamically, and creatively.

The use of augmented reality made it possible to reduce the complexity and abstraction of the covered content, utilizing the mobile device as a supplementary tool alongside the traditional teaching method.

IV. CONCLUSIONS

This study aimed to enhance the learning abilities of elementary school children, with an initial focus on the subject of Science. We sought to make a significant contribution to the educational system, overcoming challenges and simplifying the transmission and illustration of classroom-taught content, particularly in the aforementioned knowledge areas.

Within this context, we explored the feasibility of integrating technologies, such as augmented reality, into tra-
ditional teaching methods in educational institutions. One of the difficulties encountered during this process revolved around the implementation of highly detailed objects. This challenge necessitated a meticulous study for the creation of three-dimensional modeled images to ensure the quality and accuracy of virtual representations. With this perspective in mind, we acknowledge the importance of future research to expand this project, progressively incorporating subjects such as Geometry, Biology, Arts, and Geography.

ACKNOWLEDGMENT

The authors would like to thank FAPEMIG, CNPq and CAPES for their financial support.

REFERENCES

[1] M. E. Kouzi, A. Mao, and D. Zambrano, "An educational augmented reality application for elementary school students focusing on the human skeletal system," in *2019 IEEE Conference on Virtual Reality and 3D User Interfaces (VR)*, 2019, pp. 1594–1599.

[2] L. O. Lopes and V. Gonçalves, "Evaluation of the augmented reality educational application for the 2nd cycle of primary school," in *2021 16th Iberian Conference on Information Systems and Technologies (CISTI)*, 2021, pp. 1–6.

[3] T. Bratitsis, P. Bardanika, and M. Ioannou, "Science education and augmented reality content: The case of the water circle," in *2017 IEEE 17th International Conference on Advanced Learning Technologies (ICALT)*, 2017, pp. 485–489.

[4] H. kai Wu *et al.*, "Current status, opportunities and challenges of augmented reality in education," *Computers & Education*, vol. 62, no. 62, pp. 1–26, Mar/Mar. 2013.

[5] R. Tori, "A presença das tecnologias interativas na educação," *Revista de Computação e Tecnologia (ReCeT)*, vol. 2, no. 1, pp. 4–16, 2010.

[6] R. Tori, C. Kirner, and R. A. Siscoutto, "Fundamentos e tecnologia de realidade virtual e aumentada," in *Editora Sbc*, 2006, pp. 2–21.

[7] P. Sarkar, "Exploring design strategies for augmented reality learning experience in classrooms," in *2020 IEEE International Symposium on Mixed and Augmented Reality Adjunct (ISMAR-Adjunct)*, 2020, pp. 314–316.

[8] F. Leighton, "Developing mobile augmented reality with unity and vuforia," *MW20: MW*, 2020.

[9] D. Omaia and L. Machado, "Realidade aumentada com vuforia e unity," in *Tutoriais - Simpósio de Realidade Virtual e Aumentada (SVR), 22.* Sociedade Brasileira de Computação, 2020, pp. 61–6.

The Influence of Artificial Intelligence in Society

Milene Santos Moreira, Douglas de Tarso da Silva and Tales C Pimenta
UNIFEI
Itajubá - Brazil
milenesantos@gea.inatel.br

Abstract—The Artificial Intelligence – AI is continuously conquering space in our daily lives. Tasks carried out previously only by humans are now conducted by automated machines and electronics devices, including text and voice real time response. Complex activities are also being carried out by artificial intelligence to improve productivity, cut costs and assists in the development of new industrialized products, thus accelerating the time to obtain new drugs. This article makes an analysis of how the societies has been influenced by artificial intelligence, considering the most populous eastern countries, China and India; and the most populous western countries, United States of America and Brazil. Despite their great cultural differences, those nations are among the largest economies in the world. This article also shows some similarities and concerns about AI applications into several fields.

Keywords— Artificial Intelligence; Machine Learning; AI Applications; Social Influence; Brazil; China; India; United States of America

I. Introduction

Artificial intelligence is a subarea of computer technology that is popularizing through several applications in practically all areas of knowledge, influencing economical, scientific and cultural activities of humanity during the last decades. Among the areas of most evident impact are the entertainment that utilize digital marketing on a large scale, such as cinema, television, electronic games; and social media to reach as many people as possible and influence them to buy products or shift common sense related to political and social facts [1].

In the industry, artificial intelligence is shown in automation, optimization of productive systems and in the development of new final products. In the defense area, artificial intelligence is found in Unmanned Aerial Vehicles (UAV), combat drones, autonomous antiair systems, teleguided missile systems, radar signal reconnaissance, among other weapons and defense systems [2].

In the healthcare environment, artificial intelligence is found in image recognition systems to help medical diagnosis of magnetic resonance imaging, ultrasound and x-ray, medical decision support systems and healthcare management [3, 4].

In agriculture intelligent systems are used in the remote control of tractors and agricultural drones automated with GPS location guidance, weather forecasting help, genetical improvement of plants and in the energy and natural resources management [5-8].

This article addresses the consequences and social impacts in some areas of the world, caused by the implementation of Artificial Intelligence, observing more recent times.

II. Artifical Intelligence

The artificial intelligence is a branch of computer science and engineering with the objective to reproduce the cognitive characteristics of human beings into computational systems and machines. It makes use of knowledge obtained from neuroscience, with the objective of accomplishing work and solve complex problems autonomously, faster, more precise and efficient when compared to a human being [4].

Artificial Intelligence also utilizes discreet mathematics knowledge, statistics, probability, and linear algebra to represent and solve complex problems that involve logical reasoning, syllogism, deduction, information classification, optimization, and decision-making [9]. It is possible to split AI in two major subgroups. The first group is machine learning and the second is deep learning.

A. Machine Learning

Using a data classification software specific to the application, the system is trained to group the input data by similarity or proximity, rating them in groups and subgroups, and returning as output, or most probable answer, the elements of higher percentage of similarity to a reference objective. There are three categories of Machine Learning [10]:

- Supervised learning: uses a predefined database with data patterns already embedded in the data sample, so the system just makes a comparison of the obtained data with the existing data [10, 11].
- Unsupervised learning: the system qualifies and regroup the obtained data by similarity among them without external comparison references [10, 12].
- Reinforcement leaning: it is implemented a scoring or rewarding system based on a trial and error and the system learns when it obtains the highest score possible [12].

B. Deep Learning

It evolved from the Machine Learning and it is the model closest to the human brain. It uses neural networks arrays in layers to organize random unstructured data in an interactive and recursive way. Its main applications are in the cognitive areas of speech recognition, computer vision, and image recognition and interpretation [3].

The most implemented neural network topology are the ones that utilize perceptron models, multilayer perceptron, convolutional neural networks, recurring, self-recurring, regenerative adversarial networks, among others.

III. IMPACTS OF AI AROUND THE WORLD

Education, cultural habits and economical capabilities are vastly different among people and nations. Those differences reflect the needs and priorities of artificial intelligence applications in a wide range of human activities that impact the quality of life and society.

A. Brazil

In Brazil, AI is used in several branches of technology that influence society in important manners, such as public safety, judiciary system, economy, agriculture and leisure, among others. Some of the applications that influence the most daily activities are:

- Facial recognition: It can identify facial expression that show the emotional status, verify satisfaction level during interaction with products or digital services and comprehend behavioral patterns to improve these services. It can be used to control people access to restricted areas, as a biometrical authentication to authorize payments, recognize outlaws through cameras strategically placed in airports, bus stations and other public places of a high flux of people. This technology can also be applied in the recognition of plagues in plantations.
- Smart homes: In domestic applications, AI combined to Internet of Things (IoT) bring home automation to an elevated level. It is possible to manage the contents of refrigerators, in which the system informs of food close to expiration date or close to finish. Electronic equipment can be controlled by voice commands, and autonomous vacuum cleaners can adapt their operation to the moments of smaller flux of people.
- Costumer care: Automation of call centers with AI-based solutions has made a big leap. Chatbots allow people to be answered faster, reducing wait times. The systems are trained to solve nearly all types of problems, from sales to technical support and service cancellations. When the problem is out of the system capability, the call is transferred to a human and it is used as learning to update the automated system, this way it will be ready to answer a new similar call.

The growth of AI brought concerns about the negative consequences to the Brazilian society are drawing attention of government agencies that are conducting discussions about preventive and corrective measures to mitigate potential damages and losses [13].

Prejudice is a great concern. Credit analysis already uses AI and some people claim their credit was denied due to their address, or other personal detail. There are cases of people arrested by mistake due to their resemblance to an outlaw [14].

Another by concern is related to privacy. People have doubts and fears about their personal information such as spending behaviors, biometric data, economic data, and social relationships. To try and solve this problem, it was issued the Universal Data Protection Bill (law number 13,709 of August 14, 2018) [15]. This bill tries to limit the access to individual personal data that can feed artificial intelligence systems to obtain information in unauthorized and invasive ways.

B. China

In 2017, the Chinese State Council created the Plan of Development of Next Generation AI and it was established a new goal for China to become the global innovation center by 2030 [16]. The policy dictates the Chinese strategy to inject an amount of nearly 150 billion dollars in the domestic industry and to become the main AI power by 2030 [17].

In 2019, the Beijing Academy of Artificial Intelligence (BAAI) created the Beijing AI principles in which the development of AI is aligned with the future of society, humanity, and the environment. These principles present a long-term for the research, development, use, governance, and planning of AI, for the society and environment [18]. The principles also account for:

- Projecting and developing AI systems to provide development of sustainable society and nature.
- Serve society and promote human values, privacy, dignity, liberty, autonomy and human rights.
- The developers and researchers must consider the ethical, legal, and social risks, so that they are reduced and avoided.
- Continuously strive for the improvement of maturity, strength, reliability, and controllability of AI systems, ensuring safety.
- Use ethics in the development of AI systems to make them reliable, far and reduce prejudice and preconceptions.
- Reflect diversity and inclusion in AI systems, developing them to benefit the highest number of people.
- Stablish open AI platforms, to avoid data and platform monopolies, sharing the benefits of development and promoting equal development possibilities to several regions and sectors.
- Provide the necessary capacitation to operate AI systems and comprehend the possible risks.
- Provide measures to ensure the rights and duties of the stakeholders.

These principles were supported by the main universities of China (University of Tsinghua and University of Beijing), and research institutions (Institute of Automation, Chinese Academy of Science, Institute of Computer Technologies, and the Artificial Intelligence Industry Technological Innovation Strategic Alliance – AIITISA) and Chinese companies (Baidu, Alibaba and Tencent).

In 2017 it was created the Artificial Intelligence Industry Alliance – AIIA, led by the China Academy of Information and Communication Technology – CAICT of the Ministry of Industry and Information of Technology that released in 2019 a "joint vow" draft about self-discipline in the development of AI. Even though the content is broad if compared to other ethics declarations and AI governance, phrases like safe, controllable and self-discipline were included, demonstrating the tendencies of Chinese digital governance [19, 20].

979-8-3503-8083-5/23 $31.00 © 2023 IEEE

C. United States of America

In 2017 China revealed the plan to become the world leader in AI by 2025 [21, 22] an as a consequence the USA issued an Executive Order to keep the American leadership in artificial intelligence [23, 24]. It provides five guidelines to be implemented by the Committee of Artificial Intelligence of the National Science and Technology Council – NTSB:

- AI technological processes [25] in the government, in the industry, in the academy pushing scientific discoveries, economic competitiveness and national security.
- Boost in the development of technical standards appropriate to reduce the difficulties to implement AI technologies in the industries.
- Training of workers for the application of AI, preparing them to the job market.
- Ensure public trust in AI technology, protections of civil rights, privacy, and American values in the application of AI technologies.
- Support an international research and innovation environment and the opening to industrial markets of AI.

The Order also provides six strategic objectives:

- Promote research and development of AI alongside with the industries, teaching institutions, international allies, and partners, creating the progress in AI to contribute to economic and national security.
- Facilitate the access to high quality federal data, creating more value in research and development of AI, preserving safety, protection, privacy, and confidentiality.
- Reduce the barriers of use of AI technology to allow its application, protecting American technology, national and economic security, civil liberties, privacy, and values.
- Ensure the development of technical standards that are less vulnerable to malicious invasions, ensuring the public trust in AI systems and develop international technical standards for protection of innovation in AI.
- Train researchers and users of AI, ensuring that the American worker can experience all the benefits of AI.
- Create and implement a plan of action for the protection of the lead of the United States in AI against foreign competitors.

The order makes the National Institute of Standards and Technology – NIST responsible for elaborating a plan of federal involvement in the development of technical standards and support tools to trustworthy, strong, and reliable systems that use AI.

The Department of Defense of the USA issued an AI strategy according to the Executive Order [26]. It recognizes the high military investments by rival countries that threaten the USA technological lead and destabilizes the free and open international order. It will promote the vision and principles of using AI technology in an ethical and legal way and that it will share its objectives, ethical guidelines and safety procedures for the responsible development and use of AI by other countries.

Regarding the application of AI in the form of facial recognition, Joy Buolamwini, scientist, and digital activist of the Massachusetts Institute of Technology – MIT, proved that these technologies present everywhere in the world are constantly failing when it comes to woman and black people. Buolamwini tested the facial recognition systems of big companies such as Amazon, IBM and Microsoft and all them presented failures. She also worries about the systems being used also by the police [27].

D. India

AI is well received and accepted by the population of India. However, there is a fear regarding jobs [27]. A third of the interviewed believe that machine learning and the automation of industrial processes will replace the majority of jobs in the costumer care and repetitive manual jobs, thus causing a big socioeconomic impact [27].

The same study indicates that in India, not only activities of lesser complexity are being replaced by AI. CEOs of Indian companies also invest in new AI systems to replace leadership position, claiming that it is difficult to find candidates that fulfill all the needed characteristics to occupy management positions in companies. One of the reasons that people cannot fulfill the requirements at a high hierarchy level is the difficulty to reconcile personal and professional demands, which does not occur on specialized AI systems [27].

A significant impact in AI that already occur in India is related to the education sector. In universities, assistant professors have been replaced by chat bots, representing a reduction of 60% in costs of operation of those universities. These chatbots reply to questions and demands from students, but still they cannot justify the answers and decisions taken when the reasoning is questioned [30].

IV. CONCLUSIONS

The utilization of AI has contributed in significant ways to societies from around the world. National and international studies show that it can be applied in several areas of knowledge, influencing economical, scientific, and cultural activities of humanity in the last decades [1]. On the other hand, its use may have negative results due to ethical and moral hindrances, representing also the possibility of elimination of conventional jobs, potentially limiting privacy and aggravating existing social conflicts.

This article investigated if AI technologies are beneficial to the society considering economic, social, ethical, and governmental analysis. Considering the profile of each country, China seems as the most advanced, regarding ethical aspects of AI. The United States of America is making up for the time lost after the guidelines of the Executive Order. In India, even with the perception that AI is already increasing the scarcity of jobs, the majority of people support its insertion of AI in all possible sectors, as can be seen in the educational field, where nearly 80% of auxiliary college professors were replaced by AI systems, reducing costs and allowing people to have a higher education in the country.

China and USA are the big pioneers in research and development of Artificial Intelligence technology and compete between each other. This conflict is clear with the Executive

Order of the United States because it was created considering maintaining the AI lead by the USA and considering China's objective to become the AI world leader by 2030. This competition happens in the economic and security field.

In Brazil, is notable the level of AI use in big farms, changing the profile of qualification of rural workers that previously did not need a high level of qualification and now demands extremely qualified workers with knowledge of agronomy, livestock, computer programming, cartography, mechanics and robotics to be able to perform maintenance in autonomous or remote machinery such as harvesters, planters and sprayer drone. Just a single one of these machines is capable of replacing dozens of workers and reduce the time required to accomplish a task from little over a week to only a few days.

Although the Brazilian congress have approved bills such as Law 13,709, still are lacking laws that limit the levels of penetration of AI relative to jobs posts. Even AI being a lot more efficient and still demanding more qualified people to perform maintenance and design these complex and multidisciplinary systems, the time needed to capacitate competent specialists in those areas take a lot of time and might not have timely manner to train obsolete workers to place them in more modern jobs positions, increasing unemployment.

It is undeniable that AI has a great potential to speed up humanity technological advances, however, when the application of AI is directly related to people, for instance in public and financial safety, the government should find effective ways to prioritize basic right of the population such as privacy, job opportunities and eradicate all manner of prejudice that can be passed on to the machine during the training of the systems.

ACKNOWLEDGMENTS

This work was supported by CAPES, CNPq and FAPEMIG.

REFERENCES

[1] A. V. Rezaev and N. D. Tregubova, "Are sociologists ready for 'artificial sociality'? current issues and future prospects for studying artificialintelligence in the social sciences", The Monitoring of Public Opinion Economic & social Changes, no. 5 (147), 2018.

[2] H. Fatemidokht, M. K. Rafsanjani, B. B. Gupta, and C.-H. Hsu, "Efficient and secure routing protocol based on artificial intelligence algorithms with uav-assisted for vehicular ad hoc networks in intelligent transportation systems," IEEE Transactions on Intelligent Transportation Systems, pp. 1–13, 2021.

[3] F. Jiang, Y. Jiang, H. Zhi, Y. Dong, H. Li, S. Ma, Y. Wang, Q. Dong, H. Shen, and Y. Wang, "Artificial intelligence in healthcare: past, present and future," Stroke and Vascular Neurology, vol. 2, no. 4, pp. 230, 2017.

[4] L. C. Lobo, "Inteligência artificial e medicina," Revista Brasileira de Educação Médica, vol. 41, no. 2, pp. 185–193, 2017.

[5] A. Daly, T. Hagendorff, H. Li, M. Mann, V. Marda, B. Wagner, W. W. Wang, and S. Witteborn, "Artificial intelligence, governance and ethics: Global perspectives," The Chinese University of Hong Kong Faculty of Law Research Paper, no. 2019-15, 2019.

[6] C. Gonzalez Viejo, S. Fuentes, K. Howell, D. Torrico, and F. R. Dunshea, "Robotics and computer vision techniques combined with non-invasive consumer biometrics to assess quality traits from beer foamability using machine learning: A potential for artificial intelligence applications," Food Control, vol. 92, pp. 72–79, 2018.

[7] E. Svetlana, E. Oksana, B. Andrei, and S. Maria, "Directory of open access journals," Jan 2021.

[8] F. Balducci, D. Impedovo, and G. Pirlo, "Machine learning applications on agricultural datasets for smart farm enhancement," Machines, vol. 6, no. 3, 2018.

[9] P. Wang, B. Goertzel, and S. Franklin, Artificial general intelligence, Proceedings of the first AGI conference, 2008, vol. 171.

[10] T. Jameel, R. Ali, and I. Toheed, "Ethics of artificial intelligence: Research challenges and potential solutions," 3rd International Conference on Computing, Mathematics and Engineering Technologies (iCoMET). IEEE, 2020, pp. 1–6.

[11] M. Piteira, M. Aparicio, and C. J. Costa, "A ética na inteligência artificial: desafios," 14ª Conferência Ibérica de Sistemas e Tecnologias de Informação, 2019.

[12] K. A. K. Frota and R. F. Gabriel Junior, "Inteligência artificial e organização do conhecimento e da informação: revisão sistemática de literatura," Fórum de Estudos em Informação, Sociedade e Ciência, 2020

[13] T. Dwyer, "Inteligência artificial, tecnologias informacionais e seus possíveis impactos sobre as ciências sociais," Sociologias, no. 5, pp. 58–79, 2001

[14] A. F. Jr., "Reconhecimento facial no rio de janeiro erra, e mulher é detida por engano," Jul 2019.

[15] L. S. Mendes and D. Doneda, "Reflexões iniciais sobre a nova lei geral de proteção de dados," Revista de Direito do Consumidor, 2020.

[16] "Artificial intelligence development plan," Jul 2017.: https://flia.org/notice-state-council-issuing-new-generation-artificial-intelligence-development-plan/

[17] "AI policy – China." https://futureoflife.org/ai-policy-china

[18] "Beijing AI principles," May 2019. https://www.baai.ac.cn/news/beijing-ai-principles-en.htm

[19] "Chinese AI alliance drafts self-discipline 'joint pledge'," Jun 2019. https://www.newamerica.org/cybersecurity-initiative/digichina/blog/translation-chinese-ai-alliance-drafts-self-discipline-joint-pledge/

[20] "Chinese expert group offers 'governance principles' for 'responsible ai'," Jun 2019. https://www.newamerica.org/cybersecurity-initiative/digichina/blog/translation-chinese-expert-group-offers-governance-principles-responsible-ai/

[21] C. Cadell and A. Jourdan, "China aims to become world leader in ai, challenges u.s. dominance," Jul 2017. https://www.reuters.com/article/us-china-ai-idUSKBN1A5103

[22] "China unveils plan to become a world leader in ai by 2025." https://www.voanews.com/east-asia/china-unveils-plan-become-world-leader-ai-2025

[23] "President trump issues executive order to maintain american leadership in artificial intelligence," Mar 2019. https://jolt.law.harvard.edu/digest/president-trump-issues-executive-order-to-maintain-american-leadership-in-artificial-intelligence

[24] C. Metz, "Trump signs executive order pro-moting artificial intelligence," Feb 2019. https://www.nytimes.com/2019/02/11/business/ai-artificial-intelligence-trump.htm

[25] "Maintaining american leadership in artificial intelligence," Feb 2019. https://www.federalregister.gov/documents/2019/02/14/2019-02544/maintaining-american-leadership-in-artificial-intelligence

[26] S. Collins, "Summary of the 2018 department of defense artificial inteligence strategy," Department of Defense United States of America, Tech. Rep., 2018.

[27] "Google disbands ai committee before first meeting," Apr 2019. https://tech.co/news/google-disbands-ai-committee-2019-04

[28] "Bio alessandro acquisti," https://www.heinz.cmu.edu/ acquisti/bio.htm.

[29] S. Ghosh, "Artificial intelligence in india–hype or reality impact of artificial intelligence across industries and user groups," 2019. https://www.pwc.in/consulting/technology/data-and-analytics/artificial-intelligence-in-india-hype-or-reality.html

Fusing IP vendor Palmprint Biometric with Encoded Hash for Hardware IP Core Protection of Image Processing Filters

Anirban Sengupta
Computer Science and Engg.
Indian Institute of Technology Indore,
Indian Statistical Institute Kolkata,
India
asengupt@iiti.ac.in

Aditya Anshul
Computer Science and Engg.
Indian Institute of Technology Indore,
India
phd2101101007@iiti.ac.in

Sumer Thakur, Chirag Kothari
Indian Institute of Technology Indore,
India
ce190004039@iiti.ac.in,
chiragkothari2503@gmail.com

Abstract— **Image processing filters offer several significant applications, making them a crucial component of various consumer electronics and multimedia systems. These image filters are designed as dedicated reusable intellectual property (IP) cores using high level synthesis (HLS) framework. These image processing hardware IPs are essential components of several system-on-chips (SoCs) used in mission-critical applications. Therefore, protecting these image filter hardware IPs from an IP vendor's perspective is crucial against an adversary's false IP ownership claim in the context of globalized design supply chain. This paper presents a novel hardware IP protection (IPP) technique by fusing the IP vendor's palmprint biometric and his/her encoded hash to generate secret security constraints for embedding. This embedded security mark acts as a detective countermeasure during the IP conflict resolution process. The proposed approach outperforms recent IPP techniques, as evident by a lower probability of coincidence, stronger tamper tolerance, and stronger entropy value.**

Keywords— *Hardware security, Palmprint biometric, Encoded hash, image filters, HLS.*

I. INTRODUCTION

The prevalence of image processing filters has surged significantly in today's rapidly evolving technological society. This surge is primarily propelled by the synergetic effects of automation and the rapid expansion of our modern society. A set of pivotal image processing filters with broad applicability encompasses blur, sharpening, laplace edge detection (LED), vertical and horizontal embossment filters, etc. These filters have established wide-ranging applicability across diverse real-world scenarios, including multimedia processing, license plate recognition at tolls, advanced military applications, autonomous driving automation, robotic vision, etc. These filters execute distinct data and computation-intensive functions on images, encompassing edge detection, blurring, sharpening, and embossing, ultimately extracting meaningful information from the visual data. Owing to the massive usage and advanced design intricacy, these image processing filters are often designed as dedicated reusable hardware IP cores for system-on-chip (SoC) applications using the HLS framework [1].

However, designing hardware IPs for SoC applications may invite security vulnerability due to globalization in the design supply chain, leading to an illicit market willing to undercut the competition with fake components. Therefore, securing the hardware IP design from false claims of IP ownership is crucial. An adversary, potentially situated within the untrustworthy SoC integrator house and/or foundry, may attempt to fraudulently claim the design IP

ownership. The importance of safeguarding hardware IP designs from false claims of ownership cannot be overstated, driven by a multitude of factors: safeguarding IP rights, upholding a competitive edge, preserving the sanctity of trade secrets and confidential information, enabling licensing and monetization opportunities, shielding the reputation of the legitimate designer, etc. [2]. This paper presents a novel hardware IP protection (IPP) approach as a detective countermeasure for nullifying an adversary's false claim of IP ownership, using the fusion of IP vendor's palmprint biometric and encoded hash. Moreover, the paper presents the generation and embedding of secret security constraints (digital evidence) using an amalgamation of IP vendor's palmprint biometric and encoded hash.

Some prior work on signature-based hardware IPP methodologies includes [2-6]. Among these, authors in [2] introduce facial biometrics for embedding security mark to secure hardware IP cores. However, [2] incurs weaker security due to the generation of lesser hardware security constraints. Another approach [3] involves leveraging the DNA imprint of the IP vendor to safeguard DSP IP cores, providing sturdier security than [4]. However, it incurs greater computational complexity in signature generation process. Moreover, in [4], the authors use a binary variable watermarking scheme to generate security constraints. However, vulnerabilities in [4] emerge when the integrity of signature size or encoding mechanisms is compromised, compounded by weaker security owing to the generation of limited security constraints. Another approach, as described in [5], employs MD5 and SHA1 to enhance security resilience. Additionally, a hardware description language (HDL) design-level IP watermarking strategy utilizing SHA1 and RSA is presented in [6]. While [6] exhibits robust security, it becomes susceptible if the RSA key value is compromised. Additionally, [9] utilizes mathematical relations for hardware watermarking by establishing links between input data, internal computation initial values, and output data to create the final watermark. For the first time in literature, the proposed approach integrates IP vendor palmprint biometric with encoded hash to generate more robust and tamper tolerant signature for image processing filter IP cores. The proposed approach surpasses all the above-mentioned approaches [2-6] and [9] due to generation of larger security constraints and stronger entropy value.

II. PROPOSED METHODOLOGY

The overview of the proposed hardware IP protection (IPP) approach is shown in Fig. 1. The proposed approach

---

This work is technically and financially supported by Council of Scientific and Industrial Research (CSIR) grant no. 22/0856/23/EMR-II.

2023 International Conference on Microelectronics (ICM)

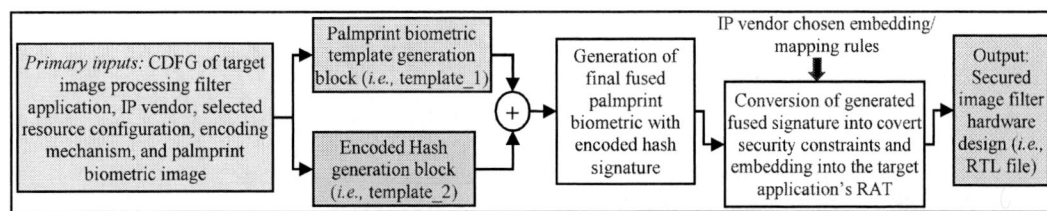

Fig.1. Overview of the proposed hardware IPP methodology

fuses two IPP paradigms by exploiting IP vendor's palmprint biometric as well as encoded hash to generate a robust and tamper tolerant security information (in the form of encoded signature), which is embedded into the design as covert security constraints for providing detective countermeasure.

A. Palmprint biometric based hardware IPP methodology

Initially, the palmprint information of an original IP vendor is captured using a high-quality digital camera, which is subsequently used to generate a unique palmprint digital template. Next, hardware security constraints are generated by converting the generated palmprint template into its corresponding covert security constraints based on IP vendor-selected embedding rules. These constraints are then implanted into the target image filter IP design during the register allocation phase of the HLS process. Fig. 2 depicts the palmprint biometric template generation process. Initially, a high-resolution camera captures the palmprint image, subjected to the IP vendor's grid size and spacing. Fig. 2. (a) shows the captured palmprint under grid size and spacing. Next, the palmprint's unique nodal points and corresponding features are identified. Each feature and corresponding nodal point are assigned distinct names for effective mapping. Fig. 2. (b) and (c) show the palmprint with nodal points and generated features, respectively. Following the generation of nodal points and feature sets, dimensions corresponding to generated features are calculated using the coordinates (x_1, y_1) and (x_2, y_2) of respective nodal points for each generated feature. The dimension is determined using the Manhattan distance. The resulting decimal dimension is then converted into its binary representation. Fig. 2. (d) illustrates each feature's dimension and corresponding binary equivalent. The binary equivalents obtained for each selected feature are concatenated according to the IP vendor's specified order of concatenation. This process yields a palmprint biometric template as the output. The generated palmprint template (*i.e.,* template 1) is fused with the obtained encoded hash (*i.e.,* template 2, discussed in II. C) to produce the final signature.

B. Advantages of palmprint biometric based hardware IPP

The template generated using the original IP vendor palmprint is inherently unique, serving as a secret mark for the target hardware IP. Extracting palmprint signature is simpler compared to facial biometrics [2]. Unlike facial biometrics [2], the palmprint biometric method exhibits more substantial feature variation, resulting in enhanced tamper tolerance. Furthermore, the palmprint-based approach boasts several advantages over contemporary techniques: it's contactless, secure from vulnerabilities, non-replicable (unlike stego-constraints and watermarking [4]), and doesn't rely on a secret key. Additionally, palmprint biometric depicts lesser complexity than DNA biometric [3].

C. Encoded hash-based hardware IPP methodology

Fig. 3 highlights the generation of encoded hash (*i.e.,* template 2). The encoded hash-based template generation

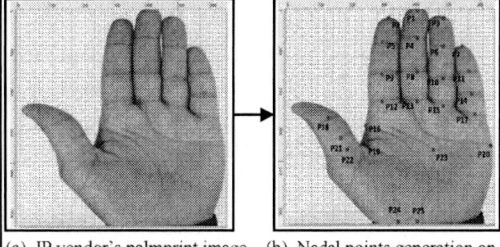

(a). IP vendor's palmprint image with grid size and spacing (b). Nodal points generation on palmprint image

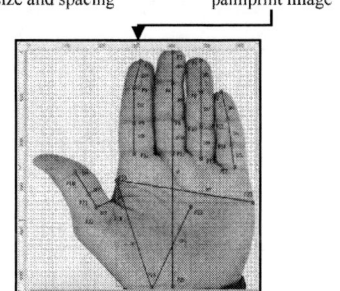

(c). Feature generation on palmprint image

Feature #	Feature name	Feature dimension	Binary representation
F1	DFT	64.03	1000000.00000111101011100001
F2	DSF	80	1010000
F3	DST	36.05	100100.0000110011001100101
F4	DSL	60.82	111100.11010010111010111
F5	DFR	70	1000110
F6	DL	260	100000100
F7	DTR	70	1000110
F8	DTF	70	1000110
F9	DSR	80	1010000
F10	DTL	50.99	110010.111111010111000101
F11	DTT	63.24	111111.0011110101110000101
F12	DHL	210.95	11010010.1111001001100110011
F13	DFM	70	1000110
F14	DFF	63.24	111111.0011110101110000101
F15	LP	300	100101100
F16	DFL	63.24	111111.0011110101110000101
F17	DSM	80	1010000
F18	DTM	80	1010000
F19	WP	393.19	110001001.00110000101000111101

(d). Features dimensions (magnitude) computation and its corresponding binary representation

(e). Generation of final palmprint biometric template post concatenation of all generated features as per IP vendor's chosen concatenation order (*i.e.,* template_1)

Fig.2. Generation of palmprint biometric template using IP vendor's palmprint biometric image

module takes CDFG of the target image filter application, IP vendor-specified resource constraints, encoding mechanism, SHA-512 algorithm, and truncation hash length as the primary input. The process of forming the encoded hash of the IP vendor is as follows: Firstly, a scheduled DFG (SDFG) is generated corresponding to the target hardware application using input DFG, resource constraints, and LIST scheduling algorithm. Next, through the obtained SDFG, an intermediate bitstream is generated based on the IP vendor's encoding rule. This rule stipulates that the output bit is '0' when the operation

979-8-3503-8083-5/23 $31.00 © 2023 IEEE   219

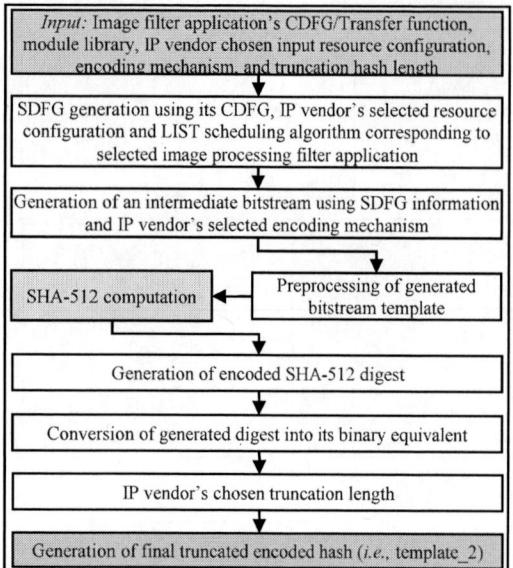

Fig. 3. Generation of encoded hash

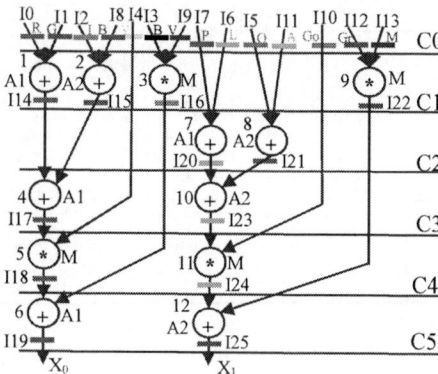

Fig. 4. Scheduled data flow graph (SDFG) of LED filter

Table I
RAT pre and post implanting security constraints corresponding to LED filter

	C0	C1	C2	C3	C4	C5
Red(R)	I0	I14/I15	I14	I17	I18	I19
Green (G)	I1	I15/I14	I15	-	-	-
Indigo (I)	I2	I16	I16	I16/I17	I16	-
Blue (BL)	I8	I22	I22	I22	I22	I25
Yellow (Y)	I4	I4	I4	I4	-	-/I19
Black (B)	I3	-/I16	-/I16	-/I16	-/I16	-
Violet (V)	I9	-/I22	-	-	-	-
Pink (I)	I7	I7	I21/I20	-	-/I18	-
Lime (LI)	I6	I6	I20/I21	I23	I24	-
Orange (O)	I5	I5	-	-	-/I18	-
Aqua (A)	I11	I11	-	-	-/I24	-
Gold (Go)	I10	I10	I10	I10	-	-
Gray (Gr)	I12	-	-	-	-	-
Maroon (M)	I13	-	-	-	-	-

and control step numbers share the same parity; otherwise, it is '1'. This intermediate bitstream is converted into a 1024-bit utilizing the IP vendor's chosen preprocessing rule. This preprocessing involves appending '1' after the intermediate bitstream and padding with '0s' up to the 896-bit mark. Additionally, a 128-bit representation corresponding to the length of the generated bitstream post the 896-bits is added, resulting in a 1024-bit. Subsequently, the generated 1024-bit bitstream is fed into the SHA-512 block to produce a 512-bit encoded hash. This hash is transformed into its binary equivalent and truncated according to the specified truncation length, generating the final encoded hash (*i.e.*, template 2). Next, the obtained palmprint template (*i.e.*, template 1) is fused (or concatenated) with the encoded hash (*i.e.*, template 2) to generate the final fused signature. The obtained signature is converted into respective security constraints based on the following IP vendor embedding rule. The embedding of the generated security constraints is performed in the register allocation phase of the HLS process.

Embedding rule: Implant an additional artificial edge within (even, even) pairs of storage variables in the register allocation table (RAT) when the bit is '0'. Conversely, an edge is integrated between (odd, odd) storage variable pairs of the RAT when the bit is '1'.

The generation of covert security constraints using the obtained signature and its embedding is explained in the next subsection using an example of LED image processing filter.

D. Demonstration of proposed approach on LED filter

The proposed approach is demonstrated on LED image processing filter application. Fig. 4 depicts the SDFG corresponding to the LED filter scheduled using two adders (+) and two (*) multipliers (obtained using the design space exploration framework). An initial register allocation table is generated from the obtained SDFG, as shown in Table I. The generated fused signature is integrated into the design in the form of covert security constraints (as digital evidence) [3], resulting in a secure hardware IP design. For the sake of demonstration, we have considered only the first 400 bits of the generated 819-bit fused signature for embedding. In Fig.

4 and Table I, *I0, I1...,I25* represent the storage variables corresponding to LED filter application. The resulting security constraints consist of pairs like *(I0, I2), ... (I22, I24), (I1, I3), ... (I19, I23)*. During embedding, two storage variable pairs of the incoming artificial edge must be allocated to different registers. This local reallocation is achieved by either swapping register colors or assigning new registers [3]. Table I shows the register allocation information before and after embedding hardware secret security constraints (Red colored registers indicate modified locations of registers). The embedded information serves as robust, imperceptible digital evidence for detective countermeasure against false IP ownership claim.

E. Resolution of IP ownership claims

In order to nullify the false IP ownership claim, the security constraints are regenerated and matched with the original embedded security constraints. Only an authentic IP vendor could successfully regenerate the exact security constraints and match them with the original embedded one; however, an adversary would fail to do so.

III. RESULTS AND ANALYSIS

The experimental analysis of the proposed hardware IPP approach was conducted on a system with a 2.30 GHz workstation and 4GB main memory (implemented in Python). The evaluation of area and latency for image filters was performed using a 15nm scale with NanGate library [7].

Entropy analysis (X_E), tamper tolerance (T_Z), and probability of coincidence (C_X) are the three parameters used for performing security analysis of the proposed approach. Entropy refers to an adversary's difficulty and uncertainty while decoding embedded covert security mark in an IP

Table II

Comparison of entropy and tamper tolerance between the proposed approach, [2], [3], [4], [5], and [6]

Security approach	Security parameters		
	Embedded constraints (q)	Entropy	Tamper tolerance
Proposed approach	400	8.27E-252	2.58E+120
Facial biometric [2]	83	1.03E-32	9.67E+24
Digital signature [5]	160	2.01E-87	1.46E+48
Watermarking [4]	240	1.66E-111	1.76E+72
DNA biometric [3]	128	2.9E-39	3.40E+38
HDL watermarking [6]	256	5.85E-99	1.15E+77

Table III

Comparison of probability of coincidence between the proposed approach, [2], [3], [4], [5], and [6]

Security approach	Benchmarks		
	Blur filter	Sharpening filter	LED filter
Proposed approach	6.05E-08	8.29E-09	2.49E-5
Facial biometric [2]	1.41E-02	2.10E-02	2.13E-03
Digital signature [5]	2.72E-04	5.85E-04	2.49E-5
Watermarking [4]	4.50E-06	1.41E-05	2.49E-5
DNA biometric [3]	1.40E-03	2.59E-03	7.59E-05
HDL watermarking [6]	1.98E-06	6.72E-06	2.49E-5

Table IV

Design cost, area and latency of proposed technique

Benchmarks	Design cost	Design Area (um^2)	Design Latency (ps)
Blur filter	0.537	147.84	927.39
Sharpening filter	0.588	243.79	794.91
LED filter	0.71	199.75	728.67
Vertical emboss ment	0.756	99.09	596.18
Horizontal embossment	0.756	99.09	596.18

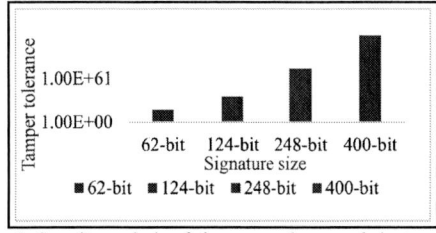

Fig. 5. Security analysis of the proposed approach in terms of varying signature sizes and its impact on tamper tolerance

design. Entropy for the proposed approach is formulated as [8] $X_E = ((1/2^d * 1/m!) * ((1/2^k)*(1/R)*(1/2^{64})))$, where '$d$' is the final generated palmprint template length and 'm' is the total number of features selected on the palmprint, 'k' is the length of truncated encoded hash, 'R' is the round computation's maximum value, and $(1/2^{64})$ is the probability of finding the exact key hash buffer initialized value in SHA-512 cryptographic module (each hash buffer is initialized with pre-defined 64-bit value). Furthermore, the tamper tolerance (T_Z) is formulated as [8] $Tz = q^t$, where 'q' denotes the total embedded constraints, and 't' represents distinctive embedding variables used in the signature formulation of the proposed approach. A larger value of T_Z results in a bigger signature space, making it challenging for an attacker to decode the precise signature combination using brute force attack, indicating more robust security. Consequently, a higher value of T_Z, therefore, enhances resilience against tampering attack. Table II reports the comparison of entropy and tamper tolerance between the proposed approach, [2], [3], [4], [5], and [6]. The proposed approach outperforms all the specified state-of-the-arts due to embedding of greater security constraints in the design, as seen in Table II. Moreover, the proposed approach provides a stronger entropy (greater uncertainty) than all above mentioned related approaches, as shown in Table II. Additionally, Fig. 5 reports the security analysis of the proposed approach in terms of varying signature sizes and its impact on tamper tolerance. The value of T_Z increases with an increase in signature size.

Next, the probability of coincidence refers to the likelihood of detecting the same covert security constraints in an unsecured design. It is a measure of the false positive of the methodology. The likelihood of another IP vendor producing the same design with security constraints must be reduced. It also indicates the presence of digital evidence inside a secured design that can be used as digital proof to handle verification of true IP vendor ownership. A stronger security technique depicts a lower value of C_X. The C_X for the proposed approach is formulated as [4,8] $C_X = (1-1/r)^q$, where 'q' denotes total embedded constraints and 'r' is the total registers required before implanting security constraints. Table III reports the comparison of C_X between proposed, [2], [3], [4], [5], and [6]. The proposed approach depicts the lowest value of C_X due to the generation and embedding of greater covert security constraints.

Further, Table IV highlights the design cost, area and latency corresponding to all selected image filter applications. The design cost metric used is as follows [8]:

$$Design\ cost = e1 * \left(\frac{(Area)}{ARmax}\right) + e2 * \left(\frac{(Latency)}{LTmax}\right) \quad (1)$$

where '$e1 = 0.5$' and '$e2 = 0.5$' represent weighing factors specified by the IP vendor. *Area* and *Latency* refer to the security embedded design area and latency, respectively. 'AR_{max}' and 'LT_{max}' denote maximum design area and latency.

IV. CONCLUSION

This paper presents a novel hardware IPP approach as a detective countermeasure for nullifying an adversary's false claim of IP ownership, using the fusion of IP vendor's palmprint biometric and encoded hash.

REFERENCES

[1] C. Pilato, S. Garg, K. Wu, R. Karri and F. Regazzoni, "Securing Hardware Accelerators: A New Challenge for High-Level Synthesis," *IEEE Embedded Systems Letters*, vol. 10, no. 3, (2018), 77-80, Sept.

[2] A. Sengupta and M. Rathor, "Facial Biometric for Securing Hardware Accelerators," *IEEE Transactions on Very Large Scale Integration (VLSI) Systems*, vol. 29, no. 1, pp. 112-123, Jan. 2021.

[3] A. Sengupta, R. Chaurasia, "Securing IP Cores for DSP Applications Using Structural Obfuscation and Chromosomal DNA Impression," *IEEE Access*, vol. 10, 2022, 50903-50913.

[4] F. Koushanfar, I. Hong, M. Potkonjak, Behavioral synthesis techniques for intellectual property protection, *ACM Trans. Des. Autom. Electron. Syst.* 10, 3 (2005), 523–545, July.

[5] E. Castillo, L. Parrilla, A. Garcia, U. Meyer-Baese, G. Botella, A. Lloris, "Automated Signature Insertion in Combinational Logic Patterns for HDL IP Core Protection," *2008 4th Southern Conference on Programmable Logic, Bariloche, Argentina*, (2008), 183-186.

[6] T. Yu and Y. Zhu, "A new watermarking method for soft IP protection," 2011 International Conference on Consumer Electronics, *Communications and Networks (CECNet)*, China, 2011, 3839-3842.

[7] Open Cell NanGate Library, 15 nm open cell library, Available: https://si2.org/open-cell-library/, last accessed on March 2023.

[8] A. Sengupta, R. Chaurasia and A. Anshul, "Robust Security of Hardware Accelerators Using Protein Molecular Biometric Signature and Facial Biometric Encryption Key," *IEEE Transactions on Very Large Scale Integration (VLSI) Systems*, vol. 31, no. 6, 826-839, 2023.

[9] B. L. Gal and L. Bossuet, Automatic low-cost IP watermarking technique based on output mark insertions. *Des. Autom. Embedded Syst.* 16, 71–92, 2012.

Emotion Recognition Based on Electroencephalogram (EEG) Signals

1st Khader Mohammad
Department of Computer Engineering
Birzeit University
khamadawwad@birzeit.edu

2nd Saleem Hamo
Department of Computer Engineering
Birzeit University
saleemhamo@gmail.com

3nd Mohammad Abbas
Department of Computer Engineering
Birzeit University
mohammed3bbas99@gmail.com

4nd Maen Mohammad
Department of Computer Engineering
Al-Quds University
Maen.khader@gmail.com

Abstract—Recently, doors have opened with a variety of contemporary applications regarding Electroencephalogram (EEG) signal analysis. These new doors can be introduced as a way to detect emotions. In this research, some models of human emotion analysis based on brain signals will be presented. This study focuses on emotion recognition based on EEG signals analysis following two approaches: Spectral and Statistical analysis. Generated feature vectors were rated with the help of several classification algorithms, bringing the accuracy of emotion detection up to 85%. Furthermore, The source of EEG signals was examined by identifying the channel from which they were recorded. To facilitate the implementation of these analytics, a software Interface (API) that provides all the required technical functionalities was designed using Python programming language. Finally, analysis results were matched with the adopted human emotions model, and by observing classification results, the feasibility of emotion recognition based on EEG signals was concluded and what circumstances need to be addressed in such a system.

I. INTRODUCTION

The field of Emotion Detection using EEG Signals is still under extensive development, and its applications are still scarce. The objective of this research is to study the feasibility of such systems, in order to provide the field with useful studies to proceed with. Thus, the research was fed with different approaches and comparisons between their results. Moreover, an API was designed to be used in future applications related to the same field of EEG signal Analysis.

Electroencephalography (EEG) is a useful and efficient model for obtaining brain signals that correlate to different states from the scalp surface area. Brain signals lie on frequencies ranging from 0.1 Hz to more than 100 Hz, divided into specified bandwidths as mentioned in table I. Each signal activates and appears when a specific state occurs whether it is an external or an internal influence [1].

A. EEG Signals

An electroencephalogram is a medical test used to read electrical activity generated by the brain. The first EEG was recorded in 1924 by a German neurologist named Hans Berger. He came up with the idea to understand the physiological basis of mental problems and physical phenomena. He came up with an assumption that electrical signals of the brain will change according to the functional status of the brain (sleep, awake, brain dysfunction) and that was later shown to be correct [2]. The test is currently mainly used in diagnosing pathological conditions such as seizures, epilepsy, and sleep disorders. It has also been used with other measurement tools to study the physiology and staging of sleep by identifying different brain waves in different sleep stages. Recently, more and more uses of EEG have started to evolve, but what we worked on, and saw was most interesting was emotion recognition.

The test starts by placing electrodes on the surface of the scalp in special locations according to a method called the 10-20 system for electrode placement. The electrodes cover all brain regions, and their placement is proportional to the size and shape of the skull. Each rod collects signals from many neurons in the area it occupies. After that, a technology called differential amplifier is used in which the differences between two electrical inputs from two rods are recorded. In a normal person, five frequency bands can be detected Delta δ, Theta θ, Alpha α, Beta β, and Gamma γ in an EEG.

How does the test work?: The test was originally made to record weak electrical currents generated in the brain without opening the skull. However, the question a person might ask is: how does the EEG read such weak signals and allow us to see clear waves on a piece of paper? The answer is that this device uses what's called an amplifier. An amplifier is a tool to amplify low amplitude (microvolts) signals and present them in a readable format. In EEG, a differential amplifier is used because these amplifiers amplify the difference in voltage of two input signals (two polarities) from two electrodes.

When measuring the potential difference between electrodes

979-8-3503-8083-5/23 $31.00 © 2023 IEEE

TABLE I
FREQUENCY BANDS AND CORRESPONDING BRAIN STATES

Identifier	Frequency Band	Brain State
Delta, δ	1-4 Hz	Primarily associated with deep sleep
Theta, θ	4-8 Hz	Appear as consciousness slips toward drowsiness
Alpha, α	8-13 Hz	Usually found over the occipital region. Indicates relaxed awareness without attention
Beta, β	13-30 Hz	Associated with active thinking and concentration
Gamma, γ	30-100 Hz	Represents binding of different populations of neurons

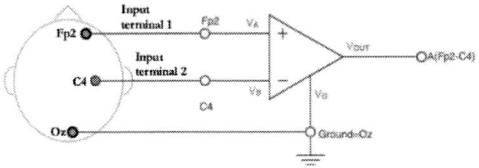

Fig. 1. Differential Amplifier

F7 and T3 for example, we cannot know the exact protentional (charge) of the input from F7, nor can we know it for the input from T3. The amplifier only shows the difference between the two inputs. Each time you take two signals from two electrodes, you get one wave on an EEG showing the difference between the two signals. To get the whole picture of brain activity, montages have been created. A montage is an orderly and logical way to connect channel electrode pairs to each amplifier. Each montage shows a different result on an EEG.

After a differential amplifier amplifies the signal, it is amplified again by normal amplifiers to increase the signal strength. The signals are then filtered to make EEG signals readable. Low-frequency filters, high-frequency filters, and notch filters are the main filters used in EEG [3].

B. Emotion Detection & EEG Signals

Emotion Detection Recognition (EDR) is a method used in the detection and recognition of human emotions. It functions by the incorporation of technological capabilities such as facial recognition, speech and voice recognition, biosensing, machine learning, and pattern recognition [5]. As EEG Signals are well known to carry brain's emotional state, it's highly probable to scan those emotions using EEG Signals Analysis.

According to the circumplex model of emotion, emotions are distributed in a two-dimensional circular space, containing arousal and valence dimensions [5]. It's possible to compare signals rising from different electrodes based on their locations. Thus, matching an EEG signal with the corresponding emotional state, along with the ability to localize an EEG signal, raises the feasibility of recognizing

the source of a particular emotional state.

Arousal-Valence Emotion Model: Arousal represents the vertical axis and valence represents the horizontal axis, while the center of the circle represents a neutral valence and a medium level of arousal. In this model, emotional states can be represented at any level of valence and arousal, or at a neutral level of one or both of these factors. Circumplex models have been used most commonly to test stimuli of emotion words, emotional facial expressions, and effective states [4].

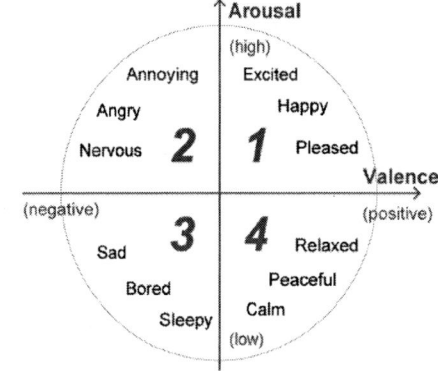

Fig. 2. Arousal-Valence Emotion Model

C. EEG Dataset

In this research, a pre-collected dataset was used. Mainly, it represents a couple of recorded EEG signals from 28 participants, who have been playing games with different emotional effects. Each sample in the raw data represents a single person playing a specific game. Hence, using this dataset was useful to examine the idea of emotion detection. More details about the dataset, participants, games, and recording conditions are shown in the Related Work section under the title of the original paper where the experiment was first done.

II. SIGNALS ANALYSIS

This section shows the used approaches in feature extraction. In each approach, the major concepts are demonstrated according to the feature's definition, the procedure applied to extract them, and how to handle each case.

Generally speaking, the feature extraction techniques take place in one of these approaches: Statistical Analysis and Spectral Analysis.

A. Preprocessing

Preprocessing was done as a first step in working with EEG signals since EEG signals have noises and artifacts from the surrounding environment. Collected EEG signals passed through a 5th order sinc filter to normalize and denoise EEG signals. In the figure below, we can see the signals from the 14 channels after and before preprocessing

B. First Approach: Spectral Analysis

Spectral Analysis Methods are concerned with extracting information about signals from their spectral domain. Hence, the spectrum of signals in a specific interval is the measure.. Since EEG signals contain specific bandwidths represented by Delta δ, Theta θ, Alpha α, Beta β and Gamma γ, it encourages the curiosity to investigate the nature and occurrence of each bandwidth, and what is the relation between them as features and other subjects.

Fourier Transform: Its a mathematical transform which transform the function from the time domain into the frequency domain by using this equation.

$$\frac{1}{2\pi} \int_0^{2\pi} 2F(\omega)e^{i\omega t}\, d\omega \tag{1}$$

And it could be as:

Fast Fourier Transform (FFT)

Fast Fourier Transform is an algorithm which computes Discrete Fourier Transform (DFT) to convert the signal from time domain to frequency domain. This method employs mathematical means or tools to EEG data analysis.

$$X(m,w) = \sum_{n=-\infty}^{\infty} x[n]w[n-m]e^{-jwn} \tag{2}$$

Where w[n] is the windowing function.

By using the spectral analysis approach, some features were defined, such as the power density of each specified bandwidth in the signal. In other words, the ratio of each bandwidth occurrence in a signal is considered as a measure.

C. Second Approach: Statistical Analysis

Statistical Moments: The used dataset is separated based on the game, participant, and the channel it was read from. According to this structure, the statistical moments were calculated for each signal to generate a feature vector per sample.

Hajroth Parameters: Hajroth parameters indicate statistical properties used in signal processing in time domain [6]. They are commonly used in the analysis of electroencephalography signals for feature extraction, the parameters are:

- Hajorth Activity: The activity parameter represents signal power, the variance of a time function.

$$(Activity = var(y(t)) \tag{3}$$

- Hajorth Mobility: The mobility parameter represents the mean frequency or the proportion of standard deviation of the power spectrum.

$$Mobility = \sqrt{\frac{var(\frac{dy}{dt})}{var(y(t)}} \tag{4}$$

Where var(f(t)) is the variance of the function f(t).

- Hajorth Complexity: The change in frequency is represented by the Complexity parameter. Provides an estimate of the Signal bandwidth (The complexity parameter shows how a signal's shape is comparable to a pure sinusoidal wave. Complexity's value converges to 1 as the signal shape is more similar to a pure sine wave).

$$Complexity = \frac{Mobility(\frac{dy(t)}{dt})}{Mobility(y(t))} \tag{5}$$

As mentioned in the previous section, for the same structure, Hajroth Parameters were calculated for each signal to generate a feature vector per sample.

III. DESIGNED API

During this research, several techniques have been used to accomplish the desired objectives. It was clear that the flow of each technique is somehow similar to others regarding its inputs and outputs. Additionally, there have been many parts where results vary according to specific input parameters. Hence, there was a need to design a general API that handles all the analysis processes according to the input parameters.

Moreover, this API is general, which means it could be used in different places where similar analysis is needed regardless of what the data represents. The following is a documentation of the functionalities included in the API.

A. Global Parameters

Applications may vary regarding their specifications. In our research, we were interested in processing data according to the discussed features, which are declared in this part of the program. Furthermore, input data formats vary among applications indeed. As mentioned earlier, this API could be customized to any format of data, and many processing requirements. This part demonstrated the main input parameter to customize the system according to data

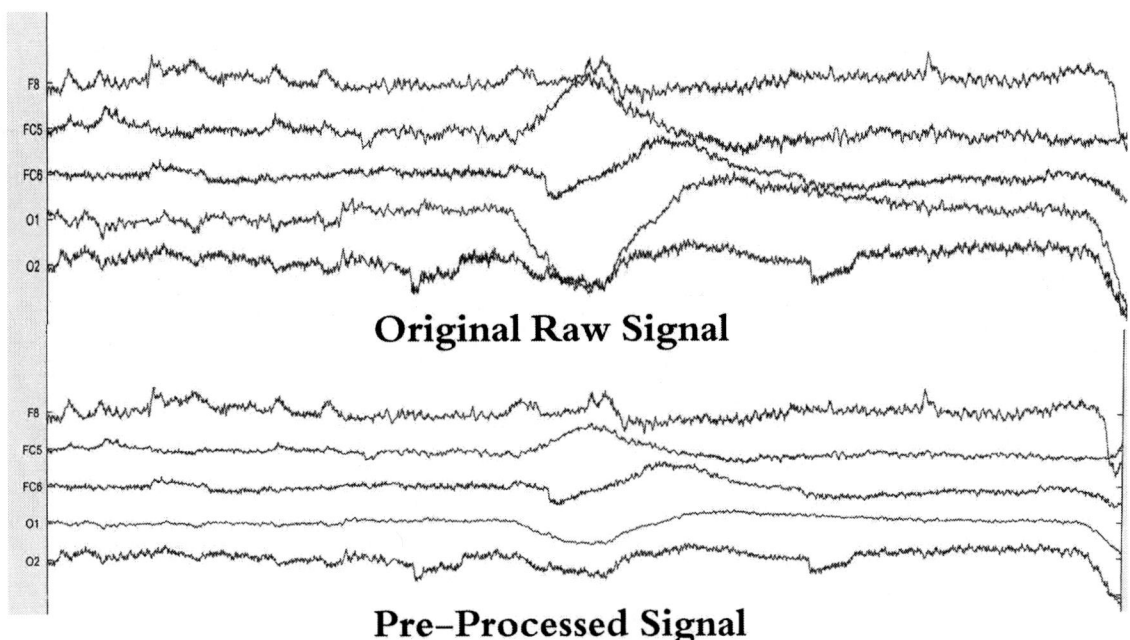

Fig. 3. Preprocessing: Noise Reduction

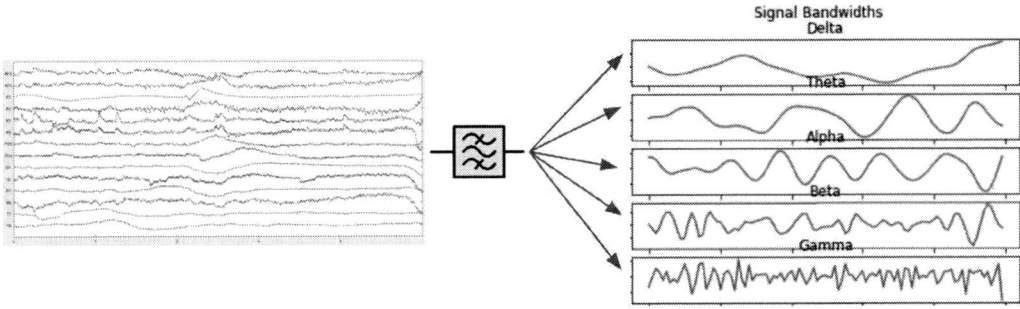

Fig. 4. Signal Bandwidths Filtering

format and specifications.

B. Functions Documentation

The following are the main functions in the designed API:

Reading Data: This function is application-dependent, which means it could vary between applications according to the input data format. In our application, data was divided into games and samples from each participant.

Find Feature Vector: This function finds feature vectors for the input data according to a given input parameter to select the analysis type. It returns feature vectors in (.csv) format.

Classification: This function receives feature vectors in (.csv) format in addition to the selected classification algorithm, so it executes the specified classification algorithm on feature vectors and prints result metrics.

Feature Extraction Per Channel: This function mainly executes filtration function features vector based on the desired channel number. In other words, the user can process data based on specific input channels. This is helpful in studying the effect of each one separately.

Features Filter: This function filters feature vectors based on selected channel numbers and specified bandwidths to return a feature vector containing the specified inputs only.

IV. RESULTS & DISCUSSION

In this paper, we will show the outcome of this research, with the help of numeric analysis results for comparing different approaches. Additionally, the main objectives will be discussed by showing the answers to research questions.

1) Comparing Different Classifiers Results: After performing different processing steps on data, each sample was labeled with the actual class it belongs to and then classified with the help of different classification algorithms, as a way to test the accuracy of those features. Following are the output metrics of the classification process among the different approaches.

TABLE II
ORDINARY STATISTICAL RESULT

Classifier	Accuracy	Precision	Recall	f1-score	Kappa
DT	0.392	0.413	0.393	0.396	0.19
Random forest	0.526	0.523	0.527	0.523	0.369
Naive Bayes	0.258	0.248	0.259	0.249	0.0119
KNN	0.3125	0.320	0.313	0.313	0.0833
NN	0.803	0.815	0.804	0.803	0.738

TABLE III
ORDINARY STATISTICAL RESULT

Classifier	Accuracy	Precision	Recall	f1-score	Kappa
DT	0.3125	0.31	0.313	0.311	0.083
Random forest	0.2768	0.283	0.277	0.279	0.0357
Naive Bayes	0.178	0.170	0.179	0.173	-0.0952
KNN	0.375	0.381	0.375	0.376	0.1667
NN	0.5178	0.517	0.518	0.517	0.3571

TABLE IV
SPECTRAL ANALYSIS RESULTS

Classifier	Accuracy	Precision	Recall	f1-score	Kappa
DT	0.42	0.442	0.429	0.430	0.2381
Random forest	0.73	0.736	0.732	0.731	0.6429
Naive Bayes	0.285	0.296	0.286	0.286	0.0476
KNN	0.723	0.735	0.723	0.726	0.631
NN	0.8482	0.852	0.848	0.847	0.7976

2) Arousal-Valence Emotion Model Evaluation: With the help of the implemented techniques, it is possible to match the used dataset and the Arousal-Valence Emotion Model. So, we used the previous methods to examine the mentioned concept. The power distribution among channels was calculated and displayed in figure 5 below.

Referring to figure 2 and figure 6 it is expected to have a clear match between electrode positions and emotions source according to. Arousal-Valence Emotion Model.

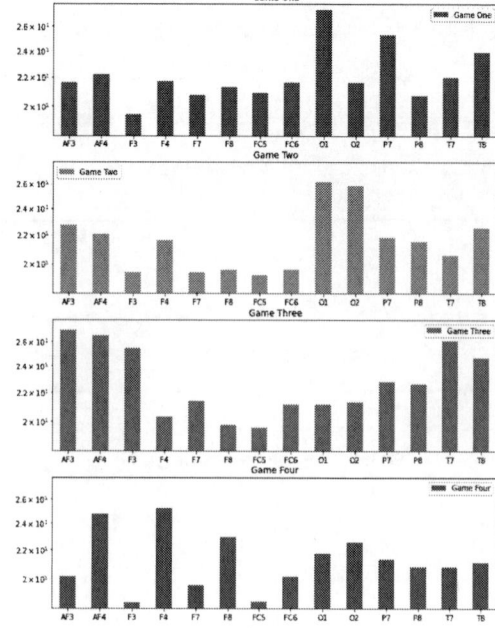

Fig. 5. Arousal-Valence Emotion Model Evaluation

According to figure 5, Firstly, results show that signals related to game one 'Boring' were mostly carrying power in 'O1' and 'P7' channels, which are located in the third quarter of the scalp. After that, game two, which is classified as 'Calm', showed the maximum power in channels 'O1' and 'O2' which are located in the third and fourth quarters. Then, game three 'Horror' has the power density concentrated in channels 'AF3', 'AF4'. 'F3' and 'T7'. Finally, the last game is classified as 'Funny' and the power is concentrated in 'AF4' and 'F4'. Figure 6 below shows the location of the electrodes with the corresponding related places in the scalp based on AVEM.

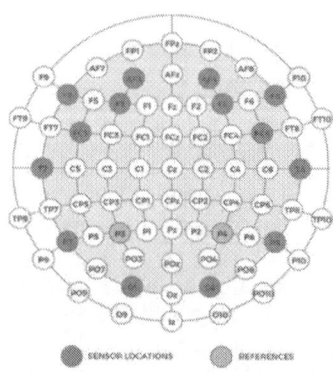

Fig. 6. Recall Electrodes Locations and Corresponding Emotions (AVEM)

3) Discussion:

Analysis Results: It was clear from the above results that there was an accuracy of 85% with classifying EEG Signals according to the emotions they carry among 4 classes as they are the played games. This result was accomplished by applying spectral analysis to extract features of bandwidth weights, with the help of WEKA tool using MLP classifier. It is important to point out that all the results above were computed using WEKA as well.

Using the API: The main advantage of the designed API was the ability to repeat the experiment with different input parameters several times, in addition to the fact that it is reusable. On the other hand, the classification process had shown fewer accuracy metrics than WEKA, due to the auto setting of classification parameters in WEKA, which was harder to implement using Python libraries.

V. Conclusion

In this study, the main objective was to study the feasibility of detecting emotions from EEG Signals and to observe the outcoming results with the electrode placed on the scalp and the power spectral density of each EEG-defined bandwidth. Additionally, the feasibility of using EEG Signals as authentication biometrics was studied.

A published dataset was used with 4 different types of games collected from 28 participants. Different approaches were used in this study. Best analysis results were observed in Spectral Analysis methods, in which the power density functions for each signal-defined bandwidth was calculated as feature vector, and classified with the help of MLP classification algorithm to come up with 85% accuracy in classifying samples into their classes. Also, the signal power was calculated for each sample channel and compared to evaluate the Arousal-Valence Emotion Mode. This result was accomplished by applying spectral analysis to extract features of bandwidth weights, with the help of WEKA tool. It should be noted that some channels are useless and cannot be used as a feature source.

After viewing our results using different features and different classification methods, we can see the different features give different values for the metrics for the same classifier. This leads us to the conclusion that there are some features better than others to represent data.

Here, we saw that Statistical Features don't give us enough information about data. Meanwhile, Spectral Analysis Features give better information about the data since each feature vector refers to a ratio of each bandwidth power (Delta δ, Theta θ, Alpha α, Beta β and Gamma γ) and each bandwidth indicates a special emotion in that part of the brain. Ultimately, the analysis of the bandwidths gives us more details about the emotions of the participant, helping us classify better.

On the other hand, different classifiers give us different values for the metrics, and not all classifiers are good for this type of data, especially data that have multiple classes to classify (in Multiclass). That being said, it's obvious and predicted that the Decision Tree and Naive Bayes are not useful for building a model for EEG Data. On the other hand, other classes (especially Neural Network) were very useful in our case because the data was not linearly separable.

A general API was developed using Python. As a first step, the provided functionalities were useful in reading data with multiple formats, generating feature vectors, and running classification algorithms.

In addition to emotion detection, using EEG signals as an authentication biometric was studied, trying to match the key concepts of biometric authentication and EEG signal biometric features. Results showed 70% accuracy in a developed system to detect participants. This was done as a secondary topic to our project.

In conclusion, to answer the proposed question, it is possible to detect emotions using EEG signals. However, it is subjected to a few circumstances that need to be addressed as discussed in the report. One of these circumstances is that there is a need to choose suitable features by taking into consideration the channel effect and its place on the scalp. Also, it is crucial to increase the accuracy to ensure the suitability of the technology. As discussed, more accuracy can be obtained by using probable preprocessing techniques with a matching classifier. Finally, all of the objectives were achieved and discussed in this report.

References

[1] Satheesh Kumar and Bhuvaneswari. "ANALYSIS OF ELECTROEN-CEPHALOGRAPHY (EEG) SIGNALS AND its categorization–a study". In: Procedia Engineering 38 (2012), pp. 2525–2536. ISSN: 1877-7058. DOI: https://doi.org/10.1016/j.proeng.2012.06.298.

[2] L F Haas. "HANS BERGER (1873–1941), RICHARD CATON (1842–1926), AND ELECTROENCEPHALOGRAPHY". In: Journal of Neurology, Neurosurgery & Psychiatry 74.1 (2003), pp. 9–9. ISSN:0022-3050. DOI: 10.1136/jnnp.74.1.9

[3] M. Teplan. "FUNDAMENTALS OF EEG MEASUREMENT". In: Journal of the Institute of Measurement Science 2 (2002).

[4] Wiem Mimoun Ben Henia and Zied Lachiri. "EMOTION CLASSIFICATION IN AROUSAL VALENCE MODEL USING MAHNOB-HCI DATABASE". In: International Conference on Engineering 8 (2017), pp. 318–323. ISSN: 2575-1328. DOI: 10.1109/icemis.2017.8272991.

[5] MOTION CLASSIFICATION. accessed: 2021-01-18. URL: en.wikipedia.org/wiki/emotion_classification .

[6] Yu-ri Lee Seung-Hyeon and Hhyoung Kim. "A NOVEL EEG FEATURE EXTRACTION METHOD USING HAJORTH PARAMETER". In: (2014)

Hyperdimensional Computing Versus Convolutional Neural Network: Architecture, Performance Analysis, and Hardware Complexity

Eman Hassan[1], Meriem Bettayeb[1], Baker Mohammad[1], Yahya Zweiri[2], and Hani Saleh[1]

[1]System-on-Chip (SoC) Lab, Electrical Engineering and Computer Science Department, Khalifa University, Abu Dhabi, UAE

[2]Department of Aerospace Engineering, and Advanced Research and Innovation Center (ARIC), Khalifa University, Abu Dhabi, UAE

Abstract—The interest in brain-inspired computing architectures has been growing, particularly in the context of edge devices with constrained resources for executing cognitive functions. One such approach is hyperdimensional computing (HDC), a novel concept that draws inspiration from the large representation of human neuronal activity. HDC has demonstrated effectiveness in one-dimensional tasks, like text identification and activity recognition, offering advantages in power consumption and response time over convolutional neural networks (CNNs). This paper compares HDC and CNN regarding architecture, accuracy, and hardware complexity, explicitly focusing on image classification tasks. Our findings indicate that CNNs generally outperform HDC in two-dimensional tasks but require significantly more computational resources. In contrast, HDC offers adequate results using just 16% of the data needed for training. Additionally, experiments conducted using a Raspberry Pi 4 show that HDC can enhance inference speed and energy efficiency by approximately 2.5 times relative to CNNs.

Index Terms—Hyperdimensional computing, Convolutional neural networks, Encoding and learning, Image classification.

I. INTRODUCTION

Brain-inspired architectures are considered a vital research topic in the current digitalized era. The advancements in deep learning algorithms such as a convolutional neural network (CNN) provide excellent accuracy in diverse applications like image classification, object segmentation, and time-series applications [1]. Nevertheless, CNN's algorithms demand several million multiply-and-accumulate (MAC) operations; for example, GoogleNet achieved a 6.7% error rate using around 7 million weights for the ImageNet dataset [1]. That makes them very expensive for power-constrained devices or impractical for particular high-speed applications [2]. Moreover, training is usually done on the cloud, which would raise security issues for sensitive data applications. Therefore, there is a significant need to rethink computing by developing algorithms engineered for performing cognitive tasks on devices with limited resources and constrained budgets [3].

In the human brain, there are around 100 billion neurons, which are interconnected by approximately 1000 trillion synapses. Consequently, the brain's possible states can be represented by vectors in an extremely high-dimensional space (HS). In that sense, the hyper-dimensional computing (HDC)

paradigm, has emerged as a form of brain-inspired computing. These high-dimension vectors, named hypervectors (HD), have holographic representation and are composed of independently identical distributed (i.i.d) elements. HDC is motivated by observing that the key aspects of human perception and cognition can be explained using mathematical properties in HS. In HDC, vectors with high but fixed dimensions (e.g. D = 10,000) are seeds (basis), representing various information distributed randomly along with the vector positions. The vectors could be binary $\{0,1\}$, bipolar $\{1,-1\}$, ternary $\{1,0,-1\}$, integer and floating numbers [4].

HDC is a potential solution for limited resources devices, as it does not include the computationally demanding algorithms found in the widely used CNN [5]. A key distinction between these two computing approaches is that HDC diverges from the traditional dimensionality reduction strategies established in machine learning. Instead, it emphasizes expanding dimensionality, mimicking neuronal activities to represent the semantic relationship between objects. Conversely, CNN focuses on feature extraction, exploiting relevant filters to build a network of learnable parameters. Capitalizing on that, HDC has been implemented widely in many applications such as text classification, image classification, biosignal processing, sensory inputs recognition and robotics [4].

This paper aims to conduct a rigorous comparative study between HDC and the widely-used CNN, Lenet-5, focusing on supervised image classification. By conducting this comparative analysis, we aim to shed light on the capabilities and limitations of HDC for image classification, thereby contributing to the broader understanding of its applicability in 2D data structures. Nonetheless, before diving into the comparative study, a brief recap the fundamentals of HDC is provided. In this paper, we aim to answer the following questions:

1) How does the HDC model perform compare to CNNs for image classification tasks in terms of accuracy and hardware efficiency (latency, power, and energy consumption) ?

2) What is the impact of training data size and vector dimensionality on the performance of HDC model?

3) What are the computational efficiencies in terms of the number of parameters and number of computations of

2023 International Conference on Microelectronics (ICM)

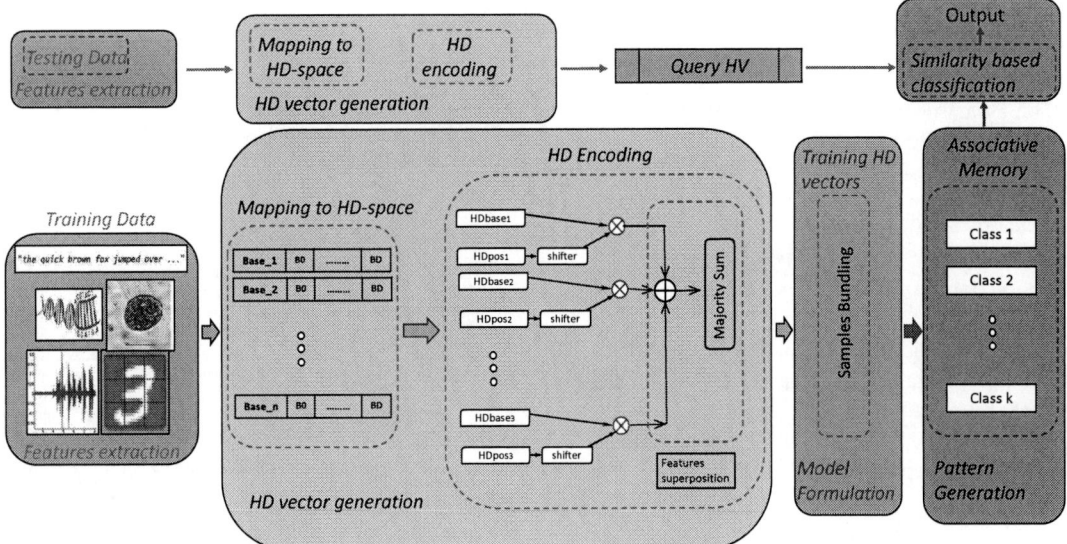

Fig. 1: Overview of the Hyperdimensional Computing Process for Classification: The pipeline starts with IM/CIM, where high-dimensional (HD) vectors, representing either integer or continuous input features, are randomly generated and saved. During the encoding phase, HD feature vector are mapped to high space using MAP operations, creating an abstract HD representation of the input. In the training phase, these HD vector samples are combined to form a class prototype, which is then stored in associative memory for later use in inference.

HDC compared to CNNs?

The scope of this study is limited to specific types of image datasets, and the evaluation metrics used for comparison will be detailed in subsequent sections. The rest of the paper is organized as follows: An overview of HDC fundamentals is covered in Section II. Section III analyzes the HDC model for the image classification task. Section IV analyzes two competing paradigms, HDC and CNN, in terms of architecture, performance, and hardware complexity for 2D applications. Finally, the conclusions and some directions for future work are presented in Section V.

II. HYPERDIMENSIONAL COMPUTING FUNDAMENTALS

HDC starts, as depicted in Fig.1, by assigning a seed vector to each symbol for a given application, storing them in an item memory (IM). A dense seed vector contains distributed elements with equal probability (P). For example, for binary representation, the number of 0s and 1s in the vector are identical with $P = 1/2$. Moreover, randomly generated vectors in high-dimension space are likely to be orthogonal, robust to variation, and easily distinguished from random noise [6].

HDC is designed to represent symbolic and compositional structures using a set of well-designed operations. The three primary operations are binding two vectors using an element-wise multiplication of vectors of the same dimension, bundling or merging a set of vectors using an element-wise addition, and permutation, a specialized form of binding that uses a specific matrix called the permutation matrix. This operation requires only one HD vector and carries out a cyclical rotation or shift. Permutation is typically used to capture temporal or spatial patterns in data. Binding and permutation operations produce different orthogonal vectors, while the bundling operation preserves the similarity and thus can be used to represent a set of HD vectors [7]. The next section investigates the HDC for image classification utilizing well-known MNIST [8] and fashion-MNIST [9] datasets. The model performance is tested and analyzed under different training dataset sizes and HD vector dimensions, as detailed in the next section.

III. HDC FOR SUPERVISED IMAGE CLASSIFICATIONS

This section explores applying HDC in supervised image classification, focusing on the MNIST dataset. In this model, every pixel in a 28×28 image corresponds to a unique HD binary vector. The system uses 784 distinct HD vectors, randomly generated and stored in the IM. These vectors serve as seed vectors and remain constant throughout the system's operation. The dimensionality (D) is set to 10,000 for these HD vectors.

To preserve the spatial relationships between pixels, each pixel's intensity is bound to its corresponding HD seed vector in the IM. For simplicity, grayscale images are transformed into black-and-white (BW) format. In this format, a pixel value of 0 indicates black, and 1 signifies white. These images are subsequently translated into HS employing either linear or orthogonal encoding methods [10]. The orthogonal mapping process is carried out through steps, summarized in Algorithm 1 and further detailed in our previous works [4].

In the following, we examine the HDC model performance as a function of training dataset size and the vector dimension.

A. Impact of Training Set Size on HDC Accuracy

For this study, we investigate the impact of training set size on the classification accuracy of the MNIST dataset. In the initial experiment, 50 images for each digit were randomly

979-8-3503-8083-5/23 $31.00 © 2023 IEEE

Algorithm 1: HDC Model for Image Classification

Input: Feature Inputs
Output: Similarity Score
Data: MNIST dataset for training and testing
/* **Encoding Module** */
  for $j = 1$ *to* n, $n = 10$ *(number of classes)* **do**
    for $x_i \in X_j$ **do**
      Set $D = 10,000$.
      Initialize 784 HD seed binary vectors
        randomly and store them in Item Memory
        (IM).
      Binarize and flatten the input image.
      Apply 1-bit shift for black pixels HD and
        0-bit shift for white pixels HD vector.
      Aggregate all pixels HD vectors to generate
        the HD representation of the digit.
    end
  end
/* **Training Module** */
  Aggregate HD vectors for similar samples into a
  single pattern HD_c using the equation:
  $HD_c = \sum_{i=1}^{n} H_i$
  Binarize each class HD vector using the majority
  sum operation: $HD_c = [\sum_{i=1}^{n}(H_i)]$, where $[.]$
  refers to the majority sum operation.
/* **AM Module** */
  Compare query vector H_q to the stored classes HD
  HD_{ci} using Hamming distance
  $\text{score} = \sum_{i=1}^{D}(HD_{ci} \oplus H_q)$.
  $Best_{(Score)} = argmin(score)$

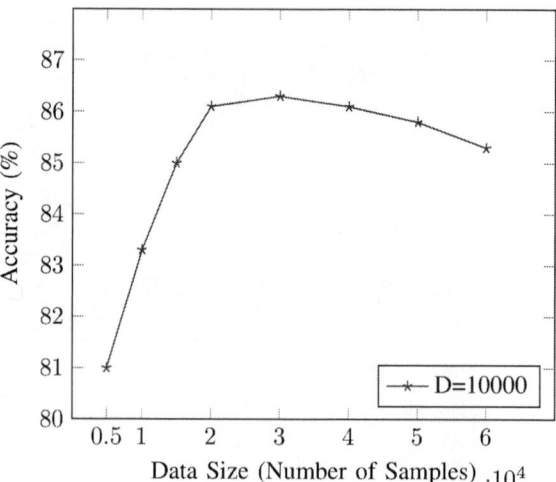

Fig. 2: the relationship between data size and accuracy for Linear HD Encoding on the MNIST dataset.

selected for training. After completing the training stage, the AM comprises 10 HD vectors, which represent all the variations of the digits. The experiment is repeated using different sample sizes (100, 500, 1000, 2000, 3000, 4400, and 5200 samples) from each class to evaluate the model's performance. To assess the encoder's accuracy, 1000-10000 unseen images were chosen, and the overall accuracy was calculated by determining the percentage of correctly classified digits in the test dataset. As shown in Fig 2, the accuracy for small sets begins at a low value and then gradually increases in a linear fashion, reaching a maximum of just below 86%. Beyond this point, the accuracy starts to decline. This decrease is attributed to the limitation of HDC's capacity, which sets a threshold on the number of bundled HD vectors [10]. Once this threshold is crossed, it becomes increasingly difficult to decode individual operands accurately.

That explains the accuracy when the pattern increases above a specific limit. Also, results show that using only 16% of the dataset for training, the value of Recall (86.5%), Precision (85%), and F1-score (86%), respectively, reflects the classifier robustness and a balanced dataset.

It is essential to underline that for HD architecture, the

hamming distance acts as a quantitative similarity metric by directly comparing distributed representations without decoding those representations. Results in [10] show that a more significant number of common elements leads to more similarity between resulting vectors. In the MNIST HD representation, certain digits (such as 4 and 9) have high resemblance resulting in significant overlap in their HD vectors. This makes it difficult to distinguish analyzable patterns accurately. In detecting patterns, the number of overlapped elements that can be detected robustly is limited. One solution to reduce errors is to use non-linear mapping for the HD space inputs instead of linear mapping. Another suggestion is to extract the main features of the image using CNN or Vision Transformer techniques (ViT) and then map those features to HD space [11]. Retraining techniques could also improve the accuracy of the system [4].

B. Effect of HD Vector Dimension on Accuracy

In the above study, a vector length of $D = 10k$ was assumed throughout the HDC simulations. However, some studies, such as [12], suggest that an HD vector with a dimension of 1k elements can still adequately represent the system. This reduction in vector size would positively impact both the execution time and the chip area required for implementation. Results in [12] show that a 1k HD vector representation can achieve 90.4% accuracy compared to 97.8% when utilizing a 10k HD vector for language recognition tasks. For the MNIST dataset employed in this work, the impact of varying vector dimensions D on the model's performance across different training dataset sizes is examined. Fig.3 illustrates that higher dimensions generally lead to better accuracy, particularly for larger training sets. This suggests the model benefits from higher-dimensional spaces, especially when trained on more extensive datasets. However, it is also worth noting that the gains in accuracy diminish beyond a certain point, indicating a trade-off between computational complexity and model performance. Specifically, the accuracy reaches around 86% when an HD of $8k$ dimension is utilized using 20% of the training dataset. Beyond that dimension, the accuracy improves slightly but at the cost of increased time and area. Nevertheless, a significant cost reduction is achievable when using a 2k HD vector, where the accuracy drops only by 5%.

IV. HDC VERSUS CNN

A. Comparison Between HDC and CNN LeNet-5 Architecture

a) Item Memory vs. Input Layer: The IM in HDC and the input layer in LeNet-5 serve as the foundational layers that initiate the processing of input data. In HDC, the IM stores HDs representing different features or items. These HDs are then used in encoding to transform the input data into a high-dimensional space. On the other hand, the input layer in LeNet-5 takes the input image and applies six feature filters, transforming the image into an intermediate layer. The similarity here is that both layers act as the entry point for input data and produce an intermediate representation used in subsequent layers [13]. In terms of mapping, the IM in HDC can be seen as analogous to the feature maps (FM) set in LeNet-5's input layer, both serving to encode the input into a form suitable for further processing.

b) Encoding Process: In HDC, the encoding process involves specific operations where the HD vector of the samples is generated by summing the HD vectors stored in the item memory, each weighted by the corresponding value of the feature V_i. This operation is mathematically equivalent to a matrix multiplication between the input feature HD vectors and the IM HD vectors for feature indices W_i. In Lenet-5, this is similar to the linear transformation $y = Wx + B$, where x and y are the input and output of the layer, and W and B are the weights and biases, respectively. In HDC, the Item Memory is the weight W, and biases are set to zero [13]. LeNet-5, on the other hand, employs 'tanh' or 'sigmoid' as its activation function after the convolution and pooling layers. On the other hand, HDC leverages activation functions such as 'cosine' or 'tanh' during encoding. These activation functions serve two purposes: introducing non-linearity to the resultant HD vector and normalizing its elements to a specific range.

Activation functions in HDC enhance the model's ability to capture complex patterns and relationships within the input data, introducing nonlinearities required for effective learning. Furthermore, the normalization of HD vectors ensures stability during training, consistent scaling across different features, and compatibility with specific operations or architectures. LeNet-5's max-pooling technique, employed for downsampling the feature size, finds an analogy in HDC's approach of reducing or selecting the appropriate HD vector dimension suitable for the corresponding task. It's important to note that in HDC, the dimensionality of the HD vectors should be predefined before storing the HD seeds in the Item Memory (IM). This comprehensive encoding process forms the foundation for HDC's unique capabilities in representing and processing high-dimensional data.

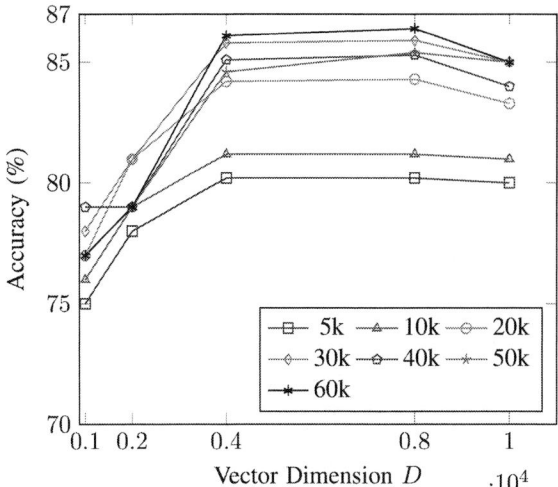

Fig. 3: The accuracy of MNIST classification correlates with the dimension of the HD vector. Dimensions of 1k, 2k, 4k, 8k, and 10k for the HD vector were chosen for accuracy analysis. The findings verify that with just 30% of the training data, a vector size between 4k and 8k is sufficient to achieve peak accuracy.

c) Associative Memory vs. Classifier Layer (CL): In HDC, the AM performs classification by calculating the similarity between the query HD and each class HD stored in memory. The cosine or hamming distance similarity can be simplified into a vector product, transforming the inference process into a form of matrix multiplication. This is similar to the linear layer in Lenet-5, specifically the classifier layer, which also employs matrix multiplication for inference. Regarding activation functions, HDC uses 'argmax' to identify the class with the highest similarity score [14]. In neural networks like LeNet-5, the 'SoftMax' activation function serves a similar purpose but is differentiable, making it more suitable for gradient-based optimization methods. The mapping here is twofold: First, HDC and LeNet-5 employ a form of matrix multiplication in their final classification layers, although for different purposes. Second, the 'SoftMax' function in LeNet-5 can be seen as analogous to the 'argmax' operation in HDC to identify the most likely class. A summarized comparison of the architectural components for HDC and LeNet-5 can be found in Table I.

B. Performance and Hardware Complexity Analysis

HDC is a promising model for edge devices as it does not include the computationally demanding training step found in the widely used CNN [5]. Although HDC has its challenges, encoding alone accounts for about 80% of the training's execution time. Additionally, some encoding algorithms may increase the size of encoded data by 20x [15].

Prior research has shown that HDC performs better than deep neural networks (DNN) for 1D datasets, such as speech recognition [16]. Nonetheless, the complexity increases when transitioning to the 2D data structure. This study evaluates and determines the computational operations and parame-

TABLE I: Summary of Similarities and Mappings between HDC and CNN-LeNet-5 Architecture. In the Learning row for LeNet-5, η represents the learning rate, and $\mathcal{L}$ represents the loss function. While for HDC, $\vec{P}_{class}$ represents the prototype vector for a class, which is updated during the learning process.

Aspect	HDC	LeNet-5
Input Layer	High-Dimensional Vectors (HDVs)	32x32 Pixel Grayscale Image
Encoding	Composite Vector Representation: $\sum_{j=1}^{n} \sum_{i=1}^{d} V_i \otimes W_i$	Linear Transformation: $y = Wx + B$
Learning	Associative Memory Updating: $\vec{P}_{class} = \vec{P}_{class} + \vec{v}_{input}$	Backpropagation: $W^{(t+1)} = W^{(t)} - \eta \frac{\partial \mathcal{L}}{\partial W},$ $b^{(t+1)} = b^{(t)} - \eta \frac{\partial \mathcal{L}}{\partial b}$
Dimensionality Management	Use of High-Dimensional Spaces	Max-Pooling
Activation Function	Binarization/Mapping to Binary or Bipolar Vectors	'tanh', 'sigmoid', or 'ReLU'
Classification Layer	Associative Memory (Prototype Vectors)	Fully Connected Layer + SoftMax
Similarity Measure	Cosine or Hamming Distance	SoftMax Output (Probability Distribution)
Activation & Optimization	Argmax for Classification	SoftMax for Differentiable Optimization

ters necessary for each paradigm. For MNIST classification through CNN, we employ the LeNet5 architecture within the Caffe framework [8]. LeNet5, recognized for its straightforward design, is comprised of seven layers. In this research, Caffe, a comprehensive deep learning framework, is utilized for simulating and assessing the effectiveness of LeNet in classifying MNIST digits [17]. Furthermore, to assess the efficiency of the models at the inference stage, we ran tests on two different platforms. The first platform had a 2.7 GHz Dual-Core Intel Core i5 processor and 8GB of memory, while the second one was a Raspberry Pi 4, which is a low-power device that provides high performance and has an ARM Cortex-A72 clocked at 1.8 GHz and 8GB of RAM.

TABLE II: Computing Complexity at inference of 2D (MNIST) using both HDC and CNN Computing Paradigms.

	CNN	HDC
Full Training Accuracy	99.04%	86%
Partial Training Accuracy (20%)	98.3%	86.4%
Partial Training Accuracy (8%)	97.7%	85.3%
No. of Parameters	431.08k	7940k
MAC operations	146.75M	-
No. of Computations	974.08k	8030.78k
CPU Latency	7.5 ms	2.54 ms
IoT RPi4 Latency	16.5 ms	6.58 ms
IoT RPi4 Energy	200 mJ	80 mJ
IoT RPi4 Power	5 W	4.3 W

The LeNet5 convolutional neural network, used for MNIST digit recognition, has a classification accuracy of approximately 99%, achieved through a learning rate of 0.01 for 10,000 iterations. On the other hand, the HDC classification accuracy for the same MNIST dataset is around 86%, achieved

through single-pass training on the full training dataset[1].

To compute the number of parameters for the HDC model during inference, we assume the following: 1) A matrix with dimensions $784 \times 10k$, utilized for random seeds, is created and stored in the IM. 2) The associative memory contains encoded classes, each occupying a space of $10k \times 10$.
Regarding computational operations, it is assumed that: 1) The quantity of bit shifts totals approximately half of 784 (0.5×784), based on the assumption that about 50% of the seed array undergoes shifting (an operation at the HD vector level). 2) The effect of the initial shifting is counteracted by an equivalent shifting operation in the opposite direction (0.5×784). 3) Column-wise addition involves $(784 - 1) \times 10k$ operations to from the HD vector for the image. 4) An XOR operation between the query and the patterns encoded in the AM involves $10 \times 10k$ calculations. 5) Post-XOR, an addition operation is carried out, requiring $10 \times (10k - 1)$ computations. 6) Finally, a total of 9 comparison operations are conducted.

Besides, the main results of both CNN and HDC implementation on CPU and Raspberry Pi are presented in Table II. From these results, CNN maintains a higher level of accuracy, especially when using the full training set and a vector dimension of D = 10k in the HDC framework. An additional analysis is conducted on the impact of training set size on CNN's accuracy, mainly training LeNet with 10k and 5k samples per class. The marked accuracies were 98% and 97.70%, in contrast to HDC's 86% and 85.3%, respectively. This reinforces the superior performance of CNN in 2D applications, with only a minimal accuracy reduction of 1.4%, which aligns with previous studies [8]. Moreover, we tested the HDC model for image classification on a more complex dataset called Fashion-MNIST [9], which is considered a new version of the original MNIST dataset with more features. It shares the same structure for training and testing datasets

[1]Datasets used in this study include MNIST [8] and Fashion-MNIST [9], both of which can be accessed publicly online. The code developed for this research is also publicly available on GitHub at: https://github.com/emfhasan/HDCvsCNN_StudyCase

and the exact image size and the number of classes. These datasets present a challenging task for HDC models, as many fail to achieve state-of-the-art performance. The HDC model accuracy for the Fashion-MNIST dataset averaged around $\sim$ 72% utilizing 20% of the data. In comparison, the CNN model accuracy for the same dataset reached $\sim$ 91.6%. For this model architecture, the number of parameters for HDC is higher than CNN. Still, for HDC design, no MAC operations are required, which would positively reflect on area, energy, and execution time. Besides, In our study, we compared the performance of HDC and CNN-Lenet5 models during the inference phase, using a 6K-dimension vector optimized for Raspberry Pi 4. While the CNN model achieved an accuracy of approximately 91.6%, the HDC model was faster and more energy-efficient. Specifically, HDC took only 6.5 ms for inference, compared to CNN's 16.5 ms. This speed advantage is attributed to HDC's use of bitwise operations and high-dimensional vector representations, eliminating the complex MAC operations required by CNN's multiple layers.

A detailed power analysis in Table II on Raspberry Pi revealed that HDC improves inference speed by 2.52x and energy efficiency by 2.5x compared to CNN. These findings confirm the computational efficiency of HDC, despite its higher parameter count. Future work will further explore optimization techniques to enhance HDC's accuracy and efficiency.

V. CONCLUSIONS AND RECOMMENDATIONS FOR FUTURE WORK

In this paper, we have conducted an in-depth comparison between two computational paradigms for image classification: CNN and HDC. While CNNs excel in low-level feature extraction and generally offer higher accuracy, they come at the cost of computational efficiency and energy consumption. On the other hand, HDC models are computationally efficient and energy-friendly but lack the capability for detailed feature extraction, which is a critical aspect of image classification tasks. HDC's limitations in low-level feature extraction make it less competitive compared to CNNs in current applications. However, its advantages in computational efficiency should be further utilized. In order to improve HDC's performance in image classification tasks, future works may explore innovative approaches that address its limitations. Additionally, efficient encoding algorithms can be analyzed to handle the capacity issue in HDC. Another important direction for future research would be to leverage HDC's performance for more complex datasets.

ACKNOWLEDGMENT

This publication is based upon work supported by the Khalifa University Competitive Internal Research Award (CIRA) under Award No. [CIRA-2019-026] and System-on-Chip Center Award No. [RC2-2018-020].

REFERENCES

[1] M. P. Véstias, "A survey of convolutional neural networks on edge with reconfigurable computing," *Algorithms*, vol. 12, no. 8, p. 154, 2019. [Online]. Available: https://doi.org/10.3390/a12080154

[2] M. Bettayeb, F. Zayer, H. Abunahla, G. Gianini, and B. Mohammad, "An efficient in-memory computing architecture for image enhancement in ai applications," *IEEE Access*, vol. 10, pp. 48 229–48 241, 2022. [Online]. Available: https://doi.org/10.1109/ACCESS.2022.3171799

[3] M. Bettayeb, H. Tesfai, B. Mohammad, and H. Saleh, "Asic-based implementation of random spray retinex algorithm for image enhancement," in *2022 IEEE 65th International Midwest Symposium on Circuits and Systems (MWSCAS)*. IEEE, 2022, pp. 1–4. [Online]. Available: https://doi.org/10.1109/MWSCAS54063.2022.9859352

[4] E. Hassan, Y. Halawani, B. Mohammad, and H. Saleh, "Hyperdimensional computing challenges and opportunities for ai applications," *IEEE Access*, 2021. [Online]. Available: https://doi.org/10.1109/ACCESS.2021.3059762

[5] Y. Halawani, B. Mohammad, M. A. Lebdeh, M. Al-Qutayri, and S. F. Al-Sarawi, "ReRAM-based in-memory computing for search engine and neural network applications," *IEEE Journal on Emerging and Selected Topics in Circuits and Systems*, vol. 9, no. 2, pp. 388–397, 2019. [Online]. Available: https://doi.org/10.1109/JETCAS.2019.2909317

[6] P. Kanerva, "Hyperdimensional Computing: An Introduction to Computing in Distributed Representation with High-Dimensional Random Vectors," *Cognitive Computation*, vol. 1, no. 2, pp. 139–159, 2009.

[7] K. Schlegel, P. Neubert, and P. Protzel, "A comparison of vector symbolic architectures," *Artificial Intelligence Review*, vol. 55, no. 6, pp. 4523–4555, 2022.

[8] Y. LeCun, L. Bottou, Y. Bengio, and P. Haffner, "Gradient-based learning applied to document recognition," *Proceedings of the IEEE*, vol. 86, no. 11, pp. 2278–2324, 1998. [Online]. Available: https://doi.org/10.1109/5.726791

[9] H. Xiao, K. Rasul, and R. Vollgraf, "Fashion-mnist: a novel image dataset for benchmarking machine learning algorithms," *arXiv preprint arXiv:1708.07747*, 2017.

[10] D. Kleyko, "Vector Symbolic Architectures and their Applications Computing with Random Vectors in a Hyperdimensional Space," Ph.D. dissertation, Luleå University of Technology Luleå, Luleå, Sweden, 2018.

[11] M. Bettayeb, E. Hassan, B. Mohammad, and H. Saleh, "Spatialhd: Spatial transformer fused with hyperdimensional computing for ai applications," in *2023 IEEE 5th International Conference on Artificial Intelligence Circuits and Systems (AICAS)*. IEEE, 2023, pp. 1–5. [Online]. Available: https://doi.org/10.1109/AICAS57966.2023.10168629

[12] M. Imani, A. Rahimi, D. Kong, T. Rosing, and J. M. Rabaey, "Exploring hyperdimensional associative memory," *IEEE International Symposium on High Performance Computer Architecture (HPCA)*, pp. 445–456, 2017.

[13] D. Ma and X. Jiao, "Hyperdimensional computing vs. neural networks: Comparing architecture and learning process," *arXiv preprint arXiv:2207.12932*, 2022.

[14] E. Hassan, H. Tesfai, B. Mohammad, and H. Saleh, "ASIC implementation of associative memory and hamming distance for hdc application," in *2021 28th IEEE International Conference on Electronics, Circuits, and Systems (ICECS)*. IEEE, 2021, pp. 1–5. [Online]. Available: https://doi.org/10.1109/ICECS53924.2021.9665633

[15] M. Imani, C. Huang, D. Kong, and T. Rosing, "Hierarchical hyperdimensional computing for energy efficient classification," *2018 55th ACM/ESDA/IEEE Design Automation Conference (DAC)*, pp. 1–6, 2018. [Online]. Available: https://doi.org/10.1145/3195970.3196060

[16] M. Imani, D. Kong, A. Rahimi, and T. Rosing, "Voicehd: Hyperdimensional computing for efficient speech recognition," in *IEEE International Conference on Rebooting Computing (ICRC)*, 2017, pp. 1–8. [Online]. Available: https://doi.org/10.1109/ICRC.2017.8123650

[17] Y. Jia, E. Shelhamer, J. Donahue, S. Karayev, J. Long, R. Girshick, S. Guadarrama, and T. Darrell, "Caffe: Convolutional architecture for fast feature embedding," *arXiv preprint arXiv:1408.5093*, 2014.

979-8-3503-8083-5/23 $31.00 © 2023 IEEE

PSO-GA-based Federated Learning for Predicting Energy Consumption in Smart Buildings

Nader Bakir
Faculty of Science
Beirut Arab University
Tripoli, Lebanon
n.bakir@bau.edu.lb

Ahmad Samrouth
Faculty of Computer Science
Lebanese University
Tripoli, Lebanon
ahmad.samrouth@st.ul.edu.lb

Khouloud Samrouth, Member, IEEE
Cybersecurity and Forensics Department
Arab Open University
Beirut, Lebanon
ksamrouth@aou.edu.lb

Abstract—Smart buildings are increasingly equipped with a multitude of sensors and IoT devices, generating vast amounts of data. Generating a common global efficient model for predicting energy consumption accurately in these environments is crucial for optimizing energy usage, reducing costs, and achieving sustainability goals. However, there are limitations in terms of communication resources and data privacy. Hence, Federated Learning (FL) has emerged as a promising solution to address the challenges of privacy, data decentralization, and scalability in such complex systems. However, FL can be challenging to optimize, as the parameters of the global model need to be tuned to achieve the best performance on a diverse set of devices (called clients). In this paper, we propose to combine the Genetic Algorithm (GA) with the Particle Swarm Optimization (PSO) algorithm to optimize FL for smart building energy consumption prediction. Experiments show that our proposed method increases the energy consumption prediction performance and reduce the Root Mean Square Error RMSE by 7.7% in comparison with the local models.

Index Terms—Energy consumption prediction, Federated learning, Genetic Algorithm, Particle Swarm Optimization, smart building.

I. Introduction

Predicting energy consumption in smart buildings through machine learning is of paramount importance as it leverages the power of data-driven insights to revolutionize energy management. Machine learning models can analyze historical patterns, weather forecasts, occupancy data, and other relevant factors to provide accurate predictions. This predictive capability empowers building operators to make real-time adjustments, optimize energy usage, and reduce costs while maintaining occupant comfort [1]. Moreover, it aids in identifying anomalies and potential faults in building systems, enabling proactive maintenance and enhancing overall energy efficiency. By harnessing the potential of machine learning, smart buildings not only save resources and reduce environmental impact but also pave the way for more sustainable and intelligent urban spaces [2].

In particular, traditional centralized machine learning techniques collect local data samples from distributed smart buildings and aggregate them in a central location for training. This arises the data privacy and data confidentiality concern for smart building operators. Particularly, smart buildings are sharing their sensitive energy data consumption with central servers which increases their attack surfaces to multiple cyber and physical threats.

In response to data privacy and confidentiality requests, Federated learning (FL) approach comes in handy. In contrast with traditional machine learning, FL enables the collaborative training of a machine learning model across multiple decentralized devices or servers without exchanging the data itself [3].

In particular, to ensure data privacy, each device performs local training, updates its model based on its own sensitive data, and then sends only the model updates to the central server, where all model updates from distributed smart buildings are aggregated to improve the global model. Currently, aggregating local model updates is an active research question aimed at achieving an optimized global model.

In this paper, we propose to combine Genetic Algorithms (GA) with Particle Swarm Optimization (PSO) to optimize federated learning for smart building energy consumption prediction. On one hand, GA algorithms excel at exploring diverse regions of the solution space. In FL, this capability is crucial for mitigating issues related to data heterogeneity among participating smart buildings. On the other hand, PSO helps the global model converge faster and more efficiently compared to traditional FL algorithms [4]. Hence, the PSO-GA hybrid algorithm combines the simplicity and convergence control of PSO with the strengths of GA for a better federated learning. In particular, we use this hybrid PSO-GA algorithm in the optimization phase at the client and server side.

The following of this paper is organized as follows. Section II provides a brief overview of GA algorithm and PSO. In Section III, we analyze some related works. Then, in Section IV, we detail our proposed PSO-GA algorithm to optimize federated learning for smart building energy consumption prediction. Next, Section V describes the experimental setup and results. Finally, Section VI concludes this paper.

II. Background

A. Federated Learning

Federated Learning [5], also known as collaborative learning, represents a novel methodology for training decentralized machine learning models, encompassing a wide array of edge devices such as smartphones, medical wearables, vehicles, and

979-8-3503-8083-5/23 $31.00 © 2023 IEEE

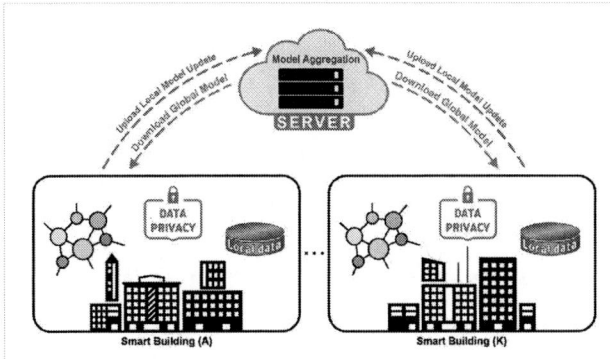

Fig. 1. Federated Learning approach

IoT devices [6]. In this collaborative paradigm, these devices collectively refine a shared global model while retaining their training data locally, eliminating the need for central data exchange. As illustrated in Figure 1, the FL workflow is as follows:

  i. First, distributed servers/devices receive initial model from the server.
 ii. Then, they train the received model with data available on the client local server.
iii. Next, they share model updates with the central server.
 iv. Finally, the central server aggregates the local model updates to get the global model.

The above process can go for rounds until a number of maximum rounds is done or an evaluation score threshold (usually Root Mean Squared Error (RMSE)) is reached. We also note that there are 2 main types of FL: Horizontal FL and Vertical FL. While the first type deals with similar data types but different instances, the latter addresses different data types with common instances.

B. Genetic Algorithms

Inspired by the process of natural selection and genetic evolution, Genetic Algorithms (GAs) are a class of optimization and search algorithms that belongs to population-based optimization approach. They are used to find approximate solutions to optimization and search problems by mimicking the mechanisms of biological evolution, such as selection, crossover (recombination), and mutation. They are particularly useful for solving complex, multidimensional, and non-linear optimization problems where traditional gradient-based methods may not be effective.

The process begins first by generating an initial population of potential solutions to the optimization problem. Then, each individual in the population is evaluated for its fitness or quality with respect to the problem's objective function. Then, individuals are selected from the current population based on their fitness scores. Next, pairs of selected individuals (parents) are combined to produce new individuals (offspring). Occasionally, random changes (mutations) are applied to some of the offspring's genetic information. The offspring replaces

some of the less fit individuals in the current population. This process continues for a fixed number of generations or until a termination condition is met. Once the algorithm terminates, the best individual in the final population is considered the approximate solution to the optimization problem.

One of the advantages of genetic algorithms is their ability to explore diverse regions of the solution space, which can help discover global optima [7].

C. Particle Swarm Optimization

Under the same category of population-based optimization, PSO is inspired by the social behavior of birds flocking [8]. The core idea behind PSO is to model a population of candidate solutions (particles) that move through a search space to find the optimal or near-optimal solution. Each particle represents a potential solution and is associated with a position and a velocity. The particles adjust their positions based on their own experiences and the experiences of their neighbors to collectively converge toward the best solution. In a machine learning context, the position and velocity of each particle are updated according to Equation 1.

$$v_l^{t+1} = \alpha*v_l^t + c_1*rand_1*(w^{pbest}-v_l^t) + c_2*rand_2*(w^{gbest}-v_l^t) \tag{1}$$

where α is the inertia coefficient, *pbest* refers to the personal best and represents the particle's best-known position, *gbest* refers to the global best solution found by any particle in the entire swarm, $c1$ is the *pbest* acceleration constant, $c2$ is the *gbest* acceleration constant controlling the positions of particles toward the optimal solution, $rand1$ and $rand2$ are random values to introduce randomness.

Then, the weights of the machine learning model are updated according to Equation 2.

$$w^{t+1} = w^t + v^t \tag{2}$$

D. PSO-GA Hybrid Algorithm

Authors in [9] combined the strengths of particle swarm optimization with genetic algorithms. Listing 24 depicts their PSO-GA hybrid optimization algorithm. In particular, they proposed a hybrid PSO-GA with a new crossover operator, Velocity Propelled Averaged Crossover (VPAC), incorporating the PSO velocity and position update rules with the ideas of selection, crossover and mutation from GAs. The PSO-GA hybrid algorithm was highly competitive and outperforming both the GA and PSO.

```
1
2 begin{Algorithm}
3 #Initialize the particle population, $S$, with $n$
      particles.
4 #Initialize the global best particle, $g$, to be the
      best particle in $S$.
5     for each particle i \in 1,....,1 do
6         Randomly initialize $x_i$
7
8 #REQUIRE A problem to be solved, $P$.
9     \Repeat
10        for each p \in S$ do
```

```
11          Evaluate the fitness of $p$.
12          Update the velocity of $p$ using the PSO
      algorithm.
13          Update the position of $p$ using the PSO
      algorithm.
14
15      Select the top $k$ particles from $S$ and
    recombine them to produce new particles using
    the GA algorithm.
16      Mutate the new particles and evaluate their
    fitness.
17      Add the new particles to $S$.
18      Update the global best particle, $g$, to be
    the best particle in $S$.
19  \until(Termination criterion is met)
20
21  return The global best particle, $g$.
22
23  end{algorithm}
```

Listing 1. PSO-GA hybrid algorithm proposed by [9].

III. RELATED WORKS

In recent years, Federated Learning demonstrated the potential in addressing privacy concerns and improving energy consumption prediction in smart buildings [10], [11].

The FL algorithm baseline used in smart building energy prediction is the Federated Stochastic Gradient Descent (FedSGD) where local smart buildings compute gradient updates on their data. These updates are then aggregated centrally to update the shared global model. A variant of FedSGD is FedAvg where local smart buildings share the updated local weights instead of the computed local gradients [12].

Alternatively, recent contributions [13], [14] presented novel applications of federated learning in the context of smart building energy prediction, highlighting its potential to mitigate privacy breaches while achieving accurate energy consumption prediction. We can categorize these contributions into 2 categories. The first category of contribution adopted the baseline FedSGD and improved by adding some weight preprocessing and/or postprocessing it while keeping the classic SGD as an optimizer. The other category of contribution consists in replacing the SGD optimizer by another kind of optimizers.

Under the first category, authors in [15] and [16] focus on speeding up the convergence process in federated learning. Authors stated that clients with similar properties tend to update their model in almost the same direction. Therefore, they applied clustering technique to group clients with similar properties (hence generating comparable updates) prior to federated learning. This resulted in a more efficient energy demand prediction model.

Liu et al. [17] implement a systematic privacy-preserving federated learning framework in the power system with an alternative encryption scheme based on the Diffie-Hellman key exchange protocol to greatly boost computational efficiency.

He et al. [18] propose a FL approach with differential privacy (DP) for smart building energy consumption prediction. Their approach enables participating buildings to train a collaborative prediction model without sharing their individual energy consumption data.

Yue et al. [19] propose a predictive coding based compression scheme for federated learning. The scheme has shared prediction functions among all devices and allows each worker to transmit a compressed residual vector derived from the reference.

Mendes et al. [20] introduce an innovative federated learning framework designed to forecast the temporal net-demand in transactive energy communities, particularly in building environments. This novel approach capitalizes on a central agent (referred to as the aggregator) that provides centralized supervision, facilitating collaborative efforts among individual clients (i.e., buildings). These buildings voluntarily participate in the collaborative process, aimed at enhancing the accuracy of their demand prediction.

Under the second category, authors in [21] replaced the SGD optimizer with GA algorithm. Specifically, the server selects the appropriate combination of clients each round with a GA. In each round, the client's combinations are evaluated anew, which are continually explored.

Recently, authors in [4] and [22] used the Particle Swarm Optimization (PSO) to optimize the hyperparameter settings for the local ML models in an FL environment.

IV. PROPOSED METHOD

In order to overcome the drawbacks of the optimization strategy using GA or PSO solely with federated learning, we propose to use the hybrid PSO-GA algorithm that combines the strength of both optimization algorithms. Hence, our proposed method falls under the second category of enhancing FL techniques introduced in Section IV. In particular, we use the hybrid PSO-GA algorithm as an optimizer on the client side and aggregator on the server side. We call our proposed method FedPSO-GA.

As a general view and as a typical FL technique, our proposed method works by training a shared model on distributed data without sharing the data itself and optimize the parameters of the shared model on each client device (each smart building). The various client devices then communicate the updated parameters to the server, which aggregates the updates and updates the shared model. This process is repeated until the convergence criterion is met.

As illustrated in Figure 2, our main contribution relies in replacing the SGD optimizer at the client and server sides by the hybrid PSO-GA population-based optimization algorithm. Particularly, we detail the process as follows:

Client side:

- Initialize a population of particles (local solutions).
- Evaluate the fitness of each particle on the local data.
- Update the particle's position and velocity based on its own position and velocity, as well as the position of the best particle in the population as previously explained in Section II.
- Select the top particles and recombine them to produce new particles (GA crossover and mutation) to obtain the optimal local solution in the current round.

979-8-3503-8083-5/23 $31.00 © 2023 IEEE

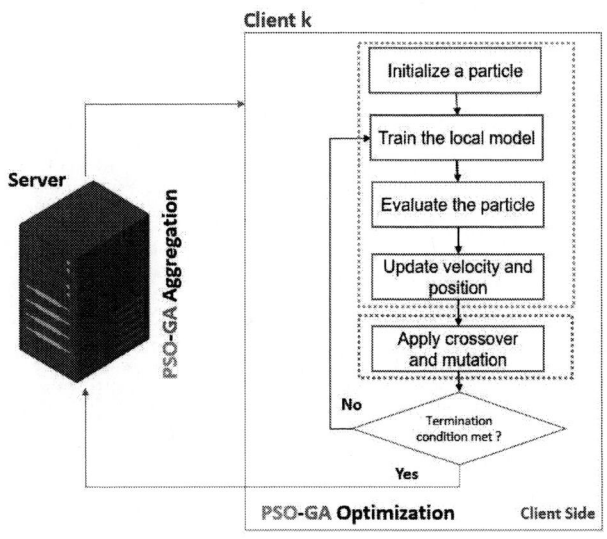

Fig. 2. Our proposed Hybrid PSO-GA FL method

- Communicate the updated particle position and the local model to the server.

Server side:

- Receive the updated local optimal particle positions and the local models from the various clients.
- Aggregate the updated particle positions to produce a new global best particle using the PSO-GA algorithm similar to the way it was applied on the client side.
- Update the global model using the updated particle positions.
- Broadcast the new global best particle and the updated global model to the clients.

After an indicated number of rounds, we the optimal global model to predict the energy consumption of the smart buildings.

V. EXPERIMENTS

A. Dataset

In our experiments, we used the CU-BEMS dataset [1], which is a smart building electricity consumption and indoor environmental sensor dataset collected from a seven-story office building in Bangkok, Thailand. The dataset contains one-minute interval data for electricity consumption of individual air conditioning units, lighting, and plug loads in each of the 33 zones of the building, as well as temperature, humidity, and ambient light measurements of the same zones. The dataset covers a period of 18 months from July 1, 2018, to December 31, 2019. In our experiments, we consider each floor as a client.

B. Model Framework and Development Environment

As the dataset contains sequential data, we adopted the LSTM (Long Short-Term Memory) deep learning model. Such

TABLE I
ADOPTED LSTM MODEL ARCHICTECTURE

Hyperparameters	
Number of LSTM Layers	128
Number of Neurons per Layer	128

TABLE II
ENERGY CONSUMPTION PREDICTION ERROR (RMSE), (MAE)

Training Scenario	RMSE	MAE
FedAvg	43.31	30.48
FedDist	44.20	31.22
FedPSO	41.2	29.7
FedPSO-GA (ours)	40.3	28.1

model is designed to capture and learn long-range dependencies in sequential data, making it effective for our energy consumption prediciton. The model architecture is depicted in Table I

We conducted our experiments using TensorFlow on a computer with Intel(R) Core(TM) i5-1235U 1.30 GHz CPU, and with 8 GB RAM. The distributed training scenarios are all simulated on this one computer.

C. Training Scenarios

To assess the efficiency of implementing our PSO-GA based Federated Learning (FL) method for energy consumption prediction, we compare our proposed method with 4 other algorithms: 1) the classic FL with averaging as an aggregation function (FedAvg), 2) Federated Distance (FedDist), 3) PSO-based Federated Learning (FedPSO), and 4) GA-based Federated Learning (FedGA).

In the FL training scenarios, we used the Horizontal Federated learning as the different clients (buildings) have the same set of features. We then configured the global model training with the number of clients = 7, number of epochs = 30, client-epoch=3, batch = 10, the inertia coefficient $\alpha = 0.3$, $pbest$ acceleration constant $c1 = 0.7$ and $gbest$ acceleration constant $c2 = 1.4$ and the learning rate = 0.0025.

D. Experimental Results

As shown in Table II, our proposed PSO-GA based FL algorithm outperforms all the other algorithms by scoring the lowest RMSE (40.3) and the lowest Mean Average Error (MAE) (28.1).

On the other hand, Figure 3 illustrates the RMSE variations in function of training epochs. Our proposed method also scores the lower RMSE through the different epochs.

In this paper, we did not cover the complexity analysis of our proposed method but it will the scope of our future work.

VI. CONCLUSION

To reduce data communication overhead and ensure data privacy, Federated Learning is advised. Several variations of FL approach were presented in the literature. In this paper, we proposed a new variant of FL by combining Genetic Algorithms (GA) with Particle Swarm Optimization PSO for local

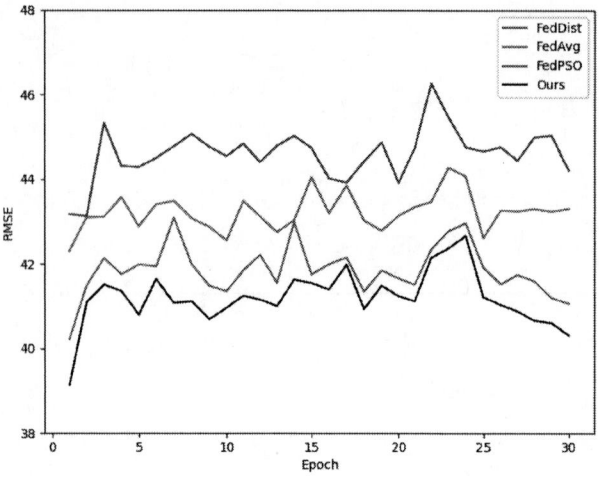

Fig. 3. RMSE Comparison of Federated Learning Algorithms

model optimization on the clients side, and for the aggregation level of the global model on the server. More particularly, the hybrid PSO-GA algorithm applied on the server side selects from the various local models updates (particles) the solution gives the highest fitness score. Experiments show that our PSO-GA based FL method, outperforms the several existing FL-based methods by reducing the Root Mean Square Error.

REFERENCES

[1] M. Pipattanasomporn, G. Chitalia, J. Songsiri, C. Aswakul, W. Pora, S. Suwankawin, K. Audomvongseree, and N. Hoonchareon, "Cu-bems, smart building electricity consumption and indoor environmental sensor datasets," *Scientific Data*, vol. 7, 07 2020.

[2] K. Alanne and S. Sierla, "An overview of machine learning applications for smart buildings," *Sustain. Cities Soc.*, vol. 76, no. 103445, p. 103445, Jan. 2022.

[3] M. Settles and T. Soule, "Breeding swarms: a GA/PSO hybrid," in *Proceedings of the 7th annual conference on genetic and evolutionary computation*, 2005, pp. 161–168.

[4] S. Park, Y. Suh, and J. Lee, "Fedpso: Federated learning using particle swarm optimization to reduce communication costs," *Sensors*, vol. 21, no. 2, p. 600, 2021.

[5] L. Li, Y. Fan, M. Tse, and K.-Y. Lin, "A review of applications in federated learning," *Computers Industrial Engineering*, vol. 149, p. 106854, 2020. [Online]. Available: https://www.sciencedirect.com/science/article/pii/S0360835220305532

[6] Q. Xia, W. Ye, Z. Tao, J. Wu, and Q. Li, "A survey of federated learning for edge computing: Research problems and solutions," *High-Confidence Computing*, vol. 1, no. 1, p. 100008, Jun. 2021.

[7] Y. Zhou, N. Zhou, L. Gong, and M. Jiang, "Prediction of photovoltaic power output based on similar day analysis, genetic algorithm and extreme learning machine," *Energy*, vol. 204, p. 117894, 2020. [Online]. Available: https://www.sciencedirect.com/science/article/pii/S036054422031001X

[8] D. Wang, D. Tan, and L. Liu, "Particle swarm optimization algorithm: An overview," *Soft Comput.*, vol. 22, no. 2, p. 387–408, jan 2018. [Online]. Available: https://doi.org/10.1007/s00500-016-2474-6

[9] K. Premalatha and A. Natarajan, "Hybrid pso and ga for global maximization," *Int. J. Open Problems Compt. Math*, vol. 2, no. 4, pp. 597–608, 2009.

[10] D. C. Nguyen, M. Ding, P. N. Pathirana, A. Seneviratne, J. Li, and H. V. Poor, "Federated learning for internet of things: A comprehensive survey," Apr. 2021.

[11] B. Li, Y. Wu, J. Song, R. Lu, T. Li, and L. Zhao, "DeepFed: Federated deep learning for intrusion detection in industrial cyber–physical systems," *IEEE Trans. Industr. Inform.*, vol. 17, no. 8, pp. 5615–5624, Aug. 2021.

[12] A. Taik and S. Cherkaoui, "Electrical load forecasting using edge computing and federated learning," p. 1–6, 2020.

[13] S. V. Dasari, K. Mittal, S. GVK, J. Bapat, and D. Das, "Privacy Enhanced Energy Prediction in Smart Building using Federated Learning," in *2021 IEEE International IOT, Electronics and Mechatronics Conference (IEMTRONICS)*. IEEE, apr 21 2021.

[14] J. Li, C. Zhang, Y. Zhao, W. Qiu, Q. Chen, and X. Zhang, "Federated learning-based short-term building energy consumption prediction method for solving the data silos problem," *Building Simulation*, vol. 15, no. 6, pp. 1145–1159, dec 10 2021.

[15] C. Briggs, Z. Fan, and P. Andras, "Federated learning for short-term residential load forecasting," 2022.

[16] Y. L. Tun, K. Thar, C. M. Thwal, and C. S. Hong, "Federated Learning based Energy Demand Prediction with Clustered Aggregation," in *2021 IEEE International Conference on Big Data and Smart Computing (BigComp)*. IEEE, 1 2021.

[17] H. Liu, X. Zhang, X. Shen, and H. Sun, "Privacy-preserving power consumption prediction based on federated learning with cross-entity data," *2022 34th Chinese Control and Decision Conference (CCDC)*, pp. 181–186, 2022. [Online]. Available: https://api.semanticscholar.org/CorpusID:256877592

[18] Q. Li, Z. Wen, Z. Wu, S. Hu, N. Wang, Y. Li, X. Liu, and B. He, "A survey on federated learning systems: Vision, hype and reality for data privacy and protection," *IEEE Trans. Knowl. Data Eng.*, vol. 35, no. 4, pp. 3347–3366, Apr. 2023.

[19] K. Yue, R. Jin, C.-W. Wong, and H. Dai, "Communication-efficient federated learning via predictive coding," *IEEE Journal of Selected Topics in Signal Processing*, vol. 16, pp. 369–380, 2021. [Online]. Available: https://api.semanticscholar.org/CorpusID:236772038

[20] N. Mendes, P. S. Moura, J. Mendes, R. A. de Salles, and J. Mohammadi, "Federated learning enabled prediction of energy consumption in transactive energy communities," *2022 IEEE PES Innovative Smart Grid Technologies Conference Europe (ISGT-Europe)*, pp. 1–5, 2022. [Online]. Available: https://api.semanticscholar.org/CorpusID:254098790

[21] D. Kang and C. W. Ahn, "Ga approach to optimize training client set in federated learning," *IEEE Access*, vol. 11, pp. 85 489–85 500, 2023.

[22] B. Qolomany, K. Ahmad, A. Al-Fuqaha, and J. Qadir, "Particle swarm optimized federated learning for industrial iot and smart city services," in *GLOBECOM 2020-2020 IEEE Global Communications Conference*. IEEE, 2020, pp. 1–6.

Optimizing Charging Schedules for WRSNs: A Multi-Criteria Decision-Making Approach with Multiple Charger Vehicles

Samah Abdel Aziz[1,2], Ammar Hawbani [1,5], Xingfu Wang [1], A.S. Ismail [1,2], Nasir Saeed[3], Saeed H. Alsamhi[4], Liang Zhao[5], and Ahmed Al-Dubai[6]

[1] School of Computer Science and Technology, University of Science and Technology of China, Hefei, 23002, China.
[2] Faculty of Science, Zagazig University, Zagazig, 44519, Egypt.
[3] Department of Electrical and Communication Engineering, United Arab Emirates University Al Ain 15551, UAE
[4] Technological University of the Shannon Midlands Midwest, Ireland
[5] Shenyang Aerospace University, Shenyang 110136, China
[6] Computing school of Edinburgh Napier University, United Kingdom.
Email: samahhabib10@yahoo.com; anmande@ustc.edu.cn; a.sami@zu.edu.eg, wangxfu@ustc.edu.cn;
nasir.saeed@uaeu.ac.ae; salsamhi@ait.ie; lzhao@sau.edu.cn; a.al-dubai@napier.ac.uk

Abstract—**Traditional development of wireless rechargeable sensor networks (WRSNs) focused on charger deployment, but recent research overlooked the efficient use of multiple chargers in 2-D or 3-D networks and multi-node energy transfer. This paper proposes a novel approach to address these challenges by modeling the charging scheduling problem in WRSNs with multiple Wireless Charging Vehicles (WCVs) as a multi-criteria decision-making problem. The approach involves two steps, including the enhanced k-means clustering for dividing tasks and traffic distribution and using a Fuzzy logic system to prioritize and optimize charging requests. Our extensive simulations demonstrate the effectiveness and competitiveness of our scheme, showing superior performance compared to prior approaches.**

Index Terms—**wireless rechargeable sensor networks, Charging scheduling, Fuzzy Scheduling, k-means clustering.**

I. INTRODUCTION

Wireless Rechargeable Sensor Networks (WRSNs) capitalize on Wireless Power Transfer (WPT) technology, enabling sensors to wirelessly harness energy from sources such as Wireless Charging Vehicles (WCVs) and UAVs as required [1]. WPT stands as a well-established technology with the capacity to extend the operational life of IoT, 5G, and 6G devices, establishing WRSNs as indispensable across an array of domains, encompassing traffic monitoring, smart homes, industrial control, environmental sensing, healthcare, agriculture, military operations, and a multitude of applications spanning robotics, underwater vehicles, induction motors, and generators. Charging scheduling within WRSNs can be categorized into two primary types: periodical and on-demand [2]–[4]. Periodical strategies entail predefined schedules, often likened to solving a Traveling Salesman Problem, though they exhibit constraints concerning scalability and adaptability. Conversely, on-demand strategies prioritize charging based on immediate requirements, accounting for the ever-evolving dynamics of the network. In our proposed system, our specific focus is on the on-demand approach, which has garnered considerable

attention within the research community. A plethora of related studies has zeroed in on on-demand methodologies in two-dimensional networks [3], [5].

Recharging scheduling refers to the process of arrangement of the sensor nodes in the charging queue to determine which sensor node is replenished first. Thus, the challenge in WRSN research is designing effective charging schedules considering battery deadlines. Single charger systems are suitable for simple networks but not practical for larger scenarios. Multiple WCVs have emerged to recharge multiple nodes simultaneously. This paper addresses key questions: (1)How can multiple WCVs be strategically deployed to avoid overlapping charging regions? (2)How can a mechanism ensure that each sensor node receives recharge from only one WCV? (3)What methods can incorporate network attributes for informed and optimal charging scheduling decisions?

II. NETWORK MODEL AND PROPOSED METHOD

In our scenario, a group of Wireless Charger Vehicles (WCVs) with travel speed V and energy capacity E_{WCV} is responsible for monitoring and recharging homogeneous sensors (SN) in an agricultural field. The field includes a fixed base station (BS) at the center, which handles data collection, maintenance, and provides energy support to the WCVs. The BS possesses complete information about all sensor nodes in the network. The sensor nodes are static, identical, and assumed to be fully charged during network setup. When a sensor node's energy falls below a threshold $(RE(SN_i) < E_{th})$, it sends a charging request to the BS in a multi-hop manner. After receiving all the requests and creating a queue, a WCV is dispatched from the BS based on a predefined charging schedule and its energy capacity E_{WCV}. If a WCV cannot fully recharge all the sensors during its tour, it returns to the BS for replenishment. The network's structure and routing paths remain constant, and a WCV can recharge

979-8-3503-8083-5/23 $31.00 © 2023 IEEE

multiple sensor nodes within its charging range. The WCV determines the charging schedule C_S, known as the charging queue R_Q, based on network information, specifying the order in which the sensor nodes are served and replenished.

A. Problem Definition

The goal is to determine a charging schedule ($C_S = \{BS \rightarrow SN_i \rightarrow SN_{i+1} \rightarrow \cdots SN_{t_r}\}$) for the Wireless Charger Vehicle (WCV) when it receives a set of charging requests ($R_Q = \{R_{QSN_i}, \cdots, R_{QSN_m}\}$). The objective is to improve the overall network lifetime by employing a charging algorithm that enhances the life-success rate and energy efficiency of the network, thus extending its lifespan (i.e. the problem is to determine the priority order in the C_S, specifically identifying which sensor should be recharged first to maximize the network's performance).

B. Proposed Method

In this section, we introduce our proposed scheme that is shown in Fig. 1 and have the following main two steps:

1) After the completion of network initialization, the network is partitioned using an enhanced k-means clustering approach. This partitioning aims to ensure a balanced distribution of tasks throughout the system, promoting fairness in resource allocation by generating traffic and requests equitably across diverse locations.

2) The second aspect of the problem involves the challenge of determining the priority order for serving the sensor nodes in the charging schedule by using a Fuzzy logic system.

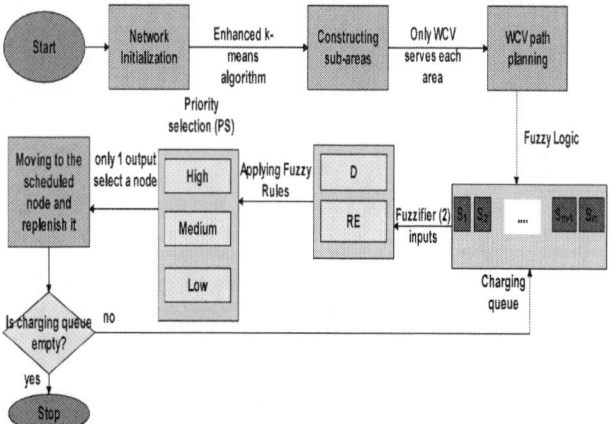

Fig. 1: Our proposed multi-criteria decision-making solution.

C. Network Partitioning

Network partitioning assigns WCVs to separate areas, with each sub-area having one WCV. The Base Station (BS) divides the network into adaptive "k" areas and assigns a working area to each WCV based on node status information. Our goal is to balance traffic load across nodes and ensure fairness in charging load among WCVs. We use the unsupervised enhanced

k-means algorithm for this partitioning task. The steps of the enhanced k-means algorithm are shown in Algorithm 1.

Algorithm 1 Cluster Centroid Initialization and Subarea Construction

---

Require: Set of sensor nodes SN, set of WCVs W
Ensure: All subareas A_n and initial centroids C_n
1: Initialize n as the number of WCVs.
2: **for** $i \leftarrow 1$ to n **do**
3:     Define $C_n(x_n, y_n)$ as the cluster center of A_n.
4:     Initialize C_n for all clusters.
5: **end for**
6: Define $selectedNodes$ as an empty list.
7: $ChangeFlag \leftarrow 1$
8: **while** $ChangeFlag = 1$ **do**
9:     $ChangeFlag \leftarrow 0$
10:     **for** $j \leftarrow 1$ to n **do**
11:         Store the current cluster center as C_j.
12:     **end for**
13:     **for all** sensor nodes SN_j in SN **do**
14:         Calculate distances between SN_j and all C_n.
15:         Choose the nearest cluster center as C_i.
16:         Add SN_j to the corresponding subarea A_i.
17:         Refresh cluster center C_i by updating x_n and y_n.
18:     **end for**
19:     **for** $k \leftarrow 1$ to n **do**
20:         **if** C_k is not equal to the stored cluster center **then**
21:             $ChangeFlag \leftarrow 1$
22:         **end if**
23:     **end for**
24: **end while**
25: **return** All subareas A_n and initial centroids C_n.

D. Fuzzy Inference Model

In our work, recharging scheduling relies on various attributes, i.e., multi-criteria, that influence the decision-making process for recharging scheduling for prioritizing sensor node recharge, which is based on multi-criteria [6]. Thus, in this section, we present two parts of the fuzzy inference model. In II-D1, we delve into the attributes and fuzzifier, while II-D2 focuses on our output priority selection.

1) Attributes and Fuzzifier: In our system, determining the priority of sensor nodes in the charging schedule is crucial. Each WCV is responsible for a cluster of sensor nodes, and we prioritize nodes based on multiple attributes. We employ fuzzy logic to assess and compare these attributes [6], establishing fuzzy rules linking them to priority. These rules define conditions for higher priority. The fuzzy inference engine computes priority values for each node. When a node's energy drops below a threshold, it sends a charging request with status information (D, RE). This information feeds into the fuzzy inference model, which selects high-priority nodes for recharging, as shown in Fig. 2. We represent all the fuzzy sets by curve fitting. Curve fitting is a process to create a

979-8-3503-8083-5/23 $31.00 © 2023 IEEE   240

curve or mathematical function that has the best fit to a set of collected data points from our experimental simulation.

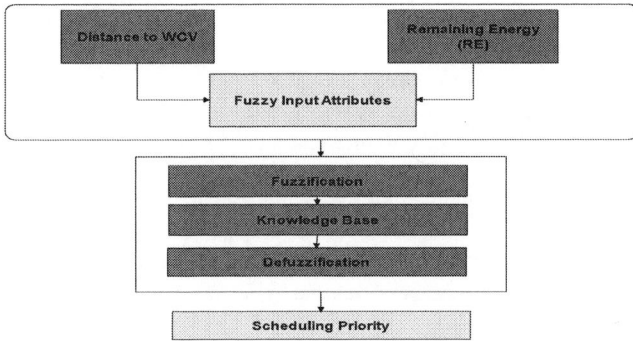

Fig. 2: Fuzzy inference model.

1) **Distance to WCVs:** We define three distance measurement categories, Near, Medium, and Far, as $T(D) = \{Near, Medium, Far\}$. Each category is represented as a fuzzy set. Specifically, we have fuzzy sets for Near, Medium, and Far distances denoted as $\tilde{D}_N$, $\tilde{D}_M$, and $\tilde{D}_F$, respectively. These fuzzy sets incorporate membership functions that describe the degree of belonging to each distance category and are shown in the left of Fig. 3.

For Near distances, we interpolate the membership function $\mu_N^D(x)$ using experimental simulation data and the Boltzmann distribution. The formula is provided as:

$$\mu_N^D(x) = \begin{cases} a_2 + \frac{a_1 - a_2}{1 + e^{(x - x_0)/dx}} \\ a_1 = 1.10758; a_2 = -6.62677e^{-4}; \\ x_0 = 0.194; dx = 0.08; \end{cases} \quad (1)$$

To interpolate the membership function $\mu_M^D(x)$ for Medium distances, we employ the Cauchy-Lorentz distribution, represented as:

$$\mu_M^D(x) = \begin{cases} y_0 + \frac{2A}{\pi} \frac{w}{4(x - x_c)^2} + w^2 \\ y_0 = -0.27853; x_c = 0.5007; \\ w = 0.50711; A = 1.03129; \end{cases} \quad (2)$$

Similarly, for Far distances, the membership function $\mu_F^D(x)$ is interpolated using an inverse polynomial peak function with a specific center:

$$\mu_F^D(x) = \begin{cases} y_0 + \frac{a}{1 + a_1\left(2\frac{x - x_c}{w}\right)^2 + a_2\left(2\frac{x - x_c}{w}\right)^4 + a_3\left(2\frac{x - x_c}{w}\right)^6} \\ y_0 = -0.00476; x_c = 0.99404; \\ w = 0.69733; a = 1.00415; \\ a_1 = 2.32565; a_2 = 0.60687; a_3 = 3.40939; \end{cases} \quad (3)$$

2) **Remaining Energy:** We categorize remaining energy levels into Low, Medium, and High, represented by $T(RE) = \{Low, Medium, High\}$. Fuzzy sets are employed to define these categories, and their membership functions are visualized in the middle of Fig.3.

For the Low category, we have the fuzzy set $\tilde{RE}_L$ characterized by the membership function $\mu_L^{RE(x)}$. This function is obtained through interpolation using the Gumbel distribution, as given:

$$\mu_L^{RE(x)} = \begin{cases} 1 - e^{-e^{(x-a)/b}} \\ a = 0.18443; b = -0.05597; \end{cases} \quad (4)$$

The Medium category is represented by the fuzzy set $\tilde{RE}_M$, with the membership function $\mu_M^{RE(x)}$. Interpolation of this function is achieved using an inverse polynomial peak function centered at specific values, as shown:

$$\mu_M^{RE(x)} = \begin{cases} y_0 + \frac{a}{1 + a_1\left(2\frac{x - x_c}{w}\right)^2 + a_2\left(2\frac{x - x_c}{w}\right)^4 + a_3\left(2\frac{x - x_c}{w}\right)^6} \\ y_0 = -0.001561; x_c = 0.50005; \\ w = 0.40616; a = 1.01693; \\ a_1 = 0.36111; a_2 = -0.49596; a_3 = 0.98967; \end{cases} \quad (5)$$

Similarly, the high category is represented by the fuzzy set $\tilde{RE}_H$, and its membership function $\mu_H^{RE(x)}$. The interpolation for this function employs the Gamma cumulative distribution function, as expressed:

$$\mu_H^{RE(x)} = \begin{cases} y_0 + \frac{a_1}{b^a \Gamma(a)} \int_0^x t^{a-1} e^{-\frac{t}{b}} dt \\ a1 = 0.99294; a = -158.3673; \\ b = 0.005; y_0 = 0.00149; \end{cases} \quad (6)$$

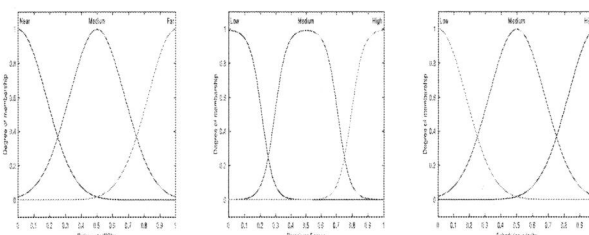

Fig. 3: (from left to right) (a) Distance to WCVs (meter) membership functions. (b) Remaining energy (joule) membership functions. (c) The crisp output of the criteria: Scheduling priority.

2) *Scheduling priority:* The output of our system is the scheduling priority is displayed in the right of Fig.3. This priority selection represents the high priority assigned to a single starved sensor, which is chosen as the first selection. The scheduling priority is determined based on three measurement items: $T(SP) = \{Low, Medium, High\}$. For low priority, we interpolate the membership function ($\mu_L^{SP(x)}$) using experimental simulation data and the Logistic distribution. The formula is provided as:

$$\mu_L^{SP(x)} = \begin{cases} a_2 + \frac{a_1 - a_2}{1 + (x/x_0)^p} \\ a_1 = 0.98525; a_2 = -0.03916 \\ x_0 = 0.21055; p = 2.88696 \end{cases} \quad (7)$$

979-8-3503-8083-5/23 $31.00 © 2023 IEEE

In a similar manner, the medium priority, the membership function $\mu_M^{SP(x)}$ is interpolated by employing an amplitude-adjusted version of the Gaussian peak function.

$$\mu_M^{SP} = \begin{cases} y_0 + ae^{-\frac{(x-x_c)^2}{2w^2}} \\ y_0 = -0.00167; x_c = 0.50133 \\ w = 0.17667; a = 1.00304 \end{cases} \quad (8)$$

Finally, for the high priority, the interpolation of the membership function $\mu_H^{SP}(x)$ is performed by using the Boltzmann distribution and can be represented as:

$$\mu_H^{SP}(x) = \begin{cases} a_2 + \frac{a_1 - a_2}{1 + e^{(x-x_0)/dx}} \\ a_1 = -0.00903; a_2 = 1.1063, \\ x_0 = 0.80676; dx = 0.08222 \end{cases} \quad (9)$$

In our scheme, the rule set comprises a total of 9 rules, which are shown in Table I.

TABLE I: Fuzzy rules for priority selection (IF-THEN).

R#	D	RE	Scheduling priority(SP)
1	Near	Low	High
2	Near	Medium	High
3	Near	High	Low
4	Medium	Low	High
5	Medium	Medium	Medium
6	Medium	High	Low
7	Far	Low	High
8	Far	Medium	Medium
9	Far	High	Low

III. EXPERIMENTAL RESULTS

We randomly distribute 300-700 nodes across an area ranging from 300×300 m^2 to 700×700 m^2 for the WRSN. The network includes a fixed Base Station (BS) located at the center, denoted as $L_0 = (0,0)$, responsible for data collection and task assignments. We use specific parameter values: 5 WCVs, each assigned to a single cluster. Sensor nodes with a maximum energy capacity of 100 J. WCVs with a capacity of 20 kJ. Random energy consumption rates range from 0.01 J/s to 0.05 J/s for various activities like sensing, receiving, transmitting, and communication. WCVs' movement speed (V) set between 2 and 5 m/s. These parameters define the characteristics of our simulation in the WRSN. The OMC_TSP approach introduces a multi-node charging method that operates on-demand and utilizes multiple MCs [7]. We have employed the OMC_TSP scheme specifically to illustrate the performance results to compare with energy usage efficiency and life success rate.

- Energy Usage Efficiency:- Energy usage efficiency ($EUE = \eta$) is characterized as the proportion between the energy acquired by nodes and the total energy conveyed from the BS to the WCVs. The essential role of the EUE comes into play when aiming for impressive charging efficiency. A higher EUE validates the timely charging of a greater quantity of nodes. By comparing our

proposed method with the OMC_TSP scheme, we found that our results are more efficient than those mentioned, as shown in Fig. 4.

- Life-Success Rate (LSR) is a metric that measures the percentage of SNs successfully charged by the WCV out of the total requested SNs. A higher LSR indicates increased network stability and connectivity, which is essential for improving charging efficiency. Our method effectively schedules nodes by considering multiple factors to handle multiple requests. Additionally, the balanced distribution of network loads allows each WCV to rapidly complete recharging tasks, resulting in higher node charging and improved LSR, as shown in Fig. 4.

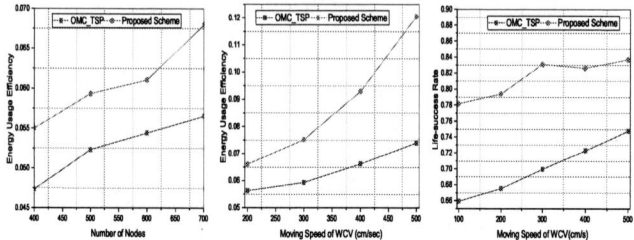

Fig. 4: (From left to right) (a) Performance comparison of EUE vs the number of nodes, (b) EUE vs moving speed of WCV, and (c) LSR vs moving speed of WCV.

CONCLUSION

This paper introduced a new approach to address the shortcomings of traditional WRSN development. It reframes the issue of scheduling charging in WRSNs with multiple Wireless Charging Vehicles (WCVs) as a multi-criteria decision-making problem. This approach incorporates enhanced k-means clustering and utilizes Fuzzy logic with two input parameters: D and RE. The output is the SP, determining which sensor node should be recharged first. Through the simulation, we have showcased the superior efficiency of our approach when compared with previous methods. This research opens up new possibilities for more efficient and effective WRSN design.

REFERENCES

[1] B. Qureshi et al., "A state-of-the-art survey on wireless rechargeable sensor networks: Perspectives and challenges," vol. 28, no. 7, pp. 3019-3043, 2022.

[2] Z. Lyu, Z. Wei, X. Wang, Y. Fan, C. Xia, and L. J. I. S. J. Shi, "A periodic multinode charging and data collection scheme with optimal traveling path in WRSNs," vol. 14, no. 3, pp. 3518-3529, 2020.

[3] C. Lin, J. Zhou, C. Guo, H. Song, G. Wu, and M. S. J. I. T. o. M. C. Obaidat, "TSCA: A temporal-spatial real-time charging scheduling algorithm for on-demand architecture in wireless rechargeable sensor networks," vol. 17, no. 1, pp. 211-224, 2017.

[4] C. Lin, D. Han, J. Deng, and G. J. I. T. o. V. T. Wu, "P ^{2} S: A primary and passer-by scheduling algorithm for on-demand charging architecture in wireless rechargeable sensor networks," vol. 66, no. 9, pp. 8047-8058, 2017.

[5] M. Yang et al., "Dynamic charging scheme problem with actor–critic reinforcement learning," vol. 8, no. 1, pp. 370-380, 2020.

[6] R. R. J. I. T. o. s. Yager, Man, and Cybernetics, "On ordered weighted averaging aggregation operators in multicriteria decisionmaking," vol. 18, no. 1, pp. 183-190, 1988.

[7] T. J. C. C. Rault, "Avoiding radiation of on-demand multi-node energy charging with multiple mobile chargers," vol. 134, pp. 42-51, 2019.

2023 International Conference on Microelectronics (ICM)

Exploring H₂S Gas Sensing with Graphene Nanoribbon Field Effect Transistors: A Semi-Empirical Simulation Approach

Asma Wasfi
*Electrical and Communication
Engineering Department
United Arab Emirates University
Al Ain, United Arab Emirates*

Mohamed Atef
*Electrical and Communication
Engineering Department
United Arab Emirates University
Al Ain, United Arab Emirates*

Falah Awwad
*Electrical and Communication
Engineering Department
United Arab Emirates University
Al Ain, United Arab Emirates
f_awwad@uaeu.ac.ae*

Abstract— In recent years, there has been a lot of interest in Graphene Nanoribbon Field Effect Transistors due to their sensitivity, selectivity and real time detection abilities. One area of interest is their potential to sense molecules. In this study we investigate the use of Graphene Nanoribbon Field Effect Transistors as sensors for detecting hydrogen sulfide molecules at various concentrations including one, two and three hydrogen sulfide molecules. To achieve this detection capability, we employ Graphene Nanoribbon Field Effect Transistors sensors with a designed configuration optimized for sensing hydrogen sulfide. These sensors exhibit responses depending on the concentration of hydrogen sulfide molecules present enabling differentiation between various concentrations. Our approach involves using an empirical modeling technique combined with non-equilibrium Greens function calculations to thoroughly analyze the sensing process for different hydrogen sulfide molecule concentrations. Through simulations and analysis our results demonstrate that these Graphene Nanoribbon Field Effect Transistors sensors can effectively detect single, double, and triple hydrogen sulfide molecules. This highlights their potential for applications in detecting hydrogen sulfide gas. To summarize our findings, we have shown that Graphene Nanoribbon Field Effect Transistors have versatility as sensors for detecting varying concentrations of molecules. This opens possibilities for their use, in gas sensing and related fields.

Keywords— field-effect transistor, graphene, nanoribbon, H₂S, semi-empirical modeling, electronic properties.

I. INTRODUCTION

Graphene Nanoribbon Field Effect Transistors (GNRFETs) have garnered significant attention in recent years due to their exceptional sensitivity, selectivity, and real-time detection capabilities, making them promising candidates for a wide range of applications, including gas sensing [1-2]. As the demand for efficient and reliable gas sensors continues to grow, GNRFETs offer a compelling avenue for exploration. This paper delves into the intriguing realm of gas sensing with GNRFETs, specifically focusing on their potential as sensors for detecting hydrogen sulfide (H₂S) molecules.

Hydrogen sulfide (H₂S) is a colorless, highly toxic gas with a distinct odor of rotten eggs. Its presence in various industrial and environmental settings necessitates the development of efficient detection methods [3]. GNRFETs, with their remarkable properties and tunable characteristics, present an exciting opportunity to address this need. In this study, we embark on an exploration of the capabilities of GNRFET sensors in detecting H₂S molecules at varying concentrations.

Our investigation involves the utilization of GNRFET sensors meticulously configured to optimize their sensitivity to H₂S [4]. We employ a semi-empirical simulation approach that combines empirical modeling techniques with non-equilibrium Greens function calculations [5]. Through comprehensive simulations and analyses, we probe the sensing process for different concentrations of H₂S molecules, ranging from single molecules to higher concentrations. Saleh et al. investigated the performance of copper and zinc co-doped zigzag graphene nanoribbon (CuZn-ZGNR) for the detection of H₂S gas using simulation based on density functional theory (DFT). The results showed that the sensitivity of the CuZn-ZGNR system towards H₂S gas detection was significantly improved compared to pristine and doped systems (ZGNR, Zn-ZGNR, and Cu-ZGNR) [6].

The results of our study underscore the versatility of GNRFET sensors in effectively detecting single, double, and triple H₂S molecules, highlighting their potential for applications in gas sensing and related fields. As we navigate through the subsequent sections of this paper, we will explore the GNRFET configuration, elucidate the computational methodology employed, present our detailed results and discussions, and ultimately draw conclusions that shed light on the remarkable promise of GNRFETs as advanced gas sensors.

The ever-evolving field of nanotechnology, along with the robust computational tools at our disposal, positions us at the forefront of innovation in gas sensing technology. As we delve deeper into the intricacies of GNRFET-based H₂S detection, we contribute to the broader landscape of sensor technology, offering insights that could have far-reaching implications for safety, environmental monitoring, and industrial applications.

The novelty of this study lies in the detailed use of Quantumwise to analyze a specially designed GNRFET sensor for detecting H₂S molecules. The sophisticated methodology, combining the Non-equilibrium Greens function with the Empirical method (NEGF+SE), sets the research apart. The key breakthrough is the revelation that each H₂S molecule imparts a distinctive signature due to its unique chemical and electronic attributes. This finding underscores the specificity and remarkable sensitivity of the GNRFET sensor.

The study's nuanced exploration demonstrates a significant advancement by revealing a variation in the sensor's response to different quantities of H₂S molecules. This not only establishes the efficacy and precision of the

979-8-3503-8083-5/23 $31.00 © 2023 IEEE

GNRFET sensor in H_2S detection but also contributes to the understanding of how these sensors interact with specific molecules at an electronic level. Overall, the research paves the way for GNRFETs to be recognized as versatile and precise sensors, offering a novel approach to gas sensing technology.

This paper is organized as follows: Section II presents graphene nanoribbon field effect transistor configuration, Section III presents computational methodology, Section IV presents results and discussion, and section V concludes the paper.

II. GRAPHENE NANORIBBON FIELD EFFECT TRANSISTOR CONFIGURATION

We conducted simulations using the Quantumwise Atomistix Toolkit (ATK) and its graphical user interface Virtual Nano Lab (VNL) to detect hydrogen sulfide (H_2S) molecules. ATK VNL offers atomic scale modeling and calculations, for nanoscale systems with various calculators to assess electron transport characteristics of quantum systems. Specifically, we used an empirical (SE) calculator based on ATK SE [7] to simulate the proposed GNR FET.

To ensure results and efficient simulation runs we utilized a high-performance computing environment (HPC). Our setup included seven computing nodes, each equipped with 36 processing units. Through these simulations we analyzed the electrical currents and device density of states in the presence of H_2S molecules. This provided us insights, into the sensors ability to detect H_2S.

The GNR FET is used to detect molecules of hydrogen sulfide (H_2S). It consists of two types of nanoribbons: armchair nanoribbons (AGNR) and zigzag graphene nanoribbons (ZGNR). To simulate the GNR FET we used ATK VNL. Employed both ZGNR and AGNR configurations. The sensor structure was modeled using the ATK builder tool. To construct the region of the FET we created AGNR with a width of four atoms using the Nanoribbon Plugin Tool. The electrodes on the other hand were made from ZGNR with a width of six atoms also designed using the Nanoribbon Plugin Tool. The unique z shaped structure of the sensor was achieved by connecting the two electrodes through AGNR material. The left and right electrodes consist exclusively of ZGNRs. Completing the FET sensor structure involved incorporating a gate beneath the part of the z shaped structure, which consisted of two layers. A dielectric layer and a metallic layer as shown in Figure 1. The permittivity of the dielectric material was set to 3.9 ε. The length of the electrodes measured 7.4 Å while the channel length extended to 19.2 Å [8, 9, 11]. This study is a proof of concept that GNRFET can be used to detect H_2S molecules.

Figure 2 provides a visual representation of the molecular structures of hydrogen sulfide (H_2S) at different concentrations, offering insights into the arrangement of atoms within the H_2S molecules. Specifically, Figure 2(a) illustrates the structure of a single H_2S molecule, Figure 2(b) displays the configuration of two H_2S molecules, and Figure 2(c) presents the arrangement of three H_2S molecules. The color key aids in distinguishing between sulfur (yellow), hydrogen (white), and carbon (gray) atoms, enhancing our understanding of the molecular compositions.

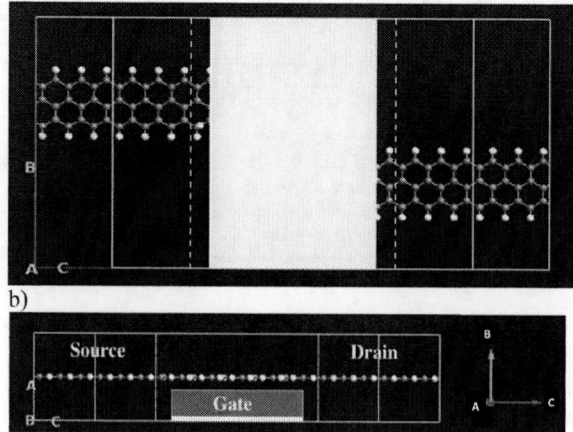

Fig. 1. The GNR FET sensor is illustrated schematically in (a) and a cross sectional view is shown in (b). It consists of two metal components (source and drain) an AGNR channel and a gate located underneath the channel. In terms of color-coding hydrogen is represented as white while carbon appears as gray.

In Figure 3, we present a cross-sectional view of the Graphene Nanoribbon Field Effect Transistor (GNRFET) device in the presence of 2 H_2S molecules. This figure offers a valuable glimpse into the interaction between H_2S molecules and the GNRFET structure, highlighting how the electrical properties of the device are affected by the presence of these molecules. This visual representation provides essential context for our discussions on the impact of H_2S on GNRFET performance in Section IV.

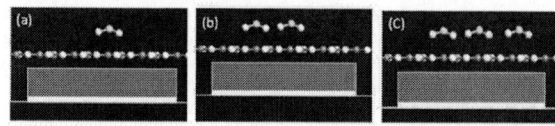

Fig. 2. H_2S Molecular Structures: (a) One molecule, (b) Two molecules, and (c) Three molecules. Color Key: Yellow for sulfur, White for hydrogen, Gray for carbon.

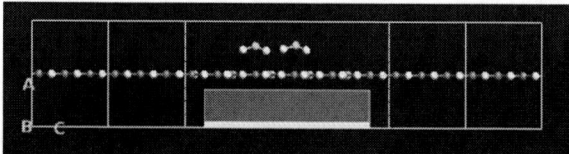

Fig.3. Cross-sectional view of the GNR-FET device with 2 H_2S molecules.

III. COMPUTATIONAL METHODOLOGY

In this study we examined a sensor using simulations based on data. To simulate the sensor accurately we employed an expanded version of the empirical model called the Extended Hückel technique. This technique incorporates a self-Hartree potential.

To optimize the geometry of the structure we utilized Generalized Gradient Approximation (GGA) with density theory (DFT). We ensured that the atomic structure reached its state by minimizing atomic forces to be lower, than 0.05

eV/Å. For sampling purposes we set up a 5 x 5 x 50 grid for k points and used a mesh cut off energy of 10 Ha [10, 11].

To analyze transport properties such as device density of states and transverse current we employed the Extended Hückel technique. The grid mesh cut off energy used was set at 10 Hartree and for producing Brillouin zone information we utilized a configuration with dimensions of $5 \times 5 \times 50$ k points [10,11].

For solving equations related to distribution we applied Neumann boundary conditions in directions A and B while using Dirichlet boundary conditions in direction C. The Neumann boundary condition ensures that there is a constant electrostatic potential, with zero derivative [10, 11].

IV. RESULTS AND DISCUSSION

A. Device Density of State

The inclusion of hydrogen sulfide (H_2S) molecules has caused a recognizable change, in the Density of States (DOS) of the GNRFET. Figure 4 presents an analysis of the DOS between the GNR FET in its state and when exposed to different concentrations of H_2S molecules clearly demonstrating significant differences. The figure shows that the GNRFET in its state has a higher number of energy states compared to when H_2S molecules are introduced. The energy spikes in the GNR FET occur at energy levels indicating alterations in the structure. For example, the absence of H_2S molecules at -1.3 eV generates energy spikes indicating changes in the DOS. These observations emphasize how sensitive the GNRFET is to detecting quantities of H_2S molecules highlighting its potential for gas sensing applications.

Furthermore, the alterations induced by the inclusion of H_2S molecules extend beyond the qualitative realm, delving into a quantitative analysis of the DOS within the GNRFET. As illustrated in Figure 4, the DOS comparison between the pristine GNRFET and its exposure to varying concentrations of H_2S molecules vividly showcases significant distinctions. Specifically, the figure elucidates that the GNRFET, in its original state, exhibits a higher density of energy states compared to its configuration when influenced by H_2S. The discernible energy spikes in the GNR FET correspond to specific energy levels, indicating structural modifications. These precise observations not only underscore the GNRFET's sensitivity but also provide a quantitative foundation for understanding the structural changes induced by varying quantities of H_2S molecules, reinforcing its potential for discerning gas sensing applications.

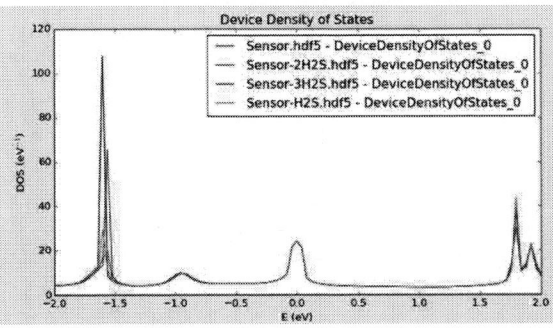

Fig. 4. The GNRFET sensor device density of state for H_2S molecules at various concentrations including one, two and three H_2S molecule.

B. The Transverse Current

a)

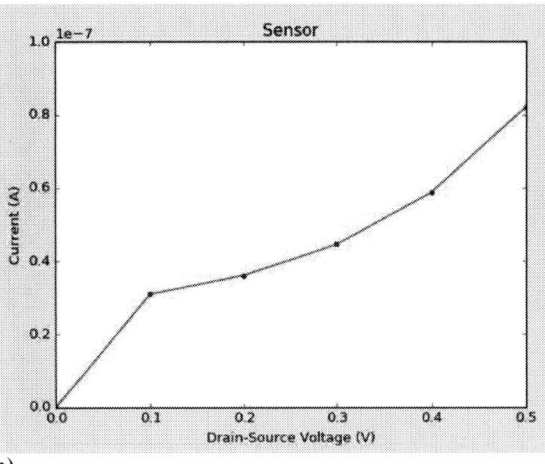

b)

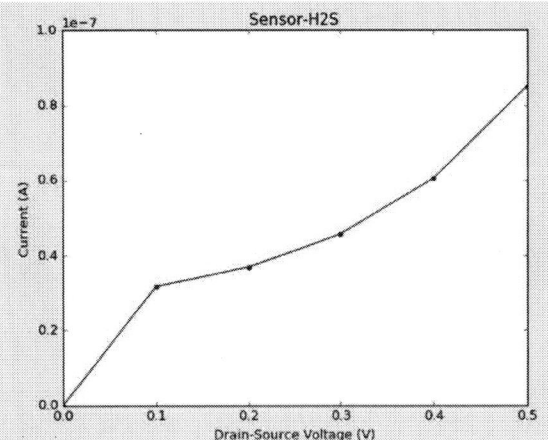

c)

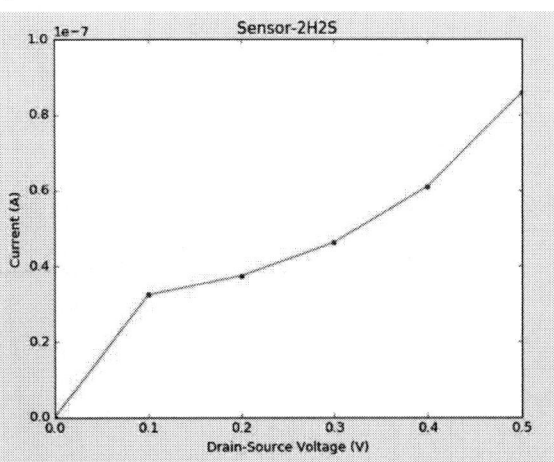

d)

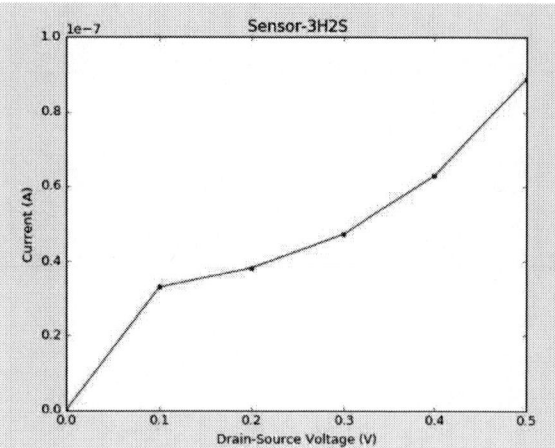

e)

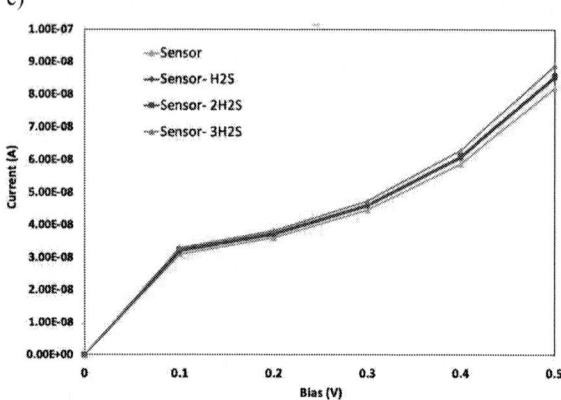

Fig. 5. The current of the GNR-FET when different concentration of H_2S is positioned on the channel: a)bare, b)one H_2S molecule, c) two H_2S molecules, d) three H_2S molecule.

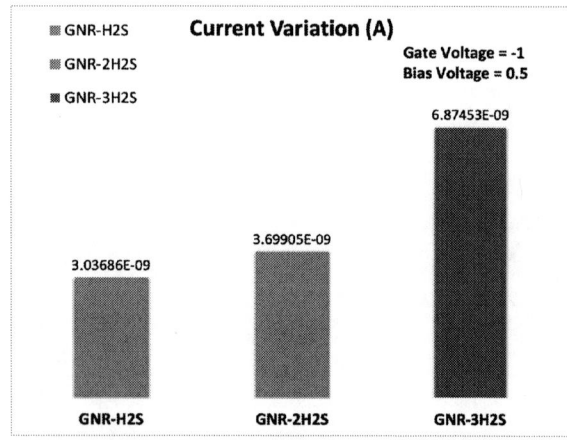

Fig. 6. The current variation of the GNR-FET when different concentration of H_2S is positioned on the channel.

In our research, on detecting hydrogen sulfide (H_2S) molecules we primarily focus on measuring the current as the parameter of interest. The results, shown in Figure 5 were obtained using a self-Extended Hückel (NEGF+SC EH) method. We placed different concentrations of molecules

along the GNR FET channel and measured the transverse current at a bias voltage of 0.5V and a gate potential of -1 V.

Due to their chemical and electronic structures each H_2S molecule produces a current signature as depicted in Figure 5. When different concentrations of H_2S molecules are present along the channel it significantly impacts the charge density of the sensor resulting in a signature for each concentration. Furthermore Figure 5 highlights that depending on the concentration of H_2S molecules there are variations in sensor readings, with levels of current observed for different concentrations. The differences, in readings can be explained by the characteristics of H_2S molecules both in terms of their chemical properties and how they affect the response of the sensor.

Figure 6 shows that the higher the concentration of the H_2S results in higher variation in current and higher sensitivity.

V. CONCLUSION

In this paper, Quantumwise has been leveraged to meticulously scrutinize the electronic transport characteristics of the specially tailored GNR FET sensor engineered for the precise detection of hydrogen sulfide (H_2S) molecules. Employing a sophisticated methodology that amalgamates the Non-equilibrium Greens function with the Empirical method (NEGF+SE), we conducted a comprehensive analysis to discern the profound influence of H_2S molecules on the electrical properties of the GNRFET.

Our discerning exploration revealed a striking revelation: each H_2S molecule imparts a distinctive signature owing to its unique chemical and electronic attributes. This distinctiveness not only underscores the specificity of the GNRFET sensor but also underscores its remarkable sensitivity. Our findings illuminate a nuanced variance in the GNRFET sensor's response to varying quantities of H_2S molecules, unequivocally establishing its efficacy and precision in the realm of H_2S molecule detection.

In summation, this study underscores the profound potential of GNRFETs as versatile and precise sensors for the detection of H_2S molecules, offering a promising avenue for advancements in gas sensing technology.

REFERENCES

[1] Tripathy, S. K., J. K. Singh, and G. M. Prasad. "Bilayer Graphene Nanoribbon Transistor for Butane Gas Detection." In International Conference on Communication, Devices and Computing, pp. 359-365. Singapore: Springer Nature Singapore, 2023.

[2] Ramaraj, Sankar Ganesh, Manoharan Muruganathan, Osazuwa G. Agbonlahor, Hisashi Maki, Yosuke Onda, Masashi Hattori, and Hiroshi Mizuta. "Carbon molecular sieve-functionalized graphene sensors for highly sensitive detection of ethanol." Carbon 190 (2022): 359-365.

[3] Hedlund, Frank Huess. "Confined space hazards: Plain seawater, an insidious source of hydrogen sulfide." Journal of Occupational and Environmental Hygiene (2023): 1-7.

[4] Tan, Guang-Lei, Dan Tang, Xiao-Min Wang, and Xi-Tao Yin. "Overview of the recent advancements in graphene-based H2S sensors." ACS Applied Nano Materials 5, no. 9 (2022): 12300-12319.

[5] Wasfi, Asma, Ahmed Al Hamarna, Omar Mohammed Hasani Al Shehhi, Hazza Fahad Muhsen Al Ameri, and Falah Awwad.

"Graphene Nanoribbon Field Effect Transistor Simulations for the Detection of Sugar Molecules: Semi-Empirical Modeling." Sensors 23, no. 6 (2023): 3010.

[6] E. Salih and A. I. Ayesh, "Co-doped zigzag graphene nanoribbon based gas sensor for sensitive detection of H2S: DFT study," *Superlattices and Microstructures,* vol. 155, p. 106900, 2021/07/01/ 2021, doi: https://doi.org/10.1016/j.spmi.2021.106900

[7] J. Cerdá and F. Soria, "Accurate and transferable extended Huckel-type tight-binding parameters," Physical Review B, vol. 61, no. 12, pp. 7965-7971, 03/15/ 2000, doi: 10.1103/PhysRevB.61.7965.

[8] Q. Yan *et al.*, "Intrinsic Current−Voltage Characteristics of Graphene Nanoribbon Transistors and Effect of Edge Doping," *Nano Letters,* vol. 7, no. 6, pp. 1469-1473, 2007/06/01 2007, doi: 10.1021/nl070133j.

[9] M. H. Rashid, A. Koel, and T. Rang, "Simulations of Graphene Nanoribbon Field Effect Transistor for the Detection of Propane and Butane Gases: A First Principles Study," *Nanomaterials,* vol. 10, no. 1, p. 98, 2020. [Online]. Available: https://www.mdpi.com/2079-4991/10/1/98.

[10] P.-H. Chang, H. Liu, and B. K. Nikolić, "First-principles versus semi-empirical modeling of global and local electronic transport properties of graphene nanopore-based sensors for DNA sequencing," *Journal of Computational Electronics,* vol. 13, no. 4, pp. 847-856, 2014/12/01 2014, doi: 10.1007/s10825-014-0614-8.

[11] M. Gupta, N. Gaur, P. Kumar, S. Singh, N. K. Jaiswal, and P. Kondekar, "Tailoring the electronic properties of a Z-shaped graphene field effect transistor via B/N doping," *Physics Letters A,* vol. 379, no. 7, pp. 710-718, 2015.

DRAM Bitline as A Delay Path for Potential PUF

Enas Abulibdeh, Leen Younes, Baker Mohammad, Hani Saleh, Mahmoud Alqutayri, Khaled Humood*

System on Chip Lab (SoC), Department of Electrical and Computer Engineering

Khalifa University, Abu Dhabi, UAE

University of Edinburgh, Edinburgh, Scotland

100059804,100049417,baker.mohammad,hani.saleh,mahmoud.alqutayri,khaledm.humood} @ku.ac.ae

Abstract—The Physical Unclonable Function (PUF) exploits the physical variation of the underlying device to generate a unique value that can be used as the device signature or a secure key. Most DRAM-based PUFs violate one or more operating rules of the commodity DRAM, which turns it into a source of entropy. This work proposes a new DRAM-based PUF approach that is based on evaluating the variation of adjacent cells and implicitly producing the response by the normal read operation. The simulation results prove the model's functionality by achieving 53.8% for bit aliasing.

Index Terms—PUF, Commodity DRAM, Access Transistor, Sense Amplifier, Activation Command, Timing Parameter, Variation

I. INTRODUCTION

PUF is a security block that inherits the physical variation of the integrated devices, which can be utilized to secure System on Chips (SoCs) [1]. Also, it presents a solution for limited resources systems such as the Internet of Things (IoT) [2]. The direction of constructing a low-cost PUF proposes exploiting the conventional elements of the system such as the memory [3]. DRAM represents an attractive candidate for a PUF because of its availability in most systems [4]. Also, the dynamic nature of DRAM cells provides the opportunity to extract unpredictable values [5]. The DRAM cell consists of a capacitor and an access transistor as shown at the right of Fig. 1. When the access transistor is activated by the wordline (WL), the charge that is stored in the cell's capacitor is shared through the bitline (BL), which is pre-charged to 0.5Vdd. The sense amplifier (SA), which is shown at the bottom of Fig. 1, locates at the end of the BL, senses the changes on the BL, and evaluates the cell's value accordingly. The DRAM cells are arranged in a two-dimensional array, where each row represents one block that is activated once by a single WL, and the cells at the same offset of each row are connected via a single BL and evaluated by a single SA. The set of SAs that evaluate and store the row's value is defined as the row buffer. This organization of DRAM cells and the periphery circuit is shown in Fig. 1.

The memory controller (MC) controls the access to DRAM rows by issuing three basic commands [6]: activation (ACT), Read/Write (RD/WR), and Pre-charge (PRE) commands. The ACT command connects the selected DRAM row to the row buffer. The RD/WR command transfers the data from/to the row buffer to/from the data bus. The PRE command clears the row buffer and pre-charges the BLs to Vdd/2 for the next read or write operation. The row address strobe (RAS) or WL is a

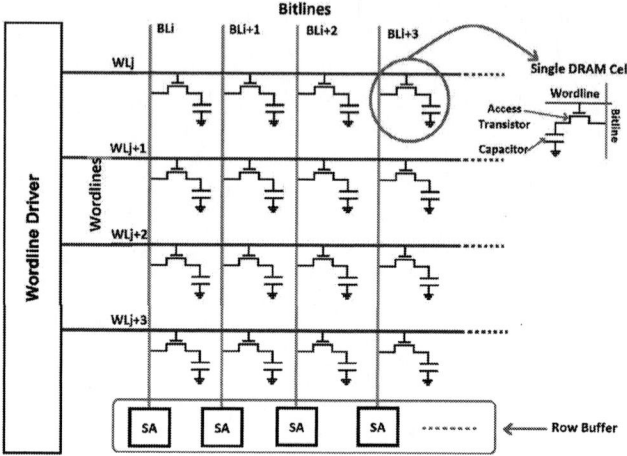

Fig. 1: Basic Organization of DRAM.

signal that activates the target row by connecting the row cells to the BLs. t_{RAS} defines the time period that the target row is being activated within. While the column address strobe (CAS) signal enables the columns within the row, and t_{CL} or t_{CAS} is the minimum waiting time to get the first data ready after sending a RD/WR command. Column-row delay (t_{RCD}) is the minimum time between row activation and column activation for a particular operation. The pre-charge time (t_{RP}) represents the minimum time duration that the PRE command is applied before sending the next ACT command. The read operation is initiated by issuing PRE command and lasts for t_{RP}. The ACT command is issued and activates WL for t_{RAS}. RD command enables the sensing and the data becomes available after t_{RAS} - (t_{RCD} + t_{CL}) [7].

When WL reaches all the row's cells, the BL signal propagates to the SA at a different rate on each cell due to the physical variation of the access transistor, cell's capacitor, and the load capacitor. Also, the variation of internal transistors of the SA contributes in BL variation. The accumulated variation of BL is different from one cell to another in the same row and represents a source of entropy that can be utilized to generate unpredictable values.

This work exploits the variation in the propagation delay between two adjacent cells in the same row to implement a delay-based PUF. Where the challenge represents the row address and each response bit is generated by evaluating the delay

979-8-3503-8083-5/23 $31.00 © 2023 IEEE

248

difference between two cells. The proposed scheme allows the signature generation during the normal read operation. The suggested approach offers two key features in comparison with the existing approaches: (1) Fast generation of the response. The response is generated by the normal read operation and the cell restoring is not required [8]. (2) More stable. The variation does not depend on a single cell individually, it depends on the variation between two cells.

This paper is organized as follows: section II lists the recent works published in the same domain, section III describes and analyzes the proposed approach, section IV shows the simulation results and analyzes them. Finally, section V summarizes what has been done so far.

II. RELATED WORK

Command enables the sensing when the column address strobe (CAS) is asserted. Latency-based DRAM PUFs exploit the standard timing parameters of commodity DRAM to extract efficient entropy from the addressed DRAM block. J. Miskelly et al. [9] set t_{RCD} to the lowest value to drive PUF response. The challenge is a set of timing parameter configurations and an offset to the target PUF section, and the response is the read value accordingly. The implementation is limited on the software level by a proper configuration of the memory controller. Due to the stability issues, the PUF reading is repeated 100 times during the enrollment stage. Thus, the post-processing of PUF reflects on the time generation of the response, which is around 3 ms and in the worst cases is 31 ms. Precharge latency-based PUF (PreLatPUF) [10] elaborates on pre-charge latency to generate the device signature. Reducing precharge time produces faulty read operations with an unpredictable value. The cells are evaluated and discovered based on the reduction of the latency. Thus, an exhaustive post manufacturing process is required to address the candidate cells that can be involved in PUF's response generation. The disabling period is long enough to ensure the uniqueness of PUF's responses. However, the response's evaluation times become an order of minutes. Also, a dedicated memory is needed to overcome the granularity issue [11]. Decay-Based DRAM PUF [12] deactivates the refreshing period to allow charge decaying. However, the scheme consumes a time period in terms of minutes to generate the device signature. Rapid Run-time DRAM PUF [13] also disables the refreshing period and introduces the decay period, after which the response's bits are evaluated. The evaluation time is less than [12] but still greater than the standard refreshing time (64ms). Both PUFs in [12] and [13] require a post manufacturing processing to address the potential cells for PUF operation.

III. THE PROPOSED APPROACH

This work proposes a new approach to implement delay-based PUF on the commodity DRAM that exploits the variation in the propagation delay of two adjacent BLs. The proposed implementation utilizes the same commercial DRAM array with a slight addition in the periphery circuit. Ideally,

upon the activation of the WL, charge sharing occurs on all the cells simultaneously.

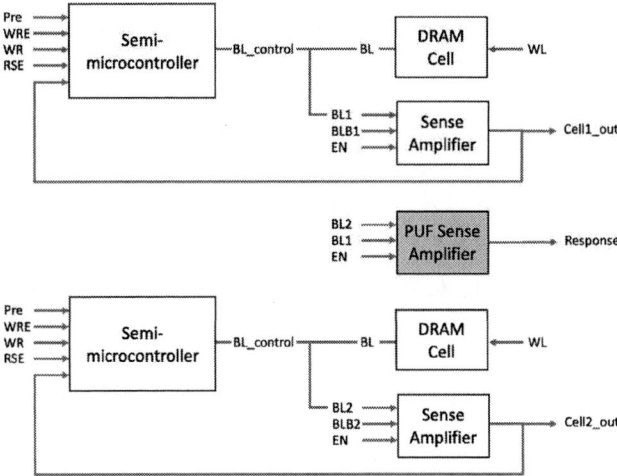

Fig. 2: The Schematic of 1-bit PUF's Response.

However, due to the process variation the BL varies from one cell to another. Consequently, the charge sharing on each cell occurs at a different rate. Thus, BLs of two adjacent cells is compared by SA (PUF's SA), and the output is PUF response. Fig. 2 shows the ordinary structure of adjacent DRAM cells and PUF's SA. Each two DRAM cells in the same row can be combined by a PUF's SA to create a row buffer for the response.

The delay variation is captured along the BL path, starting from the cell to the output of the SA. The initial voltage on BL and DRAM cell is given by Eq. 1 , where Q_{BL} and Q_C are the initial charge on the load and the cell capacitors, respectively. C_{BL} and C_C are the capacitance of the load and the cell capacitor, respectively.

$$V_{BL_{initial}} = \frac{Q_{BL}}{C_{BL}}, \; V_{C_{initial}} = \frac{Q_C}{C_C} \tag{1}$$

When the access transistor is ON (linear mode), the charge is shared and the final voltage on BL is given by Eq. 2.

$$V_{BL_{final}} = \frac{Q_{BL} + Q_C}{C_{BL} + C_C} - V_{th} \tag{2}$$

Where V_{th} is the threshold voltage of the DRAM cell's access transistor. Thus, the voltage difference that is sensed by SA can be modeled by Eq. 3

$$\Delta V_{BL} = V_{th} + \frac{C_C(V_{BL_{initial}} - V_{C_{initial}})}{C_{BL} + C_C} \tag{3}$$

Assuming the reading operation is done with V_{WL} = Vdd, the load capacitor is pre-charged to 0.5Vdd by the PRE command, and the row cells are initialized to Vdd during PUF mode activation, then $V_{BL_{initial}}$ = 0.5Vdd and $V_{C_{initial}}$ = Vdd. With substitution in Eq. 3, the final model is given by Eq. 4

$$\Delta V_{BL} = V_{th} - \frac{0.5(Vdd)C_C}{C_{BL} + C_C} \tag{4}$$

979-8-3503-8083-5/23 $31.00 © 2023 IEEE   249

Eq. 4 shows that V_{th}, C_C, and C_{BL} control the sensed voltage by SA and present the source of BL variations. The physical variation of access transistor (e.g: length), and load and cell capacitors (e.g: area of the plates) introduce accumulated variation at BL, which reflects on varying the propagation delay on all BLs. The effect of mismatching of the SA transistors is studied extensively in the literature. One of the works [14] explains it in details. The variations of SA and load capacitor create a source of entropy that varies from one system to another. The PUF's SA is enabled after t_{RCD} and disabled after the same time period that is used in the ordinary SA during reading or writing operation. Fig. 3 shows the necessary signals to generate the response bit. PUF-Enable (third to last) is enabled at the same time the WL is activated.PUF-Enable lasts for the same period as the ordinary SA's enable. The switch waveform (next to last) highlights the two case scenarios. When the switch is open, i.e. logic-0, the DRAM capacitance of the second cell is smaller than that of the first cell. While when the switch is closed, i.e. logic-1, the capacitance value is larger. Since there is no need to restore the cell's value, the destructive nature of the DRAM cell changes the value of the second cell while the value of the first cell remains the same. This is in the case of an open switch and is shown in the magnified portion of the fifth waveform during the timeline 40ns - 50ns. The response is evaluated to logic-0 (last waveform). Whereas in the second case, where the switch is closed, the DRAM cell changes the value of the first cell while the value of the second cell remains the same and the response is evaluated to logic-1. To avoid any biasing by the stored values, the cells should be initialized to logic-1 before the process starts.

IV. RESULTS

The proposed idea is implemented and evaluated using Cadence tool with Monte Carlo simulation over 50K samples.

The cell capacitance and BL parasitic capacitance are selected to be 30 fF and 85 fF, respectively [7]. The DRAM cells are initialized to 1.2 V. The voltage supply is selected to be 1.2 V and WL activation voltage is selected to be 1.5 V. The cell's access transistor and capacitor are sampled during the simulation based on the normal distribution, as shown in Fig. 4, which reflects both process variation and mismatching.

Practically, the propagation delay through the BL path involves the variation of the cell's capacitor, access transistor, load capacitor and the SA's transistors. The response is evaluated based on each source of variation separately, then the accumulated variation as shown in Fig. 5 As for the variation on the BL and DRAM cell capacitors, they yield higher uniformity in comparison to the variation on the access transistors as well as the SA. Moreover, Fig. 5 shows the distribution of the response based on the accumulated variation, which has lower uniformity than the distribution that is generated by capacitors' variation and higher uniformity than that is generated by transistors' variation. The variation of the SA and the load capacitor can not be the only elements taken into consideration, because the SA and load capacitor

are the same for all cells sharing the same BL in the same array. Therefore, the access transistors and the cell capacitor variations are the distinguishable elements between DRAM rows.

To assess the performance of the circuit for PUF utilization, the response bit is evaluated by bit-aliasing and bit-reliability. The configurations are defined in two ways: SA and C_{BL} configuration to emulate different memory arrays, and cell capacitor and access transistor configuration to emulate the same array. A set of notations are defined and used in criteria formulation, which are: (1) I: the number of SA and C_{BL} configurations, where I_i represents the ith configuration. (2) R: the number of C_{cell} and V_{th} configurations, where R_r represents r^{th} configuration. (3) $I_i(R_r)$: is the response bit of rth configuration of C_{cell} and V_{th} under i^{th} configuration of SA and C_{BL}. (4) T: the number of different environmental temperatures. (5) $T_t(I_i(R_r))$: is the response bit of rth configuration of C_{cell} and V_{th} under i^{th} configuration of SA and C_{BL} at t^{th} temperature. The bit-aliasing evaluates the variation of the bit over the devices' axis, which is given by Eq. 5. The ideal score is 50%, which means the response bit will be evaluated as logic-1 in half of the different memory arrays and logic-0 for the rest, and that guarantees uniformity and uniqueness among devices.

$$Bit - Aliasing = \frac{1}{I \cdot R} \sum_{r=1}^{R} \sum_{i=1}^{I} I_i(R_r) \qquad (5)$$

The stability of the circuit is evaluated by measuring the changes against temperature or voltage variations. Ideally, the circuit generates the same output at different environmental situations, which in an ideal case should be zero (no changes). Practically, the output may flip due to the environmental changes. Eq. 6 states the percentage of error occurrence over multiple configurations at different temperatures.

$$Bit - Reliability = \frac{2}{T \cdot (T-1) \cdot I \cdot R} \sum_{i=1}^{I} \sum_{c=r}^{R} \sum_{t=1}^{T-1} \sum_{b=t+1}^{T}$$
$$|T_t(I_i(R_r)) - T_b(I_i(R_r))| \qquad (6)$$

The circuit achieves 53.8% for bit-aliasing, which is approximately an ideal case. The bit-reliability is 14.2% over three different temperatures, which requires a selective scheme as post-process to define the bit confidence.

V. CONCLUSION

This work proposes a delay-based PUF on the commodity DRAM that evaluates the delay of two adjacent cells in the same activated row. The response is generated by comparing the BLs of two adjacent cells where the sense amplifier is the evaluation element. The simulation results proves the uniqueness and uniformity of the design, which are efficient for PUF response.

979-8-3503-8083-5/23 $31.00 © 2023 IEEE

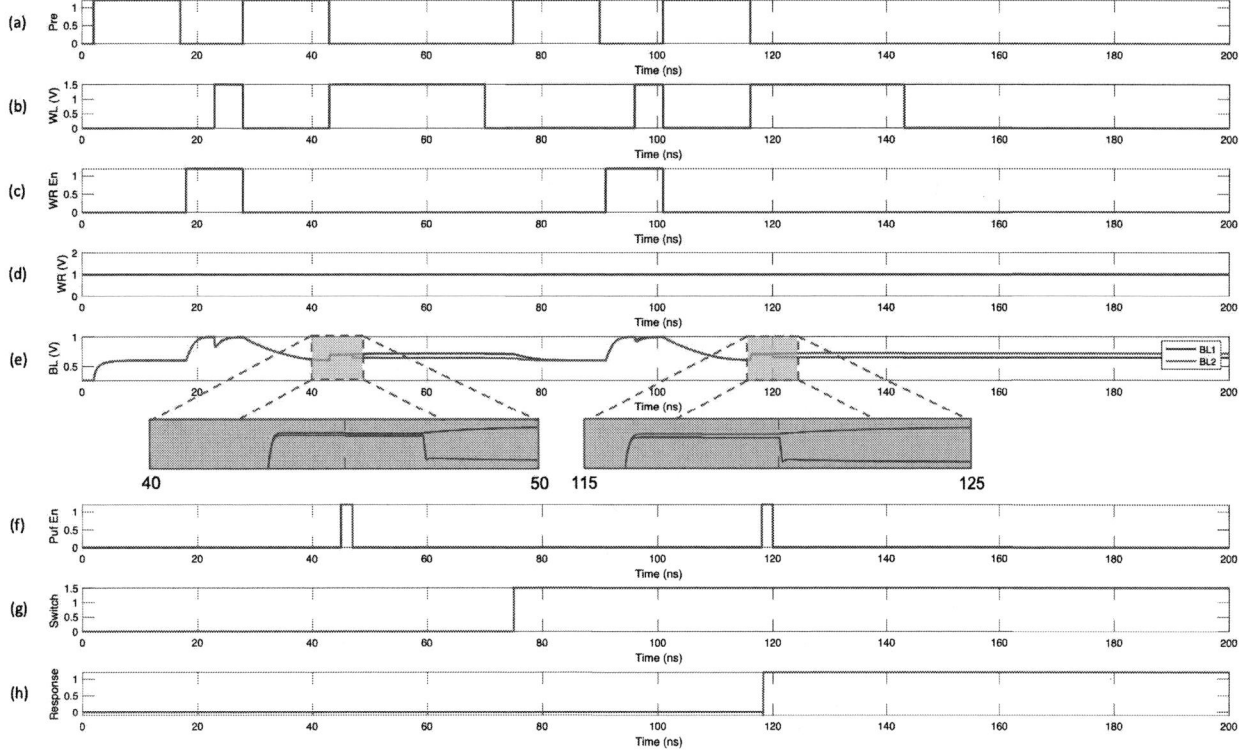

Fig. 3: The transient analysis of a 1-bit response PUF where (a) is the precharge signal, (b) is the assertion of the word line, (c) is enabling the write mode, (d) is the value written onto the dram cells, (e) is the bitlines of the two DRAM cells and the magnified portions represent the two possible outcomes of the circuit, (f) is enabling the PUF sense amplifier, (g) initially the switch is open making the dram capacitance of the 2nd cell smaller than that of the 1st, later on the situation is reversed, (h) is the PUF response.

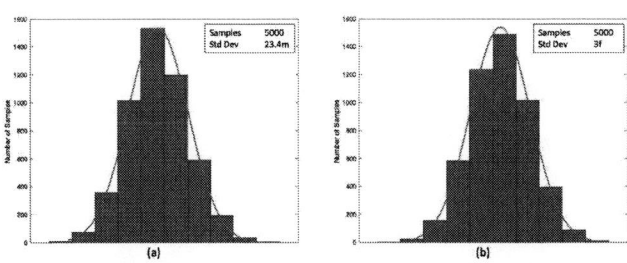

Fig. 4: The statistical distribution of (a) Vth of the access transistor and (b) the capacitance of DRAM cell, utilized by the simulation.

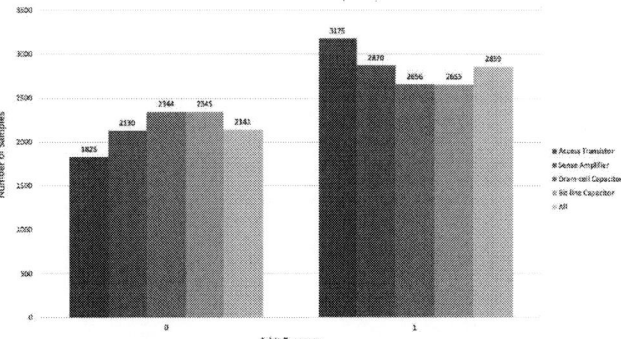

Fig. 5: 1-Bit response distribution based on the variation along the path of the BL.

REFERENCES

[1] A. Stanciu, M. N. Cirstea, and F. D. Moldoveanu, "Analysis and evaluation of puf-based soc designs for security applications," *IEEE Transactions on Industrial Electronics*, vol. 63, no. 9, pp. 5699–5708, 2016.

[2] T. Idriss, H. Idriss, and M. Bayoumi, "A puf-based paradigm for iot security," in *2016 IEEE 3rd World Forum on Internet of Things (WF-IoT)*, 2016, pp. 700–705.

[3] T.-N. Nguyen, S. Park, and D. Shin, "Extraction of device fingerprints using built-in erase-suspend operation of flash memory devices," *IEEE Access*, vol. 8, pp. 98 637–98 646, 2020.

[4] Q. Tang, C. Zhou, W. Choi, G. Kang, J. Park, K. K. Parhi, and C. H. Kim, "A dram based physical unclonable function capable of generating¿ 10 32 challenge response pairs per 1kbit array for secure chip authentication," in *2017 IEEE Custom Integrated Circuits Conference (CICC)*. IEEE, 2017, pp. 1–4.

[5] J. S. Kim, M. Patel, H. Hassan, and O. Mutlu, "The dram latency puf: Quickly evaluating physical unclonable functions by exploiting the latency-reliability tradeoff in modern commodity dram devices," in *2018 IEEE International Symposium on High Performance Computer*

Architecture (HPCA), 2018, pp. 194–207.

[6] L. Orosa, Y. Wang, I. Puddu, M. Sadrosadati, K. Razavi, J. Gómez-Luna, H. Hassan, N. Mansouri-Ghiasi, A. Tavakkol, M. Patel *et al.*, "Dataplant: Enhancing system security with low-cost in-dram value generation primitives," *arXiv preprint arXiv:1902.07344*, 2019.

[7] K. Humood, B. Mohammad, and H. Abunahla, "Dtrng: Low cost and robust true random number generator using dram weak write scheme," in *2021 IEEE International Symposium on Circuits and Systems (ISCAS)*. IEEE, 2021, pp. 1–5.

[8] S. Bavikadi, P. R. Sutradhar, K. N. Khasawneh, A. Ganguly, and S. M. Pudukotai Dinakarrao, "A review of in-memory computing architectures for machine learning applications," in *Proceedings of the 2020 on Great Lakes Symposium on VLSI*, 2020, pp. 89–94.

[9] J. Miskelly and M. O'Neill, "Fast dram pufs on commodity devices," *IEEE Transactions on Computer-Aided Design of Integrated Circuits and Systems*, vol. 39, no. 11, pp. 3566–3576, 2020.

[10] B. B. Talukder, B. Ray, D. Forte, and M. T. Rahman, "Prelatpuf: Exploiting dram latency variations for generating robust device signatures," *IEEE Access*, vol. 7, pp. 81 106–81 120, 2019.

[11] W. Xiong, A. Schaller, N. A. Anagnostopoulos, M. U. Saleem, S. Gabmeyer, S. Katzenbeisser, and J. Szefer, "Run-time accessible dram pufs in commodity devices," in *International Conference on Cryptographic Hardware and Embedded Systems*. Springer, 2016, pp. 432–453.

[12] A. Schaller, W. Xiong, N. A. Anagnostopoulos, M. U. Saleem, S. Gabmeyer, B. Škorić, S. Katzenbeisser, and J. Szefer, "Decay-based dram pufs in commodity devices," *IEEE Transactions on Dependable and Secure Computing*, vol. 16, no. 3, pp. 462–475, 2019.

[13] I. Kumari, M.-K. Oh, Y. Kang, and D. Choi, "Rapid run-time dram puf based on bit-flip position for secure iot devices," in *2018 IEEE SENSORS*. IEEE, 2018, pp. 1–4.

[14] A. Hajimiri and R. Heald, "Design issues in cross-coupled inverter sense amplifier," in *ISCAS '98. Proceedings of the 1998 IEEE International Symposium on Circuits and Systems (Cat. No.98CH36187)*, vol. 2, 1998, pp. 149–152 vol.2.

979-8-3503-8083-5/23 $31.00 © 2023 IEEE

2023 International Conference on Microelectronics (ICM)

Real-Time Switched Capacitor Based Power Side-Channel Attack Detection

Leen Younes, Baker Mohammad, Mahmoud Al-Qutayri, Hani Saleh, Dima Kilani *

System on Chip Lab (SoC), Department of Electrical and Computer Engineering
Khalifa University, Abu Dhabi, UAE
{100049417,baker.mohammad,hani.saleh,mahmoud.alqutayri}@ku.ac.ae
* School of Engineering
University of British Columbia, Kelowna, Canada
dima.kilani@ubc.ca

Abstract—**Side-channel attacks (SCAs) are regarded as significant risks to the hardware implementation of cryptographic systems. Side-channel information, such as timing, power, and electromagnetic radiation, is leaked through the system and can be exploited for secret key extraction. This work proposes a real-time and compatible detection method for power SCAs. The technique utilizes a switched capacitor DC-DC (SC-DCDC) converter in conjunction with a lightweight artificial intelligence engine for power SCA detection. The proposed system, referred to as EoH, possesses the capability to perform dynamic voltage scaling and learn the behaviors of the cryptographic system to identify potential attacks. The switching activities of the SC-DCDC converter can be viewed as measurements of the cryptographic function. Therefore, a recurrent neural network was chosen as it processes time-series data most effectively. The technique is system-specific, meaning that during the enrollment phase, the normal operation of the system is learned. Furthermore, the technique can be expanded to include other types of SCAs and is not limited to power.**

Keywords— Side Channel Attack, Detection, Power Analysis, Voltage Regulator, SC-DCDC Converter

I. INTRODUCTION

Securing on-chip data ranks as a top priority in modern integrated circuits. Side-channel attacks on complex integrated systems have been effectively used to extract sensitive data, thereby compromising the security of the chip [1]. The noninvasive side-channel attacks (SCAs) target the implementation of secure cryptographic systems, rather than the algorithms themselves. They exploit the physical characteristics and information leaked from indirect sources [2]. The side-channel information can take the form of power consumption [3], [4], timing [5], or electromagnetic radiation [6], [7]. This information is correlated with the secret key as well as the inputs. Additionally, it is dependent on the internal state of the system and the intermediate values being computed [8].

SCAs can be broadly classified into two categories, passive and active, based on the extent of control the attacker has over the device. Passive attacks revolve around merely observing the behaviour of the system without interfering with its operation. Power, electromagnetic, and timing SCAs fall under this category. In passive attacks, the attacker is only capable of injecting controlled input data and collecting side-channel information. Subsequently, the data is analyzed to establish the relationship between the collected and processed data, eventually leading to the retrieval of the secret key [2], [9]. On the other hand, active attacks involve modifying the system's operation and injecting signals to influence its behaviour. By doing so, the attacker can selectively extract information to aid in inferring the key or breaking the cryptographic module [2]. Active attacks include fault injection.

Comparing the two types in general, passive attacks are considered to be less invasive and less risky than active attacks, yet they are still very effective in extracting sensitive information from a system.

This paper describes a method for real-time power SCAs detection. The method employs the voltage regulator within the power management unit along with a lightweight artificial intelligent (AI) engine. The proposed system, shown in Fig. 1, utilizes a switched capacitor DC-DC (SC-DCDC) converter, where its switching activity serves as an indicator of the load current.

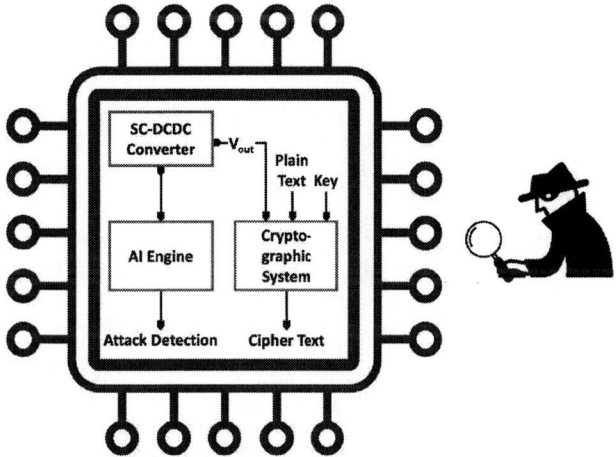

Fig. 1: The block diagram of the proposed system utilized for the detection of power side-channel attacks. A switched capacitor DC-DC (SC-DCDC) converter is employed to supply the cryptographic system. The artificial intelligence (AI) engine continually monitors the switching activities of the converter, as it is a direct measure of the load current, and indicates a compromised system upon the occurrence of any abnormalities.

Based on the switching activity, the AI learns the normal operation of the cryptographic system during the enrollment phase. Subsequently, it continuously monitors the switching activities, detects any abnormalities and classifies them as attacks. Furthermore, the AI triggers the SC-DCDC converter to provide voltage scaling as a countermeasure.

The paper is organized as follows: A brief introduction to power SCA and its various types is provided in Section II. An overview of some power SCA detection methods published in the literature is

979-8-3503-8083-5/23 $31.00 © 2023 IEEE

presented in Section III. Section IV describes the proposed system and the execution flow for detecting power SCA. The paper concludes in Section V."

II. BACKGROUND

A power analysis attack is an effective, non-invasive way of revealing secret information, typically the key, about a device by capturing its power signatures.Researchers from both academia and industry have devoted significant attention to studying power analysis

attacks as the leading form of side-channel analysis. It has been demonstrated to effectively crack the Advanced Encryption Standard (AES) within a few minutes. The susceptibility to power analysis attacks stems from the fact that systems consume varying amounts of power depending on the computations they perform. For instance, the power signatures of a distinct encryption algorithm may vary due to diverse implementations and the use of different cryptographic operations. An attacker can exploit this data to infer confidential information by monitoring and analyzing the target device's power consumption during specific operations. Since power analysis attacks are based on passive observations of power consumption, they are particularly challenging to detect using conventional security measures [2].

There are various types of power analysis attacks: simple power analysis (SPA), differential power analysis (DPA), and correlation power analysis (CPA).

- Simple Power Analysis

  SPA interprets the power consumption directly during the operations of the cryptographic system. It operates under the assumption that the system's power consumption depends on the operation being carried out. Therefore, it is essential to have basic knowledge regarding the characteristics of the employed algorithm and any potential software or hardware implementations. SPA is commonly used when there is limited access to the device and only a few power traces are available [2], [10], [11].

- Differential Power Analysis

  DPA [3] is a statistical method that does not require knowledge of the cryptographic algorithm and is therefore considered a refined form of SPA. This method takes advantage of the fact that the number of simultaneously switching logic gates affects the power profile or electromagnetic field that is radiated. As a result, a relationship exists between the state of the key bit and the power profile. The key can be inferred by applying the difference in means to the collected data. This method successfully attacks the system by acquiring a large number of power traces, even with basic countermeasures implemented [2], [10].

- Correlation Power Analysis

  Similar to DPA, CPA is also a statistical method that does not require knowledge of the cryptographic algorithm. It also requires the acquisition of a large number of power traces and then utilizes statistical techniques in the data analysis phase. To infer the key, CPA looks for the correlation factor rather than the difference in means [2].

III. RELATED WORK

The majority of the existing methodologies for detecting SCAs require the external insertion of a small resistor to the system's power supply. This external insertion impacts the power grid's effective impedance; hence, it induces abnormal voltage variations. The work presented in [12] leverages this fact and reveals those abnormal variations to detect SCAs. During the design phase, the power profiles prior to and following the connection of the resistor are captured.

During runtime, the on-chip sensors continuously measure the power signals, which are then digitized through an integrated analog-to-digital converter. Consequently, the abnormal voltage variations are sensed and subsequently analyzed through a trained linear classifier embedded within the system. Following a similar approach, [13] then analyzes the power profiles through logistic regression.

Furthermore, the proposed approach in [14] is ring oscillator based rather than machine learning based. It exploits the frequency sensitivity of the ring oscillator to the supply voltage. A voltage drop due to the resistor insertion at a node results in a small frequency shift in the ring oscillator. This frequency shift leads to a time delay (and corresponding phase difference) between the rising edges of a ring oscillator connected to a node with resistor insertion and another connected to a regular node. Consequently, the number of edges in a specific time window differs between these two ring oscillators, where it will be larger at a regular node. The difference in the number of rising edges serves as a measure of the phase difference and functions as a SCA detection metric. It is measured through a counter and a comparator, eliminating the need for phase detectors.

Moreover, in the study conducted in [15], a switched capacitor circuit was employed to sense the voltage resulting from the insertion of the resistor. The voltage drop is stored on a sampling capacitor and is then amplified by the switched capacitor. The SCA is detected by comparing the amplified voltage drop with a pre-determined threshold voltage.

IV. METHODOLOGY

The general concept of SCAs is to feed input data and observe the output while simultaneously analyzing leaked data, such as power in power analysis attacks, in response to the controlled changes in the input data. The key value is then inferred from the collected data [16].

As previously mentioned, systems are susceptible to power analysis attacks because their power consumption varies with multiple factors. The fundamental theory of CMOS power dissipation is briefly described in order to provide further clarification. Transistors, being the fundamental building blocks of all circuits, are eventually used to implement cryptographic systems. Therefore, the total power consumption of the cryptographic systems is composed of three components: dynamic, static, and leakage power. The major contributor to the power consumption is the dynamic power that is consumed during the circuit's active mode. It directly correlates with the data activity, where the switching of the states, as well as the frequency of the toggling have a direct effect on it. Therefore, the activity of the system can be identified by monitoring the power consumption.

Energy efficiency stands as the fundamental constraint in current systems-on-chip (SOCs). Server-class CPUs are thermally limited, necessitating improved energy efficiency for enhanced performance; on the other hand, IoT devices and mobiles rely heavily on low consumption of energy. As a result, power management systems are imperative, demanding swift responses to workload fluctuations. Nowadays, most SoCs feature a power management unit, comprising a voltage regulator, be it of the linear or switching type, for voltage regulation and power switching. In this paper, the focus is on the switching type, more specifically, a SC-DCDC converter. The SC-DCDC converter facilitates noninvasive load measurement. In Fig. 2, the switching activity of the SC is depicted against variations in the load current. According to the simulation results, a low toggling frequency implies a reduced load current. Similarly, an increased toggling frequency indicates a higher load current. As illustrated in the right side of the figure, the relationship between the load current and the toggling frequency is approximately linear. Thus, one may regard the switching activity as a direct measure of the data activity.

Making use of this notion, the switching activity is utilized as a tool in the detection mechanism for SCAs. The proposed system, referred to as **EoH**, leverages the SC-DCDC converter, the design of which

979-8-3503-8083-5/23 $31.00 © 2023 IEEE

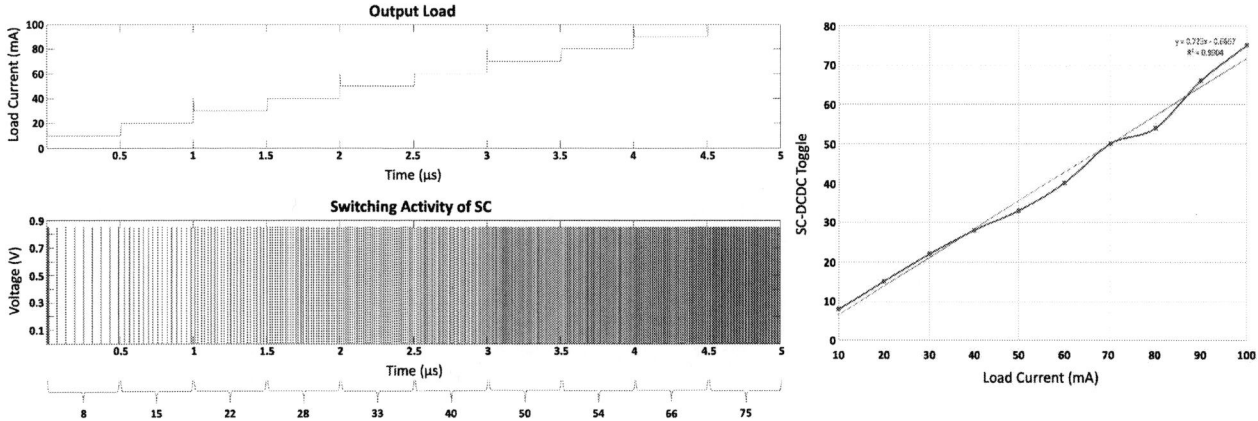

Fig. 2: The change in the switching activity of the switched capacitor DC-DC converter against the change of the load current. The toggling becomes more frequent as the load increases. The relationship between the two is approximately linear.

is elaborated upon later, to supply the cryptographic system. **EoH** is not only capable of providing dynamic voltage scaling through the voltage regulator, but also learning the behaviors of the cryptographic system and detecting potential attacks. Figure 1 illustrates the structure of **EoH**. During normal operations, the switching activity can be monitored and learned, indicating the occurrence of encryption. Any abnormality would signify a potential attack.

The lightweight AI engine uses the switching activity of the SC-DCDC converter as a measurement of the cryptographic function. During the enrollment phase, the AI engine monitors and learns the activities of the cryptographic system through extracting the changes in the switching activity over a specific time window. Various features with discriminative representations are extracted and can be used later in the classification pipeline. Now, the behavior of the system under normal operation is known and can serve as a reference across any abnormal activities. During inference, the AI engine monitors the activity of the cryptographic system and configures the features to identify whether an attack occurs, classifying the system as either secure or compromised. Upon the detection of an attack, the SC-DCDC converter is exploited as a countermeasure, with the voltage scaled to influence the power profile. The process adapted by **EoH** is presented in Fig. 3.

A. AI Engine

For the proposed attack detection technique, the switching activities collected from the SC-DCDC converter will serve as the dataset for the machine learning algorithm during the enrollment phase. As mentioned earlier, the data is in the form of a time series; therefore, it is represented as an array of numbers, commonly referred to as a 1D tensor. Conventionally, time is represented along the second axis. The enrollment process employs annotated datasets, categorizing it under supervised learning. Fig. 4 illustrates the primary steps for conducting a supervised detection approach.

All neural networks share the common trait of having no memory; every input is processed separately with no context from one input to the next. As a result, for the network to process a sequence or temporal series of data, it must be presented with the entire sequence. Recurrent neural networks (RNNs) adopt a similar principle to biological intelligence, although simplified. Biological intelligence processes information incrementally whilst creating an internal model based on of prior information and updating it as new information arrives. Similarly, RNN processes sequences by examining each element individually while also keeping track of what has been seen thus far. Consequently, the RNN is able to internally loop over

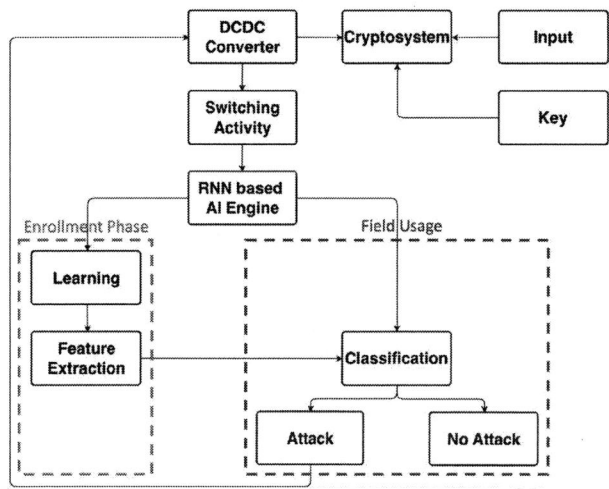

Fig. 3: The execution flow adapted by the proposed system for detecting power SCA and countering it.

sequence elements, resetting its state between each separate sequence. This property makes them the ideal choice for processing the data collected from the SC-DCDC converter.

B. Switched Capacitor DC-DC converter

The design, based on the work presented in [17], consists of five main blocks: a gain controller, a re-configurable switched capacitor circuit, a non-overlapping two phase circuit, a pulse frequency modulation (PFM) controller and additionally a switch modulation enable circuit. Fig. 5 presents the block diagram of the SC regulator.

The re-configurable SC-DCDC converter has four modes of operation that down-convert the external supply based on the gain setting. Each mode corresponds to a specific configuration of flying capacitors that provides conversion gain of 1, ½, 2/3 and ¾. Thus, when using a supply voltage of 0.8V, the generated voltages would be 0.8V (bypass mode), 0.6 (high performance mode), 0.53 (low performance mode) and 0.4 (standby mode). Additionally, the output voltage is regulated with a frequency modulation technique over a wide range of load current. Frequency modulation allows the adjustment of the frequency of the clock in response to the load current. Switch modulation is

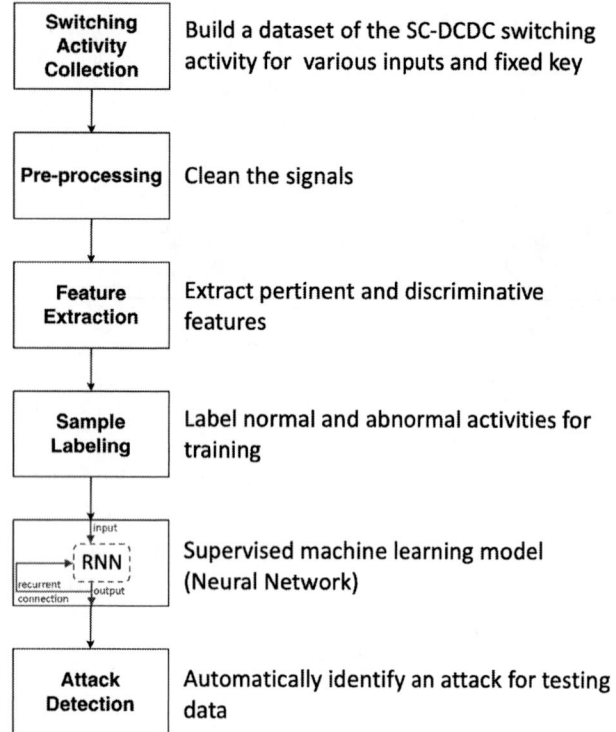

Fig. 4: The main steps for conducting a supervised attack detection adapted to the proposed technique.

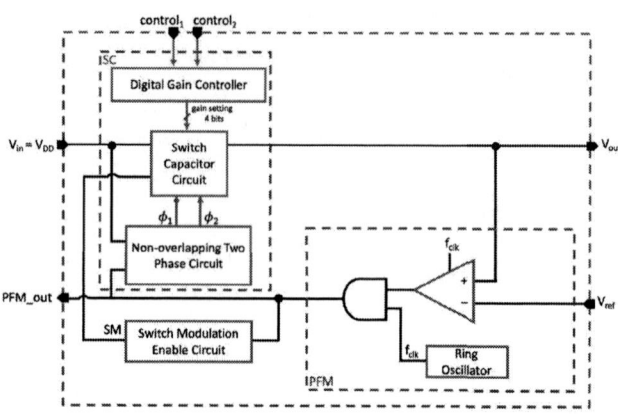

Fig. 5: Switched capacitor DC-DC converter block diagram.

employed to improve efficiency at lower loads. Furthermore, a dead time, during which all the switches are off for a short period of time, is ensured through the generation of two non-overlapping phases. This is done to avoid any short-circuit currents.

1) Switched Capacitor Circuit: To enhance the efficiency of the dc-dc converter, the bypass mode is introduced. In situations requiring high power, the SC is bypassed and the input voltage is directly supplied, resulting in no switching activity. As for the remaining gains, the charging and discharging phases control the operation. During the charging phase, the flying capacitors are charged while in the discharging phase, they discharge through the load capacitor, thereby generating the output load.

2) Switch Modulation: To further improve the efficiency, switch modulation is implemented. This is done automatically based on the activity of the SC, as there is a strong relationship between the switching activity and the load current where the higher the load the more frequent the switching occurs. The switch modulation enable circuit detects whether the load current is high or low. Then, the circuit reduces the transistor width for lower currents, consequently reducing the leakage power, which, in turn, improves the power efficiency.

V. CONCLUSION

A real-time detection technique for power side-channel attacks is proposed. It employs the voltage regulator, a crucial component of the power management units that exist in most systems-on-chip nowadays. This work utilized a switched capacitor DC-DC (SC-DCDC) converter, to be specific, along with a lightweight artificial intelligence (AI) engine. The RNN based AI learns the activities of the cryptographic system and detects an attack. Furthermore, it provides dynamic voltage scaling as a countermeasure. While this work primarily focuses on power SCAs, it's worth noting that it can also detect any misuse of the cryptographic system by other types of passive, non-invasive SCAs.

VI. ACKNOWLEDGMENT

This work is funded and supported by Technology Innovation Institute (TII) [grant: EX2021-005]. Additional support from Khalifa University, System on Chip Lab (SoC)[grant: RC2-2018-020]

REFERENCES

[1] D. Genkin, I. Pipman, and E. Tromer, "Get your hands off my laptop: physical side-channel key-extraction attacks on pcs: Extended version," *Journal of Cryptographic Engineering*, vol. 5, pp. 95–112, 2015.

[2] S. Bhunia and M. Tehranipoor, *Hardware security: a hands-on learning approach.* Morgan Kaufmann, 2018.

[3] P. Kocher, J. Jaffe, and B. Jun, "Differential power analysis," in *Annual international cryptology conference.* Springer, 1999, pp. 388–397.

[4] S. Mangard, E. Oswald, and T. Popp, *Power analysis attacks: Revealing the secrets of smart cards.* Springer Science & Business Media, 2008, vol. 31.

[5] P. C. Kocher, "Timing attacks on implementations of diffie-hellman, rsa, dss, and other systems," in *Annual International Cryptology Conference.* Springer, 1996, pp. 104–113.

[6] J.-J. Quisquater and D. Samyde, "Electromagnetic analysis (ema): Measures and counter-measures for smart cards," in *International Conference on Research in Smart Cards.* Springer, 2001, pp. 200–210.

[7] D. Agrawal, B. Archambeault, J. R. Rao, and P. Rohatgi, "The em side-channel (s)," in *CHES*, vol. 2. Springer, 2002, pp. 29–45.

[8] A. Barenghi, L. Breveglieri, I. Koren, and D. Naccache, "Fault injection attacks on cryptographic devices: Theory, practice, and countermeasures," *Proceedings of the IEEE*, vol. 100, no. 11, pp. 3056–3076, 2012.

[9] S. Guilley, L. Sauvage, J.-L. Danger, N. Selmane, and R. Pacalet, "Silicon-level solutions to counteract passive and active attacks," in *2008 5th Workshop on Fault Diagnosis and Tolerance in Cryptography.* IEEE, 2008, pp. 3–17.

[10] P. Kocher, J. Jaffe, B. Jun, and P. Rohatgi, "Introduction to differential power analysis," *Journal of Cryptographic Engineering*, vol. 1, pp. 5–27, 2011.

[11] D. Mukhopadhyay and R. S. Chakraborty, *Hardware security: design, threats, and safeguards.* CRC Press, 2014.

[12] D. Utyamishev and I. Partin-Vaisband, "Real-time detection of power analysis attacks by machine learning of power supply variations on-chip," *IEEE Transactions on Computer-Aided Design of Integrated Circuits and Systems*, vol. 39, no. 1, pp. 45–55, 2018.

[13] F. Kenarangi and I. Partin-Vaisband, "Exploiting machine learning against on-chip power analysis attacks: Tradeoffs and design considerations," *IEEE Transactions on Circuits and Systems I: Regular Papers*, vol. 66, no. 2, pp. 769–781, 2018.

[14] N. Gattu, M. N. I. Khan, A. De, and S. Ghosh, "Power side channel attack analysis and detection," in *2020 IEEE/ACM International Conference On Computer Aided Design (ICCAD).* IEEE, 2020, pp. 1–7.

979-8-3503-8083-5/23 $31.00 © 2023 IEEE

[15] N. Kaushik and J. Hu, "A switched-capacitor power side-channel attack detection circuit in 65-nm cmos," in *2021 IEEE International Symposium on Circuits and Systems (ISCAS)*. IEEE, 2021, pp. 1–5.

[16] D. Das, S. Ghosh, A. Raychowdhury, and S. Sen, "Em/power side-channel attack: White-box modeling and signature attenuation counter-measures," *IEEE Design & Test*, vol. 38, no. 3, pp. 67–75, 2021.

[17] D. Kilani, M. Alhawari, B. Mohammad, H. Saleh, and M. Ismail, "An efficient switched-capacitor dc-dc buck converter for self-powered wearable electronics," *IEEE Transactions on Circuits and Systems I: Regular Papers*, vol. 63, no. 10, pp. 1557–1566, 2016.

An Embedded Real-Time Driver Monitoring System based on AI and Computer Vision

1st Leila Sharara, *Member, IEEE*
Electrical and Computer Engineering
Wayne State University
Detroit, USA
leilasharara@wayne.edu

2nd Alexandros Politis
Electrical and Computer Engineering
Wayne State University
Detroit, USA
alexandrospolitis@wayne.edu

3rd Lubna Alazzawi
Electrical and Computer Engineering
Wayne State University
Detroit, USA
alazzawi@wayne.edu

4th Mohammed Ismail, *Fellow, IEEE*
Electrical and Computer Engineering
Wayne State University
Detroit, USA
ismail@wayne.edu

Abstract—This research aims to develop a comprehensive device that ensures both safety and comfort for car users by integrating a Driving Monitoring System (DMS) with a Seat Electronic Control Unit (ECU). The primary objective of this system is to monitor and regulate the driver's alertness level. By utilizing infrared sensors and a pre-trained Artificial Intelligence (AI) algorithm with advanced facial recognition capabilities, the DMS accurately detects the driver's head position and eye status during both daytime and nighttime driving situations. In the event of detecting signs of drowsiness or distraction, the system promptly activates a warning mechanism by sending an electrical signal to the car's Seat ECU. The Seat ECU manages the seat motor drivers and position sensors, initiating a car seat vibration and triggering an audible alert to notify the driver of any potential danger. It must satisfy specific requirements, including real-time control of various motors, fast processing speed, low power consumption, low heat dissipation, low standby current, and low Electromagnetic Interference (EMI). These two systems are combined into a compact device that also incorporates a night vision camera mounted on the steering column to record data without obstructing the driver's view. The system operates reliably in real driving conditions, unaffected by facial expressions or eye occlusions, while achieving a frame rate of up to 45 Frames Per Second (FPS).

Index Terms—Artificial Intelligence, Machine Learning, ADAS, DMS, ECU, MediaPipe, OpenCV, HPE, EAR, EOR.

I. Introduction

Safe driving requires full attention and coordination between the driver's mind and body. Impaired driving, such as drunk, distracted, and drowsy driving, is a significant cause of accidents and fatalities on the road. The National Highway Traffic Safety Administration (NHTSA) reports that 94% of vehicular accidents are due to human error. Distracted driving accounts for 15% of accidents, while drowsy driving accounts for 10%, resulting in a significant number of fatalities annually [1]. DMS, particularly those utilizing advanced AI and facial recognition technologies, and Advanced Driver Assistance System (ADAS) with features such as adaptive cruise control and lane departure warnings, can prevent accidents and fatalities by augmenting the driver's decision-making process.

Furthermore, ongoing developments in automotive seating solutions aim to integrate safety and security features, providing an additional layer of protection and thereby augmenting overall vehicular safety [1], [4].

AI-based and computer vision algorithm approaches have gained significant attention in the automotive safety field, exhibiting high detection accuracy and processing speeds. However, these methods often lack comprehensive testing and implementation in actual automotive devices, as well as validation based on general operating standards. Achieving a fully developed vision-based Advanced Driver Assistance System (ADAS) requires careful consideration of critical factors.

This research work introduces a cost-effective and robust automotive device that integrates a Seat ECU and DMS to monitor driver alertness effectively. It provides thorough details into the hardware architecture and algorithm development, presenting a hybrid model that ensures accurate real-time Head Pose Estimation (HPE) and eye status detection. A functional prototype was developed, implemented, and tested, yielding successful results [4]. The rest of this paper is structured as follows: Section II provides an overview of related work. Section III presents the Methodology of the DMS system, followed by the system implementation discussion in Section IV. In Section V, the performance evaluation is presented, while Section VI focuses on the challenges and potential future improvements. Finally, Section VII concludes the work.

II. Related work

Leading automakers such as General Motors, Mercedes-Benz, Toyota, and Ford have integrated advanced DMS and ADAS into their vehicles. These cutting-edge technologies utilize an array of sensors and cameras to ensure driver attentiveness and safety [18]. Additionally, extensive research has been dedicated to the development of DMS and ADAS within the realm of automotive safety. Academic research groups have developed real-time DMS systems using camera-based approaches to detect drowsiness and distraction. These systems

979-8-3503-8083-5/23 $31.00 © 2023 IEEE

TABLE I
COMPARISON OF VARIOUS DRIVER MONITORING SYSTEMS [4]

Parameter	This Work	[15]	[16]	[17]
Methodology	CNN	CNN + LSTM	DBN + Sliding Window	CNN + RNN
Detected Impairment	Drowsiness, Distraction	Drowsiness, Distraction	Drowsiness	Drowsiness
Mechanism	Facial Features	Facial spatiotemporal	Facial features	Facial features & vehicle dynamics
Detection Rate (FPS)	45	15	30	15
Alert Mechanism	Audio+ Seat Vibration	Audio	NO	Audio + Visual
Device Implementation	YES	NO	NO	NO
Extended Functionality	YES	NO	NO	NO
Standby Power	3.6 mw at 12 V_{dc}	Unattended	Unattended	Unattended

employ advanced neural network architectures and techniques like CNNs, LSTMs, and sliding windows to achieve high detection accuracy and real-time processing speeds. For instance, Zeng et al. achieved 96.7% accuracy at 15 FPS, while Y. Li et al [15] outperformed other methods with 92.4% accuracy at 30 FPS. Alam et al. integrated facial features with vehicle data, achieving 93.3% accuracy at 15 FPS [16]. Comparably, F. Vicente *et al* system effectively detected eyes off the road at 25 FPS [17].

While the aforementioned studies have demonstrated high accuracy rates in controlled environments, there remain potential areas for refinement and enhancement. It is imperative to evaluate the performance of these systems in real-world scenarios encompassing diverse lighting conditions and driving contexts to ensure their robustness. Moreover, there is a need for a more seamless integration of DMS with ADAS and other automotive technologies, in order to establish a comprehensive safety framework that addresses both driver behavior and vehicle automation. While the ability to detect critical events is crucial, these studies could be directed toward the development of interventions aimed at preventing accidents or promptly alerting drivers in real-time. Further insights into the current state-of-the-art technologies and techniques used in DMS and ADAS could be referred to [2]–[4]. Table I provides a detailed comparison of our system's features with those of related studies.

III. METHODOLOGY

Onboard DMS and ADAS can effectively hold great promise in enhancing road safety. However, the successful implementation of these systems requires a strong and dependable approach to ensure accurate perception and tracking capabilities [8].

A. Artificial Intelligence (AI)

AI utilizes Machine Learning (ML) techniques to mimic human abilities by analyzing vast datasets, detecting patterns, and predicting anomalies. Deep Learning (DL) is a subfield of ML that consists of a variety of supervised and unsupervised models. Among the main DL models used in computer vision are Deep Convolutional Neural Networks (CNNs) that have been widely used in object detection, facial recognition, and video classification [5]. The designed DMS detection pipeline was constructed using OpenCV for facial detection and Google's ML solution, MediaPipe, for real-time facial landmark extraction. To achieve accurate detection of drowsiness and distraction, the DMS focused on four main submodules: facial landmark detection, HPE, eye state tracking, and Eyes Off the Road (EOR) detection [8]–[10].

MediaPipe utilizes BlazeFace, a feature extraction network that approximates a mesh of the face geometry, and estimates 468 3D-facial landmarks defining the face's features in real-time. BlazeFace is based on MobileNet-V1/V2 and uses Depthwise Separable Convolution (DSC) layers. DSC splits computation into two stages - filtering and combining with an activation function in between. This approach improves network efficiency and speed while maintaining accuracy and robustness. Additionally, MediaPipe's Iris ML solution was deployed to isolate the eye area and estimate bounding boxes around the eyes. These bounding boxes are then passed to a smaller regression model for accurate iris landmarks detection. The iris landmarks are then used to determine the eye center. Subsequently, the eyelid landmarks (P_1 to P_6) and iris landmarks are employed to estimate the eye state [4].

B. Eye Status Detection

Drowsiness can be detected through eye movements, blinks, and pupil contrast reactions. Computer vision algorithms analyze eye movement patterns and head pose to differentiate between normal blinking and drowsiness. The average frequency of blinks typically ranges from 10 to 15 blinks per minute, with a duration of 100 to 150 milliseconds. However, during drowsiness, the frequency of blinks can decrease and their duration may increase, resulting in prolonged eye closures lasting more than 5 seconds [5], [6]. Detecting drowsiness involves analyzing blink time, velocity, and frequency to determine changes in eye state. This can be achieved by estimating the Eye Aspect Ratio (EAR) using eye landmark positions as illustrated in Fig. 1. The EAR quantifies the real-time openness of the eyes based on measurements of horizontal and vertical eye positions which can be computed with Equation (1) [4]:

$$EAR = \frac{\|P_2 - P_6\| + \|P_3 - P_5\|}{2\|P_1 - P_4\|} \tag{1}$$

Subsequently, the MediaPipe Iris model utilizes eight extra landmarks positioned around the iris contours to determine the eye center and estimate values for Eyes Off the Road (EOR). The model establishes a range of 0.40 to 0.60 as an

979-8-3503-8083-5/23 $31.00 © 2023 IEEE

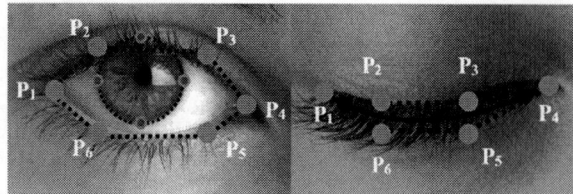

Fig. 1. The eyelid landmarks P_1 through P_6 are marked in filled green circles, whereas the iris landmarks are marked in unfilled green circles.

indicator of a driver paying attention to the road. When the ratio falls outside this range, it signifies an EOR state, implying a distracted driver [4], [6], [11]. The ratio is determined by Equation (2) as follows:

$$EOR = \frac{d}{D} \qquad (2)$$

Where d is the distance from the outer corner of the eye (P_1) to the eye center, and D is the total distance between the inner and the outer corners of the eye $\overline{P_1 P_4}$ [4].

C. Head Pose Estimation

For drivers to effectively maintain awareness of their car's surroundings, it is necessary for them to continuously move their heads and eyes. While tracking eye gaze with high accuracy in an unconstrained environment is feasible, implementing eye gaze estimation becomes challenging under constrained conditions. These limitations include variations in light conditions, occlusions, and non-frontal faces. HPE plays a crucial role in overcoming these limitations by analyzing complex head gestures.

The employed MediaPipe solution detects the 2D face area and returns (X, Y, Z) coordinates that can be used to estimate head pose angles. The estimation is achieved by solving the Perspective-n-Pose (PnP) problem, which determines translation and rotation using Euler angles. The Euler angles, representing pitch, yaw, and roll allow for the determination of the orientation of the head's movement. The rotation matrix is defined by the triplet of Euler angles (α, β, γ), which combine three rotations around the (x, y, z) axes, as depicted in Equation (3) [12], [13].

$$R = \begin{bmatrix} cos\gamma & sin\gamma & 0 \\ sin\gamma & cos\gamma & 0 \\ 0 & 0 & 1 \end{bmatrix} \begin{bmatrix} cos\beta & 0 & -sin\beta \\ 0 & 1 & 0 \\ sin\beta & 0 & cos\beta \end{bmatrix} \begin{bmatrix} 1 & 0 & 0 \\ 0 & cos\alpha & -sin\alpha \\ 0 & sin\alpha & cos\alpha \end{bmatrix} \quad (3)$$

The derivative of the rotation matrix with respect to Euler's angles can then be used for PnP estimation. The PnP problem starts by slicing the facial area of the 2D image and placing it in the 3D metric space. The relation between the 2D and the 3D coordinate systems for one point is given in Equation (4) [12], [13]:

$$\begin{bmatrix} u \\ v \\ 1 \end{bmatrix} = \begin{bmatrix} f_x & 0 & c_x \\ 0 & f_y & c_y \\ 0 & 0 & 1 \end{bmatrix} \begin{bmatrix} r_{11} & r_{12} & r_{13} & t_1 \\ r_{21} & r_{22} & r_{23} & t_2 \\ r_{31} & r_{32} & r_{33} & t_3 \end{bmatrix} \begin{bmatrix} X \\ Y \\ Z \\ 1 \end{bmatrix} \quad (4)$$

Where (u,v) are the coordinates of the projection points in pixels, $[u, v, 1]^T$ is the 2D image taken by the camera, A = [f_x 0 c_x; 0 f_y c_y; 0 0 1] is the input camera matrix while (c_x,c_y) from within the A matrix are the principal points at the image center, (f_x, f_y) from within the A matrix are the focal lengths expressed in pixel units, (X, Y, Z) are the coordinates of a 3D point in the World Coordinate Space (WCS), the Z coordinate is the depth value present while capturing the image, and [R|t] is the estimated camera matrix (perspective), i.e., 3D rotation and translation of the camera [12].

Once these parameters are initialized, the corresponding landmarks obtained in each frame to perform HPE can be found. The 3D world gaze estimation can then be obtained by substituting the rotation matrix with the PnP solution by two-sided quaternion multiplication as in Equation (5) and (6) [13]:

$$\begin{pmatrix} 0 \\ \vec{N_d} \end{pmatrix} = {}_B^N q_k \cdot \begin{pmatrix} 0 \\ \vec{B_d} \end{pmatrix} \cdot {}_B^N \dot{q}_k + \begin{pmatrix} 0 \\ \vec{N_t} \end{pmatrix} \qquad (5)$$

$$\vec{N_d} = \begin{pmatrix} x & y & z \end{pmatrix}^T \qquad (6)$$

Where $\vec{N_d}$ is the 3D gaze estimation in the WCS, the head orientation estimation ${}_B^N q_k$ is the Kalman filter output quaternion, ${}_B^N \dot{q}_k$ is the initial alignment quaternion, and $\vec{N_t}$ is the HPE transformed into the WCS [13].

IV. SYSTEM IMPLEMENTATION

An embedded DMS captures real-time images of the driver's face, which are processed using hardware and software modules. To be compatible with different automotive environments, the DMS must meet several essential requirements. It should be stable and able to operate without the need for calibration for each driver or system configuration. The system must accurately detect the driver's facial image under varying lighting conditions without obstructing their vision. Additionally, it should function effectively regardless of head movements, facial expressions, or any eye occlusions. Finally, adhering to automotive operating standards and incorporating durable hardware for data acquisition, as well as reliable software for data processing, are essential elements in the design of the proposed DMS illustrated in Fig. 2.

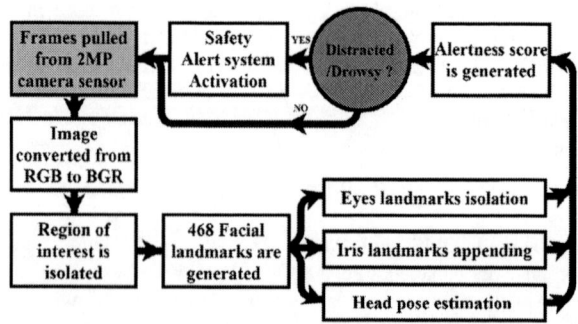

Fig. 2. Architecture of the Driver Monitoring System [4].

Fig. 3. The PCB of the seat ECU showcasing the diverse integrated components.

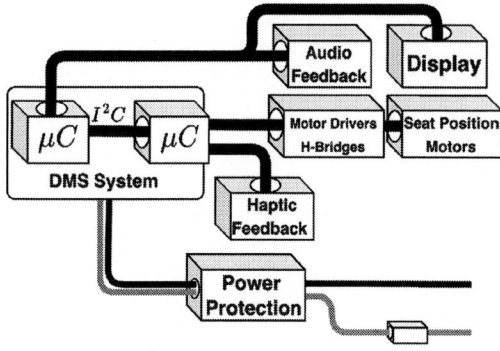

Fig. 4. Block diagram illustrating the integration of the seat ECU and DMS systems.

A. Data Acquisition

The data acquisition module employs a 2MP-CMOS camera sensor featuring an IR cut-off filter and Infrared LEDs. This setup enables the capture of high-quality images even in diverse lighting conditions, including low-light situations such as nighttime. With a resolution of 1920x1080 pixels and a wide viewing angle, the camera ensures detailed visuals while maintaining focus at an ideal working distance. It is strategically mounted on the steering column to avoid obstructing the driver's field of view.

B. Data Processing Module

The designed DMS system utilizes a low-cost Single Board Computer (SBC) integrated with the car's seat ECU for data processing. The SBC features a quad-core ARM Cortex-A73 CPU cluster running at up to 1.8 GHz and a dual-core Cortex-A53 cluster running at up to 1.9 GHz, providing powerful performance. It is equipped with 4GB DDR4 RAM running at 1320 MHz, powered by the seat ECU at 12V DC. The SBC is housed in a large metal casing designed as a heat sink to optimize CPU and RAM heat dissipation while adhering to Electromagnetic Compatibility (EMC) requirements. The system's capabilities and responsiveness were enhanced by overclocking the SBC's CPU, increasing its clock rate beyond its original design [14]. The SBC's heatsink effectively manages the additional generated heat, providing proper cooling for the system. As a result, the DMS system maintained stability, operated at an increased speed, and enhanced performance.

The seat ECU depicted in Fig. 3 is design to manage various functions for a power automotive seat. The ECU uses a Microcontroller (MC) to control seat motor drivers, display data, and monitor seat control switches. To optimize cost, the system employs six H-bridges to control five motors. The motors are equipped with integrated custom Hall sensors, allowing for the storage of individual seat positions. This integration simplifies the system, and reduces power loss and cable costs while also enabling bidirectional data transmission.

To ensure protection, the system incorporates safety measures such as overheating and reverse polarity protection. The seat ECU is connected to the DMS, providing safety and comfort. It interacts with the DMS to alert the driver of distracted or drowsy driving by initiating seat vibration. The structure of the system, its interfaces, data flow, and component communication are illustrated in Fig. 4.

C. Data Processing Software

The DMS algorithm is implemented on an SBC running the Linux OS embedded operating system. The algorithm is developed using Python-based scripting and utilizes various libraries, including MediaPipe for facial extraction, OpenCV for drawing and picture morphing, and Numpy and SciPy for auxiliary functions. For more detailed information on the implemented algorithm, the reader is referred to [[4],alg1-alg6]. The depicted Fig. 5 illustrates the system flowchart, which showcases the data path within the DMS and the decision-making process at different levels.

V. PERFORMANCE EVALUATION

Developing a computer vision algorithm is just one part of the broader product design process, with rigorous testing across various driving scenarios as a major challenge. The performance evaluation of the designed DMS involved measuring driver behaviors in a stationary car under different illumination conditions as shown in Fig. 6. Constrained conditions such as facial expressions and occlusions were also assessed. The driver followed specific actions, including varying head angles and gaze directions, while looking at different target locations during daytime and nighttime.

To accurately determine driver drowsiness, the EAR is calculated using Equation (1). A minimum EAR threshold of 0.25 is manually calibrated, considering eyes open if the EAR is above this value. If the EAR remains below this threshold for more than two seconds, the system triggers an alert, taking into account the elimination of eye blinks as potential false positives. Additionally, the iris location estimation and a refined gaze threshold contribute to a more reliable system. The EOR

2023 International Conference on Microelectronics (ICM)

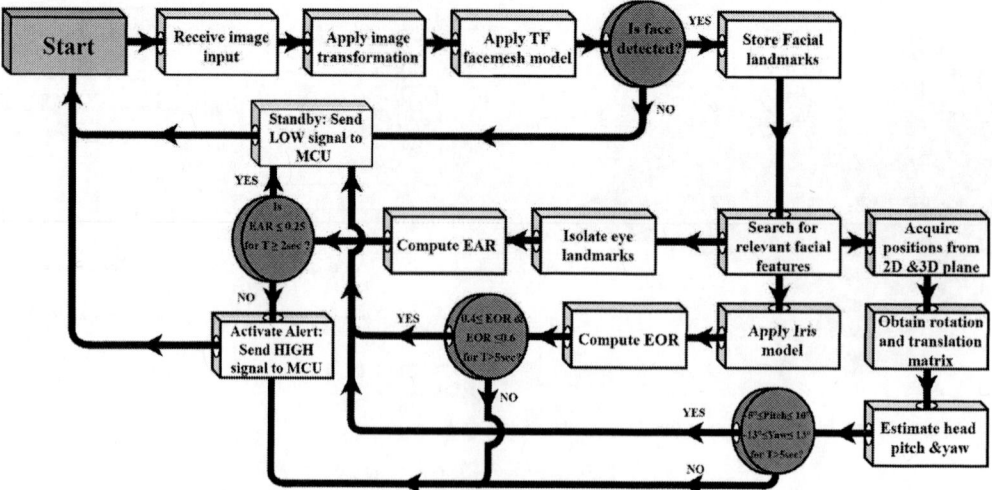

Fig. 5. Data Flow and Decision-making in the DMS system.

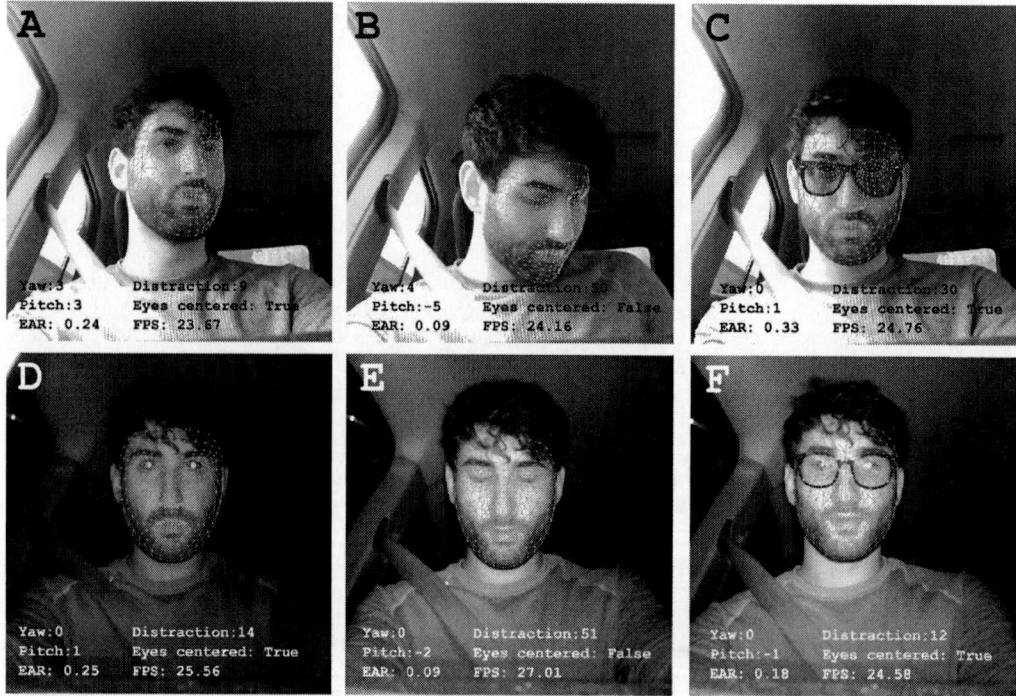

Fig. 6. Real-time HPE and Eye Status Detection. Daytime: (A) Driver looking on the road, (B) Driver distracted and (C) Driver wearing sunglasses. Nighttime: (D) Driver looking forward, (E) Driver exhibiting drowsiness , (F) Driver wearing eyeglasses.

parameter allows the system to account for deviations from the optimal eye behavior range (0.40-0.60) defined by Equation (2) where any deviation from the range activates the warning system. The head pitch range is defined as 10° upward and 5° downward, while the head yaw range spans 13° to the left and 13° to the right relative to the intrinsic camera parameters. The time frame during which the EOR and/or Head pose must stay within prescribed thresholds is configured as a five-second countdown prior to the activation of the alert system. This ensures improved accuracy and eliminates the possibility of false positive alert signals.

Ensuring the robustness of a device is of paramount importance for designers. This concern should be tackled through a range of design measures, including considerations for stress, variation effects, as well as environmental and usage conditions [19]. the physical robustness of the seat ECU was tested through drop tests, water ingress tests, and temperature tests. Integration of the Seat ECU and DMS, along with an audible warning signal, effectively restored driver alertness. Lastly, The designed safety and comfort system is compact and universal which allows for easy retrofitting in any vehicle.

VI. CHALLENGES AND FUTURE IMPROVEMENT

Onboard DMS plays a crucial role in enhancing transportation safety and management. However, monitoring driver alertness under demanding driving conditions is a challenging task.

A. Constrained Conditions

The complexity and variability of facial expressions pose challenges to the efficacy of DMS. Intense facial expressions and occlusions can result in false positive alerts. However, implementing a time limit helped filtering out most of these expressions. The presence of eye occlusion, particularly when a driver is wearing polarized sunglasses, can hinder the accuracy of eye tracking algorithms. However, the utilized hybrid model based on both eye status algorithms and HPE can still rely on head movement algorithms to detect drowsiness and distraction. Additionally, The DMS performance can decrease during low illumination conditions such as nighttime and shadowy driving. To address this, an infrared camera is used to provide efficient illumination without impacting the driver's vision. Finally, the system's accuracy during nighttime driving is consistent with the daytime.

B. Future Improvement

Future deployments of advanced DMS and ADAS hold promise for enhancing road safety. A key improvement could be deploying our algorithms within an OS container on an SBC device, enhancing consistency, resource efficiency, scalability, and security across diverse vehicle systems. To further enhance these systems, improvements should encompass integrating multiple sensors, optimizing vehicle responses, and universalizing the use of HMI and DMS technologies. Efforts must focus on improving accuracy and reducing false alerts, striking a balance between sensitivity and precision. Furthermore, integrating facial recognition could bolster theft prevention by restricting access to authorized users only. Additionally, combining DMS with ADAS for autonomous safety responses is another exciting future prospect.

VII. CONCLUSION

This research work emphasizes the importance of DMS in enhancing road safety by detecting drowsiness and distractions. The DMS and seat ECU systems were integrated into a simple, compact unit that triggers a driver warning audio signal and seat vibration if the driver is drowsy or distracted, using data from various sensor technologies. The integration of DMS with ADAS and other automotive systems, results in increased accuracy and robustness, reduced power consumption, fewer electronic components, and a smaller overall area. The developed system achieved a maximum detection rate of 45 FPS in real-time, unaffected by factors such as time of day, facial expressions, or eye occlusions. This low-cost, universal system has the potential to significantly improve road safety if implemented in most vehicles.

REFERENCES

[1] Bieber, Christy. "Distracted Driving Statistics and amp; Facts In 2023." Forbes Advice, Edited by Brette Sember, 23 Feb. 2023.

[2] Fernández, A., Usamentiaga, R., Carús, J. L., & Casado, R. (2016). Driver distraction using visual-based sensors and algorithms. Sensors, 16(11), 1805.

[3] Hermann S. Driver monitoring–new challenges for smart sensor-based systems. Stud Health Technol Inform. 2004;108:103-10. PMID: 15718636.

[4] Sharara, Leila, et al. "A Real-Time Automotive Safety System Based on Advanced AI Facial Detection Algorithms." IEEE Transactions on Intelligent Vehicles (2023).

[5] Voulodimos, Athanasios, et al. "Deep learning for computer vision: A brief review." Computational intelligence and neuroscience 2018 (2018).

[6] Vicente, Francisco, et al. "Driver gaze tracking and eyes off the road detection system." IEEE Transactions on Intelligent Transportation Systems 16.4 (2015): 2014-2027.

[7] Wang, Jian-Gang, and Eric Sung. "Study on eye gaze estimation." IEEE Transactions on Systems, Man, and Cybernetics, Part B (Cybernetics) 32.3 (2002): 332-350.

[8] Murphy-Chutorian, Erik, and Mohan Manubhai Trivedi. "Head pose estimation in computer vision: A survey." IEEE transactions on pattern analysis and machine intelligence 31.4 (2008): 607-626.

[9] Rakhmatulin, Ildar, and Andrew T. Duchowski. "Deep neural networks for low-cost eye tracking." Procedia Computer Science 176 (2020): 685-694.

[10] Voulodimos, Athanasios, et al. "Deep learning for computer vision: A brief review." Computational intelligence and neuroscience 2018 (2018).

[11] Kuwahara, Akihiro, et al. "Eye fatigue estimation using blink detection based on Eye Aspect Ratio Mapping (EARM)." Cognitive Robotics 2 (2022): 50-59.

[12] Rocca, François, Matei Mancas, and Bernard Gosselin. "Head pose estimation by perspective-n-point solution based on 2d markerless face tracking." Intelligent Technologies for Interactive Entertainment: 6th International Conference, INTETAIN 2014, Chicago, IL, USA, July 9-11, 2014. Proceedings 6. Springer International Publishing, 2014.

[13] Wöhle, Lukas, and Marion Gebhard. "Towards robust robot control in cartesian space using an infrastructureless head-and eye-gaze interface." Sensors 21.5 (2021): 1798.

[14] Hikmat, Ahmed, and Azween Abdullah. "The Impact of Overclocking the CPU to the Genetic Algorithm." Ijcsns 9.5 (2009): 175.

[15] S. Zhang, Y. Wu, Q. Zhang, and J. Liu, "Real-time driver drowsiness detection via a convolutional neural network and an attention mechanism," in Proceedings of the 10th International Conference on Intelligent Systems and Knowledge Engineering, 2015, pp. 31–38.

[16] Y. Li, Y. Wang, M. Zhao, Y. Liu, and Y. Zhang, "Real-time driver drowsiness detection based on deep belief network and sliding window technique," Neurocomputing, vol. 229, pp. 22–31, 2017.

[17] A. Alam, M. Bhuiyan, and T. R. Savarimuthu, "Real-time driver drowsiness detection using a deep learning framework," IEEE Transactions on Intelligent Transportation Systems, vol. 20, no. 3, pp. 1038–1047, 2019.

[18] C. ADAS, "Understanding ADAS: Driver attention warning and drowsiness detection," Jul 2022

[19] Sharara, L., Navidi, S.M., Al Maharmeh, H., Parekh, S., Wehbi, A., Alhawari, M. and Ismail, M., 2022, October. Analysis and Effects of Aging and Electromigration on Mixed-Signal ICs in 22nm FDSOI Technology. In 2022 29th IEEE International Conference on Electronics, Circuits and Systems (ICECS) (pp. 1-4). IEEE.

979-8-3503-8083-5/23 $31.00 © 2023 IEEE

Development, Optimization, and Application of ML based Modeling of Printed VO$_2$ RF Switch

Ahmad Khusro*, Mohammad Hashmi†, and Muhammed Akmal Chaudhary‡

*Department of Electrical Engineering, IIT Kanpur, Kanpur, India
†School of Engineering and Digital Sciences, Nazarbayev University, Astana, Kazakhstan
3 College of Engineering and Information Technology, Ajman University, Ajman, UAE
E-mail:†ahmadk@iitk.ac.in, †mohammad.hashmi@nu.edu.kz, ‡m.akmal@ajman.ac.ae

Abstract—This paper proposes a globally optimized behavioral modeling algorithm using cascaded feed-forward neural network in conjugation with Particle Swarm Optimization (PSO) for fully printed VO$_2$ based RF switches. The proposed model makes use of varied set of operating conditions (geometric dimensions and operating temperature) over a frequency range of 0.01 to 33 GHz. The modeling algorithm enables flexibility in selecting the optimized hyperparameters such as number of neurons in each hidden layer, and activation function at each layer. The developed model is tested for both interpolation and frequency extrapolation cases up to 40 GHz to establish the validity and robustness of the modeling algorithm. An excellent agreement between the measured and the modeled performance over a broad frequency range demonstrates a good generalization capability and successful model development strategy. The proposed model is then evaluated within a commercial circuit simulator (Keysight's ADS) to demonstrate usefulness in RF circuit design.

Index Terms—Neural Networks, PSO, Printed VO$_2$ RF Switch, MATLAB Cosimulation, ADS Ptolemy

I. INTRODUCTION

The rapid advancements in wireless standards have led to innovative solutions in the Radio Frequency (RF) front-end circuits, components, and systems [1]–[9]. For example, vanadium di-oxide (VO$_2$) has emerged as an alternate for the design of reconfigurable RF switches [10]. It is a Metal-Insulator-Transition (MIT) switch in which VO$_2$ behaves like a dielectric at room temperature and then metal above 68 °C. However, VO$_2$ switch's generic model is not yet available for use in commercial computer-aided design (CAD) environment. This limits the usefulness of VO$_2$ switch in the design, simulation and analysis of high frequency circuits. Recently, there is a lot of emphais on the use of machine learning (ML) in the modeling of devices [9], [11]–[17] and modeling of VO$_2$ switch can benefit from the ML based approach.

In this paper, the cascaded feed-forward (CFF) configuration of Artificial Neural Network (ANN) is used for modeling the printed VO$_2$ RF switch is explored. The inputs are length, width, thickness and temperature for the broad frequency range of 10 MHz to 40 GHz for shunt configuration. The flexibility in modeling algorithm is ensured by using case and switch statements to select the best combination of activation at each layer and number of neurons in hidden layers. Moreover, the weights and bias of the network is adjusted using backpropagation algorithm in conjugation with Particle Swarm optimization (PSO). The robustness and scalability of the

developed model is demonstrated by model testing for set of test inputs including varied dimensions and temperatures, as well as novel set of extrapolated frequency range of 33 GHz - 40 GHz. The model validation shows a good generalization capability for interpolation as well as extrapolation. Furthermore, the proposed model employs the advantage of seamless incorporation in commercial CAD tool. It is then evaluated within the CAD simulator by taking reconfigurable T-resonator band-stop filter as a case study. An excellent agreement between the modeled and measured data demonstrates the effective CAD incorporation of the model.

The subsequent sections highlight RF characterization of the printed VO$_2$ switch, development of CFF based model, training and testing of the model, validation of the proposed technique, and finally the integration of the proposed model into CAD environment for its usefulness in RF circuit design.

II. STEPS FOR THE BEHAVIORAL MODEL DEVELOPMENT OF PRINTED VO$_2$ SWITCH AND MODEL VALIDATION

A. RF Characterization of Printed VO$_2$ switch

It is designed on a co-planar waveguide (CPW) containing a centrally placed conductor which carries signal and is inserted between the two ground planes as shown in Fig. 1(a). In this work, the RF switch characterization is done for shunt configuration shown in Fig. 1(b). The fabricated RF switch on a sapphire substrate has been screen printed using silver nanoparticles and VO_2 nano-particles based inks [18]. Sapphire is chosen as a substrate owing to its low dielectric loss (ε_r = 9.3, tanδ = 0.0001). The RF characterization of the printed RF switch, under electrical and thermal stimuli is performed on a probe station using Vector Network Analyzer (VNA). The printed VO$_2$ switches are located in the probe station with a thermal chuck and connected to VNA through RF probes. Under electrical triggering, the bias current is provided by the current source through a bias tee whereas the temperature is controlled through the chuck temperature. During the ON state, VO$_2$ film acts as a dielectric and, therefore, RF signal passes through the signal trace.

B. Model Development and Optimization using PSO

The proposed electro-thermal model for the printed VO$_2$ RF switch, shown in Fig. 2, is based on cascade feed-forward (CFF) architecture. The modeling of electro-thermal behavior

979-8-3503-8083-5/23 $31.00 © 2023 IEEE

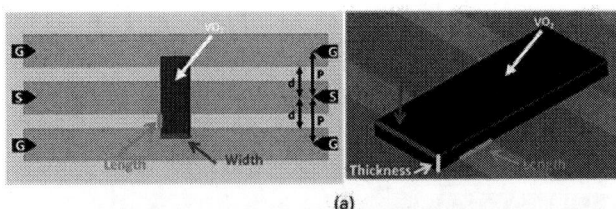

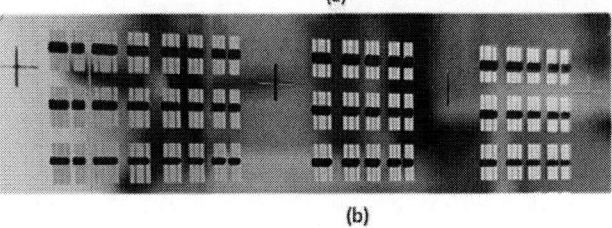

Fig. 1. (a) Design Layout using CPW structure, and (b) Shunt configuration of printed VO$_2$ RF switch

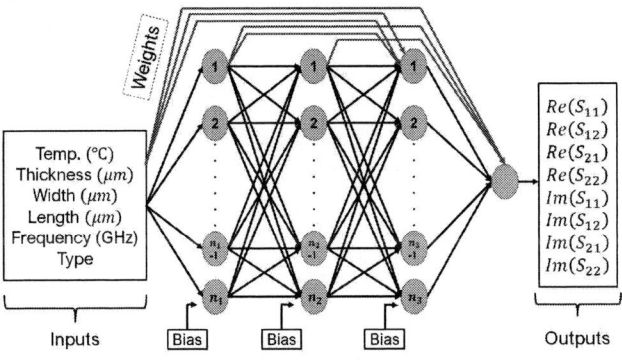

Fig. 2. Proposed CFF Neural Network Model of Switch

of printed VO$_2$ RF switch maps six predictor variables into feature space for various two-port S-parameters. The six predictor variables are: (i) width varying in the range [0.2mm-1.2mm] with a step size of 0.2mm, (ii) length of 0.1mm and 0.15mm, (iii) 0.4mm and 0.6 mm as thickness of the switch, (iv) type of CPW (narrow or wide) used to fabricate the switch, (v) temperature [20°C-100°C] with size of 10°C (excluding the measurement at 30°C), and (vi) frequency range [0.01 GHz-33 GHz]. CFF architecture has relatively extra more potential to extract the hidden insights of complex non-linear relationship very quickly. The relation through the inputs to the outputs of the CFF neural network architecture based printed RF switch model can be described using (1)-(4), where x_i is the input of m signals, w_{ji}^h represents the weight linked to node j for input i of layer (h), b_j^h represents the bias weight linked to node j of layer (h), f_j^h is the activation function of node j of layer h and O_j^h represents the output of node j for layer h.

$$\varphi_j^{(1)} = b_j^{(1)} + \sum_{i=1}^{n} w_{ji}^{(1)}.x_i \quad (1)$$

$$O_j^{(1)} = f_j^{(1)}(\varphi_j^{(1)}) \quad (2)$$

$$\varphi_j^{(h)} = b_j^{(h)} + \sum_{p=1}^{h-1}\sum_{k=1}^{n} O_{jk}^{(p)}.w_{jk}^p + \sum_{i=1}^{m} w_{ji}^{(h)}.x_i \quad (3)$$

TABLE I
PARAMETER INITIALIZATION

PSO Parameters		Particle's Range	
Parameters	Value	Parameters	Range
Population Size	n =100	Weights and bias	[-3,3]
Acceleration	$c_1 = 1.8$	Activation Function	Tan-sigmoid (TS), Log-sigmoid (LS),
Coefficients	$c_2 = 2.1$		Elliott Sigmoid (ES), Pure Linear (PL)
Inertia Weight	$\omega = 1.5$	No. of neurons	[15,30]

TABLE II
OPTIMIZED HYPERPARAMETERS AND PREDICTION ACCURACY

Parameters/ Response	Optimized Parameters and Test Error		
	NNs (n1, n2, n3)	AFs (1, 2, 3, 4)	Test Error
$Re(S_{11})$	(20, 20, 21)	(T-S,T-S,T-S,PL)	6.997e-5
$Im(S_{11})$	(22, 18, 17)	(L-S,T-S,L-S,PL)	6.996e-5
$Re(S_{12})$	(19, 19, 21)	(T-S,T-S,T-S,PL)	6.992e-5
$Im(S_{12})$	(20, 20, 20)	(L-S,T-S,L-S,PL)	6.997e-5
$Re(S_{21})$	(18, 19, 18)	(T-S,T-S,T-S,PL)	6.943e-5
$Im(S_{21})$	(20, 20, 20)	(L-S,T-S,L-S,PL)	7.171e-5
$Re(S_{22})$	(21, 21, 22)	(T-S,T-S,T-S,PL)	6.995e-5
$Im(S_{22})$	(23, 22, 22)	(L-S,T-S,T-S,PL)	6.985e-5

$$O_j^{(h)} = f_j^{(h)}(\varphi_j^{(h)}) \quad (4)$$

For an accurate modeling of the RF switch using ANN, exact number of neurons (NNs) in the hidden layer and activation functions (AFs) at each stage still remains an open question. Apparently, numerous published literature for RF component modeling uses hit and trial process to determine the hyperparameters which often results in underfit or overfit models [19]. We propose a new method where hyperparameters such as NNs and AFs for each stage along with weights and bias are optimized using particle swarm optimization (PSO) algorithm [20]. PSO is preferred due to its easy implementation, excellent interpretability and its ability to provide the global optimum along with the fast convergence.

The proposed model utilize 3/5th of data for training and the remaining 2/5th is reserved for testing and validation excluding the inputs corresponding to frequency extrapolation range. The parameters that needs to be optimized (particles) and generated in feature space are shown in Table I along with its defined range. The optimization refers to minimizing the error function $E(w)$ i.e. mean square error (MSE). The final output of the CFF based printed RF switch model is evaluated using (5)-(6) which includes the back-propagation algorithm functioning in conjugation with PSO. Eventually, the optimal hyperparameters of CFF based neural network model, given in Table II, of the printed VO$_2$ RF switch are obtained and this completes the development stage.

$$w_{jpn}^{(h)} = w_{jp}^{(h)} + E(w).w_{jp}^{(h)} \quad (5)$$

$$y = f^{(o)}\left(b^{(o)} + \sum_{p=1}^{h-1}\sum_{k=1}^{n} O_{jk}^{(p)}.w_{jk}^{(p)} + \sum_{i=1}^{m} .w_{jin}^{(o)}.x_i\right) \quad (6)$$

C. Model Validation

This section measures the reliability, extrapolation accuracy and prediction ability by calculating $E(w)$ on new set of inputs

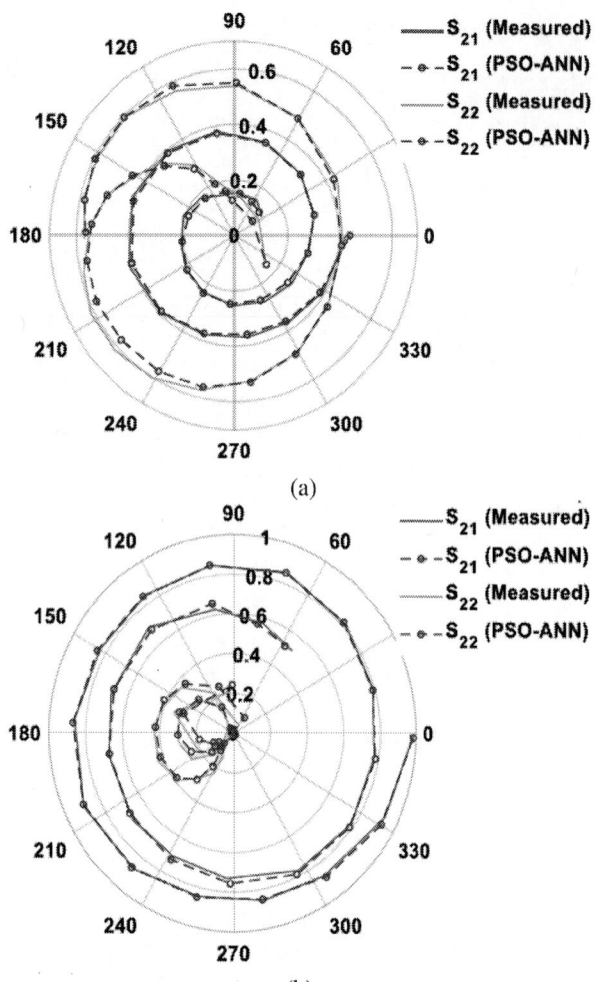

(a)

(b)

Fig. 3. Comparison between the measured data and the proposed CFF-PSO model for VO$_2$ RF switch for a set of test inputs including frequency (33-40 GHz) (a) OFF sate [Temp.=80°C; Wide CPW; Thickness=0.4; Length=0.1; Width=0.8], (b) ON state [Temp.=40°C; Narrow CPW; Thickness=0.6; Length=0.15; Width=0.6]

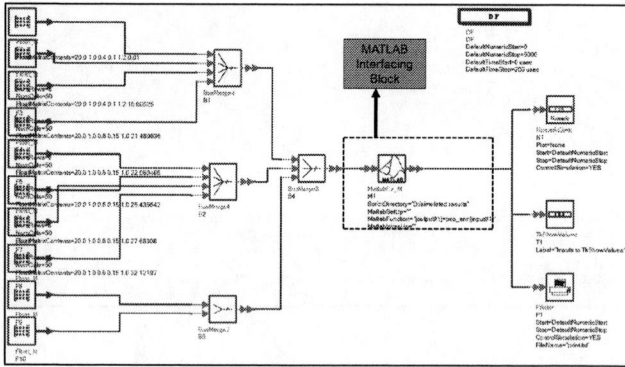

Fig. 4. Demonstration of establishing interface between MATLAB and ADS

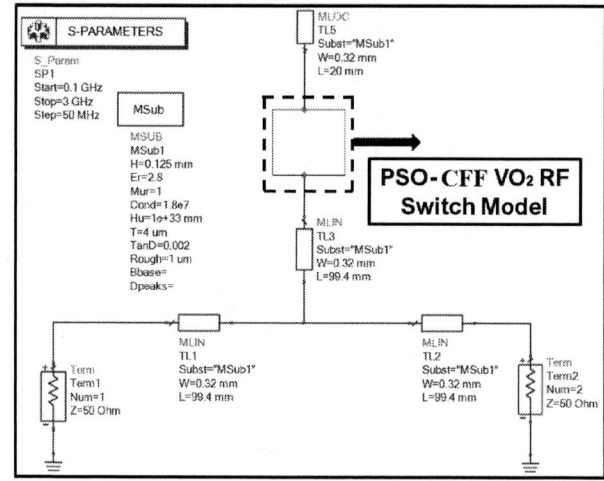

Fig. 5. Utility of the Proposed Model in Bandwidth Reconfigurable T-resonator Circuit Design

for extrapolation frequency range from 33 GHz-40 GHz. The model shows a very good agreement with promising $E(w)$ of approximately 7e-5 on test set (including broad extrapolation frequency range of (33-40 GHz). The model is tested for OFF and ON state with different geometric dimensions. An excellent agreement between the measured and modelled S-parameters for complete frequency range of 0.1-40 GHz is depicted in Fig. 3.

III. CAD INCORPORATION OF THE DEVELOPED MODEL

This section demonstrates the seamless incorporation of the proposed PSO-CFF model developed in MATLAB into circuit simulator (Keysight's ADS) and discusses the utility of model in RF filters. The MATLAB cosimulation of ADS Ptolemy in Advanced Design System (ADS) aids to establish the interface between MATLAB and ADS. To establish the interface, as shown in Fig. 4, MATLAB Interfacing block calls a function developed in MATLAB to which the input is provided to the

trained model and response is recorded using sink. The bus merger is used to combine the large size inputs in a single go as a feeder to MATLAB interfacing block. Numeric start and numeric stop in data flow controller is fixed at 0 and 5000 respectively, since the size of one set of test input samples is 5001. The float matrix block is used to insert the inputs in ADS environment and the response is recorded using the numeric sink.

Taking an example of RF filter, employing a single behavioral model of VO$_2$ RF switch, as shown in Fig. 5, the substrate is characterized with a thickness of 0.125 mm, relative permittivity of 2.8 and loss tangent of 0.002. Two sets of test inputs are provided to the proposed model representing ON (Temp. = 50°C) and OFF (Temp. = 80°C) states of the switch keeping same set of geometric parameters (Thickness = 0.4, Length = 0.1, Width = 1). It can be inferred from the Fig. 6 that the considered T-resonator operates at two different set of frequencies (1.7 GHz and 2.2 GHz) with the same passband insertion loss of approximately 2.55 dB. The operating frequency of the filter is shifted to lower value under OFF state and it is due to the length of filter getting extended and vice-versa in the ON state of the switch. It is pertinent to mention

2023 International Conference on Microelectronics (ICM)

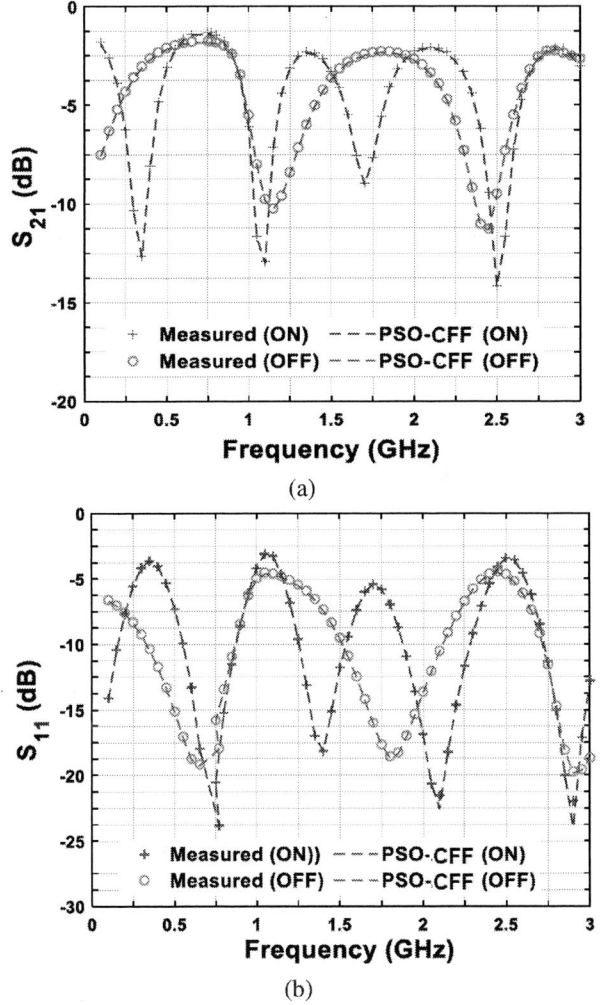

(a)

(b)

Fig. 6. Filter characteristics deployed with the VO_2 switch model for the test set inputs (Temp. =50°C; Thickness=0.4; Length=0.1; Width=1) in ON state and (Temp. =100°C; Thickness=0.6; Length=0.15; Width=0.8) OFF state

that the simulation and analysis of the design using proposed model takes only few seconds which is relatively low when compared to the EM simulation that takes around 8-9 minutes to generate the response.

IV. CONCLUSION

In this paper, PSO optimized flexible CFF neural network model is presented for printed VO_2 based switch. This research work overcomes the earlier hit and trial based method of assigning the values to hyperparameters. Instead, the proposed modeling technique backed by PSO chooses the optimum set of hyperparameters from a vast library in a single run, making the approach faster, less complex and accurate. The robustness of the proposed model is evaluated by testing it under various input for frequency extrapolation cases. The proposed model is robust as it shows a good extrapolation accuracy for a range of 7 GHz under ON and OFF state. Finally, the paper demonstrates the easy deployment of the proposed VO_2 based

RF switch model in CAD tool environment for the design and analysis of bandwidth reconfigurable T-resonators.

V. ACKNOWLEDGEMENT

This work was funded by grant # AP19677597 from MHES, Republic of Kazakhstan. The authors thank the IMPACT lab, KAUST for providing the VO_2 characterization data.

REFERENCES

[1] M. S. Hashmi *et al.*, "Electronic multi-harmonic load-pull system for experimentally driven power amplifier design optimization," *IEEE MTT-S Int. Microw. Symp., Boston, USA*, pp. 1549-1552, June 2009.
[2] Z. Zhao *et al.*, "A fast small signal modeling method for GaN HEMTs," *Solid-State Electronics*, vol. 175, pp. 107946, 2021.
[3] Y. S. Noh and I. B. Yom, "A Linear GaN High Power Amplifier MMIC for Ka-Band Satellite Communications, *IEEE Microwave and Wireless Component Letters*, vol. 26, no. 8, pp. 619-621, Aug. 2016.
[4] K. Dautov, et al, "Compact multi-frequency system design for SWIPT applications," *International Journal of RF and Microwave Computer-Aided Engineering*, vol. 31, no. 6, e22632, June 2021.
[5] M. S. Hashmi and F. M. Ghannouchi, "A flexible dual-inflection point RF predistortion linearizer for microwave power amplifiers," *Progress in Electromagnetics Research C*, vol. 13, pp. 1-18, 2010.
[6] M. A. Maktoomi and M. S. Hashmi, "A coupled-line based L-section DC-isolated dual-band real to real impedance transformer and its application to a dual-band T-junction power divider," *Progress in Electromagnetics Research C*, vol. 55, pp. 95-104, 2014.
[7] A. M. Zaidi, et al., "Dual-Band Design Techniques for Microwave Passive Circuits: A Review and Applications," *IEEE Microwave Magazine*, vol. 23, Iss. 7, pp. 61-77, July 2022.
[8] M. A. Maktoomi and M. S. Hashmi, "A Performance Enhanced Port Extended Dual-Band Wilkinson Power Divider," *IEEE Access*, vol. 5, pp. 11832-11840, June 2017.
[9] P. Zhao and K. Wu, "Homotopy Optimization of Microwave and Millimeter-Wave Filters Based on Neural Network Model," *IEEE Transactions on Microwave Theory & Techniques*, pp. 1–11, 2020.
[10] E. A. Casu et al., "Shunt capacitive switches based on VO2 metal insulator transition for RF phase shifter applications," *47th European Solid-State Device Research Conference*, Leuven, pp. 232-235, 2017.
[11] A. Khusro, M. S. Hashmi, and A. Q. Ansari, "Enabling the development of accurate intrinsic parameter extraction model for GaN HEMT using support vector regression (SVR)," *IET Microwaves, Antennas & Propagation*, vol. 13, no. 9, pp. 1457-1466, Jul. 2019.
[12] J. Cai et al., "Bayesian Inference-Based Behavioral Modeling Technique for GaN HEMTs," *IEEE Transactions on Microwave Theory and Techniques*, vol. 67, no. 6, pp. 2291-2301, June 2019.
[13] S. Husain, M. Hashmi and F. M. Ghannouchi, "Comprehensive Investigation and Comparative Analysis of Machine Learning-Based Small-Signal Modelling Techniques for GaN HEMTs," *IEEE J. Electron Devices Soc.*, vol. 10, pp. 1015-1032, 2022.
[14] A. Khusro *et al.*, "A Generic and Efficient Globalized Kernel Mapping-Based Small-Signal Behavioral Modeling for GaN HEMT," *IEEE Access*, vol. 8, pp. 195046-195061, 2020.
[15] A. Majumder, et al., "Optimization of small-signal model of GaN HEMT by using evolutionary algorithms," *IEEE Microw. Wireless Compon. Lett.*, Vol. 27, No. 4, pp. 362-364, Apr. 2017.
[16] L. Zhai et al, "A reliable parameter extraction method for the augmented GaN high electron mobility transistor small-signal model," *Int. J. RF Microw. Comput.-Aided Eng.*, Dvol. 32, no. 8, pp. e23210, 2022.
[17] A. Khusro, et al, "An accurate and simplified small signal parameter extraction method for GaN HEMT," *International Journal of Circuit Theory and Applications*, Vol. 47, Issue 6, pp. 941-953, June 2019.
[18] S. Yang, M. Vaseem, and A. Shamim, "Fully Inkjet-Printed VO2-Based Radio-Frequency Switches for Flexible Reconfigurable Components," *Adv. Mater. Technol.*, vol. 4, no. 1, p. 1800276, 2019.
[19] H. T. Nguyen and A. F. Peterson, "Machine Learning for Automating the Design of Millimeter-Wave Baluns," *IEEE Transactions on Circuits and Systems I: Regular Papers*, vol. 68, no. 6, pp. 2329-2340, June 2021.
[20] Y. Sun, et al., "A Particle Swarm Optimization-Based Flexible Convolutional Autoencoder for Image Classification," *IEEE Trans. Neural Nets. and Learning Systems*, vol. 30, no. 8, pp. 2295-2309, Aug. 2019.

979-8-3503-8083-5/23 $31.00 © 2023 IEEE

2.45GHz Low-Power Diode Bridge Rectifier Design

KOUBAR Gabriel
Aix-Marseille Université CNRS,
Université de Toulon, IM2NP
UMR7334, 13397
Marseille, France
Lebanese University
Beirut, Lebanon
gabriel.koubar@im2np.fr

HADDAD Fayrouz
Aix-Marseille Université CNRS,
Université de Toulon, IM2NP
UMR7334, 13397
Marseille, France
fayrouz.haddad@im2np.fr

BOUCHRA Nessakh
Aix-Marseille Université CNRS,
Université de Toulon, IM2NP
UMR7334, 13397
Marseille, France
bouchra.nessakh@im2np.fr

RAHAJANDRAIBE Wenceslas
Aix-Marseille Université CNRS,
Université de Toulon, IM2NP
UMR7334, 13397
Marseille, France
wenceslas.rahajandraibe@im2np.fr

SADEK Sawsan
Lebanese University
Doctoral school of science and
technologies Lebanese University
Beirut, Lebanon
sawsansadek70@gmail.com

Abstract—This article focuses on the design of full-wave bridge rectifiers and the comparative evaluation of two Schottky diodes' performance (HSMS2850 and SMS7630) for energy harvesting under low-power conditions. It also explores various impedance matching configurations. The system operates at a frequency of 2.45 GHz, commonly used in Wi-Fi applications, and investigates two scenarios: one with an input power of -5 dBm and the other at -15 dBm. The findings indicate that the highest achieved Power Conversion Efficiency (PCE) is 57% for the -5 dBm scenario and 33% for the -15 dBm scenario, both exhibiting S11 values below -30dB.

Keywords—Rectenna, RF energy harvesting, Rectifier, Full wave bridge, Schottky diodes.

I. INTRODUCTION

In an increasingly connected world, wireless communication has become an integral part of our daily lives. From smartphones and laptops to smart home devices and IoT sensors, the demand for seamless connectivity is ever-present. However, powering these wireless devices with conventional batteries poses challenges in terms of efficiency, environmental impact, and maintenance costs. To address these challenges, researchers and engineers have turned their attention to a promising solution known as Wi-Fi energy harvesting. This innovative approach aims via rectennas to harness the ambient radio frequency (RF) energy present in our environment and convert it into usable electrical power [1]. By tapping into the existing Wi-Fi networks that permeate our urban and industrial areas, a sustainable and cost-effective source of energy can be potentially unlocked. Wi-Fi energy harvesting reduces our reliance on disposable batteries, eases the need for frequent replacements.

Moreover, Wi-Fi energy harvesting aligns with the growing global emphasis on energy efficiency and sustainability. By converting unused RF energy into usable power, we can reduce our carbon footprint and contribute to the development of greener and smarter cities.

Additionally, it opens up possibilities for powering low-energy devices, such as wireless sensors, in remote or hard-to-reach locations where traditional power sources are limited or impractical.

In this work, different full-wave bridge rectifiers are designed and compared using ADS (Advanced Design System). They use Schottky diodes (HSMS2850 and SMS7630) with diverse impedance matching topologies. In section 2, the state-of-the-art of rectenna systems is presented. The design of the bridge rectifier and the obtained results are discussed and compared in section 3. Conclusions are given in section 4.

II. RECTENNA SYSTEM

The primary components of a rectenna system, as illustrated in "Fig. 1", consist of five main blocks.

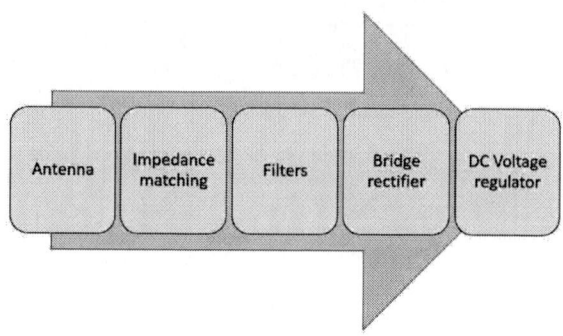

Fig. 1. Rectenna system overview.

Firstly, there is an antenna designed to capture and collect power with a significant gain level. This serves as the power collector for the system.

Secondly, a matching impedance network is employed to connect the output of the antenna to the input of the remaining blocks of the rectenna. The purpose of this network is to ensure optimal power transmission.

Next, a series of filters is incorporated into the system if needed, to eliminate any unwanted direct current (DC) and alternative current (AC) signals that could potentially disrupt the functionality of the bridge elements. These filters help to maintain the integrity of the signals within the rectenna system.

The fourth block consists of a full-wave bridge rectifier, responsible for converting the incoming AC signal into a double-alternance signal.

Finally, a voltage regulator is employed stabilize the converted signal, transforming it into a DC signal. This voltage regulator ensures a consistent and reliable power output from the rectenna system.

Overall, the combination of these five blocks forms the fundamental structure of a rectenna system, allowing for efficient energy collection and voltage regulation for various applications.

Recent works have focused on enhancing the performance of rectifiers through employing relatively higher input power (Pin). In [2], for Pin of 13 dBm, the Power Conversion Efficiency (PCE) is 81%. In [3], for Pin of 27 dBm, the PCE is 75% and in [4], for Pin of 6 dBm, the PCE is only 45%. Furthermore, for low input power of 0 dBm, in [5], they obtained a PCE of 46%. In [6], a PCE of 35% is achieved for Pin of -10 dBm and in [4], 20% is achieved for -6 dBm.

The main objective of this research work is to investigate the performance of a rectenna system employing Schottky diodes (excluding the antenna). The study specifically focuses on comparing the power energy conversion efficiency of the rectifier operating at a frequency of 2.45 GHz, in order to position our study at power levels lower than the references mentioned earlier, specifically below 0 dBm. Two low power levels of -5 dBm and -15 dBm have been selected. By conducting a thorough analysis, we have identified HSMS2850 and SMS7630 diodes as the most suitable choices for this study due to their low threshold voltage in order to achieve optimal performance.

III. SIMULATION RESULTS

The design and study process of the bridge rectifier consisted of the following steps:

1) The SPICE Model parameters of the diode components HSMS2850 and SMS7630 were incorporated into ADS (Advanced Design System) by referring to their respective datasheets, as shown in Table 1.

TABLE 1. SMS7630 AND HSMS2850 SPICE MODEL PARAMETERS

	SMS7630	HSMS2850
Rs (ohm)	20	25
Vf (V)	0.240	0.250
Cj0 (pF)	0.14	0.18
Vb (V)	2	3.8

2) A simulation was set up in ADS (Advanced Design System), starting with a bridge rectifier configuration and voltage regulator. This served as the baseline for subsequent tests (cf. "Fig.2").

3) Simulations were conducted to evaluate the performance of the Bridge rectifier configuration.

4) In order to replicate the circuit for lab measurements, impedance matching solutions were introduced into the circuit. Two approaches were explored using either discrete components or microstrip lines.

5) Additional simulations were performed to assess the effectiveness of the impedance matching solutions and compare the results with the baseline configuration.

A. Bridge rectifier baseline simulation.

The Power Conversion Efficiency is commonly calculated by dividing the DC power output by the AC power input and expressing it as a percentage.

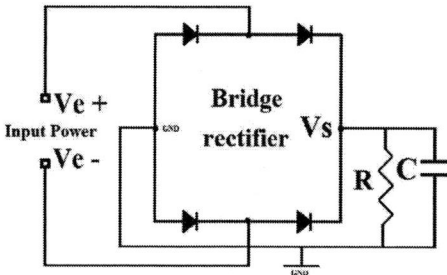

Fig. 2. Bridge rectifier baseline topology.

Following a series of simulations aimed at maximizing efficiency, the results indicated an optimal load resistance of 3000 Ohms when using diode SMS7630 (cf. "Fig. 3") and 2900 Ohms with diode HSMS2850, both with an input power of -15 dBm. Consequently, a load resistance of 3000 Ohms was chosen as the optimal configuration.

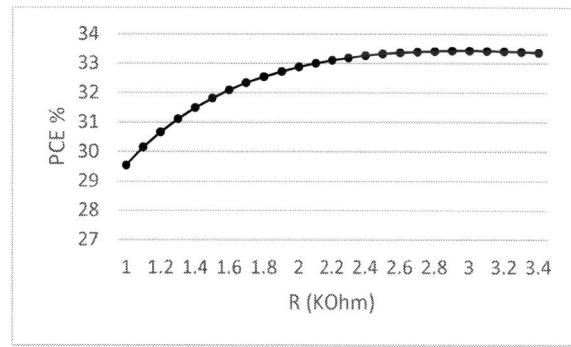

Fig. 3. The PCE as a function of the load resistance (R) using SMS7630.

To rectify and stabilize the voltage at the output, a capacitance of 100 pF was added, the results are presented in Table 2.

TABLE 2. PCE OF BASELINE TOPOLOGY

	HSMS2850		SMS7630	
Input Power	-15 dBm	-5 dBm	- 15 dBm	-5 dBm
PCE	24%	53%	33%	60%

B. Bridge rectifier with discrete components impedance matching simulation.

Initially, the impedance matching is achieved using discrete components.

Matching simulations were performed for both LC and CL configurations using ADS (cf "Fig. 4"). The observed impact on the PCE was less than 1% difference between the two types.

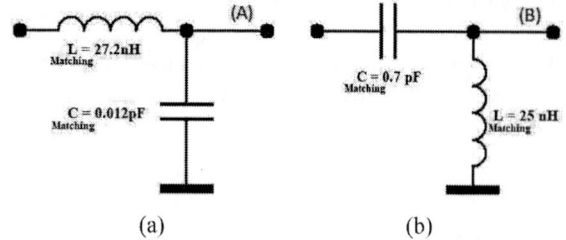

Fig. 4. Impedance matching topologies: (a) LC for SMS7630 and (b) CL for HSMS2850.

For the HSMS2850 Schottky diode, the CL configuration has exhibited better performance, while for the SMS7630, the LC configuration (cf. "Fig. 5") demonstrated better results.

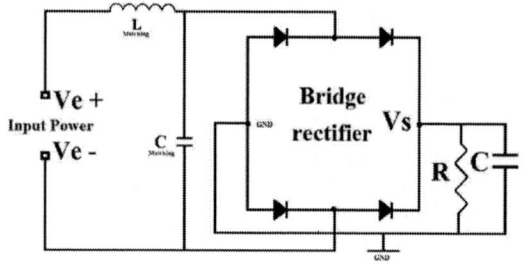

Fig. 5. Rectifier system using LC for impedance matching.

Table 3 illustrates the obtained outcomes.

TABLE 3.   PCE AND S11 COMPARISON BETWEEN THE TWO DIODES USING DISCRETE COMPONENTS FOR IMPEDANCE MATCHING

	HSMS2850		SMS7630	
Input Power (dBm)	-15	-5	- 15	-5
Discrete components	CL		LC	
PCE	23%	49%	33%	57%
S11	< -30 dB			

C. Bridge rectifier Microstrip impedance matching simulation.

The five key factors that come into play when choosing a substrate are:

1) Dielectric Constant (ε_r): It is an important parameter to consider when selecting a high-frequency circuit substrate. For high-frequency circuits like 2.45 GHz, a substrate with a dielectric constant of at least 3.5 is recommended.

2) Loss Tangent (tan δ): It is a measure of the energy lost when passing through a substrate. Substrates with low loss tangent exhibit better performance at high frequencies. For 2.45 GHz applications, it is advisable to choose a substrate with a loss tangent below 0.02.

3) Thermal Stability: It is another important parameter to consider when selecting a substrate for high-frequency circuits. Substrates capable of withstanding high temperatures are more suitable for high-frequency circuits as they help maintain stable electrical characteristics.

4) Board Thickness: It should be chosen based on the operating frequency of the circuit and the characteristic impedance of the transmission line used in the circuit. For operation at 2.45 GHz, a board thickness ranging from 0.8 to 1.6 mm is recommended.

5) Cost: It is an important factor to consider when selecting a substrate for high-frequency circuits. Higher-performing substrates tend to be more expensive, so it is crucial to find a balance between performance and cost.

TABLE 4.   RF POOL RO4350B SUBSTRATE PARAMETERS

Substrate	ε_r	tan δ	Thickness (mm)	Frequency
RO4350B	3.66	0.0031	1.54	> 500 MHz

After analyzing various substrates and considering the above-mentioned recommendations, we have opted for RO4350B as our selected choice (cf. Table 4).

Table 5 presents the results obtained for both diode types using ideal microstrip lines and microstrip lines utilizing the Roger RO4350B substrate. The data reveals an average difference of 3 to 4 % between the two lines types.

TABLE 5.   PCE AND Microstrip line

	HSMS2850		SMS7630	
Input Power (dBm)	- 15	- 5	-15	- 5
PCE with Ideal line	18%	43%	33%	56%
PCE with RO4350B line	16%	40%	29%	52%
S11 with RO4350B line	< -40 dB			

D. Impedance matching types comparison

Table 6 presents a summary of the power conversion efficiency for both diodes using the two different types of impedance matching.

As presented in Table 6 and "Fig. 6" it is evident that the SMS7630 diode exhibits superior performance compared to the HSMS2850 diode. Additionally, the discrete components utilized in the impedance matching circuit demonstrate lower losses and higher PCE in comparison to the microstrip lines configuration.

TABLE 6.   OVERALL SUMMARY OF THE DIFFERENT IMPEDANCE MATCHING OPTIONS VERSUS THE BASELINE CONFIGURATION

	HSMS2850		SMS7630	
Input Power (dBm)	-15	-5	- 15	-5
PCE (Baseline)	25%	53%	33%	60%
PCE (Baseline matched by discrete components)	23%	49%	**33%**	**57%**
PCE (Baseline matched by Microstrip lines)	16%	40%	29%	52%

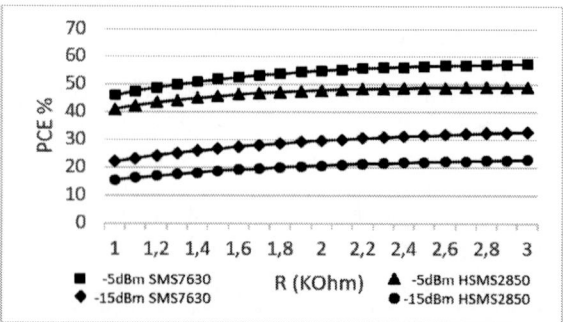

Fig. 6. PCE of both Schottky diodes with Pin of -5 dBm and -15 dBm matched with discrete components.

Table 7 provides a comparison of the achieved power conversion efficiency results with those reported in other literature works for different input power levels. It can be noted that obtained results are very competitive to the actual state-of-the-art.

TABLE 7. OUR CIRCUIT PCE VS OTHER REFERENCES

Ref	Pin	PCE
[5]	-5 dBm	41 %
	-15 dBm	27 %
[4]	-6 dBm	20 %
This work	**-5 dBm**	**57 %**
	-15 dBm	**33 %**

E. Transistors Bridge rectifier.

To increase the bridge rectifier efficiency, other technologies were explored [7]. A comprehensive primary examination led to the choice of the 130nm CMOS technology based on its dimensional characteristics, intended operating frequency, and minimal power consumption, all aimed at optimizing the efficiency of the bridge.

Using the Cadence Virtuoso tool, a series of circuit simulations were carried out. This process initiated with the characterization of transistors and proceeded with an investigation into half-wave rectifiers, followed by a comparison of three distinct full-wave bridge rectifier configurations:

- Configuration 1: Four transistors type NMOS, all connected in diode mode as in "Fig.7".

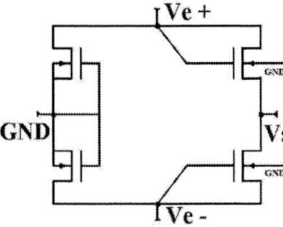

Fig. 7. Four transistors in diode mode.

- Configuration 2: Four transistors type NMOS, with two of them operating in diode mode and the other two in switch mode as in "Fig. 8".

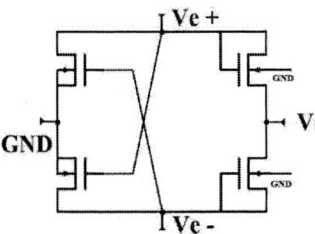

Fig. 8. Four transistors, two in diode mode and two in switch mode.

- Configuration 3: Two NMOS and two PMOS, all of which are in switch mode as in "Fig. 9"

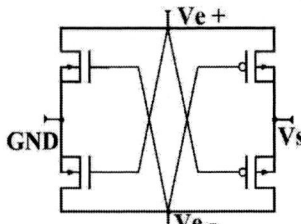

Fig. 9. Four transistors in switch mode.

Configuration 3 demonstrates that transistors operating in switch mode exhibit reduced power consumption and enhanced efficiency at low input power levels, achieving a PCE of 69% at -5 dBm and 43% at -26 dBm. This represents a notable efficiency improvement when compared to discrete components (Schottky diode), as illustrated in Table 8.

TABLE 8. CMOS BRIDGE PCE VS OTHER REFERENCES

Ref	Pin	PCE
[5]	-5 dBm	41 %
	-15 dBm	27 %
[4]	-6 dBm	20 %
This work	-5 dBm	57 %
	-15 dBm	33 %
CMOS bridge rectifier [7]	**-5 dBm**	**69%**
	-26 dBm	**43%**

IV. CONCLUSION

This paper introduces a promising bridge rectifier design that incorporates Schottky diodes, exhibiting satisfactory conversion performance at power levels below 0 dBm suitable to energy harvesting applications. The rectifier achieves a power conversion efficiency of 57% (and 33%) at -5 dBm (and -15 dBm) respectively. The rectifier is now under fabrication and will be measured in the laboratory to validate the simulation results.

It is worth emphasizing the importance of the utilization of MMIC CMOS technology with low threshold voltage to enhance the performance of the bridge rectifier and minimize power consumption reaching PCE of 69% at -5 dBm.

REFERENCES

[1] I. D. Bougas, M. S. Papadopoulou, K. Psannis, P. Sarigiannidis, and S. K. Goudos, *"State-of-the-Art Technologies in RF Energy Harvesting Circuits - A Review,"* Symposium on communication Engineering, 2020. DOI: 10.1109/WSCE51339.2020.9275507.

[2] M. A. Halimi, D. Surender, and T. Khan, *"Design of a 2.45 GHz operated Rectifier with 81.5% PCE at 13 dBm Input Power for RFEH/WPT Applications,"* IEEE Indian Conference on Antennas and Propagation, 2021. DOI: 10.1109/InCAP52216.2021.9726346.

[3] S. A. Rotenberg, S. K. Podilchak, P. D. H. Re, C. Mateo-Segura, G. Goussetis, and J. Lee, *"Efficient Rectifier for Wireless Power Transmission Systems,"* IEEE Trans. Microw. Theory Tech., vol. 68, no. 5, 2020. DOI: 10.1109/TMTT.2020.2968055.

[4] D. Khan *et al.*, *"A 2.45 GHZ high efficiency CMOS RF energy harvester with adaptive path control,"* Electronics, vol. 9, no. 7, 2020. DOI: 10.3390/electronics9071107.

[5] E. Coskuner and J. J. Garcia-Garcia, *"Metamaterial impedance matching network for ambient rf-energy harvesting operating at 2.4 GHz and 5 GHz,"* Electronics, vol. 10, no. 10, 2021. DOI: 10.3390/electronics10101196.

[6] D. Vital, S. Bhardwaj, and J. L. Volakis, *"Textile-Based Large Area RF-Power Harvesting System for Wearable Applications,"* IEEE Transaction on Antennas and Propagation, vol. 68, no. 3, 2020. DOI: 10.1109/TAP.2019.2948521.

[7] G. Koubar, F. Haddad, S. Sadek, R. W. Rahajandraibe, "Low power CMOS bridge Rectenna," *2022 IEEE Conference on Antenna Measurements and Applications (CAMA),* Guangzhou, China, 2022, pp. 1-5. DOI: 10.1109/CAMA56352.2022.10002549.

A 0.3-V 10-nW CMOS OTA with Feedforward Body-Driven Structure

Hirokazu Yoshizawa and Naoki Inoue
Graduate School of Information Systems
Saitama Institute of Technology
Fukaya, Saitama 369-0293, Japan
yoshiz_h@sit.ac.jp

Abstract—**A 0.3-V 10-nW rail-to-rail CMOS operational transconductance amplifier (OTA) with a feedforward body-driven structure is proposed. It has a two-stage structure with a class-AB output stage. To achieve operations with a supply voltage of 0.3 V, the first-stage differential amplifier circuit does not use a tail current source but uses two PMOS current sources to provide a constant current to the body-driven NMOS transistors of each input pair. Simulation results show that the proposed OTA has a DC gain of 53 dB, a unity-gain frequency of 2.9 kHz, and a phase margin of 51° for a supply voltage of 0.3 V with a capacitive load of 30 pF using SPICE parameters for a 0.18-μm CMOS process. For common-mode input voltages ranging from 0.036 to 0.230 V, the DC gain is higher than 40 dB. The silicon area is 0.0048 mm². The power consumption of the circuit is 10 nW, resulting in a high figure of merit.**

Keywords—operational transconductance amplifier, low voltage, low power consumption, feedforward body-driven

I. INTRODUCTION

The number of battery-powered portable electronic devices, such as smartphones, has increased in recent years, making it crucial to reduce the power consumption of their circuits. Furthermore, energy harvesting circuits, such as solar power and thermal power generation, are becoming more common. Some biomedical applications also require low power operation.

To respond to this global trend, operational transconductance amplifiers (OTAs) are also required to operate with low power consumption. OTAs operating with supply voltages below 1 V have been investigated for over 20 years [1-11]. Particularly in the last decade, there have

been numerous reports of OTAs that operate with a power supply voltage of 0.5 V or less [3-11].

In this study, we report a power-efficient CMOS OTA operating at a supply voltage of 0.3 V with a feedforward body-driven structure.

II. PROPOSED STRUCTURE

Fig. 1 depicts the proposed OTA circuit for a 0.3-V supply operation. In this circuit, the tail current source is removed from the differential amplifier used in the first stage of the commonly used two-stage OTAs.

Instead of a tail current source, M3 and M4 provide a constant bias current to the input devices M1 and M2. M1 and M2 are in a current mirror configuration, and differential input voltages are applied to their body terminals. Consequently, M1–M4 constitute the first gain stage. M5–M7 form a class-AB output stage with M8. They also form a feedforward path from the input to the output. Note that the body terminals of M7 and M8 in the output stage are also connected to the input voltages V_{inp} and V_{inn}, respectively, to increase the gain of the proposed OTA. We call this a feedforward body-driven structure.

Fig. 2 shows a block diagram of the proposed OTA. Similar to the block diagram shown in [10], we adopted the multipath zero compensation [12] to remove the unfavorable right-half-plane zero. g_{mf} denotes the feedforward transconductance and consists of M5–M8. g_{mb1} denotes the body transconductance of M1, and g_{m8} denotes the transconductance of M8. R_1 and R_2 denote the output resistances of the input and output stages, respectively. C_1 denotes the output capacitance of the input stage, and C_L denotes the load capacitance.

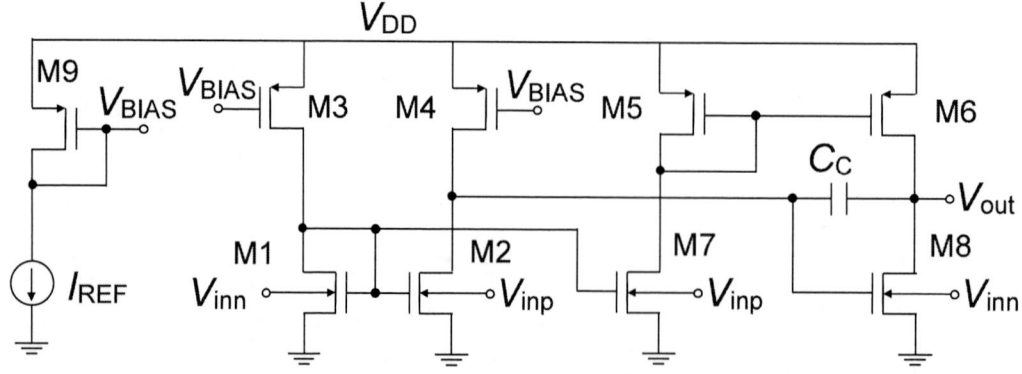

Fig. 1 Proposed low-voltage OTA.

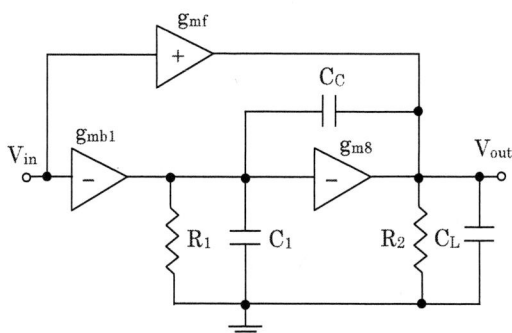

Fig. 2   Block diagram of the proposed OTA.

TABLE I.   TRANSISTOR SIZES

MOS transistors	W/L sizes
M1–M4, M5, M7	15 μm/1 μm
M6, M8	10 × 15 μm/1 μm

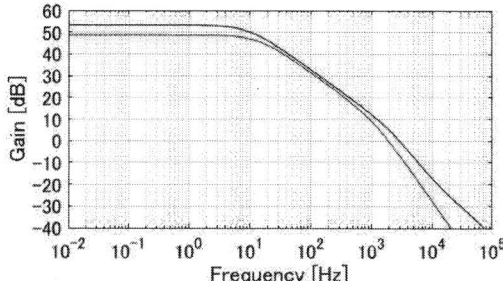

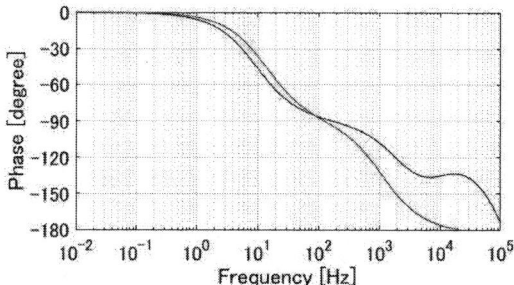

Fig. 3   Gain (above) and phase (below).

The small signal open-loop transfer function of the proposed OTA is given by

$$\frac{V_{out}}{V_{in}} \cong \frac{A_{DC}\left[1 + s\dfrac{(g_{mf} - g_{mb1})C_C}{g_{mb1}g_{m8}}\right]}{(1 + sg_{m8}R_1R_2C_C)\left(1 + s\dfrac{C_L}{g_{m8}}\right)}, \quad (1)$$

where $A_{DC} = g_{mb1}g_{m8}R_1R_2 + (g_{mf} - g_{mb1})R_2$, $R_1 = 1/(g_{d2} + g_{d4})$, $R_2 = 1/(g_{d6} + g_{d8})$, and g_{mf} is given by

$$g_{mf} = \frac{1}{2}\left(g_{mb1}\frac{g_{m7}g_{m6}}{g_{m1}g_{m5}} + g_{mb7}\frac{g_{m6}}{g_{m5}} + g_{mb8}\right). \quad (2)$$

Two poles and a left-half-plane (LHP) zero in (1) are given as follows:

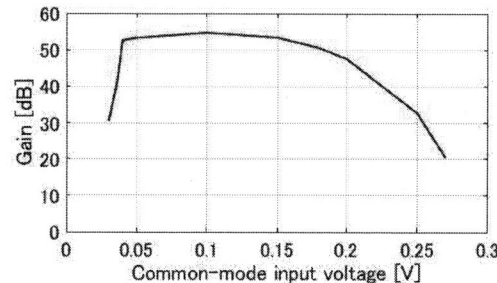

Fig. 4   DC gain for common-mode input voltage.

$$p_1 \cong -\frac{1}{g_{m8}R_1R_2C_C}, \quad (3)$$

$$p_2 \cong -\frac{g_{m8}}{C_L}, \quad (4)$$

$$z_1 \cong -\frac{g_{mb1}g_{m8}}{(g_{mf} - g_{mb1})C_C}. \quad (5)$$

Because the LHP zero frequency becomes lower than that in [10], it is easier to bring back the phase shift caused by the poles, which improves the phase margin. Furthermore, the number of current paths in the proposed circuit is less than that in [10]. Therefore, the total current becomes smaller for the same unity-gain frequency.

III.   SIMULATIONS

To verify the performance of the proposed circuit shown in Fig. 1, we ran HSPICE simulations using parameters for a 0.18-μm deep n-well CMOS process. The threshold voltages of the NMOS and PMOS transistors were 0.46 and –0.39 V, respectively. The MOS transistor sizes are presented in Table I. The supply voltage was 0.3 V and the compensation capacitor Cc was 0.4 pF. A bias current I_{REF} of 2.5 nA was applied externally.

In Fig. 3, the blue line indicates the AC analysis result of the proposed circuit shown in Fig. 1 (using the feedforward body-driven structure); the red line indicates the AC analysis result when the body terminals of the two NMOS transistors M7 and M8 in Fig. 1 are connected to the ground instead of the input voltages. As shown in Fig. 3, the former circuit connection (the blue line) provides a higher DC gain than the latter (the red line). The former also has a better phase margin than the latter because of its higher g_{mf} and lower LHP zero frequency. The DC gain was 53 dB, the unity-gain frequency was 2.9 kHz, and the phase margin was 51° with a load capacitance of 30 pF. The total current consumption of the circuit was 34 nA.

Fig. 4 illustrates the effect of the common-mode input voltage on the DC gain. A DC gain of more than 40 dB was obtained for common-mode input voltages ranging from 0.036 to 0.230 V.

Fig. 5 shows the difference between the input and output voltages of the proposed OTA in a unity-gain configuration. It is within ±5 mV for the entire input range. Fig. 6 shows the current consumption of the proposed OTA versus common-mode input voltage. When the common-mode voltage exceeds 0.25 V, the current consumption of the OTA significantly decreases, because M6 in the output stage enters its triode region and the current in M6

979-8-3503-8083-5/23 $31.00 © 2023 IEEE

2023 International Conference on Microelectronics (ICM)

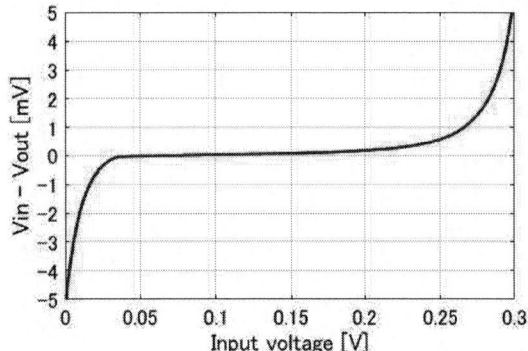

Fig. 5 Difference between the input and output voltages of the proposed OTA in a unity-gain configuration.

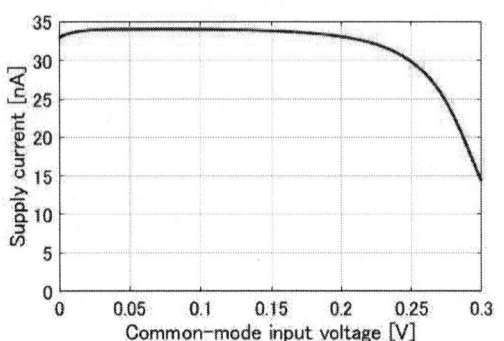

Fig. 6 Supply current of the proposed circuit versus common-mode input voltage.

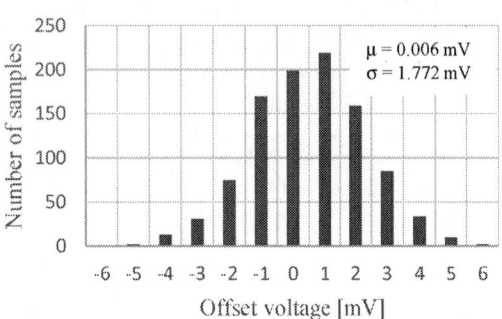

Fig. 7 Offset voltage variation over 1000 Monte Carlo simulations.

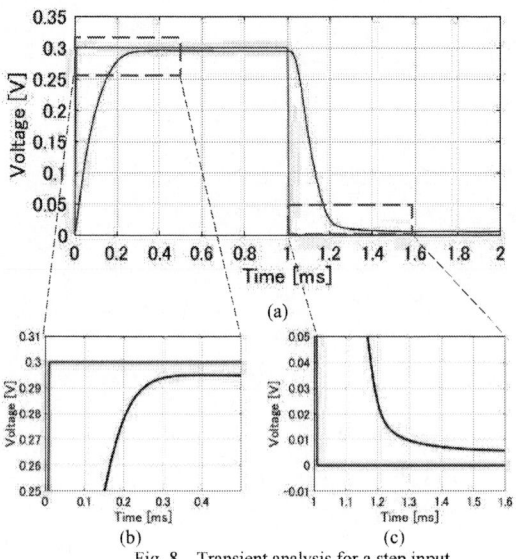

Fig. 8 Transient analysis for a step input.

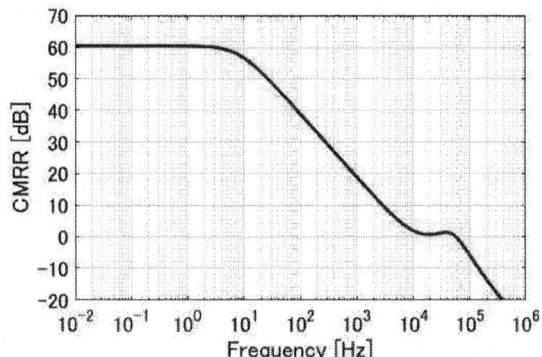

Fig. 9 Common-mode rejection ratio (CMRR).

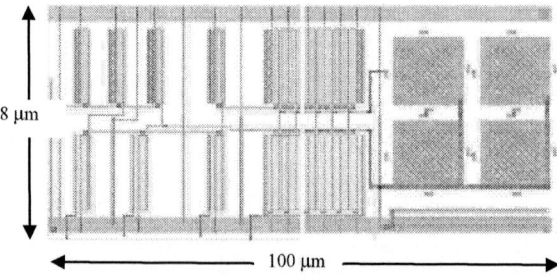

Fig. 10 Layout of the proposed OTA.

TABLE II. PROCESS VARIATION RESULTS

	TT	FF	FS	SF	SS
DC gain [dB]	53	54	51	48	48
Unity-gain frequency [kHz]	2.9	3.0	2.4	2.9	2.3
Phase margin [degrees]	51	52	39	54	40

decreases. On the other hand, M1–M4 in the input stage have a nearly constant current regardless of the common-mode input voltage.

Fig. 7 shows the variation of the offset voltage over 1000 Monte Carlo simulations. We assumed a 3 sigma variation in the threshold voltage difference of the input pair MOS transistors M1 and M2 to be 1 mV. The mean offset voltage was 0.006 mV and its standard deviation was 1.772 mV.

Fig. 8 illustrates the transient analysis for a step input in a unity-gain configuration. The red and blue lines indicate the input and output voltages, respectively. The output swing was from 5.2 to 294.3 mV. The slew rate was 1.4 V/ms for the rise and 1.6 V/ms for the fall. Fig. 9 shows the common-mode rejection ratio (CMRR) of the proposed OTA. The CMRR at DC was 60 dB. Fig. 10 depicts a chip layout of the proposed circuit; the active area is 0.0048 mm².

Table II summarizes the four-corner process variation results for the AC analysis at 25°C. In Table II, the notation F means fast and S means slow. For example, FS indicates a combination of fast NMOS and slow PMOS.

979-8-3503-8083-5/23 $31.00 © 2023 IEEE

TABLE III. TEMPERATURE VARIATION RESULTS

Temperature [°C]	−20	25	70
DC gain [dB]	48	53	48
Unity-gain frequency [kHz]	2.5	2.9	2.6
Phase margin [degrees]	38	51	53

Table III presents the results of temperature variations. The results are almost stable, ranging from −20°C to 70°C. In Table IV, the performance of the proposed OTA is compared with that of state-of-the-art OTAs operating at supply voltages equal to or below 0.5 V. The figure of merit (FoM) of the proposed OTA was calculated using the following equations:

$$FoM_S = \frac{UGF \cdot C_L}{V_{DD} \cdot I_{total}}, \tag{6}$$

$$FoM_L = \frac{SR_{AVG} \cdot C_L}{V_{DD} \cdot I_{total}}, \tag{7}$$

$$FoM_{AS} = \frac{UGF \cdot C_L}{V_{DD} \cdot I_{total} \cdot Area}, \tag{8}$$

$$FoM_{AL} = \frac{SR_{AVG} \cdot C_L}{V_{DD} \cdot I_{total} \cdot Area}, \tag{9}$$

where UGF is the unity-gain frequency, C_L is the load capacitance, SR_{AVG} is the average slew rate, V_{DD} is the supply voltage, I_{total} is the current consumed in the entire circuit, and $Area$ is the chip area of the OTA. The FoM$_{AS}$ for this study, obtained using (8), is higher than other reported results with a DC gain of 50 dB or more.

IV. CONCLUSIONS

We have proposed a 0.3-V 10-nW OTA with a feedforward body-driven structure. The simulation results show that the proposed OTA has a DC gain of 53 dB and a unity-gain frequency of 2.9 kHz for a supply voltage of 0.3 V with a capacitive load of 30 pF. The feedforward body-driven structure is simple and efficient to improve the DC gain and unity-gain frequency.

ACKNOWLEDGMENT

This work was supported by VLSI Design and Education Center (VDEC), the University of Tokyo in collaboration with Synopsys, Inc., Cadence Design Systems, and Rohm Corporation.

REFERENCES

[1] B. J. Blalock, P. E. Allen, and G. A. Rincon-Mora, "Designing 1-V opamps using standard digital CMOS technology," *IEEE Trans. Circuits Syst. II*, vol. 45, no. 7, pp. 769–780, Jul. 1998.

[2] A. D. Grasso, S. Pennisi, G. Scotti, and A. Trifiletti, "0.9-V class-AB miller OTA in 0.35-μm CMOS with threshold-lowered non-tailed differential pair," *IEEE Trans. Circuits Syst. I, Reg. Papers*, vol. 64, no. 7, pp. 1740–1747, Jul. 2017.

[3] T. Kulej, "0.5-V bulk-driven CMOS operational amplifier," *IET Circuits, Devices Syst.*, vol. 7, no. 6, pp. 352–360, Nov. 2013.

[4] E. K. F. Lee, "A Sub-0.5V, 1.5 μW Rail-to-Rail Constant gm Opamp and Its Filter Application," in *Proc. IEEE Int. Symp. on Circuits and Systems*, pp. 197-200, 2012.

[5] Z. Qin, A. Tanaka, N. Takaya, and H. Yoshizawa, "0.5-V 70-nW rail-to-rail operational amplifier using a cross-coupled output stage," *IEEE Trans. Circuits Syst. II, Exp. Briefs*, vol. 63, no. 11, pp. 1009–1013, Nov. 2016.

[6] L. Magnelli, F. A. Amoroso, F. Crupi, G. Cappuccino, and G. Iannaccone, "Design of a 75-nW, 0.5-V subthreshold complementary metal–oxide–semiconductor operational amplifier," *Int. J. Circuit Theory Appl.*, vol. 42, no. 9, pp. 967–977, Sep. 2014.

[7] L. H. C. Ferreira and S. R. Sonkusale, "A 60-dB gain OTA operating at 0.25-V power supply in 130-nm digital CMOS process," *IEEE Trans. Circuits Syst. I, Reg. Papers*, vol. 61, no. 6, pp. 1609–1617, Jun. 2014.

[8] T. Kulej and F. Khateb, "Design and implementation of sub 0.5-V OTAs in 0.18-μm CMOS," *Int. J. Circuit Theory Appl.*, vol. 46, no. 6, pp. 1129–1143, Jun. 2018.

[9] K.-C. Woo and B.-D. Yang, "A 0.25-V Rail-to-Rail Three-Stage OTA With an Enhanced DC Gain," *IEEE Trans. Circuits Syst. II, Exp. Briefs*, vol. 67, no. 7, pp. 1179-1183, Jul. 2020.

[10] T. Kulej and F. Khateb, "A Compact 0.3-V Class AB Bulk-Driven OTA," *IEEE Trans. Very Large Scale Integr. (VLSI) Syst.*, vol. 28, no. 1, pp. 224-232, Jan. 2020.

[11] A. Ballo, A. D. Grasso, S. Pennisi, and G Susinni, "A 0.3-V 8.5-μA Bulk-Driven OTA," *IEEE Trans. Very Large Scale Integr. (VLSI) Syst.*, VOL. 31, no. 9, pp. 1444-1448, Sep. 2023.

[12] R. G. H. Eschauzier and J. H. Huijsing, "An operational amplifier with multipath Miller zero cancellation for RHP zero removal," *Proc. Eur. Solid-State Circuits Conf.*, pp. 122-125, 1993.

TABLE IV. COMPARISON OF OTA PEFORMANCES

	[5]* 2016	[6]* 2014	[7]* 2014	[9]* 2020	[10]* 2020	[11]* 2023	This work**
Supply voltage [V]	0.5	0.5	0.25	0.25	0.3	0.3	0.3
Power [nW]	70	75	18	26	12.6	2550	10
Supply current [nA]	140	150	72	104	42	8500	34
Load capacitance C_L [pF]	40	30	15	15	30	150	30
DC gain [dB]	77	70	60	70	> 64.7	38	53
Unity-gain frequency [kHz]	4.0	18	1.88	9.5	2.96	810	2.9
Phase margin [degrees]	56	55	53	88	52	71.3	51
Slew rate (rise) [V/ms]	2	3	0.6	2	1.9	120	1.4
Slew rate (fall) [V/ms]	2	3	0.8	2	6.4	120	1.6
CMRR (at DC) [dB]	55	N/A	N/A	62.5	110	39.8	60
Silicon area [mm² x10⁻³]	36	57	83	2	8.5	1.06	4.8
FoM$_S$ [MHz · pF/μW]	2.29	7.20	1.57	5.48	7.05	47.65	8.53
FoM$_L$ [V · pF/(μs·μW)]	1.142	1.360	0.584	1.154	9.880	7.06	4.500
FoM$_{AS}$ [MHz · pF/(μW·mm²)]	64	126	19	2740	829	44950	1777
FoM$_{AL}$ [V · pF/(μs· μW·mm²)]	32	24	7	577	1162	6659	938

*experimental results, **simulation results.

A High PSRR CMOS Voltage and Current Reference in One Circuit Without Amplifier for Low Power Applications

Ashutosh Pathy[1], Andleeb Zahra[1], Amir Ahmad[2], and Zia Abbas[1]

[1]CVEST, International Institute of Information Technology, Hyderabad, India, [2]College of Information Technology, United Arab Emirates University, UAE

ashutosh.pathy@research.iiit.ac.in, amirahmad@uaeu.ac.ae, zia.abbas@iiit.ac.in

Abstract— This paper presents a low power voltage reference and current reference in one circuit. The reference voltage is the sum of the complementary to absolute temperature (CTAT) gate-source voltage of a MOSFET operating in the subthreshold region and a scaled version of the proportional to absolute temperature (PTAT) ΔV_{GS} of two MOSFETs operating in subthreshold region. The generated temperature independent reference current is the ratio of the reference voltage and a resistance. The proposed circuit does not use an operational transconductance amplifier (OTA) to generate the reference current, thereby eliminates the variation in current due to any offset of the OTA. Furthermore, the circuit uses a native MOSFET for ameliorating the DC PSRR. The proposed circuit is designed in a 130nm CMOS process. The circuit works with a minimum supply voltage of 1.2V while generating a reference voltage of around 0.7V and a reference current of around 100nA. A TC of 100 ppm/°C for the reference voltage and a TC of 108 ppm/°C for the reference current are noted for mismatch and process variations from Monte-Carlo simulations for 300 samples. The circuit consumes only 240nW of power, which makes it suitable for ultra-low power applications.

Keywords— CTAT, Line sensitivity, Native MOSFET, PTAT, PSRR, Subthreshold region

I. INTRODUCTION

Wireless sensor networks (WSN) are often deployed in uninhabited locations for the collection and transmission of different types of sensitive information. Such situations are characterized by limited resources leading these networks to be powered by energy harvesters. Solar energy, piezoelectric energy, and thermal energy are a few examples of energy that can be harvested. The harvested energy is stored in a rechargeable battery that provides a very small amount of power for low energy applications. Such low power systems employ duty cycling technique in which the active mode is short-term and the sleep mode is long-term. Still, a stable reference signal (current, voltage, frequency) is required for both the modes. Therefore, it is highly desired to reduce the power consumption of the reference circuit down to nano-watt level.

Recently, various types of voltage reference and current reference circuits operating in the subthreshold region were proposed [1-8]. The CTAT voltage required for the reference voltage is generated with the gate-source voltage of a MOSFET. To generate the required PTAT voltage, either a NMOS composite pair is used [1] or the $\Delta V_{GS}/R$ PTAT current of a beta-multiplier is allowed to flow through a larger resistor which is a scaled version of the resistor used in the beta-multiplier circuit

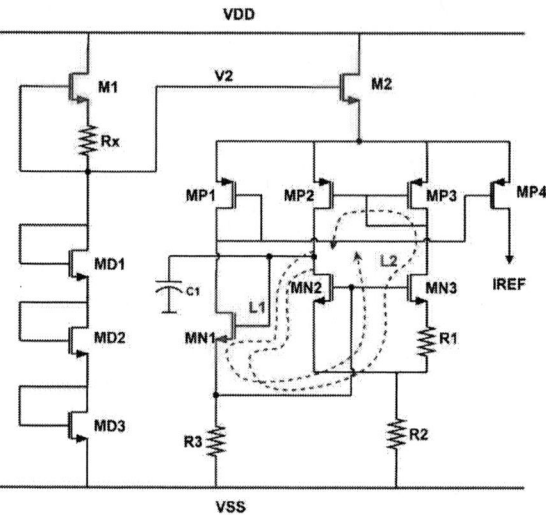

Fig. 1. Simplified Schematic of the proposed all-in-one reference circuit

[4]. Furthermore, there are two mainstream methods to generate a reference current for low power applications.

In the first method, an OTA is used to generate a voltage drop across a resistor [9-10]. In the second method, PTAT currents and CTAT currents are generated separately and then added up to generate a temperature independent current [11]. The former method results in a wider variance of the reference current due to OTA's offset and the latter method is power hungry.

The proposed circuit uses the second method to generate the PTAT voltage while avoiding the mentioned mainstream methods of generating the reference current. Instead, the circuit uses a negative feedback loop inside the circuit's core to generate the reference current. Therefore, the reference current achieves low variance and low power operation. For improving the supply regulation, the proposed circuit makes use of native NMOS transistors. One significant benefit of using native MOSFET is its zero-threshold voltage which enables the circuit to always start. Furthermore, zero threshold voltage enables the circuit to work closer to the supply as it avoids the V_{GS} drop of a standard V_T device if used otherwise.

The rest of the paper is detailed as follows. Section II explains the proposed architecture. Section III presents the simulated results. Finally, conclusions are drawn in section IV.

979-8-3503-8083-5/23 $31.00 © 2023 IEEE

II. PROPOSED ALL-IN-ONE REFERENCE

The proposed reference circuit (Fig. 1) consists of a PSRR enhancement circuit and the reference core. The operating principle of the circuit is explained as follows:

A. MOSFET's V_{GS} in Subthreshold Region

In subthreshold region, a MOSFET can work analogously to a BJT and its current in the subthreshold region is given by [1]:

$$I_D = Is \left(\frac{W}{L}\right) exp \left(\frac{V_{GS}-V_{th}}{\eta V_T}\right) \left\{1 - exp\left(\frac{-V_{DS}}{V_T}\right)\right\} \quad (1)$$

For $V_{DS} > 4V_T$, I_D is almost independent of V_{DS} and is given by:

$$I_D = Is \left(\frac{W}{L}\right) exp \left(\frac{V_{GS}-V_{th}}{\eta V_T}\right) \quad (2)$$

From (2), gate-source voltage (V_{GS}) can be written as:

$$V_{GS} = V_{th} + \eta V_T \, ln \left(\frac{I_D}{Is\frac{W}{L}}\right) \quad (3)$$

where, $Is = \mu_n(\eta - 1)C_{ox}V_T^2$; V_{th} is the threshold voltage of the transistor which has a CTAT nature, V_T is the thermal voltage , μ_n is mobility, η is the subthreshold slope factor.

B. Operating Principle of Reference Circuit

The proposed voltage reference circuit is a modified Brokaw cell [12] where the BJTs have been replaced with the MOSFETs (MN2 and MN3). All the MOSFETs in the circuit operate in the subthreshold region. From Fig. 1, the voltage across resistor R1 is given by:

$$V_{R1} = V_{GS}^{MN2} - V_{GS}^{MN3} \quad (4)$$

V_{GS}^{MN2} and V_{GS}^{MN3} are the gate-source voltage of MN2 and MN3. Using (3) and (4), V_{R1} can be written as:

$$V_{R1} = \eta V_T \, ln \left(\frac{I_{MN2} \, S_{MN3}}{I_{MN3} S_{MN2}}\right) \quad (5)$$

$$V_T = \frac{KT}{q} \quad (6)$$

Where, S_{MN2} and S_{MN3} are the aspect ratio of MN2 and MN3. From equation (5-6) it can be noted that the voltage across the resistor R1 is PTAT in nature. Also, the voltage (V_{R2}) across the resistor R2 can be written as:

$$V_{R2} = 2 \frac{VR1}{R1} R2 \quad (7)$$

The required reference voltage (*VREF)* is the sum of VR2 and the gate-source voltage of the MOSFET MN2. VR2 is PTAT in nature and V_{GS}^{MN2} is CTAT in nature. The size of the resistor R2 is adjusted such that VREF is temperature independent (Fig. 3).

The proposed all-in-one reference circuit does not follow the mainstream method of using an OTA to generate the reference current. Instead, it employs a feedback loop consisting of MN1 to drop the reference voltage across a resistor (R3). Such technique is power and area efficient. Furthermore, one can avoid the variation of the generated current due to the offset of an OTA. The operation of the feedback loops can be explained as follows. The circuit (Fig. 1.) consists

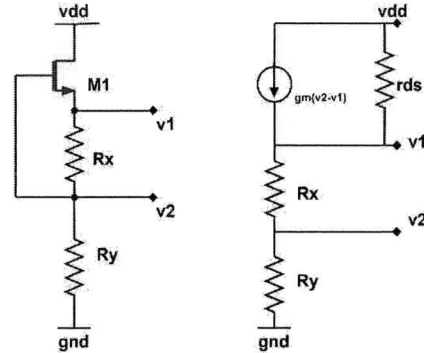

Fig. 2. Equivalent small signal circuit for the biasing circuit

of a positive feedback loop (L2) and a negative feedback loop (L1). The gain of the positive feedback loop is rather small compared to the gain of the negative feedback loop due to degeneration at the source of MN3. Also, there is only one high impedance node (drain of MN2). Therefore, a small capacitor (C1) at the drain of MN2 is good enough to ensure the stability of the loop.

C. PSRR Enhancement Circuit

PSRR is one of the most critical specs for a voltage reference since it provides the reference signal to a low drop-out regulator under which many critical blocks operate in a power management IC. If the supply of the voltage reference is connected directly to the source of MP2 and MP3, any disturbance at the supply will directly affect the reference voltage. Therefore, it is necessary to isolate the supply from the source of PMOS transistors. A native MOSFET (M2) is used to isolate the supply in the proposed architecture.

In our proposed circuit, the gate of the native MOSFET (M2) is biased with an auxiliary circuit consisting of a native MOSFET (M1) and the diode connected devices (MD1-MD3). The native MOSFET M1 acts as a current source when a resistor is connected between its gate and source. The biasing circuit provides a voltage which is equal to $3V_{GS}$. For the reference core to have good PSRR, the biasing circuit must be immune to changes in the supply voltage. Since the gate of the MOSFET M1 is isolated from the supply, the biasing circuit has a high PSRR of 50dB. The DC PSRR (ratio of v2 and supply) of the biasing circuit can be derived from its small signal equivalent circuit (Fig. 2.) as follows:

$$PSRR = 20 \, log \left(\frac{R_y}{R_x + R_y + rds + gm*rds*R_x}\right) \quad (8)$$

Where, Ry is the resistance offered by the diode connected MOSFETs (MD1-MD3).

D. Minimum Value of Supply Voltage

The primary advantage of using a native device (M2) is that it does not incur the unnecessary gate-source voltage drop like that of a higher threshold voltage device [13]. Therefore, the voltage supplied to the reference circuit is closer to the main supply voltage. The minimum supply voltage required for proper operation of the proposed circuit can be expressed as:

$$VDD,min \approx V_{REF} + V_{GS} + 2*V_{dsat} \quad (9)$$

979-8-3503-8083-5/23 $31.00 © 2023 IEEE

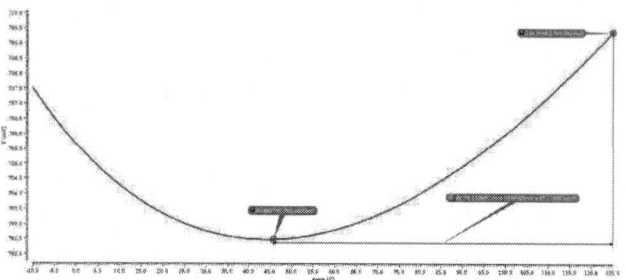

Fig. 3. Reference voltage (VREF) across temperature in nominal corner

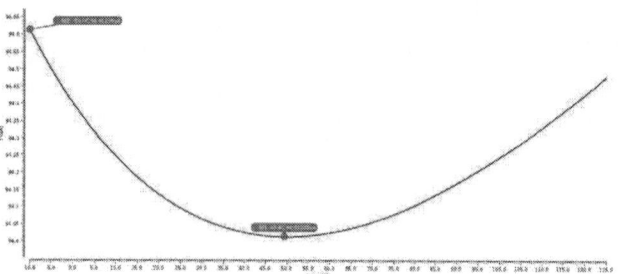

Fig. 4. Reference current (IREF) across temperature in nominal corner

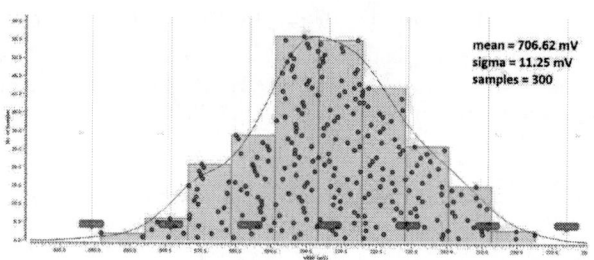

Fig. 5. Reference voltage (VREF) as noted in the Monte-Carlo simulations

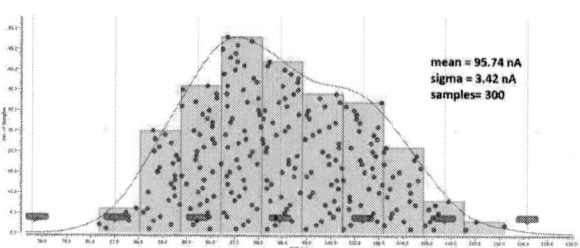

Fig. 6. Reference voltage (VREF) as noted in the Monte-Carlo simulations

E. Sources of Process Variation and Mismatch

From equation (7), the reference voltage does not vary with mismatch and process variations in the resistors. Furthermore, the reference voltage is a strong function of the gate-source voltage of the MOSFET MN2. Therefore, process variations in the threshold voltage of the MOSFET MN2 induce process variations in the reference voltage. Mismatch error in the reference voltage is due to the mismatch in the PMOS current mirrors, which is minimized with choosing large dimensions for the current mirror. The reference current is affected by the process variation of resistor (R3) and the VGS of MN2. A trimming arrangement [4] could be used for the resistor R3 to minimize the spread of the reference current.

III. RESULTS AND DISCUSSIONS

The proposed all-in-one reference is designed in a 130nm CMOS process. Monte-Carlo simulation was performed for the proposed circuit to check its immunity to process and mismatch variations. The reference voltage has a mean value of 706.62mV and a spread (sigma/mean) of 1.6% (Fig. 5). The reference current has a mean value of 95.74nA and a spread of 3.6% (Fig. 6). Furthermore, the reference voltage has an average TC of 100ppm/°C and the reference current has an average TC of 108 ppm/°C (Fig. 7 and Fig. 8). The proposed circuit achieves a PSRR of 70dB at DC (Fig. 9). The feedback loop is frequency compensated by placing a 1pF capacitor at the high impedance node. The DC loop gain is 35dB and the phase margin for the feedback is 90 degrees (Fig. 10). The branch consisting of MP1 consumes a current of 96nA.

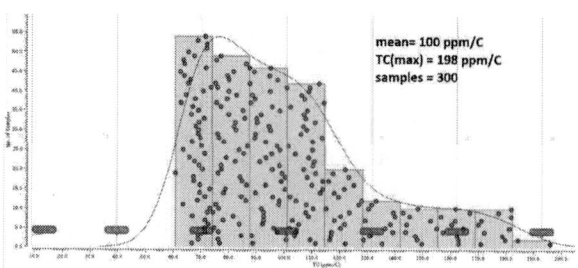

Fig. 7. TC of the reference voltage as noted in the Monte-Carlo simulations

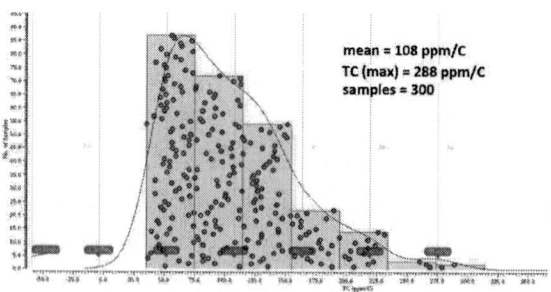

Fig. 8. TC of the reference current as noted in the Monte-Carlo simulations

The reference core consumes 80nA (40nA in each branch). The bias generation circuit consumes 20nA at nominal condition. The proposed circuit is compared with the state of the art in Table 1.

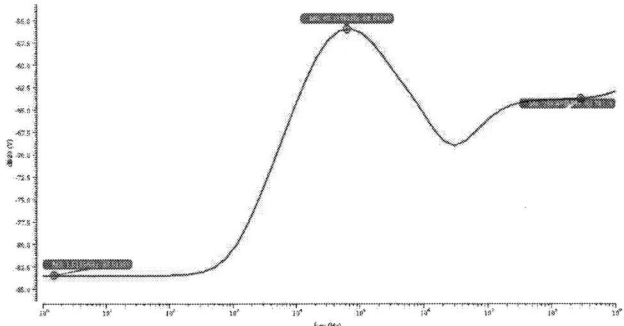

Fig. 9. PSRR of the voltage reference as noted in nominal corner

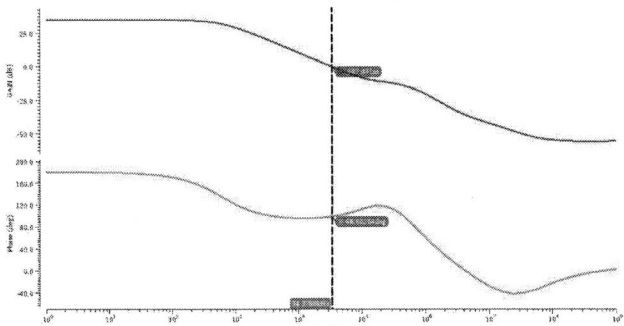

Fig. 10. Stability analysis (loop gain and phase of the feedback loop)

IV. CONCLUSIONS

An all-in-one current and voltage reference circuit which achieves a high PSRR of 83dB at DC is presented in this paper. A feedback arrangement is used instead of a traditional OTA to generate the temperature independent reference current. The circuit includes an auxiliary circuit consisting of native MOSFET to isolate its supply from the core supply. The circuit consumes only 240nW at 1.2V supply making it suitable for low power applications.

REFERENCES

[1] A. Pathy, B. Adithya and Z. Abbas, "A 0.85V Supply, High PSRR CMOS Voltage Reference without Resistor and Amplifier for Ultra-Low Power Applications," *2021 IEEE International Midwest Symposium on Circuits and Systems (MWSCAS)*, 2021, pp. 995-998, doi: 10.1109/MWSCAS47672.2021.9531826.

[2] J. Jiang, W. Shu, J. Chang and J. Liu, "A novel subthreshold voltage reference featuring 17ppm/°C TC within −40°C to 125°C and 75dB PSRR," 2015 IEEE International Symposium on Circuits and Systems (ISCAS), 2015, pp. 501-504, doi: 10.1109/ISCAS.2015.7168680.

[3] L. Wang, C. Zhan, J. Tang, Y. Liu and G. Li, "A 0.9-V 33.7-ppm/°C 85-nW Sub-Bandgap Voltage Reference Consisting of Subthreshold MOSFETs and Single BJT," in *IEEE Transactions on Very Large Scale Integration (VLSI) Systems*, vol. 26, no. 10, pp. 2190-2194, Oct. 2018, doi: 10.1109/TVLSI.2018.2836331.

[4] H. Zhuang, Z. Zhu and Y. Yang, "A 19-nW 0.7-V CMOS Voltage Reference With No Amplifiers and No Clock Circuits," in *IEEE*

Table 1: Comparison with the state of the art

SPECIFICATIONS	This Work	[2] ISCAS	[8] TCAS-II	[9] TCAS-II	[13] TCAS-II
Technology (nm)	130	65	180	180	180
V_{REF} (mV)	706	140	NA	368	1.2
I_{REF} (nA)	95.74	NA	1500	10.3	51
Supply (V)	>1.2	>0.5	>1.8	>0.7	>2
Temperature Range (°C)	-10 to 125	-40 to 125	-40 to 100	-40 to 125	-40 to125
Power (nW)	240	32	2000	28	192
VREF TC (max) TC (avg)	198 100	NA 17	NA NA	NA 43.1	NA 32.7
IREF TC (max) TC (avg)	288 108	NA NA	NA 1457	180 160.8	NA 89
PSRR (dB) @10Hz	-83	-75	NA	-65	-85

Transactions on Circuits and Systems II: Express Briefs, vol. 61, no. 11, pp. 830-834, Nov. 2014, doi: 10.1109/TCSII.2014.2345298.

[5] D. Osipov and S. Paul, "Simulation-based design procedure for sub 1V CMOS current reference," *Design, Automation & Test in Europe Conference & Exhibition (DATE), 2017*, 2017, pp. 1663-1666, doi: 10.23919/DATE.2017.7927261.

[6] S. Agarwal, A. Pathy and Z. Abbas, "A 9.5nW, 0.55V Supply, CMOS Current Reference for Low Power Biomedical Applications," in *IEEE Transactions on Circuits and Systems II: Express Briefs*, vol. 69, no. 9, pp. 3650-3654, Sept. 2022, doi: 10.1109/TCSII.2022.3183146.

[7] D. Cordova, A. C. de Oliveira, P. Toledo, H. Klimach, S. Bampi and E. Fabris, "A sub-1 V, nanopower, ZTC based zero-VT temperature-compensated current reference," *2017 IEEE International Symposium on Circuits and Systems (ISCAS)*, 2017, pp. 1-4, doi: 10.1109/ISCAS.2017.8050289.

[8] M. S. Eslampanah Sendi, S. Kananian, M. Sharifkhani and A. M. Sodagar, "Temperature Compensation in CMOS Peaking Current References," in *IEEE Transactions on Circuits and Systems II: Express Briefs*, vol. 65, no. 9, pp. 1139-1143, Sept. 2018, doi: 10.1109/TCSII.2018.2805832.

[9] L. Wang and C. Zhan, "A 0.7-V 28-nW CMOS Subthreshold Voltage and Current Reference in One Simple Circuit," in *IEEE Transactions on Circuits and Systems I: Regular Papers*, vol. 66, no. 9, pp. 3457-3466, Sept. 2019, doi: 10.1109/TCSI.2019.2927240.

[10] J. Lee and S. Cho, "A 1.4-µW 24.9-ppm/°C Current Reference With Process-Insensitive Temperature Compensation in 0.18-µm CMOS," in *IEEE Journal of Solid-State Circuits*, vol. 47, no. 10, pp. 2527-2533, Oct. 2012, doi: 10.1109/JSSC.2012.2204475.

[11] B. B. Yadav, K. Mounika, A. Bathi and Z. Abbas, "67ppm/°C, 66nA PVT Invariant Curvature Compensated Current Reference for Ultra-Low Power Applications," *2020 IEEE International Symposium on Circuits and Systems (ISCAS)*, 2020, pp. 1-5, doi: 10.1109/ISCAS45731.2020.9180857.

[12] A. P. Brokaw, "A simple three-terminal IC bandgap reference," in *IEEE Journal of Solid-State Circuits*, vol. 9, no. 6, pp. 388-393, Dec. 1974, doi: 10.1109/JSSC.1974.1050532.

[13] W. Huang, L. Liu and Z. Zhu, "A Sub-200nW All-in-One Bandgap Voltage and Current Reference Without Amplifiers," in *IEEE Transactions on Circuits and Systems II: Express Briefs*, vol. 68, no. 1, pp. 121-125, Jan. 2021, doi: 10.1109/TCSII.2020.3007195.

A Low-Power Analog Integrated Deep Spatio-Temporal Inference Network with Application to Digit Classification

Vassilis Alimisis, Nikolaos P. Eleftheriou and Paul P. Sotiriadis

Department of Electrical and Computer Engineering
National Technical University of Athens, Greece
E-mail: alimisisv@gmail.com, eleftheriou_nikos@hotmail.com , pps@ieee.org

Abstract—In the realm of advanced machine learning, a burgeoning paradigm, known as deep machine learning, has emerged to address intricate, high-dimensional data in a structured manner, taking cues from biological inspirations. This study contributes novel findings utilizing a recently introduced deep learning framework, termed the Deep Spatio-Temporal Inference Network. It is a discriminative architecture in deep learning, amalgamates elements from unsupervised learning to establish dynamic pattern representation, concurrently incorporating Bayesian model. The proposed inference is composed of parallel-connected Mahalanobis distance circuits and a distance comparator circuit. As a result, this work proposes a novel low power ($2.43\mu W$), low voltage (0.6V) analog architecture of a Deep Spatio-Temporal Inference Network with application to digit classification. Confirmation of the analog classifier's effective functioning is achieved through validation with a real-world dataset (93.15% accuracy). The implementation of the proposed architecture is executed within the TSMC 90nm CMOS process, and its behavior is simulated utilizing the Cadence IC Suite.

Index Terms—Deep Spatio-Temporal Inference Network, digit classification, low-power design, analog VLSI implementation

I. Introduction

Digital imaging technology has revolutionized the conventional use of film by transitioning to a realm of bits and bytes [1]. In this paradigm, the quality of an image is gauged by the pixel count it possesses. The heightened resolution of an image corresponds to an increased abundance of these diminutive yet vividly coloured dots [2]. Unlike the traditional camera, which relies on lenses to focus light onto film for image formation, the digital camera employs an image sensor. This sensor, often a CMOS or a charge coupled device (CCD), undertakes the task of translating light into electric charges [2].

The CMOS image sensor, notably found in smartphones, employs color-filter layers to impart hues, while photodiodes perform the crucial role of converting light into electrical signals [2], [3]. This amalgamation culminates in the creation of a digital image, further refined through on-chip image processing in certain applications such as artificial vision and image recognition [3]. On the other hand, CCD image sensors, a preferred choice in machine-vision systems, embody transistorized light sensors on an integrated circuit [4]. These sensors meticulously integrate received light, transmuting elec-

trons into the electrical signals that ultimately manifest as video or still images in various formats [4]. This diversity in sensor technology underscores their respective contributions to distinct domains of visual technology.

Deep-learning neural networks excel across various applications, from speech recognition to self-driving cars [5]. They succeed at deciphering complex patterns in datasets, often surpassing human capabilities. In camera-related applications, diverse neural network variants enhance image quality by addressing blurriness, enhancing colours, and rectifying pixel issues [6]. They also excel at specific tasks like isolating regions of interest. For instance, in surveillance, these networks create feature maps that highlight crucial parts of an image, such as facial details or pedestrian counts [7]. This focused approach reduces memory and computational demands, crucial for resource-efficient edge applications.

The motivation is based on the power and area efficiency requirements of image sensors [8], [9], this paper proposes a new, power-efficient and analog hardware architecture for deep learning that integrates principles from unsupervised learning for dynamic pattern representation alongside Bayes inference. This is called Deep Spatio-Temporal Inference Network. The implemented network is a promising classifier appropriate for battery dependent image smart sensor classification systems, since it achieves 93.15% accuracy. It is implemented and confirmed on a measured digit recognition dataset [10]. The accuracy of the proposed implementation is confirmed through post-layout simulation results obtained in a TSMC 90nm CMOS process and simulated using Cadence IC Suite. This validation involves a comparison with a software-based implementation. Furthermore, a comprehensive comparison study between the proposed classifier and analog classifiers is included for the sake of thoroughness.

The rest of this work is ordered in the following manner. In Section II the the characteristics of the implemented network are explained and a clarification of its mathematical foundations is provided. The suggested structure and the fundamental components of the proposed classifier are outlined in Section III. The desired behavior of the implemented network is verified via a digit classification dataset and a comparison

---

979-8-3503-8083-5/23 $31.00 © 2023 IEEE

with the software-based counterpart is presented in Section IV. Section V provides a comparison study with related analog classifiers. Section VI concludes this work.

II. DEEP SPATIO-TEMPORAL INFERENCE NETWORK MATHEMATICAL MODEL

Deep Spatio-Temporal Infernce Network consists of multiple instances of an identical cortical circuit, referred to as nodes [11]. Each node is a parameterized model that learns through unsupervised learning. These nodes exist across all hierarchy layers, aiming to grasp significant spatiotemporal patterns shown in presented data [11]. The lowest layer nodes take raw sensory input, e.g., image pixels, and continually develop a belief state to characterize observed sequences. Higher layers receive belief states from lower corresponding layers. These beliefs across the hierarchy are utilized as valuable features given to a classifier or regression learner, which can be trained through supervised learning.

Firstly, the selected winning centroid relies exclusively on the Euclidean distance [11], [12]. The distance d_x between a centroid x and an observed input o is represented as follows:

$$d_x = ||o - \mu_x||\psi_x. \tag{1}$$

The clustering algorithm uses the starvation trace ψ_x to involve centroids initially positioned far from dense areas in the observation space. This helps centroids that might otherwise never be chosen for updates due to their distant location. This enables inactive or starved centroids to gradually adjust their perceived distance to input vectors over time. When not selected as the centroid, they decrease this apparent distance; conversely, their apparent distance increases when they are the selected centroid. The mean estimate of the winning centroid, μ_x, is adjusted towards the present input along with the estimated variance σ_x^2 in a combined manner so that:

$$\mu_x \leftarrow \mu_x + \alpha(o - \mu_x), \tag{2}$$

$$\sigma_x^2 \leftarrow \sigma_x^2 + \beta[(o - \mu_x)^2 - \sigma_x^2], \tag{3}$$

where α and β are positive numbers close to 0. Then, the posterior distribution $Pr(o|s)$ is derived by normalizing the Euclidean distances between the input and every centroid s, such that :

$$n_s = \sum_{i=1}^{d} \frac{(o_i - \mu_{i,s})^2}{\sigma_{i,s}^2}, \tag{4}$$

$$p_s = \frac{n_s^{-1}}{\sum_{s' \in S}(n_{s'}^{-1})}, \tag{5}$$

III. PROPOSED ARCHITECTURE

In this section, we analyze the proposed analog implementation of the Network. The architecture presented is versatile, accommodating various numbers of classes, centroids, and input dimensions. For the implementation of the Euclidean distance function in equation (1), we employ a current-mode Mahalanobis distance circuit [13], shown in Fig. 1. This

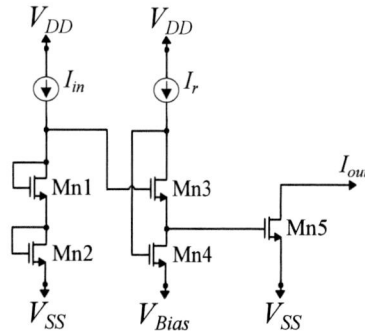

Fig. 1: The Mahalanobis distance is approximated by the translinear circuit which computes the $\frac{I_{in}^2}{I_r}$.

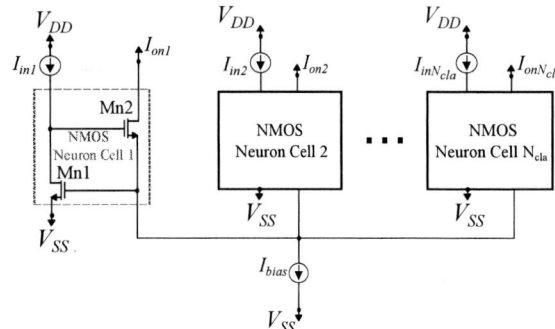

Fig. 2: A N_{cla}-neuron Standard Lazzaro NMOS Winner-Take-All (WTA) circuit.

circuit, which operates in a translinear fashion, embodies the mathematical expression: I_{in}^2 divided by I_r. Summation within the current domain is straightforward, achieved by connecting wires containing the currents to be summed.

Moving forward, the nearest centroid is determined through a distance comparison circuit, specifically a Winner-Take-All (WTA) circuit [14]. In a classification problem with N_{cla} classes, the typical Lazzaro WTA circuit consists of N_{cla} neurons. These neurons share a shared bias current, as depicted in Fig. 2. Each sub-circuit in the WTA circuit corresponds to an individual class. The WTA circuit effectively identifies the class with the highest input current and assigns a non-zero output current to the corresponding neuron. Simultaneously, the remaining neurons receive an output current of zero.

The architecture of the proposed network, as depicted in Fig. 3, is developed for a classification problem involving N_{cla} classes and N_d features (input dimensions). The quantity of centroids within each class is a hyperparameter, typically determined through exploratory data analysis. In this generalized schematic, each class comprises one centroid and N_d input dimensions, illustrated in Fig. 3. The output of each Mahalanobis distance circuit (MDC) describes each input dimension. The mathematical model, as described by equations (4) and (5), involves the summation of Euclidean distances. This summation is executed within each circuit, leveraging current mirrors (CM) to minimize potential distortions in cal-

979-8-3503-8083-5/23 $31.00 © 2023 IEEE

culations that might arise from undesired effects on the output currents of the Mahalanobis circuits. The classifier's prediction is denoted by the resulting output currents, characterized by high or low values. The dimensions of the transistors are equal to $W/L = 3.2\mu m/1.6\mu m$. Notably, all transistors in the mentioned designs operate in the sub-threshold region, with voltage source rails set as $V_{DD} = -V_{SS} = 0.3$ V and $I_r = 3nA$.

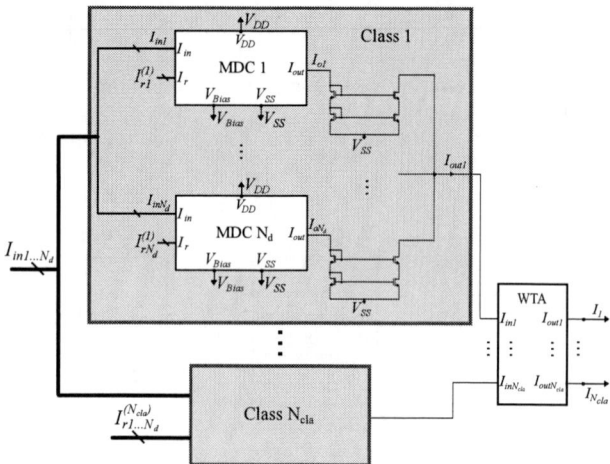

Fig. 3: Block diagram of a generic Analog Deep Spatio-Temporal Inference Network with N_{cla} classes and N_d features. These classes perform the summation of currents generated by the N_d MDC circuits associated with each input. The final output is determined through a WTA circuit, employing a current-mode representation.

IV. DIGIT RECOGNITION AND SIMULATION OUTCOMES

In this section, the proposed network is validated through a testing on a digit recognition problem [10]. The architecture presented herein is realized utilizing the TSMC $90nm$ CMOS process in conjunction with the Cadence IC suite. The power supply rails for the entire classifier are established at $V_{DD} = -V_{SS} = 0.3V$. All simulation results are derived from the layout, as depicted in Fig. 4, through post-layout simulations. This classification problem considers a handwritten digit recognition task, which consists of $N_{cla} = 10$ classes and $N_d = 64$ inputs. For comparison purposes, we have reduced the number of classes to $N_{cla} = 2$, which consists of 5 centroids per class (binary classifier, odd/even). The dataset used is provided by Python's Sklearn package [15] and consists of $8x8$ pixel images of digits. Each pixel consists of a grayscale value between 0 and 16. The classifier receives all relevant metrics directly. The required parameters for the system are determined through the calculation of the mean value, variance, and prior probability for each class.

To test the proposed classifier both in terms of classification specificity and circuit's behavior over PVT variations, two separate tests are conducted on the layout. To address the experimental variability, the results from 20 different training-test iterations are presented in Fig. 5. The sensitivity of the circuit is further validated through a Monte Carlo analysis.

More specifically, Fig. 6 illustrates the Monte Carlo Histogram for N = 100 points.

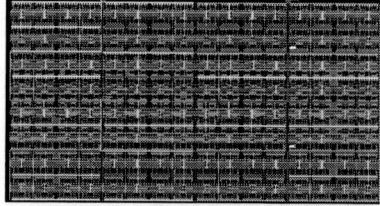

Fig. 4: Layout of the proposed Deep Spatio-Temporal Inference Network architecture based on the design methodology.

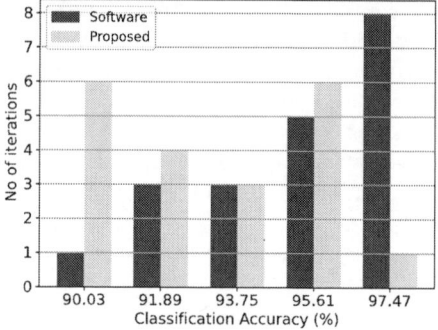

Fig. 5: Classification results of the proposed architecture and the equivalent software model on the digit recognition dataset over 20 iterations.

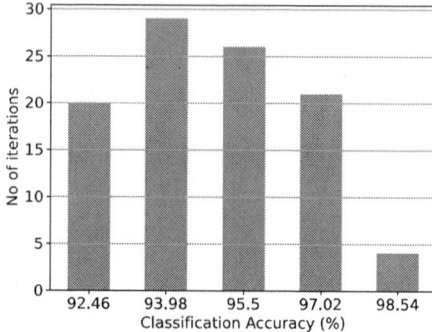

Fig. 6: Post-layout Monte-Carlo simulation results of the proposed architecture on the digit recognition dataset with $\mu_M = 94.87\%$ and a standard deviation of $\sigma_M = 1.73\%$

V. PERFORMANCE SUMMARY AND DISCUSSION

This section aims to present a comparative analysis of various analog classifiers developed by our research team. By adjusting these classifiers to the same application as the one tested in this work a fair and unbiased comparison can be conducted. In Table I a performance summary is illustrated for a Fuzzy [16], a Gaussian Mixture Model (GMM) [17], a Bayesian [18], a Threshold [19], a Support Vector Machine (SVM) [20] and a centroid-based [21] classifier.

Firstly, our work surpasses the performance of the related analog classifiers in mean accuracy, processing speed, and

979-8-3503-8083-5/23 $31.00 © 2023 IEEE

TABLE I: Analog classifiers' comparison on the Digit Recognition

	Classifier	Min accuracy	Mean accuracy	Max accuracy	Power consumption	Processing speed	Energy per classification	No. of Dimensions
This work	Spatio	89.10%	93.15%	97.20%	$2.43\mu W$	$430K \frac{\text{classifications}}{s}$	$\frac{5.65 \text{ pJ}}{\text{classification}}$	64
[16]	Fuzzy	85.20%	90.82%	95.10%	$2.71\mu W$	$4.55K \frac{\text{classifications}}{s}$	$\frac{595.6 \text{ pJ}}{\text{classification}}$	13
[17]	GMM	77.70%	83.72%	88.90%	$3.38\mu W$	$100K \frac{\text{classifications}}{s}$	$\frac{33.8 \text{ pJ}}{\text{classification}}$	13
[18]	Bayes	73.40%	81.75%	84.20%	$2.08\mu W$	$100K \frac{\text{classifications}}{s}$	$\frac{20.8 \text{ pJ}}{\text{classification}}$	13
[19]	Threshold	78.60%	82.55%	86.40%	$1.21\mu W$	$100K \frac{\text{classifications}}{s}$	$\frac{12.1 \text{ pJ}}{\text{classification}}$	13
[20]	SVM	84.40%	85.74%	86.90%	$82.12\mu W$	$140K \frac{\text{classifications}}{s}$	$\frac{586.57 \text{ pJ}}{\text{classification}}$	13
[21]	Centroid	86.30%	91.32%	95.40%	$3.42\mu W$	$100K \frac{\text{classifications}}{s}$	$\frac{34.2 \text{ pJ}}{\text{classification}}$	13

energy consumption per classification. It is important to emphasize that, for this specific application, we have a high input dimension number. The proposed topology offers a distinct advantage in that it obviates the need for Principal Component Analysis (PCA), enabling the utilization of all 64 input dimensions without any loss of information. To attain optimal accuracy, the remaining topologies should truncate the dimensions to 13. This is the main limitation of the previous related works (specific number of input dimensions). While our network demonstrates the ability to accurately classify all 10 classes, we transformed the problem into a binary classification scenario to facilitate a meaningful comparison with binary analog classifiers [16], [19], [20].

VI. Conclusion

In this work, a new, power-efficient ($2.43\mu W$), low supply (0.6V) architecture of an analog Deep Spatio-Temporal Inference Network for digit recognition was proposed. The presented architecture consists of Mahalanobis distance circuits and a distance comparator circuit. All post-layout simulation results were obtained using the TSMC 90nm CMOS process and were compared with a software-based implementation and a variety of analog classifiers. The implemented architecture achieves 93.15% classification accuracy and reasonable sensitivity characteristics. It can serve as a fundamental building block in intelligent sensor systems designed for image classification.

References

[1] H. J. Trussell and M. J. Vrhel, *Fundamentals of digital imaging*. Cambridge University Press, 2008.

[2] G. Petrie and A. S. Walker, "Airborne digital imaging technology: a new overview," *The Photogrammetric Record*, vol. 22, no. 119, pp. 203–225, 2007.

[3] A. El Gamal and H. Eltoukhy, "Cmos image sensors," *IEEE Circuits and Devices Magazine*, vol. 21, no. 3, pp. 6–20, 2005.

[4] M. Bigas, E. Cabruja, J. Forest, and J. Salvi, "Review of cmos image sensors," *Microelectronics journal*, vol. 37, no. 5, pp. 433–451, 2006.

[5] H. Osipyan, B. I. Edwards, and A. D. Cheok, *Deep neural network applications*. CRC Press, 2022.

[6] M. Gruosso, N. Capece, and U. Erra, "Human segmentation in surveillance video with deep learning," *Multimedia Tools and Applications*, vol. 80, pp. 1175–1199, 2021.

[7] M. Di Benedetto, F. Carrara, L. Ciampi, F. Falchi, C. Gennaro, and G. Amato, "An embedded toolset for human activity monitoring in critical environments," *Expert Systems with Applications*, vol. 199, p. 117125, 2022.

[8] F. Tang, D. G. Chen, B. Wang, and A. Bermak, "Low-power cmos image sensor based on column-parallel single-slope/sar quantization scheme," *IEEE Transactions on Electron Devices*, vol. 60, no. 8, pp. 2561–2566, 2013.

[9] M. Maheepala, M. A. Joordens, and A. Z. Kouzani, "Low power processors and image sensors for vision-based iot devices: a review," *IEEE Sensors Journal*, vol. 21, no. 2, pp. 1172–1186, 2020.

[10] F. Nelli and F. Nelli, "Recognizing handwritten digits," *Python Data Analytics: With Pandas, NumPy, and Matplotlib*, pp. 473–486, 2018.

[11] D. George and J. Hawkins, "Towards a mathematical theory of cortical micro-circuits," *PLoS computational biology*, vol. 5, no. 10, p. e1000532, 2009.

[12] C. M. Bishop and N. M. Nasrabadi, *Pattern recognition and machine learning*. Springer, 2006, vol. 4, no. 4.

[13] R. J. Wiegerink, *Analysis and synthesis of MOS translinear circuits*. Springer Science & Business Media, 2012, vol. 246.

[14] J. Lazzaro, S. Ryckebusch, M. A. Mahowald, and C. A. Mead, "Winner-take-all networks of o (n) complexity," *Advances in neural information processing systems*, vol. 1, 1988.

[15] F. Pedregosa, G. Varoquaux, A. Gramfort, V. Michel, B. Thirion, O. Grisel, M. Blondel, P. Prettenhofer, R. Weiss, V. Dubourg *et al.*, "Scikit-learn: Machine learning in python," *the Journal of machine Learning research*, vol. 12, pp. 2825–2830, 2011.

[16] E. Georgakilas, V. Alimisis, G. Gennis, C. Aletraris, C. Dimas, and P. P. Sotiriadis, "An ultra-low power fully-programmable analog general purpose type-2 fuzzy inference system," *AEU-International Journal of Electronics and Communications*, vol. 170, p. 154824, 2023.

[17] V. Alimisis, G. Gennis, K. Touloupas, C. Dimas, M. Gourdouparis, and P. P. Sotiriadis, "Gaussian mixture model classifier analog integrated low-power implementation with applications in fault management detection," *Microelectronics Journal*, vol. 126, p. 105510, 2022.

[18] V. Alimisis, G. Gennis, C. Dimas, and P. P. Sotiriadis, "An analog bayesian classifier implementation, for thyroid disease detection, based on a low-power, current-mode gaussian function circuit," in *2021 International conference on microelectronics (ICM)*. IEEE, 2021, pp. 153–156.

[19] V. Alimisis, G. Gennis, E. Tsouvalas, C. Dimas, and P. P. Sotiriadis, "An analog, low-power threshold classifier tested on a bank note authentication dataset," in *2022 International Conference on Microelectronics (ICM)*. IEEE, 2022, pp. 66–69.

[20] V. Alimisis, G. Gennis, M. Gourdouparis, C. Dimas, and P. P. Sotiriadis, "A low-power analog integrated implementation of the support vector machine algorithm with on-chip learning tested on a bearing fault application," *Sensors*, vol. 23, no. 8, p. 3978, 2023.

[21] V. Alimisis, V. Mouzakis, G. Gennis, E. Tsouvalas, C. Dimas, and P. P. Sotiriadis, "A hand gesture recognition circuit utilizing an analog voting classifier," *Electronics*, vol. 11, no. 23, p. 3915, 2022.

Regional CubeSat Communication and Constellation Design Evaluation

Khaled Mohammed, Hamzeh Abu Qamar, Ruhul Amin Khalil, and Nasir Saeed
Department of Electrical and Communication Engineering, United Arab Emirates University, Al Ain 15551, UAE
Email: {201950180; 202050889; ruhulamin; nasir.saeed}@uaeu.ac.ae

Abstract—In the rapidly evolving landscape of space networks, CubeSats have emerged as pivotal components facilitating cost-effective networks, supporting many applications, including monitoring, surveillance, and in-space backhauling. Besides, the forthcoming Emirati Interplanetary Mission in 2028 underscores the imperative of efficient constellation designs, particularly in remote and challenging areas. However, developing such intricate systems from the ground up is beset with formidable costs and technical complexities. In response to these challenges, our project aims to create a simulator of small satellite constellations that will, in advance of their deployment, serve as a proactive measure to mitigate potential issues and curtail the risk of costly satellite failures. We perform the link budget analysis and provide simulations of various regional constellations, showcasing their dynamic orbital mobility and providing a cost-effective platform for comprehensive performance testing, focusing on vulnerability assessment. The proposed simulator will validate the optimality of satellite constellation design and highlight the manifold terrestrial benefits of CubeSats, substantially elevating performance metrics such as the link budget. The transformative potential of our methodology lies in its capacity to significantly reduce production costs while concurrently bolstering the efficiency of CubeSat communication systems. This, in turn, positions our approach as a formidable contender in the competitive market landscape, poised to reshape the future of satellite technology.

Index Terms—CubeSats, constellations, link budget analysis, backhauling.

I. INTRODUCTION

Ensuring equitable access to the digital world is paramount in our increasingly interconnected global society. Unfortunately, a significant disparity persists between regions with robust internet connectivity and those grappling with limited or no access. To promote resilience and inclusivity globally, we must make concerted efforts to bridge this digital divide. As of the close of 2019, projections from the International Telecommunication Union (ITU) indicate that approximately 3.6 billion individuals still lacked access to the internet, underscoring the magnitude of this challenge [1]. This issue is particularly acute in the least developed countries, where only two out of every ten people are privileged to be online. Space-related technology has become increasingly captivating, with space research offering new frontiers. Among these frontiers, miniature satellite constellations, commonly called CubeSat constellations, have emerged as a desirable prospect [2]. What distinguishes them is their versatility,

This work is supported by IEEE Antennas and Propagation Society Undergraduate Summer Research Scholarship.

design, deployment cost-effectiveness, and capacity to support a wide range of applications [3]. These applications encompass everything from remote sensing, aiding in the understanding and monitoring of the Earth's surface and atmosphere, to space exploration missions that expand our knowledge of the cosmos and even provide internet connectivity in remote and underserved areas. However, the practical deployment of CubeSat mega-constellations to meet the anticipated demand presents formidable real-world challenges [4]. This challenge is partly due to the substantial time and effort required for conceiving, designing, and assembling these satellite systems. Moreover, it is essential to ensure a robust and reliable design while rigorously assessing the feasibility of a mission before deploying these costly satellites. This prudent approach is critical for mitigating the significant risks of mission failure and the potential degradation of expensive satellite assets [5]. Besides, despite the growing interest in CubeSats to perform different missions, research on designing their digital twin or a simulator remains limited due to the lack of feasibility of the essential metrics of CubeSat design. Also, open-source design applications – like STK and GMAT – lack critical design parameters, such as revisit time and inter-satellite links. Moreover, it is near impossible to design a perfect digital twin as there is always a possibility of micrometeoroids or orbital debris collisions and noise and crosstalk from other neighboring satellites [6]. Moreover, no studies investigate real-time link budget analysis using digital twins. With real-time link analysis, we can continuously monitor the bit error rate using various mathematical models in [7]. Our approach stands poised to improve the small satellite constellation landscape by delivering substantial reductions in production costs, all while elevating the operational efficiency of communication systems to new heights. Our seamless integration of virtual and real-world techniques via the proposed simulator empowers us to proactively harness data insights, ensuring continuous system monitoring, early issue mitigation, minimal downtime, and discovery of novel applications. In this way, our mission is to profoundly contribute to the ongoing growth and evolution of the space technology sector in the UAE while concurrently narrowing the digital divide problem and performing environmental monitoring.

II. PROPOSED SIMULATOR DESIGN

The design of a simulator for mega-constellations of Cube-Sats is critical, ensuring the reliable transmission of data between space and Earth, Earth to space, or through intermediary communication relays. International entities govern the allocation of frequency bands for specific functions, such as space-to-Earth, Earth-to-space, and inter-satellite communication [8]. In the case of CubeSats, uplink frequencies are typically kept higher than downlink frequencies to minimize signal attenuation and reduce power requirements [9]. Existing studies have predominantly focused on the analysis of VHF (144 - 146 MHz) and UHF (435 - 438 MHz) amateur bands, which are commonly employed in different missions [10]. Nevertheless, some investigations have delved into Ka-band, X-band, S-band, and L-band frequencies [11], [12]. In the design of these communication links, factors such as range, throughput, and received signal quality assume paramount importance for communication engineers. Hence, the comprehensive analysis provided by a link budget is indispensable, as it accounts for all gains and losses within the communication link. According to [13], employing the free-space propagation model, we can evaluate the normalized signal-to-noise ratio (γ) at the ground station to determine the link budget for the satellite duplex link:

$$\gamma = \frac{E_b}{N_o} = \frac{P_t G_t G_r}{kTLR_b} \qquad (1)$$

Here, E_b represents the energy-per-bit, N_o denotes the noise spectral density, P_t signifies the transmit power, G_t corresponds to the transmitter antenna gain, G_r represents the receiver antenna gain, k stands for the Boltzmann constant, T reflects the system temperature noise, L encompasses the total loss, and R_b relates to the desired data rate. The total path loss L is equal to:

$$L = L_{FS}L_{atm}L_{POL}L_{FTx}L_{FRx}L_{DTx}L_{DRx}. \qquad (2)$$

All these different types of losses are defined in Table II. By simplifying (2), we consider $L_{FTx} = L_{FRx} = L_F$ and $L_{DTx} = L_{DRx} = L_D$.

$$\begin{aligned} L(dB) = {}& L_{FS}(dB) + L_{atm}(dB) + L_{POL}(dB) \\ & + 2L_F(dB) + 2L_D(dB). \end{aligned} \qquad (3)$$

With path loss in (3), link reliability is assessed by calculating the Bit Error Rate (BER) based on the signal-to-noise ratio. The BER calculation based on SNR is given in Table I for some of the well-known digital modulation schemes.

Moreover, for effective CubeSat utilization in LEO, distinct parameters necessitate design considerations due to their smaller footprint compared to medium Earth orbit (MEO) and geostationary orbit (GEO) satellites. This difference is primarily influenced by satellite altitude, with higher altitudes yielding larger footprints. To achieve global coverage, we adapt in this paper the Walker Constellation model with symmetric CubeSats (having the same inclination angle and altitude) for the local coverage in UAE as shown in Fig. 1

TABLE I: BER formulas for different modulation techniques.

Coherent Modulation Technique	BER (P_b)
BPSK	$0.5(\sqrt{\frac{E_b}{N_o}})$
M-PSK	$\frac{1}{m} \times (\sqrt{\frac{mE_b}{N_o}} \times \sin(\frac{\pi}{M}))$
M-QAM	$\frac{2}{m}(1 - \frac{1}{\sqrt{M}}) \times (\sqrt{\frac{3m}{2(M-1)} \times \frac{E_b}{N_o}})$

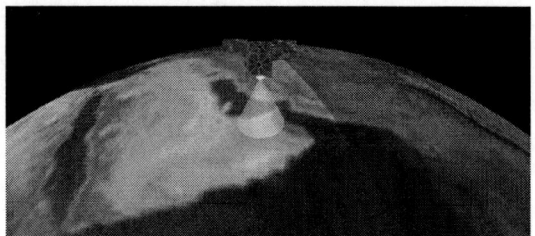

Fig. 1: Satellite orbital trajectory.

Using (4), we can calculate the orbital period T:

$$T = \sqrt{\frac{4\pi^2 R^3}{GM}}, \qquad (4)$$

where R is the average radius of the orbit, the universal gravitational constant $G = 6.6726 \times 10^{-11} Nm^2/kg^2$, $M_{earth} = 5.98 \times 10^{24} kg$, $R_{earth} = 6.37 \times 10^6 m$. Then, we can find the six initial Keplerian elements of the CubeSat:

1) Semi-major axis (a):

$$a = \sqrt[3]{\frac{P^2 GM}{4\pi^2}}, \qquad (5)$$

where P is the orbital period, and M is the combined mass of primary and secondary bodies.

2) Eccentricity (e):

$$\vec{e} = \frac{\vec{v} \times \vec{h}}{\mu} - \frac{\vec{r}}{r}, \qquad (6)$$

where v is orbital speed, $\vec{h}$ is angular momentum vector, μ is standard gravitational parameter, and r is orbit radius.

3) Inclination angle (i):

$$\cos(i) = \frac{\hat{Z} \cdot \vec{h}}{|\hat{Z}||\vec{h}|}. \qquad (7)$$

where Z is the angular momentum vector.

4) Right Ascension of Ascending Node (Ω):

$$\cos(\Omega) = \frac{\hat{I} \cdot \vec{n}}{|\hat{I}||\vec{n}|}, \qquad (8)$$

where Ω is the angle from the vernal equinox to the ascending node. The satellite passes through the equatorial plane through the ascending node, moving from south to north.

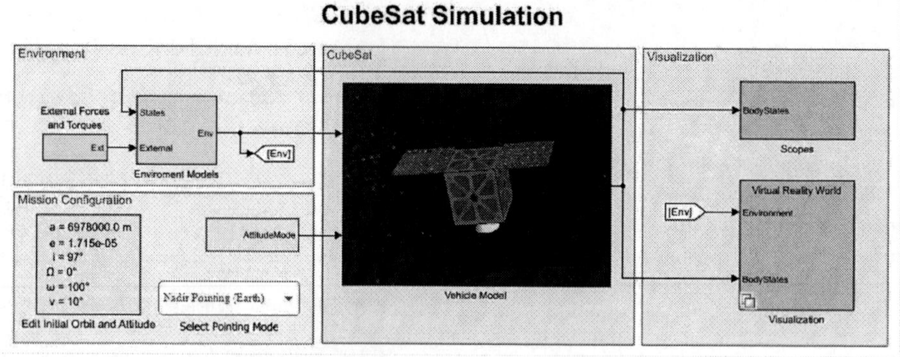

Fig. 2: Block diagram on Simulink.

5) Argument of Perigee (ω):

$$\cos(\Omega) = \frac{\vec{n} \cdot \vec{e}}{|\vec{e}||\vec{n}|}, \tag{9}$$

where Ω is the angle from the ascending node to the eccentricity vector, n is a vector pointing toward the ascending node.

6) True mean anomaly (M):

$$\cos(v) = \frac{\vec{e} \cdot \vec{r}}{|\vec{e}||\vec{r}|} \tag{10a}$$

$$\tan(\frac{E}{2}) = \sqrt{\frac{1-e}{1+e}} \tan(\frac{v}{2}) \tag{10b}$$

$$M = E - e\sin(E) \tag{10c}$$

where E is an eccentric anomaly. Table 2 shows the Keplerian elements found using equations (2-8) for a simple orbit design for our target location. Next, we will simulate the satellite-to-ground link's constellation design and communication link budget.

III. NUMERICAL RESULTS

We utilized the "Aerospace Block set" Simulink Toolbox to analyze Expo City's mission geometry for Dubai's coverage, defining the CubeSat vehicle model and its mission parameters aligned with the initial design parameters. The Simulink block diagram is depicted in Fig. 2, and the satellite trajectory over a 3D globe is shown in Fig. 1. This analysis used the initial parameters in Table II. Additionally, we simulated the orbit design in MATLAB Simulink through the CubeSat Simulation Project Toolbox, which offers a pre-configured CubeSat for simulation and visualization via Simulink 3D Animation. A MATLAB graphical user interface (GUI) was also employed to simulate the three-dimensional orbit of the satellite, requiring input for defining the orbit trajectory and CubeSat attitude. Given that a majority of Low Earth Orbit (LEO) satellites operate within the L-band frequency range of 1-2 GHz [14], we can reasonably assume that the Signal-to-Noise Ratio (SNR) at a typical ground station falls within the range of up to 20 dBs. Therefore, our simulation encompasses an SNR range from -5 to 20 dBs, in line with established guidelines [15]. Utilizing

TABLE II: Initial input parameters.

Initial Parameters	Value
a: semi-major axis (km)	6978
e: eccentricity	0.00001715
i: inclination (deg)	97°
Ω: right ascension of ascending node (deg)	0°
ω: argument of periapsis (deg)	100°
v: true anomaly (deg)	10°
h: altitude (km)	500
Latitude (deg)	25.5°
Longitude (deg)	55.3°
L_{FS}: Free Space Loss (dB)	138.48
L_{atm}: Atmospheric Loss (dB)	0.2802
L_{POL}: Polarization Loss (dB)	2.31
L_D: Denoising Loss (dB)	0.0012
L_F: Feeder Loss (dB)	1

these SNR values, we conducted a comprehensive performance analysis of the CubeSat-to-ground link, depicted in Fig. 3 across various modulation schemes. Furthermore, to assess the practicality of our findings, we considered a specific scenario using the Simulink block diagram presented in Fig. 1. In this scenario, we evaluated the performance of a single CubeSat tasked with providing coverage to Expo City in Dubai. In Fig. 4, we present the Geodetic Latitude (depicted in Blue) and Geodetic Longitude (depicted in Red) of the CubeSat. The geodetic latitude of the CubeSat represents the angle formed between the equatorial plane and the perpendicular line intersecting the CubeSat's position on the Earth's surface. On the other hand, the Geodetic Longitude signifies the angle within the equatorial plane, defined by line 'a' connecting the

979-8-3503-8083-5/23 $31.00 © 2023 IEEE

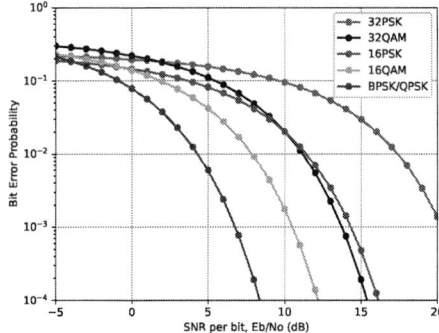

Fig. 3: BER vs. SNR per bit for different modulation techniques.

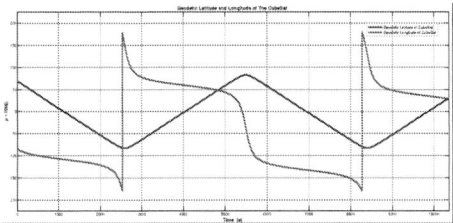

Fig. 4: Geodetic Latitude and Longitude of the CubeSat.

Earth's center to the prime meridian and line 'b' connecting the center to the meridian along which the CubeSat is situated. This information provides crucial insights into the CubeSat's position and orientation relative to the Earth's surface. In

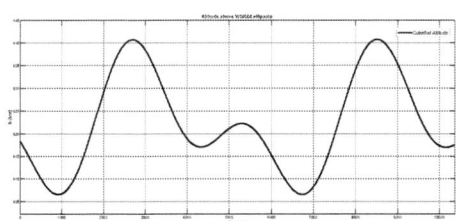

Fig. 5: Altitude above WGS84 of the CubeSat.

Fig. 5, we present the height above the CubeSat's WGS84 ellipsoid (referred to as HAE), representing a mathematical model of the Earth—an ellipsoid. To provide context, it's important to note that Fig. 4 and Fig. 5 together depict periodic signals of the CubeSat's Geodetic Latitude, Geodetic Longitude, and Altitude. This periodicity aligns closely with a period of approximately 5663 seconds. This period coincides with the orbital period of the CubeSat, a relationship that can be precisely calculated using (4).

IV. CONCLUSION

In conclusion, as we peer into the future of communication networks, we observe a shift toward fast broadband with the promise of 6G technology. However, we must embrace advanced modeling and simulation to address the ever-expanding coverage challenges of tomorrow's non-terrestrial networks. Developing a simulator for CubeSat constellations is a necessity and a strategic imperative. This tool enables us to overcome challenges and ensure satellite networks' cost-effective and successful deployment. Our study unveiled a sophisticated simulation platform tailored for advanced satellite networks. Through precise modeling of communication links and constellations and digital prototyping, we showcased CubeSat performance with real-time orbital motion and communication link analysis, focusing on regional coverage in the UAE. Future directions for research in this field involve fine-tuning the simulator for diverse geographical areas, designing mega-constellations, exploring adaptive communication protocols, and delving deeper into the optimization of CubeSat constellations to meet the evolving demands of global connectivity.

REFERENCES

[1] N. Saeed, A. Elzanaty *et al.*, "CubeSat Communications: Recent Advances and Future Challenges," *IEEE Commun. Sur. & Tuts.*, vol. 22, no. 3, pp. 1839–1862, 2020.

[2] W. Marshall, "Space technology is improving our lives and making the world a better place: here's how," in *World Economic Forum*, 2017.

[3] S. Malisuwan and B. Kanchanarat, "Small Satellites for Low-Cost Space Access: Launch, Deployment, Integration, and In-Space Logistics," *American J. of Indus. and Business Manag.*, vol. 12, no. 10, pp. 1480–1497, 2022.

[4] M. Xia, S. Hu *et al.*, "Data Naming Mechanism of LEO Satellite Mega-Constellations for the Internet of Things," *Applied Sci.*, vol. 12, no. 14, p. 7083, 2022.

[5] A. Salehi, M. Fakoor *et al.*, "Conceptual Design Process for LEO Satellite Constellations Based on System Engineering Disciplines." *CMES-Computer Modeling in Eng. & Sci.*, vol. 131, no. 2, 2022.

[6] R. Onrubia, D. Pascual *et al.*, "Satellite cross-talk impact analysis in airborne interferometric global navigation satellite system-reflectometry with the microwave interferometric reflectometer," *Remote Sensing*, vol. 11, no. 9, p. 1120, 2019.

[7] A. Goldsmith, *Wireless Communications*. Cambridge University Press, 2008.

[8] N. Saeed, H. Almorad *et al.*, "Point-to-point communication in integrated satellite-aerial 6G networks: State-of-the-art and future challenges," *IEEE Open J. Commun. Soc.*, vol. 2, pp. 1505–1525, 2021.

[9] O. Popescu, "Power budgets for CubeSat Radios to support ground communications and inter-satellite links," *IEEE Access*, vol. 5, pp. 12 618–12 625, 2017.

[10] D. Barbarić, J. Vuković, and D. Babic, "Link budget analysis for a proposed CubeSat Earth observation mission," in *41st Int. Conven. on Info. and Commun. Techno., Electro. and Micro., (MIPRO)*. IEEE, 2018, pp. 0133–0138.

[11] O. Kegege, Y. F. Wong, and S. Altunc, "Advances in Ka-band Communication System for CubeSats and SmallSats," in *Aerospace Systems Conference of the National Society of Black Engineers*, no. GSFC-E-DAA-TN33991, 2016.

[12] J. Issler, A. Gaboriaud, F. Apper *et al.*, "CCSDS Communication products in S and X Band for CubeSats," 2014.

[13] I. R. P. Series, "Specific attenuation model for rain for use in prediction methods," *Rec. P. 838-3, ITU-R*, 2005.

[14] David Daly, "The Basics of LEO Satellite Systems," https://www.basecampconnect.com/leo-satellite-systems/, online; accessed 13 August 2023.

[15] Veripos, "How do I check the L-band reception status of my LD2/LD2S demodulator?" https://veripos.com/support/legacy-products/ld2-ld2s-equipment-eol/how-do-i-check-the-l-band-reception-status-of-my-ld2-ld2s-demodulator, online; accessed 13 August 2023.

Innovative Hardware Architecture for Zero Emission Sea Drones

Ilya Kavalchuk
Faculty of Military and Security
Higher Colleges of Technology
Abu Dhabi, UAE
ikavalchuk@hct.ac.ae

Saad Zafar
Faculty of Military and Security
Higher Colleges of Technology
Abu Dhabi, UAE
szafar1@hct.ac.ae

Shahid Islam
Faculty of Military and Security
Higher Colleges of Technology
Abu Dhabi, UAE
sislam@hct.ac.ae

Abstract—Sea drones have become an essential part of marine and naval processes due to their cost, possibility to operate without the need for services and personnel allocation and functions, they provide. Despite having different purposes, sea drones fulfill the need of the cheap, relatively slow and, for some applications, almost undetectable solutions. However, the problem with pollution from internal combustion engines, created noise and vibrations limit the use of such systems in restricted waters and natural parks. This paper discusses the architecture of the zero emission fully electric unmanned drone of the small size powered by hydrogen fuel cell. The paper demonstrates layout of the drone, interconnection of the drivetrain components and evaluates operational parameters of it.

Keywords—half submerged vessels, hydrogen fuel cell, propulsion system, sea drones

I. INTRODUCTION

Naval exploration is always linked with great potential danger for the explorers due to the unpredictable nature of the oceans and related weather conditions [1]. In addition, modern naval platform combine systems of various nature which leads to the extended number of personnel required shipboard to operate with all relevant systems. In combination, it leads to high operational costs of such explorational missions and potential loss of people due to the weather and sea conditions.

Simultaneously, large crew size for the vessel require vessel to grow in size, which leads to the increased power requirements. The common propulsion systems for medium and large size vessels are based on diesel engines or, for some military operation, on nuclear power. Chin et al. [2] proposed a hybrid propulsion with two or more sources of propulsion such as diesel, batteries, and other renewable energy for lower emissions. The operation of commonly used diesel engines create significant air pollution, noise generation and vibrations. Such footprints have negative impact on the marine environment, demask the vessel and lead to the extended requirements for the air quality, on-board fuel reserves and related consumables. Furthermore, diesel engines require certain service operations, which further increase running costs.

Alternatives to the diesel engines were proposed in [3], and it is a battery powered system. Despite being fully electric and providing zero emissions, such systems possess significant technical limitations for the vessels [4]. First of all, such systems have significant weight, which is a negative factor for naval platforms. Higher weight of the propulsion system leads to lower potential payload capacity and decreases usefulness of the system. In addition, higher weight leads to the increase submerged volume, which increases drag and related power resistance, thus limiting potential top speed and longevity of a vessel. Secondly, battery-based systems do not hold significant potential range due to the limited capacity of the batteries, and they are linked with the long recharging time, while onshore. Thirdly, hard sea environments relate to the increased degradation of the batteries limiting the lifetime of the system. Fourthly, batteries create fire hazard and might damage the vessel and, even, lead to the sinking of it.

Another trend in sea platform development is linked with the unmanned solutions. The presence of people on board sea platforms creates certain challenges and hazards both for the nature and for the involved people. Unmanned platforms do not require certain life support systems, food and drink storage, thus can contain more useful payload and have longer at-sea time in comparison with human operated systems. Modern communication technologies, unmanned sea drones are used by marine biologists [5], military [6] and some transportation users [7]. Despite being reliable and cheap, most of these solutions are still powered by internal combustion engines and still possess all the issues associated with this type of propulsion system.

Fuel cells have become a common direction of propulsion systems for the transportation applications in 21st century. They have better energy density, as a system, in comparison with batteries, have no self-discharge processes and reasonable efficiency, as were discussed in [8]. The commonly identified issues, in applications to the road vehicles include absence of reverse power flow on board of the vehicle, lower power rating and issues with transitional operation, as the reaction time of chemical reaction within fuel cells is significantly higher than the reaction time of batteries or supercapacitors. As a result, certain systems were proposed with hybrid energy storage systems with higher reliance on HFCs for powering electrical loads with constant or slowly varying power consumption [9].

Hydrogen powered vessels were also discussed in [10], but the system was never applied to the unmanned solutions and for exploration and military conditions. No noise and no emission of the fuel cell combined with fast refueling time and capability to run longer, make such solution a potential answer to most common issues of the existing unmanned naval drones.

The operation of sea drones and their architecture was discussed in several papers, with the key focuses being put towards either remote control solutions [11], sensors and vision systems [12] and other control-related problems [13]. Another pull of research works is related to the design of specific drones for very unique environments, especially in the area of biological observations [14]. The current gap does not address development of the universal solution with a wider range of needs and links in architecture solutions between biological research and military vessels.

This paper discussed power consumption, requirements and proposes architectural solution of the small size unmanned naval platform to be used for long term observations of natural environments or military applications. The second part of the paper discusses the hull structure of such half-submerged platform, speed and distance requirements and provides power requirements for the propulsion system. The third part discussed auxiliary loads of unmanned systems and power requirements of related systems. The fourth part presents proposed architecture of the platform with the use of fuel cell and discusses selection of the storage system. While the fifth part evaluates performance of the proposed solution and its operational benefits and drawbacks. The final part provides the conclusion of the conducted study and discusses the potential areas of further improvements of the system.

II. PROPULSION SYSTEM REQUIREMENT

A. Hull Structure

Half-submerged unmanned sea drones possess significant benefits for the end-user in terms of overall performance. Hulls of such systems have smaller dimensions in comparison with conventional vessels, as the vessel is allowed to be submerged and does not need to carry life-support systems. Design includes lowering the drag, created by the vessel.

Drag, being a function of the hull design and dimensions, is heavily impacted by the cross sectional area, shape of the nose and transom, and the shape of the general superstructure. The specific shapes of the hull may be adjusted based on the mission, requirements and overall choices of the team. For the presented solution, the following design intents were used:

a) Streamlined shape:

Streamlined layout of the hull helps to reduce hydrodynamic drag with subsequent improvements of the efficiency of the system. For the design used for unmanned solution, torpedo-like hull shape was selected to ensure longer travelled distance on a single set of fuel.

b) Hull construction materials:

Material selection for the sea vessel includes certain concepts, which, sometimes, contradict one another. The key requirements for the unmanned solution include lightweight, corrosion resistance and impact resistance. Price is another factor to be considered during the material selection process. Resistance to biofouling is also an important factor. Polymer-based composites, fiberglass and Kevlar, were chosen for the proposed solution.

c) Sealing and pressure resistance

A critical aspect of the hull design is ensuring it can withstand the pressure of the operating depth. As no personnel are present onboard, additional seals should be introduced to protect installed equipment and drivetrain electrical components from the potential water leakage. In the developed hull, 30 meters depth was selected for the pressure protection and additional water ingress resistance system with internal walls, separating drone's system was introduced.

d) Payload bays and access hatches

As designed platform can be used for various missions, hull design includes dedicated payload compartments for adding or removing certain systems or components. Additional devices might include various configuration of sensors, cameras, data processing units, or other specific

equipment. Designed compartments, being integral part of the hull, can withstand the same pressures and depths limits as the other parts of the hull. The same approach was used to add services access hutches for drivetrain equipment maintenance.

e) Propulsion system mounts and fuel cell placement

The mounting point for the propulsion systems considered required maneuvering and speed performance for the given conditions. Back side of the hull was chosen to install two propulsion engines, to ensure weight balance, stability and controllability of the vessel. The steering system is positioned towards the middle part of the hull, as it provides not only controllability and extra maneuverability, but also give stability component and decrease the capsizing probability. Fuel cells, relevant hydrogen storage solutions and capacitors are distributed along the hull for the balance purposes.

f) Buoyancy control system

The design of the drone includes a buoyancy control system to adjust vertical positioning of the vessel with various payloads and fuel levels. In the developed solution, adjustable ballast system is foam-filled compartment, which can be emptied easily and fine-tune buoyance during the missions.

With all considerations listed above, the final design shape and target speed to identify power requirements for the propulsion part of the design, were completed using the Elizabeth Swann hull design curve, as depicted in Fig. 1 [15].

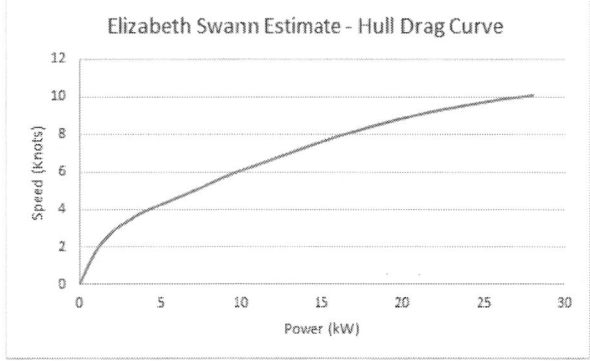

Fig. 1. Hull Power Requirements Analysis

For the selected design, 70% efficiency was assumed to identify the final required power. During optimization process, as the speed of the hull can be sacrificed to improve travelled distance, cruising speed of 7.5 knots was chosen with the power consumption of 14.5 kW continuous. For the higher speed, as 10 knots were considered, 27 kW of power would be required, which would significantly limit destination distance and reduce usability of the solution. Smaller power requirements also led to the selection of smaller motors and overall weight reduction.

B. Hull Drag Analysis

The power analysis was done for the hull design, discussed above for the cruising speed of 3.9 m/s and for seawater normal conditions with the water density of 1025 kg/m³. The dimensions of the hull were chosen based on the common architectures of unmanned streamlined sea drones, with the length of 7 meters, width of 0.95 meters and height of 0.85 meters. To reduce the visibility of the drone and cause destruction for the environment, it was assumed that 80% of the hull is submerged and should be calculated for buoyancy thrust. Weight of the hull with relevant systems, was assumed to be adjustable to meet the need of hiding the hull and it might vary from 2230 kg for the minimal water displacement, empty fuel storage and empty ballast system, with the weight being going up to 7065 kg with fully filled buoyancy system.

Based on the construction findings, it is possible to calculate thrust requirement for the given conditions. Thrust of the propeller and propulsion motors depends on drag resistance Fd and buoyancy thrust requirements for the highest submerging conditions Fb. They can be calculated as [16]:

$$F_d = 0.5 C_d \rho A V^2 = 0.5*0.1*1025*0.5074*3.9^2 = 395.5 \ N \quad (1)$$

$$F_b = m_{submerged}*g = 0.8*7065*9.81 = 55446 \ N \quad (2)$$

The total thrust requirements for the conditions mentioned, sum of two thrust should be supplied [16]:

$$F_{thrust} = F_d + F_b = 55446 + 395.5 = 55841.5 \ N \quad (3)$$

As the power is related to the speed of motion, the total power of the propulsion system can be calculated by multiplying thrust force with the cruising velocity [16].

$$P_{drivetrain} = F_{thrust}*V_{cruise} = 55841.5*3.9 = 217.8 \ kW \quad (4)$$

C. Propeller Analysis

Propeller is a crucial element of the propulsion system. Selection of the propeller was conducted based on the several factors, including: efficiency maps, cavitation impact, safety factors, aspect ratio and total blade area, pitch value and other geometrical and material performances. Ideal solution compatible with the drone architecture yields efficiency of 74%. It means, that the output propulsion motor power is [16]:

$$P_{motor} = P_{drivetrain}*\eta_{Propeller} = 217.8/0.74 = 294.324 \ kW \quad (5)$$

Geometry of the propeller is defined by its diameter. Fig. 2 demonstrates the effects on varying the diameter of the propeller on the drag force of the vessel. Due to the initially chosen dimensions, diameter was investigated in the region from 0.3 m to 0.35 m, as it can be observed on Fig. 2 as the force optimized value.

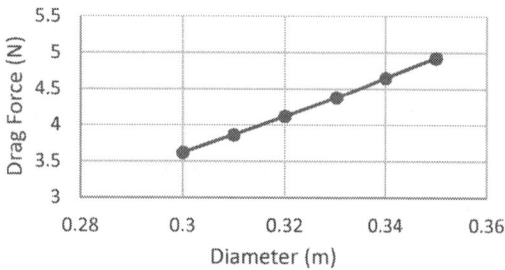

.Fig. 2. Effect of Propeller Diameter on Drag Force

The further optimization was conducted with respect to the power of the propeller. Same diameters were used, as above with the results summarized on Fig. 3 demonstrating impact of the diameter on the power requirements. The figure supports the design intent and chosen diameter for the propeller for optimization purposes.

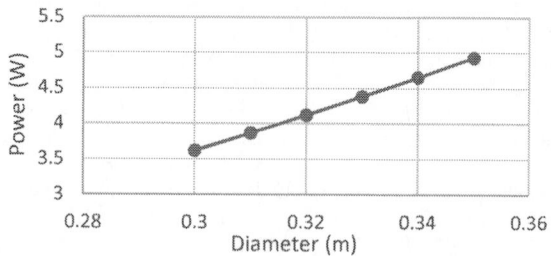

Fig. 3. Effect of Varioation of Propeller Diameter on Power

The effect of variation of the propeller velocity on the drag force is represented in Figure 4. The velocity varies from 1 m/s to 1.5 m/s. The drag force is increased from 3.63 W to 5.45 W.

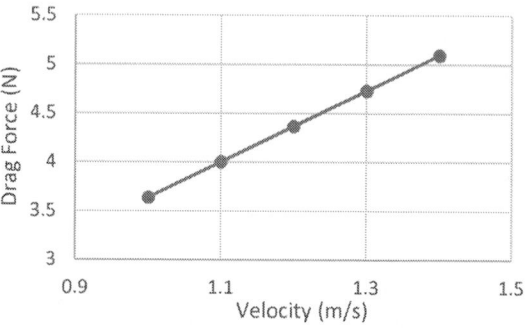

Fig. 4. Effect Of Variation Of Propeller Velocity On Drag Force

Finally, analysis of the propeller velocity and its impact on power was investigated, as shown on Fig. 5.

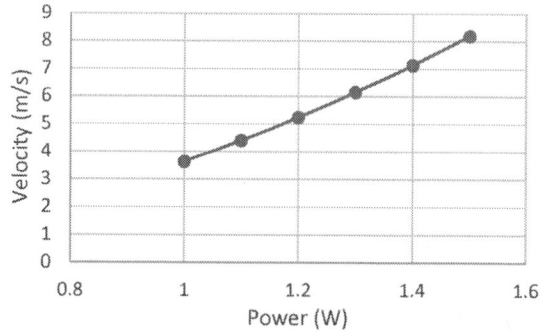

Fig. 5. Effect Of Variation Of Propeller Velocity On Power

Using the need to reach the speed, 300 kW motors were selected with the two motors architecture being used. Each motor has rating of 150 kW and ensures operation state in case of one motor failure. Smaller motors allow to use smaller individual propellers, which has positive impact on maneuverability of the solution and efficiency of the combine propulsion.

III. System Architecture

Developed platform supposed to fulfill several functions, which defines elements of the platform and their interconnection. Based on the safety requirements and the need to reach safely narrow places and complete required maneuvering, while keeping in view possibility of high-speed motion, architecture includes two types of driving electric motors – two high-speed driving motors, which are used for open sea navigation; and maneuvering and steering motors of significantly lower power but with enhanced torque capabilities. As it was shown above, the power output of the propulsion motors are chosen as 150 kW each with maneuvering motor having peak power of 10 kW – sufficient to have minor course corrective measures for the given drone design.

As the water environment has high resistance to motion and provides nearly no opportunity to recover used energy, drivetrain power electronic converters are designed with unidirectional power flow, in comparison with bi-directional converters commonly used in on-land vehicles. Two separate inverters are used for maneuvering motors and for driving motors, based on the power and frequency requirements of them. Two driving motors are added with the following purposes – to ensure system operation with the failure of one of the motors, and to enhance speed capability. In case of double motors failure, maneuvering motor has capacity to slowly move the vessel to the safe position.

Inverters, to drive motors are used as common DC/AC frequency change inverters with three phase output to operate installed induction squirrel cage motors on the propulsion side. IGBT-based system was chosen due to its speed and efficiency with multilevel inverter architecture being installed. Speed control is linked with the communication and positioning systems with predictive definition of the speed. The peak current level is set with the safety factor of 3 to resist build up of the sea weeds on the propellers during operations in shallow waters and with high presence of the weeds in the close proximity to the propellers. Based on the efficiency of 92%, the output power rating is set to:

$$P_{inverters} = P_{motors}/\eta_{Inverter} = 300/0.92 = 326 \ kW \quad (6)$$

For safety reasons, 350 kW inverter was chosen for the propulsion application. For cost saving reasons, chosen motors have brushless design with 3 phase primary operation. It means, that 24-switch, two level DC/AC inverter with variable frequency 3 phase output can be used in the developed architecture.

For starting operations and to ensure running of the motors on the low speed or when motion of the propellers is limited by external factors, ultracapacitor is added for rapid adjustments of torque in challenging environments. As was discussed above, sea vessels have relatively constant power consumption, thus any solution larger than the capacitor, would be irrelevant. Interaction between the capacitor and FC is arranged through bi-directional converter to allow FC to charge capacitor, when excessive energy is available or when the state of charge of the capacitor falls below pre-defined threshold.

The key challenge in the design of the water drone is insurance of the pay load capacity, while maintaining required range. Common solutions require up to 1000 km of the operating range, meaning that the main weight of the propulsion system should be dedicated towards energy storage system. Hydrogen fuel, having very good energy density from the fuel perspective, has an explosive nature meaning extra security measures to be taken with the storage solutions. One of the safest storage systems is based on the metal hydride chemicals, with hydrogen being stored in solid form. Such storage system cannot be used for the on-water systems due to its extensive weight, limiting overall operational capabilities of the sea drone.

Alternative to the hydride include liquid storage systems and high pressure hydrogen bottles. Liquid system operation is connected with low temperature conversion during discharge process. As the heat should be supplied from dedicated system, such solution is also not practical for the drone application. In order to maintain safety, as well as meet the weight requirements, the proposed solution includes high pressure gas storage system.

As the platform is designed for research and data collection purposes, it includes set of auxiliary systems for such operations. They are all combined in a single low voltage bus line and interconnected with the main fuel cell through unidirectional DC/DC inverter. Auxiliary systems include different types of proximity and obstacle detection sensors, positioning and navigation systems and communication systems.

Proximity and obstacle detection system is presented as a fusion of various sensors, combined in one system, which controls maneuvering motors as a primary objective. Small size radar is used to detect obstacles over long distances and adjust global passage plan, pre-defined before the beginning of the vessel mission. For the shorter-range detection, electro-optical sensors and set of cameras is used. Cameras are installed to develop 360-degree view for the mission, as well as for motion control with short proximity. Main aim is to detect type of the object and certain performances. Used cameras function in greyscale mode, as the main one, with data being processed onshore and the feed being sent over the communication system. Color feed is used for the selected areas surveillance with data being save on board of the vessel to minimize bandwidth, sent to the control center.

Electro-optical sensors are used for obstacle and range detection due to their performance with the sea environment. Sonar system with the water-based operation was not integrated in the architecture due to the dimensional and weight restrictions. However, for specific operations, it can be added as a replacement of the cameras to get additional research and surveillance data, specially related to seabed research, profiling and detection of in-water species.

Long-range navigation and control is organized through advanced satellite-based communication system, similar to Star Link with direct connection to the central control unit. Positioning is determined in real time using GPS sensors, and the data is sent in line with the surveillance and other sensors data to the central unit with required latency of 50-60 ms, which can be achieved with modern satellite systems. In case of the bad weather conditions and limited satellites signal stability, small data storage device will be utilized to save necessary data and pre-programmed GPS track will be used for location control.

Developed architecture is summarized in Fig. 6.

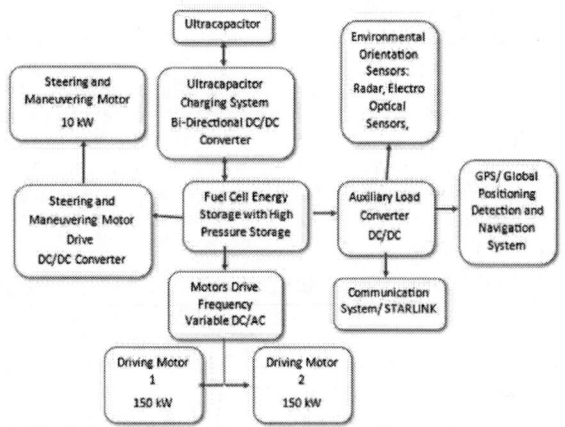

Fig. 6. Summary of Developed Sea Drone Architecture

Developed architecture includes set of power converters, linking low power loads with the main storage system. As can be seen, there are three converters being added – DC/AC high power inverter with 350 kW inverter was chosen, steering mechanism is powered through a stepper motor with two directional rotation and 70 steps per turn accuracy, thus the inverter installed is utilized only to operate with the steering motor and has power of 30 kW with low operating speed, but with excess of the torque; and the final converter supplies power to the control and other auxiliary systems. As the drone includes high computational power to process navigation and obstacle detection tasks, communication system and other related auxiliary system, the power of a converter is set on 30 kW to fulfill all power requirements of the systems.

Size of the storage defines the driving range. For the developed platform high autonomy with extended travelling range up to 850 km is the key benefit. Based on the fuel cell efficiency and storage capacity, the following estimations were done:

$$P_{total}=P_{propulsion}+P_{maneuvering}+P_{auxilliary}=380\ kW \quad (7)$$

$$t_{operating}=d_{travelled}/V_{cruise}=850/14=60.7\ h \quad (8)$$

$$E_{required}=P_{total}*t_{operating}=23180\ kWh \quad (9)$$

Based on the selected storage solution, the energy density of the hydrogen is approximated on 33 kWh/kg. Using requirements, the required storage mass of hydrogen is:

$$m_{hydrogen}=E_{required}/E_{density\ of\ H}=23180/33=700\ kg \quad (10)$$

with such mass, the submerging status of the vessel will be changing, thus, its visibility will be compensated by the buoyancy tanks to ensure lower visibility for the environment.

Target design specification, including the hall is presented in Table 1 below. It includes storage system performance as well as overall parameters of the vessel. Parameters like mass and power can be corrected based on the requirements for the platform and can be further optimized.

TABLE I.   SEA DRONE SPECIFICATION

S. No	Parameter	Units	Values
1	Dimensions (L/W/H)	m	7/0.95/0.85
2	Total Mass (Dry/Full)	Kg	2230/7065
3	Water Displacement	Kg	5652
4	Cruisig Speed	m/s (knots)	3.9 (7.5)
5	Peak Power	kW	300
6	Desired Travelled Distance	Km	850
7	Noize Level	dB	75
8	Fuel Cell Type	-	PEM
9	Output Power of a Fuel Cell	kW	400
10	Hydrogen storage capacity	Kg	700

Developed architecture was evaluated with respect to the dynamic and distance performance of the vessel and its operating time. Analysis was done using seawater conditions and it included reduction of the weight with consumption of the hydrogen and its weight reduction. As the weight reduces, the buoyancy resistance force is decreasing, subsequently reducing power consumption. It allows to create a safety operating margin for the energy storage.

Power consumption of the auxiliary systems was assumed to remain constant throughout the simulation. Operation and power consumption of the maneuvering motors was assumed also constant, as it is used for stabilization.

Results of the operating simulation, as energy consumption and remaining weight of the hydrogen are shown on Figure 7.

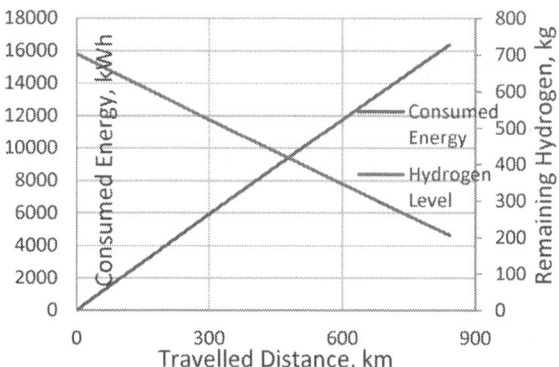

Fig 7. Developed System Performance Evaluation

As can be seen from Figure 7, targeted performance was fully achieved and certain safety margins exist for potential additional losses on the motion in presence of high waves. As expected, such performance is superiod in comparison with conventional battery drones with the travelled distance of 300-500 km.

IV. CONCLUSION AND FUTURE WORK

The paper presents the innovative concept of a low noise and low visibility unmanned sea platform. Such system can be used for surveillance, natural research or military needs. Developed solutions include low refueling time and low noise pollution through the utilization of the hydrogen fuel cell, as the main concept and benefits in comparison with alternative solutions. Developed innovative solution ensures longevity through extended energy storage capacity and use of two motors.

For the future work, thermal balance of such vehicles and fuel cell operation optimization should be performed. If combined with the intensive research on propellers design and hull shape optimization, smaller platform with similar range can be developed.

References

[1] H. P. Huntington *et al.*, "Vessels , risks , and rules : Planning for safe shipping in Bering Strait," *Mar. Policy*, vol. 51, pp. 119–127, 2015, doi: 10.1016/j.marpol.2014.07.027.

[2] C. S. Chin, Y. Tan, and M. V. Kumar, "Study of Hybrid Propulsion Systems for Lower Emissions and Fuel Saving on Merchant Ship during Voyage," *Jounal Mar. Sci. Eng.*, 2022.

[3] G. Eason, B. Noble, and I. N. Sneddon, "On certain integrals of Lipschitz-Hankel type involving products of Bessel functions," *Phil Trans Roy Soc Lond.*, vol. A247, pp. 529–551, 1955.

[4] C. A. Reusser and J. R. Perez Osses, "Challenges for Zero-Emissions Ship," *J. Mar. Sci. Eng.*, vol. 9, 2021.

[5] J. C. Maxwell, *A Treatise on Electricity and Magnetism*, 3rd ed. Oxford: Claredon, 1892.

[6] I. S. Jacobs and C. P. Bean, "Fine particles, thin films and exchange anisotropy," in *Magnetism*," New York, USA, 1963.

[7] R. Kellermann, T. Biehle, and L. Fischer, "Transportation Research Interdisciplinary Perspectives Drones for parcel and passenger transportation : A literature review," *Transp. Res. Interdiscip. Perspect.*, vol. 4, p. 100088, 2020, doi: 10.1016/j.trip.2019.100088.

[8] S. Brynolf, J. Hansson, E. James, O. Gr, H. A. Gasteiger, and J. Suchsland, "Review —

Electromobility : Batteries or Fuel Cells ?," *J. Electromechanical Soc.*, 2015, doi: 10.1149/2.0211514jes.

[9] T. Zimmermann, P. Keil, M. Hofmann, M. Horsche, and A. Jossen, "Review of system topologies for hybrid electrical energy storage systems," *J. Energy Storage*, vol. 8, 2016, doi: 10.1016/j.est.2016.09.006.

[10] M. Pagliaro, "Hydrogen-powered boats and ships," *Curr. Trends Future Dev. Bio- Membr.*, pp. 411–419, 2020, doi: 10.1016/B978-0-12-817384-8.00018-2.

[11] V. Dobref, I. Popa, P. Popov, and I. Scurtu, "Unmanned Surface Vessel for Marine Data Acquisition," presented at the IOP Conference Series: Earth and Environmental Science, IOP Publishing, 2018, p. 012034.

[12] A. Miller, B. Miller, and G. Miller, "Navigation of Underwater Drones and Integration of Acoustic Sensing with Onboard Inertial Navigation System," *Drones*, vol. 5, no. 3, p. 83, 2021.

[13] F. J. Velez *et al.*, "Wireless sensor and networking technologies for swarms of aquatic surface drones," presented at the 2015 IEEE 82nd Vehicular Technology Conference (VTC2015-Fall), IEEE, 2015, pp. 1–2.

[14] G. Schofield, N. Esteban, K. A. Katselidis, and G. C. Hays, "Drones for research on sea turtles and other marine vertebrates–A review," *Biol. Conserv.*, vol. 238, p. 108214, 2019.

[15] D. Pulvertaft, *Figureheads of the Royal Navy*. Pen and Sword, 2011.

[16] W. Shi, H. Grimmelius, and D. Stapersma, "Analysis of ship propulsion system behaviour and the impact on fuel consumption," *Int. Shipbuild. Prog.*, vol. 57, no. 1–2, pp. 35–64, 2010.

2023 International Conference on Microelectronics (ICM)

Layout-based reliability analysis of openMSP430 register file under external radiations

Vivek Bansal*, Otmane Ait Mohamed*, Fakhreddine Ghaffari+

*Department of Electrical and Computer Engineering Concordia University Montreal, Canada
+CY Cergy Paris University
vivek.bansal@mail.concordia.ca, otmane.aitmohamed@concordia.ca, fakhreddine.ghaffari@cyu.fr

Abstract—This paper presents an application specific-approach to estimate the soft error failure rate of a circuit due to the effect of external radiations. The proposed approach utilizes the Satisfiability modulo theory (SMT) to evaluate the impact of soft errors on the microprocessor's register file. We investigate the vulnerable factors at the gate level using the layout information of the circuit. A case study is conducted by performing an analysis on the register file of the openMSP430 core for benchmark programs. Exhaustive analyses focused on single and multiple event transients were performed. These analyses have been used to measure the vulnerability factor of different registers, identifying their application-specific criticality.

Index Terms—soft-error, criticality, transient, register-file

I. INTRODUCTION

With the advancements in the semiconductor industry and the aggressive downscaling of devices, electronic circuits such as semiconductor memories and microprocessors have become more prone to transient or soft errors [1]. A soft error occurs when a logic value of a logic element is erroneously changed due to the effects of radiation particle bombardment on the device. The radiation particles travel through the substrate of the device's transistor and may cause a transient voltage glitch i.e. single event transient (SET). Due to charge sharing between neighbor transistors, SETs can cause multiple transistors to be affected i.e. single event multiple transients (SEMT) [2].

Several research papers have been published to study SET or SEMT of the digital circuits [3] [4] [5]. Effect of single and multiple event transients [5] is studied using simulation-based fault injection into the model of the LEON3 processor. Faults are injected into flip-flops, register file, and cache memories and their behavior against transient fault have been studied. Therefore, transient faults in any part of the circuit have become a significant threat to its reliability. The processor is one of the main components in the design of digital circuits, thus, an accurate and efficient model is needed to analyze the transient faults in the processor.

Also, register files in a processor are frequently accessed part throughout the execution of a program, so any soft error that occurs in the register file will propagate to the other parts of the processor and will change the original functionality of the program thus leading to the incorrect output of the program. Several error handling techniques such as error correction code (ECC) and recovery through exiting copy (REC) [6] [7] are used to protect the register file against soft errors. All these techniques are often based on data redundancy

which represent a very huge architecture overhead for the processor. Therefore, it is very important to check the soft error vulnerability of the register file that is being used throughout the program and then decide which part of the processor should be more protected.

In this paper, we propose an application-based approach to calculate the soft error vulnerability of the program and its associated register file based on the block-based dependency graph and register data flow graphs. This approach relies on the vulnerability factor of the register file block in the core that is being used throughout the program. The vulnerability of the register file block is calculated based on SETs and SEMTs propagation using the Satisfiability Modulo Theory (SMT). Unlike related techniques that deals with the RTL of the design, our approach is much accurate as it considers not only the synthesised design but also the application. To demonstrate the proposed approach, we present the analysis of the openMSP430 core as a case study.

The remainder of this paper is organized as follows: Section II describes the overview of openMSP430 and section III lists preliminaries for our methodology. The detailed proposed methodology is described in section IV. The experiment on the selected benchmark is presented in section V. Finally, we conclude the analysis based on the experiment in Section VI.

II. ARCHITECTURE

OpenMSP430 is a synthesizable 16-bit microcontroller core, compatible with Texas Instrument's MSP430 core family [8]. It is based on Von Neumann's architecture. The microcontroller has a single dedicated address space for instruction as well as for data. The core is composed of 16 different registers (R0-R15), each register contains 16-bits. *R0- R3* have dedicated functions, *R0* as a program counter (PC), R1 as a stack pointer (SP), R2 as status register and *R2/R3* as constant generator registers. *R4-R15* are general purpose registers.

As Fig. 1 shows, MSP430 architecture incorporates 16-bit RISC CPU, peripherals, and a flexible clock system that interconnect using a von-Neumann common memory address bus (MAB) and memory data bus (MDB). In our case study, we use register transfer level (RTL) description of the open-MSP430, freely available for academic research purposes [8].

979-8-3503-8083-5/23 $31.00 © 2023 IEEE

2023 International Conference on Microelectronics (ICM)

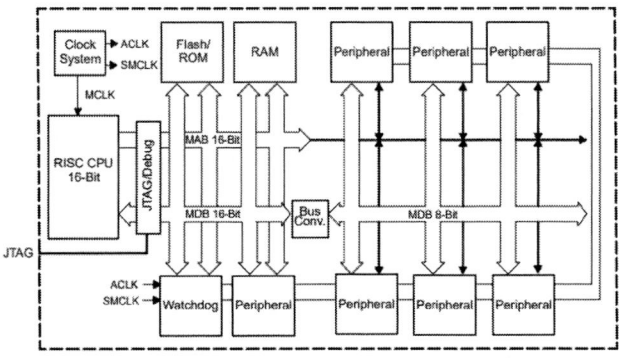

Fig. 1. MSP430 Architecture

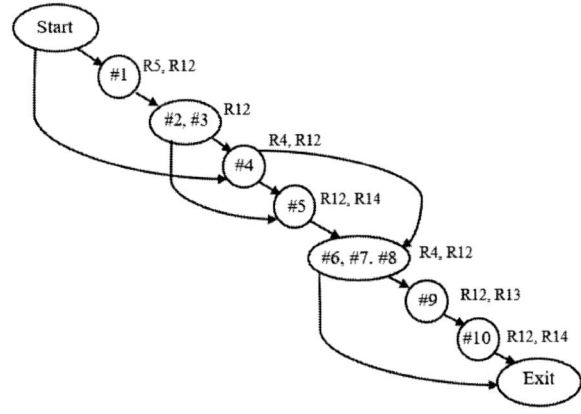

Fig. 3. Register Data Flow Graph.

III. PRELIMINARIES

A sample program as shown in Fig. 2 having multiple blocks of executable instructions [9] is represented by a block-based dependency graph (BDG). Each block is a set of statements whose execution begins with the first instruction and ends with the last. A Register data flow graph (RDFG) for each block is generated based on its sequence of statements.

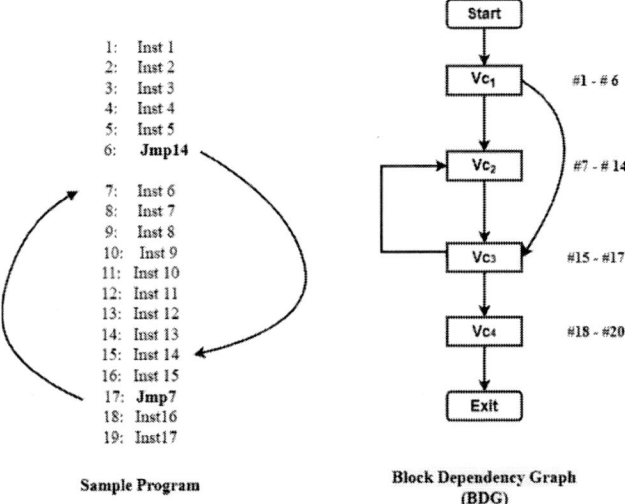

Fig. 2. A typical Sample Program with its Block-dependency Graph (BDG).

Block-based Dependency Graph (BDG): A program can be defined as a block-based dependency flow graph $Gc = \{Vc, Ec\}$, where Vc represents the set of vertices $\{Vc1, Vc2, Vc3.....\}$ denoting each block in the program and Ec represents a set of edges to possible flows of control from one block to another. A basic block terminates at either an instruction that has a branch to another basic block or the instruction is receiving the transfer of control flow from two or more places in the program. Figure 2 shows an example of a block-based dependency graph for the example program. It consists of four nodes (Vc_1, Vc_2, Vc_3, Vc_4) excluding start and exit node. Nodes represent the different blocks of the sample program. Each node has a set of instructions as

mentioned in the BDG. For example, Inst 1 - Inst5 and Jmp14 are executed in a block denoted as $Vc1$, and instructions Inst 14, Inst 15, and Jmp7 are executed in another block denoted as $Vc3$).

Register Data Flow Graph (RDFG): A RDGF can be defined as $Gd= \{Vd, Ed\}$. Vd is a set of nodes $\{Vd1, Vd2, Vd3.....\}$, each node is associated with an operation (s) that is being performed in that block. If execution of a RDFG node $vd1$ has data dependency to another RDFG node $vd2$, then the RDFG has edge between them. Generally, the directed edge represents how the data stored in one register is being used in the execution of the next instruction. An example of a register data flow graph for one block is shown in Fig. 3. Each node represents one or more instructions depending upon the register data dependency between them. All the registers associated with the execution of particular instruction (s) are also denoted in the graph. .

IV. PROPOSED METHODOLOGY

We present a brief overview of the proposed methodology in Fig. 4. The proposed technique starts by compiling the design program to get assembly language instructions with the help of the GCC compiler. Based on the assembly language description of the program, the block-based dependency graph and register data flow graphs are obtained based on the mode of operation and data flow throughout the program. Further, the Satisfiability modulo theory (SMT) is applied to each node in every register data flow graph to estimate soft error vulnerability factors. Additionally, we calculate the vulnerability of the register file and program reliability after the complete execution of the program.

A. Satisfiability Modulo Theory (SMT) Analysis

To perform Satisfiability Modulo Theory (SMT) analysis, we use the Register-Transfer Level (RTL) design of open-MSP430 core as it is freely available [8]. Using a design synthesis tool (i.e. Synopsys Design Vision), we synthesize the complete RTL design to obtain a gate-level netlist having

979-8-3503-8083-5/23 $31.00 © 2023 IEEE

vulnerability factor (i.e. soft error probability) is calculated for the register file is calculated using an SMT solver.

B. Register File Vulnerability

To calculate the soft error rate of the register file, we define a register vulnerability factor (RVF) for each register in the openMSP430 core. The vulnerability factor is defined as the probability that a transient fault or soft error may produce the wrong output. The vulnerability factor of any register is considered in the execution of the program if the register is read by the processor in the next instruction. The vulnerability factor of the register is overwritten when it is loaded with new data. Basically, we measure each register utilization in each operation through run-time tracing of its read and write access.

To compute the RVF of any register during the execution of any instruction, the first step is to identify the type of operation and flow of its execution in the microprocessor. Secondly, we determine all registers that are being used in the execution of the instruction. After that, the vulnerability factor of each register used in the operation is recalculated using the vulnerability factor of the register file. After each instruction of the block, all the values are passed to the next instruction. If the next instruction reads the same register used in a previous instruction, then it is vulnerable that has the vulnerability factor calculated in the previous step. This value will be updated after the execution of the current instruction.

To compute the vulnerability of the register file, the RVF of each register is calculated for the complete execution trace of the program. Then, the vulnerability of the register file is computed as the sum of all registers' RVFs divided by the number of registers (N) in the register file.

$$Vulnerability\ of\ Register\ File = \frac{\sum_{i=0}^{i=N} RVFi}{N} \quad (1)$$

V. Analytical Experiment

To demonstrate the scalability of the proposed method, we have performed several analytical experiments on different benchmark programs written in C language: (1) QuickSort Algorithm, (2) Matrix Multiplication, and (3) Fibonacci Series Number Calculation. Based on the instruction set of MSP430 instruction set, we compile the source code written in C-Language into assembly language using GCC compiler. The execution path of all the programs depends on the input parameters. For each benchmark, we extract the execution path and the block-based dependency graph (BDG) is build based on this path as described in section IV. With python scripting, register data flow graphs (RDFGs) for each node in the BDG are constructed to check how the data is flowing from one register to another register depending on the mode of operation.

We present a glimpse of the Fibonacci series in Fig. 6, showing four different blocks of a program. Each block has a set of instructions that denotes the one node in the BDG. In the Fibonacci series, different types of operations like MOV, ADD, PUSH, etc. are utilized. For each kind of operation, the

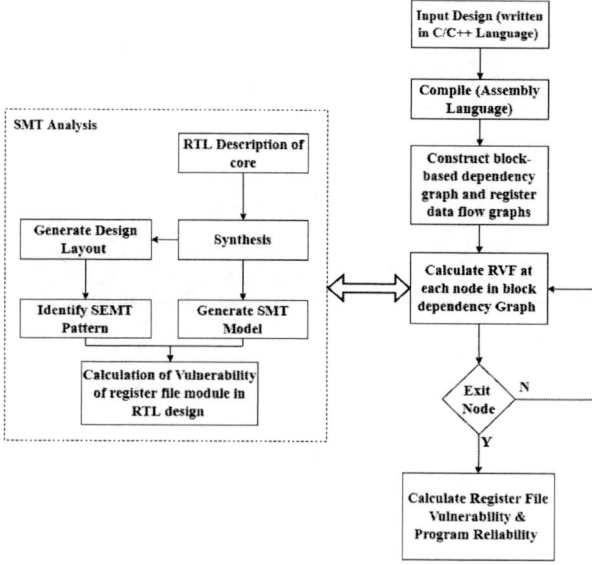

Fig. 4. Basic Steps of Proposed Methodology.

multiple module declarations. With Python scripting and network graph theory, we obtain a module hierarchy that helps us to create an SMT model for each module in the synthesized openMSP430. Fig. 5 shows the openMSP430 processor's hierarchy view, seen after the synthesis step. Additionally, we generate the layout for modules in the core using a layout tool (i.e. Cadence Encounter). After the generation of the layout, we route the design and extract the timing information in the standard delay file (.sdf).

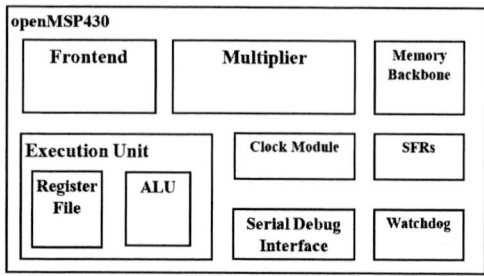

Fig. 5. Design Hierarchy after synthesis.

To evaluate the impact of soft errors on the circuit, we utilized Multiple Event Transient Analysis (META) tool [10]. META is an automated EDA layout based tool to estimate the vulnerability of the circuits against single event transient (SET) and single event multiple transient (SEMT) fault. To calculate the vulnerability factor of the register file module in the openMSP430 core, we analyze the SEMT propagation at the gate level. An SMT model of the design is built using the synthesized gate-level netlist and the timing reports (.sdf file) available for its generated layout. At the same time, SEMT patterns i.e. error sites due to radiations, are also extracted. Using SEMT patterns and the SMT model of the design, the

```
main:
    I1      SUB.W #8, R1
    I2      MOV.W #0, @R1
    I3      MOV.W #1, 4(R1)
    I4      MOV.W @R1, R12
    I5      ADD.W 4(R1), R12
    I6      MOV.W R12, 2(R1)
    I7      MOV.W #3, 6(R1)
    I8      BR #.L2
.L3:
    I9      MOV.W 4(R1), @R1
    I10     MOV.W 2(R1), 4(R1)
    I11     MOV.W @R1, R12
    I12     ADD.W 4(R1), R12
    I13     MOV.W R12, 2(R1)
    I14     ADD.W #1, 6(R1)
.L2:
    I15     MOV.B #10, R12
    I16     CMP.W 6(R1), R12 { JGE .L3
    I17     NOP
    I18     ADD.W #8, R1
    I19     RET
```

Fig. 6. Glimpse of Fibonacci Series Program

processor has a unique path and it utilizes a set of modules in the processor to execute the instruction. At the first node of BDG, all the registers do not have any vulnerability values. The vulnerability values (RVF) of the register will start to change while the processor is running some operation.

According to the RDFG of any block of the algorithm, the first node *"START"* receives RVF values for all registers from the *"EXIT"* node of the previous RDFG node. If any instruction reads the vulnerable register while its execution and writes the data back to another register then the RVF of the reading register remains the same and the RVF of the write register is overwritten over the previous value. The soft error probability of registers corresponding to the operation is calculated and stored in a list. The list containing soft error probabilities of all registers is passed to the next instruction in the block. At the *"EXIT"* node of the RDFG, all updated values will be passed to the next node of BDG. Similarly, after execution of all instructions in the program, RVFs of all registers are extracted from *"EXIT"* node of the BDG to calculate the vulnerability of the register file using equation 1.

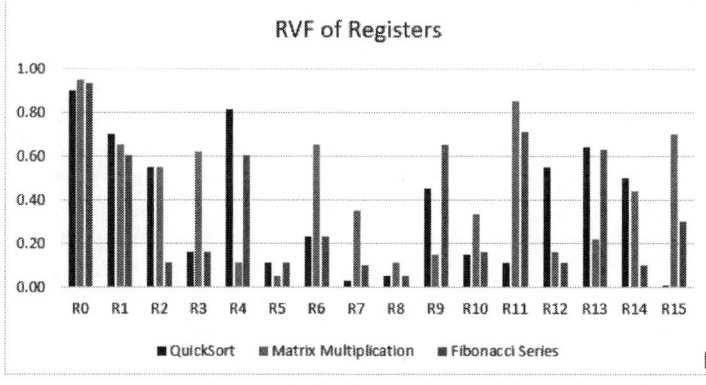

Fig. 7. Register Vulnerabilities.

We have computed register vulnerability factors for all three

mentioned benchmarks. Fig. 7 illustrates the comparison of three benchmark programs with their register vulnerability factors. It shows soft error vulnerability factor values of all registers. As seen from the graph, the program counter register *(R0/PC)* has a higher probability of having a soft error because this register is used in the execution of every instruction.

TABLE I
EXPERIMENTAL RESULTS

Benchmarks	Vulnerability of Register File
QuickSort	0.3718
Matrix Multiplication	0.4306
Fibonacci Series	0.3386

Table I indicates the vulnerability of the register file for the benchmark programs. The register file for the matrix multiplication algorithm has the maximum vulnerable value.

VI. CONCLUSION

In this paper, we have proposed a methodology to estimate the vulnerability factors of registers against transient faults post synthesis considering the design layout. We tested our method on various benchmarks with an openMSP430 micro-controller core and extracted the most vulnerable registers after their execution. Besides the program counter *(PC)* and stack pointer registers *(SP)* which are the most vulnerable for all benchmarks applications, our methodology highlights that the most commonly used register throughout the execution are more vulnerable compared to other general-purpose registers.

REFERENCES

[1] I. Chatterjee et al., "Impact of Technology Scaling on SRAM Soft Error Rates," in IEEE Transactions on Nuclear Science, vol. 61, no. 6, pp. 3512-3518, Dec. 2014.

[2] P. Huang et al., "Heavy-Ion-Induced Charge Sharing Measurement With a Novel Uniform Vertical Inverter Chains (UniVIC) SEMT Test Structure," in IEEE Transactions on Nuclear Science, vol. 62, no. 6, pp. 3330-3338, Dec. 2015.

[3] D. Munteanu et al., "Modeling and Simulation of Single-Event Effects in Digital Devices and ICs," IEEE Transaction on Nuclear Science, VOL. 55, NO. 4, August 2008.

[4] Akira Mochizukil et al., "Multiple-Event-Transient Soft-Error Gate-Level Simulator for Harsh Radiation Environments", published in TEN-CON 2015 - 2015 IEEE Region 10 Conference.

[5] H. Abbasitabar et al., "Susceptibility Analysis of LEON3 Embedded Processor against Multiple Event Transients and Upsets," 2012 IEEE 15th International Conference on Computational Science and Engineer-ing, Nicosia, 2012, pp. 548-553.

[6] H. Kanbara et al., "Probability of calculation failures by soft errors in an embedded processor core," 2009 IEEE 8th International Conference on ASIC, Changsha, Hunan, 2009, pp. 601-604.

[7] A. Eker et al., "Exploiting Existing Copies in Register File for Soft Error Correction," in IEEE Computer Architecture Letters, vol. 15, no. 1, pp. 17-20, 1 Jan.-June 2016.

[8] "OpenMSP430 project", [Online]. Available at "https://opencores.org/projects/openmsp430".

[9] Jianli Li et al., "Reconstructing control flow graph for control flow checking," 2010 IEEE International Conference on Progress in Infor-matics and Computing, Shanghai, 2010, pp. 527-531.

[10] V. Bansal et al., "META: A Layout Based Tool to Estimate the Vulnerability of Digital Circuits to Multiple Event Transient," 2022 20th IEEE Interregional NEWCAS Conference (NEWCAS), Quebec City, QC, Canada, 2022, pp. 450-454.

A Novel Architecture of CXL Protocol Data Link Layer for Low Latency Memory Access

Basma H. Mohamed
Computer and Systems Engineering Dept.
Ain Shams University
Cairo, Egypt
basmahesham648@gmail.com

Esmail Hany
Siemens EDA, Cairo, Egypt
Cairo, Egypt
esmail.hany@siemens.com

Mahmoud El-Tahawy
Siemens EDA, Cairo, Egypt
Cairo, Egypt
mahmoud.eltahawy@siemens.com

Mohamed Abdel-Salam
Siemens EDA, Cairo, Egypt
Cairo, Egypt
mohamed.abd_el_salam_ahmed@siemens.com

M. Watheq El-Kharashi
Computer and Systems Engineering Dept.
Ain Shams University
Cairo, Egypt
watheq.elkharashi@eng.asu.edu.eg

Mona Safar
Computer and Systems Engineering Dept.
Ain Shams University
Cairo, Egypt
mona.safar@eng.asu.edu.eg

Abstract—With the increase in memory-intensive applications, there is a need to provide low latency paths for memory access and coherent caching between host processors and storage devices. This paper proposes a novel architecture for the Data Link Layer of Compute Express Link (CXL) protocol, as defined in CXL Specification Revision 2.0. The architecture presented here contains the transmission and reception chains, which ensures the reliable conveyance of the packets between two CXL components using the CXL protocol. This paper explains how the proposed architecture ensures reliable data conveyance, responses and requests, through the addition of a 16-bit Cyclic Redundancy Check (CRC) to each 512-bit FLow control unIT (FLIT) in the transmitter side, and how the packets are processed in receiver side. The architecture is intended to have simple implementation in addition to having minimum datapath delays. The whole architecture is designed using HDL then verified using test-benches for different protocol scenarios to make sure that the functionality of the system is achieved. In addition, the whole design is synthesized using United Microelectronics Corporation (UMC) 130 nm [1] and Taiwan Semicomductors Manufactor Company (TSMC) 28 nm [2] technologies to get the maximum achieved frequency and calculate the corresponding power and area. The proposed hardware implementation manages to work up to 1 GHz using TSMC 28 nm technology and 438.59 MHz using UMC 130 nm technology. Its estimated dynamic and leakage power consumption by UMC 130 nm technology at 1 GHz are 95.3698 mW and 171.17 mW, respectively. In addition, its area cost extracted after synthesis is $242\,124.8\,\mu m^2$.

Index Terms—Computational Storage Devices (CSDs), Compute Express Link (CXL) protocol, Near memory computing, Peripheral Component Interface Express (PCIe) protocol

I. INTRODUCTION

The amount of data created in any application is rising exponentially. Initially, these data were kept in storage systems. Application servers must retrieve data from storage devices in order to process it, which imposes a cost for transferring data to the system. To that end, devising new system architectures that move data near the processor is key. By integrating processing engines into storage devices to process data, "move process to data" paradigm is pushed to its farthest limits by

Computational Storage Devices (CSDs) [3]–[5]. However, the growth of memory bandwidth and computing capabilities per silicon die presents an intriguing scaling issue for CSDs. In order to overcome scaling issues, many chip suppliers and data centre operators are resorting to parallelism, which causes the compute and memory capacities to continue growing exponentially. Memory designs may, in theory, keep up with this rising need for bandwidth as memory capacity increase. However, the interfaces to the memory subsystems are the restriction.

Cache-coherent "far" memory protocols [6]–[11] have been introduced as a solution, opening up a variety of options for memory subsystems. By connecting memory to ports that would often be used for storage, they offer a way for cloud and High-performance computing (HPC) providers to increase memory capacity and bandwidth [12], [13]. These interfaces provide potential for subsystem-level acceleration and with the possible inclusion of both memory and storage behind this interface. Consequently, they enable data transfer inside the integrated storage subsystem, reducing the bandwidth bottleneck into the CPU [14], [15].

Compute Express Link (CXL) 2.0 standard provides an interface that leverages the PCIe 5.0 physical layer and electricals. It uses extremely low latency pathways for memory access and coherent caching between host processors and devices like accelerators and memory expanders that must share memory resources [9], [12], [13], [16]. It offers the option to support either PCIe or CXL configuration. Achieving low latency data access helps in supporting newer Artificial Intelligence (AI) applications and processing larger data sizes. In addition, coherent caching reduces latency for memory access, which makes the system more efficient.

CXL is a dynamic multi-protocol technology made to work with memory and accelerators. It offers a wide range of protocols, including memory access semantics (CXL.mem), caching protocol semantics (CXL.cache), and I/O semantics

979-8-3503-8083-5/23 $31.00 © 2023 IEEE

that are similar to PCIe (i.e., CXL.io) [17], [18] across discrete or on-package links. For enumeration and discovery, error reporting and host physical address (HPA) search, CXL.io is necessary [9], [13], [19].

The purpose of this paper is to introduce a hardware architecture proposal for data link layer of universal CXL protocol according to CXL standard 2.0. It assumes that only CXL.mem part is implemented in the designed layer. The algorithm of each introduced block is described as well to achieve the required functionality. Finally, the expected and resulted timing diagrams in addition to the performance evaluation results are provided.

The paper contributions are summarized in the list below:

1) A low latency paths for memory access between host processors and devices are guaranteed in different operations.
2) A modular strategy is used to address the design's intrinsic complexity of link layer.
3) High bandwidth and low energy interface protocol design is proposed and implemented.

This paper is organized as follows, Section II introduces CXL standard architecture and specifications based on revision 2.0. Section III introduces the link layer block diagram and the design implementation of each individual block. Section IV shows the full chain simulation on funcional simulator [20]. Section V discusses the resulted design performance and cost generated by synthesis tools [21], [22]. Finally, a conclusion section summarizes what is introduced.

II. CXL Protocol Specification Overview

An overview of the CXL protocol specification 2.0 is provided in this section. CXL protocol provides a flexible (Flex) high-speed port that helps the designs in deciding whether to offer CXL or native PCIe protocol. Alternate protocol negotiation is used to make the choice during link training and it is dependent on the plugged-in device [9].

Flex bus architecture is formed as a number of layers, shown in [9]. The controller layer of CXL, which composes of transaction and link layers, is segmented into logic for CXL.io and another logic for CXL.mem and CXL.cache. Additionally, the transaction layer and link layer merge the CXL.mem and CXL.cache logic. Afterwards, CXL link layer is connected to an arbitration multiplexer (ARB/MUX), which by means accepts the traffic from the two logic paths; PCIe and CXL streams [9], [23]. Finally, the physical layer receives the FLIT, a unit amount of data for the message in link-level, and prepares its transmission on the line [9], [17], [18].

The transaction layer is the top CXL layer and it interfaces with the application layer through CXL Protocol Interface (CPI) to read the request information or respond with the request feedback [23]. It is accountable for the preparation and processing of Transaction Layer Packets (TLPs) [9].

Link layer is the middle CXL layer, which deals with both transaction layer and physical layer. Its fundamental role is the assembling and disassembling of TLPs in a packet called FLIT. It has the responsibility of data integrity, including

discovery and correction of errors. Additionally, it is in charge of managing the FLITs' credit-based flow [9].

The specification's four architectural CXL layers [9], [23] combine the packets into a complete CXL FLIT to carry out the information transaction. Each layer of the transmitting CXL device adds its own information as the transaction progresses. On the contrary, when it reaches the receiver side of the link mate, this design completely reverses.

III. Link Layer Implementation

This section provides the implementation details for link layer in the host side. An overview of the overall architecture is presented. Then, it is divided into couple of modules according to protocol functionality and a description of each module is given. The CXL.mem link layer is responsible generating or receiving a 528-bit FLIT. The data link layer packs multiple of TLPs in the generated FLIT and protects the contents of this FLIT by adding the CRC bits. It calculates the CRC value based on the packed TLPs from the transaction layer and the FLIT header it has just added. The CRC algorithm is explained in [9].

In addition, link layer is responsible for FLIT error handling by recalculating the CRC at the receiver side and initiate FLIT re-transmission in case of error detection at the remote side. An additional block should be responsible for link layer initialization after the physical layer getting into a reset state. It is responsible also for traffic management through credit information exchange between the host and device [24].

Link layer has four main blocks to perform the above functionalities. These blocks are shown in the proposed link layer block architecture in Fig. 1. Each of them is implemented separately then integrated together to perform the protocol operations. An overview of each of them is introduced as well as follows.

A. FLIT Packing and Un-packing

This block is responsible for packing multiple of protocol TLPs to help improving link efficiency. In addition, it packs the control FLITs, as regulated by the standard [9].

B. Retry Mechanism and Cyclic Redundancy Check (CRC)

Re-transmission, or Link Layer Retry (LLR), is a feature of the link layer that allows for recovery from faulty transmission. The terms "local" and "remote" are used to describe the entity that discovers a FLIT error and the entity that sends the FLIT incorrectly, respectively [25].

The LLR algorithm is achieved using two finite state machines, which both exist in each entity and the transmitter and receiver overall state is determined by these state machines. The first one is Remote Retry State Machine (RRSM), which is triggered at the entity receiving the error-FLIT. The other one is Local Retry State Machine (LRSM), which is activated at the remote entity transmitting a corrupted FLIT after being detected by the local receiver. CXL standard [9] shows the retry sequence between local and remote devices.

979-8-3503-8083-5/23 $31.00 © 2023 IEEE

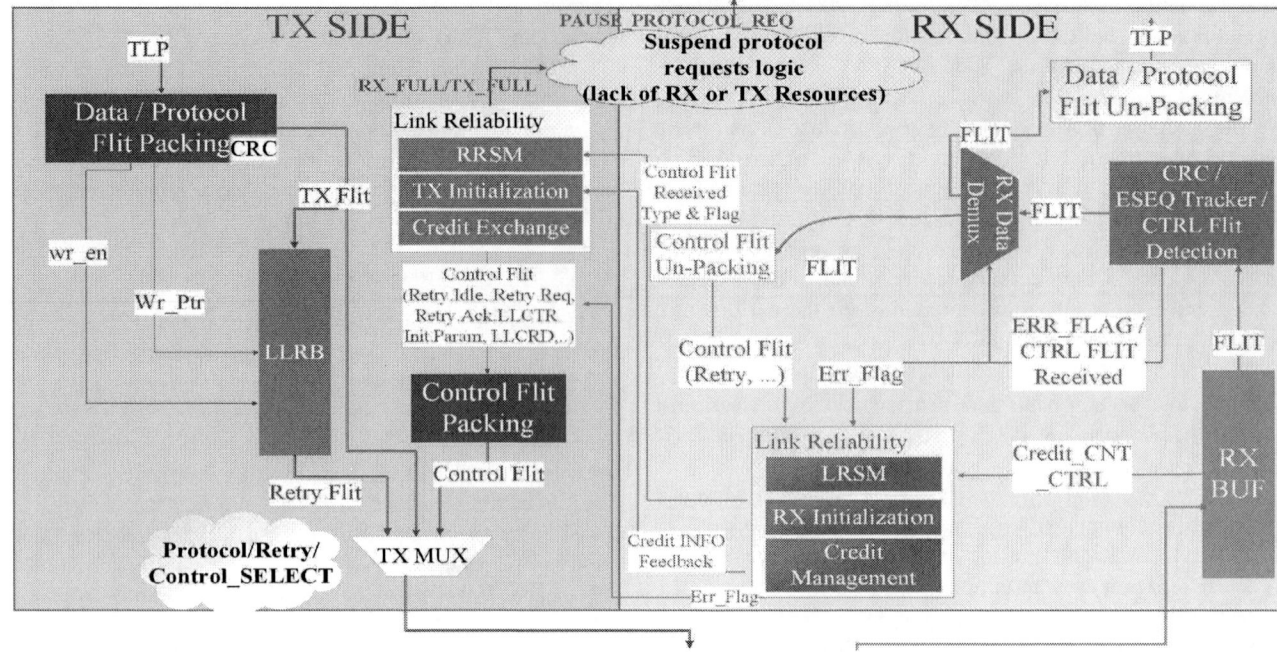

Fig. 1. Proposed data link layer architecture.

1) LRSM: LRSM algorithm in the RX side is composed of a CRC calculator, a state machine controller block (LRSM) and couple of counters. These counters help in determining the overall state of the transmission and reception state [9].

If a TimeOut counter reaches its maximum waiting for Retry.Ack FLIT reception, LRSM re-sends the Retry.Req. It keeps re-requesting the FLIT re-transmission, if Retry.Ack is not received successfully, until reaching the maximum retry counts (MAX_NUM_RETRY) then it triggers physical layer re-initialization. After exiting the re-initialization state, the LRSM starts retry request mechanism again until either having a successfull Retry.Ack or ending up in a retry abortion state.

2) RRSM: RRSM algorithm in the TX side is composed of a local link layer retry buffer (LLRB), LLRB controller, LLRB counters, and state machine controller block (RRSM) [9].

LLRB module stores all retriable FLITs. When it is necessary, it re-transmits the unacknowledged stored FLITs. There are couple of counters that are used to track the write and read position in the buffer, which are described in [9].

C. Link Layer Initialization

After physical layer link transition from down to up, link layer initialization must start. It is the phase, where the link parameters are initialized and credits are exchanged between host and device before transmission. A timing diagram example shown in Fig. 2 describes how the link layer is initialized and credits are exchanged as stated in [9].

The exchanged Link Layer Control Initialization Parameter (LLCTR.Init.Param) FLIT in Fig. 2 carries some initialization information, listed in [9]. If a CRC error is detected during this

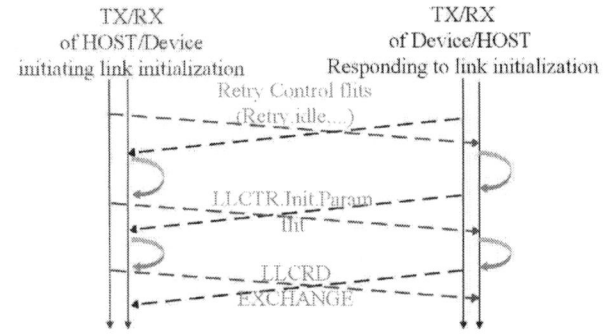

Fig. 2. CXL Standard link layer initialization sequence.

sequence, a retry sequence is initiated until an error-free FLIT is received. At the end of this sequence, both entities exchange their available resources in a Link Layer Credit (LLCRD) FLIT to exit the initialization phase successfully.

There are two state machine blocks, which control the described sequence in both TX and RX sides. Both of them are responsible for determining the overall initialization state of the link layer. These state machines trigger the retry blocks, if necessary.

D. Credit Management and RX Buffer

CXL protocol defines a credit value in order to track the resources available inside the RX buffer in the collocated receiver. It indicates whether the receiver is able to receive a new FLIT to manage the traffic between the host and device. The transmitter sends the credit return fields for use by the remote transmitter. Messages sent on data channels require a single data credit for the entire messages, including the header of the message. Moreover, credit is exchanged in the LLCRD FLIT in the link layer initialization sequence and in the forced LLCRD FLIT in case of asymmetric traffic. The RX buffer has the same size as the retry buffer, which is assumed to be 50 entries in this design.

IV. FUNCTIONAL VERIFICATION

This section discusses the approaches used to verify the functionality of the proposed hardware architecture. Different functional scenarios, used in the testing, will be explained as well. Verification for the full architecture is carried out by building a test-bench generating different test scenarios and compared using automated checkers to verify the output samples correctness.

A. Assumptions and Limitations

Couple of assumptions are taken into consideration while implementing the proposed layer to provide initial simulation results:

- Link width is ×16 with 32 GT/s data rate, therefore clock rate used is 1 GHz.
- Simple TLP packing rules are used, where only 1 TLP is packed per FLIT.
- There is only one header per FLIT.

B. Test Scenarios

There are different test scenarios implemented for both TX and RX sides to cover different operations. The simulation results of the covered scenarios are performed on the top-level of the transaction and link layer. In the transmission testing, write and read operations are introduced and verified. Similarly, the reception side of the host is verified in both reading and writing completion. Retry and initialization mechanisms' verification results are shown as well.

1) Transmission Verification Results: Write and read operation requests are tested. A valid FLIT output signal is high when a FLIT is ready to be passed to the ARB/MUX, as shown in Fig. 3 and Fig. 4.

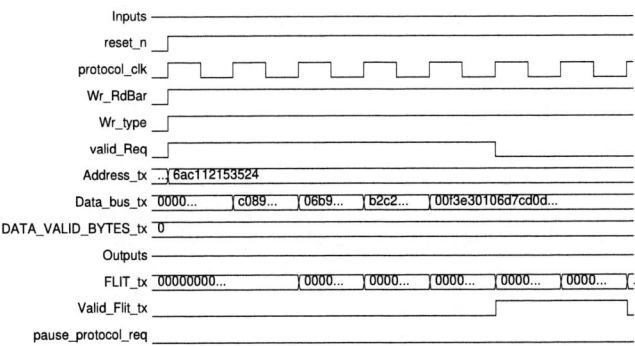

Fig. 3. TX write operation simulation results.

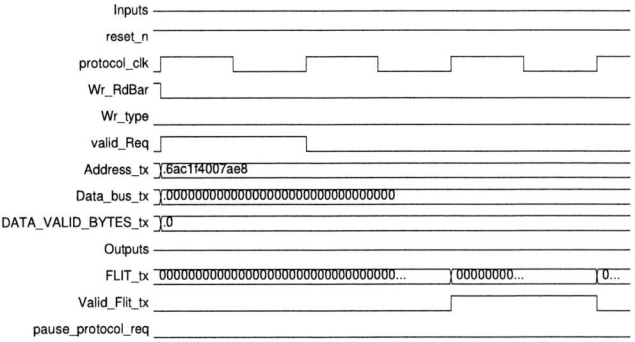

Fig. 4. TX read operation simulation results.

2) Reception Verification Results: Read and write operations responses are tested as well. However, read operation reception supports two different data read modes; a full cacheline read mode, which is 64 bytes (4 data slots), and half cacheline, which is 32 bytes (2 data slots). Therefore, the received data could need two FLITs similar to the transmission write mode or only one FLIT as in Fig. 5.

A valid response output signal is high when the FLIT is processed and the feedback is ready to be delivered to the application layer, as shown Fig. 5. In addition, a valid data signal is asserted in the read operation whenever the data chunks are ready to be read from least to most data chunk order, as shown in Fig. 5. Write completion simulation result is similar to the half cacheline read reception operation but with having write acknowledgement signal asserted instead of receiving data chunks.

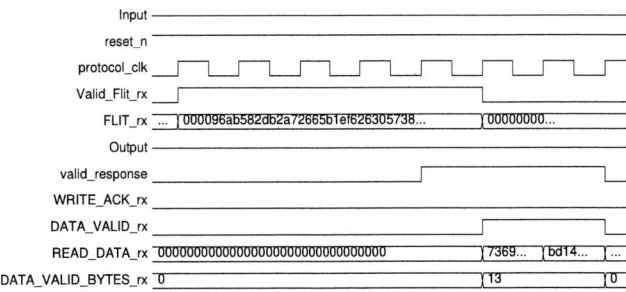

Fig. 5. RX half cacheline read operation simulation results.

3) Retry Mechanism Verification Results: In order to validate the retry mechanism sequence, two retry scenarios should be covered. The first scenario tests the host initiating a retry request as a result of receiving an erroneous FLIT. The other scenario tests the host responding to a retry request initiated from the remote entity. The implementation functionality for both scenarios, shown in Fig. 6 and Fig. 7, are verified by comparing the transmitted FLITs with the sequence illustrated in [9].

4) Initialization Mechanism Verification Results: Link layer initialization sequence is tested by having a communicating host and device after having the physical layer link powered

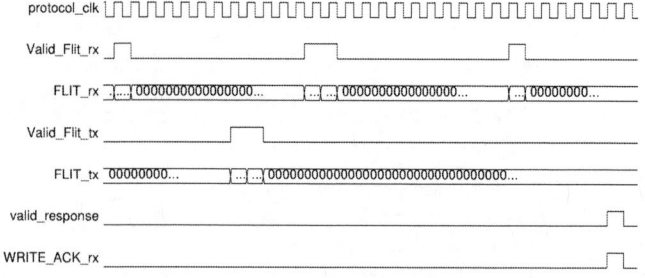

Fig. 6. Host initiating retry sequence simulation results.

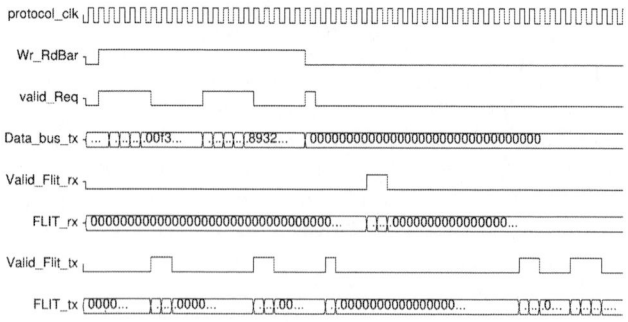

Fig. 7. Host responding to retry sequence simulation results.

up. It is assumed that an error free sequence is being received by the host to test the normal initialization scenario from the host perspective. The implementation functionality in Fig. 8 is verified by comparing the transmitted FLITs with the standard sequence in Fig. 2.

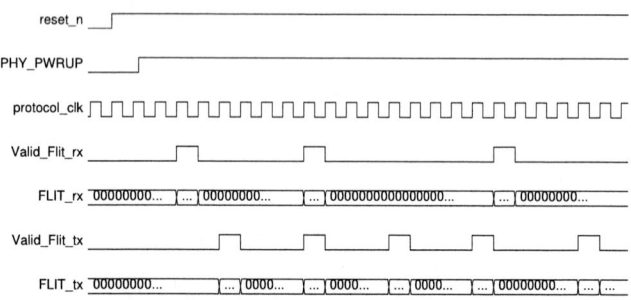

Fig. 8. Normal (error-free) initialization sequence simulation results.

V. PERFORMANCE RESULTS

The performance metrics for the realized CXL controller layers are introduced in this section. UMC 130 nm and TSMC 28 nm technologies are used to check the maximum frequency that can be achieved, as shown in Table I [21], [22]. UMC 130 nm technology is used as well to calculate power at 1 GHz frequency and area, as shown in Table II.

Table II shows that the implementation consumes significant amount of power. This is due to having TX and RX paths, which operate together to handshake and feedback some

information to each other. Some modules are gated using enable signals, however others keep tracking and counting or either ready to process any required data. Moreover, having large RX buffer and LLR buffer increases the design power consumption. In addition, synthesis using large technology, as in UMC 130 nm, results in consuming more power than newer technologies. This is due to having larger gates. This reason results in having large area as well as shown in the Table.

Design enhancements are performed to improve design timing, which are discussed below. In addition, the controller layer maximum processing time in each path is calculated.

TABLE I
ACHIEVED FREQUENCY BY THE REALIZED IMPLEMENTATION ACROSS
DIFFERENT ASIC TECHNOLOGIES.

Technology	Frequency before timing improvements	Frequency after timing improvements
TSMC 28 nm	859.8 MHz	1 GHz
UMC 130 nm	354.14 MHz	438.59 MHz

TABLE II
DESIGN COST USING 130 NM UMC TECHNOLOGY AT 1 GHZ FREQUENCY.

Estimated dynamic power	95.3698 mW
Leakage power	171.17 mW
Area	242 124.8 μm^2

A. Timing Improvements

The challenge in achieving higher frequencies is that the used technologies are TSMC 28 nm and UMC 130 nm, which will not be able to meet a 1 GHz frequency in digital implementation. However, some timing improvement effort could be carried out to address some of the critical paths in the design. One of these paths is the long CRC combinational logic. Sampling the output of this path significantly improves the timing results especially using TSMC 28 nm technology, as shown in Table I.

B. Latency and Processing Time Calculation

As latency is a big concern in this protocol, it is required to minimize the design processing time in the designed controller layers. The maximum calculated processing time in both transmission and reception operations are listed below. These calculations show how the design has negligible processing cycles compared to the specification requirement [9].

1) Transmission Path: Maximum processing time in transmission path is the write mode in full cacheline. It has total 6 cycles of the operating clock. They are divided as follows; one cycle is consumed in TLP generation, then link layer packing operations takes five cycles.

2) Reception Path: Maximum processing time in reception path is the full cacheline read response mode. It has total 12 cycles of the operating clock. Three of the total cycles are consumed while writing and reading the FLIT inside the buffer. Afterwards, error detection checks consume one cycle. Un-packing the FLIT inside the link layer two cycles. Finally, one more cycle is consumed in the TLP processing in the transaction layer. Another FLIT should be processed for the remaining data slot, which consumes one cycle to be read from the buffer. Similar cycles are consumed as in the first FLIT.

C. Comparison

As a result of not having open access performance data of similar work, it is not manageable to evaluate and compare the resulted hardware implementation performance with other similar implementations. In addition, using older tool versions and different technologies will result in having different values than other similar work. However, the implementation could achieve a negligible processing time compared to the specification requirement [9]. In addition, the proposed implementation manages to meet timing at 1 GHz on TSMC 28 nm technology.

VI. CONCLUSION

Goals underlying this research is to offer a reduced latency memory access paths between host processors and devices through CXL memory protocol controller simple implementation. It provides data link layer architecture taking into consideration minimizing the delays in the datapaths and following CXL specification 2.0 [9]. Furthermore, this paper gives an open source architecture for the proposed CXL layer and implementation algorithms of each block inside.

The proposed architecture has been verified by functional simulation. It is verified in different functional modes mentioned by the specification. Its output is checked against the reference sequence described by the standard using internally developed verilog checkers.

Finally, its performance is improved to achieve 1 GHz frequency using TSMC 28 nm technology and 438.59 MHz frequency using UMC 130 nm technology. In addition, the design cost is estimated using UMC 130 nm technology, showing that the dynamic and leakage power consumption are 95.3698 mW and 171.17 mW, respectively. Furthermore, its area cost extracted after synthesis is $242\,124.8\,\mu m^2$.

For the near future, it is planned to map the system design on a FPGA and integrate with an existing logical layer. The communication between the implemented host and a CXL device is planned to be tested as well. Afterwards, a higher link performance can be achieved by modifying some of the blocks to apply more complex rules and assumptions.

REFERENCES

[1] Robert Tsao. United Microelectronics Corporation, https://www.umc.com/, 1980. (Accessed: November 10th, 2023.).

[2] Taiwan Industrial Technology Research Institute, Hsinchu. Taiwan Semicomductors Manufactor Company, https://www.tsmc.com/, February 21st, 1987. (Accessed: November 10th, 2023.).

[3] A. HeydariGorji, M. Torabzadehkashi, S. Rezaei, H. Bobarshad, V. Alves, and P. H. Chou. Stannis: Low-Power Acceleration of DNN Training Using Computational Storage Devices. In *2020 57th ACM/IEEE Design Automation Conference (DAC)*, 2020.

[4] M. Torabzadehkashi, S. Rezaei, A. HeydariGorji, H. Bobarshad, V. Alves, and N. Bagherzadeh. Computational storage: An Efficient and Scalable Platform for Big Data and HPC Applications. *Journal of Big Data*, 6(1), 2019.

[5] S. H. Lee. The Rise of Memory in the Ever-Changing AI Era – From Memory to More-Than-Memory. In *2022 IEEE Symposium on VLSI Technology and Circuits (VLSI Technology and Circuits)*, pages 272–275, 2022.

[6] Gen-Z Consortium. Gen-Z Specifications, https://genzconsortium.org/, 2017. (Accessed: November 10th, 2023.).

[7] Open-CAPI Consortium. Open-CAPI Specifications, https://opencapi.org/, 2017. (Accessed: November 10th, 2023.).

[8] CCIX Consortium. CCIX Specifications, https://ccixconsortium.org/, 2017. (Accessed: November 10th, 2023.).

[9] CXL Consortium. Compute Express Link Specification, Revision 2.0, https://www.computeexpresslink.org/post/compute-express-link-2-0-specification-now-available, November, 2020. (Accessed: November 10th, 2023.).

[10] CCIX, GEN-Z, OpenCAPI: Overview and Comparison - Youtube, https://www.youtube.com/watch?v=w-rAj1liqrw, 2017. (Accessed: November 10th, 2023.).

[11] CXL Vs. CCIX - Youtube, https://www.youtube.com/watch?v=MB2xF96HU$_0t = 21s$, 2019. *(Accessed : November10th, 2023.)*.

[12] S. J. Park, H. Kim, K.-S. Kim, J. So, J. Ahn, W.-J. Lee, D. Kim, Y.-J. Kim, J. Seok, J.-G. Lee, H.-Y. Ryu, C. Y. Lee, J. Prout, K.-C. Ryoo, S.-J. Han, M.-K. Kook, J. S. Choi, J. Gim, Y. S. Ki, S. Ryu, C. Park, D.-G. Lee, J. Cho, H. Song, and J. Y. Lee. Scaling of Memory Performance and Capacity with CXL Memory Expander. In *2022 IEEE Hot Chips 34 Symposium (HCS)*, 2022.

[13] D. Das Sharma, S. Tavallaei, and CXL Consortium. Compute Express Link™ 2.0 White Paper, https://www.computeexpresslink.org/resource-library, November, 2020. (Accessed: November 10th, 2023.).

[14] D. Fakhry, M. Abdelsalam, M. W. El-Kharashi, and M. Safar. A Review on Computational Storage Devices and Near Memory Computing for High Performance Applications. In *Memories, Materials, Devices, Circuits and Systems, 4:100051*, July, 2023.

[15] D. Fakhry, M. Abdelsalam, M. W. El-Kharashi, and M. Safar. An HBM3 Processing-in-Memory Architecture for Security and Data Integrity: Case study. In *Proceedings of Green Sustainability: Towards Innovative Digital Transformation (ITAF 2023), ser. Lecture Notes in Networks and Systems (LNNS), Springer International Publishing*, 2023.

[16] D. Das Sharma. Compute Express Link®: An Open Industry-Standard Interconnect Enabling Heterogeneous Data-Centric Computing. In *2022 IEEE Symposium on High-Performance Interconnects (HOTI)*, pages 5–12, 2022.

[17] PCI-SIG. PCI Express® Base Specification Revision 5.0 Version 1.0, https://pcisig.com/specifications, May, 2019. (Accessed: November 10th, 2023.).

[18] PCI-SIG. PCI express® Base Specification Revision 6.0 Version 0.7, https://pcisig.com/specifications, October, 2020. (Accessed: November 10th, 2023.).

[19] Synopsys Inc. Compute Express Link (CXL), https://www.synopsys.com/designware-ip/technical-bulletin/compute-express-link-standard-2019q3.html, 2019. (Accessed: November 10th, 2023.).

[20] Siemens EDA. https://eda.sw.siemens.com/en-US/ic/questa/simulation/, 1981. (Accessed: November 10th, 2023.).

[21] Synopsys, Inc. Design Compiler tool, https://www.synopsys.com/implementation-and-signoff/rtl-synthesis-test/dc-ultra.html, 1986. (Accessed: November 10th, 2023.).

[22] Siemens EDA. Oasys, https://eda.sw.siemens.com/en-US/ic/oasys-rtl/, 1981. (Accessed: November 10th, 2023.).

[23] Intel. CXL- Cache/Mem Protocol Interface (CPI), Revision 0.7, https://www.intel.ca/content/www/ca/en/io/cxl-cache-mem-protocol-interface-cpi.html, March, 2020. (Accessed: November 10th, 2023.).

[24] M. Aguilar, A. Veloz, and M. Guzman. Proposal of Implementation of the 'Data Link Layer' of PCI-Express. In *(ICEEE). 1st International Conference on Electrical and Electronics Engineering*, pages 64–69, 2004.

[25] C. Chou, S. Chittor, A. Khan, A. Kimar, P. K. Mannava, R. S. Ram, S. Sen, S. Venkatesan, and K. Padwekar. LINK LEVEL RETRY SCHEME, U.S. Patent 7,016,304 b2. Google Patents, https://patentimages.storage.googleapis.com/a2/7f/0d/318d58caea8ffb/US7016304.pdf, Mar. 21, 2006. (Accessed: November 10th, 2023.).

979-8-3503-8083-5/23 $31.00 © 2023 IEEE

The Electric Crisis of Brazil in 2021
Origins and a Solution Proposal

Milene Santos Moreira
Universidade Federal de Itajubá
Itajubá - Brazil
milenesantos@gea.inatel.br

Tales C Pimenta
Universidade Federal de Itajubá
Itajubá - Brazil
tales@unifei.edu.br

Abstract—**The purpose of this article is to analyze the origins and propose solutions to the 2021 electricity crisis in Brazil. The intention is to raise subsidies for reflections on social issues and issues related to the pandemic, which contribute in a way to decisions of the Brazilian electricity sector, in order to solve problems that contribute to the instability of the voltage of the electricity networks, the that can lead to blackouts. The study method is classified in terms of the approach to problems as qualitative research, and in terms of its nature, as a theoretical study. The study concluded that the main origins of the 2021 electricity crisis in the country are: the low average rainfall recorded in Brazil, the lack of strategic planning for the Brazilian electricity sector and the excessive dependence on the water source for electricity generation in the country. As for the proposals for solutions to the electrical crisis, the research identified the need to create a distributed and renewable generation in the country in the medium to long term.**

Keywords— Electric Crisis, Renewable Energy, Hydroelectric Power, Planning.

I. INTRODUCTION

An electricity crisis is determined by many issues and it is associated with the ability to guarantee the supply of electricity. Those issues are due to economic, social, environmental, technological and political nature. An electrical crisis originates from a perspective where there are serious obstacles to the availability of electrical to meet the demands of the population. Electric energy is essential for human life, encompassing the human development index and also other development indices, such as health, education, sanitation, transport, industry, commerce, etc. So, as an effect of the crisis, there is a deceleration in the well-being of the Brazilian population and the economy.

The electrical crisis of 2001, known as the "2001 Blackout", resulted in major economic impacts (consequence of rationalization) such as reduction in tax collection, large increase in unemployment, increase in the commercial balance deficit. It was also a great nuisance to the population. The expansion of electricity installation capacity grew by 35% between 1990 and 2000, but energy consumption grew approximately 49% [1].

The 2021 electrical crisis is very similar, in many aspects, to the 2001 electrical crisis. The great need for hydroelectric plants, the low rainfall and the lack of long-term public planning for the construction of renewable plants that sustainable and not harmful to the environment, make up these similarities. The short-term results are mainly the need to run thermoelectric plants as a secondary source of electricity and the consequently the increase in electricity cost throughout the country.

Energy generation is a process of conversion or transformation. Electric energy is "generated" from the transformation of primary energy sources, which come from nature.

Primary energy sources can be wind, river water, coal, natural gas, etc. Generally, in electric power plants, the primary energy sources are transformed into electricity by means of turbines, which are equipments that rotate, transforming the primary source into electrical energy. What makes the turbines turn are wind, steam or water, so that those plants are respectively named as eolic, thermoelectric and hydroelectric plants [2], as represented in Fig. 1. Depending on how the conversion of primary energy into electrical energy is carried out, it can generate impacts of economic, social, technological and environmental nature. In this regard, this study analyzes the causes and solutions proposals for the 2021 electricity crisis in Brazil.

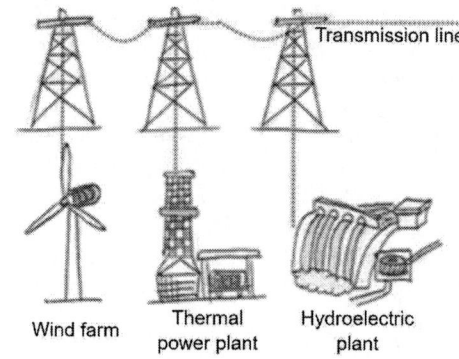

Fig. 1. Main types of energy generation in Brazil.

In addition to this introduction, the article is divided into 6 parts, namely: methodology, main points in the history of planning in the Brazilian electric sector, the Brazilian electricity sector, origins, the proposed solutions and conclusion. Figure 1 demonstrates the electrical transmission lines.

II. METHODOLOGY

The method used to approach the electrical crisis in Brazil is qualitative and its nature was classified as a theoretical study. Qualitative, because it presents a better vision and understanding of the problem, and theoretical, because it analyzes the origins of the problem and proposes solutions through a theoretical basis, building explanatory aspects and discussing issues relevant to the research topic.

In order to obtain the necessary data for the article, bibliographic research was carried out through books and periodicals that address the electrical crisis in Brazil. Data collection was carried out on two fronts, namely: the origin of the electrical crisis in Brazil and the proposed solutions for the electrical crisis. Economic, social, environmental and technological aspects were considered, starting from a general reflection on the electric crisis in Brazil, stating its limits and potentials. Data analysis was necessary to obtain proposals for solutions to the electrical crisis in Brazil.

III. MAIN POINTS IN THE HISTORY OF PLANNING OF THE ELECTRIC SECTOR IN BRAZIL

In 1990 the government was no longer able to invest in the electricity sector. The electricity sector then entered a very delicate situation, with voltage instability and, consequently, the occurrence of blackouts. The problem was solved through the privatization of these state-owned companies. Thus, for privatization to proceed correctly and to last, fundamental conditions were defined, the so-called regulation or restructuring of the Brazilian electrical system. Key points were defined such as: the energy cost, the relationship between energy generators and distributors, market participation rules, minimum investment obligations, etc [3].

Until the end of the 90's, the energy sector planning was conducted by Eletrobras, and after the restructuring, the energy sector planning was "orphaned" (without centralized planning) until 2004 [4].

In 2004, the Empresa de Pesquisa Energética – EPE (Energy Research Company) was created, in order to provide study and research services that are intended to subsidize the planning of the energy sector to the Ministry of Mines and Energy (MME), which remains until today [5].

IV. THE BRAZILIAN ELECTRICAL SECTOR

Electric energy is essential to our lives, since it provides quality of life, which is translated into economic and social development. The main source of energy in Brazil is hydroelectric power (electricity generated from running water in rivers), which corresponds to 62% of the Brazilian capacity. It is a clean and renewable energy source. In second place comes the thermoelectric plants (electricity generated from the burning of natural gas, coal, fossil fuels, biomass, nuclear, etc.), with a percentage of 28%. The remaining 10% comes from other sources including wind farms (electricity generated from wind energy) and photovoltaic generation [6].

Electric energy travels a long way before reaching the final user. It is produced by the plants, transmitted by the transmission lines, which form a basic network. It passes through substations to reach the local distribution networks, which them reach the industry, commerce, houses and public structures. Energy may be delivered also to large consumers by transmission lines. Figure 2 demonstrates the path taken by energy since its production.

More than 300 companies, including generators, transmitters, distributors and free consumers, are part of the National Interconnected System – SIN [7]. There are hundreds of plants and more than 100,000 km of transmission lines and to operate this complex system.

The National Electric System Operator – ONS, created in 1998, is a non-profit, private organization governed by civil law, which is responsible for coordinating and controlling the generation and the transmission of electric energy of the SIN, which is regulated by National Electric Energy Agency – ANEEL [6][7][8].

In order to know where the problem comes from, initially it is necessary to understand that a planning must be made for the electric sector with the objective of having a forecast of the electric energy demand. Later it is defined "who" will supply the electric energy, what will be the plants, how it will be work to have the offer available at a pre-defined quality and criteria.

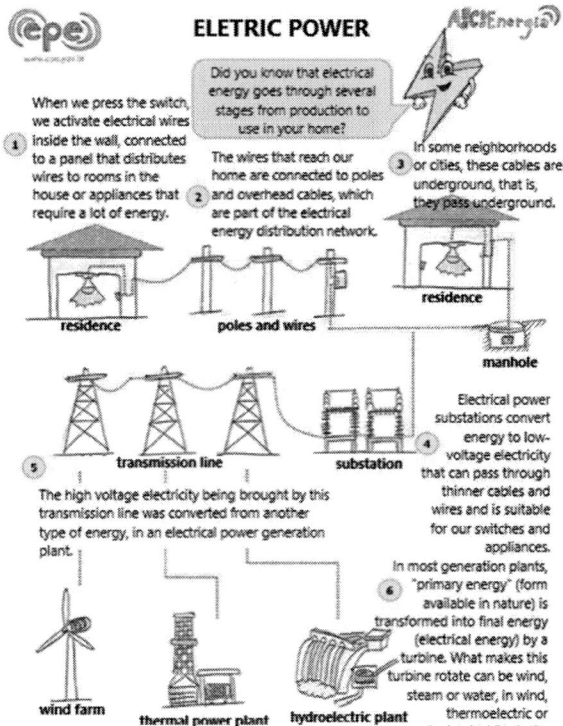

Fig. 2. The path taken from energy to homes.

A. The ANEEL

The ANEEL was created in 1996. It is the regulatory agency of the electricity sector, is linked to the Ministry of Mines and Energy - MME.

The ANEEL regulates the policies and guidelines of the electric sector that are governed by the Federal Government, inspecting the energy supply to consumers and also mediating conflicts between agents in the electric sector. ANEEL also grants the right to explore energy generation, transmission, distribution and trading services. ANEEL also defines energy tariffs, as established by law and through contracts that are granted and signed with companies [6].

B. The EPE

The Energy Research Company - EPE conducts studies to assist the government's choices. It aims to provide efficient public services, to promote social well-being, to meet the demands of society and support sustainable development [9].

Based on the definitions and guidelines of the energy policies, studies and researches are carried out to guide the development of the electric sector. Studies and researches establish energy planning and prepare important documents for actions and monitoring, namely: ten-year plans, long-term plans, periodic bulletins and reviews, conjuncture analysis and specific studies [9].

Those studies and researches generate results capable of being used in other analyses, by considering the formulation and evaluation of alternatives to serve the country's energy services. As the main objective, projects are carried out and commercial action plans are also prepared. Auctions are conducted for generation and transmission of electric energy, which enables the expansion of energy supply, according to the guidelines, principles and goals established, which are made by the Union, through the competent agencies, such as the National Energy Policy Council – CNPE the MME [9].

In addition to the mentioned organizations, there are also other institutions active in the energy sector in Brazil, such as the Electricity Sector Monitoring Committee (CMSE) and the Electric Energy Commercialization Chamber (CCEE) [6].

Faced with all this restructuring and adaptation of the energy sector, the following questions are presented: What went wrong to cause this electrical crisis? Have not been built enough power plants? The private sector had been encouraged efficiently?

V. ORIGINS

A. The Hydropower Plants

As previously stated, 62% of the energy is produced by the water that moves the turbines of the hydroelectric plants. However, the long-term climate change, became a serious problem for the country. Reduced rainwater means no water in hydroelectric plants, and the consequences are devastating.

Climate changes alters the water availability and distribution, which affects the entire electric sector in different ways, especially for the hydroelectric generation. The lack of rain in certain regions leads to significantly lower energy generation and greater complexity in the operation of reservoirs. In addition, the more extreme weather events increase, the less use of the inflowing waters may occur. However, other regions may receive greater availability of water and, as a consequence, present a greater capacity for generating energy [10].

B. Climate Changes

Climate change also has consequences for energy production forecast. Hydroelectric production forecast is based on the assessment of water availability in each river basin. When the flow regime is changed, there is an impact on the availability of water and, as a consequence, there is a serious mismatch between the conditions of real operations in relation to the energies that are guaranteed in those basins, leading the system to voltage instability [10].

Voltage stability is one of the characteristics that must be considered when talking about an electrical system. When the value of electrical voltage is decreased and increased, it can lead to the malfunction of machines, equipment and even huge systems, which can cause financial losses to energy consumers. Voltage stability is the ability of the electrical system to maintain voltage at an adequate level, both under normal operating conditions and after disturbances occur. Stability depends on maintaining the balance between generation and consumption of an electrical system [11,12].

Therefore, the load forecast for the following years is substantially considerable. ONS, CCEE and EPE publish load forecast data for the Annual Energy Operation Planning. In 2022, there is a forecast of a load increase of 2.7% in the SIN, reaching an average of 71,373 MW. For the period from 2022 to 2026, the forecast has an average annual increase of 3.4%, reaching 81,604 MW at the end of the period [13]. According to the forecast for the coming years, which is significantly high, important actions must be taken for the Brazilian electrical system, in order to remedy unwanted results when it comes to an electrical crisis.

C. Pandemic

The pandemic also has recently impacted all sectors of the economy, and thus impacted the electrical system. In April 2020, at the beginning of the pandemic in Brazil, the electricity consumption suffered a 6.6% retraction, as compared to April 2019. The main impacts were on the commerce (-17.9%) and industry (-12.4%) [14]. In other words, there was a decrease in the demand for electricity, therefore, the pandemic did not promote the electricity crisis. Figure 3 shows the EPE review of April 2020 [14].

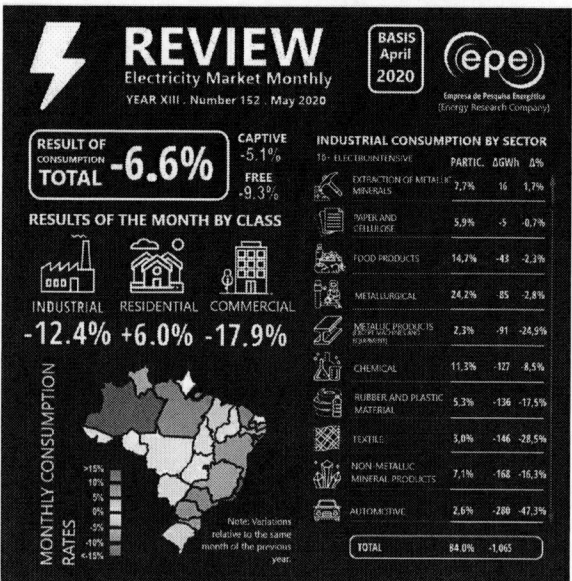

Fig. 3. April 2020 monthly review of the Brazilian electricity market.

D. Lack of Proper Public Management

Great debates about the electricity crisis appear when questions are raised about the ability of public management to achieve goals that are effective for the sustainable energy supply. Frequently there are interferences from the general public, such as ideological interferences and interferences from people who has great influence over the decision-making authorities, through relationships in the electricity segments [15].

Electricity crises usually reflect the lack of a long-term strategy about the use of energy resources. When it comes to electrical crisis, they are also considered points of greatest concern, such as moral and ethical.

As aforementioned, the electricity crisis in Brazil is a consequence of periods of severe difficulties to supply electrical energy to the society. These supply difficulties are directly related to the structural realities of the disposition of energy resources and to the strategies used in the short, medium and long term, to guarantee electricity in competitive and sustainable standards.

The absence of long-term planning in the Brazilian electricity sector, to strategically promote the diversification of sources in the 2000s and 2010s, made the scenario even worse at each crisis. The dependence on hydroelectric plants caused severe socio-environmental impacts since there was a drastic reduction in the levels of reservoirs.

The electrical crisis of 2001, or the 2001 blackout, was very similar to the current crisis. It was observed that the decisions about investment on the energy sector could not be managed by private groups alone. Transactional groups began to motivate themselves and incorporate the hydroelectric system as assets to produce electricity at low cost, thus causing a lack of investment in the expansion of the electrical system, and thus the rationing.

VI. Solution Proposal

As a solution proposal, we must consider the immense range of alternative sources to expand the production of electric energy in Brazil, such as biomass, photovoltaic and wind power. We have enough sustainable resources to make Brazil's generation a distributed generation. However, even with this range of alternative sources, we still have the problem of lack of energy production and consequently, blackouts. It is therefore, the omission in the public and private management of energy resources, since the climate changes are not included in the country's climate plans [16].

There is a lack of faster strategic planning for the electricity sector in Brazil, in the short, medium and long term, since the population demands are faster than planning. More generation plants, from diversified sources, can be created, so that we may be not dependent only on a single source, as it is the case of hydroelectric plants, which always leads to a lack of sufficient energy, thus promoting energy crises. Demands for electricity are increasingly which forces the government to maintain investments in hydro generation and does not prioritize the creation of more distributed generation, such as solar, wind and biomass energy sources, which are renewable energy sources.

The creation of sustainable plants, with greater diversification, more competitive and renewable as compared to other countries, is a consequence of government planning. If the government does not commit itself to the urgent creation of a distributed generation, considering the medium and long term instead of just short term will force the creation of new hydroelectric plants in Amazon area, thus compromising the distributed generation as a long-term solution [17].

The proposal is to invest in other renewable energy sources, such as wind and solar, to reduce the high dependence on hydroelectric plants in order to promote more sustainability to the Brazilian energy matrix.

VII. Conclusions

Based on the research related to the origins and proposals for solutions for the 2021 Brazilian electricity crisis, this article concluded that the most relevant origins are: the lack of rains for a long time, the lack of proper planning for the Brazilian electric sector and the great dependence on the water source as the main generation of electric energy in the country. As for the proposed solutions for the electrical crisis, the surveys found the need to create a distributed and renewable generation in the country in the medium to long term. It is also considered that short-term actions may bring harm to the population, but they are necessary due to the emergency.

Acknowledgment

This work was supported by CAPES, CNPq and FAPEMIG.

References

[1] Tolmasquim, M.T., Guerreiro, A. e Gorini, R., "Prospective vision of the Brazilian energy matrix: energizing the country's sustainable development", Brazilian Energy Magazine, v. 13, n. 1, 2007.

[2] Forms of Energy, EPE, 2021, Available in: https://www.epe.gov.br/pt/abcdenergia/formas-de-energia#PRIMARIA-UTIL/, Access at: Aug. 29, 2022.

[3] Lorenzo, H.C.d., "The brazilian electricity sector: past and future", Institutional repository UNESP, pag. 147–170, 2001-2002.

[4] History of Eletrobrás, Eletrobrás, 2021, Available in: https://eletrobras.com/pt/Paginas/Historia.aspx, Access at: Aug. 29, 2022.

[5] EPE – who we are, 2021, Available in: https://www.epe.gov.br/pt/a-epe/quem-somos, Access at: Aug. 29, 2022.

[6] Find out more about the Brazilian electricity sector, ANEEL, 2021, Available in: https://www.aneel.gov.br/home?p_p_id=101&p_p_lifecycle=0&p_p_state=maximized&p_p_mode=view&_101_struts_action=\%2Fasset_publisher\%2Fview_content&_101_returnToFullPageURL=\%2F&_101_assetEntryId=14476909&_101_type=content&_101_groupId=654800&_101_urlTitle=faq&inheritRedirect=true, Access at: Dec. 01, 2021.

[7] What is ONS, ONS, 2021, Available in: http://www.ons.org.br/paginas/sobre-o-ons/o-que-e-ons, Access at: Aug. 29, 2022.

[8] The electrical system, USP, 2021, Available in: https://sistemas.eel.usp.br/docentes/arquivos/5840834/59/SistemaEletrico1.pdf, Access at: Aug. 29, 2022.

[9] The electrical system, EPE, 2021, Available in: https://www.epe.gov.br/pt/areas-de-atuacao/planejamento-energetico, Access at: Aug. 29, 2022.

[10] Matsumura, E.H. e Ferreira, T.V.B., "Climate change and developments on energy planning studies: Initial considerations", EPE, december, 2008.

[11] Kundur, P., Paserba, J., Ajjarapu, V., Andersson, G., Bose, A., Canizares, C., Hatziargyriou, N., Hill, D., Stankovic, A., Taylor, C., Van Cutsem, T., e Vittal, V., "Definition and classification of power system stability IEEE/CIGRE joint task force on stability terms and definitions", IEEE Transactions on Power Systems, v. 19, n. 3, p. 1387–1401, 2004.

[12] Carvalho, C.S., "Comparison of voltage instability identification methods based on synchronized measurements", 2016, Ph.D. thesis, UFRJ, march, 2016.

[13] Load forecast for annual energy operation planning cycle 2022 (2022-2026), Available in: http://www.ons.org.br/paginas/energia-no-futuro/suprimento-energetico, Access at: Aug. 29, 2022.

[14] May review reflects impacts of the COVID-19 pandemic. Available in: https://www.epe.gov.br/pt/imprensa/noticias/resenha-de-maio-reflete-impactos-da-pandemia-da-covid-19, Access at: Aug. 29, 2022.

[15] Lisboa, J., Coelho, A., Coelho, F. and Almeida, F., "Introduction to organization management", Barcelos: Vida Econômica, 2004.

[16] Risk of blackout and electricity rationing: back to the future?, Available in: https://www.cepea.esalq.usp.br/br/opiniao-cepea/risco-de-apagao-e-racionamento-de-energia-eletrica-de-volta-para-o-futuro.aspx, Access at: Aug. 29, 2022.

[17] Borges, F.Q. e Zouain, D.M., "The electrical matrix in the state of Pará and its position in promoting sustainable development", Planning and public policies, v. 2, n. 35, 2021

A Fully-Differential Low-Noise Instrumentation Amplifier for Electrical Impedance Tomography

Ibrahim Alkhalifa
Electrical Engineering Department
King Fahd University of Petroleum & Minerals
Dhahran, Saudi Arabia
King Abdullah University of Science and Technology
Thuwal, Saudi Arabia
g201682380@kfupm.edu.sa

Yaqub Mahnashi
Electrical Engineering Department
Bioengineering Department
Center for Communication Systems and Sensing
King Fahd University of Petroleum & Minerals
Dhahran, Saudi Arabia
ymahnashi@kfupm.edu.sa

Abstract—**Electrical impedance tomography (EIT) is a promising technology for healthcare. It enables patient bedside monitoring, thus establishing an early warning capability. Lung ventilation monitoring is an example of EIT applications. At the core of this system is an instrumentation amplifier needed to detect the small signals at the electrodes attached to the body. This paper proposes a fully-differential low-noise instrumentation amplifier for lung ventilation monitoring. The proposed design is implemented using 0.18μm CMOS TSMC technology and validated using simulations in the Cadence Virtuoso environment. The proposed design achieves a closed-loop gain of 20 dB, a 100 Hz - 200 kHz bandwidth, and a noise floor of $10nV/\sqrt{Hz}$ at a power consumption of 242 μW. Thanks to its fully differential nature, the proposed design achieves a common-mode rejection ratio (CMRR) of 200. In comparison with prior art, the proposed design achieves improved performance in terms of noise and CMRR.**

Index Terms—**CMOS amplifier, electrical impedance tomography, fully differential circuits, impedance spectroscopy, instrumentation amplifier, low-noise, lung ventilation monitoring.**

I. INTRODUCTION

Several ailments that might affect humans such as lung disease, and various types of cancerous cells are traditionally investigated and discovered using well-established imaging techniques such as CT scan and MRI. Recently, a different imaging method called electrical impedance tomography (EIT) has gained attention due to its portability and radiation-free features [1], [2]. This technique exploits the resistivity variations of the tissue and provides insightful information about the tissue. For example, EIT is used for the lungs [3], classification of hand gestures [4], cell culture [5] and even as a remote-sensing technique demonstrated using fish detection [6].

As shown in Fig. 1, EIT system works simply by applying an AC current, $i(t)$, through a pair of electrodes that are attached to a tissue, and then measuring the voltage generated at each of the remaining electrodes using an instrumentation amplifier (IA). Then, the measured voltage is processed to generate a color map that indicates the level of resistivity of the tissue. An EIT system usually requires multiple channels to be able to provide a high-quality image. For example, 16 channels are used in [4], where the electrodes are attached to a bracelet to generate a resistivity colormap of the user's wrist.

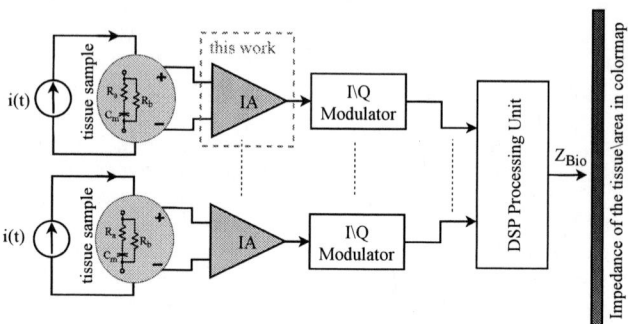

Fig. 1. Conceptual block diagram of the mutli-channel EIT system with simplified impedance model of the tissue sample.

Focusing on one channel of the EIT system, the analog readout circuit comprises a current stimulator, an IA, and an I/Q modulator. For the current stimulator, a Wein-bridge fully-differential oscillator is usually opted for. This oscillator generates a voltage signal which is then converted to current through a current driver. Due to their capacitive nature, some body tissue exhibits different impedances at different frequencies. Therefore, the oscillator has to provide a programmable and wide range frequency operation in the range of 0 Hz to 10 MHz [7].

In [8], a readout circuit consisting of transconductor (TC) and transimpedance (TI)-based IA, I/Q demodulator, and G_m-C filters is proposed and implemented using 0.18μm CMOS technology. The design offers a programmable gain through a TC degenerated with a bank of resistors achieving a 0 dB – 40 dB gain, and a wide bandwidth of 87 MHz. An AC coupling capacitor and a MOS resistor are used to couple the input and set the common-mode input level to V_{cm}. G_m-C filters are used at the output to obtain the DC component, which contains information about the impedance of the tissue.

On the other hand, a 100 Hz – 10 MHz programmable gain readout circuit is proposed in [9]. The circuit consists of an IA with a current conveyer architecture (CCIA) whose gain is set by a programmable ratio of resistors, an ADC and its driver, and a digital control circuitry. Programmable gain was used to achieve a high dynamic range (DR), the control is done through an automatic gain control (AGC) feedback loop. Since a frequency range of 100 Hz – 1 MHz is not needed for DC signals, the bias current of the IA is reduced to save power (0.47 mW, 3.3 mW). The use of an ADC is done to demodulate the I/Q outputs digitally. It is noteworthy that the majority of power dissipation is due to the ADC driver.

In [10], a 13-channel, 1.53 mW EIT system is proposed. Here, we make a remark regarding the relation between the number of channels and the number of electrodes required. Given a target number of channels (N), the required number of electrodes is $N + 3$ [11]. Since there are 13 channels, this implies that 13 ADCs are required! Such a large number would entail a heavy penalty in area and power. Thus, in this paper frequency division multiplexing (FDM) is used to combine the data generated by all 13 channels into one stream that feeds into a single ADC. In addition, I/Q demodulation is done before IA to relax the bandwidth requirements of the remaining circuitry. By exploiting the constant inverted U-shape of the amplitude of the measured voltages, fixed gains could be used, which eliminates the need for an AGC. All of these techniques, in addition to the use of a single 1 V supply, resulted in a state-of-the-art Power/Channel of 0.118 mW.

In [7], an 8-channel 10 MHz EIT system is proposed for early breast cancer detection. Notably, a phase locked loop (PLL) is used to reduce errors in phase measurement. A wideband CCIA is chosen to allow high-frequency tissue interrogation. Thus, a state-of-the-art $1.6 \ m\Omega/\sqrt{Hz}$ impedance resolution is achieved. Positive feedback is used to generate a large-value resistor required at the input of IA to couple the input and reject the DC electrode offset.

It can be seen form previous works that the IA is an essential block in an EIT system since it amplifies a weak signal in the presence of large common-mode signals. Furthermore, it must achieve this while adding minimal noise. In this paper, a low-noise, fully differential IA is proposed. Simulation results shows that the proposed design is improved in terms of noise and common-mode rejection ratio (CMRR) in comparison with previous art. The remainder of this paper is organized as follows. In Section II, the proposed design is presented. Section III provides the results of the proposed design and a comparison with prior works. Section V concludes the paper and suggests future research directions.

II. PROPOSED DESIGN

In instrumentation applications, the amplifier must have low noise, high input impedance, high CMRR, offset rejection, and wide input common-mode range (ICMR); therefore, in this work, an amplifier with capacitive feedback is proposed as shown in Fig. 2. The transfer function of the amplifier with negative feedback is given by:

$$\frac{v_{out}}{v_{in}} = \frac{-sR_FC_1}{sR_FC_F + 1} \quad (1)$$

From (1), the midband gain is $-\frac{C_1}{C_F}$, with a zero at DC, and a pole at $\frac{1}{R_FC_F}$.

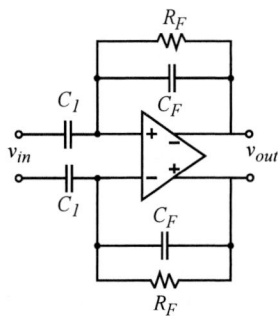

Fig. 2. The closed-loop configuration of the proposed amplifier.

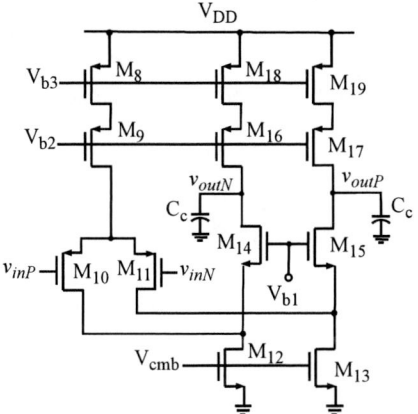

Fig. 3. Circuit realization for the folded cascode OTA.

The amplifier design consists mainly of three parts. The core amplifier, a voltage reference circuit, and a common-mode feedback (CMFB) circuit. The core is constructed using a folded cascode architecture depicted in Fig. 3. ICMR can be quantified as following:

$$V_{th10} + V_{dsat12} \leq V_{in} \leq V_{DD} - V_{dsat8} - V_{dsat9} - V_{gs10} \quad (2)$$

where V_{th} is the threshold voltage of the MOSFET, V_{dsat} is the MOSFET pinch-off voltage, and V_{gs} is the voltage between the gate and the source of the MOSFET.

The input voltage can swing below ground as seen in (2). Also, by adding a parallel NMOS input pair, the input could swing above V_{DD}; however, this has not been considered in this design, as it requires additional circuitry to maintain g_m constant.

A. Noise Analysis

The noise of the proposed amplifier is analysed considering the input-referred noise (IRN). The IRN of a folded cascode is given by:

$$IRN = 2\left[4kT\gamma\left(\frac{1}{g_{m10}} + \frac{g_{m12}}{g_{m10}^2} + \frac{g_{m18}}{g_{m10}^2}\right) + \left(\frac{1}{C_{ox}g_{m10}^2 f}\right)\left(\frac{K_{fp}I_{10}}{L_{10}^2} + \frac{K_{fp}I_{18}}{L_{18}^2} + \frac{K_{fn}I_{12}}{L_{12}^2}\right)\right] \quad (3)$$

979-8-3503-8083-5/23 $31.00 © 2023 IEEE

where k is the Boltzman constant, T is the temperature in kelvin, γ is process-dependent constant for the thermal noise (not to be confused with the body-effect coefficient), C_{ox} is the oxide capacitance density, K_f is a process-dependent parameter specific to the flicker noise effect. The length of the MOSFET is denoted by L and its transconductance g_m.

By choosing to make the noise dominated by the input pair, there are two options that can be considered with regard to the thermal noise. First, if the transistors are biased with the same current, then their inversion level can be different [12]. Second, if the transistors carry different currents, then choosing the same inversion level suffices. In the proposed design, M_{18} carries $1/10^{th}$ of the current passing M_{10}, while M_{12} has roughly the same current as M_{10}. Therefore, M_{10} and M_{18} are baised in the weak inversion, and M_{12} in the strong inversion. The flicker noise can be minimized by the appropriate choice of the gate length (L). In addition, it is well-known that the flicker noise of PMOS is lower than NMOS. Therefore, the flicker noise is dominated by M_{12}. When negative feedback is applied, IRN of the OTA gets referred to the input with gain of $(\frac{C_{in}+C_F+C_1}{C_1})^2$, where C_{in} is the capacitance seen at the input of OTA [13]. As a result, the noise referred to input signal of the IA block can be quantified as following [14]:

$$\frac{V_{in}}{V_{R,n}} = \frac{1}{sR_FC_1}. \tag{4}$$

Therefore, the total noise seen at the input of the IA block can be optimized mainly by R_F and C_1.

B. Voltage References Circuit

In this work, a supply-independent voltage reference circuit, also known as the bandgap circuit, is also designed for the proposed IA as shown in Fig. 4. M_1-M_4 constitute the core of the voltage reference circuit, M_5-M_7 are employed to generate the DC voltage V_{b3} which is used to bias part of the PMOS transistors. Note that NMOS transistors in the folded cascode OTA, see Fig. 3, are biased by the CMFB circuit, which will be discussed shortly. The current reference (I_{ref}) passing through the resistance R can be determined by:

$$V_{GS4} = V_{GS3} + I_{ref}R \tag{5}$$

Therefore, given a desired I_{ref}, one can start by selecting V_{GS4} to obtain desired V_{b2}, then choosing V_{GS3} and R to arrive at the desired I_{ref}. Note that, the desired V_{b1} is generated by appropriately sizing M_2. Very large gate length, L, is chosen for M_1-M_4 transistors in order to minimize the modulation effect on the channel length. The transistor sizes are summrized in Table I.

C. Common-mode Feedback Circuit (CMFB)

To accurately define the DC level at the output, CMFB is employed. To avoid using resistors to sense the common-mode level, and since clocks are not available, an active scheme is used as shown in Fig. 5. This circuit is, by design, insensitive to differential-mode signals. It forms a negative feedback loop to force the common-mode level of the output to be equal

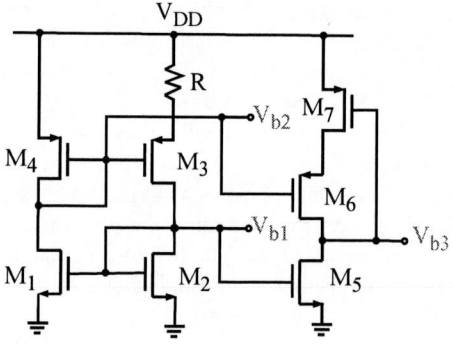

Fig. 4. Circuit realization for voltage references: V_{b1}, V_{b2} and V_{b3}.

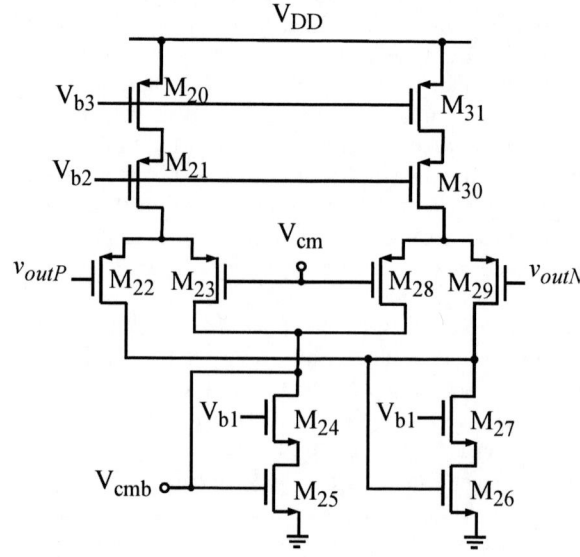

Fig. 5. Circuit realization of the active common-mode sensing circuit.

to V_{cm}. The loop is stabilized by the same compensation capacitor (C_c) used by the folded cascode OTA in Fig. 3. Noise of CMFB circuitry appears as a common mode and is thus rejected.

TABLE I
TRANSISTOR SIZING OF THE PROPOSED CIRCUITS.

Transistors	W (μm)	L (μm)	I_D (μA)
M_1, M_2, M_5	0.7	19	1
M_3	355	19	1
M_4	57	19	1
M_6, M_7	18	0.18	1
M_8, M_9	1080	0.18	60
M_{10}, M_{11}	1080	0.18	30
M_{12}, M_{13}, M_{24}-M_{27}	33	3.5	66
M_{14}, M_{15}	3	3.5	3
M_{16}-M_{19}	54	0.18	3
M_{20}, M_{21}, M_{30}, M_{31}	594	0.18	33
M_{22}, M_{23}, M_{28}, M_{29}	297	0.18	16.5

TABLE II
PERFORMANCE OF THE PROPOSED DESIGN COMPARED TO THE RECENT LITERATURE.

Reference	TCASII [8]	TBCAS [9]	A-SSCC [15]	ICICM [17]	IEEE Access [18]	This work
Bandwidth	87 MHz	10 MHz	190 kHz	777.5 kHz	3 MHz	200 kHz
IRN ($nV/\sqrt{Hz}$)	16.7	10	42	NA	27.8	10
Topology	Fully-differential	Fully-differential	Fully-differential	Single-ended	Single-ended	Fully-differential
CMRR (dB)	164	66	NA	148	86	> 200
Number of Channel	1	2	6	1	1	1
Power/Channel (mW)	0.292	1.05 - 10.85	1.73	0.039	0.268	0.242
Results	Simulation	Measurement	Measurement	Simulation	Measurement	Simulation

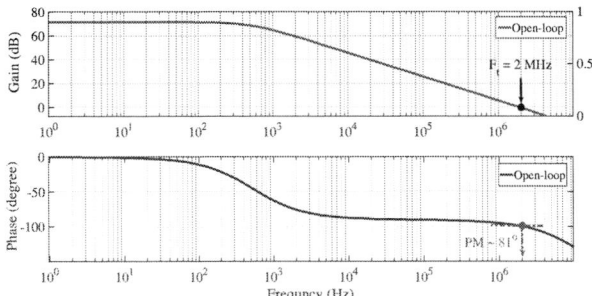

Fig. 6. The gain and phase vs frequency for open-loop configuration of the proposed amplifier.

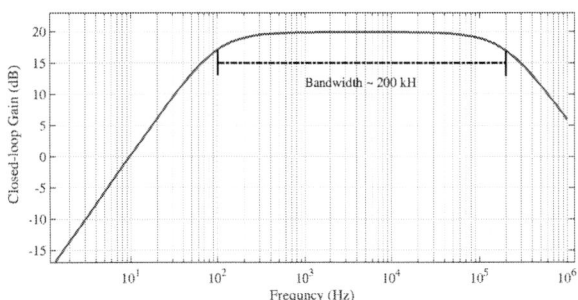

Fig. 7. The gain vs frequency for closed-loop configuration of the proposed amplifier.

III. RESULTS AND DISCUSSION

The complete circuit of the proposed design is implemented in 0.180μm CMOS TSMC technology and verified using simulations in Cadence Virtuoso environment. Extensive runs have been conducted to optimize the design, and with the help (g_m/I_D) methodology, transistor sizes are chosen as shown in Table I along with the drain current for each transistor. The supply-independent voltage reference circuit shown in Fig. 4 is tested first against the variation of the supply voltage. Therefore, V_{DD} is swept between 1.1- 1.9 V, and a reference current (I_{REF}) of $0.98 - 1.01$ μA is obtained. Therefore, a maximum deviation of 2% is achieved in I_{ref}.

The OTA is designed to have an open loop gain of > 60 dB, and a gain-bandwidth product (F_t) of 2 MHz. This is suitable for the frequency band required for lung ventilation monitoring, which is 10-200 kHz [15]. The amplifier is simulated in open-loop configuration as shown in Fig. 6, a phase margin of 81° is achieved using $C_c = 32$ pF. For the CMRR, a very low common-mode gain is obtained at DC of about -70 dB for a single-ended setup, which results in CMRR of 145 dB. On the other hand, the common-mode gain reaches almost -246 dB when the output signal is measured differentially. Therfore, for differential-ended setup, the CMRR maintains 200 dB over the open-loop bandwidth, i.e. 2 MHz.

Considering the lung ventilation system, the input signal ranges from 0.01-40 mV_{pp} [16]. Therefore, the proposed circuit is designed to accommodate the largest input with a gain of 20 dB to avoid saturating the readout chain while also ensuring that the noise of the whole system is still dominated by IA. C_1 is chosen to be 40 pF to enhance the IRN as seen in (4). To obtain a very large resistor for R_F with a reasonable area, an NMOS biased in weak inversion with the maximum L

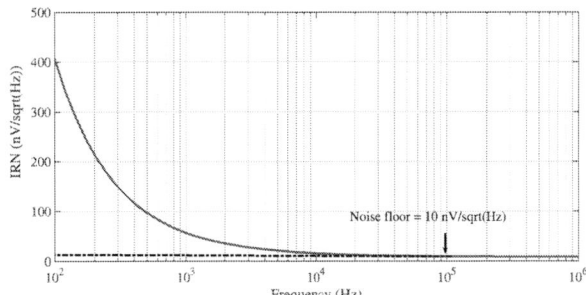

Fig. 8. Input referred noise of the proposed amplifier.

and minimum W allowable in technology is used. The closed-loop gain is shown in Fig. 7, where a 20 dB gain and a bandwidth of 200 kHz is achieved. The noise performance is illustrated using the IRN as depicted in Fig. 8. It can be seen that the amplifier achieves a noise floor of $10nV/\sqrt{Hz}$.

A comparison with previously reported designs is presented in Table II. It can be seen that our work achieves the highest CMRR thanks to a fully differential scheme. In addition, the achieved IRN compares well with previous works while consuming a favorable power consumption. Although [17] achieved a lower power consumption, IRN is not reported. As reported earlier, 200 kHz bandwidth would be sufficient for some EIT applications such as lungs ventilation monitoring. Therefore, the proposed design achieves the best compromise between power consumption and IRN among the reported works.

IV. CONCLUSION

This paper presented a fully-differential IA for monitoring lung ventilation. A noise floor of $10nV/\sqrt{Hz}$ at a power consumption of 242 μW strikes a compromise, as illustrated by comparison with previous works. Future work should target minimizing power consumption while ensuring a stable CMFB loop. Also, the bandwidth can be extended to cover wide range of EIT applications.

REFERENCES

[1] S. Mansouri, Y. Alharbi, F. Haddad, S. Chabcoub, A. Alshrouf, and A. A. Abd-Elghany, "Electrical impedance tomography - recent applications and developments," *J Electr Bioimpedance*, vol. Nov 20;12(1), pp. 50–62, 2021.

[2] P. Kassanos, "Bioimpedance sensors: A tutorial," *IEEE Sensors Journal*, vol. 21, no. 20, pp. 22 190–22 219, 2021.

[3] M. Zheng and B. Ibrahim, "Performance prediction, sensitivity analysis and parametric optimization of electrical impedance tomography using a bioelectrical tissue simulation platform," in *2021 43rd Annual International Conference of the IEEE Engineering in Medicine Biology Society (EMBC)*, 2021, pp. 2864–2870.

[4] M. Nawaz, R. W. Chan, A. Malik, T. Khan, and P. Cao, "Hand gestures classification using electrical impedance tomography images," *IEEE Sensors Journal*, vol. 22, no. 19, pp. 18 922–18 932, 2022.

[5] P. Linderholm, L. Marescot, M. H. Loke, and P. Renaud, "Cell culture imaging using microimpedance tomography," *IEEE Transactions on Biomedical Engineering*, vol. 55, no. 1, pp. 138–146, 2008.

[6] L. J. Nowak and M. Lankheet, "Fish detection using electrical impedance spectroscopy," *IEEE Sensors Journal*, vol. 22, no. 21, pp. 20 855–20 865, 2022.

[7] J. Lee, S. Gweon, K. Lee, S. Um, K.-R. Lee, and H.-J. Yoo, "A 9.6-mw/ch 10-mhz wide-bandwidth electrical impedance tomography ic with accurate phase compensation for early breast cancer detection," *IEEE Journal of Solid-State Circuits*, vol. 56, no. 3, pp. 887–898, 2021.

[8] J. Pérez-Bailón, M. T. Sanz-Pascual, B. Calvo, and N. Medrano, "Wideband compact 1.8 v-0.18 m cmos analog front-end for impedance spectroscopy," *IEEE Transactions on Circuits and Systems II: Express Briefs*, vol. 69, no. 3, pp. 764–768, 2022.

[9] M. Takhti and K. Odame, "A power adaptive, 1.22-pw/hz, 10-mhz readout front-end for bio-impedance measurement," *IEEE Transactions on Biomedical Circuits and Systems*, vol. 13, no. 4, pp. 725–734, 2019.

[10] B. Liu, G. Wang, Y. Li, L. Zeng, H. Li, Y. Gao, Y. Ma, Y. Lian, and C.-H. Heng, "A 13-channel 1.53-mw 11.28-mm2 electrical impedance tomography soc based on frequency division multiplexing for lung physiological imaging," *IEEE Transactions on Biomedical Circuits and Systems*, vol. 13, no. 5, pp. 938–949, 2019.

[11] H. Rajaguru, P. Rathinam, and R. Singaravelu, "Electrical impedance tomography (eit) and its medical applications: a review," *Int J Soft Comp Eng*, vol. 3, pp. 193–8, 01 2013.

[12] R. Harrison and C. Charles, "A low-power low-noise cmos amplifier for neural recording applications," *IEEE Journal of Solid-State Circuits*, vol. 38, no. 6, pp. 958–965, 2003.

[13] M. Steyaert, Z. Y. Chang, and W. Sansen, "Low-noise monolithic amplifier design: Bipolar versus cmos," *Analog Integr. Circuits Signal Process.*, vol. 1, no. 1, p. 9–19, sep 1991. [Online]. Available: https://doi.org/10.1007/BF02151022

[14] H. Wang and P. P. Mercier, "A current-mode capacitively-coupled chopper instrumentation amplifier for biopotential recording with resistive or capacitive electrodes," *IEEE Transactions on Circuits and Systems II: Express Briefs*, vol. 65, no. 6, pp. 699–703, 2018.

[15] S. Hong, J. Lee, J. Bae, and H.-J. Yoo, "A 10.4 mw electrical impedance tomography soc for portable real-time lung ventilation monitoring system," in *2014 IEEE Asian Solid-State Circuits Conference (A-SSCC)*, 2014, pp. 193–196.

[16] M. Kim, J. Jang, H. Kim, J. Lee, J. Lee, J. Lee, K.-R. Lee, K. Kim, Y. Lee, K. J. Lee, and H.-J. Yoo, "A 1.4-m ω -sensitivity 94-db dynamic-range electrical impedance tomography soc and 48-channel hub-soc for 3-d lung ventilation monitoring system," *IEEE Journal of Solid-State Circuits*, vol. 52, no. 11, pp. 2829–2842, 2017.

[17] Y.-Y. Qian, Z.-G. Wang, Y.-K. Liu, and Z.-J. Zhou, "A high gain and high cmrr instrumentation amplifier for biomedical applications," in *2019 IEEE 4th International Conference on Integrated Circuits and Microsystems (ICICM)*, 2019, pp. 61–64.

[18] I. Corbacho, J. M. Carrillo, J. L. Ausín, M. Domínguez, R. Pérez-Aloe, and J. F. Duque-Carrillo, "Wide-bandwidth electronically programmable cmos instrumentation amplifier for bioimpedance spectroscopy," *IEEE Access*, vol. 10, pp. 95 604–95 612, 2022.

Radiation-hardened stabilized power supply unit based on bipolar transistors

Takato Tanizawa
Electrical and Electronic Engineering
Shizuoka University
Hamamatsu, Japan

Minoru Watanabe
Graduate School of Natural Science and Technology
Okayama University
Okayama, Japan
minoru-watanabe@okayama-u.ac.jp

Abstract—**For work to decommission melted reactors in nuclear power plants after an accident, a radiation-hardened stabilized power supply unit is demanded for use with any system. However, currently available stabilized power supply units is invariably vulnerable to radiation. The radiation tolerance is always less than 1 kGy. Since the radiation intensity in the Fukushima Daiichi nuclear power plant is estimated as greater than 1000 Sv/h, the expected lifetime of power supply units is therefore limited to less than one hour. This paper presents a new design for radiation-hardened stabilized power supply units. The power supply units never use field effect transistors: they only use bipolar transistors. The experimentally assessed power supply unit has achieved 2 MGy total-ionizing dose tolerance, which is sufficiently higher than currently available radiation-hardened power supply units.**

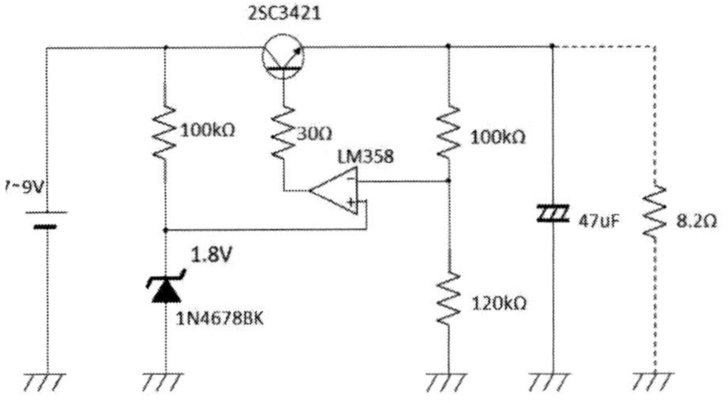

Fig. 1. Circuit diagram of a radiation-hardened stabilized power supply unit using an operational amplifier, a Zener diode, and a bipolar transistor.

I. INTRODUCTION

At the Fukushima Daiichi nuclear power plant, stabilized power supply units with high radiation tolerance are demanded for radiation-hardened robots and measurement systems [1][2]. However, the radiation intensity around the melted nuclear fuel and broken reactors in the Fukushima Daiichi nuclear power plant is estimated as higher than 1000 Sv/h. Since commercially available stabilized power supply units is too vulnerable to radiation or the total-ionizing-dose tolerance is less than 1 kGy, the lifetime of commercially available stabilized power supply units is limited to less than one hour. Therefore, commercially available stabilized power supply units are not adequate for application to such radiation-hardened robots and measurement systems [3]–[6]. To realize battery-based robots and battery-based measurement systems, a radiation-hardened stabilized power supply unit must be implemented.

This paper therefore presents a new design for a radiation-hardened stabilized power supply unit. The power supply units use no field effect transistor and use only bipolar transistors. Results of experimentation have demonstrated that the power supply unit has achieved 2 MGy total-ionizing dose tolerance, which is sufficiently higher than that of currently available radiation-hardened power supply units.

II. RADIATION TOLERANCE OF LITHIUM-ION BATTERY CELLS

Our earlier research has confirmed that lithium-ion battery cells are robust against radiation. Six lithium-ion battery cells were prepared for total-ionizing-dose tolerance experimentation. Their output voltage is 3.785 V. Their capacity is 2600 mAh. The lithium-ion battery cells have been exposed to gamma rays which were emitted by a Co60 gamma radiation source. The total ionizing dose reached 301.7–328.6 kGy with no explosion or fire accident. Any degradation has never been confirmed. Therefore, it could be estimated that the lithium-ion battery cell lifetime is longer than a period of 300 hr under the 1000 Sv/h intense radiation condition at the Fukushima Daiichi nuclear power plant. A new radiation-hardened stabilized power supply unit using full bipolar transistors has been designed based on the radiation experiment results obtained for lithium-ion battery cells.

III. DESIGN OF A RADIATION-HARDENED STABILIZED POWER SUPPLY UNIT USING AN OPERATIONAL AMPLIFIER AND BIPOLAR TRANSISTORS

First, a radiation-hardened stabilized power supply unit using an operational amplifier, a Zener diode, and a power

979-8-3503-8083-5/23 $31.00 © 2023 IEEE

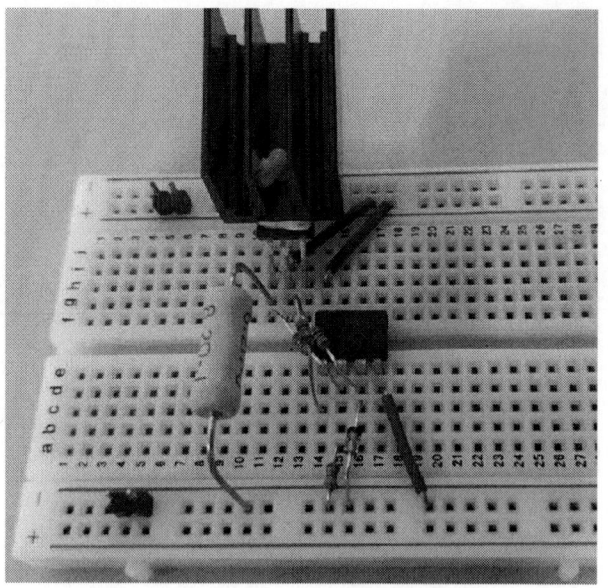

Fig. 2. Photograph of the radiation-hardened stabilized power supply unit using an operational amplifier, a Zener diode, and a bipolar transistor presented in Fig. 1.

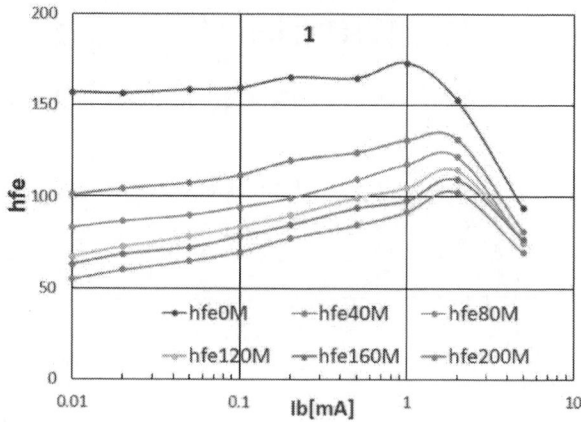

Fig. 4. Current gain or amplification factor of the power bipolar transistor (2SC3421; Toshiba Corp.) for the total ionizing dose.

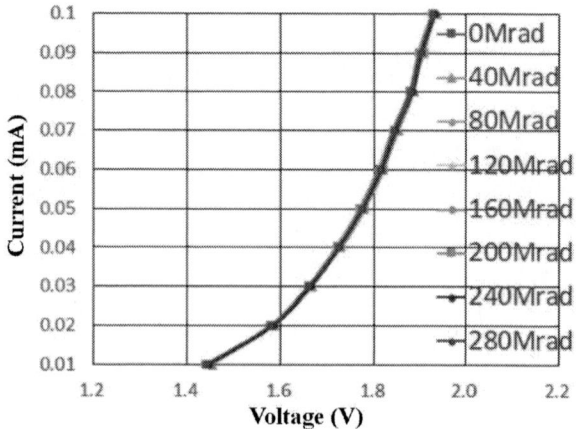

Fig. 3. V-I characteristic variation of the Zener diode (1N4678BK; Central Semiconductor LLC) for the total ionizing dose.

bipolar transistor was designed as presented in Figs. 1 and 2. Compared with field effect transistors (FETs), bipolar transistors are more robust against radiation under a strong radiation environment like that at the Fukushima Daiichi nuclear power plant. Therefore, the radiation-hardened stabilized power supply unit consists of an operational amplifier with bipolar transistors and a bipolar power transistor. The radiation tolerance of the radiation-hardened stabilized power supply unit was measured using a Cobalt 60 gamma radiation source. Findings indicate that an approximately 100 Mrad total-ionizing-dose tolerance can be achieved. However, our goal is to realize a higher than 200 Mrad total-ionizing-dose tolerance for radiation-hardened stabilized power supply units. Among the semiconductor components (an operational amplifier, a Zener diode, and a bipolar transistor), the one most vulnerable to radiation is the operational amplifier. The V-I characteristic variation of the Zener diode (1N4678BK; Central Semiconductor LLC) for the total-ionizing-dose is depicted in Fig. 3. Even if the total ionizing dose exceeds 280 Mrad, the V-I characteristic remains unchanged. In addition, the current gain or amplification factor of the power bipolar transistor (2SC3421; Toshiba Corp.) deteriorates depending on the total ionizing dose, as portrayed in Fig. 4. However, even if the total ionizing dose of the transistor reaches 200 Mrad, the current gain or amplification factor of the power bipolar transistor can be maintained as higher than 50. Therefore, by designing the minimum current gain to satisfy the specifications of the radiation-hardened stabilized power supply unit, the power bipolar transistor degradation can be disregarded. However, the total-ionizing-dose tolerance of operational amplifiers was limited to approximately 100 Mrad.

IV. DESIGN OF THE RADIATION-HARDENED STABILIZED POWER SUPPLY UNIT CONSTRUCTED WITH FULL BIPOLAR TRANSISTORS

Therefore, a new radiation-hardened stabilized power supply unit that does not use any operational amplifier was designed as portrayed in Figs. 5 and 6. According to our evaluation, the total-ionizing-dose tolerance of bipolar transistors depends on the transistor size. Even if the operational amplifier is constructed using bipolar transistors, the size of each transistor inside the operational amplifier is small or inadequate. Therefore, as described herein, the amplifier itself was constructed using discrete bipolar transistors: 2SA1015 and 2SC1815. Current gain degradation or amplification factor of the bipolar

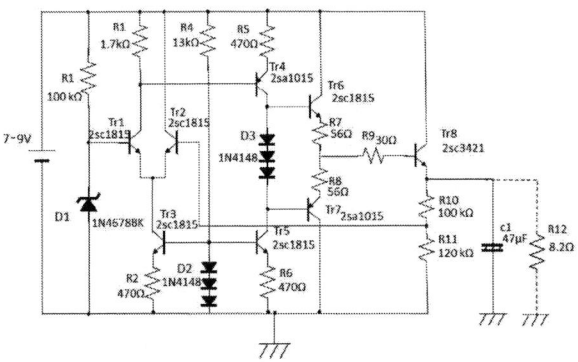

Fig. 5. Circuit diagram of the radiation-hardened stabilized power supply unit using bipolar transistors.

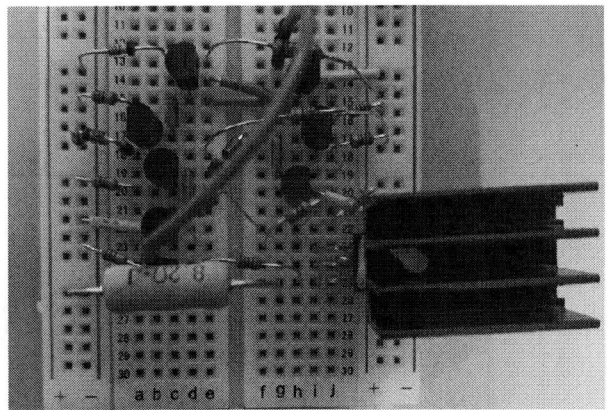

Fig. 6. Photograph of the radiation-hardened stabilized power supply unit using bipolar transistors presented in Fig. 5.

Fig. 7. Photograph of the radiation experiment using a Cobalt 60 gamma radiation source.

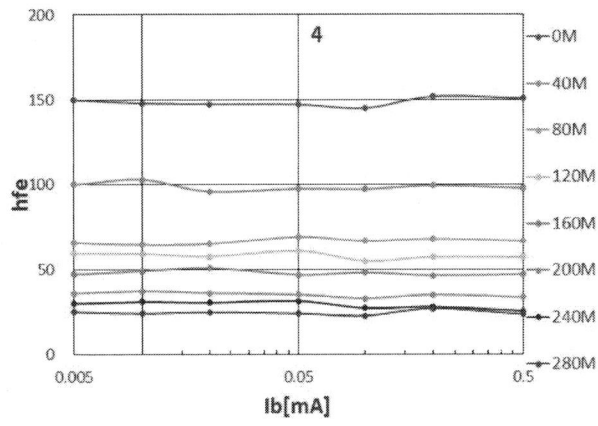

Fig. 8. Current gain or amplification factor of the bipolar transistor (2SA1015; Toshiba Corp.) for the total ionizing dose.

transistors 2SA1015 and 2SC1815, as shown respectively in Figs. 8 and 9. Results show that, even if the total ionizing dose of the transistors reaches 200 Mrad, the current gain or amplification factor of the power bipolar transistor can be maintained as greater than 25. Based on those findings, the operational amplifier was designed with a margin. The radiation tolerance of the new radiation-hardened stabilized power supply unit with no operational amplifier and using bipolar transistors and diodes was measured using a Cobalt 60 gamma radiation source, as shown in Figs. 10 and 11. Four radiation-hardened stabilized power supply units were used for this experiment. Results show that a greater than 200 Mrad total-ionizing-dose tolerance was achieved. Even after a 200 Mrad total-ionizing-dose is applied, the radiation-hardened stabilized power supply units were able to provide 3.3 V with 300 mA. By increasing the degradation margin using large discrete transistors instead of small transistors inside an operational amplifier, the total-ionizing-dose tolerance of

the radiation-hardened stabilized power supply unit can be improved.

V. CONCLUSION

For work to decommission the melted reactors and to extract melted nuclear fuel at the Fukushima Daiichi nuclear power plant, a radiation-hardened stabilized power supply unit is demanded for use with all systems. However, commercially available stabilized power supply units is invariably vulnerable to radiation. The radiation tolerance is less than 1 kGy. Therefore, the total-ionizing-dose tolerance of commercially available stabilized power supply units remains inadequate. This paper has presented a proposal of a radiation-hardened stabilized power supply unit using bipolar transistors. The radiation tolerance of the radiation-hardened stabilized power

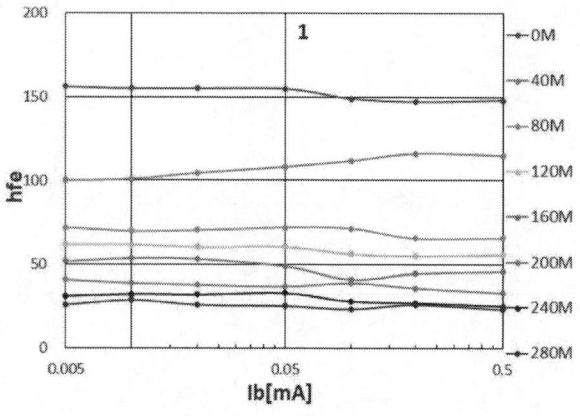

Fig. 9. Current gain or amplification factor of the bipolar transistor (2SC1815; Toshiba Corp.) for the total ionizing dose.

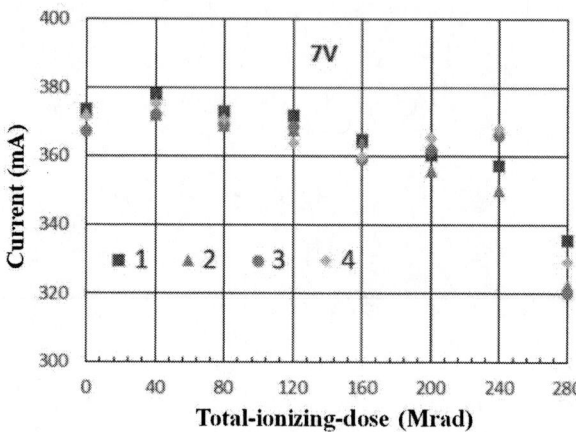

Fig. 11. Output current of the radiation-hardened stabilized power supply unit using bipolar transistors.

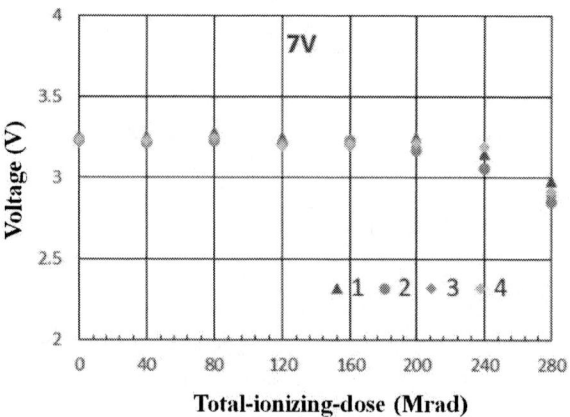

Fig. 10. Output voltage of the radiation-hardened stabilized power supply unit using bipolar transistors.

supply unit has been confirmed as 2 MGy by experimentation using a Co60 gamma radiation source. Currently, radiation-hardened field programmable gate arrays have been developed [7]–[10]. This radiation-hardened stabilized power supply unit is applicable to such radiation-hardened field programmable gate arrays.

ACKNOWLEDGEMENTS

This research was partly supported by the Initiatives for Atomic Energy Basic and Generic Strategic Research No. JPJA22F22683756, the Ministry of Education, Science, Sports and Culture, Grant-in-Aid for Scientific Research(B), No. 21H03407, and the Ministry of Education, Science, Sports and Culture, Grant-in-Aid for Challenging Research (Pioneering), No. 22K18415. The VLSI chip in this study was fabricated in the chip fabrication program of VLSI Design and Education

Center (VDEC), the University of Tokyo in collaboration with Rohm Co. Ltd. and Toppan Printing Co. Ltd.

REFERENCES

[1] K. Ouyang, W. Chen, Z. He, "Analysis of dose rate around molten corium deposited on the seabed after a severe bottom of the ship melt-through nuclear accident," IEEE International Conference on Computer and Communications, 2017.

[2] T. Ishikawa, T. Ohba, A. Hasegawa, K. Akahane, S. Yasumura, K. Kamiya, G. Suzuki, "Comparison between external and internal doses to the thyroid after the Fukushima Daiichi Nuclear Power Plant accident," Journal of Radiation Research, Vol. 64, Issue 2, 2023.

[3] R. D. Kulkarni, V. Agarwal, "Reliability analysis of a modern power supply under nuclear radiation effects," The Fifth International Conference on Power Electronics and Drive Systems, 2003.

[4] J. Deming, E. Krause, A. Kirby, B. Sandoval, D. Beckman, Z. Miller, C. A. Maldonado, "A Compact Modular High Voltage Power Supply for Space Applications," IEEE Aerospace Conference, 2023.

[5] J. C. Cubillos, N. Trikoupis, J. Mekki, "Radiation Tolerance of Programmable Voltage Supply and High Galvanic Insulation Readout Electronics Used by CERN's LHC Cryogenics," IEEE Transactions on Nuclear Science, Vol. 63, Issue 4, pp. 2022 - 2028, 2016.

[6] N. Stojadinovic, S. D. Veljkovic, V. Davidovic, I. Manic, S. Golubovic, Gamma-irradiation effects in power MOSFETs for application in communication satellites 5th International Conference on Telecommunications in Modern Satellite, Cable and Broadcasting Service, No.01EX517, 2001.

[7] K. Ando, M. Watanabe, N. Watanabe, "Optical multi-context scrubbing operation on a redundant system," Optics Express, Vol. 31, Issue 23, pp. 38529-38539, 2023.

[8] H. Shinba, M. Watanabe, "Radiation-hardened configuration-context realization for field programmable gate arrays," Applied Optics, Vol. 59, Issue 19, pp. 5680-5686, 2020.

[9] T. Fujimori, M. Watanabe, "Optically reconfigurable gate array using a colored configuration," Applied Optics, Vol. 57, Issue 29, pp. 8625-8631, 2018.

[10] R. Moriwaki, M. Watanabe, "Optical configuration acceleration on a new optically reconfigurable gate array VLSI using a negative logic implementation," Applied Optics, Vol. 52, No. 9, pp. 1939–1946, 2013.

An Analog Integrated, Low-Power, Area-Efficient, Gilbert, Modulo-based Classifier with Application to Lung-Cancer Classification

Vassilis Alimisis, Nikolaos P. Eleftheriou, Savvas Leventikidis and Paul P. Sotiriadis

Department of Electrical and Computer Engineering
National Technical University of Athens, Greece
E-mail: alimisisv@gmail.com, eleftheriou_nikos@hotmail.com, savvas01@yahoo.gr, pps@ieee.org

Abstract—**This study presents an alternative approach to develop low-power (744nW) analog classifiers capable of efficiently handling multiple input features while maintaining high levels of accuracy and minimizing power consumption. The proposed classifier relies on Voting and Bayes mathematical models, incorporating Gilbert two-signal four-quadrant multipliers and current comparators. The analog classifier is validated through testing with a real-world lung-cancer surgery dataset, achieving an accuracy of 75.45%. It predicts all testset samples of patients suffering from lung-cancer. Additionally, a comparison with related analog classifiers using the same dataset is conducted. The models are trained via a software-based implementation. The proposed architecture is realized using the TSMC 90nm CMOS process and simulated using the Cadence IC Suite.**

Index Terms—**Modulo-based classifier, Lung-cancer classification, low-power design, analog VLSI implementation**

I. INTRODUCTION

The rapid expansion of the Internet of Things (IoT) has given rise to a variety of devices and sensors, many of which operate solely on batteries, making efficient power management crucial [1]. IoT devices find application in various consumer and industrial sectors, some of which lack online recharging capabilities. To address this, hardware designers are increasingly turning to innovative power management solutions.

A notable emerging trend involves integrating IoT applications with Machine Learning (ML) algorithms to extract valuable insights from real-time data [2]. In pursuit of real-time computation, a new domain is emerging, leveraging advanced computation methods like edge computing and analog computing. Edge computing [3] processes data as close to the source as possible, enhancing speed and efficiency. Analog computing [4] aligns more closely with the continuous nature of physical laws, often requiring fewer components compared to digital circuits. Additionally, analog circuits, by operating in the sub-threshold region [5], significantly reduce power consumption.

Recent advancements in wireless remote medical devices have sparked interest in monitoring various physiological parameters related to human health conditions with a focus on portability, particularly through wearable architectures [6].

Motivated by the demand for low-power and low-area solutions in analog computing for ML and IoT applications, an efficient, high-speed analog Bayesian classifier designed for lung-cancer classification is introduced. The proposed classifier has been rigorously tested on a real-world lung-cancer surgery dataset [7].

The remainder of this paper is organized as follows. Section II refers to a brief presentation of classifier's mathematical model. The proposed architecture and the basic building blocks of the proposed classifier are described in Section III. The proper behavior of the proposed classifier is confirmed via a real-world lung-cancer classification dataset and compared with the software-based implementation in Section IV. Section V provides a comparison study with related analog classifiers. Some concluding remarks are given in Section VI.

II. MATHEMATICAL MODELLING

The Naive Bayes classifier is a straightforward probabilistic classification method that applies Bayes' theorem while assuming independence between input features [8]. Even with this assumption, it can achieve impressive accuracy when combined with kernel density estimation. By employing Bayes' theorem, the conditional probability of a vector input X belonging to a class C_k is expressed as:

$$p(C_k|X) = \frac{p(C_k)p(X|C_k)}{p(X)}. \tag{1}$$

In this context, $p(C_k)$ represents the prior probability of class k, $p(X)$ denotes the evidence probability of the input X, and $p(X|C_k)$ signifies the value of the probability density function (PDF) of class k for the input X. Specifically, for a multivariate Gaussian PDF with a diagonal covariance matrix, as assumed by the Bayesian model, $p(X|C_k)$ is defined as:

$$p(X|C_k) = \prod_{n=1}^{N} \frac{1}{\sqrt{(2\pi) \cdot \sigma_{kn}^2}} e^{-\frac{1}{2} \cdot \frac{(x_n - \mu_{kn})^2}{\sigma_{kn}^2}}. \tag{2}$$

In this context, N represents the number of features, leading to the generation of $N-d$ Gaussian functions. Parameters μ_{kn} and σ_{kn} denote the mean value and variance corresponding to the n-th feature of class k respectively, while x_n stands for the

979-8-3503-8083-5/23 $31.00 © 2023 IEEE

n-th feature of the input vector X. The final decision for the winning class is taken by applying the argmax operator to the probabilities $p(C_k|X)$ for all classes. In practical application, the evidence probability is often disregarded, and the output of the classifier can be described as:

$$y = \operatorname{argmax}\{p(C_k|X)\} = \operatorname{argmax}\{p(C_k)p(X|C_k)\}. \quad (3)$$

for $k \in \{1, 2, ..., K\}$.

In this study, to represent each sub-class using a single feature (a $1 - D$ cell), the mathematical model is articulated through a voting classifier [9], which can be approximated as:

$$y = \operatorname{mod}\{(C_1(X), C_2(X), C_3(X), ... C_K(X)\}. \quad (4)$$

In this context, $C_k(x)$ signifies the output of each 1-D Gilbert decision cell (GDC), essentially representing each sub-classifier. To further illustrate this concept, let's consider a scenario with five features involved in a binary classification task, each carrying equal weight. The functions $C_1(x)$, $C_2(x)$, $C_3(x)$ collectively yield the output for the first class (class 1), while $C_4(x)$, $C_5(x)$ produces the output for the second class (class 0). Consequently, the result is calculated as $y = \operatorname{mod}(1, 1, 1, 0, 0) = 1$ (indicating class 1).

III. PROPOSED ARCHITECTURE

In this section, the proposed architecture of the analog classifier along with its basic building blocks is presented. Since it can accommodate various numbers of classes and input dimensions, it is scalable and provides high versatility. Firstly, for the realization of GDC in equation (4), Gilbert two signal four-quadrant multipliers (Gilbert cells) [5] along with current comparators (Winner-take-all circuit) [10] are employed, as shown in Fig. 1. Transistors M_{n1}-M_{n4} and M_{n7}-M_{n10} implement the two Gilbert cells and transistors M_{n5}, M_{n6}, M_{n11} and M_{n12} implement the Winner-take-all (WTA) circuit. The GDC circuit operates in a translinear fashion, producing two decision output currents that signify the decisions for each feature in both classes. In this context, a higher current indicates the winning class.

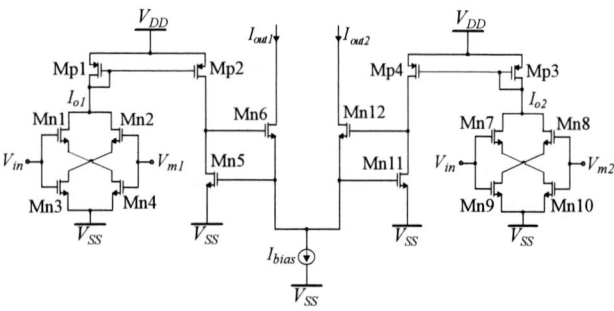

Fig. 1: The implementation of the GDC circuit. It consists of Gilbert two signal four-quadrant multipliers (Gilbert cells) along with current comparators. Here V_{in} is the input voltage and V_{m1} and V_{m2} are parameters describing the mean values of each function. The output current represents the decision according to a specific feature.

The architecture of the proposed classifier, as shown in Fig. 2, is designed for a classification problem involving

$N_{cla} = 2$ classes and $N_d = 16$ features (input dimensions). This illustration consists of 16 GDC circuits (input dimensions) and one WTA circuit (modulo implementation), shown in Fig. 2. Each GDC circuit describes the voting strength for each class regarding a specific feature. It produces two output currents, each one represent a feature's decision. All the currents related to one class are summed via current mirrors (CMs) to minimize potential distortions in calculations that might arise from undesired effects on the output currents of the GDC. The resulting output currents, distinguished by their high or low values, indicate the classifier's final prediction. All transistor dimensions are set to $W/L = 1.6\mu m/1.6\mu m$. The power supply rails set as $V_{DD} = -V_{SS} = 0.3$ V and all transistors operate in the sub-threshold region in order to achieve low power consumption.

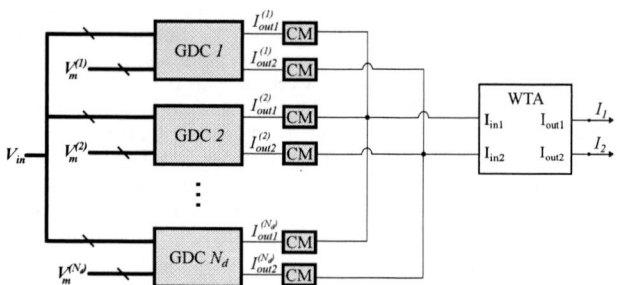

Fig. 2: The proposed classifier's top-level architecture.

IV. LUNG-CANCER DATASET AND SIMULATION RESULTS

In this section, the effectiveness of the proposed classifier using a real-world dataset related to lung-cancer [7] is challenged. The proposed architecture has been implemented in the TSMC $90nm$ CMOS process, employing the Cadence IC suite. All the simulation tests are conducted on the layout (post-layout simulations) illustrated in Fig. 3. This classification task revolves around lung-cancer, encompassing $N_{cla} = 2$ distinct classes and $N_d = 16$ inputs. As for the classifier's training, a software-based implementation is employed to tune the required parameters. All related metrics are directly fed into the hardware classifier. The necessary parameters for the system are computed by evaluating the mean value and prior probability of each class.

The Thoracic Surgery Data is a dataset available from the UCI Machine Learning Repository [7]. It encompasses clinical information about patients who underwent thoracic surgery for lung-cancer treatment. This dataset is valuable for research in medical and healthcare fields, as it includes attributes such as age, performance status, tumor size, and other relevant factors. The aim is to predict the survival status of patients after surgery based on these features. This dataset serves as a valuable resource for machine learning practitioners and researchers aiming to develop predictive models in the context of thoracic surgery outcomes.

To evaluate the proposed classifier's performance in terms of classification specificity and the circuit's behaviour under

Process-Voltage-Temperature (PVT) variations, two distinct tests were carried out on the layout. To account for random effects, the outcomes of 20 different training-testing iterations are depicted in Fig. 4. It predicts all the patients who have cancer but it has false-positive alarms for patients who have not cancer. As a result, it can be used as a wake-up engine for a digital back-end. Subsequently, the circuit's sensitivity on random variations is affirmed through a Monte Carlo analysis. Specifically, Fig. 5 displays the Monte Carlo Histogram for N = 100 data points.

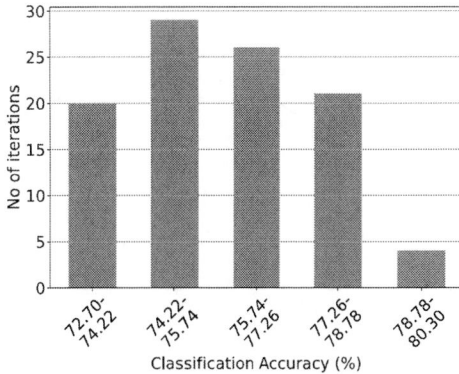

Fig. 5: Post-layout Monte-Carlo simulation results of the proposed architecture on the lung-cancer classification dataset with $\mu_M = 75.87\%$ and a standard deviation of $\sigma_M = 1.73\%$.

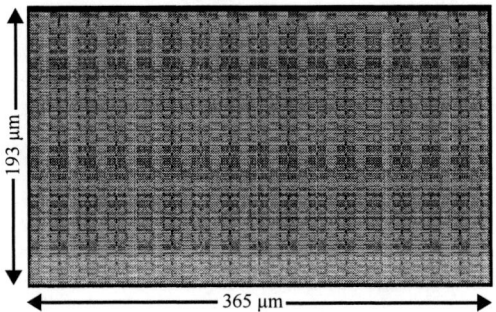

Fig. 3: Layout of the proposed classifier's architecture based on the design methodology (extra dummy transistors are used).

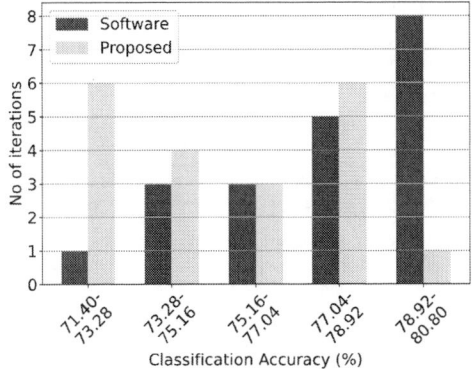

Fig. 4: Classification results of the proposed architecture and the equivalent software model on the lung-cancer classification dataset over 20 iterations.

V. COMPARISON STUDY AND DISCUSSION

In related literature, it is clear that the majority of analog classifiers are typically tailored to specific applications. It is a great challenge to compare different ML models or hardware implementations on the same application and deduce fair results. However, this challenge enables for adapting analog classifiers to serve a common application, thereby simplifying the process of evaluating performance that encompasses both ML models and alternative methodologies. Table I offers an overview of the performance comparison along a variety of related classifiers. Here, Gaussian Mixture Model (GMM) [11], Radial Basis Function [12], Long Short-Term Memory (LSTM) [13], K-means [14], Bayesian [17], ANN (Artificial

Neural Network) [15], Fuzzy [16], Support Vector Machine (SVM) [18], Threshold [19], Multilayer Perceptron (MLP) [20] and centroid-based [21] classifiers, all within the context of lung-cancer disease classification, are summarized.

The presented research introduces a solution, offering a balance between accuracy, power efficiency, and energy consumption per classification when compared to equivalent classifiers in the field. It's imperative to highlight that in this specific application, the design deal with a high input dimensionality. The proposed configuration holds a significant edge by obviating the necessity for Principal Component Analysis (PCA), allowing for the incorporation of all 16 input dimensions without any loss of crucial information. In contrast, many alternative topologies must reduce the dimensions to 11 to achieve optimal accuracy, representing a noteworthy constraint in prior similar studies [13]–[15], [20]. While the proposed classifier demonstrates its proficiency in accurately classifying a broader range of classes, we opt for a binary classification scenario to ensure a fair comparison. This adjustment facilitates a more meaningful assessment in relation to binary analog classifiers [16], [18], [19].

In terms of classification accuracy, the proposed architecture outperforms all its counterparts, except for MLP [20], LSTM [13] and K-means [14]. While these models achieve higher accuracy, they come at the cost of increased complexity and power consumption along with a larger silicon area due to their components number. On the other end of the spectrum, the Threshold classifier achieves the lowest power consumption in comparison with the other classifiers, albeit with a trade-off in accuracy and processing speed, attributed to its simple model design [19]. It's important to note that in biomedical applications of this kind, swift processing speed isn't of paramount importance, primarily due to their infrequent occurrence. Therefore, in the analysed approach, processing speed is decreased to enhance accuracy and optimize power consumption. Additionally, it touts lower energy consumption per classification compared to all classifiers, except for ANN [15], which achieves a lower classification accuracy.

TABLE I: Analog classifiers' comparison on the Lung-Cancer Disease Classification

	Classifier	Worst accuracy	Mean accuracy	Best accuracy	Power consumption	Processing speed	Energy per classification	No. of Dimensions
This work	Modulo	71.40%	75.45%	79.50%	$744nW$	$320k\frac{classifications}{s}$	$\frac{2.33 \text{ pJ}}{classification}$	16
[11]	GMM	68.40%	71.27%	73.80%	$2.97\mu W$	$100K\frac{classifications}{s}$	$\frac{29.70 \text{ pJ}}{classification}$	11
[12]	RBF	66.70%	70.41%	72.70%	$27.87\mu W$	$200k\frac{classifications}{s}$	$\frac{139.35 \text{ pJ}}{classification}$	11
[13]	LSTM	94.10%	97.54%	100.00%	$22.54mW$	$870M\frac{classifications}{s}$	$\frac{25.91 \text{ pJ}}{classification}$	16
[14]	K-means	88.30%	91.41%	95.10%	$111.12\mu W$	$5M\frac{classifications}{s}$	$\frac{22.22 \text{ pJ}}{classification}$	16
[15]	ANN	68.90%	72.43%	76.50%	$2.63\mu W$	$14M\frac{classifications}{s}$	$\frac{0.19 \text{ pJ}}{classification}$	16
[16]	Fuzzy	73.80%	78.65%	81.60%	$3.67\mu W$	$4.55K\frac{classifications}{s}$	$\frac{806.59 \text{ pJ}}{classification}$	11
[17]	Bayes	63.70%	68.72%	71.30%	$1.79\mu W$	$100K\frac{classifications}{s}$	$\frac{17.90 \text{ pJ}}{classification}$	11
[18]	SVM	70.10%	72.37%	74.70%	$67.63\mu W$	$140K\frac{classifications}{s}$	$\frac{483.07 \text{ pJ}}{classification}$	11
[19]	Threshold	67.60%	70.77%	75.90%	$920nW$	$100K\frac{classifications}{s}$	$\frac{9.20 \text{ pJ}}{classification}$	11
[20]	MLP	86.10%	87.56%	89.40%	$354.18\mu W$	$930k\frac{classifications}{s}$	$\frac{380.84 \text{ pJ}}{classification}$	16
[21]	Centroid	71.40%	73.87%	76.30%	$2.98\mu W$	$100K\frac{classifications}{s}$	$\frac{29.80 \text{ pJ}}{classification}$	11

VI. CONCLUSION

In this work, an alternative approach for a power-efficient (744nW), low voltage (0.6V), analog classifier for lung-cancer surgery classification was proposed. The presented architecture consists of Gilbert two-signal four-quadrant multipliers and current comparators. The circuit's parameters were adjusted through offline training of a Bayes software classifier. The post-layout simulation was conducted through a TSMC 90nm CMOS process and the results were assessed in comparison with both a software-based implementation and a variety of related analog classifiers. The realized architecture demonstrates a decent classification accuracy of 75.45% with notable sensitivity properties.

REFERENCES

[1] K. Gulati, R. S. K. Boddu, D. Kapila, S. L. Bangare, N. Chandnani, and G. Saravanan, "A review paper on wireless sensor network techniques in internet of things (iot)," *Materials Today: Proceedings*, vol. 51, pp. 161–165, 2022.

[2] J. P. Bharadiya, "Leveraging machine learning for enhanced business intelligence," *INTERNATIONAL JOURNAL OF COMPUTER SCIENCE AND TECHNOLOGY*, vol. 7, no. 1, pp. 1–19, 2023.

[3] Y. Mao, C. You, J. Zhang, K. Huang, and K. B. Letaief, "A survey on mobile edge computing: The communication perspective," *IEEE communications surveys & tutorials*, vol. 19, no. 4, pp. 2322–2358, 2017.

[4] W. Haensch, T. Gokmen, and R. Puri, "The next generation of deep learning hardware: Analog computing," *Proceedings of the IEEE*, vol. 107, no. 1, pp. 108–122, 2018.

[5] S.-C. Liu, *Analog VLSI: circuits and principles*. MIT press, 2002.

[6] V. Custodio, F. J. Herrera, G. López, and J. I. Moreno, "A review on architectures and communications technologies for wearable health-monitoring systems," *Sensors*, vol. 12, no. 10, pp. 13 907–13 946, 2012.

[7] [Online]. Available: https://archive.ics.uci.edu/dataset/277/thoracic+surgery+data

[8] C. M. Bishop and N. M. Nasrabadi, *Pattern recognition and machine learning*. Springer, 2006, vol. 4, no. 4.

[9] M. A. Khan, M. A. Khan Khattk, S. Latif, A. A. Shah, M. Ur Rehman, W. Boulila, M. Driss, and J. Ahmad, "Voting classifier-based intrusion detection for iot networks," in *Advances on Smart and Soft Computing: Proceedings of ICACIn 2021*. Springer, 2022, pp. 313–328.

[10] J. Lazzaro, S. Ryckebusch, M. A. Mahowald, and C. A. Mead, "Winner-take-all networks of o (n) complexity," *Advances in neural information processing systems*, vol. 1, 1988.

[11] V. Alimisis, G. Gennis, K. Touloupas, C. Dimas, M. Gourdouparis, and P. P. Sotiriadis, "Gaussian mixture model classifier analog integrated low-power implementation with applications in fault management detection," *Microelectronics Journal*, vol. 126, p. 105510, 2022.

[12] S.-Y. Peng, P. E. Hasler, and D. V. Anderson, "An analog programmable multidimensional radial basis function based classifier," *IEEE Transactions on Circuits and Systems I: Regular Papers*, vol. 54, no. 10, pp. 2148–2158, 2007.

[13] Z. Zhao, A. Srivastava, L. Peng, and Q. Chen, "Long short-term memory network design for analog computing," *ACM Journal on Emerging Technologies in Computing Systems (JETC)*, vol. 15, no. 1, pp. 1–27, 2019.

[14] R. Zhang and T. Shibata, "An analog on-line-learning k-means processor employing fully parallel self-converging circuitry," *Analog Integrated Circuits and Signal Processing*, vol. 75, pp. 267–277, 2013.

[15] S. T. Chandrasekaran, R. Hua, I. Banerjee, and A. Sanyal, "A fully-integrated analog machine learning classifier for breast cancer classification," *Electronics*, vol. 9, no. 3, p. 515, 2020.

[16] E. Georgakilas, V. Alimisis, G. Gennis, C. Aletraris, C. Dimas, and P. P. Sotiriadis, "An ultra-low power fully-programmable analog general purpose type-2 fuzzy inference system," *AEU-International Journal of Electronics and Communications*, vol. 170, p. 154824, 2023.

[17] V. Alimisis, G. Gennis, C. Dimas, and P. P. Sotiriadis, "An analog bayesian classifier implementation, for thyroid disease detection, based on a low-power, current-mode gaussian function circuit," in *2021 International conference on microelectronics (ICM)*. IEEE, 2021, pp. 153–156.

[18] V. Alimisis, G. Gennis, M. Gourdouparis, C. Dimas, and P. P. Sotiriadis, "A low-power analog integrated implementation of the support vector machine algorithm with on-chip learning tested on a bearing fault application," *Sensors*, vol. 23, no. 8, p. 3978, 2023.

[19] V. Alimisis, G. Gennis, E. Tsouvalas, C. Dimas, and P. P. Sotiriadis, "An analog, low-power threshold classifier tested on a bank note authentication dataset," in *2022 International Conference on Microelectronics (ICM)*. IEEE, 2022, pp. 66–69.

[20] K. Lee, J. Park, and H.-J. Yoo, "A low-power, mixed-mode neural network classifier for robust scene classification," *Journal of Semiconductor Technology and Science*, vol. 19, no. 1, pp. 129–136, 2019.

[21] V. Alimisis, V. Mouzakis, G. Gennis, E. Tsouvalas, C. Dimas, and P. P. Sotiriadis, "A hand gesture recognition circuit utilizing an analog voting classifier," *Electronics*, vol. 11, no. 23, p. 3915, 2022.

Ultra-Low Power Self-polarized Dynamic Threshold Telescopic OTAs Circuits for Biomedical Applications

Dalila Laouej, Houda Daoud and Mourad Loulou

Information Technologies and Electronics Laboratory
National Engineering School of Sfax
Sfax, Tunisia
dalila.laouej@enis.tn, houda.daoud@enetcom.usf.tn, mourad.loulou@ieee.org

Abstract—Very low voltage and ultra low power consumption fully differential telescopic operational transconductance amplifier (OTA) circuits based on Dynamic Threshold MOS (DTMOS) technique were proposed in this paper. The first proposed DT-OTA (OTA-I) was designed using the conventional structure of a telescopic OTA. While 'the second one (OTA-II) was designed using the bulk-driven technique at all amplifier levels. These two circuits are self-polarized and were designed for biomedical applications. The proposed two DT telescopic OTAs has good performances especially in terms of power consumption. The first circuit has a gain of 54.7 dB, a GBW of 312 kHz while the power consumption is only 270 nW. The DT telescopic OTA-II has a gain of 67 dB, a GBW of 400 kHz while the power consumption is only 250 nW. The DT-OTA circuits are implemented with the TSMC 0.18μm CMOS technology.

Keywords—Dynamic Threshold MOS (DTMOS); Telescopic OTA; bulk-driven technique; self-polarized; very low voltage; ultra-low power consumption; Biomedical devices; TSMC 0.18μm CMOS technology.

I. INTRODUCTION

Nowadays, continuous and progressive reduction in the size of CMOS devices follows also a continuous decrease in the size of wearable devices. In fact, the embedded systems are polarized by batteries while the power consumption must be reduced as possible. One of the most appreciated areas by these advances is the biomedical field such as the Wireless Body Area Network (WBAN). The WBAN, which its general communication architecture is presented in Fig.1.a, was standardized to IEEE 802.15.6 [1, 2] for short distances and low power applications. As shown, this network can't be empty of embedded micro-sensors; the general structure is presented in Fig.1.b [3]. In fact, the micro-sensors are the key devices in this network. These devices can be implemented inside, outside or used around the human body. The major challenge of these components is the power consumption. Therefore, the implantable devices are polarized by batteries, which are neither changeable nor chargeable. Thus, to have a long service life, all the electronics blocks must consume as little as possible of power, around of hundreds of micro Watts (μW). One of critical and primordial

block used for designing a variety of analog integrated circuits such as Gm-C filters, a comparator, an integrator, an oscillator, and so on, was the operational transconductance amplifier (OTA). Telescopic OTA is one of the most suitable structures investigated for low voltage and low power consumption applications. It was the most powerful structure in terms of noise and power consumption. From the literature, the OTA circuits used for this field consume a little power such as 59nW in [4], 145nW in [5], 11.9μW in [6], 20μW in [7] and so on. Generally, in electronics, implementing an analogue integrated circuit at low-voltage was always affected by the limitation in the threshold voltage (V_{th}) since an NMOS transistor is conductor just when the gate-source voltage (V_{gs}) is larger than V_{th}. Therefore, in order to overcome this limitation there are different useful methods and techniques, such as polarizing the MOSFET transistor in the weak inversion using the bulk-driven (BD) or the dynamic threshold techniques. Also we found the using of the floating gate MOSFET or the quasi-floating gate approaches. These methods allow us to scull down the design complexity and increase the supply voltage to reach the threshold voltage of the MOSFET transistor. The dynamic threshold technique was used for designing many OTA structures as presented in [8-12]. Thus, the objective of this work is to use the dynamic threshold approach to design a two different telescopic OTA structures dedicated to biomedical applications.

In this paper, we first present the two proposed DT telescopic OTA circuits design. Next, we present the dynamic threshold approach. In section IV, the simulations results of the two DT OTA circuits' performances are presented. In section V a performances comparison is introduced. Finally, Section VI details the conclusion.

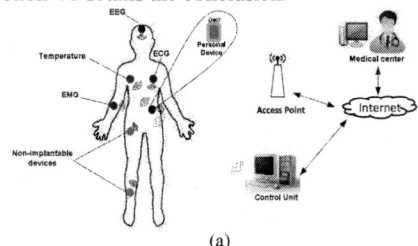

(a)

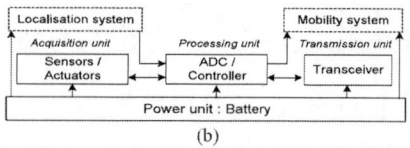

(b)

Fig. 1. Wireless Body Area Network (WBAN) (a) general communication architecture (b) general structure of a medical micro-sensor

II. THE PROPOSED DT TELESCOPIC OTA CIRCUITS DESIGN

A. DT telescopic OTA-I Circuit analysis

The structure of the first proposed DT telescopic OTA (DT-OTA-I) is presented in Fig. 2. As shown in this figure, the input signals are injected into the tied gate-bulk terminals of the differential input stage (NM1-NM2) which is polarized by a simple current mirror (NM5-NM6). Cascodes transistors stages (PM1-PM3) and (PM2-PM4) are used to increase the static gain. The cascode stage (PM5-NM7), composed of connected-diode PMOS (PM5) with simple NMOS (NM7), is used to polarize the transistors (NM3-NM4).

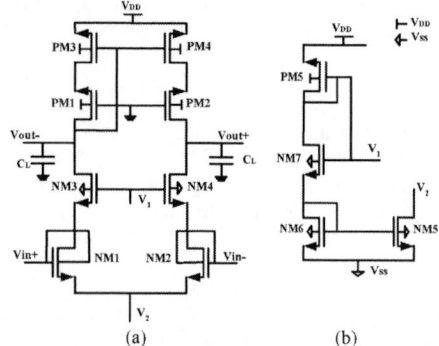

(a)   (b)

Fig. 2. First proposed DT Telescopic OTA a)DT-OTA-II main structure b) circuit polarization

The equivalent small signal of the DT telescopic OTA-I is given by Fig. 3 below.

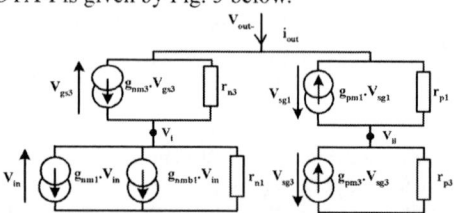

Fig. 3. Equivalent small-signal model of DT Telescopic OTA-I

For this circuit model;

$V_{sg1}=V_{s1}-V_{g1}=V_{s1}= V_{out-}$, because $V_{g1}=0$.

$V_{sg3}= V_{s3}-V_{g3}= V_{ii}$.

According to this model, the open-loop voltage gain and gain bandwidth (GBW) expressions of the DT Telescopic OTA are given by (1) and (2):

$$A_V = G_{nm1}.\frac{(g_{nm3}r_{n1}r_{n3})(g_{pm1}r_{p1}r_{p3})}{(g_{nm3}r_{n1}r_{n3})+(g_{pm1}r_{p1}r_{p3})} \quad (1)$$

$$GBW = \frac{G_{nm1}}{2\pi C_L} \quad (2)$$

Where, G_{nm1} ($G_{nm1}=g_{nm1}+g_{nmb1}$), g_{nm3} and g_{pm1} are respectively the transconductances of transistors NM1, NM3 and PM1. r_{n1}, r_{n3}, r_{p1} and r_{p3} are respectively the drain-source resistances of

transistors NM1, NM3, PM1 and PM3. C_L is the capacitance at the output node.

B. DT telescopic OTA-II Circuit analysis

The structure of the second proposed DT telescopic OTA (DT-OTA-II) is presented in Fig. 4. For this circuit, the input signals are injected into the tied gate-bulk terminals of the differential input stage (NM1-NM2) which is polarized by a bulk-driven current mirror (NM5-NM6). Cascodes transistors stages (PM1-PM2) and (PM3-PM4) are used to increase the static gain. All these stages are implemented using the bulk-driven technique. The transistor connected-diode NMOS (NM7) is used to polarize the (PM1-PM2) and the bulk-driven current mirror (NM5-NM6) stages while the NM8 transistor is used to polarize the (PM3-PM4) stage.

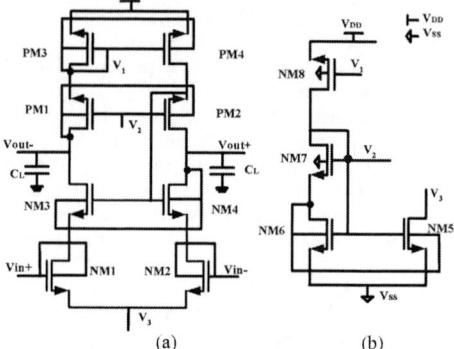

(a)   (b)

Fig. 4. second proposed DT Telescopic OTA a)DT-OTA–II main structure b) circuit polarization

According to the circuit small signal model, the expressions of the open-loop gain and the GBW of the circuit are given below:

$$A_V=G_{nm1}.R_L=G_{nm1}.(R_{out1}//R_{out2}) \quad (3)$$

Where:

$$R_{out1}=g_{pm1}.r_{p1}r_{p3}$$

$$R_{out2}=(1+r_{n3}(1+g_{nm3}))\left[1+\frac{1}{(1+r_{n3}g_{nm3})(1+r_{n3}r_{n1}g_{nm3})}\right] \quad (4)$$

$$GBW = \frac{G_{nm1}}{2\pi C_L} \quad (5)$$

III. THE DYNAMIC THRESHOLD APPROACH:METHOD DESCRIPTION

Dynamic threshold MOS (DTMOS) technique was first introduced in 1994 [13]. This technique is mostly used in digital applications [10]. Contrary to the bulk-driven method where the operational amplifier input signal is applied into the bulk terminal while the gate terminal is biased by a fixed voltage V_{gs} ($>V_{th}$ for NMOS) to form an inversion layer and the drain current I_D is controlled by V_{gs} (Fig.5.a), using the dynamic threshold approach, the input signal is injected in the same time into the bulk and the gate terminals as shown in Fig.5.b. The main idea of this technique is to connect the gate and the bulk of the MOS device together.

Dynamic threshold approach is a method proposed to overcome the MOSFET transistor voltages limitation such as the supply voltage V_{DD} and the threshold voltage V_{th}. In fact, it permits to operate with a very low supply voltage and a low threshold voltage. In fact, in this case the threshold voltage will be reduced and therefore the transistor will be conductor at a very low voltage as shown in figure 6. Indeed, using this method we can note that the threshold voltage can be scale down for the same CMOS technology.

The threshold voltage of a MOS transistor is the gate to source (V_{gs}) voltage needed to create the conduction channel. This voltage of an NMOS transistor can be expressed by Eq.6. From this equation, it is seen that V_{th} can be controlled by varying the substrate bias voltage (V_{bs}). In fact, according to this expression, the DTMOS technique reduces the transistor off state leakage current and also reduces the threshold voltage as the transistor is on state ($V_{bs}>0$) [9-11].

$$V_{th} = V_{T0} + \gamma\left(\sqrt{2|\varphi_F| - V_{bs}} - \sqrt{2|\varphi_F|}\right) \quad (6)$$

Where, V_{T0}, γ, V_{bs} and φ_F are respectively the zero bias threshold voltage, the body effect coefficient, the body bias voltage and the Fermi potential.

Using this technique, the transconductance G_m is larger than a conventional gate or a bulk driven transistor. In fact, it increased by g_{mb} and is given by (7) [9, 12].

$$G_m = g_m + g_{mb} \quad (7)$$

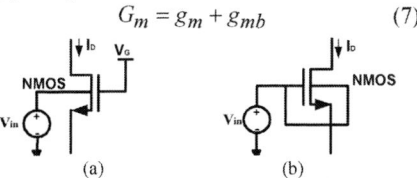

(a)   (b)

Fig. 5. (a) bulk-driven and (b) dynamic threshold transistor

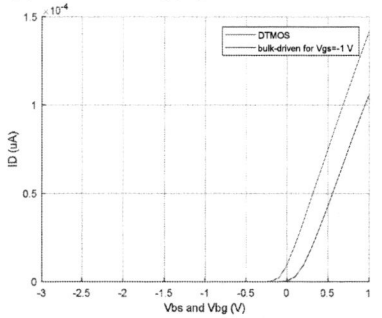

Fig. 6. Drain currents I_D versus bulk-source (V_{bs}) and bulk-gate (V_{bg}) of the bulk-driven and the dynamic threshold NMOS

IV. SIMULATIONS RESULTS

A. DT telescopic OTA-I simulations results

An OTA is characterized by many features such as open-loop voltage gain, unity-gain bandwidth, SR, output voltage swing, CMRR, PSRR, etc. To evaluate the proposed OTA structure, we simulate its features and characteristics. For this type of circuit, the DC gain is one of the most important parameters. It indicates the capacitance of circuit to amplify the differential inputs signals. Thus, to prove the stability of the circuit, the phase marge (PM) must be more than 45 degrees. In fact, the first designed DT Telescopic OTA has a gain of 54.7dB, a GBW of 312 kHz and a phase margin of 89.7 degrees as shown in Fig.7. Table1 summaries the first proposed DT telescopic OTA performances.

This circuit achieves good performances especially in terms of power consumption; such it consumes only 270nw, and input referred noise as shown in Fig.8.

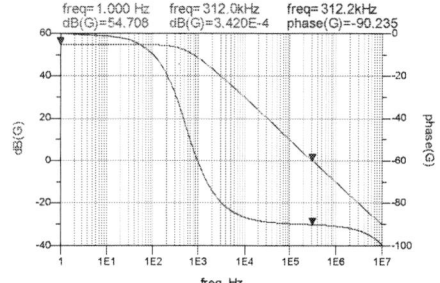

Fig. 7. Gain, GBW and phase curves of DT-OTA-I

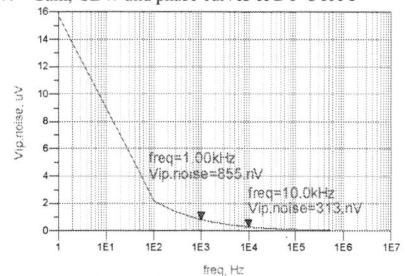

Fig. 8. Input referred noise of the proposed DT-OTA-I

TABLE I. TELESCOPIC DT-OTA-I PERFORMANCES

Performances (C_L=1 pF)	Simulations results
DC Gain (dB)	54.7
GBW (kHz)	312
CMRR(dB)	57.8
PSRR (dB)	72.6
Output swing (V)	[-0.4, 0.23]
Input referred noise voltage at f=1 kHz((nV/√Hz)	855
Pc (nW)	270

B. DT telescopic OTA-II simulations results

The second designed DT Telescopic OTA has good performances. In fact, it achieves a gain of 67dB, a GBW of 400 kHz and a phase margin of 89.6 degrees as shown in Fig. 9.

Table 2 summaries the second proposed DT telescopic OTA performances. Also this second circuit achieves good performances especially in terms of power consumption; such it consumes only 250nw, and input referred noise which is very low as shown in Fig.10. Hence, the input referred noise voltage is low due to the high transconductance value of the input stage.

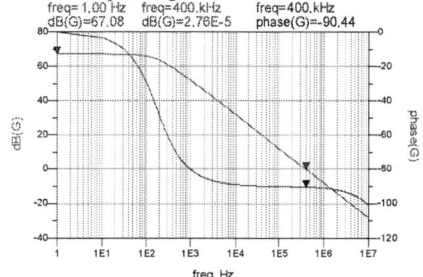

Fig. 9. Gain, GBW and phase curves of the DT-OTA-II

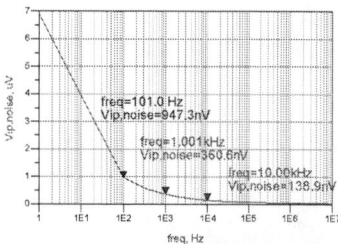

Fig. 10. Input referred noise of the proposed DT-OTA-II

TABLE II. TELESCOPIC DT-OTA-II PERFORMANCES

Performances (C_L=1 pF)	Simulations results
DC Gain (dB)	67
GBW (kHz)	400
CMRR(dB)	77.1
PSRR (dB)	62
Output swing (V)	[-0.46, 0.393]
Input referred noise voltage at f=1 kHz (nV/√Hz)	360nV
Pc (nW)	250

V. PERFORMANCES COMPARISONS

The two Telescopic OTAs circuits are implemented using TSMC 0.18μm CMOS technology. These two circuits can be suitable for low power biomedical applications. This section presents the performances summary and comparison of the two OTAs circuits performances with other published works. Firstly, we start by comparing our two circuits. From the finding simulations results we can notice that the second structure have good performances. In fact, we obtain an increasing in the value of all the parameters. This augmentation appears especially in the value of the power consumption and the input referred noise, which is less than of the half of the first one. That means, the BD technique improves more the OTA performances. Secondly, as mentioned in table 3, comparing with other works, the two proposed DT telescopic OTAs achieves the lowest power consumption except [10] and the good phase margin except [14]. Thus, our circuits have a good stability where their phase margins are almost 90 degrees. Furthermore, our OTA circuits are fast, they have a good DC gain. The CMRR describes the sensitivity of the OTA to reject a common mode signal. It is relatively high; that means that our OTAs are less sensitive to the noise and the common mode voltage.

TABLE III. PERFORMANCES SUMMARY OF THE OTAS AND COMPARISON WITH PUBLISHED WORKS

References	[10]	[14]	[15]	This work DT-OTA-I	This work DT-OTA-II
CMOS Technology (μm)	0.18	0.18	0.18	0.18	0.18
Supply Voltage (V)	0.4	±0.6	±0.5	±0.5	±0.5
DC gain (dB)	91	23.05	50.2	54.7	67
GBW (kHz)	111	379.7	156800	312	400
CMRR (dB)	106	-	61.89	57.8	77.1
PSRR (dB)	-	-	-	72.6	62
Phase (degree)	66	93.8	72.4	89.7	89.56
C_L(pf)	15	1	0.1	1	1
P_C(nW)	386	1397	2800	270	250

VI. CONCLUSION

In this paper a very low power two Telescopic OTA circuits using the DT technique have been designed to be used for very low power biomedical devices. The two proposed OTA circuits using the DT approach operate in the sub-threshold region to enable low supply voltage circuit operation at ±0.5V. Under a load condition of 1 pF, the first OTA circuit provides a DC gain of 54.7 dB, a GBW of 312 kHz and consumes 270nW while the second one provides a DC gain of 67 dB, a GBW of 400 kHz and consumes only 250nW. Therefore, the simulation results of the OTA circuits prove that these designs can be satisfies the required performance of WBAN applications.

REFERENCES

[1] D. Laouej, H. Daoud and M. Loulou, "A Very Low Power Delta Sigma Modulator Using Optimized Bulk Driven Telescopic OTA for Biomedical Devices", IEEE I. Conference on Design and Test of integrated and micro & non-Systems (DTS), pp. 1-6, 2020.

[2] J. Negraa, I. Jemilia, A. Belghith,"Wireless Body Area Networks: Applications and technologies", Procedia Computer Science, vol. 83, pp. 1274–1281, 2016.

[3] D. Laouej, H. Daoud and M. Loulou, "Design of sixth order Butterworth Gm-C Filter using Particle Swarm Optimization program for Biomedical Application. IEEE I. Conference in Microeltronics (ICM), pp. 1-5, 2017.

[4] S. Bano, G. B. Narejo and S. M. U. Ali shah, "Power Efficient Fully Differential Bulk Driven OTA for Portable Biomedical Application", journal of Electronics, 2018.

[5] A. Ghaemnia and O. Hashemipour, "An ultra-low power high gain CMOS OTA for biomedical applications", Analog Integrated Circuits and Signal Processing, 2019.

[6] F. Moulahcene, N.E Bouguechal, I. Benacer and S. Hanfoug, "Design of CMOS Two-stage Operational Amplifier for ECG Monitoring System Using 90nm Technology", International Journal of Bio-Science and Bio-Technology Vol. 6, No. 5 , pp. 55-66, 2019.

[7] E.Lohitha and Dr.E.John Alex, "Ultra low power and area efficient OTA for portable and wearable applications", Dogo Rangsang Research Journal, Vol. 12, pp. 29-41, 2022

[8] H.F.Achigui, C.J. Fayomi, M, Sawan, "1 V DTMOS Based Class AB Operational Amplifier: Implementation and Experimental Results", IEEE Journal of Solid-State Circuits, Vol. 41, No. 11, pp. 2440-2448, 2006.

[9] Madhulika, S. Kumari and M. Gupta, "Design and Analysis of DTMOS based Low Voltage OTA and its Filter Application", International Conference on Recent Trends on Electronics, Information, Communication & Technology (RTEICT), pp. 685-690, 2021.

[10] E. Kargaran , M. Sawan , K. Mafinezhad and H. Nabovati, "Design of 0.4V, 386nW OTA Using DTMOS Technique for Biomedical Applications", IEEE 55th International Midwest Symposium on Circuits and Systems (MWSCAS), pp. 270-273, 2012.

[11] A. Yazdani-Nejad and S. Hossein Pishgar, "Design and Simulation of OTA using DTMOS Technique in 180 nm CMOS Process", International Journal of Computer Applications, pp. 20-22, 2016.

[12] M. Mahendra, Sh. Kumari and M. Gupta, "DTMOS Based Low Power Adaptively Biased Fully Differential Transconductance Amplifier with Enhanced Slew-Rate and its Filter Application", IETE Journal of Research, pp.1-20, 2021.

[13] F. Assaderaghi, D. Sinitsky, S. Parke, J. Bokor, P. K. Ko, and C. Hu, "A dynamic threshold voltage MOSFET (DTMOS) for ultra-low voltage operation," in Int. Electron Devices Meeting, Techn. Digest, pp. 809–812, 1994.

[14] A. Yazdani-Nejad, S. H. Pishgar, "Design and Simulation of OTA using DTMOS Technique in 180 nm CMOS Process," International Journal of Computer Applications (1975-8887), vol. 139, no.7, April 2016.

[15] D. Laouej, H. Daoud and M. Loulou, "An ultra-low power hybrid 2nd order Feed Forward ΔΣ modulator design for implantable medical devices", Analog Integrated Circuits and Signal Processing, pp. 277-289, 2021.

Energy-efficient Computation-In-Memory Architecture using Emerging Technologies

Rajendra Bishnoi[*], Sumit Diware[*], Anteneh Gebregiorgis[*], Simon Thomann[***], Sara Mannaa[**],
Bastien Deveautour[**], Cédric Marchand[**], Alberto Bosio[**], Damien Deleruyelle[**],
Ian O'Connor[**], Hussam Amrouch[***], Said Hamdioui[*]

[*]Computer Engineering Laboratory, Delft University of Technology, The Netherlands
[**]University Lyon, ECL, INSA Lyon, CPE Lyon, Institut de Nanotechnologies de Lyon, France
[***]Chair of AI Processor Design; Munich Institute of Robotics and Machine Intelligence,
Technical University of Munich, Germany

Abstract—**Deep Learning (DL) has recently led to remarkable advancements, however, it faces severe computation related challenges. Existing Von-Neumann-based solutions are dealing with issues such as memory bandwidth limitations and energy inefficiency. Computation-In-Memory (CIM) has the potential to address this problem by integrating processing elements directly into the memory architecture, reducing data movement and enhancing the overall efficiency of the system. In this work, we propose CIM architecture using three distinct emerging technologies. Firstly, a CIM architecture utilizing Ferroelectric Field-Effect Transistors (FeFET) is shown and the resulting errors from the analog compute scheme are injected into the emerging algorithm of Hyperdimensional Computing. Subsequently, we explore Vertical Nanowire Field-Effect Transistors (VNWFETs) based CIM within a 3D computing architecture, demonstrating improved energy efficiency and reconfigurability for CIM. Additionally, we improve the accuracy of the Resistive Random Access Memories (RRAM) based CIM architecture using two mapping-based solutions. These three technologies exhibit non-volatile characteristics, and when integrated into the CIM architecture, they yield significant advantages, including enhanced energy efficiency, reliability, and accuracy in computing processes.**

I. INTRODUCTION

Deep neural networks (DNNs) have demonstrated significant advancements across a variety of applications, including image recognition, speech recognition, healthcare and natural language processing [1]–[3]. In general, DNNs are configured using several layers with many inputs that are connected through weights to their outputs and learn useful representations by adjusting their weights algorithmically. This architectural design allows DNNs to effectively capture complex patterns and relationships within data, enabling them to excel in tasks requiring sophisticated decision-making capabilities. However, they face challenges for computation efficiency, especially as they grow in complexity and require substantial computational resources, that hinder their deployment in resource-constrained environments like IoT devices or mobile platforms.

Current computing systems, including Central Processing Units (CPUs), Graphics Processing Units (GPUs), Field-Programmable Gate Arrays (FPGAs), and Tensor Processing Units (TPUs), are developed using CMOS-based von-Neumann architecture [4]–[6]. In these systems, the physical separation of memory and compute units results in a

large number of data transfers necessary to execute vector-matrix multiplication (VMM) operations for neural network applications, leading to a degradation in performance and energy efficiency. Additionally, CMOS technology is struggling with challenges such as excessive sub-threshold leakage and scalability issues. Computation-in-memory, where data processing occurs directly within the memory, demonstrates potential for enhancing computation efficiency and addressing the data transfer bottleneck. The CIM architecture, combined with emerging non-volatile technologies, delivers various advantages, including leakage-free storage, high density, high scalability, and faster accesses. These features have the potential to further enhance the computation efficiency of the system.

In this paper, we exploit three emerging technologies, namely Ferroelectric Field-Effect Transistors (FeFET), Vertical Nanowire Field-Effect Transistors (VNWFETs), and Resistive Random Access Memories (RRAM), to enhance Computational-In-Memory (CIM) capabilities for neural network development. The contributions for this paper are as follows:

- We address ferroelectric stochasticity and temperature effects in FeFET-based IMC, proposing innovative strategies like on-chip cooling using thermoelectric devices to ensure reliable computing.

- We highlight the potential of VNWFETs in 3D computing architecture, demonstrating enhanced energy efficiency and reconfigurability advantage for computing-in-memory and approximate computing.

- We present two mapping-based solutions to enhance the accuracy of the RRAM-based CIM architecture for Neural Networks, namely the unbalanced bit-slicing scheme and the mapping-aware biased training methodology.

The rest of this paper is organized as follows. Section II presents the fundamentals of CIM architecture. CIM Using FeFET Technology is discussed in Section III. Section IV and Section V provides details of the VNWFET- and RRAM-based CIM architectures, respectively. Finally, Section VII concludes the paper.

979-8-3503-8083-5/23 $31.00 © 2023 IEEE

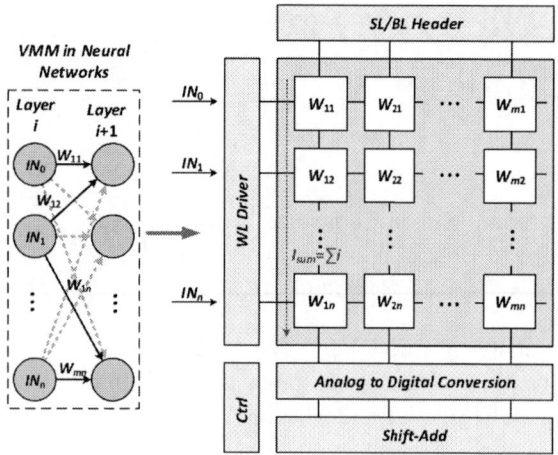

Fig. 1: Computation-In-Memory architecture [7].

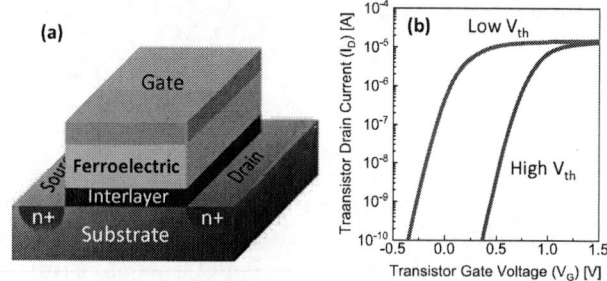

Fig. 2: (a) FeFET where the high-κ layer is replaced by a thick (e.g., 10 nm) layer of ferroelectric material. (b) Two distinguishable states are created by polarizing the ferroelectric layer, i.e., low-V_{TH} and high-V_{TH}, corresponding to high and low currents, respectively.

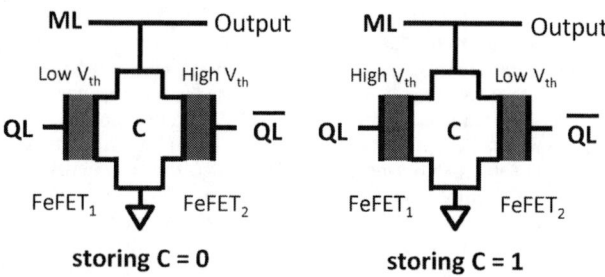

Fig. 3: Circuit of the FeFET-based Ternary Content Addressable Memory (TCAM) cell implementation. A complementary write scheme stores logic 0 and 1 in a cell.

II. COMPUTATION-IN-MEMORY (CIM) ARCHITECTURE

Computation-in-memory (CIM) proves to be a highly efficient alternative to the conventional von-Neumann architecture for the implementation of Vector-Matrix Multiplication (VMM) in neural network hardware. The mapping of VMM operation between layers onto the CIM architecture is illustrated in Figure 1. In the CIM configuration, data is stored as conductance in memory elements arranged in a grid-like structure called crossbar [8]–[12]. This crossbar operates in the analog domain and interfaces with other digital components in the system through peripheral circuits such as digital-to-analog converters (DACs) and analog-to-digital converters (ADCs). The weight matrix is translated into conductances within the crossbar, and input voltages (IN's) are applied using DACs. This sets off a current flow through all conductances, following Ohm's law and simulating element-wise multiplication of voltages and conductances. In the same column, currents from conductances accumulate according to Kirchhoff's law, generating output currents (I's). Each column, consequently, performs a multiply-and-accumulate operation in the analog domain. Simultaneous output production from all columns allows the VMM operation to be executed with an O(1) time complexity. Following this, the column currents are converted into digital outputs using ADCs and transmitted to other system components for subsequent processing or storage. This process demonstrates the potential advantages of CIM for the neural network computations.

III. CIM USING FERROELECTRIC FIELD-EFFECT TRANSISTOR

Applications like Deep Neural Networks (DNNs) and big data are heavily data-centric, demanding excessive amounts of on-chip memory, while traditional CMOS-based SRAM is power and area-hungry. At the same time, the slow-down in classical CMOS technology scaling boosted the attention of emerging non-volatile memory (NVM) technologies. The literature has proposed several different technologies like, Spin-Transfer Torque Magnetic RAM (STT-MRAM) [13], Resistive RAM (ReRAM) [14] or Ferroelectric FET (FET) [15]. Classically SRAM needs at least six transistors, whereas NVMs needs only one memory device plus typically one access transistor [16]. Additionally, NVMs technologies retain their stored information even when powered off, drastically reducing static power consumption. These advantages enable larger on-chip memories compared to conventional SRAM. However, tight integration with logic is challenging for most NVMs due to their lacking CMOS compatibility [17].

Since 2007, hafnium oxide (HfO_2) has been adopted as a high-κ material in existing manufacturing processes. Later, in 2011, ferroelectricity was discovered in $Hf_{0.5}Zr_{0.5}O_2$. Replacing the high-κ layer of a traditional CMOS transistor with a thick HfO_2-based ferroelectric (FE) layer creates a FeFET (see Figure 2(a)). Except for the addition of two masks, no new process steps or materials are necessary [15], [18], making FeFET a fully CMOS-compatible emerging NVM [18], [19]. Further, as the FE layer only increases the height of the device, the footprint is unchanged, enabling high-density memory.

The non-volatility of FeFET comes from the polarization of the FE layer itself. By applying strong gate voltage biases (e.g., $\pm 4\,V$), the polarization can be changed, i.e., a state is *written* into the FE layer. In turn, the polarization of FE layer affects the characteristics of the underlying transistor, where a positive polarization decreases the transistor's V_{TH} (low-V_{TH} state), while a negative polarization increases V_{TH} (high-V_{TH} state). With these two states, shown in Figure 2(b), FeFET can be used to store binary information. To *read* a FeFET, a much smaller probing gate bias is used ($\approx 0.7\,V$), and the flowing current indicates the stored state within the FeFET.

Using FeFET TCAM cells can be implemented extremely area-efficient, as only two FeFETs are necessary. In compari-

979-8-3503-8083-5/23 $31.00 © 2023 IEEE

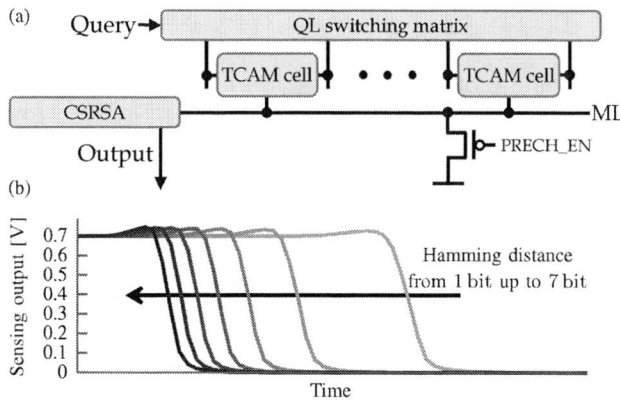

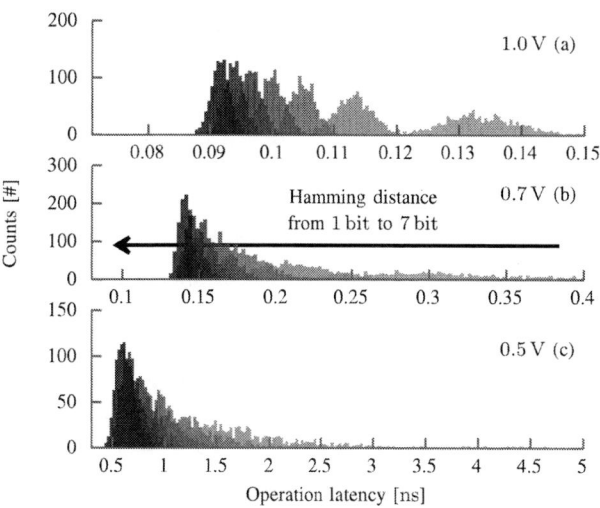

Fig. 4: (a) Circuit of TCAM block. The number of cells (block size) is variable. (b) Example of the output voltage waveforms of the CSRSA for a block size of 15 bit. Data from [20].

Fig. 5: The operation latencies of FeFET TCAM under process variation at three different supply voltages. The variation for one miss is particularly high at 0.5 V ranging from 0.8 ns to 5.0 ns (not recognizable in the plot). The block size is 15 bit. The results are based on 1000 Monte Carlo SPICE simulations per voltage level and Hamming distance. Data from [20].

son, 16 transistors are needed using conventional technology. Figure 3 shows the circuit of a FeFET-based TCAM cell and how data is stored using a complementary write scheme of the two FeFET. On an abstract level, a TCAM cell will compare the provided query data with the internally stored data (i.e., $Q = C$). The Match Line (ML) (the cell's output) is precharged to the supply voltage during the initialization. When the two data are the same, a match occurs, and the cell will respond with logic 1 and block the path from ML to ground. In case of a miss, when the two data are different, the cell will have *one* conducting path from ML to ground, and the result will be a logic 0.

Using several of such TCAM cells, a block can be constructed by connecting the cells to a shared ML shown in Figure 4(a). The ML now depends on the miss/match state of all the connected cells, and the number of cells that report a mismatch is proportional to the discharge current of the ML. This is due to the parallel discharge paths that are formed in the cells reporting a mismatch. By using a clocked self-referencing sense amplifier, the discharge current is translated to the temporal domain shown in Figure 4(b). As the discharge current is proportional to the number of cells reporting a miss, the degree of mismatch is equal to the Hamming distance of the bit string stored in the block and an applied query string. From the moment of applying the query data until the sharp output transition of the sense amplifier, an operation delay can be measured, representing the different Hamming distances.

Analog computing schemes are very sensitive to noise and process variation, leading to errors in the final result of the calculation. Using a TCAM block to calculate the Hamming distance faces similar issues. The fact that emerging technologies like FeFET exhibit even more process variation compared to mature technologies exacerbates the problem further. Figure 5 shows the operation latency distributions at different supply voltages of the TCAM block under process variation. With increasing Hamming distances, the margins between the distributions shrink until they start to overlap. However, with the overlap to the neighboring distributions, the operation latency will be interpreted wrongly, leading to

errors in the calculation. This can be abstracted and modeled using the concept of a confusion matrix. Each row represents a true Hamming distance value, and the columns represent the potential output value. The cells denote the probability for a specific value X to result in a value of Y; ideally, all the probability mass is on the main diagonal. Evidently, the error probability of the circuit is very high, which will break error-sensitive algorithms like deep neural networks and other conventional machine learning algorithms [21]. Thus, to make use of such unreliable hardware, an emerging computing paradigm that is able to tolerate errors becomes a must.

A. Hyperdimensional Computing on Analog Hardware

Hyperdimensional Computing (HDC) is an emerging brain-inspired algorithm that uses large vectors with elements in the thousands to represent real-world objects in the hyperspace. By using these large-scaled vectors, HDC does not store information in the conventional sense, where each element represents a unique piece of data, but rather relies on patterns encoded in the vectors [22]. To retrieve data, these patterns are recognized, giving HDC strong error and noise resistance [22]. After the concept was proposed by Kanerva in 2009 [23], HDC has been employed in a wide range of applications. Amongst others, language recognition [22], image classification [24], EMG signal processing [25], wafer map defect pattern classification [26], and voice recognition [27] have been showcased in literature.

In order to encode the complex real-world entities from the example applications, only the three basic HDC operations are necessary. Further, the implementation of these operations depends on the data type that is used for the elements of the vectors, which can be only binary, integers, real, or even complex numbers. The binary case, in particular, is

979-8-3503-8083-5/23 $31.00 © 2023 IEEE

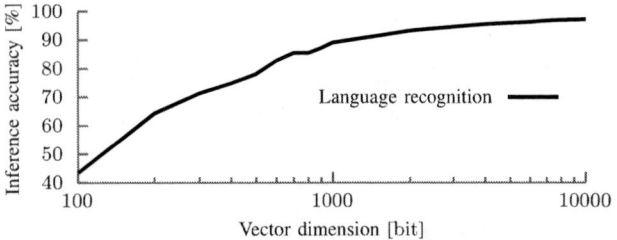

Fig. 6: Relation of inference accuracy and hypervector dimension. Data from [20].

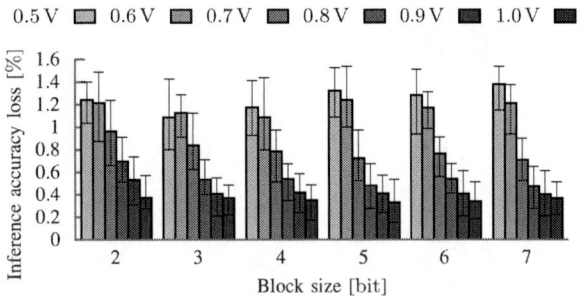

Fig. 7: Inference accuracy loss due to process variation on FeFET-based TCAM block for language recognition. Data from [20].

very computationally lightweight, making HDC very energy efficient. HDC follows a similar flow to other machine learning approaches, where a model is first trained on data and then tested. To infer the model with a query, all the class hypervectors are stored in the associative memory. The result of an inference is the class vector with the highest similarity to the query vector, and the respective class label is selected. Figure 6 shows the accuracy of the language recognition task using binary HDC across different vector dimensions. With increasing vector dimensions, the pattern storage capacity improves, and the accuracy grows higher, peaking around 97 % with a dimension of 10 000 bit.

For binary vectors, the similarity metric is implemented through the Hamming distance. As HDC is very error-resilient, it is a perfect candidate to employ unreliable analog compute schemes whilst maintaining reasonable performance. To test this, we have injected the errors coming from the previously described TCAM block into the inference algorithm of HDC via the abstracted confusion matrices. Figure 7 shows the inference accuracy loss on the test set for the different supply voltages we have tested. As the probabilistic-based error model introduces additional randomness, we have repeated the inference of the whole test set 100 times for each configuration plotted to have sufficient statistical data. Thus, the bars represent the average accuracy loss, and the error bars indicate the standard deviation. From the histograms in Figure 5, we could already estimate the increase in error probability with lower supply voltages. This carries forward, and the peak average accuracy loss is around 1.3 % for 0.5 V V_{DD}. Increasing V_{DD} decreases the average accuracy loss throughout below 0.4 %, showing the trade-off between energy and inference accuracy.

IV. CIM USING VERTICAL NANOWIRE FIELD-EFFECT TRANSISTORS

Vertical Nanowire Field Effect Transistors (VNWFETs) represent a promising emerging technology that holds substantial promise for minimizing footprint and, consequently, reducing interconnect capacitance. This has the potential to enhance energy efficiency and seamlessly integrate with advanced 3D integration strategies. In this perspective, the VNWFET emerging technology that may fulfill the energy efficiency, performance, and compact design prerequisites for CIM, serving as viable alternatives to conventional von Neumann machines. Indeed, adding ferroelectric material to the VNWFET gatestack [28] enables nonvolatile logic as well as nonvolatile reconfigurability.

In previous works such as [29] and [30] we presented a design methodology aimed at bridging the gap between an established (laboratory-scale) VNWFET technology and its corresponding compact model, progressing towards standard static logic cell design and characterization, and ultimately extending to logic synthesis.

A. Vertical Nanowire Field Effect Transistors (VNWFET) : Device & Compact Model

In this study, we employ a VNWFET with a junction-less Gate-All-Around (GAA) structure [31], illustrated in Figure 8. The nanowire channel, homogeneously highly doped and patterned into a boron-doped silicon substrate, controls the current between drain and source contacts. The GAA structure ensures an effective channel length of 14 nm, improving electrostatic control and overcoming scaling issues in conventional planar transistors.

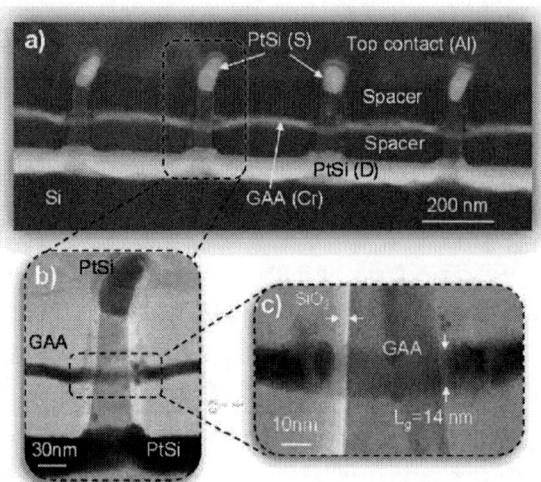

Fig. 8: VNWFET device from [31]: (a) STEM image in cross section of the vertical transistor implemented in nanowire arrays, (b) single VNWFET showing its (c) gate formation

The VNWFET device modeling incorporates carrier transport physics in the junctionless architecture. A SPICE simulation methodology, founded on a unified charge-based control model [32], accurately captures the 3D technology. The compact model addresses depletion and accumulation charges, short channel effects, velocity saturation, Drain-Induced Barrier Lowering (DIBL), Band-To-Band Tunneling (BTBT),

Gate-Induced Drain Leakage (GIDL), and Schottky contact formation [33], [34]. To enhance model accuracy at low drain bias, a semi-empirical field-dependent mobility model is implemented, demonstrating good agreement between measurement and compact model.

B. Logic Cell Characterization: Simulation Flow and resulting delays & Power Consumption

For this study, we implemented 4 basic complementary static: INV1X1, NAND2X1, NOR2X1 and XOR2X1 and a D Flip-Flop. SPICE simulations utilize the compact model implemented as a Verilog-A executable model making use of the vertical nanowire technology and exploring the impact of the variation of critical parameters. Gate physical length (L_g) and nanowire (NW) diameter (d_{nw}) are fabrication-dependent parameters set to $L_g = 18nm$ and $d_{nw} = 22nm$ = 22nm based on experimental devices. This study explores the number of nanowires for p-type and n-type transistors as a key design parameter. Initial steps involve verifying n-type and p-type VNWFET device functionality through DC-sweep simulation, ensuring expected I_{DS}/V_{GS} behavior. Subsequent simulations focus on an elementary inverter gate to determine the optimal ratio between n-type and p-type nanowires for balanced noise margins and well-matched rise and fall times. The number of nanowires are defined as shown in table I

TABLE I
Number of NWs used for each version of a logic gate

Logic Cell	n-type NW values	p-type NW values
INV	4, 24, 44, 64	4, 24, 44, 64
NAND	8, 48, 88, 128	4, 24, 44, 64
NOR	4, 24, 44, 64	8, 48, 88, 128
XOR	8, 48, 88, 128	8, 48, 88, 128

Accurate measurement of latency in individual logic gates is essential. This study evaluates delay in output transition compared to input transition(s), which was measured for both rise and fall transitions according to equation 1.

$$delay = t(V_{out} = 0.5V_{dd}) - t(V_{in} = 0.5V_{dd}) \qquad (1)$$

Output rising and falling times are also measured, representing the time for the output voltage to rise/fall from $0.1V_{dd}$ to $0.9V_{dd}$.

Logic cell power consumption comprises static (leakage) and dynamic power. Leakage power arises from cell leakage current in a static state, defined as $P_{leak} = I_{dd}V_{dd}$. Whereas dynamic power includes switching power from load capacitor (C_l) charging and short circuit power during pull-up and pull-down network conduction. Total energy (E_t) consumption per transition is calculated, considering switching energy (E_s) and internal energy as shown in equation 2. In our case, we considered that $V_i = 0.1V_{dd}$ and $V_f = 0.9V_{dd}$ whereas t_i corresponds to the time at which $V_{out} = V_i$ and t_f corresponds to the time at which $V_{out} = V_f$.

$$E_{int} = E_t - E_s = V_{dd}|\int_{t_i}^{t_f} I_{dd}\, dt| - |\int_{V_i}^{V_f} C_l V_{out}\, dv_{out}| \quad (2)$$

C. Logic Synthesis: from Standard Cell Library Characterization to Synthesis Experiment

This study adopts the standard liberty file format for standard cell library characterization. This hierarchical file includes information on delay models, unit attributes, operating conditions, and lookup tables (LUTs) for timing and power consumption based on input slew and load capacitance. The nonlinear delay model, incorporating input slew rate and load capacitance, necessitates LUTs to store values for synthesis optimization.

The liberty file details timing and power consumption at the output pins of each cell. Matrices reflect values related to output transitions affected by a single input transition, considering the input's effect on the output transition. A time sensing parameter identifies how input transitions affect output transitions, classified as negative unate, positive unate, or non unate.

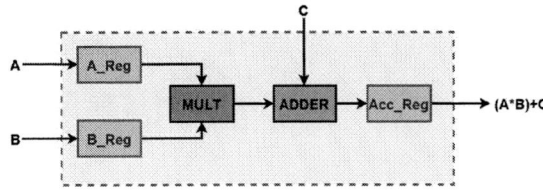

Fig. 9: Processing Element of a systolic array dedicated to MAC operations

We chose to compare the number of cells, delay and power performance of the VNWFET library generated, with two technology libraries: a free 45nm pdk [35] and a 65nm industrial library. We chose to retrain these library equivalent logic cells (INV, NAND, NOR, XOR and D FLip-Flop) to realize a fair comparison. Table II shows how many versions of the same combinational cell is available for each technology library.

TABLE II
Number of employed cells per logic function for each technology library

Library	INV	NAND	NOR	XOR
VNWFET	4	3	3	3
45nm CMOS	4	1	1	1
65nm CMOS	20	15	15	10

The targeted circuit used as benchmark is a Processing Element (PE) within a Systolic Array accelerator designed for matrix-to-matrix multiplication through Multiply and Accumulate operations (MAC). Illustrated in Figure 9 is the PE architecture, comprising a multiplier, an adder, and three D flip-flops responsible for storing and synchronizing operands, weights, and the accumulated result of the operation. Each library is utilized to synthesize PE versions with data sizes of 4, 8, 16, and 32 bits, aiming to assess the scalability of VNWFET in comparison to traditional CMOS technology.

D. Preliminary results on VNWFET performance

The synthesis results reveal two primary aspects: logic optimization and power/timing trade-off. Concerning logic optimization, Table III indicates that the VNWFET exhibits strong performance in comparison to the 45nm library, potentially attributed to the broader selection of cells within the VNWFET library. However, when compared to the 65nm library, the VNWFET library still performs admirably despite having a less optimal pool of cells. Although the cell count may not explicitly indicate area performance, a reduced number of cells inherently contributes to improved place and route solutions, thereby limiting the overall footprint.

TABLE III
Number cells per circuit based on VNWFET and 45/65 nm CMOS technology library

Circuit	NVWFET	45nm CMOS	65nm CMOS
4-bits PE	230	530	221
8-bits PE	737	1674	705
16-bits PE	2553	5697	2321
32-bits PE	9395	20608	8492

The timing outcomes, as illustrated in Figure 10, indicate that the VNWFET library exhibits commendable performance compared to the 45nm CMOS library, particularly for smaller circuits. However, the trend suggests that the advantage diminishes as circuits grow in size. Additionally, when compared to the 65nm CMOS, which performs optimally, the disadvantage becomes more pronounced. Nevertheless, the minimal timing differences are promising in the context of these preliminary results.

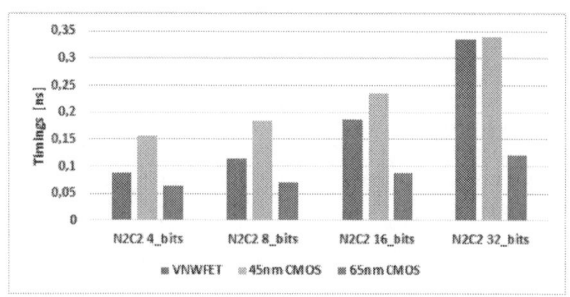

Fig. 10: VNWFET timing results vs 45/65 nm CMOS libraries

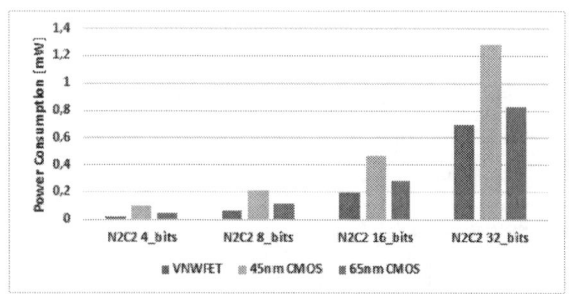

Fig. 11: VNWFET power consumption results vs 45/65 nm CMOS libraries

An equal comparison is made for power consumption results obtained. It is important to emphasize that these results are derived from estimates obtained through the synthesis tool. The power consumption estimates are illustrated in Figure 11. The VNWFET technology library exhibits lower power consumption across each size version of the PE. The scaling trend of VNWFET aligns with that of the CMOS libraries,

indicating that these results are likely to remain favorable at higher integration rates.

To push further, the next step will be to add Ferroelectric layer (see Section III) to to VNFET to obtain VNFeFET. We thus plan to have Non-Volatile Boolean gates [36] and operators meaning that we can permanently store a given input data (i.e., an operand) in order to reduce the data transfer and obtain a hybrid approach between CIM and the classical von Neumann paradigm

V. CIM USING RESISTIVE RADOM-ACCESS MEMORY

A. RRAM technology

RRAM devices function as storage elements where data is stored in resistance states. The operational mechanism of RRAM devices is centered around the reversible creation or disruption of a Conductive Filament (CF) within a resistive layer, leading to high and low resistance states. The structural composition of RRAM devices is depicted in Figure 13. It comprises of a metallic oxide that is sandwiched between a Top Electrode (TE) and a Bottom Electrode (BE) as described in the figure. The magnitude of the CF determines the resistance state of the device, which, in turn, is dependent upon the polarity of the applied supply voltage [37]–[41]. To elaborate, when a sufficiently high positive voltage (exceeding the set threshold voltage, V_{set}) is applied, bonds between ions break, creating a conductive filament (CF) comprising vacancies capable of conducting current. This amplifies the CF's size, corresponding to the low resistance state. Conversely, when a negative voltage (below the reset threshold voltage, V_{reset}) is applied, some ions migrate back into the oxide region, thereby diminishing the CF's size. Consequently, the device attains a high resistance state. These RRAM devices are non-volatile in nature, scalable, compact, with reduces access latency and energy consumption as well as compatible with the CMOS technology.

B. CIM architecture using bit-slicing

CIM architectures encounter a challenge in meeting the high bit-precision requirements of neural network applications. This is primarily because the bit capacity of RRAM devices is typically insufficient compared to the bit size needed for neural network weights. Additionally, achieving full-precision inputs and outputs in neural networks necessitates the use of high-resolution digital-to-analog converters (DACs) and analog-to-digital converters (ADCs). However, these components are costly in terms of both energy consumption and physical space, significantly diminishing the hardware efficiency advantages of CIM [42]–[44]. To address these issues, bit-slicing techniques [8], [9], [45] are implemented within CIM architectures.

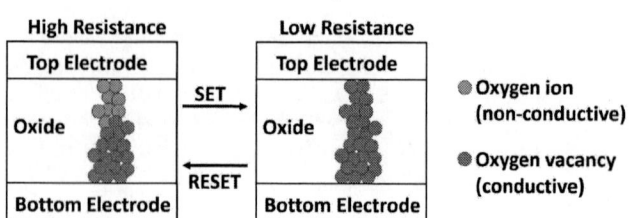

Fig. 12: RRAM technology.

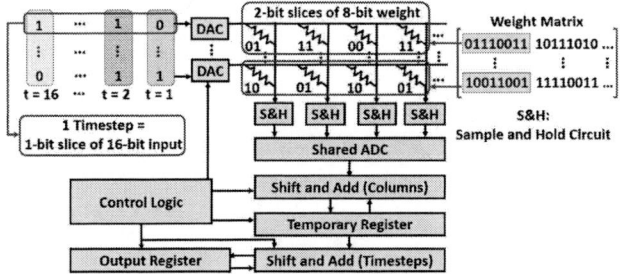

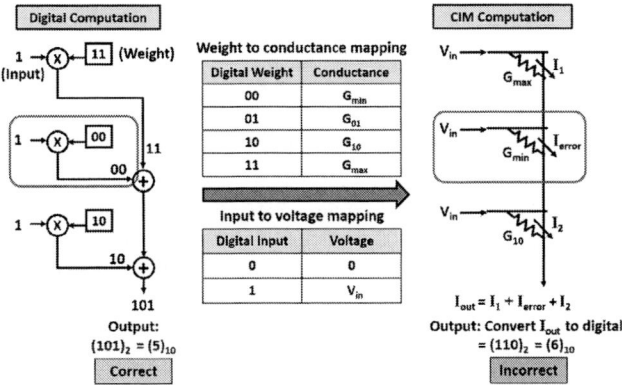

Fig. 13: RRAM-based CIM architecture with bit-slicing.

Fig. 14: Demonstration of non-zero G_{min} error.

Bit-slicing involves breaking down the full-precision neural network weights and inputs into smaller bit-size segments known as slices, illustrated in Figure. For example, 2-bit slices of an 8-bit weight are transformed into conductances and assigned to distinct columns in the RRAM crossbar. Simultaneously, 1-bit slices of a 16-bit input are converted into voltages and distributed across different time steps, sequentially applied to the crossbar. Column currents resulting from time-multiplexed voltage inputs are then converted to digital values. Through shift-and-add operations across both columns and time steps, the complete full-precision output is obtained. Nevertheless, such architecture faces the issue of finite conductance states (Non-zero G_{min} error) as well as conductance variations.

Non-Zero G_{min} error: In Bit-slicing CIM architectures, a zero weight-slice in the neural network is represented by an RRAM device with non-zero conductance equal to the minimum device conductance value (G_{min}). In the digital domain, multiplying any non-zero input with a zero weight must result in a zero output. However, when it mapped to CIM architecture (see Figure 14), a non-zero output current is produced when a non-zero input in the form of a voltage is applied to an RRAM with G_{min} conductance. This is known as non-zero Gmin error, which creates a functional mismatch between digital output and CIM output resulting in erroneous VMM operation and degraded neural network accuracy.

Conductance variations: The programmed conductance of a RRAM deviates from its target value due to the stochastic nature of oxygen vacancy creation/depletion and fabrication imperfections like variable oxide thickness [18]. This phe-

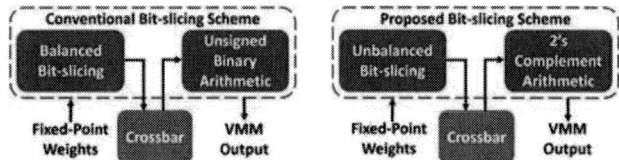

Fig. 15: Overview of conventional and proposed bit-slicing schemes [46].

nomenon is called conductance variation, shown in Fig. 3. It leads to incorrect weight storage as RRAM conductance, resulting in poor accuracy. In this paper, we improve the accuracy of RRAM-based neural network architectures in the presence of conductance variation.

C. Unbalanced bit-slicing scheme

A bit-slicing scheme consists of two fundamental components: 1) bit-slicing logic which determines how the slices are created, and 2) arithmetic which determines how the partial outputs from sliced columns are combined. The *balanced* bit-slicing (BBS) logic in state-of-the-art bit-slicing schemes provides low sensing margin resulting in significant impact of non-zero G_{min} errors, while unsigned binary arithmetic in these schemes leads to high accumulative non-zero G_{min} error on combining the partial outputs as shown in Figure 15.

We have proposed an unbalanced bit-slicing (UBS) scheme in [47] which changes the way in which neural network weights are mapped to bit-slicing CIM crossbar to mitigate the impact of non-zero G_{min} error. The proposed UBS scheme provides high sensing margin for important bits (MSBs) by using an RRAM with n-bit maximum capacity as an m-bit memory-cell (slice) where $m<n$. This provides sufficient sensing margin to the MSB column output and make them immune to non-zero G_{min} error as shown in Figure 16. This suffices for good accuracy due to the robustness of neural networks to minor computational fluctuations i.e. errors in less important bits (LSBs). Moreover, use of 2's complement arithmetic in [47] leads to reduction in accumulative non-zero G_{min} error after combining the partial outputs due to weighted subtraction as shown in Eq. 3a and Eq. 3b obtained using 8-bit weights with maximum 2 bits/RRAM as an example. The conductance subscripts indicate binary slice value. The digital outputs of the columns are denoted by D_i, while the accumulated digital outputs are indicated by D_f. $D_i = T_i + E_i$, where T_i is the ideal output value and E_i the error due to non-zero G_{min}. Similarly, $D_f = T_f + E_f$, where T_f is the ideal accumulated output value and E_f the accumulated error due to non-zero G_{min}.

$$E_f = 64 \cdot E_1 + 16 \cdot E_2 + 4 \cdot E_3 + E_4 \qquad (3a)$$
$$E_f = (-128) \cdot E_1 + 64 \cdot E_2 + 16 \cdot E_3 + 4 \cdot E_4 + E_5 \qquad (3b)$$

Owing to the cumulative effect of high MSB sensing margin and 2's complement arithmetic on non-zero G_{min} error, unbalanced bit-slicing scheme achieves up to 8.8× and 1.8× classification accuracy compared to state-of-the-art CIM architectures for single-bit memristors and two-bit memristors respectively, at reasonable energy overheads arising due to extra columns.

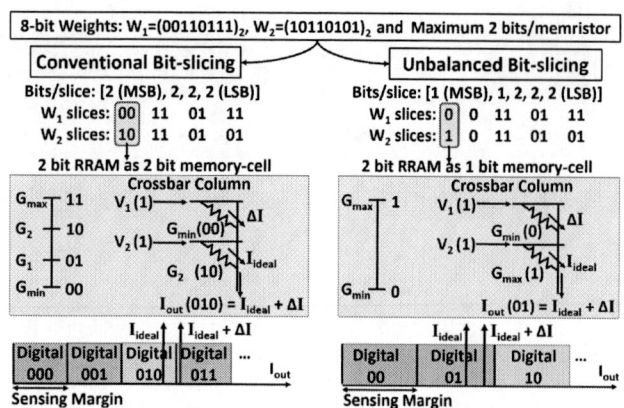

Fig. 16: Impact of sensing margin on bit-slicing schemes.

Fig. 18: Overview of the conventional and proposed biased training methodologies

D. Mapping-aware biased training methodology

Figure 17 shows a CIM-based multiply-accumulate operation, where I_{error} is the error current in a single RRAM device due to conductance variation. As small I_{error} is desirable, the preference order of states in Fig. 4 is: G00 (best), G01, G11, G10 (worst). Despite having a higher variation percentage, G00 and G01 are preferred over G11 and G10 as their small mean values result in small I_{error}. Hence, the preference order of conductance states must be based on I_{error} contribution instead of the variation percentage.

The deployment of a neural network on CIM hardware for inference involves two phases as shown in Figure 18: i) Training the neural network weights to obtain high classification accuracy. ii) Mapping the trained weights to RRAM device conductances for inference on CIM hardware. Conventional training can result in the mapping of weights to conductance states having a high variation impact (unfavorable states). This can lead to low hardware accuracy despite high software accuracy. A mapping-aware biased training, as proposed in [49], restricts the neural network weights during training, so that their post-training values directly get mapped to conductance states having a low variation impact (favorable states). However, restricting too many weights hinders backpropagation and leads to low software accuracy. This in turn results in low

hardware accuracy, as it is upper bounded by software accuracy. Conversely, if too few weights are restricted, the hardware accuracy will be poor as many RRAM devices can get mapped to unfavorable states. Hence, our proposed mapping-aware biased training only restricts the important weights. This leads to high software accuracy due to the adaptability of non-important weights and also provides high hardware accuracy as important weights get mapped to favorable RRAM states.

We first train the neural network in a standard (hardware-unaware) manner. These weights are used as initial weights for mapping-aware biased training for faster convergence. We then determine a favorability constraint on the weights to ensure the mapping of desired weight bits to favorable conductance states. It depends on RRAM device bit capacity and CIM mapping scheme details like fixed-point format, underlying CIM architecture, etc. For example, consider 2-bit RRAM devices (slices), 8-bit fixed-point weights (6-bit fraction), and CIM architecture in [9]. The mapping scheme first converts trained weights to 2's complement fixed-point format. It then shifts the 2's complement weight range by 2^7 to overcome the difficulty in isolating the sign contribution from a multi-bit slice [9]. Figure 19 then shows the favorability constraint to map the most significant 2-bit slice to favorable states (G00, G01, and G10) for this example. We now perform a new epoch of backpropagation using training data and then determine

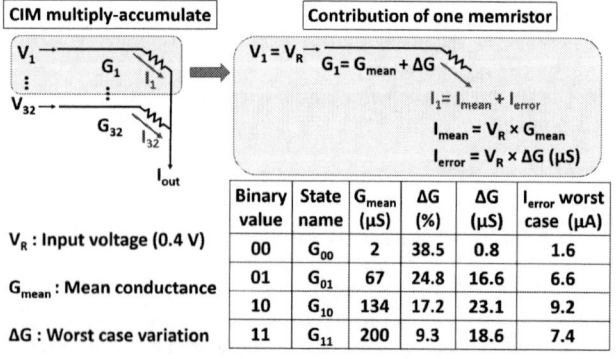

Fig. 17: Favorable conductance states analysis for a 2-bit memristor (four conductance states). The used conductance variation data is obtained from [48].

Binary value	State name	G_{mean} (µS)	ΔG (%)	ΔG (µS)	I_{error} worst case (µA)
00	G_{00}	2	38.5	0.8	1.6
01	G_{01}	67	24.8	16.6	6.6
10	G_{10}	134	17.2	23.1	9.2
11	G_{11}	200	9.3	18.6	7.4

V_R : Input voltage (0.4 V)

G_{mean} : Mean conductance

ΔG : Worst case variation

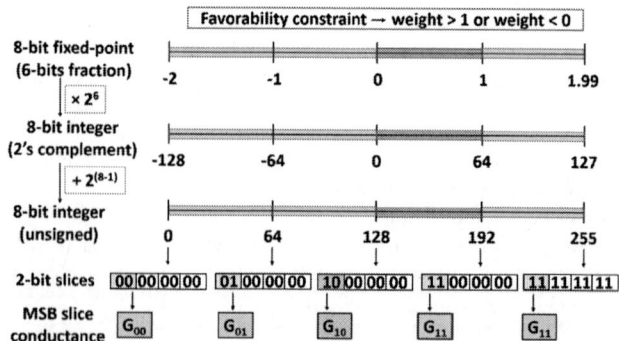

Fig. 19: Illustration of favorability constraint derivation for mapping MSB slice of 8-bit weight to favorable conductance states in a 2-bit RRAM device

which weights are important for high hardware accuracy. In CIM hardware design for the same network, instead of individual weights, some crossbar columns (groups of weights) are more important than others for high hardware accuracy. Weights in these columns are restricted as per the favorability constraint and test accuracy is evaluated. This process is repeated for a given number of biased training epochs and weights with the best post restriction test accuracy are mapped to CIM hardware. The proposed biased training achieves up to 2.4× hardware accuracy and up to 2.4× correct operations per unit energy compared to the conventional training, without incurring any hardware overhead. Such high accuracy and energy efficiency can facilitate the deployment of CIM-based neural networks for edge-AI.

VI. CONCLUSIONS

Deep neural networks demand a lot of computation efficiency, and the existing hardware is not suitable as they have separated processor units and memory units that can impose significant latency and bandwidth constraints, limiting the overall speed and effectiveness of deep neural network computations. Computation-in-memory has the potential to considerably enhance deep neural network efficiency by integrating processing and storage within the same module, thereby reducing latency, minimizing data transfer bottlenecks. In this paper, we have presented CIM architecture using three emerging technologies. Initially, a CIM structure incorporating Ferroelectric Field-Effect Transistors (FeFET) is introduced to calculate the Hamming distance function using an analog computing scheme. To test the error resiliency of the emerging HDC algorithm, the errors from the unreliable hardware have been modeled and injected into the inference step of HDC. Following this, we investigated a CIM system employing Vertical Nanowire Field-Effect Transistors (VNWFETs) within a 3D computing framework, demonstrating enhanced energy efficiency and reliability through CIM and approximate computing. Furthermore, we have presented the improvement in the precision of the Resistive Random Access Memories (RRAM) based CIM architecture by implementing two mapping-based solutions, namely the unbalanced bit-slicing scheme and the mapping-aware biased training methodology. These findings underscore the transformative potential of CIM architectures in overcoming current hardware limitations and advancing the capabilities of deep neural network computations.

ACKNOWLEDGMENT

We thank Paul R. Genssler for his work on Hyperdimensional Computing. This work was partly supported by the EU H2020 grant "DAIS" (grant agreement No. 101007273) and the EU HORIZON-JU-RIA grant "NEUROKIT2E" (grant agreement No. 101112268), and by the FVLLMONTI European Union's Horizon 2020 research and innovation programme under grant agreement No 101016776.

REFERENCES

[1] C. Szegedy et al., "Going deeper with convolutions," in *IEEE conference on computer vision and pattern recognition*, 2015, pp. 1–9.

[2] Google, Facebook and microsoft are remaking themselves around AI. [Online]. Available: https://www.wired.com/2016/11/google-facebook-microsoft-remaking-around-ai/

[3] S. Diware et al., "Severity-based hierarchical ecg classification using neural networks," *TBioCAS*, vol. 17, no. 1, pp. 77–91, 2023.

[4] Intel processor family. [Online]. Available: https://www.intel.in/content/www/in/en/products/processors/core.html

[5] NVIDIA turing architecture GPUs. [Online]. Available: https://www.nvidia.com/en-in/geforce/turing/

[6] N. Jouppi et al., "TPU v4: An Optically Reconfigurable Supercomputer for Machine Learning with Hardware Support for Embeddings," in *International Symposium on Computer Architecture*, 2023.

[7] M. A. Yaldagard and other, "Read-disturb detection methodology for rram-based computation-in-memory architecture," in *International Conference on Artificial Intelligence Circuits and Systems (AICAS)*, 2023, pp. 1–5.

[8] A. Ankit et al., "PUMA: A Programmable Ultra-efficient Memristor-based Accelerator for Machine Learning Inference," in *International Conference on Architectural Support for Programming Languages and Operating Systems*, 2019.

[9] A. Shafiee et al., "ISAAC: A Convolutional Neural Network Accelerator with In-Situ Analog Arithmetic in Crossbars," in *International Symposium on Computer Architecture*, 2016.

[10] A. Singh et al., "A 115.1 tops/w, 12.1 tops/mm 2 computation-in-memory using ring-oscillator based adc for edge ai," in *International Conference on Artificial Intelligence Circuits and Systems (AICAS)*, 2023, pp. 1–5.

[11] A. Singh, M. Fieback et al., "Accelerating rram testing with a low-cost computation-in-memory based dft," in *International Test Conference (ITC)*, 2022, pp. 400–409.

[12] A. Singh et al., "Referencing-in-array scheme for rram-based cim architecture," in *DATE*, 2022, pp. 1413–1418.

[13] A. V. Khvalkovskiy, D. Apalkov, S. Watts, R. Chepulskii, R. S. Beach, A. Ong, X. Tang, A. Driskill-Smith, W. H. Butler, P. B. Visscher, D. Lottis, E. Chen, V. Nikitin, and M. Krounbi, "Basic principles of STT-MRAM cell operation in memory arrays," *Journal of Physics D: Applied Physics*, vol. 46, no. 7, p. 074001, feb 2013. [Online]. Available: https://iopscience.iop.org/article/10.1088/0022-3727/46/7/074001

[14] H. Akinaga and H. Shima, "Resistive random access memory (reram) based on metal oxides," *Proceedings of the IEEE*, vol. 98, no. 12, pp. 2237–2251, Dec 2010.

[15] S. Dünkel, M. Trentzsch, R. Richter, P. Moll, C. Fuchs, O. Gehring, M. Majer, S. Wittek, B. Müller, T. Melde, H. Mulaosmanovic, S. Slesazeck, S. Müller, J. Ocker, M. Noack, D. . Löhr, P. Polakowski, J. Müller, T. Mikolajick, J. Höntschel, B. Rice, J. Pellerin, and S. Beyer, "A fefet based super-low-power ultra-fast embedded nvm technology for 22nm fdsoi and beyond," in *2017 IEEE International Electron Devices Meeting (IEDM)*, Dec 2017, pp. 19.7.1–19.7.4.

[16] D. Reis, M. Niemier, and X. S. Hu, "A computing-in-memory engine for searching on homomorphically encrypted data," *IEEE Journal on Exploratory Solid-State Computational Devices and Circuits*, pp. 1–1, 2019.

[17] J. G. Alzate, U. Arslan, P. Bai, J. Brockman, Y. J. Chen, N. Das, K. Fischer, T. Ghani, P. Heil, P. Hentges, R. Jahan, A. Littlejohn, M. Mainuddin, D. Ouellette, J. Pellegren, T. Pramanik, C. Puls, P. Quintero, T. Rahman, M. Sekhar, B. Sell, M. Seth, A. J. Smith, A. K. Smith, L. Wei, C. Wiegand, O. Golonzka, and F. Hamzaoglu, "2 mb array-level demonstration of stt-mram process and performance towards 14 cache applications," in *2019 IEEE International Electron Devices Meeting (IEDM)*, Dec 2019, pp. 2.4.1–2.4.4.

[18] S. Beyer, S. Dünkel, M. Trentzsch, J. Müller, A. Hellmich, D. Utess, J. Paul, D. Kleimaier, J. Pellerin, S. Müller, J. Ocker, A. Benoist, H. Zhou, M. Mennenga, M. Schuster, F. Tassan, M. Noack, A. Pourkeramati, F. Müller, M. Lederer, T. Ali, R. Hoffmann, T. Kämpfe, K. Seidel, H. Mulaosmanovic, E. T. Breyer, T. Mikolajick, and S. Slesazeck, "Fefet: A versatile cmos compatible device with game-changing potential," in *2020 IEEE International Memory Workshop (IMW)*, 2020, pp. 1–4.

[19] T. Mikolajick, S. Slesazeck, M. H. Park, and U. Schroeder, "Ferroelectric hafnium oxide for ferroelectric random-access memories and ferroelectric field-effect transistors," *MRS Bulletin*, vol. 43, no. 5, p. 340–346, 2018.

[20] S. Thomann, P. R. Genssler, and H. Amrouch, "Hw/sw co-design for reliable tcam-based in-memory brain-inspired hyperdimensional computing," *IEEE Transactions on Computers*, vol. 72, no. 8, pp. 2404–2417, 2023.

979-8-3503-8083-5/23 $31.00 © 2023 IEEE

[21] L. Liu, Y. Guo, Y. Cheng, Y. Zhang, and J. Yang, "Generating robust dnn with resistance to bit-flip based adversarial weight attack," *IEEE Transactions on Computers*, vol. 72, no. 2, pp. 401–413, 2022.

[22] A. Rahimi, P. Kanerva, and J. M. Rabaey, "A Robust and Energy-Efficient Classifier Using Brain-Inspired Hyperdimensional Computing," in *Proceedings of the International Symposium on Low Power Electronics and Design*, 2016.

[23] P. Kanerva, "Hyperdimensional computing: An introduction to computing in distributed representation with high-dimensional random vectors," *Cognitive computation*, vol. 1, pp. 139–159, 2009.

[24] D. Kleyko, E. Osipov, A. Senior, A. I. Khan, and Y. A. Şekercioğğlu, "Holographic graph neuron: A bioinspired architecture for pattern processing," *IEEE Transactions on Neural Networks and Learning Systems*, vol. 28, no. 6, pp. 1250–1262, 2017.

[25] G. Karunaratne, M. Le Gallo, G. Cherubini, L. Benini, A. Rahimi, and A. Sebastian, "In-memory hyperdimensional computing," *Nature Electronics*, vol. 3, 2020.

[26] P. R. Genssler and H. Amrouch, "Brain-inspired computing for wafer map defect pattern classification," in *IEEE International Test Conference*, 2021.

[27] M. Imani, D. Kong, A. Rahimi, and T. Rosing, "Voicehd: Hyperdimensional computing for efficient speech recognition," in *2017 IEEE International Conference on Rebooting Computing (ICRC)*, 2017.

[28] H. Fujisawa, K. Ikeda, and S. Nakashima, "Nonvolatile operation of vertical ferroelectric gate-all-around nanowire transistors," *Japanese Journal of Applied Physics*, vol. 60, no. SF, p. SFFB10, 2021.

[29] S. Mannaa, A. Poittevin, C. Marchand, D. Deleruyelle, B. Deveautour, A. Bosio, I. O'Connor, C. Mukherjee, Y. Wang, H. Rezgui, M. Deng, C. Maneux, J. Müller, S. Pelloquin, K. Moustakas, and G. Larrieu, "3-d logic circuit design-oriented electrothermal modeling of vertical junctionless nanowire fets," *IEEE Journal on Exploratory Solid-State Computational Devices and Circuits*, vol. 9, no. 2, pp. 116–123, 2023.

[30] S. Mannaa, C. Marchand, D. Deleruyelle, B. Deveautour, I. O'Connor, and A. Bosio, "Vnwfet-based technology: From device modelling to standard cell library," in *2023 IEEE 23rd International Conference on Nanotechnology (NANO)*, 2023, pp. 576–581.

[31] G. Larrieu and X.-L. Han, "Vertical nanowire array-based field effect transistors for ultimate scaling," *Nanoscale*, vol. 5, no. 6, pp. 2437–2441, 2013.

[32] A. Hamzah, R. Ismail, N. E. Alias, M. L. P. Tan, and A. Poorasl, "Explicit continuous models of drain current, terminal charges and intrinsic capacitance for a long-channel junctionless nanowire transistor," *Physica Scripta*, vol. 94, no. 10, p. 105813, 2019.

[33] C. Mukherjee, M. Deng, F. Marc, C. Maneux, A. Poittevin, I. O'Connor, S. Le Beux, C. Marchand, A. Kumar, A. Lecestre *et al.*, "3d logic cells design and results based on vertical nwfet technology including tied compact model," in *2020 IFIP/IEEE 28th International Conference on Very Large Scale Integration (VLSI-SOC)*. IEEE, 2020, pp. 76–81.

[34] C. Mukherjee, A. Poittevin, I. O'Connor, G. Larrieu, and C. Maneux, "Compact modeling of 3d vertical junctionless gate-all-around silicon nanowire transistors towards 3d logic design," *Solid-State Electronics*, vol. 183, p. 108125, 2021.

[35] A. B. Kahng, H. Lee, and J. Li, "Horizontal benchmark extension for improved assessment of physical cad research." Association for Computing Machinery, 2014, p. 27–32.

[36] A. Bosio, M. Cantan, C. Marchand, I. O'Connor, P. Fiser, A. Poittevin, and M. Traiola, "Emerging technologies: Challenges and opportunities for logic synthesis," in *2021 24th International Symposium on Design and Diagnostics of Electronic Circuits & Systems (DDECS)*, 2021, pp. 93–98.

[37] W. Kim, A. Chattopadhyay, A. Siemon, E. Linn, R. Waser, and V. Rana, "Multistate memristive tantalum oxide devices for ternary arithmetic," *Scientific reports*, vol. 6, no. 1, p. 36652, 2016.

[38] M. Fieback *et al.*, "Testing scouting logic-based computation-in-memory architectures," in *ETS*, 2020, pp. 1–6.

[39] R. Bishnoi *et al.*, "Special session–emerging memristor based memory and cim architecture: Test, repair and yield analysis," in *VTS*, 2020, pp. 1–10.

[40] C. Bengel *et al.*, "Reliability aspects of binary vector-matrix-multiplications using reram devices," *Neuromorphic computing and engineering*, vol. 2, no. 3, p. 034001, 2022.

[41] M. Fieback *et al.*, "Defects, fault modeling, and test development framework for rrams," *JETC*, vol. 18, no. 3, pp. 1–26, 2022.

[42] A. Singh *et al.*, "Srif: Scalable and reliable integrate and fire circuit adc for memristor-based cim architectures," *TCAS-I*, vol. 68, no. 5, pp. 1917–1930, 2021.

[43] M. Mayahinia *et al.*, "A voltage-controlled, oscillation-based adc design for computation-in-memory architectures using emerging rerams," *JETC*, vol. 18, no. 2, pp. 1–25, 2022.

[44] A. Singh *et al.*, "Low-power memristor-based computing for edge-ai applications," in *ISCAS*, 2021, pp. 1–5.

[45] A. Gebregiorgis *et al.*, "Dealing with non-idealities in memristor based computation-in-memory designs," in *VLSI-SoC*, 2022, pp. 1–6.

[46] S. Diware, A. Gebregiorgis *et al.*, "Unbalanced bit-slicing scheme for accurate memristor-based neural network architecture," in *AICAS*, 2021.

[47] S. Diware *et al.*, "Accurate and energy-efficient bit-slicing for rram-based neural networks," *Transactions on Emerging Topics in Computational Intelligence*, vol. 7, no. 1, pp. 164–177, 2022.

[48] A. Prakash and H. Hwang, "Multilevel cell storage and resistance variability in resistive random access memory," *Physical Sciences Reviews*, vol. 1, no. 6, 2016.

[49] S. Diware *et al.*, "Mapping-aware biased training for accurate memristor-based neural networks," in *International Conference on Artificial Intelligence Circuits and Systems (AICAS)*, 2023, pp. 1–5.

AUTHOR INDEX

Abbas, Mohammad 222
Abbas, Zia 276
Abdallah, Abdelrahman Amgad 196
Abdel-Raheem, Esam 96, 113
Abdel-Salam, Mohamed 298
Abdleraheem, Mohamed 107
Abdulrahman, Maram 74
Abu-Issa, Abdellatif 162
Abulibdeh, Enas 64, 248
Afifi, Shereen Moataz 1, 46
Afifi, Shereen 16, 196
Ahmad, Amir 276
Ahmed, Faheem 141
Ahmed, Islam Osama 158
Alazzawi, Lubna 258
Aldein, Elham Alaa 107
Al-Dubai, Ahmed 239
Alhawari, Mohammad 11, 91
Ali, Aser Ashraf 46
Ali, Noor Faris 41
Alimisis, Vassilis 172, 280, 317
Alkhalifa, Ibrahim 308
Alkhammash, Hend I. 137
Alqutayri, Mahmoud 64, 248
Al-Qutayri, Mahmoud 7, 202, 253
Alsamhi, Saeed H. 239
Al-Sheikh, Nirmeen 131
Amrouch, Hussam 325
Anshul, Aditya 149, 218
Areibi, S. 178
Arivazhagan, L 153
Atef, Mohamed 41, 107, 243
Atia, Maryam M. 21
Awwad, Falah 41, 243
Aydoghmishi, Faezeh Mohammadi 113
Azab, Eman 74
Aziz, Samah Abdel 239
Azizan, Azizul 26
Bakir, Nader 234
Bansal, Vivek 294
Barn, C. 178
Bauer, Lars 184
Bayoumi, Ahmed H. A. 190
Bayoumi, Magdy A. 36
Bettayeb, Meriem 228
Bishnoi, Rajendra 325
Bosio, Alberto 325
Boudargham, Nadine 58

Chaudhary, Muhammed Akmal 264
Da Silva, Douglas De Tarso 214
Daoud, Houda 321
Das, G. 145
De Carvalho, Jaqueline Corrêa Silva 210
De Carvalho, Marcos Alberto 210
Deleruyelle, Damien 325
Deveautour, Bastien 325
Diware, Sumit 325
Edwards, Melvin D. 11
Ehsanfar, Shahab 119
El Ghany, Mohamed A. Abd 74
Eladawy, Mohamed Ahmed 158
El-Din, Omar Hossam 158
Eleftheriou, Nikolaos P. 172, 280, 317
Elghany, Mohamed A. Abd 21, 52
El-Kharashi, M. Watheq 298
Elshafie, Mohamed Ahmed 158
El-Tahawy, Mahmoud 298
Fahmy, Hossam A. H. 190
Fahmy, Omar M. 1
Fahmy, Omar 16
Fathi, Dalia A. 36
Fayez, Mina Hanna 158
Fayrouz, Haddad 268
Fernandes, Miller Henrique Lúcio 210
Fetteha, Marwan A. 32
Gabriel, Koubar 268
Garg, S. 145
Garg, Supriya 145
Gaspar, Danilo 119
Gebregiorgis, Anteneh 325
Ghaffari, Fakhreddine 294
Ghoneim, Mohamed S. 36
Gouda, Amr I. 21
Grewal, G. 178
Gupta, N. 145
Hamad, Mustapha 58
Hamdioui, Said 325
Hammad, Hany 206
Hamo, Saleem 222
Hamza, Husna 168
Hany, Esmail 298
Hashmi, Mohammad 125, 264
Hassan, Eman 228
Hawbani, Ammar 239
Hegazy, Ali 80
Hegazy, Laila 16

Henkel, Jörg 184
Hroub, Ayman 7, 162
Humood, Khaled 248
Husain, Saddam 125
Husi, Géza 86
Hussein, Mousa 41
Ibrahim, Sameh A. 190
Inoue, Naoki 272
Islam, Rumana 96
Islam, Shahid 288
Ismael, Mohammad 162
Ismail, A. S. 239
Ismail, Mahmoud G. 21
Ismail, Mohammed 258
Izam, Tengku Faiz Tengku Mohmed Noor 26
Jarndal, Anwar 125, 153, 168
Kadirbay, Bagylan 125
Kasjoo, S. R. 145
Kassem, Abdallah 58
Kaur, Ranpreet 46
Kavalchuk, Ilya 288
Khairuddin, Anis Salwa Mohd 26
Khalil, Ruhul Amin 284
Khusro, Ahmad 264
Kilani, Dima 253
Kothari, Chirag 149, 218
Kumar, S. 145
Laouej, Dalila 321
Leventikidis, Savvas 317
Loan, Sajad A. 137
Loulou, Mourad 321
Maciel, Eder Costa 210
Madian, Ahmed H. 32, 36
Mahdy, Ahmed 70
Mahnashi, Yaqub 308
Mannaa, Sara 325
Marchand, Cédric 325
Mashaly, Maggie 74
Medlin, Donald 86
Mendes, Luciano Leonel 119
Meribout, Mahmoud 202
Modak, Sudipta 113
Mohamad, Mahazani 26
Mohamed, Basma H. 298
Mohamed, Otmane Ait 294
Mohamed, Usama Sayed 107
Mohammad, Baker 64, 228, 248, 253
Mohammad, Khader 131, 222
Mohammad, Maen 222
Mohammed, Khaled 284
Moreira, Milene Santos 214, 304
Mustafa, Basmala 1

Nassar, Hassan 184
Nessakh, Bouchra 268
O'Connor, Ian 325
Olivares, Cristian Y. 101
Pathy, Ashutosh 276
Pimenta, Tales 119, 210, 214, 304
Politis, Alexandros 258
Qamar, Hamzeh Abu 284
Rahman, Rihat 91
Razak, Mohd Zulhakimi Ab 26
Rennó, Vanessa Mendes 119
Rueda, Luis 113
Saadah, Alaa 86
Saeed, Nasir 239, 284
Safar, Mona 298
Sahyon, Micheal Safwat 158
Said, Lobna A. 32, 36
Salah, Khaled 80
Saleh, Hani 7, 64, 202, 228, 248, 253
Salem, Mohammed A.-M 21, 52
Samrouth, Ahmad 234
Samrouth, Khouloud 234
Sanduleanu, Mihai 70
Sarhan, Nabil J. 11, 91
Saud, Jad 86
Sawsan, Sadek 268
Sayed, Wafaa S. 32
Schettini, Norelli 101
Sengupta, Anirban 149, 218
Sharara, Leila 258
Sharma, B. 145
Sharma, D. K. 145
Shiblee, M. 137
Singh, A. K. 145
Singh, P. 145
Soomro, Afaque Manzoor 141
Sotiriadis, Paul P. 172, 280, 317
Stouraitis, Thanos 7, 202
Taha, Mohamed Ayman 158
Taha, Radwa Essam 46
Taha, Radwa 1, 52, 196
Talaat, Mohamed Gamal 158
Talip, Mohamad Sofian Abu 26
Tanizawa, Takato 313
Tarique, Mohammed 96
Tesfai, Huruy Tekle 202
Thakur, Sumer 149, 218
Thomann, Simon 325
Tolba, Mohammed F. 7
Varnosfaderani, Shiva Maleki 91
Vermeulen, S. 178
Wael, Miran 74

Waly, Sarah M...52
Wang, Xingfu...239
Wani, Mohammad Salim 137
Waqas, Muhammad 141
Wasfi, Asma...243
Wasif, Sandy 74
Watanabe, Minoru..313
Wenceslas, Rahajandraibe.................................268
Yoshizawa, Hirokazu 272
Younes, Leen 248, 253
Youssef, Rafik...184
Zafar, Saad...288
Zaghloul, Yasmine Abdalla................................206
Zahra, Andleeb...276
Zhao, Liang...239
Zheng, Xiao Ran...86
Zweiri, Yahya...228

IEEE
445 Hoes Lane
Piscataway, NJ 08854-4141

ISBN 979-8-3503-8083-5